Evolutionary Biology and Conservation of Titis, Sakis and Uacaris

The Neotropical primate family Pitheciidae consists of four genera – *Cacajao* (uacaris), *Callicebus* (titis), *Chiropotes* (bearded sakis) and *Pithecia* (sakis) – whose 40+ species display a range of sizes, social organizations, ecologies and habitat uses. Though few are well known and the future survival of many is threatened, pitheciids have been little studied.

This book is the first to review the biology of this fascinating and diverse group in full. It includes fossil history, reviews of the biology of each genus, and, among other studies, specific treatments of vocalizations and foraging ecology. These studies are integrated into considerations of current status and future conservation requirements on a country-by-country basis for each species. Designed to be a state-of-the-art summary of current knowledge, *Evolutionary Biology and Conservation of Titis, Sakis and Uacaris* is a collective effort of researchers currently (and historically) working on these remarkable animals.

Liza M. Veiga spent most of her working career based at the Federal University of Pará and the Emílio Goeldi Museum, both in Belém, Brazil. Passionate as she was about the conservation biology of Brazilian primates, this volume represents a lasting legacy of her academic contributions.

Adrian A. Barnett is an Honorary Research Fellow at the Instituto National de Pesquisas da Amazonas in Manaus, Brazil, and at the Centre for Research in Evolutionary and Environmental Anthropology at Roehampton University. He is a tropical biologist whose research focuses on rare, little-known and hard-to-find species and has spent 15 years studying tropical primates, particularly the conservation and ecology of uacaris.

Stephen F. Ferrari is a Professor of Zoology in the Biology department of the Federal University of Sergipe, São Cristovão, Brazil. While his research covers a wide variety of mammalian taxa, he has a particular focus on the primate genera titis and bearded sakis, their ecology and conservation.

Marilyn A. Norconk is Professor of Anthropology and Graduate Faculty in the School of Biomedical Sciences at Kent State University, Kent, OH, USA. Her research focuses on the behavioral ecology of South American monkeys, particularly on the feeding ecology and social behavior of white-faced sakis and bearded sakis in Venezuela and Suriname.

Cambridge Studies in Biological and Evolutionary Anthropology

Series editors

Also available in the series

47. *The First Boat People* Steve Webb 0 521 85656 6
48. *Feeding Ecology in Apes and Other Primates* Gottfried Hohmann, Martha Robbins & Christophe Boesch (eds.) 0 521 85837 2
49. *Measuring Stress in Humans: A Practical Guide for the Field* Gillian Ice & Gary lames (eds.) 0 521 84479 7
50. *The Bioarchaeology of Children: Perspectives from Biological and Forensic Anthropology* Mary Lewis 0 521 83602 6
51. *Monkeys of the Taï Forest* W. Scott McGraw, Klaus Zuberbühler & Ronald Noe (eds.) 0 521 81633 5
52. *Health Change in the Asia-Pacific Region: Biocultural and Epidemiological Approaches* Ryutaro Ohtsuka & Stanley I. Ulijaszek (eds.) 978 0 521 83792 7
53. *Technique and Application in Dental Anthropology* Joel D. Irish & Greg C. Nelson (eds.) 978 0 521 870 610
54. *Western Diseases: An Evolutionary Perspective* Tessa M. Pollard 978 0 521 61737 6

55. *Spider Monkeys: The Biology, Behavior and Ecology of the genus Ateles* Christina J. Campbell 978 0 521 86750 4
56. *Between Biology and Culture* Holger Schutkowski (ed.) 978 0 521 85936 3
57. *Primate Parasite Ecology: The Dynamics and Study of Host-Parasite Relationships* Michael A. Huffman & Colin A. Chapman (eds.) 978 0 521 87246 1
58. *The Evolutionary Biology of Human Body Fatness: Thrift and Control* Jonathan C. K. Wells 978 0 521 88420 4
59. *Reproduction and Adaptation: Topics in Human Reproductive Ecology* C. G. Nicholas Mascie-Taylor & Lyliane Rosetta (eds.) 978 0 521 50963 3
60. *Monkeys on the Edge: Ecology and Management of Long-Tailed Macaques and their Interface with Humans* Michael D. Gumert, Agustín Fuentes & Lisa Jones-Engel (eds.) 978 0 521 76433 9
61. *The Monkeys of Stormy Mountain: 60 Years of Primatological Research on the Japanese Macaques of Arashiyama* Jean-Baptiste Leca, Michael A. Huffman & Paul L. Vasey (eds.) 978 0 521 76185 7
62. *African Genesis: Perspectives on Hominin Evolution* Sally C. Reynolds & Andrew Gallagher (eds.) 978 1 107 01995 9
63. *Consanguinity in Context* Alan H. Bittles 978 0 521 78186 2
64. *Evolving Human Nutrition: Implications for Public Health* Stanley Ulijaszek, Neil Mann & Sarah Elton (eds.) 978 0 521 86916 4

Evolutionary Biology and Conservation of Titis, Sakis and Uacaris

Edited by

Liza M. Veiga
Federal University of Pará and the Emílio Goeldi Museum, Belém, Brazil

Adrian A. Barnett
Instituto National de Pesquisas da Amazonas, Manaus, Brazil and Roehampton University, UK

Stephen F. Ferrari
Federal University of Sergipe, São Cristovão, Brazil

Marilyn A. Norconk
Kent State University, Kent, OH, USA

CAMBRIDGE
UNIVERSITY PRESS

CAMBRIDGE UNIVERSITY PRESS
Cambridge, New York, Melbourne, Madrid, Cape Town,
Singapore, São Paulo, Delhi, Mexico City

Cambridge University Press
The Edinburgh Building, Cambridge CB2 8RU, UK

Published in the United States of America
by Cambridge University Press, New York

www.cambridge.org
Information on this title: www.cambridge.org/9780521881586

First published 2013

Printed and bound in the United Kingdom by the MPG Books Group

A catalogue record for this publication is available from the British Library

Library of Congress Cataloguing in Publication data

Evolutionary biology and conservation of titis, sakis and uacaris / edited by Liza M. Veiga, Federal University of Pará and the Emílio Goeldi Museum, Belém, Brazil, Adrian A. Barnett, Instituto National de Pesquisas da Amazonas, Manaus, Brazil and Roehampton University, UK, Stephen F. Ferrari, Federal University of Sergipe, São Cristovão, Brazil, Marilyn A. Norconk, Kent State University, Kent OH, USA.
pages cm. – (Cambridge studies in biological and evolutionary anthropology)
ISBN 978-0-521-88158-6 (Hardback)
1. Pitheciidae. I. Barnett, Adrian.
QL737.P959E96 2013
599.8–dc23

2012027073

ISBN 978-0-521-88158-6 Hardback

Liza Veiga came to Primatology late, but made a resounding impact, and left a lasting contribution. After degrees in economics and international development in her native Britain, and a number of years working in business management, she decided to take her doctorate in experimental psychology at the University of Pará in the Amazonian city of Belém. She had sought out new challenges throughout her life, but found her greatest inspiration in the enigmatic pitheciine monkeys of the Amazon forest. After defending her thesis, she took up a postdoctoral position at the Goeldi Museum, also in Belém, where she dedicated the final years of her tragically short life to a variety of projects in research and conservation, as well as the supervision of students. For more than a decade, Liza overcame hardships and dangers to conduct her research in the field, while working incessantly towards the protection of the monkeys and the forests they inhabit. During this time, she published a dozen scientific papers and as many book chapters, as well as producing more than fifty species accounts for the IUCN Red Data Book, including those of all the pitheciid monkeys. She was an editor of Neotropical Primates, and a consultant to numerous governmental and non-governmental bodies, including Conservation International and the Brazilian federal environment institute. But more important than any academic achievement, Liza was a dedicated colleague and a faithful friend, always ready to help out, encourage, providing inspiration and enthusiasm with an unfailing smile. She was soft-spoken but strong-willed, untiring and unerring, and extraordinarily unselfish, forever happy to share and lend a hand, rather than make demands, and was especially committed to the next generation of primatologists, eager to ensure continuity. Liza has left an important legacy, an example to follow, an inspiration to us all, but we hope, above all, that her work and dedication will help ensure the survival of her beloved primates over many generations to come.

Contents

Contributors

Krista Adamek
Department of Wildlife and Fisheries Sciences, Texas A&M University, College Station, TX, USA

Ana Luisa K. Albernaz
Coordenação de Ciências de terra e Ecologia, Museu Paraense Emilio Goeldi, Belém, PA, Brazil

J. Marcio Ayres†
(Wildlife Conservation Society, New York, NY, USA)

Andrew J. Baker
Philadelphia Zoo, Philadelphia, PA, USA

Karen L. Bales
Department of Psychology and California National Primate Research Center, University of California, Davis, CA, USA

Adrian A. Barnett
Centre for Research in Evolutionary and Ecological Anthropology, University of Roehampton, London, UK and *Núcleo de Biodiversidade, Instituto Nacional de Pesquisas da Amazônia, Manaus, AM, Brazil*

Christopher Barton
Durrell Institute of Conservation and Ecology, University of Kent, Canterbury, UK

John M. Bates
Department of Zoology, Field Museum of Natural History, Chicago, IL, USA

Jennie Becker
Curator of Mammals, Los Angeles Zoo and Botanical Gardens, Los Angeles, CA, USA

Bruna M. Bezerra
Departamento de Zoologia, Universidade Federal de Pernambuco, Recife, PE, Brazil

Júlio César Bicca-Marques
Departamento de Biodiversidade e Ecologia, Pontifícia Universidade Católica do Rio Grande do Sul, Porto Alegre, RS, Brazil

Richard Bodmer
Durrell Institute of Conservation and Ecology, University of Kent, Canterbury, UK

Jean P. Boubli
Wildlife Conservation Society Brasil, Rio de Janeiro, RJ, Brazil

Mark Bowler
San Diego Zoo Institute for Conservation Research, Escondido, CA 92027-700, USA and *School of Psychology, University of St Andrews, St. Andrews, Fife, Scotland*

Sarah A. Boyle
Department of Biology, Rhodes College, Memphis, TN, USA

Christini Barbosa Caselli
Department of Animal Biology, Biology Institute State University of Campinas, Campinas, SP, Brazil

Janice Chism
Department of Biology, Winthrop University, Rock Hill, SC, USA

Elena P. Cunningham
Department of Basic Sciences and Craniofacial Biology, New York University College of Dentistry, New York, NY, USA

José Maria C. da Silva
Zoological Museum, University of Copenhagen, Universitetsparke, Copenhagen, Denmark

Lesa C. Davies
Department of Anthropology, Northeastern Illinois University, Chicago, IL, USA

Nayara de Alcântara Cardoso
Instituto de Desenvolvimento Sustentável Mamirauá, Tefé, AM, Brazil

Manuella A. de Souza
Instituto Brasileira do Meio Ambiente e dos Recursos Naturais Renováveis (IBAMA), Pará, Belém, Brazil

Stella de la Torre
College of Biological and Environmental Sciences, Department of Ecology, Universidad San Francisco de Quito, Cumbayá-Quito, Ecuador

Ana Gabriela de Luna
Departmento de Ciencias Biologicas, Universidad de Los Andes, Colombia

Thomas R. Defler
Departamento de Biología, Universidad Nacional de Colombia, Bogotá, Cundinamarca, Colombia

Anthony Di Fiore
Department of Anthropology, University of Texas at Austin, Austin, TX, USA

Eduardo Fernandez-Duque
Department of Anthropology, University of Pennsylvania, Philadelphia, PA, USA and *Centre for Applied Littoral Ecology (Conicet), Argentina*

Stephen F. Ferrari
Department of Ecology, Federal University of Sergipe, São Cristóvão, SE, Brazil

Wilsea M.B. Figueiredo-Ready
Instituto de Estudos Costeiros, Universidade Federal do Pará, Campus de Bragança, Bragança, PA, Brazil

Tracy Frampton
Brevard Zoo, Melbourne, FL, USA

Paul A. Garber
Department of Anthropology, Program in Ecology, Evolution, and Conservation Biology, University of Illinois, Urbana, IL, USA

Brian W. Grafton
Department of Biological Sciences, Kent State University, Kent, OH, USA

L. Tremaine Gregory
Smithsonian Conservation Biology Institute, Center for Conservation Education and Sustainability, National Zoological Park, Washington, DC, USA

Maria L. Harada
Departamento de Genética, Universidade Federal de Pará, Belém, PA, Brazil

Amy Harrison-Levine
Department of Anthropology, University of Colorado – Boulder, Boulder, CO, USA and *Conservation Biology, Denver Zoo, Denver, CO, USA*

Walter C. Hartwig
Department of Clinical Education, Touro University College of Osteopathic Medicine, Vallejo, CA, USA

Stefanie Heiduck
Deutsches Primatenzentrum, Abteilung Verhaltensökologie & Soziobiologie, Göttingen, Germany

Eckhard W. Heymann
Deutsches Primatenzentrum, Abteilung Verhaltensökologie & Soziobiologie, Göttingen, Germany

André Hirsch
Institutional Program of Bioengineering, Federal University of São João del-Rei, Campus Sete Lagoas, Sete Lagoas, MG, Brazil

Leandro Jerusalinsky
Centro Nacional de Pesquisa e Conservação de Primatas Brasileiros, ICMBio; João Pessoa, PB, Brazil

Gareth Jones
Department of Biology, Bristol University, Bristol, Avon, UK

Richard F. Kay
Department of Evolutionary Anthropology, Duke University, Durham, NC, USA

Martin M. Kowalewski
Museo Argentino de Ciencia Naturales, Estacacion Biologico de Corrientes, San Cayetano, Corrientes, Argentina

Shawn M. Lehman
Department of Anthropology, University of Toronto, Toronto, Ontario, Canada

Laura Marsh
Global Conservation Institute, Santa Fe, NM, USA

Jesús Martinez
Wildife Conservation Society Bolivia, La Paz, Bolivia

William A. Mason
Department of Psychology and *California National Primate Research Center University of California, Davis, CA, USA*

Hope Matthews
Department of Biology, Winthrop University, Rock Hill, SC, USA and *Culture and Heritage Museums of York County, Rock Hill, SC, USA*

Wynlyn McBride
Department of Natural and Applied Sciences, University of Dubuque, Dubuque, IA, USA

Shona McCann-Wood
Durrell Institute of Conservation and Ecology, University of Kent, Canterbury, UK

W. Scott McGraw
Department of Anthropology, The Ohio State University, Columbus, OH 43210, USA

D. Jeffrey Meldrum
Department of Biological Sciences, Idaho State University, Pocatello, ID, USA

Sally P. Mendoza
Department of Psychology and California National Primate Research Center, University of California, Davis, CA, USA

Nohelia Mercado
Asociación para la Conservación e Investigación de Ecosistemas Andino Amazónicos (ACEAA – Bolivia). Sopocachi, La Paz, Bolivia

Russell A. Mittermeier
Conservation International Arlington, VA, USA

Mirjam N. Nadjafzadeh
Deutsches Primatenzentrum, Abteilung Verhaltensökologie & Soziobiologie, Göttingen, Germany

Marilyn A. Norconk
Department of Anthropology, Kent State University, Kent, OH, USA

Robert Gary Norman
New York University College of Dentistry, New York, NY, USA

Marcela Oliveira
Instituto de Desenvolvimento Sustentável Mamirauá, Tefé, AM, Brazil

Marcelo M. Oliveira
Instituto Chico Mendes de Conservação da Biodiversidade (ICMBio), Brasilia, DF, Brazil

Maria Juliana Ospina Rodríguez
Departamento de Biologia, Universidad Antonio Nariño, Bogotá, Colombia

Erwin Palacios
Conservation International Colombia, Bogotá DC, Colombia

Suzanne Palminteri
Centre for Ecology, Evolution, and Conservation, School of Environmental Sciences, University of East Anglia, Norwich, Norfolk, UK and *WWF: World Wildlife Fund Conservation Science Program, 1250 24th Street, N.W., Washington, DC, USA*

Liliam P. Pinto
Postgraduate Ecology Course, Institute of Biology, Campinas State University, Campinas, SP, Brazil and *National Center for Amazonian Biodiversity Research and Conservation, Chico Institute for Biodiversity Conservation, Manaus, AM, Brazil*

Marcio Port-Carvalho
São Paulo Forestry Institute, Bauru Experimental Station, SP, Brazil

Leila Porter
Department of Anthropology, Northern Illinois University, DeKalb, IL, USA

Carlos Portillo-Quintero
Centro de Estudios Botánicos y Agroforestales, Instituto Venezolano de Investigaciones Científicas, Maracaibo, Venezuela

George Powell
Conservation Science Program, World Wildlife Fund, Washington DC, USA

Ghillean T. Prance
Royal Botanic Gardens, Kew, Surrey, UK

Rodrigo C. Printes
Departamento de Biologia, Universidade Estadual do Rio Grande do Sul, São Francisco de Paula, RG, Brazil

Pablo Puertas
Wildlife Conservation Society Peru, Lima, Peru

P. Kirsten Pullen
Paignton Zoo Environmental Park, Paignton, Devon, UK

Helder L. Queiroz
Instituto de Desenvolvimento Sustentável Mamirauá, Tefé, AM, Brazil

Luis Reginaldo R. Rodrigues
Departamento de Genética, Universidade Federal de Pará, Belém, Pará, Brazil

Adriana Rodríguez
External Consultant, Conservation International Colombia, Bogota, Colombia

Alfred L. Rosenberger
Department of Anthropology, City University of New York, New York, NY, USA

Anthony B. Rylands
Center for Applied Biodiversity Science, Conservation International, Arlington, VA, USA

Ricardo R. Santos
Centre of Agrarian and Environmental Sciences, Federal University of Maranhão, Chapadinha, MA, Brazil

Horacio Schneider
Instituto de Estudos Costeiros, Universidade Federal do Pará, Campus de Bragança, Bragança, PA, Brazil

Eleonore Z.F. Setz
Animal Biology Department, State University of Campinas, Campinas, SP, Brazil

Suleima S.B. Silva
Wild Animal Reproduction Laboratory, Institute of Biological Sciences, Federal University of Pará, PA, Brazil

José S. Silva Júnior
Departamento de Zoologia, Museu Paraense Emilio Goeldi, Belém, PA, Brazil

Andrew T. Smith
School of Life Sciences, Arizona State University, Tempe, AZ, USA

Marcelo C. Sousa
PRODEMA, Universidade federal de Sergipe, São Cristóvão, SE, Brazil

Antonio S. Souto
Departamento de Zoologia, Universidade Federal de Pernambuco, Recife, PE, Brazil

Wilson R. Spironello
Núcleo de Biodiversidade, Instituto Nacional de Pesquisas da Amazônia, Manaus, AM, Brazil

Masanaru Takai
Primate Research Institute, Kyoto University, Inuyama, Japan

Marcelo F. Tejedor
Laboratorio de Investigaciones en Evolución y Biodiversidad, Facultad de Ciencias Naturales, Sede Esquel, Universidad Nacional de la Patagonia "San Juan Bosco", Esquel, Argentina

Cynthia L. Thompson
Department of Anatomy & Neurobiology, Northeast Ohio Medical University, Rootstown, OH, USA

Diego G. Tirira
Museo Ecuatoriano de Ciencias Naturales, Quito, Ecuador and Fundación Mamíferos y Conservación, Quito, Ecuador

Raul Tupayachi
WWF Peru, Lima, Peru

Bernardo Urbani
Centro de Antropologia, Instituto Venezuelano de Investigaciones Cientificas, Caracas, Venezuela

Liza M. Veiga
Post-graduate Program in Zoology, Museu Paraense Emilio Goeldi, Belém, PA, Brazil and *Department of Zoology, Museu Paraense Emilio Goeldi, Belém, PA, Brazil*

Marianela Velilla
Department of Wildlife Ecology and Conservation, Gainesville, FL, USA

João Valsecchi
Instituto de Desenvolvimento Sustentável Mamirauá, Tefé, AM, Brazil

Jean-Christophe Vié
Association Kwata, Study and Conservation of French Guianan Wildlife, Cayenne, France

Tatiana M. Vieira
Instituto de Desenvolvimento Sustentável Mamirauá, Tefé, AM, Brazil

Suzanne E. Walker-Pacheco
Department of Sociology and Anthropology, Missouri State University, Springfield, MO, USA

Rob Wallace
Wildife Conservation Society Bolivia, La Paz, Bolivia

Patricia C. Wright
Ecology and Evolution, Stony Brook University, Stony Brook, NY, USA

Charles E. Zartman
Departamento Botânica, Instituto Nacional de Pesquisas da Amazonia, Manaus, AM, Brazil

Foreword

In a nutshell: a brief introduction to early field exploration on the pitheciids

Young male *Pithecia pithecia chrysocephala* with Russ Mittermeier (left) and *Saguinus bicolor* with Marcio Ayres (right).

My interest in the pitheciids can be traced back to a single individual. Not a person, but rather a female white uacari (*Cacajao calvus*), a resident of the Bronx Zoo from November 28, 1956 to March 31, 1980. For a long time, she was the only member of the species in captivity anywhere in the world, and to this day one of only a small handful of specimens ever kept in captivity outside Brazil. I grew up in the Bronx, and first encountered this amazing animal as a 7-year-old. I loved the Bronx Zoo and would go there every few weeks. My first stop would either be the reptile house or the small mammal house, where this white uacari was being kept. From the first moment I saw it, I was intrigued by its long, shaggy white fur, its bright red face and odd little tail. It conjured up images of the "abominable snowman", the legendary Yeti, which was then all the vogue, and inspiring expeditions to the Himalayas. To be sure, a *miniature* "abominable snowman", but something that contributed to my fascination with this animal, so different from its other monkey and ape relatives. I must have visited this monkey on dozens of occasions over the next 24 years, and I vowed to try and observe the species in the wild some day.

Years later, my reading of Henry Walter Bates's classic book, *The Naturalist on the River Amazons* (John Murray, London, 1863), further fuelled my interest in the white uacari. Bates was quite intrigued by this animal and discussed it at length, although he never actually observed it in the wild, having only come across a captive specimen at a site called Fonte Boa, on the Rio Solimões, in Brazilian Amazonia.

At the beginning of the 1970s, while doing a year-long senior project reviewing the available literature on Neotropical primates, I also became interested in the bearded sakis. At that time, the only captive bearded sakis were held at the Cologne Zoo in Germany, under the care of pioneering primatologist Uta Hick. She was later recognized for her work by Philip Hershkovitz, when he named a new bearded saki *Chiropotes satanas utahicki* in 1985. The astonishing animals also captured my imagination, with the odd, heavily muscled lumps on their heads, their comical faces, and the thickly furred tails.

Needless to say, at that time, virtually nothing was known of these monkeys in the wild. The three recognized types of uacaris – red, white and black – were all listed in the IUCN's *Red Data Book*, as was the white-nosed bearded saki. However, a careful reading of the literature revealed that virtually everything known about these animals was conjecture and the mere repetition of a few old references. What was obvious was that few scientists had ever seen them, let alone studied them, in the wild. This stimulated my imagination even further and I resolved to do my thesis on the genera *Cacajao* and *Chiropotes*.

After two years of postgraduate courses at Harvard, I started on my thesis research. In early 1973, I began a brief sojourn at Monkey Jungle in Florida, as the guest of owner Frank Dumond. Frank had several red uacaris living in his open-air enclosure, which offered a unique opportunity to observe these animals under semi-natural conditions. Seeing them in a forest fuelled my fascination more than ever, and I was aching to get out into the field. While at Monkey Jungle, I met Roy Fontaine, a postgraduate student from Bucknell University, who was also fascinated by the uacaris and hoped to eventually study them in the field. His work at Monkey

Jungle represents the first systematic observations of uacari behavior (Fontaine & Dumond 1977; Fontaine 1981).

As no one had conducted fieldwork on any uacari or saki species, I had to start from scratch. Luckily, I could count on the help of Ernest Williams, a herpetologist at Harvard's Museum of Comparative Zoology, and his Brazilian colleague, Paulo Vanzolini of the São Paulo Museum of Zoology, to organize a 4-month expedition to search for uacaris and sakis in the Brazilian Amazon basin, which is home to the vast majority of their species. Vanzolini made the Museum's two boats available to me for the 4 months of my stay (Mittermeier & Coimbra-Filho 1977).

I began my expedition in June 1973, and surveyed the Tapajós, Negro and Içá rivers, major tributaries of the Amazon, before exploring the blackwater lakes along the Solimões – the Brazilian name for the upper Amazon where the great whitewater river joins the Rio Negro near Manaus. On the Tapajós, I was able to observe the white-nosed bearded saki *Chiropotes albinasus* (a species that should really be called the *red-nosed* saki), while I found black uacaris on a number of tributaries of the Negro, and red uacaris in the Rio Jacurapá, a small blackwater tributary of the Rio Içá. My greatest reward came towards the end of the expedition, when we steamed through the whitewater *várzea* forests of the Auatí–Paraná channel, which connects the Solimões and the Japurá rivers, to finally observe the white uacari.

While finding all these species was a pleasant surprise, I was even more encouraged to see how apparently common they were, despite their IUCN listing. In fact, in the flooded forests along parts of the Rio Negro, the black uacari was *the* most abundant monkey, even though they were hunted occasionally for food or even fishing bait.

One further discovery about the uacaris was that all of them seemed to be inhabitants of Amazonian flooded forests – the black and the red uacaris in the *igapó* (blackwater), and the white uacari in the whitewater *várzea* forests. They were also typically seen in quite large groups of many dozens of individuals, and seemed to be especially partial to seeds. While we now know that the habitat preferences of these monkeys are more complex (Aquino 1998), their relationship with flooded forests is still the most prominent aspect of their ecology.

As far as I know, I had become the first non-Brazilian ever to see all four of these pitheciid species – the white-nosed bearded saki, and the white, red and black uacaris – in the wild. While this was indicative of how little was known of Amazonian primates only 40 years ago, it also reflected the logistic difficulties of surveying such vast and relatively inaccessible areas of forest. So, while I had obtained some preliminary data (Mittermeier & Coimbra-Filho 1977), it was clear that the logistics of doing research in the Brazilian Amazon at that time were incompatible with the limitations of a doctoral project, and I made the difficult decision to look elsewhere for a field site.

I subsequently visited Peru, Colombia and Panama, and it was while I was visiting Barro Colorado Island in Panama in early 1974, that I met a Dutch botanist who had just been to Suriname, then known as Dutch Guiana. He told me of a wonderful site, the Raleighvallen-Voltzberg Nature Reserve, where he had seen good populations of the northern bearded saki, *Chiropotes sagulatus*. He said it was a relatively straightforward task to reach the site and the animals were easy to find, so I decided to give it a try.

After returning to the US and making excuses to my advisor, Professor Irven DeVore, for my complete lack of data after a year and a half in the field, I headed down to Suriname in April, 1975, accompanied by howler monkey expert Katherine Milton, who was also interested in the bearded saki. We found eight primate species living in a beautiful mosaic of forest habitats at Raleighvallen-Voltzberg, including two species of pitheciines, the northern bearded saki and the white-faced saki (*Pithecia pithecia pithecia*). I also encountered a second pitheciine researcher, Drew Buchanan, who was carrying out the first-ever study of the white-faced saki in the nearby Brownsberg Nature Park. Like so many other pioneering pitheciid researchers, he was having a difficult time finding these elusive animals, but he was able to collect some data (Buchanan *et al.* 1981).

I returned in January 1976, to carry out my thesis research. By some good fortune, I was soon joined by Dutch primatologist Marc van Roosmalen, who was interested in the local spider monkeys, *Ateles paniscus*. Marc had prepared for his fieldwork by studying the fruits of the region, and his profound knowledge of the Guianan flora was of immense value to my own research.

After some 3 years of fieldwork, we had learned a good deal about the ecology of the primates of Suriname (Mittermeier & van Roosmalen 1978, 1979, 1981, 1982, 1983; Mittermeier *et al.*, 1983; van Roosmalen *et al.* 1981, 1988). In particular, we were able to determine that both sakis, but especially *Chiropotes*, were classic seed predators, with specialized dentition that enables them to open even the hard-shelled fruits of the Lecythidaceae, the family that includes Brazil nuts. I returned to the US to defend my doctoral thesis the following year (Mittermeier 1977), while van Roosmalen (1985) stayed on to conduct further research on the black spider monkey.

Our work in Suriname stimulated the interest of several other primatologists, notably John Fleagle and Art Skopec (Fleagle & Mittermeier 1980; Fleagle *et al.* 1981; Fleagle & Meldrum 1988), and then Warren Kinzey and Marilyn Norconk who were interested in seed predation (Kinzey & Norconk 1990, 1993). To this day, an old, rusty pressure gauge, brought over by the late Warren Kinzey to measure the durability of seed casings, can be found in the research station at the foot of the Voltzberg dome, where it persists as a memorial to the man and his work. I think of him every time I go there (Norconk 1994; Rosenberger 1994).

At around the same time as I was searching for my field site, a number of primatologists, including Warren Kinzey, Bill Mason, John Robinson and Martin Moynihan, were conducting the first field studies of the fourth pitheciid genus, *Callicebus* (the titis), in Colombia, Ecuador and Peru (Mason 1968; Moynihan 1976; Kinzey *et al.* 1977; Kinzey & Gentry 1979; Robinson 1979; Kinzey & Wright 1982; Kinzey & Robinson 1983). Some years later, Janis Crandelmire-Sacco (1988) also conducted a short study in southeastern Peru. Titis may not be quite as exotic in appearance as the other pitheciids, but they proved to be almost as elusive, although the hard work and dedication of these pioneers resulted in a useful body of data on ecological and behavioral parameters.

After his *Callicebus torquatus* study group on the Rio Nanáy in Peru disintegrated, Kinzey was able to conduct a short study of *Callicebus personatus* in the Atlantic Forest of Brazil (Kinzey & Becker 1983), providing the first field data on this distinct species group. To this day, Warren is the only primatologist to have conducted long-term field studies of three genera of pitheciids (*Callicebus*, *Chiropotes* and *Pithecia*).

Meanwhile, Roy Fontaine finally got his chance to observe the red uacari in the wild, when he travelled to the Rio Tapiche in the Peruvian Amazon in 1978. He joined up with Pekka Soini, the Finnish researcher who spent some 40 years working in Peru (Mittermeier *et al.* 2004; Heymann 2004), but the red uacaris in the blackwater-flooded forests there were particularly difficult to find.

At about this same time, Brazilian primatologist José Márcio Ayres was collecting data (Ayres 1981) on bearded sakis (*C. albinasus* and *Chiropotes sagulatus*), which together with our records from Suriname provided a much broader picture of the ecology of this genus. In 1983, Márcio began his groundbreaking study of the white uacari at Lago Mamirauá, between the Solimões and the Japurá, just a little downriver from where I had first seen them a decade earlier. While he produced some amazingly detailed data on the ecology of this species (Ayres 1986, 1989), Márcio's most important legacy is his contribution to the development of the concept of the sustainable development reserve, which has had such a major role in Amazonian conservation over the past 20 years. Márcio, who died in 2003, was a great pioneer of primatology and of Amazonian conservation in general, and he is sorely missed (Fonseca 2003; Mittermeier 2003; Valladares-Pádua 2003). I will always remember him as the person who had the drive and the determination to carry out the first long-term study of a uacari, something that I myself was not willing to do in the early 1970s.

Márcio's earlier study of *C. sagulatus* took place in what is now the Biological Dynamics of Forest Fragments Project, a joint venture of INPA and the Smithsonian Institution, was followed up in 1987 by Edson Frazão, who had a frustrating experience chasing these agile denizens of the upper canopy over a vast area of forest, passing many days without so much as glimpsing his study subjects (Frazão 1992). On a different scale, while Eleonore Setz had the good sense to choose a *Pithecia pithecia* group in an isolated 10-ha fragment of forest at this site when she began her behavioral study in 1985 (Setz 1993), she soon discovered that these sakis were as shy and elusive as those Buchanan had studied in Suriname.

Shortly after this, Tom Defler began a series of primate studies on Colombia's Apaporis River, which included ground-breaking observations of the black-headed golden-backed uacari's social ecology and habitat use (Defler 1991, 1999, 2001), before the work was tragically cut short by guerrilla action. Shortly after that, Jean-Philippe Boubli began the arduous task of trying to gain data on black uacaris in white sand campinas (Boubli 1997), initiating a series of studies which culminated in the description of a new uacari taxon. In an elegant gesture he named it after the father of modern uacari research, Marcio Ayres.

While the first two decades of pitheciid fieldwork were marked by the difficulties and frustrations typical of any pioneering research, the foundation laid during this period of discovery provided both the inspiration and the practical support necessary to ensure the success of subsequent researchers. Nowadays, fieldworkers have a much better idea of where to find pitheciids in the wild, what to expect of them, and even how to habituate them. In fact, many field studies now involve the monitoring of established populations, which allows the collection of more detailed data more efficiently, although there is still a long way to go, considering the diversity of the pitheciids and their ecological characteristics, and the vast, unexplored areas of the Amazon basin that have yet to reveal their secrets.

Suffice to say that I am really delighted to see the level of interest that now exists in the Pitheciidae, which were truly mysterious animals when I began my career in the early 1970s. All of this attention bodes well for their survival, and I hope that this outstanding book will serve to stimulate further research and conservation efforts on behalf of these truly unusual primates. I extend my congratulations to Liza Veiga and Adrian Barnett for organizing the symposium at the XXI Congress of the International Primatological Society (IPS) in Entebbe, Uganda, in June 2006, and for editing this book, something for which all of us who work on the Pitheciidae are most grateful. What is especially significant is the number of researchers from habitat countries (especially Brazil, Colombia, Peru, Venezuela and Ecuador) who are conducting these studies, something that was not the case when I began my fieldwork nearly 40 years ago. This too bodes well for the future.

In closing, I would like to pay tribute to Marcio Ayres, Pekka Soini and Warren Kinzey, true pioneers of primatology who loved the Pitheciidae as we all do, but who are, sadly, no longer with us.

Russell A. Mittermeier
President, Conservation International
and
Chairman, IUCN/SSC Primate, Specialist Group

References

Aquino, R. (1998). Some observations on the ecology of *Cacajao calvus ucayalii* in the Peruvian Amazon. *Primate Conservation*, **18**, 21–24.

Ayres, J.M. (1989). Comparative feeding ecology of the uakari and bearded saki, *Cacajao* and *Chiropotes*. *Journal of Human Evolution*, **18**, 697–716.

Ayres, J.M.C. (1981). Observações sobre a ecologia e o comportamento dos cuxiús (*Chiropotes albinasus* e *Chiropotes satanas* Cebidae, Primates). Unpublished Master's dissertation, Instituto Nacional de Pesquisas da Amazônia (INPA), Fundação Universidade do Amazonas (FUA), Manaus.

Ayres, J.M.C. (1986). Uakaris and the flooded forest. Unpublished PhD thesis, University Cambridge.

Boubli, J. (1997). A study of the black uakari, *Cacajao melanocephalus melanocephalus*, in the Pico da Neblina National Park, Brazil. *Neotropical Primates*, **5**, 113–115.

Buchanan, D.B. Mittermeier, R.A. & van Roosmalen, M.G.M. (1981). The saki monkeys, genus *Pithecia*. In *Ecology and Behavior of Neotropical Primates, Vol. 1*, ed. A.F. Coimbra-Filho & R.A. Mittermeier. Rio de Janeiro: Academia Brasileira de Ciências, pp. 391–417.

Crandelmire-Sacco, J. (1988). An ecological comparison of two sympatric primates: *Saguinus fuscicollis* and *Callicebus moloch* of Amazonian Peru. *Primates*, **29**, 465–475.

Defler, T.R. (1991). Preliminary observations of *Cacajao melanocephalus* Humboldt 1811 (Primates, Cebidae) in Colombia. *Trianea*, **4**, 557–558.

Defler, T.R. (1999). Fission–fusion in the black-headed uacari (*Cacajao melanocephalus*) in eastern Colombia. *Neotropical Primate*, 7, 5–8.

Defler, T.R. (2001). *Cacajao melanocephalus ouakary* densities on the lower Apaporis River, Colombian Amazon. *Primate Report*, **61**, 31–36.

Fleagle, J.G. & Meldrum, D.J. (1988). Locomotor behavior and skeletal morphology of two sympatric pitheciine monkeys, *Pithecia pithecia* and *Chiropotes satanas*. *American Journal of Primatology*, **16**, 227–249.

Fleagle, J.G. & Mittermeier, R.A. (1980). Locomotor behavior, body size, and comparative ecology of seven Surinam monkeys. *American Journal of Physical Anthropology*, **52**, 301–314.

Fleagle, J.G., Mittermeier, R.A. & Skopec, A. L. (1981). Differential habitat use by *Cebus apella* and *Saimiri sciureus* in Central Surinam. *Primates*, **22**, 361–367.

Fonseca, G.A.B. da. (2003). The grand vision of Márcio Ayres. *Neotropical Primates*, **11**, 40–41.

Fontaine, R. (1981). The uakaris, genus *Cacajao*. In *Ecology and Behavior of Neotropical Primates, Vol. 1*, ed. A.F. Coimbra-Filho & R.A. Mittermeier. Rio de Janeiro: Academia Brasileira de Ciências, pp. 443–449.

Fontaine, R. & Dumond, F.V. (1977). The red ouakari in a seminatural environment: potential for propagation and study. In *Primate Conservation*, ed. H.S.H. Prince Rainier of Monaco & G.H. Bourne. New York: Academic Press, pp. 168–236.

Frazão, E.R. (1992). Dieta e estratégia de forragear de *Chiropotes satanas chiropotes* (Cebidae: Primates) na Amazônia Central Brasileira. Unpublished Master's dissertation, Instituto Nacional de Pesquisas da Amazônia (INPA), Fundação Universidade do Amazonas (FUA), Manaus.

Hershkovitz, P. (1985). A preliminary taxonomic review of the South American bearded saki monkeys genus *Chiropotes* (Cebidae, Platyrrhini), with the description of a new subspecies. *Fieldiana, Zoology, New Series*, **27**, iii + 46.

Heymann, E. (2004). Pekka Soini: a dedicated and brilliant naturalist. *Neotropical Primates*, **12**, 89–90.

Kinzey, W.G. & Becker, M. (1983). Activity pattern of the masked titi monkey, *Callicebus personatus*. *Primates*, **24**, 337–343.

Kinzey, W.G. & Gentry, A.H. (1979). Habitat utilization in two species of *Callicebus*. In: *Primate Ecology: Problem Oriented Field Studies*, ed. R.W. Sussman. New York, NY: Wiley and Sons, pp. 89–100.

Kinzey, W.G. & Norconk, M.A. (1990). Hardness as basis of fruit choice in two sympatric frugivorous primates. *American Journal of Physical Anthropology*, **81**, 5–15.

Kinzey, W.G. & Norconk, M.A. (1993). Physical and chemical properties of fruit and seeds eaten by *Pithecia* and *Chiropotes* in Surinam and Venezuela. *International Journal of Primatology*, **14**, 207–227.

Kinzey, W.G. & Robinson, J.G. (1983). Intergroup loud calls, range size and spacing in *Callicebus torquatus. American Journal of Physical Anthropology*, **60**, 539–544.

Kinzey, W.G. & Wright, P.C. (1982). Grooming behavior in the titi monkey, *Callicebus torquatus. American Journal of Primatology*, **3**, 267–275.

Kinzey, W.G., Rosenberger, A.L., Heisler, P. S., *et al.* (1977). A preliminary field investigation of the yellow handed titi monkey, *Callicebus torquatus torquatus*, in northern Peru. *Primates*, **18**, 159–181.

Mason, W.A. (1968). Use of space by *Callicebus* groups. In: *Primates. Studies in Adaptation and Variability*, ed. P.C. Jay. New York, NY: Holt, Rinehart and Winston, pp. 200–216.

Mittermeier, R.A. (1977). The distribution, synecology and conservation of Surinam monkeys. Unpublished PhD thesis, Harvard University.

Mittermeier, R.A. (2003). José Márcio Ayres. *Neotropical Primates*, **11**, 40.

Mittermeier, R.A. & Coimbra-Filho, A.-F. (1977). Primate conservation in Brazilian Amazonia. In *Primate Conservation*, ed. H.S.H. Prince Rainier of Monaco & G.H. Bourne. New York, NY: Academic Press, pp. 117–166.

Mittermeier, R.A. & van Roosmalen, M.G.M. (1978). De Surinaamse apen/The monkeys of Suriname. *SURALCO Magazine*, **10**, 13–21.

Mittermeier, R.A. & van Roosmalen, M.G.M. (1979). Synecology of Surinam monkeys. In *Tropical Ecology and Development, Proceedings of the Vth International Symposium of Tropical Ecology, 16–21 April 1979, Kuala Lumpur, Malaysia*, ed. J.I. Furtado. Kuala Lumpur, Malaysia: The International Society of Tropical Ecology, pp. 383–392.

Mittermeier, R.A. & van Roosmalen, M.G.M. (1981). Preliminary observations on habitat utilization and diet in eight Surinam monkeys. *Folia Primatologica*, **36**, 1–39.

Mittermeier, R.A. & van Roosmalen, M.G.M. (1982). Conservation of primates in Surinam. *International Zoo Yearbook*, **22**, 59–69.

Mittermeier, R.A. & van Roosmalen, M.G.M. (1983). A synecological study of Surinam monkeys. In *Advances in Herpetology and Evolutionary Biology. Essays in Honor of Ernest E. Williams*, ed. A.G.J. Rhodin & K. Miyata. Cambridge, MA: Museum of Comparative Zoology, Harvard University, pp. 521–534.

Mittermeier, R.A., Heymann, E.W., Salo, J. & Pyhälä, M. (2004). In Memoriam: Pekka Soini: 1941–2004. *Neotropical Primates*, **12**, 89–92.

Mittermeier, R.A., Konstant, W.R., Ginsberg, H., *et al.* (1983). Further evidence of insect consumption in the bearded saki monkey, *Chiropotes satanas chiropotes*. *Primates*, **24**, 602–605.

Moynihan, M. (1976). *The New World Primates: Adaptive Radiation and the Evolution of social Behavior, Languages, and Intelligence.* Princeton, NJ: Princeton University Press.

Norconk, M.A. (1994). Warren G. Kinzey – a personal remembrance by Marilyn Norconk. *Neotropical Primates*, **2**, 19–23.

Robinson, J.G. (1979). Vocal regulation of use of space by groups of titi monkeys, *Callicebus moloch*. *Behavioural Ecology and Sociobiology*, **5**, 1–15.

Rosenberger, A.L. (1994). Warren G. Kinzey – a founding father of platyrrhinology. *Neotropical Primates*, **2**, 18–19.

Setz, E.Z.F. (1993). Ecologia alimentar de um grupo de parauacús (*Pithecia pithecia chrysocephala*) em um fragmento florestal na Amazônia central. Unpublished PhD thesis, Instituto de Biologia, Universidade Estadual de Campinas, Campinas, São Paulo.

Valladares-Padua, C. (2003). José Márcio Ayres 1954–2003. A primatologist who liked to create parks. *Neotropical Primates*, **11**, 39–40.

van Roosmalen, M.G.M. (1985). Habitat preferences, diet, feeding strategy and social organization of the black spider monkey (*Ateles paniscus paniscus* Linnaeus 1758) in Surinam. *Acta Amazonica*, **15**(3/4, suppl.), 1–238.

van Roosmalen, M.G.M., Mittermeier, R.A. & Fleagle, J.G. (1988). Diet of the northern bearded saki (*Chiropotes satanas chiropotes*): a Neotropical seed predator. *American Journal of Primatology*, **14**, 11–35.

van Roosmalen, M.G.M., Mittermeier, R.A. & Milton, K. (1981). The bearded sakis, genus *Chiropotes*. In *Ecology and Behavior of Neotropical Primates, Vol. 1*, ed. A.F. Coimbra-Filho & R.A Mittermeier. Rio de Janeiro: Academia Brasileira de Ciências, pp. 419–441.

Preface

The family Pitheciidae includes 4 genera (*Cacajao*, *Chiropotes*, *Pithecia* and *Callicebus*), and some 50 taxa distributed over a wide area of tropical South America, although the majority are endemic to the Amazon or Orinoco basins. With a 20-million year history of independent evolution, this distinctive group includes some of the world's most unusual-looking monkeys, which have equally unusual behavior and many remarkable ecological adaptations.

Present-day pitheciids range from the small titis (*Callicebus*), arguably the most primitive of New World monkeys, to the larger, cat-sized bearded sakis (*Chiropotes*) and uacaris (*Cacajao*), uniquely specialized for the exploitation of hard-husked fruits and the predation of seeds. Concomitant variation in behavior includes monogamous nuclear families in *Callicebus* and (possibly) *Pithecia* and multimale–multifemale complex fission–fusion societies in *Cacajao* and *Chiropotes*. Pitheciids are also found in a variety of habitats, including rainforest, swampland and dry scrub, where specific ecological challenges are reflected in distinct behavioral and morphological adaptations. As a whole, the group presents a fascinating evolutionary benchmark against which many predictions of theoretical biological anthropology can be tested. At the same time, many of these intriguing primates face increasing risks of extinction, and conservation issues are a major concern, considering not only the primates themselves, but also the forests they inhabit.

For a long time, wild pitheciids were known only as fast-moving inhabitants of the high canopy or shy denizens of the understorey. They were, and still are, very difficult to study, but the past two decades have seen a surge in interest and a growing number of field studies. As a result, knowledge of pitheciid ecology has now grown to the point where it is possible to draw systematic comparisons with better-known primate taxa, such as the New World capuchins and spider monkeys and Old World groups such as the guenons and colobines. In short, the time is ripe to provide a comprehensive overview of the biology of this fascinating group of primates and to plot a course of study for future research. That is the goal of *Evolutionary Biology and Conservation of Titis, Sakis and Uacaris.*

This book aims to do what no previous publication has done: uniting information on pitheciids in a coordinated and structured manner that explores the evolutionary biology of this unique primate radiation, and deals pragmatically with current conservation problems. It comes at a key moment in the study of the pitheciids, when knowledge of their ecology and behaviour is sufficiently robust to provide testable hypotheses for future studies and guide research into coming decades, but still manageable enough for a single volume. New developments in conservation biology also make this a most apposite time for a book that provides both biological backgrounds and conservation priorities for the various pitheciid taxa.

The impetus for the book emerged during planning for a symposium on the pitheciids at the XXI Congress of the International Primatological Society that took place in Uganda in June 2006. However, it is not a simple volume of symposium proceedings; its scope extends far beyond this format, in terms of topics, content, structure and contributing authors. In fact, only 15 of the 36 chapters are based on symposium presentations, and their content has been adapted to the book's integrated format. A particularly important addition is the genus *Callicebus*, which was not included in the symposium due to limited space.

In putting together this volume we sought to provide a well-balanced and fully integrated overview of pitheciid biology that would touch on all relevant topics over a wide range of interests, while maintaining an essential coherence and clarity of vision. No other book is devoted exclusively to the pitheciids, and the majority of the coverage in more general volumes on New World primates pre-dates the important advances made in pitheciid research over the last decade. As such, we expect this volume to become the definitive reference on the pitheciids, not only for the growing number of primatologists interested specifically in this family, but also other specialists seeking comparative data and the wider community of researchers in primate ecology, conservation, behavior and taxonomy.

Editors' Acknowledgments

The editors would like to thank all at Cambridge University Press whose dedication, talent and patient kindness has resulted in this book being published. In particular, we would like to thank Sara Brunton, Martin Griffiths, Dominic Lewis, Christopher Miller, and Lynette Talbot. We are also grateful to Paola Cardias Soares, Sabine Garcia de Oliveira and Ana Guimarãoes who provided editorial assistance to Liza Veiga, as well as to Stephen D. Nash (Conservation International) for cover art and to Paulo R.K. Goulart (Federal University of Pará, Brazil) for section heading representations of titis, sakis, and uacaris. We should also like to thank all chapter authors for their cooperative spirit, willingness to share their knowledge and for their desire to profile these amazing primates. With their collective erudition, the authors have highlighted what we know and what we still have to learn about this remarkable group of animals, and how their conservation might most effectively be achieved.

The following people kindly reviewed chapters or helped in other ways: Brooke Aldrich, Donald J. Brightsmith, Rogerio Grassetto Teixeira da Cunha, John Fleagle, Agustín Fuentes, Patrícia Izar, Sergio Lucena Mendes, Katherine Milton, Charles L. Nunn, Luke Parry, Jennifer Rehg, Andrew Richie, Jo Setchell, Bjorn Schulte-Herbruggen, Rebecca Shapley, Tamaini Snaith, Mike Steiper, and Karen Strier. We thank them all.

Abbreviations

BDFFP, Biological Dynamics of Forest Fragments Project

CDI, canine dimorphism index

dbh, diameter breast height

DICE, Durrell Institute of Conservation and Ecology

EER, energetic equivalent rule

FARC, Fuerzas Armadas Revolucionarias de Colombia

HHH, humeral head height

IAC, intergroup agonistic contests

IC, independent contrast

IMI, intermembral index

ITCZ, Intertropical Convergence Zone

ITE, intertroop encounter

LCA, last common ancestor

MC^3, metropolis-coupled Markov chain Monte Carlo algorithm

ML, Maximum Likelihood

MP, Maximum Parsimony

MYA, million years ago

NWM, New World monkey

OFT, Optimal Foraging Theory

PCA, Principal Components Analysis

PCR, polymerase chain reaction

PSEV, phylogenetically structured environmental variation

RPPN, *Reservas Particulares de Patrimônio Natural*

SALMA, South American Land Mammal Age

SINE, short interspersed nuclear element

SQ, shearing quotient

WCS, Wildlife Conservation Society

SYNONOMIES FOR LIVING SPECIES OF PRIMATES MENTIONED IN TEXT

Adrian A. Barnett

It was not possible to secure unanimity over all names to be used for all primate taxa mentioned in the various chapters and appendices of this book. We therefore append the following list of taxa, giving recent synonymies where they exist. Synonymies older than a decade were considered to be supplanted if no recent publications were encountered which used them.

Acknowledgments: AAB thanks Julio C. Bizca-Marques, Sarah Boyle and Janice Chism for their help in checking synonymies.

NAME IN TEXT	SYNONOMIES
Alouatta caraya	
Alouatta palliata	
Alouatta seniculus	
Ateles belzebuth	
Ateles chamek	*A. paniscus chamek*
Ateles paniscus	
Brachyteles hypoxanthus	
Cacajao ayresi	*C. melanocephalus ayresi*
Cacajao calvus calvus	
Cacajao calvus novaesi	
Cacajao calvus rubicundus	*C. rubicundus*
Cacajao calvus ucayalii	
Cacajao melanocephalus	*C. m. melanocephalus; C. honsomei*
Cacajao ouakary	*C. m. ouakary; C. melanocephalus*
Cacajao roosevelti	*C. albinasus*
Callicebus aureipalatii	
Callicebus baptista	*C. hoffmannsi baptista*
Callicebus barbarabrownae	*C. personatus barbarabrownae*
Callicebus bernhardi	
Callicebus brunneus	
Callicebus caligatus	
Callicebus caquetensis	
Callicebus cinerascens	
Callicebus coimbrai	
Callicebus cupreus	*C. cupreus cupreus*
Callicebus discolor	*C. cupreus discolor*
Callicebus donacophilus	*C. donacophilus donacophilus*
Callicebus dubius	
Callicebus hoffmannsi	*C.* hoffmannsi hoffmannsi
Callicebus lucifer	*C. torquatus lucifer*
Callicebus lugens	*C. torquatus lugens*
Callicebus medemi	*C. torquatus medemi*
Callicebus melanochir	*C. personatus melanochir*
Callicebus modestus	
Callicebus moloch	
Callicebus nigrifrons	*C. personatus nigrifrons*
Callicebus oenanthe	
Callicebus olallae	
Callicebus ornatus	*C. cupreus ornatus*
Callicebus pallescens	*C. donacophilus pallescens*
Callicebus personatus	*C. personatus personatus*
Callicebus purinus	*C. torquatus purinus*
Callicebus regulus	*C. torquatus regulus*
Callicebus stephennashi	
Callicebus torquatus	*C. torquatus torquatus*
Callithrix jacchus	
Cebus albifrons	
Cebus apella	*Sapajus apella*
Cebus olivaceus	
Cebus robustus	*Sapajus nigritus robustus*
Cebus xanthosternos	*Sapajus xanthosternos*
Cercocebus atys	

NAME IN TEXT	SYNONOMIES
Cercocebus galeritus	
Cercocebus torquatus	
Cercopithecus nictitans	
Cercopithecus pogonias	
Cercopithecus wolfi	
Chiropotes albinasus	
Chiropotes chiropotes	*C. satanas chiropotes*
Chiropotes israelita	*C. chiropotes; C. s. chiropotes*
Chiropotes sagulatus	*C. satanas sagulatus; C. chiropotes; C. s. chiropotes*
Chiropotes satanas	*C. satanas satanas*
Chiropotes satanas chiropotes	*C. chiropotes*
Chiropotes satanas utahickae	*C. utahicki*
Colobus angolensis	
Colobus polykomos	
Daubentonia madagascariensis	
Lagothrix lagotricha	
Lophocebus albigena	*Cercocebus albigena*
Macaca fuscata	
Macaca sylvanus	
Macaca thibetana	
Mandrillus sphinx	

NAME IN TEXT	SYNONOMIES
Nasalis larvatus	
Papio anubis	
Papio cynocephalus	
Pithecia aequatorialis	
Pithecia albicans	
Pithecia hirsuta	*P. monachus monachus*
Pithecia irrorata irrorata	
Pithecia irrorata vanzolinii	
Pithecia monachus	
Pithecia monachus milleri	
Pithecia monachus napensis	
Pithecia p. chrysocephala	
Pithecia p. pithecia	
Pithecia pithecia	
Presbytis melalophos	
Presbytis rubicunda	
Procolobus badius	
Propithecus diadema	
Propithecus diadema edwardsi	*P. edwardsi*
Theropithecus gelada	
Trachypithecus auratus	
Trachypithecus phayrei	

Part I

Fossil History, Zoogeography and Taxonomy of the Pitheciids

Walter C. Hartwig & Adrian Barnett

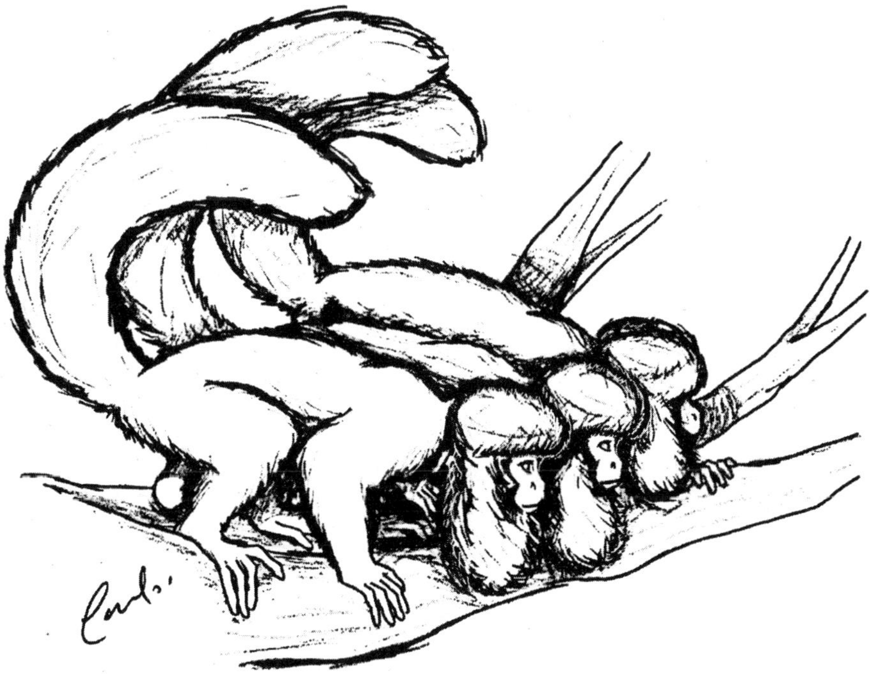

The five chapters in this section review current knowledge of the fossil record, evolutionary history, adaptive radiations, taxonomy and zoogeography of the closely related saki and uacari monkeys as well as the potentially closely related titi and owl monkeys. In the generation since *Living New World Monkeys* (*Platyrrhini*) (Philip Hershkovitz, University of Chicago Press, 1977), genetic investigations of primate evolutionary history have accelerated and now are the primary means of understanding interrelationships of the living species. In keeping with good scientific progress, the wealth of information we now have about pitheciines compared to 30 years ago opens new lines of inquiry and new needs for more detailed field and laboratory studies.

Titis, sakis and uacaris occupy a well-defined niche from a morphological perspective. They specialize in foraging on fruits with prohibitively hard pericarps. Using dental anatomy interpreted to be well-designed for puncturing and crushing the pericarp, the pitheciines are then able to access what most other species cannot – the soft inner pulp and seeds. The degree to which each living genus emphasizes sclerocarpy may underlie their evolutionary differentiation from one another within the pitheciine clade (see Kay *et al.* (Chapter 1) and Rosenberger and Tejedor (Chapter 2)).

Evolutionary Biology and Conservation of Titis, Sakis and Uacaris, eds. Liza M. Veiga, Adrian A. Barnett, Stephen F. Ferrari and Marilyn A. Norconk.
Published by Cambridge University Press.

Given the nature of the fossil record to preserve hard tissues, it is no surprise that dental indicators of sclerocarpy have influenced paleontologists to identify some fossil taxa as ancestors of modern pitheciines. Some of the geologically earliest fossil New World monkeys display mandibular trends toward seed predation and indicate the existence of fossil pitheciids as of 15 million years ago. Kay *et al.* (Chapter 1) compare the competing opinions and conclude that other Patagonian early fossil platyrrhines resemble but are not related to the radiation of saki monkeys.

The fossil record of New World monkeys is sparsely distributed in areas where fortuitous geology makes ancient deposits available for survey. Paleontologists have long recognized that essentially tropical, Amazonian species have evolutionary histories largely invisible to the known South American fossil record. Pitheciines are no exception to this disconnect, because the living taxa are distributed tightly within equatorial and Neotropical riverine environments. And yet they are known in the fossil record. Indeed, apparent direct ancestors of sakis and uacaris have been found in 12-million-year-old deposits, and convergent if not distantly related taxa have been found in even older deposits. This strongly suggests that the adaptive radiation of pitheciines was a major vector of New World monkey evolutionary history.

The intrigue of New World monkey evolutionary history is such that the two living genera whose affinities are most debated – *Aotus* and *Callicebus* – are most often compared to the lineage whose relationships are most "secure" – the pitheciines. Rosenberger and Tejedor (Chapter 2) continue the story of *Aotus* and *Callicebus*, in apt tribute to the late Warren Kinzey, by detailing how the growing molecular evolution data bear on their standing interpretation of owl and titi monkeys as pitheciines. Because the molecular, morphological and ecological data are incongruent, they argue, we should anchor hypotheses of titi and owl monkey evolutionary history to what they do in the wild.

Figueiredo *et al.* (Chapter 3) demonstrate how far technology has come in their analysis of *Cacajao* molecular data. To obtain enough specimens for a competent genetic analysis they acquired data from living animals, cadaver, and museum skins and skulls. Their results further substantiate the monophyletic status of the genus, and probe more deeply into the relatedness of populations to their geographies. Pelage similarities compelled earlier morphologists to classify populations of uacaries into the same species. This practice may mask true phylogenetic differences driven by riverine isolation and panmictic population distributions as captured by genetic analysis.

Watersheds have influenced the zoogeography of *Callicebus* and *Chiropotes* species groups as well (see Silva Júnior *et al.*, Chapter 4). Primary field censuses of pitheciine habitats are logistically complicated, and this has led to a not unwelcome dilemma for taxonomists. Pitheciine diversity and population ranges are probably greater than estimated by the field surveys and museum analyses performed to date. Further detail would require greater intrusion into these poorly known habitats, which inevitably would reduce their "nativity" even as it would improve our sense of pitheciine ecological distribution.

Within the widely distributed *Callicebus personatus* species group are populations that exhibit the expected effects of reproductive isolation. Pelage traits have influenced the use of subspecies as population boundary names that often correspond to river systems or ecosystem transition zones at the edges of the Mata Atlantica (see Printes *et al.*, Chapter 5). Genetic sampling can now evaluate how well these taxonomies align with intrinsic biological markers, but is only now getting under way.

Threats to the biodiversity and conservation of pitheciines are recurring themes in this and other sections of this volume. From a taxonomic standpoint, recognizing more and more individual populations as distinct taxonomic entities – such as subspecies – helps to focus attention on what is most easily lost by anthropogenic habitat destruction. At the same time, micro-taxonomies may exaggerate the breadth of the *adaptive radiation* of the pitheciine lineage.

The chapters in this section exhibit the significant advances in understanding that have been made over the last 20 years. We are breaking away from divining taxonomy from skins and skulls. We understand more about pitheciine evolutionary history with every decade of fossil discoveries. And we are asking different questions about adaptive radiations and dispersal mechanisms because of more intensive fieldwork and preliminary comparative genetics. Students of the next decade are likely to be drawn to saki and uacari research less because they are exotic primates and more because of the myriad of specific and micro-evolutionary questions raised by the research reported here.

Chapter 1

Pitheciidae and other platyrrhine seed predators

Richard F. Kay, D. Jeffrey Meldrum & Masanaru Takai

Phylogenetic and geochronologic background

Although a commonly agreed upon phylogeny for the platyrrhines generally, and the specific relationship of the pitheciid clade among other platyrrhines, has long been elusive, a growing consensus based on both molecules and morphology now appears to have emerged (Schneider *et al.* 1996, 2001; Ray *et al.* 2005; Horovitz 1999; Singer *et al.* 2002; Canavez *et al.* 1999; Barroso *et al.* 1997; Hodgson *et al.* 2009; Wildman *et al.* 2009). Figure 1.1A depicts a platyrrhine phylogeny based on nucleotide sequence data and *Alu* data.

Three family-rank clades supported by the molecular evidence are Cebidae (capuchins, squirrel monkeys, owl monkeys, tamarins, and marmosets), Atelidae (howlers, woolly, spider and woolly spider monkeys), and Pitheciidae (titis (*Callicebus*), sakis (*Pithecia* and *Chiropotes*) and uacaris (*Cacajao*)). The *Alu* data indicate that pitheciids are sister to a cebid–atelid clade. Some scientists consider the owl monkey *Aotus* to be a possible relative of *Callicebus*, and therefore a pitheciid *sensu lato* (Rosenberger & Tejedor, Chapter 2). However, all the recent molecular evidence confirms that *Aotus* is in fact a cebid with no close relationship to the pitheciids (Ray *et al.* 2005; Schneider *et al.* 2001; Hodgson *et al.* 2009; Wildman *et al.* 2009).

Recent work also has established a rough temporal framework for platyrrhine evolution (Schrago 2007). The estimated 20.1 Ma (Early Miocene) time of separation of Pitheciidae from other extant platyrrhine families and for the 15.6 Ma (Middle Miocene) separation of *Callicebus* from the stem lineage leading to Pitheciinae are consistent with the known, albeit sparse, fossil record. This evidence is not especially useful for palaeontologists, however. Assuming that the calibration of the branches is correct, the 95% probability intervals for these branch times are extremely broad. For the origin of the Pitheciidae, it is between 15.6 and 28.3 Ma and for the *Callicebus* branch from pitheciines, it is between 11.9 and 23.2 Ma. In most cases, fossil platyrrhines cannot be excluded from the pitheciid or pitheciine clade based on geologic age alone. Therefore, we must rely on the evidence of morphology and the fossil record itself to reconstruct more precisely the branch times related to pitheciid evolution.

Herein we make a few observations pertinent to the fossil evidence for the evolution of the Pitheciidae. We examine the anatomical attributes that correlate with the distinctive pattern of pitheciid behavior, especially with respect to the seed-predator niche. We survey the anatomical evidence of the fossils to get clues about the evolution of this behavior. We illustrate a remarkable instance of convergence of pitheciid-like anatomy and associated large-seed predation in a group of Early Miocene platyrrhines. Because emphasis is placed on the evolution of a sclerocarpic foraging niche (defined below), we have chosen to omit discussion of one proposed clade of extinct species (*Antillothrix*, *Xenothrix* and *Paralouatta*) that inhabited the Greater Antilles and that may be related to *Callicebus* (MacPhee & Horovitz 2004; Horovitz & MacPhee 1999; MacPhee *et al.* 1995; MacPhee & Fleagle 1991; MacPhee & Woods 1982; Rosenberger 1975). Of the three, only *Xenothrix* exhibits any morphology suggesting seed predation (Rosenberger 1975).

A word about systematic terminology

Systematists use the term "clade" to refer to a cluster of taxa (species, genera or families), the members of which trace back to a most recent (or last) common ancestor or LCA – Figure 1.1B (see Williams & Kay 1995 and Williams *et al.* 2010 for a discussion). The clade includes all the descendant species of the LCA, whether living or extinct. A "crown clade" is a clade that has living members. For example, the subfamily Pitheciinae is a crown clade consisting of species of the extant genera *Pithecia*, *Chiropotes* and *Cacajao*, their LCA, and all of the extinct species also descended from that LCA, and so on.

A stem taxon is one that is more closely related to one crown clade than another but is not a descendant of the LCA of any living clade. An example of a stem pitheciine is *Cebupithecia* from the Miocene of Colombia (see below). *Cebupithecia* is more closely related to living pitheciines than it is to Callicebinae, but it branched from the lineage leading to the LCA of extant pitheciines.

Evolutionary Biology and Conservation of Titis, Sakis and Uacaris, eds. Liza M. Veiga, Adrian A. Barnett, Stephen F. Ferrari and Marilyn A. Norconk. Published by Cambridge University Press.

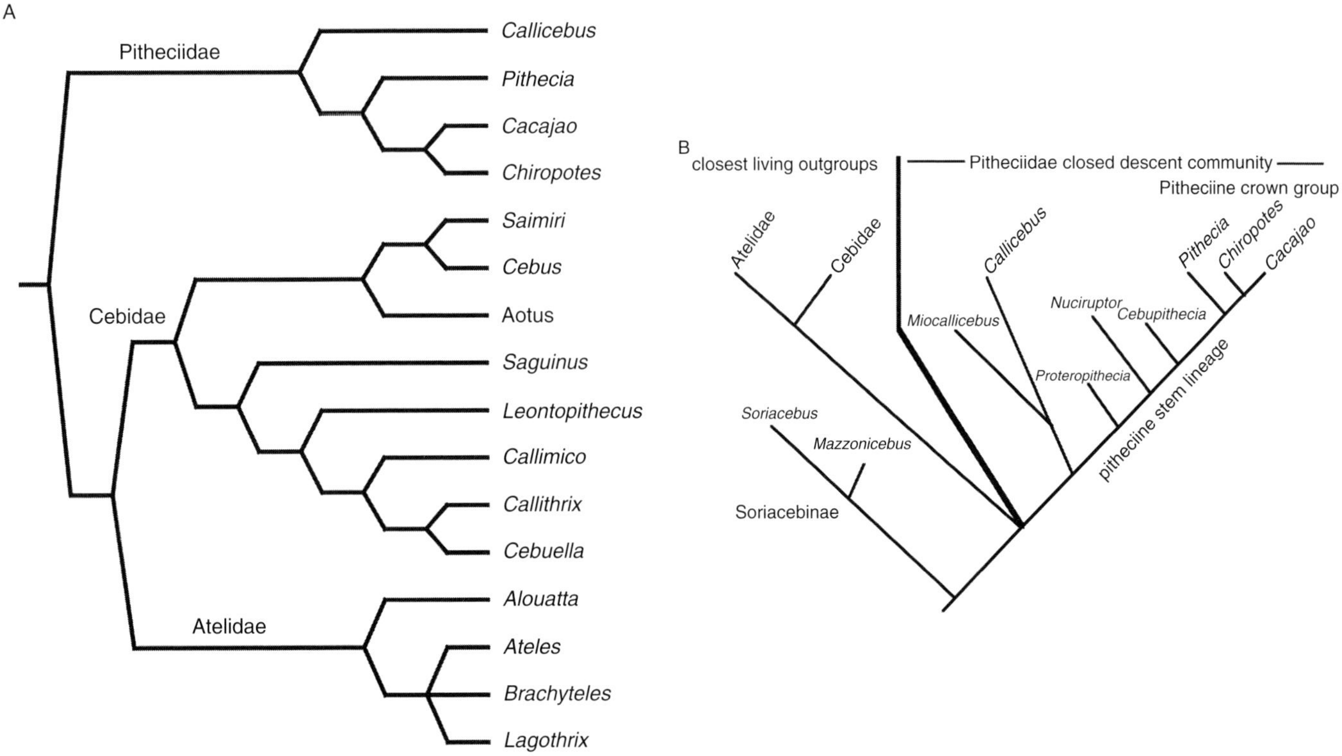

Figure 1.1 A. Cladogram of extant platyrrhine genera based on molecular sequences and *Alu* data. *Alu* elements (short interspersed nuclear elements or SINEs) complement molecular sequences because their mode of evolution is predominantly unidirectional and virtually homoplasy-free (Hillis 1999; Bashir *et al.* 2005). Molecular data strongly support the monophyly of the Platyrrhini and recognition of three clades: Atelidae, Cebidae and Pitheciidae (Harada *et al.* 1995; Barroso *et al.* 1997; Schneider *et al.* 1996, 2001; Singer *et al.* 2002). Platyrrhine monophyly is supported by 87 *Alu* elements (Singer *et al.* 2002; Ray *et al.* 2005). The tree is resolved at nodes with bootstrap values ≥ 90%, or when one or more *Alu* supports it. The *Alu* data indicate that Atelidae and Cebidae are sister taxa to the exclusion of Pitheciidae (Ray *et al.* 2005). Linkage of callitrichines with *Aotus* and a *Saimiri/Cebus* group is strongly supported by molecular data and three *Alus*, and one *Alu* links cebines with *Aotus*. *Callicebus* is supported strongly as a basal pitheciid by sequences and three *Alus*. *Aotus* is excluded consistently from the *Callicebus*–pitheciine clade (*contra* Rosenberger 1981; Rosenberger & Tejedor, Chapter 2). Atelidae has strong molecular support (including six *Alus*), with *Alouatta* as the sister to a clade consisting of *Ateles*, *Brachyteles*, and *Lagothrix* (Meireles *et al.* 1999a, 1999b). B. Simplified cladogram of platyrrhine phylogeny illustrating the systematic terms used in the text and the proposed phylogenetic position of key fossil taxa.

Pitheciid adaptations for sclerocarpy

Fruit is a major element in the diet of platyrrhines. Typically, the soft, outer layers of the fruit (the pericarp) are the parts preferred. The soft parts are obtained by swallowing the fruit whole, or by removing the edible portions with teeth and/or hands and then dropping the seed. Whole seeds ingested together with the soft outer layers can germinate if passed intact through the gastrointestinal tract. Pitheciines follow a different pattern: they extract the seed from the fruit and chew it before swallowing, and are thus known as "seed predators". A number of primate species in South America are occasional seed predators, but pitheciines are specialized to varying degrees in seed predation (van Roosmalen *et al.* 1988).

Kinzey & Norconk (1990, 1993) examined the hardness of the foods ingested by *Pithecia pithecia* and *Chiropotes satanas*. Compared to the frugivore *Ateles paniscus*, these specialist seed predators select soft seeds from fruits with very hard pericarps. Data on the hardness of the pericarp of these fruits is provided by Kinzey & Norconk (1990). The maximum hardness of the pericarp ingested by *Pithecia* is nearly five times that of fruits ingested by *Ateles*. *Chiropotes* breaks open even harder fruit pericarps to access and eat seeds – the maximum recorded fruit hardness was 27 times that of the hardest fruit opened by *Ateles* (Kinzey & Norconk 1990). Once the fruit is opened, the seed itself is soft and easily chewed (Kinzey & Norconk 1990). Kinzey and Norconk refer to this foraging adaptation as sclerocarpic foraging or sclerocarpic harvesting, which is comparable to that of avian seed predators such as parrots and macaws (Boubli 1999).

Van Roosmalen *et al.* (1988) offer the best description of how pitheciines eat these fruits. For example, with *Eschweilera* fruits from the Brazil nut family (Lecythidaceae), *Chiropotes* bites a hole into the fruit at the edge of the operculum, then uses its procumbent incisors rather like a can opener to pop it off and gain access to the seeds inside. When feeding on very hard seed pods of larger Lecythidaceae such as *Lecythis davisii* and other physically similar seed pods, the saki uses its powerful wedge-shaped canines rather than its incisors (van Roosmalen *et al.* 1988, pp. 14–15).

Titi monkeys share the seed-eating habits of pitheciines but to a lesser degree, although almost half the diet of *Callicebus lugens* may be immature seeds (Palacios *et al.* 1997). Similarly,

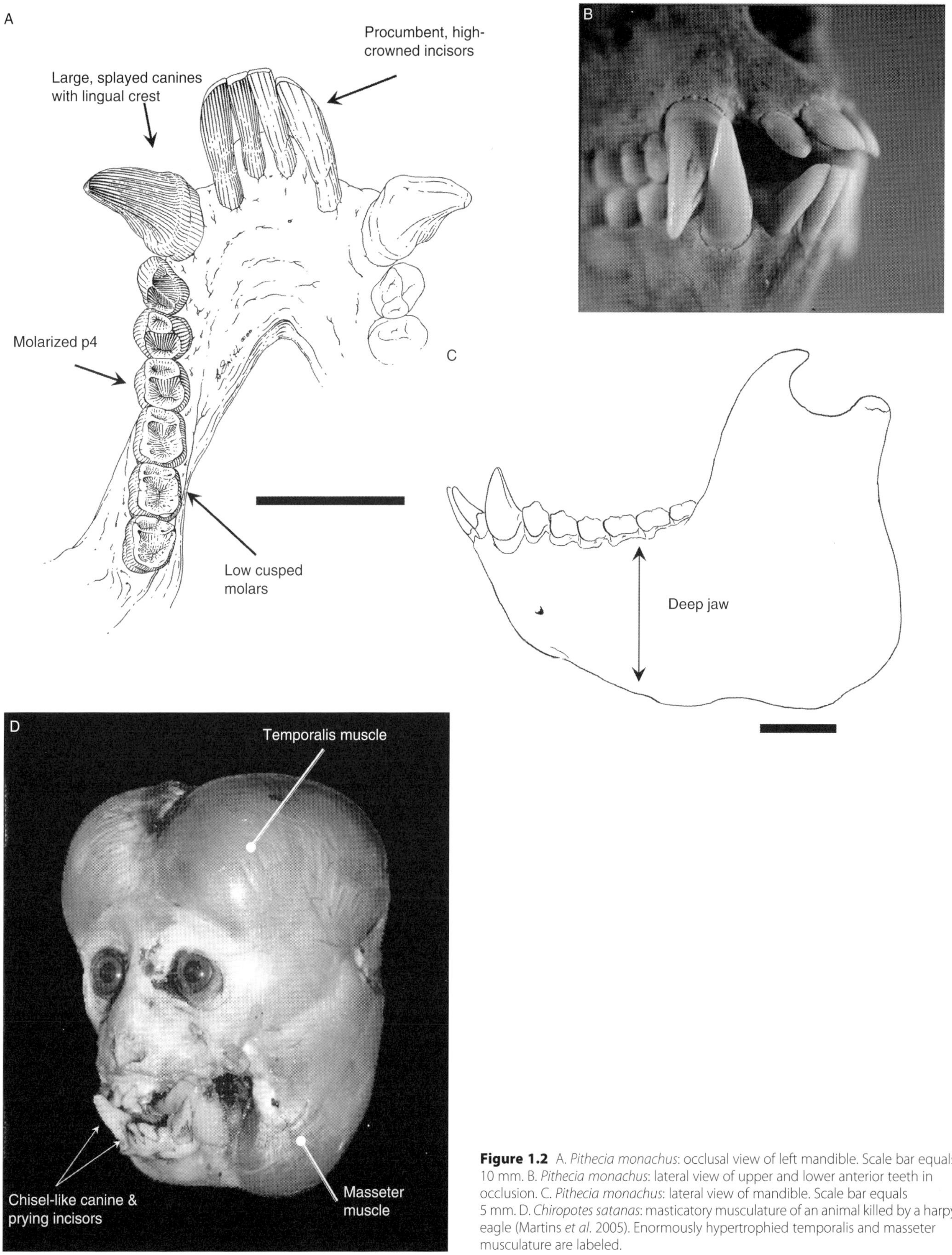

Figure 1.2 A. *Pithecia monachus*: occlusal view of left mandible. Scale bar equals 10 mm. B. *Pithecia monachus*: lateral view of upper and lower anterior teeth in occlusion. C. *Pithecia monachus*: lateral view of mandible. Scale bar equals 5 mm. D. *Chiropotes satanas*: masticatory musculature of an animal killed by a harpy eagle (Martins *et al.* 2005). Enormously hypertrophied temporalis and masseter musculature are labeled.

seeds may make up at least a quarter of the diet of *Callicebus melanochir*, but the hardness of these items is not recorded (Müller 1996).

A pattern of derived features related to sclerocarpic harvesting using the anterior dentition distinguish the living pitheciids from other extant platyrrhines (Kay 1990; Rosenberger 1992). In the pitheciines, the incisor–canine complex forms a specialized puncturing and prying mechanism (Figure 1.2A,B). The lower incisors are procumbent, narrow and styliform, forming a gouge. The canines are enlarged, laterally splayed, and have a sharp lingual crest (entocristid), producing a triangular cross-section. The specialized large-seed scraping and splitting is powered by enormously hypertrophied adductor musculature with attendant jaw deepening posteriorly (Figure 1.2C,D). The first lower molars are enlarged as well. Further specializations of the post-canine dentition include a molarized last premolar and low-cusped molars with low relief and blunt, weakly developed cutting edges (Figure 1.3A). There is virtually no disparity of height between the trigonid and talonid, with the intervening protocristid represented by a low, often indistinct ridge. Molar enamel is relatively thin and often crenulated.

One method of quantifying dietary adaptations is to measure the relative development of the shearing crests on the molar crowns. A "shearing quotient" (or SQ) can be calculated by regressing the log total crest length (y) on log tooth length (x). Deviations away from the isometric line derived from mean log x and mean log y are expressed as a percent difference: SQ = 100 (observed − expected)/expected. The molar shearing crests are poorly developed in pitheciines. In Figure 1.3B, the SQs for living platyrrhines are clustered by dietary preference. Species that eat considerable amounts of fibrous foods such as cellulose-rich leaves (*Alouatta*), or insects (chitinous exoskeletons) and fungi, which are rich in structural carbohydrate (*Callimico*), have large shearing quotients. In contrast, species that feed on less-fibrous, soft fruits (*Ateles*) or tree gum (*Callithrix*) have relatively flatter teeth, with shorter, more rounded crests. The teeth of species that specialize in eating hard seeds or splitting open tough, hard fruits (*Cebus* and the pitheciines) tend to have even less shearing crest development (Kay *et al.* 2002; Anthony & Kay 1993; Fleagle *et al.* 1997),

Pitheciidae as a morphocline

Callicebus–Pithecia–Chiropotes–Cacajao represent a morphocline of increasingly specialized dental features for sclerocarpic foraging (Kinzey 1992; Meldrum & Kay 1997a; Kay 1990; Rosenberger 1992). *Callicebus* is least specialized for seed predation among the pitheciids, but does possess some dental and mandibular morphologies associated with sclerocarpy, including posterior deepening of the mandible, and narrow, elongate incisors (Figure 1.4). However, the canines are not enlarged and the premolars are not molarized, nor is the molar structure specialized in the manner described above (cresting is more evident and the trigonid is slightly elevated above the talonid).

Pithecia, *Chiropotes* and *Cacajao* demonstrate adaptations for sclerocarpy in increasing degrees. All these species have procumbent, laterally compressed lower incisors arranged like a shovel. In a stout mandibular symphysis, the canines are progressively larger and more flared, and the chisel-like cross-sectional profile more pronounced. The last premolar looks more like a molar and the molars themselves are flatter with more poorly developed cutting crests and a tendency for the surfaces to become crenulated.

The extant pitheciines occupy an intermediate position in the morphospace defined by the shape of the platyrrhine ankle bone, the talus (Meldrum 1990). They fall between the cebids, characterized generally as small-bodied quadrupedal runners and leapers, and the atelids, which are generalized as prehensile-tailed suspensors. These distinctions are further reflected in other postcranial features such as limb

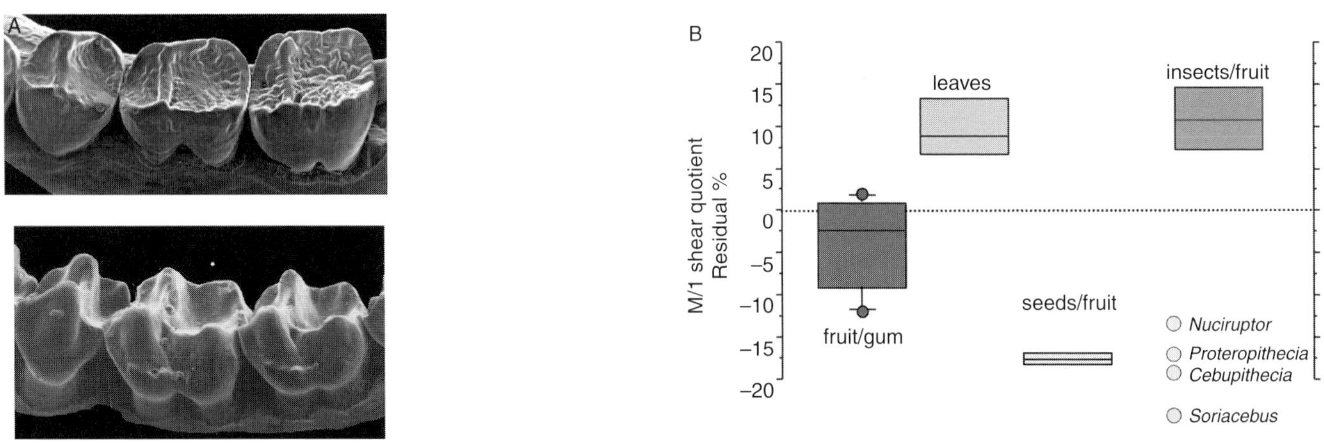

Figure 1.3 A. Occlusolateral scanning electron micrographs of the mandibular cheek teeth P_4–M_2 of *Cacajao calvus*, a seed eater (above), and *Saimiri sciureus*, a mixed insect and soft fruit eater (below), illustrating the range of shearing crest development among extant platyrrhines. The two images are bnot to the same scale. B. Shearing quotients (SQs) for living platyrrhines clustered by dietary preference. *Nuciruptor*, *Cebupithecia* and *Soriacebus* have very poorly developed shearing surfaces on their molars suggesting a diet of fruit, seeds, or gum.

proportions and the morphology of the joints of the extremities. Among the pitheciines there exists some differentiation of the details of locomotor adaptation. Fleagle & Meldrum (1988) have shown that *Pithecia* displays postcranial adaptations for vertical clinging and leaping correlated with its habit of occupying the understory. *Chiropotes* and *Cacajao*, on the other hand, show adaptations for quadrupedalism and hindlimb suspension. Hindlimb suspension may be a correlate of below-branch foraging, which provides access to suspended fruits (Meldrum 1998).

The fossil record of platyrrhine sclerocarpy

The fossil record of platyrrhines begins in the latest Oligocene (26 Ma) (Kay *et al.* 1998). For the most part, the specimens documenting platyrrhine history come from outside the current distribution of living platyrrhines. This is especially evident in the great southward extension of the range in early Miocene times, related to the "Mid-Miocene climatic optimum" between ~17 and 15 million years ago (Figure 1.5). Two distinct clades of platyrrhines demonstrate remarkable convergence in the masticatory apparatus. One group, the extant Pitheciidae and their fossil relatives, extend back to the early Middle Miocene, about 15.5 million years ago, and seem to have occupied all or most of the tropical zone. A second, older and more geographically restricted group, the *Soriacebus* clade, despite their resemblance to living pitheciines, are more probably stem platyrrhines.

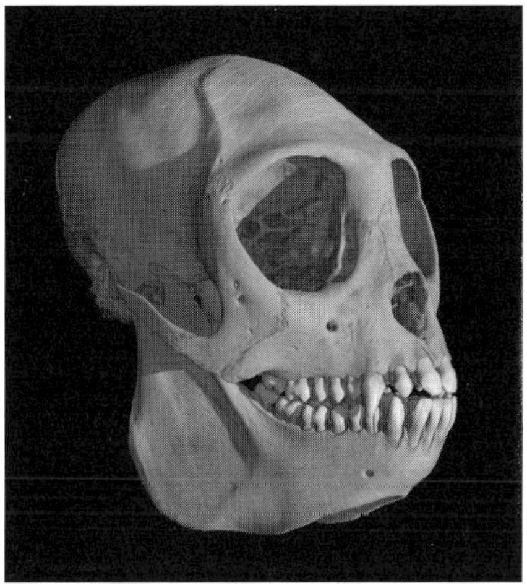

Figure 1.4 *Callicebus moloch*: right lateral view of skull. *Callicebus* is least-specialized for seed predation among the pitheciids, but does possess incipient dental and mandibular morphologies associated with sclerocarpy, including posterior deepening of the mandible and narrow incisors. Skull length equals ~6 cm.

Pitheciidae

Patagonian Argentina has a rich assemblage of fossil primates of Early Miocene age. From Patagonian Argentina south of 40° S comes a single, and oldest, fossil pitheciine species. It also represents the youngest occurrence of a fossil primate in Argentina, after which climatic conditions became more arid and cooler, less suitable for occupation by forest-dwelling monkeys. *Proteropithecia* consists of teeth, jaw fragments, and a talus (Kay *et al.* 1998, 1999). This meagre material establishes only that the front-tooth specializations of the pitheciines (sakis and uacaris), associated with seed predation, were present (Figure 1.6A). The lower incisor has a stout root and the crown is projecting. Incisor crowns and roots are compressed sideways, conforming to the gouging mechanism of extant Pitheciinae. The molars of *Proteropithecia* are also flattened, although a low, rather distinct cross crest (a protocristid)

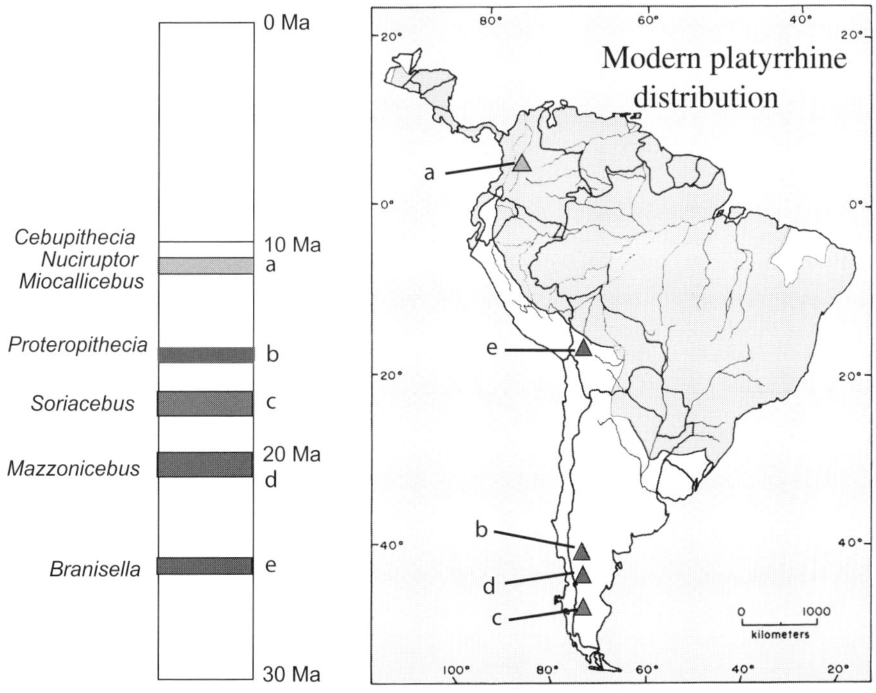

Figure 1.5 A map and timescale of the fossil platyrrhines discussed in the text. The temporal positions of the taxa are marked on the timescale at left. The modern range of platyrrhines in South America is indicated by the shaded area on the map. The positions of the localities of the fossil platyrrhines discussed in the text are indicated by triangles on the map. Symbols: a, La Venta, Neiva, Colombia; b, Cañadon del Tordillo, Neuquen, Argentina; c, Río Pinturas, Santa Cruz, Argentina; d, Gran Barranca, Chubut, Argentina; e, Salla, Bolivia.

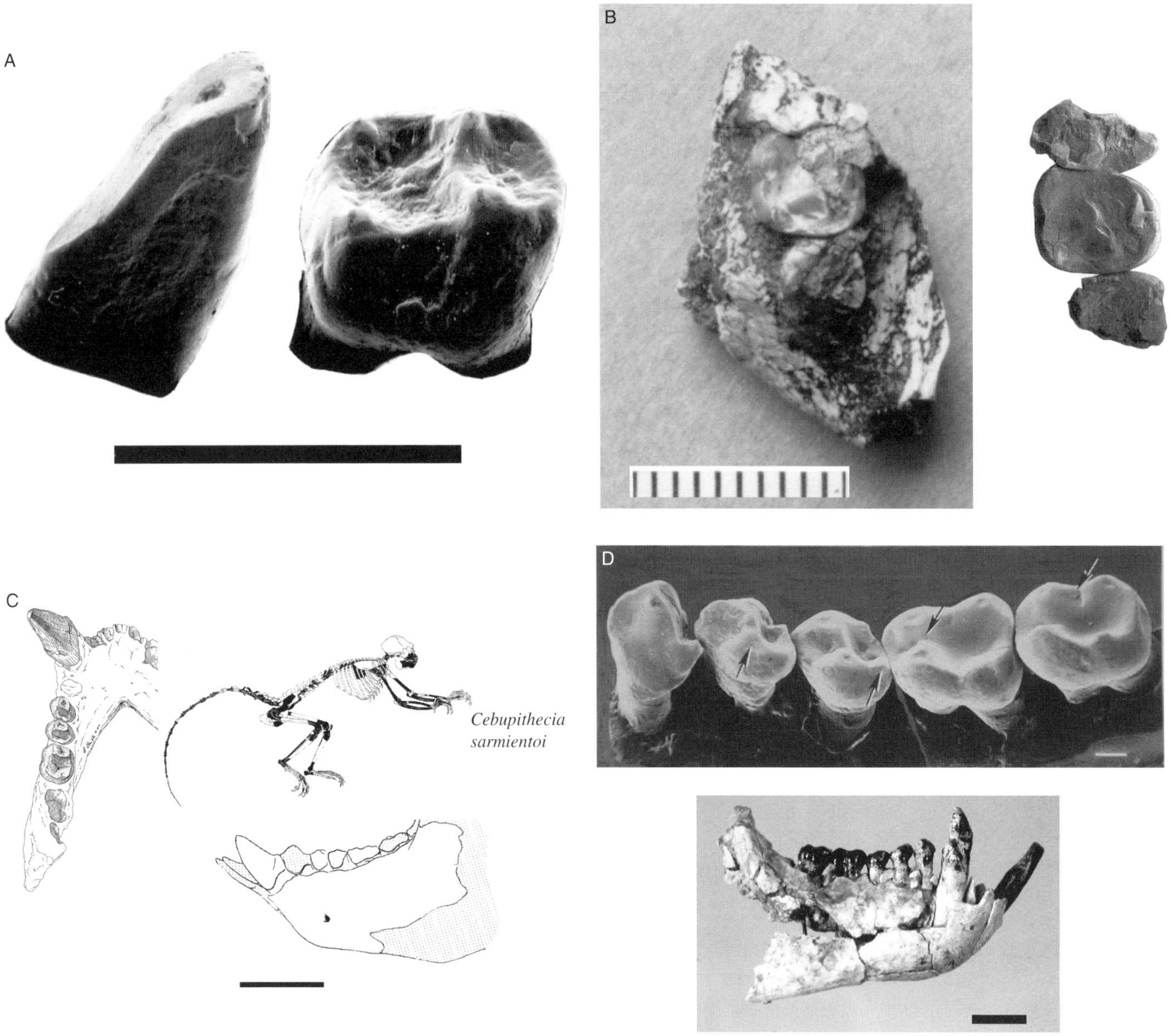

Figure 1.6 A. *Proteropithecia*, Middle Miocene, northern Patagonia, an isolated lower molar and lower incisor. Scale bar equals 5 mm. B. *Miocallicebus villaviejai*, middle Miocene, Colombia. Maxilla with right upper cheek teeth. Scale bar equals 10 mm. C. *Cebupithecia sarmientoi*, Middle Miocene, Colombia. Jaws, teeth and partial skeleton. Scale bar for lateral view of jaw equals 10 mm. D. *Nuciruptor*, Middle Miocene, Colombia. Lower jaw in lateral view and view of the cheek teeth. Scale bar for mandible equals 5 mm.

remains. The associated ankle bone resembles that of *Callicebus*, suggesting that *Proteropithecia* was a generalized arboreal quadruped.

Because *Proteropithecia* is surely a pitheciine, by inference, this species establishes that the titi-monkey (*Callicebus*) clade must already have separated from pitheciines by 15.5 Ma. Another slightly older Patagonian species, *Homunculus* (about 16.5 million years old), has been related to *Callicebus* (Rosenberger 2000, 2002; Fleagle & Tejedor 2002). However, it appears more likely that the similarities between *Homunculus* and *Callicebus* are merely shared primitive characters and evince no special affinity between the two. Several new, exceptionally well-preserved crania of *Homunculus* retain many primitive platyrrhine features and have a relatively small brain (Kay *et al.* 2006a, 2006b, 2012).

The next record of fossil pitheciids comes from 11–12-million-year-old Miocene rocks in the Tatacoa Desert, Colombia. The sites are situated at 5°N in the modern-day valley of the Magdalena River, which drains northward into the Caribbean and is separated from the Amazon and Orinoco basins by the eastern Andean Cordillera. In Miocene times, the cordillera had not yet arisen and the faunas of the Magdalena valley were in continuity with present day Orinoco and Amazon basins (Hoorn *et al.* 1995; Lundberg *et al.* 1986). Thus, the fossil monkeys from Colombia give us a Middle Miocene snapshot of the northwestern Amazon/Orinoco Basin. The Colombian fossil record documents the first definite record of titi monkeys – *Miocallicebus villaviejai* (Takai *et al.* 2001) and two stem pitheciines, *Nuciruptor* and *Cebupithecia*. We have no fossil record of any crown pitheciine species. Another Middle Miocene primate from Colombia, *Mohanamico hershkovitzi*, has been related to pitheciids (Luchterhand *et al.* 1986). Others have since suggested its affinities lie with callitriichines (Rosenberger *et al.* 1990) or express uncertainty about its phylogenetic position (Horovitz 1999; Meldrum & Kay 1997a).

Not much is known about *Miocallicebus*. The type specimen (Figure 1.6B) consists of a maxillary fragment preserving a partial root of the first upper molar, the complete second molar and a damaged third molar (Takai *et al.* 2001). Except for its much larger size, the upper second molar of *Miocallicebus* is very similar to that of extant *Callicebus*, with a nearly rectangular occlusal outline, a distinct hypocone, and a distinctive well-developed mesiolingual cingulum. The dental evidence of *Miocallicebus* confirms what we knew from the existence of *Proteropithecia*: callicebines and pitheciines had diverged from each other by the Middle Miocene.

Cebupithecia (Figure 1.6C) shares many diagnostic similarities of the jaws and teeth with living pitheciines (Orlosky 1973; Rosenberger 1979; Kay 1990). The alveolar process of the mandible deepens posteriorly as in living pitheciines, and the two sides of the lower jaw are joined at a robust symphysis. Like extant pitheciines, the incisors (as observed from the preserved roots) are modified into a gouging mechanism, being mesiodistally compressed and procumbent. A wide space separates the incisors from the robust, splayed canines, which have a chisel-like crest at the apex, lending the tooth a triangular cross-section. The first lower premolar is quite large but the others are small and do not resemble molars (i.e. they are not molariform). The molars have low relief like living pitheciines, but with smooth rather than crenulated enamel (Meldrum & Kay 1997a).

Nuciruptor is another Colombian Miocene pitheciine (Meldrum & Kay 1997a) (Figure 1.6D). A virtual contemporary of *Cebupithecia*, *Nuciruptor* has the same distinctive incisor specialization, but does not display the same degree of specialization of the canine evident in the former. The anterior premolar is not enlarged and the posterior premolar is not molariform. The molars of both *Nuciruptor* and *Cebupithecia* have very poorly developed shearing crests, with even lower SQs than any of the living pitheciines (Figure 1.3B). This strongly suggests that they were masticating hard seeds, although the molar enamel is not crenulated. In short, the mosaic of features displayed by *Nuciruptor* would seem to bear out Kinzey's (1992) prediction that specializations of the anterior dentition preceded those of the post-canine dentition in the evolution of sclerocarpy.

Given the rarity of postcranial remains for fossil pitheciines, limited conclusions can be drawn about correlations between locomotor adaptations and dietary adaptations. There is simply too little representation of the diversity of postcranial derivations, or lack thereof, to confidently correlate adaptations of limbs to those of the dentition, even if such a correlation were expected to exist. The most extensive fossil skeleton represents a single specimen of *Cebupithecia* (Figure 1.6C) (Stirton & Savage 1951; Stirton 1951). The skeleton is structurally similar to *Pithecia*, but lacks many of the derived traits of the extant pitheciine postcranium (Fleagle & Meldrum 1988; Meldrum 1993; Meldrum & Kay 1997b). For example, the distal femur (knee) has deep condyles, and the patellar groove, across which the kneecap slides, is narrow with a raised lateral margin. This suggests locomotion dominated by above-branch quadrupedal running with considerable leaping ability, whereas living pitheciines engage more frequently in climbing and suspensory behaviors as well as leaping. Similarly, extant pitheciines exhibit relatively longer limbs compared to the length of the trunk and configurations of other joints that bear some resemblances to the ateline condition. These correlates of climbing and suspensory behaviors are lacking in *Cebupithecia*. Furthermore, *Cebupithecia* has an elongate tail, unlike living, short-tailed sakis and uacaris (Meldrum & Lemelin 1991).

The *Soriacebus* clade

A second clade of platyrrhine sclerocarpic foragers consists of two genera, *Soriacebus* (Figure 1.7) and *Mazzonicebus* (Kay, 2010). The two are found in the Early Miocene of Patagonian Argentina (at > 45°S). *Soriacebus* occurs in the early part of the Santacrucian South American Land Mammal Age (SALMA) at about 17 Ma. *Mazzonicebus* appears to be its ancestor or sister taxon, and appears at 20.0 and 20.2 Ma in the Colhuehuapian SALMA (Kay 2010). Most of what we have to say about this clade is based on the published specimens of *Soriacebus* but applies equally to *Mazzonicebus*.

Soriacebus is considered by some scientists to be an extinct tribe of pitheciines (Fleagle & Tejedor, 2002; Tejedor 2005a,b), but others argue that it represents a distinct clade of early platyrrhines, unrelated to pitheciines (Meldrum & Kay 1997a; Kay *et al.* 1998). We return to these conflicting hypotheses below. *Soriacebus* resembles pitheciids in a constellation of dental and jaw anatomy associated with sclerocarpic harvesting. It possesses

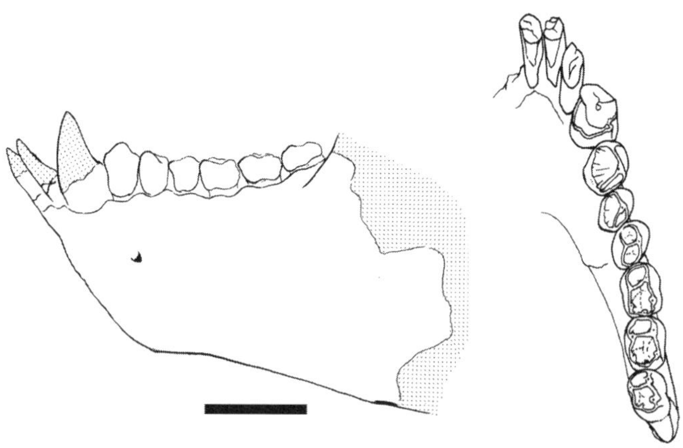

Figure 1.7 Left mandible of *Soriacebus sarmiento*, an early Miocene platyrrhine from Patagonia with convergent specializations for seed predation. Occlusal (left) and lateral (right) views. Scale bar equals 10 mm.

procumbent styliform incisors that must have served as a gouging mechanism. The lower canine is extremely robust. A large and projecting anterior premolar served as a puncturing device. A swelling on the outside of the tooth seems to have provided a sharpening edge for the large splayed pitheciine-like upper canine (Fleagle *et al.* 1997; Fleagle 1990; Tejedor 2005a,b). The lower jaw has a robust symphysis allowing recruitment of muscle forces from both sides of the jaw during powerful incisal and canine biting. The mandibular ramus deepens posteriorly, increasing its strength against powerful bending forces engendered by the action of the front teeth, and serving as an expanded surface for large chewing adductor muscles. The front teeth are extremely large in proportion to the cheek teeth behind the anterior premolar. The cutting edges of the cheek teeth are weakly developed, as in pitheciines.

Colhuehuapian-age *Mazzonicebus* from the Patagonian locality of Gran Barranca (~45°S) resembles *Soriacebus* in having robust jaws that deepen posteriorly, gouging incisors, enlarged canines and anterior premolars, and poorly developed cutting edges on its molars. However, the size disparity between the front teeth and cheek teeth is not as pronounced as in *Soriacebus* and the molar structure is markedly different in several details of the cusps.

The structural details of the premolars and molars of *Soriacebus* appear to be those of a stem platyrrhine (Kay 2010). The lower premolars have weakly developed medial cusps (metaconids), the anterior upper premolars have small hypocones and the last premolar is not molariform. Advanced soriacebine molars have elongate trigonids, versus short ones in pitheciines. This combination of clear specializations for sclerocarpic harvesting on the one hand, but differences in the details of how this adaptation is achieved, on the other, leave little doubt that the *Soriacebus* clade acquired their sclerocarpous feeding adaptations independently from pitheciines.

Concluding observations

The evolutionary history of pitheciids is characterized by an increased reliance upon large-seed predation. Pitheciids engage in "sclerocarpic harvesting", involving the removal with the incisors and canines of a hard or tough outer pericarp to obtain and chew up relatively soft nutrient-rich seeds. Although precise data about *Callicebus* are not available from the literature, *Callicebus* shows a more limited version of this adaptation with some but not all of its morphological features.

No pitheciine fossils are known from continental South America between the Recent and the middle Miocene. Stem pitheciines are represented between 15.5 and 11.5 Ma by three genera ranging from Patagonia to Colombia. Only one specimen, from the middle Miocene of Colombia, documents the *Callicebus* lineage.

In spite of this impoverished record, a few constraints can be identified with regard to the otherwise broad branch times for the separation of pitheciids from other platyrrhines and the cladogenesis between the *Callicebus* lineage and that of the saki–uacaris is provided by the fossil record. The earliest definite appearance of the *Callicebus* clade (*Miocallicebus*) is dated to the Laventan SALMA, about 12 Ma. Pitheciines present a more pronounced version of the sclerocarpic foraging jaw adaptations. They first appear as fragmentary fossil remains described as *Proteropithecia* in the later parts of the Santacrucian SALMA at about 15.5 Ma. These two species establish that the branching times for the origin of Pitheciidae and *Callicebus*–pitheciine cladogenesis must have occurred before 15.5 Ma.

The Early Miocene *Soriacebus* clade is a remarkable parallel to pitheciines in terms of anterior dental and jaw structure. *Soriacebus* offers another example from the primate fossil record of distinct clades arriving at similar solutions to common adaptive challenges, in this case, the exploitation of seeds encased within hard or tough pericarps.

Acknowledgments

We thank Adrian Barnett and Liza Veiga for the opportunity to present this paper at the Uganda meetings of the International Primatological Society. Supported by National Science Foundation grants BCS-0090255 to RFK and by the Global COE program to MT (A06 to Kyoto University).

References

Anthony, M.R.L. & Kay, R.F. (1993). Tooth form and diet in ateline and alouattine primates: reflections on the comparative method. *American Journal of Science*, **283**A, 356–382.

Barroso, C.M.L., Schneider, H., Schneider, M.P.C., *et al.* (1997). Update on the phylogenetic systematics of New World monkeys: further DNA evidence for placing pygmy marmoset (*Cebuella*) within the genus *Callithrix*. *International Journal of Primatology*, **18**, 651–674.

Bashir, A., Ye, C., Price, A.L., *et al.* (2005). Orthologous repeats and mammalian phylogenetic inference. *Genome Research*, **15**, 998–1006.

Boubli, J.P. (1999). Feeding ecology of Black-headed Uacaris (*Cacajao melanocephalus melanocephalus*) in Pico da Neblina National Park, Brazil. *International Journal of Primatology*, **20**, 719–749.

Canavez, F.C., Moreira, M.A.M., Ladasky, J.J., *et al.* (1999). Molecular phylogeny of New World primates (Platyrrhini) based on β2-microglobulin DNA sequences. *Molecular Phylogenetics and Evolution*, **12**, 74–82.

Fleagle, J.G. (1990). New fossil platyrrhines from the Pinturas Formation, southern Argentina. *Journal of Human Evolution*, **19**, 61–85.

Fleagle, J.G. & Meldrum, D.J. (1988). Locomotor behavior and skeletal morphology of two sympatric pitheciine monkeys, *Pithecia pithecia* and *Chiropotes satanas*. *American Journal of Primatology*, **16**, 227–249.

Fleagle, J.G. & Tejedor, M.F. (2002). Early platyrrhines of southern South America.

In *The Primate Fossil Record*, ed. W.C. Hartwig. Cambridge: Cambridge University Press, pp. 161–173.

Fleagle, J.G., Kay, R.F. & Anthony, M.R.L. (1997). Fossil New World monkeys. In *Mammalian Evolution in the Neotropics*, ed. R.F. Kay, R.H. Madden, R.L. Cifelli & J.J. Flynn. Washington, DC: Smithsonian Institution Press, pp. 473–495.

Harada, M.L., Schneider, H., Schneider, M.P. C., *et al.* (1995). DNA evidence on the phylogenetic systematics of New World monkeys: support for the sister grouping of *Cebus* and *Saimiri* from two unlinked nuclear genes. *Molecular Phylogenetics and Evolution*, **4**, 331–349.

Heiduck, S. (1997). Food choice in masked titi monkeys (*Callicebus personatus melanochir*): selectivity or opportunism? *International Journal of Primatology*, **18**, 487–502.

Hillis, D.M. (1999). SINEs of the perfect character. *Proceedings of the National Academy of Sciences, USA*, **96**, 9979–9981.

Hodgson, J.A., Sterner, K.N., Matthews, L.J., *et al.* (2009). Successive radiations, not stasis, in the South American primate fauna. *Proceedings of the National Academy of Sciences of the United States of America*, **106**: 5534–5539.

Hoorn, C., Guerrero, J., Sarmiento, G.A., *et al.* (1995). Andean tectonics as a cause for changing drainage patterns in Miocene northern South America. *Geology*, **23**, 237–240.

Horovitz, I. (1999). A phylogenetic study of living and fossil platyrrhines. *American Museum Novitates. New York NY*, **3269**, 1–40.

Horovitz, I. & MacPhee, R.D.E. (1999). The Quaternary Cuban platyrrhine *Paralouatta varonai* and the origin of Antillean monkeys. *Journal of Human Evolution*, **36**, 33–68.

Kay, R.F. (1990). The phyletic relationships of extant and fossil Pitheciinae (Platyrrhini, Anthropoidea). *Journal of Human Evolution*, **19**, 175–208.

Kay, R.F. (2010) A new primate from the early Miocene of Gran Barranca, Chubut Province, Argentina: paleoecological implications. In *Paleontology of the Gran Barranca*, ed. R.H. Madden, G. Vucetich, A.A. Carlini & R.F. Kay. Cambridge: Cambridge University Press, pp. 220–239.

Kay, R.F., Fleagle, J.G., Mitchell, T.R.T., *et al.* (2008). The anatomy of *Dolichocebus gaimanensis*, a primitive platyrrhine monkey from Argentina. *Journal of Human Evolution*, **54**, 323–382.

Kay, R.F., Johnson, D. & Meldrum, D.J. (1998). A new pitheciin primate from the middle Miocene of Argentina. *American Journal of Primatology*, **45**, 317–336.

Kay, R.F., Johnson, D. & Meldrum, D.J. (1999). *Proteropithecia*, new name for *Propithecia* Kay, Johnson and Meldrum, 1998 non Vojnits 1985. *American Journal of Primatology*, **47**, 347.

Kay, R.F., Kirk, E.C., Malinzak, M., *et al.* (2006a). Brain size, activity pattern, and visual acuity in *Homunculus patagonicus*, an early miocene stem platyrrhine: the mosaic evolution of brain size and visual acuity in Anthropoidea. *Journal of Vertebrate Paleontology*, **26**, 83A–84A.

Kay, R.F., Perry, J.M.G., Malinzak, M.D., *et al.* (2012). The paleobiology of Santacrucian primates. In *Early Miocene Paleobiology in Patagonia: High-Latitude Paleocommunities of the Santa Cruz Formation*, ed. S. Vizcaino, R.F. Kay & M. Bargo. Cambridge: Cambridge University Press, pp. 306–330.

Kay, R.F., Rae, T.C., Koppe, T., *et al.* (2006b). Paranasal pneumatization in the early Miocene platyrrhine *Homunculus patagonicus*. *American Journal of Physical Anthropology*, **Suppl. 42**, 112.

Kay, R.F., Williams, B.A. & Anaya, F. (2002). The adaptations of *Branisella boliviana*, the earliest South American monkey. In *Reconstructing Behavior in the Primate Fossil Record*, ed. J.M. Plavcan, C. van Schaik, R.F. Kay & W.L. Jungers. New York, NY: Kluwer Academic/Plenum Publishers, pp. 339–370.

Kinzey, W.G. (1992). Dietary and dental adaptations in the Pitheciinae. *American Journal Physical Anthropology*, **88**, 499–514.

Kinzey, W.G. & Norconk, M.A. (1990). Hardness as a basis of fruit choice in two sympatric primates. *American Journal of Physical Anthropology*, **81**, 5–15.

Kinzey, W.G. & Norconk, M.A. (1993). Physical and chemical properties of fruit and seeds eaten by *Pithecia* and *Chiropotes* in Surinam and Venezuela. *International Journal of Primatology*, **14**(2), 207–227.

Luchterhand, K., Kay, R.F. & Madden, R.H. (1986). *Mohanamico hershkovitzi*, gen et sp. nov., un primate du Miocène moyen d'Amérique du Sud. *Comptes Rendus de l'Academie de Sciences, Paris*, **Ser. II, 303**, 1753–1758.

Lundberg, J.G., Machado-Allison, A. & Kay, R.F. (1986). Miocene Characid fishes from Colombia: evolutionary stasis and extirpation. *Science*, **234**, 208–209.

MacPhee, R.D.E. & Fleagle, J.G. (1991). Postcranial remains of *Xenothrix macgregori* (Primates, Xenotrichidae) and other Late Quaternary mammals from Long Mile Cave, Jamaica. *Bulletin of the American Museum of Natural History*, **206**, 287–321.

MacPhee, R.D.E. & Horovitz, I. (2004). New craniodental remains of the Quaternary Jamaican monkey *Xenothrix mcgregori* (Xenotrichini, Callicebinae, Pitheciidae), with a reconsideration of the *Aotus* hypothesis. *American Museum Novitates. New York NY*, **3434**, 1–51.

MacPhee, R.D.E. & Woods, C.A. (1982). A new fossil cebine from Hispaniola. *American Journal of Physical Anthropology*, **58**, 419–436.

MacPhee, R.D.E., Horovitz, I., Arredondo, O., *et al.* (1995). A new genus for the extinct Hispaniolian monkey *Saimiri bernensis* Rímoli, 1977, with notes on its systematic position. *American Museum Novitates. New York NY*, **3134**, 1–21.

Martins, S.S., Lima, E.M. & Sousa e Silva Jr., J. (2005). Predation of a bearded saki (*Chiropotes utahicki*) by a Harpy Eagle (*Harpia harpyja*). *Neotropical Primates*, **13**, 7–10.

Meireles, C.M., Czelusniak, J., Ferreira, H.S., *et al.* (1999a). Phylogenetic relationships among Brazilian howler monkeys, genus *Alouatta* (Platyrrhini, Anthropoidea), based on g1-globin pseudogene sequences. *Genetics and Molecular Biology*, **22**, 337–344.

Meireles, C.M., Czelusniak, J., Schneider, M.P.C., *et al.* (1999b). Molecular phylogeny of ateline New World monkeys (Platyrrhini, Atelinae) based on γ-globin gene sequences: evidence that *Brachyteles* is the sister group of *Lagothrix*. *Molecular Phylogenetics and Evolution*, **12**, 10–30.

Meldrum, D.J. (1990). New fossil platyrrhine tali from the Early Miocene of Argentina. *American Journal of Physical Anthropology*, **83**, 403–418.

Meldrum, D.J. (1993). Postcranial adaptations and positional behavior in fossil platyrrhines. In *Postcranial Adaptations in Nonhuman Primates*, ed. D.L. Gebo. DeKalb, IL: Northern Illinois University Press, pp. 235–251.

Meldrum, D.J. (1998). Suspension as a transitional behavior in the evolution of the platyrrhine prehensile tail. In *Primate Locomotion: Recent Advances*, ed. E. Strasser, J.G. Fleagle, A.L. Rosenberger & H. McHenry. New York, NY: Plenum Press, pp. 145–156.

Meldrum, D.J. & Kay, R.F. (1997a). *Nuciruptor rubricae*, a new pitheciin seed predator from the Miocene of Colombia. *American Journal of Physical Anthropology*, **102**, 407–427.

Meldrum, D.J. & Kay, R.F. (1997b). The postcranial skeleton of Miocene platyrrhine primates. In *Vertebrate Paleontology in the Neotropics*, ed. R.F. Kay, R.H. Madden, R.L. Cifelli & J.J. Flynn. Washington, DC: Smithsonian Institution Press, pp. 459–472.

Meldrum, D.J. & Lemelin, P. (1991). Axial skeleton of *Cebupithecia sarmientoi* (Pitheciinae, Platyrrhini) from the middle Miocene of La Venta, Colombia. *American Journal of Primatology*, **25**, 69–90.

Müller, K.-H. (1996). Diet and feeding ecology of masked titis (*Callicebus personatus*). In *Adaptive Radiations of Neotropical Primates*, ed. M.A. Norconk, A.L. Rosenberger & P.A. Garber. New York, NY: Plenum Press, pp. 383–401.

Orlosky, F.J. (1973). Comparative dental morphology of the extant and extinct Cebidae. Unpublished PhD thesis, University of Washington.

Palacios, E., Rodriguez, A. & Defler, T.R. (1997). Diet of a group of *Callicebus torquatus lugens* (Humboldt, 1812) during the annual resource bottleneck in Amazonian Colombia. *International Journal of Primatology*, **18**, 503–522.

Ray, D.A., Xing, J., Hedges, D.J., *et al.* (2005). *Alu* insertion loci and platyrrhine primate phylogeny. *Molecular Phylogenetics and Evolution*, **35**, 117–126.

Rosenberger, A.L. (1975). On the distinctiveness of *Xenothrix*. *American Journal of Physical Anthropology*, **42**, 326.

Rosenberger, A.L. (1979). Phylogeny, evolution and classification of New World monkeys (Platyrrhini, Primates). Unpublished PhD thesis, City University of New York.

Rosenberger, A.L. (1981). Systematics: the higher taxa. In *Ecology and Behavior of Neotropical Primates, Vol. I*, ed. A.F. Coimbra-Filho & R.A. Mittermeier. Rio de Janeiro: Academia Brasileira de Ciências, pp. 9–27.

Rosenberger, A.L. (1992). The evolution of feeding niches in New World monkeys. *American Journal of Physical Anthropology*, **88**, 525–562.

Rosenberger, A.L. (2000). Pitheciinae. In *Encyclopedia of Human Evolution and Prehistory, Second Edition*, ed. E. Delson, I. Tattersal & J. van Couvering. New York, NY: Garland Publishing Co., pp. 562–563.

Rosenberger, A.L. (2002). Platyrrhine paleontology and systematics: the paradigm shifts. In *The Primate Fossil Record*, ed. W.C. Hartwig. Cambridge: Cambridge University Press, pp. 151–159.

Rosenberger, A.L., Setoguchi, T. & Shigehara, N. (1990). The fossil record of callitrichine primates. *Journal of Human Evolution*, **19**, 209–236.

Schneider, H., Canavez, F.C., Sampaio, I., *et al.* (2001). Can molecular data place each Neotropical monkey in its own branch? *Chromosoma*, **109**, 515–523.

Schneider, H., Sampaio, I., Harada, M.L., *et al.* (1996). Molecular phylogeny of the New World monkeys (Platyrrhini, Primates) based on two unlinked nuclear genes: IRBP interon 1 and ε-globin sequences. *American Journal of Physical Anthropology*, **100**, 153–179.

Schrago, C.G. (2007). On the time scale of New World primate diversification. *American Journal of Physical Anthropology*, **132**, 344–354.

Singer, S.S., Schmitz, J., Schwiegk, C., *et al.* (2002). Molecular cladistic markers in New World monkey phylogeny (Platyrrhini, Primates). *Molecular Phylogenetics and Evolution*, **26**, 490–501.

Stirton, R.A. (1951). Ceboid monkeys from the Miocene of Colombia. *University of California Publications in Geological Sciences*, **20**, 315–356.

Stirton, R.A. & Savage, D.E. (1951). A new monkey from the La Venta late Miocene of Colombia. *Compilaciones de los Estudios Geológicos Oficiales en Colombia, Bogotá*, **7**, 345–346.

Takai, M., Anaya, F., Suzuki, H., *et al.* (2001). A new platyrrhine from the middle Miocene of La Venta, Colombia, and the phyletic position of Callicebinae. *Anthropological Science (Japan)*, **109**, 289–307.

Tejedor, M.F. (2005a). New fossil platyrrhine from Argentina. *Folia Primatologica*, **76**, 146–150.

Tejedor, M.F. (2005b). New specimens of *Soriacebus adrianae* Fleagle, 1990, with comments on pitheciin primates from the Miocene of Patagonia. *Ameghiniana*, **41**, 249–251.

van Roosmalen, M.G.M., Mittermeier, R.A. & Fleagle, J.G. (1988). Diet of the northern bearded saki (*Chiropotes satanas chiropotes*): a neotropical seed predator. *American Journal of Primatology*, **14**, 11–35.

Wildman, D.E., Jameson, N.M, Opazo, J.C., *et al.* (2009). A fully resolved genus level phylogeny of neotropical primates (Platyrrhini). *Molecular Phylogenetics and Evolution*, **3**, 694–702.

Williams, B.A. & Kay, R.F. (1995). The taxon Anthropoidea and the crown clade concept. *Evolutionary Anthropology*, **3**, 188–190.

Williams B.A., Kay R.F. & Kirk E.C. (2010). New perspectives on anthropoid origins. *Proceedings of the National Academy (USA)*, **107**, 4797–4804.

Chapter 2

The misbegotten: long lineages, long branches and the interrelationships of *Aotus*, *Callicebus* and the saki–uacaris*

Alfred L. Rosenberger & Marcelo F. Tejedor

Introduction

An important shift in thinking has become cause for renewed scrutiny concerning the course of platyrrhine evolution and the shape of New World monkey (NWM) classification. For the first time in nearly 200 years, *Aotus* is being moved across a major taxonomic divide. It is being considered as a genus aligned with marmosets and tamarins rather than titi monkeys and saki–uacaris and, more generally, pitheciids. The emergence of this debate reflects the impact of molecular cladistics since the early 1990s. The conversation has turned from a prolonged controversy (e.g. Rosenberger 1981, 2002) over marmosets and tamarins, a dispute that was fundamental to modernizing our views of NWM evolution and appears to be resolved *for the moment*. Now, beginning with the successful molecular cladistic analysis of Schneider *et al.* (1993), the deliberation is over *Aotus*: is it a pitheciid, an atelid, a stem pitheciid or a cebid? The prevailing opinions that warrant close examination are the last and the first – *Aotus* is either a cebid or a pitheciid. However, the crux of the matter is that the hypothesis of *Aotus* as a "cebid" is based almost entirely on genes; the idea that it is a pitheciid is based entirely on morphology.

While some morphologists align *Aotus* more closely with the molecular trees (e.g. Kay 1990; Horovitz 1999; see also Meldrum & Kay 1997; Kay *et al.* 1998), we believe this assessment does not adequately account for anatomical evidence bearing on *Aotus*, *Callicebus*, *Pithecia*, *Chiropotes* and *Cacajao*, no less their fossil relatives (Rosenberger 2002). How is the *Aotus* matter different? *Aotus* has rested comfortably near *Callicebus* in morpho-space ever since higher level classifications of the platyrrhines were developed in the early 1800s (Rosenberger 1981). There was not a hint that *Aotus* could be related to anything but a pitheciid or ateline until the 1990s (see Tejedor 2001). As a consequence, morphologists challenged by the molecular evidence regarding *Aotus* have no fallback position from which our information can be reinterpreted. The molecules conflict with the morphology rather directly. The discord goes beyond that: the molecules clash with ecology and behavior. *Aotus* and *Callicebus* are bound together by a unique combination of attributes: social monogamy, biparental care with extensive input by males, no sibling care, long call advertising, territoriality, locomotion and feeding (e.g. Robinson *et al.* 1987; Wright 1996; Fernandez-Duque 2007; Norconk 2011). One is hard-pressed to find any two genera of modern NWM more alike than *Aotus* and *Callicebus*, except for the obvious dyads that split cladistically relatively recently, and only arguably into distinct genera – *Callithrix* and *Cebuella*, and *Cacajao* and *Chiropotes*.

Given the narrow scope of this chapter, a complete analysis of the problem is impossible. For one, it would require a full explication of the fossil record pertaining to *Aotus*, *Callicebus* and the saki–uacaris. Instead, to introduce these taxa in condensed form, and to clarify our use of taxonomic terms, we present a classification of pitheciids (Table 2.1), extending the scheme of Rosenberger *et al.* (1990). In an effort to summarize our assessment, we also advocate a stance rather than illuminate the conjectures and refutations: *Aotus* is a pitheciid, not a cebid. Overall, our aim is to present a synopsis of three aspects of the problem that must be accounted for in order to unravel the *Aotus* puzzle: (1) the morphological evidence linking *Aotus*, *Callicebus* and saki–uacaris cladistically; (2) a critical assessment of the molecular evidence; and (3) a synthesis of the evolution of pitheciid feeding adaptations which we further promote as heuristic evidence that *Aotus* is, in fact, pitheciid – phylogeny and adaptation are two sides of the same evolutionary coin, to paraphrase Fred Szalay.

Pithecia, *Chiropotes* and *Cacajao* – at the end of a morphocline

The craniodental morphology, our focus, leads to the following: *Aotus* is most closely related to *Callicebus*, and *Callicebus* (via the ancestral morphotype of *Aotus* and *Callicebus*) is linked with saki–uacaris. Dentally, *Pithecia*, *Chiropotes* and *Cacajao* have effectively defined pitheciids because the characters of saki-uacaris are striking structurally and adaptively while also being cladistically informative (Figure 2.1).

Evolutionary Biology and Conservation of Titis, Sakis and Uacaris, eds. Liza M. Veiga, Adrian A. Barnett, Stephen F. Ferrari and Marilyn A. Norconk. Published by Cambridge University Press.

Pithecia is the most primitive craniodentally on the whole, and thus the foundation for comparisons with other forms. *Chiropotes* and *Cacajao* present an exaggerated version of the pattern. But the greater challenge is to sort out how their anatomy evolved transformationally: how and why did platyrrhines arrive at a *Pithecia*-like pattern? Kinzey (1992), Rosenberger (1992) and Meldrum & Kay (1997) presented workable models of this transition, blending the moderns with information from the fossil record, which we extend here.

Table 2.1 A partial classification of living and fossil pitheciids, mostly to the genus level, based on Rosenberger *et al.* (1990) and Rosenberger (2002). Dagger symbols mark the fossils.

Family Pitheciidae Gray, 1849 (Mivart, 1865)
- Subfamily Pitheciinae Gray, 1849 [pitheciins]
  - *Pithecia* Desmarest, 1820
  - *Chiropotes* Lesson, 1840
  - *Cacajao* Lesson, 1840
  - †*Cebupithecia* Stirton & Savage, 1951
  - †*Nuciruptor* Meldrum & Kay, 1997
  - †*Proteropithecia* Kay *et al.*, 1998
  - Tribe Soriacebina, Rosenberger *et al.*, 1990
    - †*Soriacebus* Fleagle *et al.*, 1987
    - †*Mazzonicebus* Kay, 2010
- Subfamily Homunculinae [homunculins]
  - *Callicebus* Thomas, 1903
  - †*Homunculus* Ameghino, 1891
  - †Miocallicebus Takai *et al.*, 2001
  - †*Aotus dindensis* Setoguchi & Rosenberger, 1987
  - †*Tremacebus* Hershkovitz, 1974
- Subfamily indet.
  - †*Xenothrix* Williams & Koopman, 1952
  - †*Lagonimico* Kay, 1994
  - †*Carlocebus* Fleagle, 1990

Pithecia (Figures 2.1, 2.2) has a procumbent, wedge-like lower incisor battery, piercing canines and rugose cheek teeth, a combination that has been explained cogently as mechanical adaptations to hard-fruit harvesting and seed-eating (see Rosenberger & Kinzey 1976; Kay 1990; Kinzey 1992; Rosenberger 1992; Martin *et al.* 2003; Norconk *et al.*, Chapter 6).

The most obvious morphological link between the saki–uacaris and *Callicebus* and *Aotus* involve incisor and canine morphology (Figures 2.1 and 2.3). Although the gross anatomy of the canine and postcanine teeth of *Callicebus* bears little direct resemblance to pitheciids, the lower incisors demonstrate an uncanny likeness. They are tall, narrow and compressed together in an arch, but they do not jut out and are not shaped into the chisel-like apical edge of saki–uacaris. Upper central incisors of *Callicebus* and *Pithecia* also have an unusual lingual tubercle on the cingulum, which is not found in other platyrrhines. These features formed the beginnings of the cladistic link between *Callicebus* and pitheciids (Rosenberger 1977, 1981; Ford, 1986; Kinzey 1992; Meldrum & Kay 1997; Kay *et al.* 1998; but see Kay 1990 for a different view). The hypothesis was extended and confirmed by behavior and ecology (see Robinson *et al.* 1987; Kinzey 1992; Norconk 2011) and a host of molecular studies (see below).

For additional perspective on the functional significance of the incisor–canine complex of *Callicebus* as a cladistic link to pitheciins, the comments of Kinzey (1977, p. 140) concerning the feeding behavior of wild *Callicebus* are especially pertinent:

> Although most fruit appeared to be placed in the corner of the mouth where canine or premolars apparently tore off the husk or removed the edible pulp, a different method was used to obtain the edible portion of palm fruit [the second ranked food source]. The fruit was held between the two hands and the upper and lower incisors were used together to scrape the thin layer of hard pericarp from the pith. This behaviour very well may have accounted for the characteristic wear previously noted on *C. torquatus* incisors ...

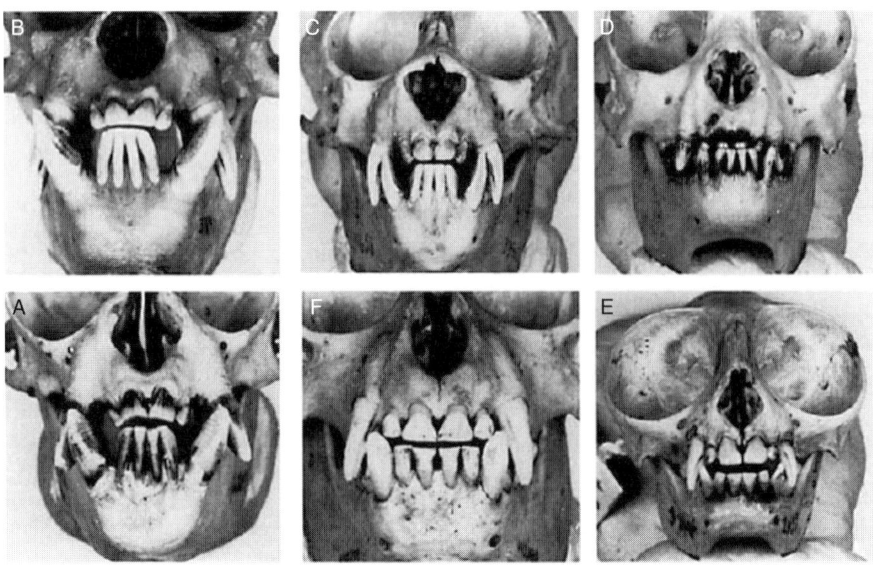

Figure 2.1 Anterior teeth of modern pitheciids and *Ateles*. Clockwise from bottom left: (a) *Cacajao melanocephalus*; (b) *Chiropotes satanas*; (c) *Pithecia pithecia*; (d) *Callicebus torquatus lugens*; (e) *Aotus grisimembra*; (f) *Ateles belzebuth hybridus*. Compared to *Ateles*, notice the reduction in the second upper incisor relative to the central incisor in *Aotus*, *Callicebus* and saki–uakaris, also the everted lower canines of *Aotus* (essentially vertical in *Ateles*), and the high-crowned upper central incisors, resembling pitheciines. The scoop-like, compressed lower incisor battery of all pitheciids, which is more proclivous in pitheciines than in homunculines, is produced by the "non-verticality" of the lateral lower incisor in this view. The face of *Pithecia* may be the most primitive form among saki–uacaris, in general (Kinzey, 1992).

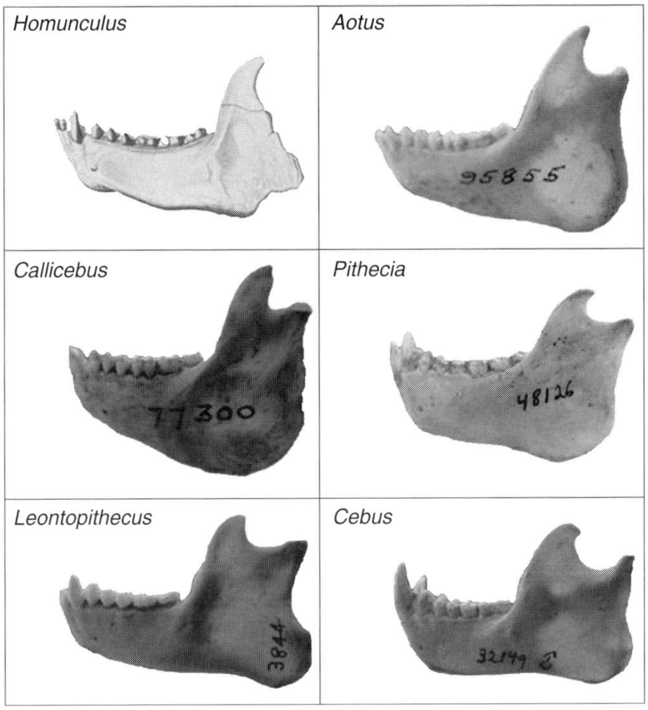

Figure 2.2 Lateral jaw profiles of selected platyrrhines brought to approximately the same length. Clockwise from top left: the Miocene fossil *Homunculus patagonicus* (from Bluntschli, 1931); *Aotus* sp., *Pithecia pithecia*, *Cebus capucinus*, *Leontopithecus rosalia*, *Callicebus* sp. Cebids typically have jaws that do not deepen much or at all, nor do they flare out posteriorly, as in pitheciids and atelids, which is the derived condition for platyrrhines. *Homunculus* and *Aotus* may represent the morphotype pitheciid condition, while the gonial inflation of *Callicebus* is derived (but see Figure 2.4) in one direction; the anteriorly deep and robust jaw of *Pithecia* (and other saki–uacaris) is derived in a different direction. The unusually elevated mandibular condyle in *Aotus* and *Callicebus* (likely in *Homunculus* also) is evident.

Kinzey implies here (and ALR, who worked with Kinzey in the field on the *Callicebus* project, confirms) that it was the hardness of the substrate beneath the pericarp that wore the incisors of *Callicebus* so heavily. Thus Kinzey's remarks (1977) anticipated the preadaptive, ecophylogenetic nature of *Callicebus* morphology and behavior as a prelude to the highly specialized, prying, gouging and stripping activities of pitheciines. While the incisors of *Aotus* are superficially different in some details, their battery is also well designed for gouging and stripping hard husks (Figure 2.3) in the same manner, as Kinzey (1974) pointed out. *Aotus* incisors are modestly high-crowned and somewhat inclined. The principal difference from *Callicebus* and pitheciines is that the incisors of some forms of *Aotus* are relatively wide at the apical edge, although this variation is not so impressive in all owl monkey taxa (Figure 2.3).

Callicebus and *Aotus* – novelties among the nondescript

Methodologically, there are two bodies of morphological evidence that speak directly to the narrow affinities of *Aotus*: studies employing parsimony algorithms and studies using non-algorithmic character analyses. The solutions of some of the older parsimony studies may (e.g. Ford 1986) or may not (e.g. Kay 1990) resemble the results from conventional character analysis, but the recent ones do not (e.g. Horovitz 1999; Horovitz *et al.* 1998). Some of the possible reasons for these discrepancies have been discussed elsewhere (Rosenberger 2002). Additional insight into the limitations of the parsimony method has highlighted the inherent potential for taxonomic sampling artifacts to bias results, whether the evidence is molecules or morphology (e.g. Rosenberger & Kearney 1995; Collins 2004; Sargis 2007; Silcox 2007; Matthews & Rosenberger 2008). In any event, the morphology-based parsimony studies that resemble molecular results offer only tepid support for the *Aotus*-cebid hypothesis.

Our character analysis relies on the morphology of the mandible, incisors, canine, face and auditory bulla. The lateral profile of mandible in NWM discriminates cebids from atelids and pitheciids (Figure 2.2; Rosenberger 1977, 1979). Cebids have a relatively horizontal body that does not expand inferiorly and posteriorly at the angle of the mandible. Widespread among early anthropoids and other primates, this condition is probably primitive among platyrrhines. The atelid–pitheciid state, a posteriorly deepening corpus with an inflated, rounded mandibular angle is very likely derived. The *Pithecia* mandible, while preserving the dilation posteriorly, is derived relative to the ancestral atelid–pitheciid state in being much deeper and thicker anteriorly in connection with their advanced incisor–canine morphology and its derived, U-shaped jaws.

Homunculus, an early Miocene pitheciid, closely resembles the typical pattern of *Aotus*. We take this pattern as the morphotypic condition of pitheciids and atelids, evidence that *Aotus* is related to pitheciids and not to cebids. However, a deeper set of resemblances is also shared by *Aotus* and *Callicebus*. Figure 2.4 shows individual mandibles belonging to three species of *Aotus* and *Callicebus*. It illustrates a variation in *Aotus* that overlaps a generic hallmark of *Callicebus*, enormous inflation of the mandibular angle.

Aotus and *Callicebus* also share a high temporomandibular joint, produced by a tall, anteroposteriorly short mandibular ramus that rises well above the tooth row. The combination of a high jaw joint and deep gonial region, where the superficial masseter muscle inserts, indicates a relatively vertical orientation of the muscle, long fibers and a relatively vertically oriented adductor force generated by them during jaw closing. Differences from cebids are evident (Figure 2.2). There the ramus tends to be low and long in the anteroposterior axis. The relatively squat ramus is especially typical of cebines. This pattern is more consistent with a temporalis-dominated feeding system applying forces in a relatively horizontal direction (Anapol & Lee 1994). Thus, the similarities of *Aotus* and *Callicebus* in form and function are themselves unique and distinguished from patterns found among cebids.

The morphology of the auditory bulla in *Aotus* and *Callicebus* is highly distinctive and unmatched by other

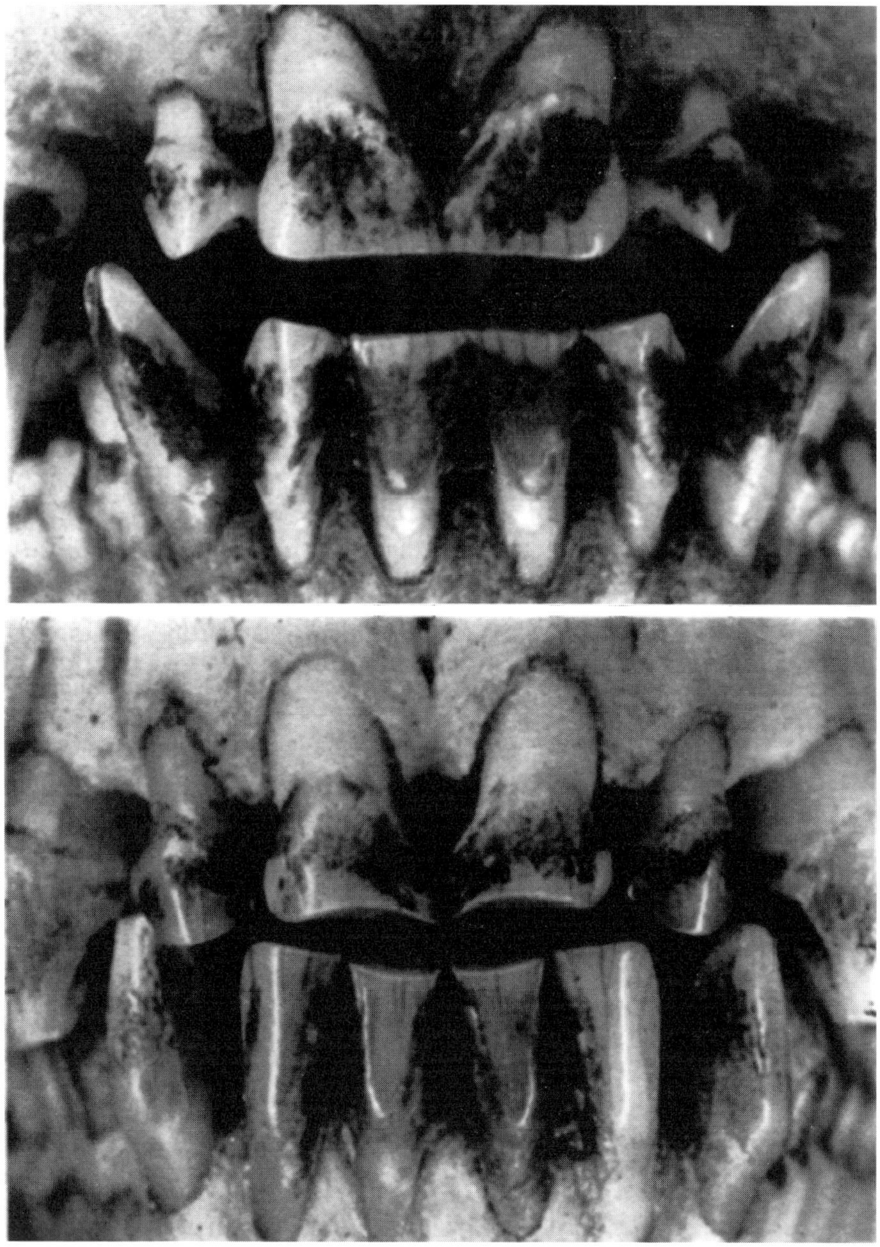

Figure 2.3 Close-ups of the anterior teeth of *Aotus trivirgatus* (top) and *Callicebus torquatus* (bottom), brought to same approximate bi-canine width. Note the relative narrowness of the lower incisor span of *Aotus* and *Callicebus* and the everted lower canines of *Aotus*, essentially absent in *Callicebus* due to extreme crown reduction. Compare the incisor proportions with the example of *Aotus* in Figure 2.1, and also the moderately everted canines of *Callicebus* in Figure 2.1.

platyrrhines (Figure 2.5). Their bullae are quite inflated and composed of a broadly distributed field of densely cancellous bone. There is an unusual, enlarged anterolateral compartment, a lobe-like extension in front of the acoustic meatus that encroaches on the temporomandibular joint. Unlike the teardrop outline of cebids, which is probably primitive for platyrrhines, the bullae of *Aotus* and *Callicebus*, like *Pithecia*, are also irregularly shaped and broad posteriorly. So, while pitheciids may be derived in overall bullar shape, the rare details found exclusively in *Aotus* and *Callicebus* are probably joint synapomorphies. Little is known about bullar functional morphology, but here the spongy bone may help dampen vibration, perhaps insulating the middle ear from bone conducted sound. An adaptive connection with the stentorian vocalizations of *Aotus* and *Callicebus*, which are prodigious especially in relation to their small body size, may have been a selective factor.

In addition to basic phenetic similarities in the crania of *Aotus* and *Callicebus*, including many features probably primitive for platyrrhines, their joint canine and facial morphologies are distinctive and probably derived. Canines are moderate (*Aotus*) and very small (*Callicebus*) in size. In all cebid genera, male canines are large and projecting, and in callitrichines even female canines are large tusks. The cebid pattern may be derived among NWM, which is not consistent with a placement of *Aotus* within the clade. *Aotus* also shares no derived cebine features; no vaulted frontal bone, no narrow

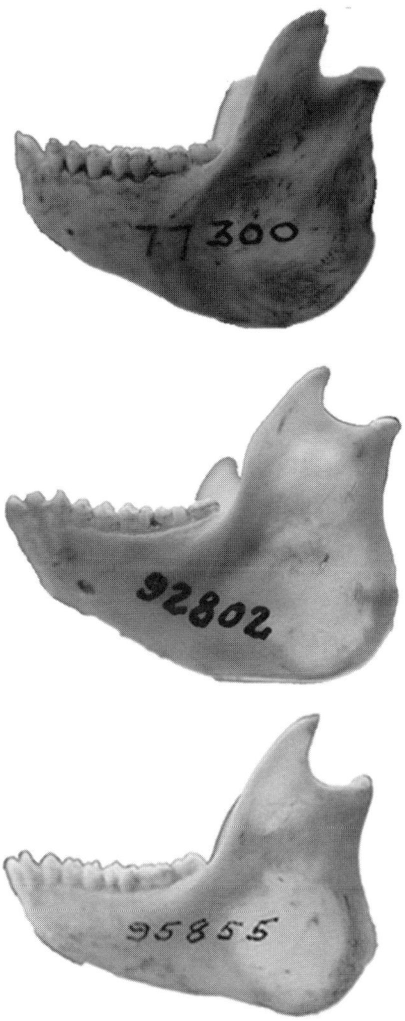

Figure 2.4
A comparison of the "typical" jaw profile in genus *Callicebus* (top, *C. torquatus*) with individual variations found in *Aotus* (middle, *A. nigriceps*; bottom, *A. infulatus*), brought to the same approximate length. The middle image is cropped slightly at the base, where it was embedded in clay.

nasals from base to tip, no wide snout, no anteroposteriorly long mandibular ramus.

It is striking that *Aotus* and *Callicebus* have a combination of moderate-to-minuscule canines (see Kay *et al.* 1988), correspondingly reduced faces with abbreviated premaxillae, relatively tall incisors, compact incisor–canine batteries and parabolic jaws. This picture also differs from our interpretation of *Homunculus* (see Tejedor & Rosenberger 2008), which is in many respects primitive for pitheciids. It had more V-shaped jaws, a precanine diastema, staggered incisors and a large snout. The distribution of characters suggests that homunculines (e.g. *Aotus*, *Callicebus* and allies; Table 2.1) and pitheciines evolved from a pattern like this in two distinct directions.

The best explanation we have for the *Aotus*–*Callicebus* pattern is that it reflects a structural compromise between adaptations for feeding and mating (Rosenberger *et al.* 1990). The incisor battery is tuned to fruit harvesting while the canine complex has been selected for a low-crowned form of monomorphism that evolved in connection with a pair-bonded monogamous mating system, a source of selection that compromises, or constrains, the dietary imperative. Metrically, the canines of *Aotus* and *Callicebus* are the least dimorphic among modern platyrrhines in their body size class (Kay *et al.* 1988), and they are clearly distinguished from other modern species that have monomorphic canines by their anatomy and biological roles. Callitrichines, for example, have large, same-sized canines in males and females and use them in agonistic situations, manifesting an altogether different socio-sexual context. This makes it highly likely that the contrasting pattern shared by *Aotus* and *Callicebus* is homologously derived (Rosenberger *et al.* 1990).

The molecular evidence – the long lineage hypothesis meets long branch attraction

We are cognizant of the impressive number of molecular cladistic studies since the 1990s, which have favored a linkage between *Aotus* and cebids. Our reading of the molecular support for this hypothesis is that it presents several problematic outcomes and contingencies. (1) The precise location of *Aotus* within the cebid branching sequence is not often replicated, and polytomies involving its position are not unusual. (2) The *Aotus* linkage within the cebid clade occurs with quantifiably low levels of support. (3) Rooting *Aotus* with the cebids often coincides with a reduced level of support for more distal clades that are very strongly supported by morphology and molecules alike. (4) The tendency is for rooting *Aotus* within the cebids adjacent to taxa that share a particular evolutionary history that may make them prone to skewed molecular results – they are long-lived lineages (Rosenberger 1979, *et seq.*) susceptible to a methodological artifact known as long branch attraction. A speciation-level process, such as reticulation, is another source of low resolution in phylogeny reconstruction (e.g. Doolittle 1999) that may have to be considered here.

Figure 2.6 summarizes the quantitative evidence backing platyrrhine clades in an array of molecular studies. They do not produce symmetrical cladograms: relationships differ; polytomies appear in different combinations, at different nodes and in different proportions relative to dichotomies; and higher-level linkages often differ. The studies from which these data were generated also tend to provide several alternative results, concluding with or without a final, "preferred" cladogram. Thus, in the absence of across-the-board consensus within and among these reports, the chart is but one way to quantitatively assess how well the *Aotus*-cebid hypothesis fares relative to other platyrrhine groupings, while the qualitative points mentioned above suggest additional reasons why caution is called for.

The *Aotus*-cebid clade ranks lowest overall in node support by comparison with the other four groups (Figure 2.6). Callitrichines and pitheciids are the clades whose monophyly is most consistently supported. The atelids vary somewhat. But surprisingly, the cebines are not linked with high

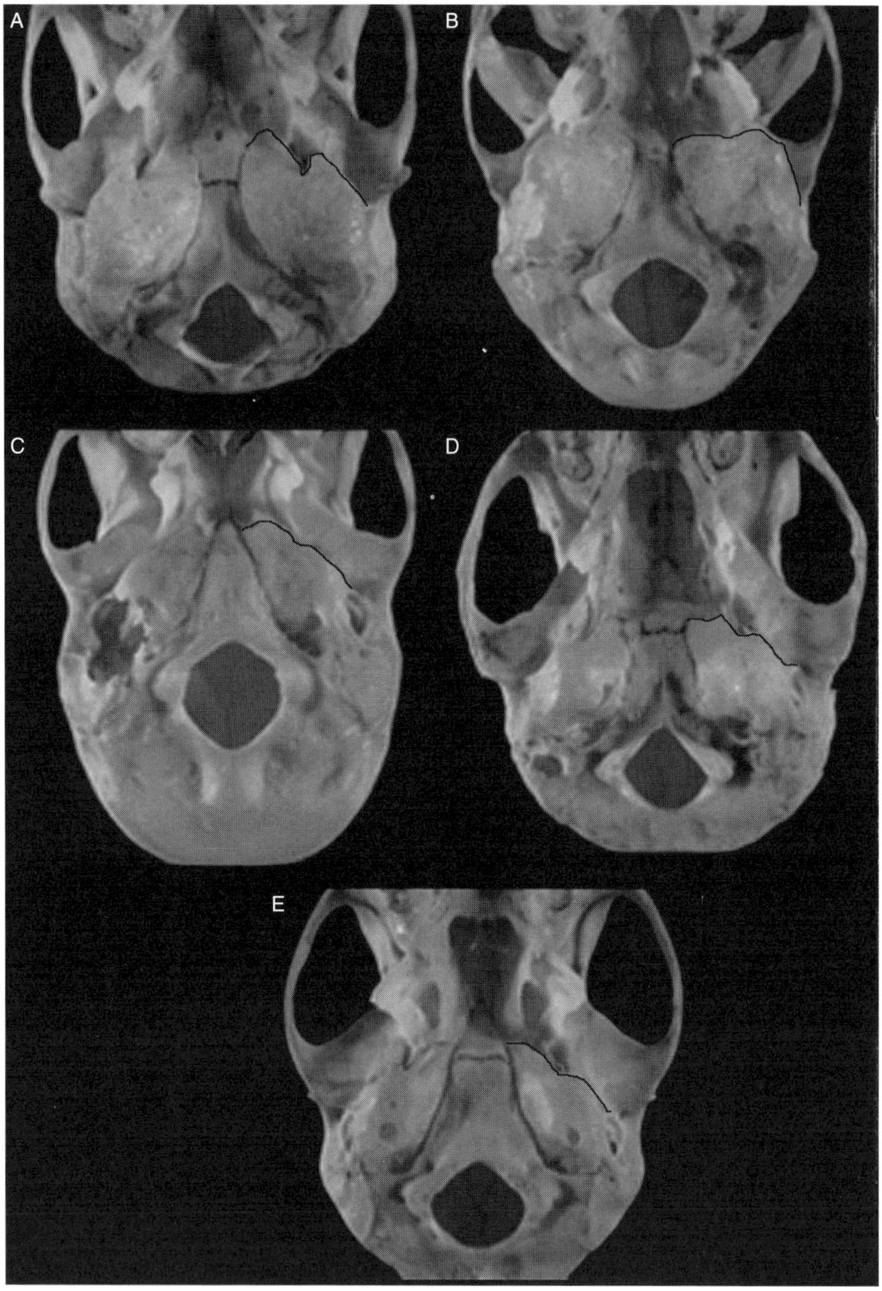

Figure 2.5 The auditory regions of selected platyrrhines, brought to approximately the same skull lengths. The anterolateral margin is outlined from the auditory meatus to the anteromedial pole. (a) *Aotus*, (b) *Callicebus*, (c) *Saimiri*, (d) *Pithecia*, (e) *Cebus*. The teardrop-shaped auditory bullae of cebids resembles archaic Old World anthropoids and is probably primitive for NWM. The irregular shape of the pitheciids, which is very wide posteriorly at the level of the eardrum, is a derived pattern.

reliability, although this is a very securely established node by morphology and molecules. Actually, the *Saimiri–Cebus* link is a triumph of modern phylogenetic reasoning, for these animals are dramatically different in so many ways. We suggest that their depressed level of support in the molecular studies is a local artifact, directly influenced by misclassification of *Aotus*.

An explanation for the persistence of a low-resolved solution for *Aotus* is that its position is simply an error that repeatedly affects the same combination of platyrrhine genera due to the long branch attraction phenomenon. Felsenstein (1978) showed that there is a high likelihood that the terminal taxa of relatively long branches will come to resemble one another due to convergence when the time interval separating their initial differentiation is relatively short (Figure 2.7). Under a random model of nucleotide evolution there is a high probability this can occur, as in theory there are only four possible character state changes, and fewer still in practise. The more time available for evolution following a limited amount of genetic separation at the origin, the more likely the character states of lineages will converge. And, as the chemistry is the same, there is no way of knowing, say, if two Gs in the same position are homologous. Bergsten (2005) showed empirically that long branch attraction is a real phenomenon, arguing that

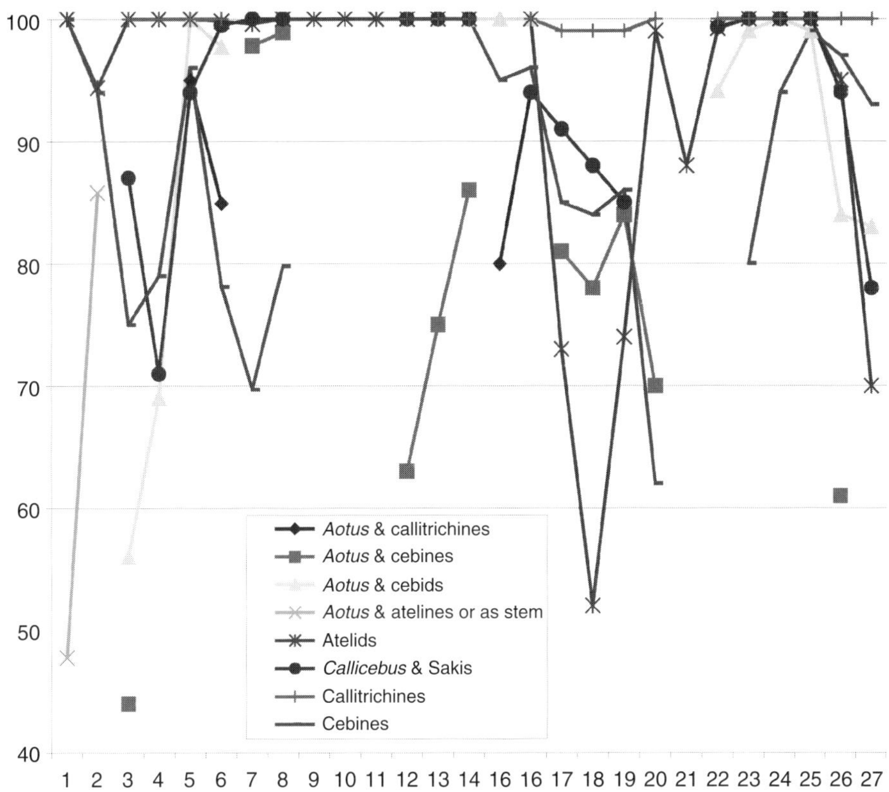

Figure 2.6 Line chart showing measures of bootstrap support for monophyletic groups returned in molecular studies of platyrrhine interrelationships based on parsimony analyses. See also text. The various "*Aotus* clades" are shown, defined in the inset legend, as in one placement that posits *Aotus* as the stem lineage of crown platyrrhines. The *Aotus* links have consistently lower and more variable support values across these studies. Sources: Canavez *et al.* (1999), Goodman *et al.* (1998), Harada *et al.* (1995), Opazo *et al.* (2006), Porter *et al.* (1997), Prychitko *et al.* (2005), Ruiz-Garcia & Alvarez (2003), Schneider *et al.* (1993, 1996, 2001), Steiper & Ruvolo (2003), and von Dornum & Ruvolo (1999).

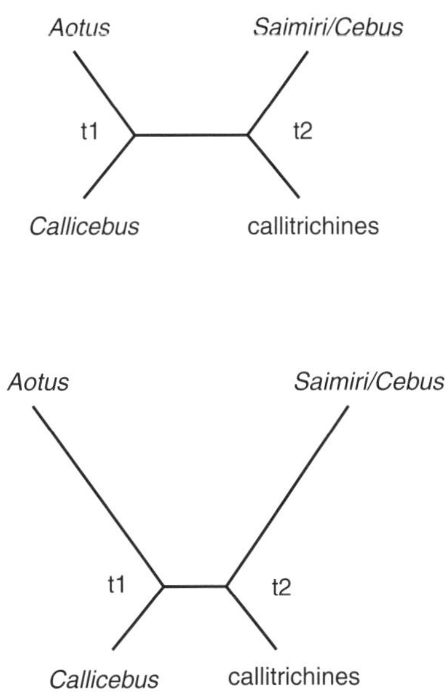

Figure 2.7 The long branch attraction artifact, modified from Felsenstein (1978). Two unrooted tree models of relationships are shown. If the temporal separation (t1–t2) between the splitting times of clades is relatively short while the descendant lineages in question have evolved for a long period of time, random selection of nucleotide substitutions will perforce result in proportionately large amounts of convergence because the pool of potential states changes remains small. Parsimony trees are therefore prone to mistake convergent similarities as homologous-derived features.

it is commonplace when the in-group also contains relatively short branches.

There are several long generic lineages, as demonstrated by the fossil record (e.g. Rosenberger 1979; Delson & Rosenberger 1984; Setoguchi & Rosenberger 1987; Rosenberger *et al.* 2009) and also the molecules (e.g. Opazo *et al.* 2006), as well as some that are very likely short (*Callithrix* and *Cebuella*; *Chiropotes* and *Cacajao*). The *Aotus* lineage, represented by a congeneric species in the middle Miocene La Venta fauna of Colombia and by *Tremacebus* in the early Miocene of Patagonia, is an established long lineage. The *Saimiri* lineage is represented by another La Ventan species and also appears to be represented in the early Miocene in Patagonia by *Dolichocebus*. In more general terms, the cebine lineage is certainly confirmed in the late–early Miocene by *Killikaike* (Tejedor *et al.* 2006). Thus the genera most closely aligned with *Aotus* in molecular studies are the most basal branches of the cebid clade and are each long-lived, thus increasing the potential for them to converge with the equally long-lived *Aotus* as an artifact: the algorithm is mistaking new analogies for derived homologies. Steiper & Ruvolo (2003), among others, have emphasized the possibility of a rapid differentiation of the early platyrrhine lineages. In other words, all the conditions Felsenstein (1978) predicted as potentially troublesome may align here. Long branch attraction may also explain the cladistic noise vexing Schneider *et al.* (2001) in their assessment of molecular studies, wherein they state: "… two major points regarding the branching pattern of the most ancient lineages remain to be clarified: (1) what is the exact branching pattern of *Aotus*, *Cebus*, *Saimiri* and the small callitrichines? (2) Which two of the three main lineages (pitheciids, atelids and cebids) are more closely related to one another?" *Aotus* may be the muddle in the middle.

Pitheciid evolution – an ecophylogenetic scenario

While the problem of *Aotus* cladistics deserves special attention as a decisive datum regarding the history of two large branches of the platyrrhines, the question of saki–uacari origins is also important. They are the only obligate seed-eaters among the living primates (Norconk *et al.*, Chapter 6). How did seed-eating evolve here?

It seems clear that the anterior teeth came first, but not in the radical configuration of modern pitheciins (Rosenberger *et al.* 1990; Kinzey 1992; Rosenberger 1992). As Kinzey, Rosenberger and colleagues inferred some years ago, there are probably related seed-eaters awaiting discovery in the fossil record that exhibit primitive "stages" of the mosaic rather than the primary tier features, e.g. occlusal flattening of P4s and the adjacent molars, crenulation of the cheek teeth, eversion of the canines, etc. *Soriacebus, Mazzonicebus* and *Homunculus* are three important examples that fit the prediction. All have saki-like attributes in the lower incisors and, in addition, *Soriacebus* and *Mazzonicebus* demonstrate robust canines and a wedge-shaped anterior lower premolar. A reinterpretation of *Homunculus* expands this argument with more anatomical detail on the anterior teeth and jaws (Tejedor & Rosenberger 2008). Discovery of *Proteropithecia* in Patagonia and *Nuciruptor* in Colombia (Meldrum & Kay 1997; Kay *et al.* 1998) also demonstrates later "stages", with more modern canine and postcanine teeth, but not the full-blown *Cebupithecia*- and saki-like low, corrugated crowns. This body of evidence represents one of the few cases among the primates where a model evolutionary sequence can be reconstructed from fossils towards the emergence of a new dietary adaptive zone.

What does this transformation mean in terms of feeding? Sclerocarpic foraging for hard, unripe fruits and for arils that coat large, hard seeds like a palm nut probably preceded obligate seed eating, perhaps as a way of minimizing competition with other sympatric platyrrhine frugivores that prefer juicy ripe fruit (Kinzey 1992; Rosenberger 1992; Norconk *et al.*, Chapter 6; see also Kinzey & Norconk 1990). As reported by Kinzey (1974, 1977), *Callicebus* uses its incisor teeth (and canines) as a rasp to remove aril from large, hard palm nuts, which also results in heavy tooth wear. We may logically interpret this as an anatomical–behavioral pattern more primitive than the pitheciid pattern.

Thus one can imagine the breaching impetus of obligate seed predators evolving as a new "processing image", extending the propensity to gouge and scrape. It could have begun with a *Callicebus*-like species – *Homunculus* would be the paleontological example – a relatively generalized fruit-feeder that finds ecological advantage by focusing on tough fruits with large seeds. At a medium-to-small body size, without exhibiting an acrobatic locomotor habit and a large home range socioecological strategy, they could perhaps have afforded (or been constrained) to eat – or were competitively advantaged by being able to eat – less ripe, less sugar-rich fruits than animals like *Ateles*. These pre-seed-eating pitheciids would have been cognitively disposed to finding and treating food objects with the anterior teeth, rather than seeking fruits of a smaller size and softer consistency that can be masticated by the postcanines without investing much energy in anterior-tooth processing. Eventually the lineage would evolve taller, more inclined incisors, better able to resist wear and to wedge leguminous pods apart with greater mechanical advantage. Adding modified canines and anterior premolars (and jaws, etc.), incorporating them as specialized puncturing and prising devices, would "complete" the transition to build a seed extraction platform.

Pitheciines may also have been preadapted to seed-eating because their digestive systems appear to be capable of processing low-quality foods, in a manner analogous to folivores (Norconk *et al.*, Chapter 6). Rosenberger *et al.* (2009) suggested that in addition to morphology, the geographical distribution of modern pitheciids, the feeding habits of *Aotus* and *Callicebus*, and especially the occurrence of homunculines, pitheciines and soriacebines in the remote, early and middle Miocene of Patagonia, suggests that the earliest pitheciids may have been "junk food" feeders living in low-productivity habitats unlike the lowland rainforests of Amazonia. They would have been adapted to eating leaves for a protein supplement, but most of the dietary needs probably came from unripe fruit or hard-husked fruit. This scenario emphasizes a year-round selective regime different from the fruit- and insect-rich environment of Amazonia.

Postscript – where does this leave us?

Aotus was once thought to be utterly nocturnal, *Callicebus* was thought to be tedious and saki–uacaris were thought to be way out there – rare and bizarre. None of these suppositions are correct. *Aotus* can be cathemeral, *Callicebus* is exciting and saki–uacaris are the living remains of a large radiation at the centre of platyrrhine evolution. Neither morphology nor molecules per se are what made these animals intriguing and interpretable. The revelations came from ecology, behavior and palaeontology. Learning why some answers diverge may be more interesting than the question that initially exposed their asymmetries. The *Aotus* conundrum may lead there. Meanwhile, as morphologists, we say to those who know the animals best – if you are interested in knowing who, phylogenetically, *Aotus* and *Callicebus* are, follow what Darwin did and watch what they do.

Acknowledgments

Thanks to the editors of this volume for inviting us to participate. A special thanks to Marilyn Norconk. We owe the point about reticulation to one of the reviewers. We thank the staffs and Departments of Mammalogy and Vertebrate Palaeontology, American Museum of Natural History, for supporting our research. MFT gratefully acknowledges fellowship awards given by the Wenner-Gren

Foundation for Anthropological Research and the New York Consortium in Evolutionary Primatology (NYCEP, funded by NSF DGE 0333415). ALR was supported by PSC-CUNY grants and a Tow Research Fellowship from Brooklyn College. Thanks also to Teresa Quaranta for assistance with illustrations.

Endnote

* This chapter is dedicated to Warren G. Kinzey, who put pitheciids on the map.

References

Anapol, F. & Lee, S. (1994). Morphological adaptations to diet in platyrrhine primates. *American Journal of Physical Anthropology*, **94**, 239–261.

Bergsten, J. (2005). A review of long branch attraction. *Cladistics*, **21**, 163–193.

Bluntschli, H. (1931). *Homunculus patagonicus* und die ihm zugereihten fossilfunde aus den Santa Cruz Schichten Patagoniens. Eine morphologische revision an hand der originaltücke in der sammlung Ameghino zu La Plata. *Morphologisches Jehrbuch LXVII (Goppert-Festschrift II)*, 811–892.

Canavez, F.C., Moreira, M.A.M., Ladasky, J.J., *et al.* (1999). Molecular phylogeny of new world primates (Platyrrhini) based on β2-microglobulin DNA sequences. *Molecular Phylogenetics and Evolution*, **12**, 74–82.

Collins, A.C. (2004). Atelinae phylogenetic relationships: the trichotomy revived? *American Journal of Physical Anthropology*, **124**, 285–296.

Delson, E. & Rosenberger, A.L. (1984). Are there any anthropoid primate "living fossils"? In *Living Fossils*, ed. N. Eldredge & S. Stanley. New York, NY: Springer-Verlag, pp. 50–61.

Doolittle, W.F. (1999). Phylogenetic classification and the universal tree. *Science*, **284**, 2124–2128.

Felsenstein, J. (1978). Cases in which parsimony or compatibility methods will be positively misleading. *Systematic Zoology*, **27**, 401–410.

Fernandez-Duque, E. (2007) Aotinae: social monogamy in the only nocturnal haplorhines. In *Primates in Perspective*, ed. C.J. Campbell, A. Fuentes, K.C. MacKinnon, M. Prager & S.K. Bearder. New York, NY: Oxford University Press, pp. 139–154.

Ford, S.M. (1986). Comment on the evolution of claw-like nails in callitrichids (Marmosets/Tamarins). *American Journal of Physical Anthropology*, **70**, 25–28.

Goodman, M., Porter, C.A., Czelusniak, J., *et al.* (1998). Toward a phylogenetic classification of primates based on DNA evidence complemented by fossil evidence. *Molecular Phylogenetics and Evolution*, **9**, 585–598.

Harada, M.L., Schneider, H., Schneider, M.P.C., *et al.* (1995). DNA evidence on the phylogenetic systematics of the new world monkeys: support for the sister-grouping of *Cebus* and *Saimiri* from two unlinked nuclear genes. *Molecular Phylogenetics and Evolution*, **4**, 331–349.

Horovitz, I. (1999). A phylogenetic study of the living and fossil platyrrhines. *American Museum Novitates*, **3269**, 40 pp.

Horovitz, I., Zardoya, R. & Meyer, A. (1998). Platyrrhine systematics: a simultaneous analysis of molecular and morphological data. *American Journal of Physical Anthropology*, **106**, 261–287.

Kay, R.F. (1990). The phyletic relationships of extant and extinct fossil Pitheciinae (Platyrrhini, Anthropoidea). *Journal of Human Evolution*, **19**, 175–208.

Kay, R.F., Johnson, D. & Meldrum, D.J. (1998). A new pitheciin primate from the middle Miocene of Argentina. *American Journal of Primatology*, **45**, 317–336.

Kay, R.F., Plavcan, J.M., Glander, K.E., *et al.* (1988). Sexual selection and canine dimorphism in New World monkeys. *American Journal of Physical Anthropology*, **77**, 385–397.

Kinzey, W.G. (1974). Ceboid models for the evolution of hominoid dentition. *Journal of Human Evolution*, **3**, 193–203.

Kinzey, W.G. (1977). Diet and feeding behavior of *Callicebus torquatus*. In *Primate Ecology: Studies of Feeding and Ranging Behaviour in Lemurs, Monkeys, and Apes*, ed. T.H. Clutton-Brock. New York, NY: Academic Press, pp. 127–151.

Kinzey, W.G. (1992). Dietary and dental adaptations in the Pitheciinae. *American Journal of Physical Anthropology*, **88**, 499–514.

Kinzey, W.G. & Norconk, M. (1990). Hardness as a basis of fruit choice in two sympatric primates. *American Journal of Physical Anthropology*, **81**, 5–15.

Martin, L.B., Olejniczak, A.J. & Maas, M.C. (2003). Enamel thickness and microstructure in pitheciin primates, with comments on dietary adaptations of the middle Miocene hominoid *Kenyapithecus*. *Journal of Human Evolution*, **45**, 351–367.

Matthews L.J. & Rosenberger A.L. (2008). Taxon combinations, parsimony analysis (PAUP*), and the taxonomy of the yellow-tailed woolly monkey, *Lagothrix flavicauda*. *American Journal of Physical Anthropology*, **137**, 245–255.

Meldrum, D.J. & Kay, R.F. (1997). *Nuciruptor rubricae*, a new pitheciin seed predator from the Miocene of Colombia. *American Journal of Physical Anthropology*, **102**, 407–427.

Norconk, M.A. (2011). Sakis, uakaris, and titi monkeys: behavioral diversity in a radiation of primate seed predators. In *Primates in Perspective* (2nd edn), ed. C.J. Campbell, A. Fuentes, K.C. MacKinnon, S.K. Bearder & R.M. Stumpf. New York, NY: Oxford University Press, pp. 123–139.

Opazo, J.C., Wildmana, D.E., Prychitko, T., *et al.* (2006). Phylogenetic relationships and divergence times among New World monkeys (Platyrrhini, Primates). *Molecular Phylogenetics and Evolution*, **40**, 274–280.

Porter, C.A., Czelusniak, J., Schneider, H., *et al.* (1997). Sequence from the 5′ flanking region of the ε-globin gene support the relationship of *Callicebus* with the Pitheciins. *American Journal of Primatology*, **48**, 69–75.

Prychitko, T., Johnson, R.M., Wildman, D.E., *et al.* (2005). The phylogenetic history of new world monkey β globin reveals a platyrrhine β to δ gene conversion in the atelid ancestry. *Molecular Phylogenetics and Evolution*, **35**, 225–234.

Robinson, J.G., Wright, P.C. & Kinzey, W.G. (1987). Monogamous cebids and their relatives: intergroup calls and spacing. In *Primate Societies*, ed. B.B. Smuts, D.L. Cheney, R.M. Seyfarth & T. Struhsaker. Chicago, IL: University of Chicago Press, pp. 44–53.

Rosenberger, A.L. (1977). *Xenothrix* and ceboid phylogeny. *Journal of Human Evolution*, **6**, 461–481.

Rosenberger, A.L. (1979). *Phylogeny, evolution and classification of new world monkeys (Platyrrhini, Primates).* Unpublished PhD thesis. City University of New York.

Rosenberger, A.L. (1981). Systematics: the higher taxa. In *Ecology and Behavior of Neotropical Primates*, ed. A.F. Coimbra & R.A. Mittermeier. Rio de Janeiro: Academia Brasileira de Ciencias, **1**, 9–28.

Rosenberger, A.L. (1992). Evolution of feeding niches in new world monkeys. *American Journal of Physical Anthropology*, **88**, 525–562.

Rosenberger, A.L. (2002). Platyrrhine paleontology and systematics: the paradigm shifts. In *The Primate Fossil Record*, ed. W.C. Hartwig. Cambridge: Cambridge University Press, pp. 151–159.

Rosenberger, A.L. & Kearney, M. (1995). The power of fossils, the pitfalls of parsimony – platyrrhine phylogeny. *American Journal of Physical Anthropology*, **38(S20)**, 130.

Rosenberger, A.L. & Kinzey, W.G. (1976). Functional patterns of molar occlusion in platyrrhine primates. *American Journal of Physical Anthropology*, **45**, 281–298.

Rosenberger, A.L., Setoguchi, T. & Shigehara, N. (1990). The fossil record of the callitrichines primates. *Journal of Human Evolution*, **19**, 209–236.

Rosenberger, A.L., Tejedor, M., Cooke, S.B., *et al.* (2009). Platyrrhine ecophylogenetics in space and time. In *South American Primates: Comparative Perspectives in the Study of Behavior, Ecology and Conservation*, ed. P. Garber, A. Estrada, J. Bicca-Marques, E. Heymann & K. Strier New York, NY: Springer, pp. 69–113.

Ruiz-Garçia, M. & Alvarez, D. (2003). RFLP analysis of mtDNA from six platyrrhine genera: phylogenetic inference. *Folia Primatologica*, **74**, 59–70.

Sargis, E.J. (2007). The postcranial morphology of *Ptilocercus lowii* (Scandentia, Tupaiidae) and its implications for primate supraordinal relationships. In *Primate Origins: Adaptations and Evolution*, ed. M.J. Ravosa & M. Dagosto. New York, NY: Springer, pp. 51–82.

Schneider, H., Canavez, F.C., Sampaio, I., *et al.* (2001). Can molecular data place each neotropical monkey in its own branch? *Chromosoma*, **109**, 515–523.

Schneider, H., Sampaio, I., Harada, M.L., *et al.* (1996). Molecular phylogeny of the New World Monkeys (Platyrrhini, Primates) based on two unlinked nuclear sequences: IRBP intron 1 and ε-globin sequences. *American Journal of Physical Anthropology*, **100**, 153–179.

Schneider, H., Schneider, M.P., Sampaio, I., *et al.* (1993). Molecular phylogeny of the New World monkeys (Platyrrhini, primates). *Molecular Phylogenetics and Evolution*, **2**, 225–242.

Setoguchi, T. & Rosenberger, A.L. (1987). A fossil owl monkey from La Venta, Colombia. *Nature*, **326**, 692–694.

Silcox, M.T. (2007). Primate taxonomy, plesiadapiforms, and approaches to primate origins. In *Primate Origins: Adaptations and Evolution*, ed. M.J. Ravosa & M. Dagosto. New York, NY: Springer, pp. 143–178.

Steiper, M.E. & Ruvolo, M. (2003). New world monkey phylogeny based on X-linked G6PD DNA sequences. *Molecular Phylogenetics and Evolution*, **27**, 121–130.

Tejedor, M.F. (2001). *Aotus* y los Atelinae: nuevas evidencias en la sistemática de los primates platirrinos. *Mastozoología Neotropical*, **8**, 41–57.

Tejedor, M. & Rosenberger, A.L. (2008). A neotype for *Homunculus patagonicus* Ameghino, 1891, and a new interpretation of the taxon. *PaleoAnthropology*, **2008**, 67–82.

Tejedor, M.F., Tauber, A.A., Rosenberger, A.L., *et al.* (2006). New primate genus from the Miocene of Argentina. *Proceedings of the National Academy of Science USA*, **103**, 5437–5441.

von Dornum, M. & Ruvolo, M. (1999). Phylogenetic relationships of the new world monkeys (Primates, Platyrrhini) based on nuclear G6PD DNA sequences. *Molecular Phylogenetics and Evolution*, **11**, 459–476.

Wright, P.C. (1996). The neotropical primate adaptation to nocturnality: feeding in the night monkey (*Aotus nigriceps* and *A.azarae*). In *Primates in Perspective*, ed. C.J. Campbell, A. Fuentes, K. MacKinnon, M. Panger & S. Bearder. New York, NY: Oxford University Press, pp. 369–382.

Chapter 3

A molecular phylogeography of the uacaris (*Cacajao*)

Wilsea M.B. Figueiredo-Ready, Horacio Schneider, Stephen F. Ferrari, Maria L. Harada, José Maria C. da Silva, José de Sousa e Silva Júnior & John M. Bates

Introduction

Traditionally, the phylogenetic relationships of platyrrhine taxa have been interpreted almost exclusively on the basis of morphological traits, and the classification of intrageneric forms in particular has relied heavily on the analysis of variation in the coloration of the pelage (Hershkovitz 1977, 1987; van Roosmalen *et al.* 2002). In the case of the Amazonian primates – of which the Pitheciini is arguably the most prominent group – the zoogeographical role of the region's rivers has been recognized since the nineteenth century (Wallace 1876). The implicit assumption is that Amazonian rivers function primarily as barriers to dispersal and gene flow. While this may be only partly true (Ayres & Clutton-Brock 1992; Peres *et al.* 1996; Ferrari 2004), the vast majority of species ranges are delimited by rivers to some extent.

Molecular techniques have brought new perspectives to both genetic and phylogenetic studies of many groups of organisms, and the primates are no exception. In fact, this approach has provided important insights into the biological diversity of the primates at all levels of organisation, ranging from phylogeny and taxonomy (Schneider *et al.* 2001; Boubli *et al.* 2008), to demography and breeding patterns (Fischer *et al.* 2004; Carnahan & Jensen-Seaman 2008; Miller-Butterworth *et al.* 2008). In some cases, molecular analyses have resulted in phylogenetic arrangements quite distinct from those produced by more traditional, morphology-oriented studies (e.g. Schneider *et al.* 2001). While controversial, such studies highlight the limitations of the more traditional, morphology-based methods, but more importantly, emphasize the need for a multidisciplinary approach, based on the systematic, integrated analysis of genetic, morphological, ecological and zoogeographic data, in the vein of the pioneering review of Rylands *et al.* (2000).

Despite the considerable recent scientific advances outlined in this book, *Cacajao* is still one of the least well-known platyrrhines, and the interpretation of intrageneric diversity has changed very little since the classic study of Hershkovitz (1987). This is due primarily to the marked paucity of specimens, localities and other data, especially considering the relatively ample distribution of the genus in the western Amazon basin, although Boubli *et al.* (2008) have recently reinterpreted the diversity of the black uacaris (*Cacajao melanocephalus* group) and described a new form.

In this chapter, a detailed analysis of the 1137 base pair-long sequence of the mitochondrial cytochrome *b* gene provides an additional baseline for the understanding of the phylogenetic and geographic relationships of the uacaris. The results not only provide new insights into the taxonomic relations of the different forms, but also have important implications for the conservation of their populations.

Methods

Tissue samples were obtained from 26 individuals collected at 15 localities (Figure 3.1), representing 6 forms of *Cacajao*: *Cacajao calvus calvus, Cacajao calvus novaesi, Cacajao calvus rubicundus, Cacajao calvus ucayalii, Cacajao melanocephalus* and *Cacajao ouakary* (Appendix 3.1). In most cases, samples were obtained from the dried skins and skulls of museum specimens, which were kindly provided by three institutions: the Goeldi Museum, the Field Museum of Natural History, and the University of São Paulo's Museum of Zoology (MZUSP). Fresh or frozen material was also obtained from the Federal University of Pará, the Museum of Vertebrate Zoology at UC-Berkeley and the American Museum of Natural History. Fifteen additional sequences of *Cacajao ayresi, C. calvus, C. melanocephalus* and *C. ouakary* from eight sites were obtained from Boubli *et al.* (2008) and included in the analyses. One specimen of *Chiropotes sagulatus* and sequences of *Pithecia monachus* were included as the outgroups.

The protocol for the extraction of DNA from museum specimens was developed by Figueiredo (2006: Appendix 3.2). For fresh specimens, DNA was extracted using the standard phenol–chloroform protocol of Hillis *et al.* (1996). For these specimens, the cytochrome *b* gene was amplified by polymerase chain reaction (PCR) using the primers L14727 and H15915 (Kocher *et al.* 1989) and a standard reaction protocol (Figueiredo 2006). However, degradation of the DNA of museum specimens impeded the amplification of the whole

Evolutionary Biology and Conservation of Titis, Sakis and Uacaris, eds. Liza M. Veiga, Adrian A. Barnett, Stephen F. Ferrari and Marilyn A. Norconk. Published by Cambridge University Press.

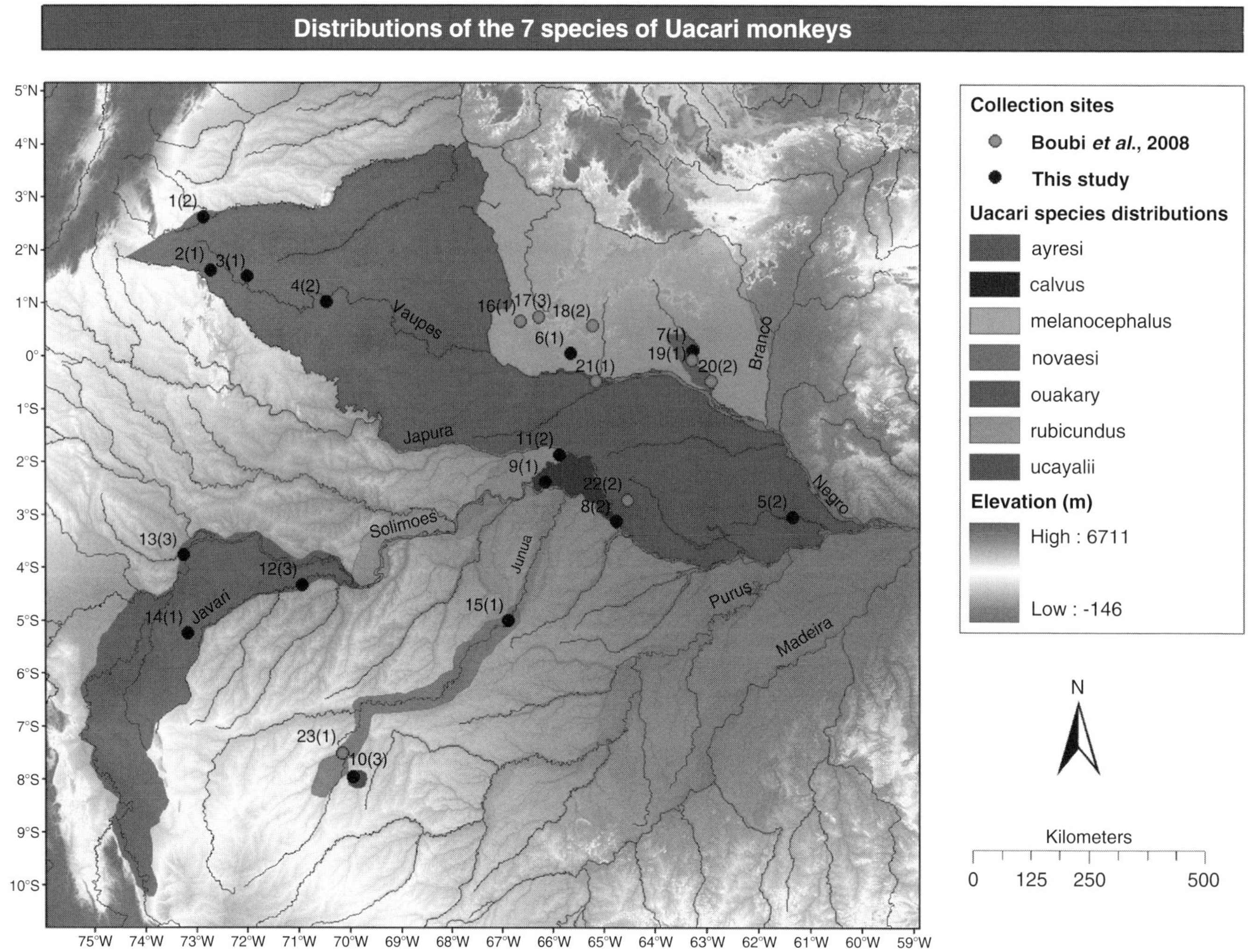

Figure 3.1 Distribution of the collecting localities of the *Cacajao* specimens included in the present analysis, and approximate limits of the geographic ranges of the different species and subspecies. (See color plate section.)

sequence of the gene using these primers. In this case, 16 new primers were developed for the amplification of shorter sequences, based on that obtained from the fresh samples. These primers were designed so that each partial sequence overlapped with neighboring sequences by at least 50 base pairs. The PCR products were visualized in 0.5% low melting agarose gel dyed with ethidium bromide. A recombinant PCR protocol was adapted from that of Nasidze & Stoneking (1999) for the combined analysis of these partial sequences (for details, see Figueiredo 2006).

Sequencing was carried out using a Big-Dye Terminator kit, according to the instructions of the manufacturer (Perkin Elmer), and was conducted in both directions for all specimens. The electropherograms were produced using the ABI 377 and ABI 3100 automatic sequencers. Version 4.1 of the Sequencer program (Genecodes) was used for the alignment and comparison of the sequences. Special care was taken to ensure the correct alignment of the subsequences of the recombinant PCRs. Sequencing in both directions, and a careful analysis of insertions, deletions, stop codons and the relative proportion of transitions and transversions (following Bates *et al.* 1999) helped confirm that the nucleotide sequences were mitochondrial, rather than nuclear pseudogenes. All procedures were repeated for a random sample of 10% of the museum samples to ensure the absence of exogenous contamination.

The genealogical relationships among the different haplotypes were reconstructed using a Metropolis-coupled Markov chain Monte Carlo algorithm (MC^3) for a Bayesian analysis run in MrBayes, version 3.1 (Huelsenbeck & Ronquist 2001). The model of molecular evolution used for each analysis was chosen based on a maximum-likelihood hierarchy test run in Modeltest, version 3.7 (Posada & Crandall 1998), which revealed model HKY85 (Hasegawa *et al.* 1985) to be the most appropriate for *Cacajao*.

The model was used to calculate genetic distances among haplotypes. Phylogenetic trees were reconstructed using Maximum Parsimony (MP) and Maximum Likelihood (ML) methods implemented in PAUP*, version 4.0b10 (Swofford 2002). The significance of groupings was evaluated by bootstrap analyses with 1000 replicates. Estimates of divergence times were based on the confidence interval of 6.3–12.9 million years ago (MYA) for the separation of *Cacajao* and *Chiropotes* (Goodman *et al.* 1998; von Dornum & Ruvolo 1999; Schneider *et al.* 2001; Figueiredo 2006).

Results

A total of 33 different haplotypes were obtained. As expected, *Cacajao* is a clearly monophyletic group, separated from *Chiropotes* by divergence values of 24.3–30.6% (Table 3.1), and 100% bootstrap probability (Figure 3.2). Divergence between *Cacajao* haplotypes, corrected by the HKY85 model of molecular evolution varied between 0.1% and 10.9%. The basic division between the black and bald uacaris is supported emphatically by a mean genetic divergence of approximately 10% and high bootstrap values. According to these values, the separation of the two lineages is estimated to have taken place at 5.8 MYA, that is, at some time during the Pliocene (Figure 3.2).

There is a quite different pattern of diversity within each group, however. In the black-headed uacaris, there is a clear

Table 3.1 Range of nucleotide divergence values (%) for pairwise comparisons of individuals representing different pitheciin taxa, corrected according to the HKY85 model (see text). The values for the comparison of the northern and southern populations of *Cacajao calvus calvus* are shown in bold type.

	Minimum–maximum nucleotide divergence (%)								
Population	**1**	**2**	**3**	**4**	**5**	**6**	**7**	**8**	**9**
1 *ouakary*	0.00–1.11								
2 *melanocephalus*	2.01–3.70	0.00–0.68							
3 *ayresi*	2.68–3.80	0.04–0.78	0.09–0.59						
4 *calvus**	6.78–9.12	7.99–9.00	8.50–9.10	0.00–0.09					
5 *ucayalii*	6.78–9.91	7.98–9.41	8.48–9.50	0.00–0.54	0.00–0.64				
6 *rubicundus*	6.78–8.93	7.99–8.79	8.50–9.09	0.00–0.09	0.00–0.45	0.00			
7 *calvus*†	6.34–9.28	7.93–10.38	8.42–10.37	**2.85–3.59**	2.85–4.19	2.85–3.43	0.09–0.51		
8 *novaesi*	6.31–9.17	7.74–10.47	8.20–9.85	2.72–3.65	2.71–4.21	2.72–3.51	0.27–0.73	0.57	
9 *Chiropotes*	24.33–30.55	27.33–28.34	28.90–29.30	27.92–28.27	27.91–29.68	27.92	25.77–26.61	24.68–26.17	–
10 *Pithecia*	47.16–52.40	50.40–51.49	51.13–51.71	49.29–49.78	49.23–50.78	49.29	48.56–50.18	48.81–51.49	63.50

* Northern population (Solimões River); †southern population (Juruá River).

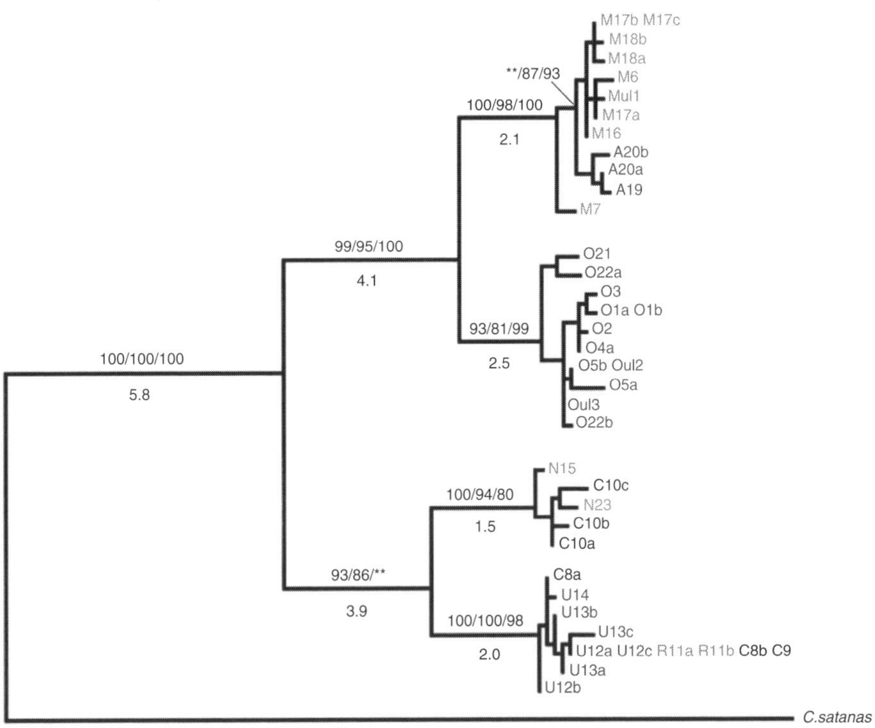

Figure 3.2 Phylogenetic tree constructed in the MrBayes program from cytochrome *b* in *Cacajao*. Numbers above branches are MP and ML bootstrap values, and Bayesian posterior probabilities, respectively, and those below the branches are estimated divergence times. Specimen codes follow Appendix 3.1, and are color-coded according to their geographic distribution (see Figure 3.1). (See color plate section.)

distinction between the two main lineages – *melanocephalus* + *ayresi* and *ouakary* – that occur on opposite banks of the Rio Negro (Figure 3.1), indicating that they represent well-defined, reciprocally monophyletic clades, and supporting their classification as true species, as proposed by Groves (2005) and Boubli *et al.* (2008). Whereas intralineage distance values were invariably lower than 1%, those between the two lineages ranged between 2.5% and 3.8%. The splitting of these two lineages is thus estimated to have taken place in the early Pliocene, around 4.1 MYA.

By contrast, genetic divergence among the bald uacaris is more closely related to geographic distribution than to morphological characteristics or their current taxonomic arrangement. In fact, specimens representing three different subspecies, *calvus*, *rubicundus* and *ucayalii* (U12a, U12c, R11a, R11b, C8b, and C9: see Figure 3.2) share an identical haplotype. This and other evidence indicates that the populations distributed along the Solimões–Ucayali/Javari river system form a monophyletic group, characterized by divergence values of no more than 0.7%. This group is clearly separated from that located in the Juruá river basin, with intergroup divergence values of 2.7–4.2%, in other words, values highly similar to those recorded between the two species of black uacari, despite the fact that the *C. c. calvus* morphotype is present in both groups. The Juruá group is also well-defined, with divergence values of no more than 0.5%.

On the basis of the molecular data alone, then, the bald uacaris present what appears to be a species-level division between two geographically distinct populations, which probably diverged at around the same time as the black uacari lineages, i.e. in the early Pliocene (3.9 MYA). This division contradicts the traditional classification, based on external morphology, given that the typical *calvus* form is present in both populations.

Discussion

This molecular phylogeny of the uacaris has revealed a number of different patterns of diversity. While predictable, the first of these is the confirmation of the monophyletic status of the genus *Cacajao*, which apparently separated from the closely related *Chiropotes* more than 5 million years ago. This coincides with the Miocene period of intense uplifting of the Andes, when the Amazon basin began to take on its present geomorphological characteristics (Campbell *et al.* 2006). Concomitantly, the basin's flooded forest ecosystems (*várzeas* and *igapós*), to which all uacaris are so well adapted, began to form. The divergence and radiation of the uacaris thus appears to have coincided with the emergence of ecosystems distinct from those occupied typically by the bearded sakis, and, presumably, the common ancestor of the two genera.

The second pattern re-emphasized clearly by this analysis is the basic division between the two main geographical and morphological groups of uacaris: the bald and the black forms. The timing of this divergence, towards the end of the Miocene, coincides with the formation of the incipient Amazon river system, and possible zoogeographic barriers within this system, in particular rivers, but possibly also lacustrine systems, which may have contributed to the speciation process.

The patterns observed within each group have widely contrasting implications for the interpretation of the diversity of the genus, however. In the case of the black uacaris, the analysis supports not only the basic division between the black- (*melanocephalus*) and golden-backed (*ouakary*) forms recognized by Hershkovitz (1987), but also their reclassification as true species, as suggested by Boubli *et al.* (2008). This classification would also reinforce the role of the Negro River as the principal zoogeographic barrier for this group. However, our results did not support the validity of the recently described *C. ayresi* (Boubli *et al.* 2008). While monophyletic in itself, this clade is nested within that of *C. melanocephalus* (Figure 3.2).

In the case of the bald uacaris, by contrast, the findings contradict traditional classifications into four distinct taxa, based on external morphology. Rather, they point to the existence of two geographic lineages, distributed along the Solimões–Ucayali/Javari and Juruá river systems, respectively. Both these lineages include populations of the *C. c. calvus* morphotype. The genetic evidence supporting this division is at least as strong as that for the *melanocephalus–ouakary* split (Table 3.1, Figure 3.2), in terms of both intralineage similarity and interlineage divergence. On the basis of this molecular evidence alone, then, it would seem reasonable to classify these two lineages as distinct species, although such a controversial decision would obviously require a more detailed investigation of morphological, ecological and zoogeographic parameters.

Other molecular studies have also contradicted traditional taxonomies. Collins & Dubach (2000), for example, were unable to detect genetic differences between the spider monkeys *Ateles chamek* and *Ateles belzebuth*. Similarly, molecular phylogenies (e.g. Schneider *et al.* 2001) have consistently placed *Callimico* in a central position within the Callitrichidae, whereas all morphological studies have identified it as an outlier, to the extent that Hershkovitz (1977) allocated the genus to its own family. Whereas most modern classifications of platyrrhines have been based – explicitly or implicitly – on the biological species concept of Mayr (1966), few have provided any direct evidence of reproductive isolation, relying on the indirect implications of morphological differentiation, primarily pelage coloration.

The black uacaris present a classic pattern of divergent coloration on opposite sides of a major zoogeographical barrier – the Negro River – which is further reinforced by the molecular data presented here. In this case, the evidence indicates clearly that the physical barrier has reinforced the divergence of the two populations. By contrast, while the molecular evidence indicates a similar basic division within the bald uacaris, this division is consistent with geographic but not morphological patterns. The distribution of this group is unusual in being predominantly *along* rather than between river systems (Figure 3.2). To a certain extent, this reflects the configuration of the extensive *várzea* systems in this region, which also confer a number of specific characteristics on the uacari populations that inhabit them, such as linear, rather than panmictic gene flow, and the constant isolation and interconnecting of relatively small subpopulations through continuous shifts in the geomorphology of the floodplain landscape. This scenario would be ideal for the fixation of alleles – e.g. for pelage coloration – in local populations, without necessarily establishing reproductive isolation, which would depend on specific selection pressures over long periods.

Unfortunately, sample sizes in the present study were too small to permit a more systematic investigation of population-level genetic variation, which might provide more conclusive insights into diversity patterns. However, the presence of the most common haplotype throughout the Solimões–Ucayali/Javari populations indicates a relatively recent process or processes of population expansion and diversification in this area (Donnelly & Tavaré 1995), related to changes in the topology of the rivers and the configuration of their *várzea* ecosystems.

Overall, then, the findings of this study re-emphasize the need for a more systematic and multidisciplinary approach to the understanding of the phylogenetic relationships among populations. They also demand caution with regard to the evaluation of morphological criteria, not only in *Cacajao*, but most if not all other platyrrhine genera. The unique ecological characteristics of the uacaris and the ecosystems they inhabit appear to have contributed to unusually complex phylogeographic patterns in this genus. Ultimately, the satisfactory resolution of these questions will be fundamental to the development of effective conservation strategies.

Acknowledgments

The following institutions and people helped with the loan of both fresh and museum preserved samples: Museu Paraense Emilio Goeldi, The Field Museum of Natural History (Chicago), Museu de Zoologia da Universidade de São Paulo, Museum of Vertebrate Zoology of the University of California at Berkeley, American Museum of Natural History, Universidade Federal do Pará, Instituto Mamirauá, James Patton, David Fleck, José M. Ayres, Helder Queiroz, Luis Fábio Silveira and Rob Voss. This work was funded by a WFF grant to WMBF, who has also been supported by funds from the Universidade Federal do Pará and from the Brazilian agencies CAPES and CNPq. Molecular data were gathered in the Pritzker Laboratory for Molecular Systematics and Evolution in The Field Museum, Chicago. Finally, we thank Jonathan Ready for helping with the geographic distribution map.

References

Ayres, J.M. & Clutton-Brock, T.H. (1992). River boundaries and species range size in Amazonian primates. *American Naturalist*, **140**, 531–537.

Bates, J.M., Hackett, S.J. & Goerck, J.M. (1999). High levels of mitochondrial DNA differentiation in two lineages of antbirds (*Drymophila* and *Hypocnemis*). *Auk*, **116**, 1093–1106.

Boubli, J.P., da Silva, M.N.F., Amado, M.V., *et al.* (2008). A taxonomic reassessment of *Cacajao melanocephalus* Humboldt (1811), with the description of two new species. *International Journal of Primatology*, **29**, 723–741.

Campbell, K.E. Jr., Frailey, C.D. & Romero-Pittman, L. (2006). The pan-Amazonian Ucayalii peneplain, late Neogene sedimentation in Amazonia, and the birth of the modern Amazon River system. *Palaeogeography, Palaeoclimatology, Palaeoecology*, **239**, 166–219.

Carnahan, S.J. & Jensen-Seaman, M.I. (2008). Hominoid seminal protein evolution and ancestral mating behavior. *American Journal of Primatology*, **70**, 939–948.

Collins, A.C. & Dubach, J.M. (2000). Phylogenetic relationships of spider monkeys (*Ateles*) based on mitochondrial DNA variation. *International Journal of Primatology*, **21**, 381–420.

Donnelly, P. & Tavaré, S. (1995). Coalescents and genealogical structure under neutrality. *Annual Review of Genetics*, **29**, 401–421.

Ferrari, S.F. (2004). Biogeography of Amazonian primates. In *A Primatologia no Brasil – 8*, ed. S.L. Mendes & A.G. Chiarello. Santa Teresa: Sociedade Brasileira de Primatologia, pp. 101–122.

Figueiredo, W.M.B. (2006). Estimativas de tempos de divergência em platirrinos e filogenia molecular e filogeografia dos uacaris, parauacus e cuxiús. Unpublished PhD thesis, Universidade Federal do Pará.

Fischer, A., Wiebe, V., Pääbo, S., *et al.* (2004). Evidence for a complex demographic history of chimpanzees. *Molecular Biology and Evolution*, **21**, 799–808.

Goodman, M., Porter, C.A., Czelusniak, J., *et al.* (1998). Toward a phylogenetic classification of primates based on DNA evidence complemented by fossil evidence. *Molecular Phylogenetics and Evolution*, **9**, 585–598.

Groves, C.P. (2005). Order Primates. In *Mammal Species of the World: A Taxonomic and Geographic Reference, 3rd edition*, ed. D.E. Wilson & D.M. Reeder. Baltimore, MD: The Johns Hopkins University Press, pp. 111–184.

Hasegawa, M., Kishino, H. & Yano, T. (1985). Dating of the human-ape splitting by a molecular clock of mitochondrial DNA. *Journal of Molecular Evolution*, **22**, 160–174.

Hershkovitz, P. (1977). *Living New World monkeys (Platyrrhini) with an Introduction to Primates*. Vol. **1**. Chicago, IL: University of Chicago Press.

Hershkovitz, P. (1987). Uakaris, New World monkeys of the genus *Cacajao* (Cebidae, Platyrrhini): a preliminary review with the description of a new subspecies. *American Journal of Primatology*, **12**, 1–57.

Hillis, D.M., Mable, B.K., Larson, A., *et al.* (1996). Nucleic acids IV: sequencing and cloning. In *Molecular Systematics, 2nd edition*, ed. D.M. Hillis, C. Moritz & B. Mable. Sunderland, MA: Sinauer, pp. 321–381.

Huelsenbeck, J.P. & Ronquist, F. (2001). MRBAYES: Bayesian inference of phylogenetic trees. *Bioinformatics*, **17**, 754–755.

Kocher, T.D., Thomas, W.K., Meyer, A., *et al.* (1989). Dynamics of mitochondrial DNA evolution in animals: amplification and sequencing with conserved primers. *Proceedings of the National Academy of Sciences USA*, **86**, 6196–6200.

Mayr, E.W. (1966) *Animal Species and Evolution*. Cambridge, MA: Belknap Press.

Miller-Butterworth, C.M., Kaplan, J.R., Schaffer, J., *et al.* (2008). Sequence variation in the primate dopamine transporter gene and its relationship to social dominance. *Molecular Biology and Evolution*, **25**, 18–28.

Nasidze, I. & Stoneking, M. (1999). Construction of larger-size sequencing templates from degraded DNA. *BioTechniques*, **27**, 480–488.

Peres, C.A., Patton, J.L. & da Silva, M.N.F. (1996). Riverine barriers and gene flow in Amazonian saddle-back tamarins. *Folia Primatologica*, **67**, 113–124.

Posada, D. & Crandall, K.A. (1998). MODELTEST: testing the model of DNA substitution. *Bioinformatics*, **14**, 817–818.

Rylands, A.B., Schneider, H., Langguth, A., *et al* . (2000). An assessment of the diversity of New World primates. *Neotropical Primates*, **8**, 61–93.

Schneider, H., Canavez, F.C., Sampaio, I., *et al.* (2001). Can molecular data place each Neotropical monkey in its own branch? *Chromosoma*, **109**, 515–523.

Swofford, D.L. (2002). *PAUP*: Phylogenetic Analysis using Parsimony (* and other methods)*. Sunderland, MA: Sinauer.

van Roosmalen, M.G.M., van Roosmalen, T. & Mittermeier, R.A. (2002). A taxonomic review of the titi monkeys, genus *Callicebus* Thomas, 1903, with the description of two new species, *Callicebus bernhardi* and *Callicebus stephennashi*, from Brazilian Amazonia. *Neotropical Primates*, **10** (Suppl.), 1–52.

von Dornum, M. & Ruvolo, M. (1999). Phylogenetic relationships of the New World monkeys (Primates, Platyrrhini) based on nuclear G6PD DNA sequences. *Molecular Phylogenetics and Evolution*, **11**, 459–476.

Wallace, A.R. (1876). *The Geographic Distribution of Animals*. London: MacMillan & Co.

Appendix 3.1 Specimens of taxa analyzed in the present study. Except where stated, localities are in the Brazilian state of Amazonas

Taxon	Specimens	Original code	Collection locality	Genbank accession number
C. ouakary	01a, b	FMNH88250–1	Upper Rio Inirida, Colombia	FJ531640–1
	02	FMNH89470	Barracon, Upper Cano Itilla, Colombia	FJ531642
	03	FMNH89471	Cano Miraflores, Rio Vaupes, Colombia	FJ531643
	04a, 04b	FMNH89468–9	Lago El Dorado, Rio Vaupes, Colombia	FJ531644–5
	05a, b	MVZ016–7	Rio Manacapuru, Brazil	FJ531646–7
	021	MN68616*	Serraria, Rio Negro, Brazil	EU560422
	022a, b	INPA5238–9*	Lago Amanã, Rio Solimões, Brazil	EU560419–20
	Oul1	UFPA-Cmo1	Unknown locality	FJ531648
	Oul3	INPA5240*	Unknown locality, Rio Solimões, Brazil	EU560421
C. melanocephalus	M6	MVZCmm1	Pico da Neblina National Park, Brazil	FJ531649
	M7	MPEG8991	Rio Curuduri, Brazil	FJ531650
	M16	INPA5252*	Serra do Demiti, Brazil	EU560417
	M17a, b, c	INPA5249–51*	Iqarapé Waputa, Brazil	EU560414–6
	M18a, b	MN68611, 68613*	Serra do Imeri, Brazil	EU560412–3
	Mul2	INPA5242*	São Gabriel da Cachoeiram, Brazil	EU560418
C. ayresi	A19	INPA5246*	Igarapé Madixi, Brazil	EU560409
	A20a, b	INPA5247–8*	Lower Rio Araçá, Brazil	EU560410–1
C. c. calvus	C8a	UFPA-Ccc1, MZUSP17537	Mamirauá, Mouth of the Rio, Brazil	FJ531651
	C8b		Japurá, Brazil	FJ531655
	C9	MPEG8990	Paraná do Marauí, São José, Brazil	FJ531656
	C10a, b, c	MPEG21861–3	Rio Jurupari, Brazil	FJ531657–9
C. c. rubicundus	R11a, b	MZUSP17552–3	Buiuçu, Auatí-Paraná channel, Brazil	FJ531652–3
C. c. ucayalii	U12a, b, c	MPEG1848–50	Estirão do Equador, Rio Javari, Brazil	FJ531660–2
	U13a, b, c	UFPA-Ccu4957–9	Rio Tapiche, Peru	FJ531654, FJ531663–4
	U14	MUSM13300	Rio Galvez, Nuevo San Juan, Peru	FJ531665
C. c. novaesi	N15	UFPA-Ccn1	Carauari, Rio Juruá, Brazil	FJ531666
	N23	INPA5241*	Sacado do Tarauacá, Brazil	EU560408
Chiropotes sagulatus		UFPA-Csa3056	Rio Trombetas, Pará, Brazil	FJ531667
Pithecia monachus		DWF156	Rio Galvez, Nuevo San Juan, Brazil	FJ531668

* sequences from Boubli *et al.* (2008)

Appendix 3.2 Protocol for the extraction of DNA from museum specimens (adapted from Figueiredo 2006)

Skin or bone fragments no bigger than 2 mm^2 were washed three consecutive times in a 5% sodium hypochlorite solution and then three times with ultrapure distilled water. DNA was extracted using the DNeasy kit (QIAGEN), with the following changes in the original protocol. (a) Samples were incubated in tissue lysis buffer and Proteinase K for at least 5 times longer than suggested by the maker. (b) Elution buffer was heated to 70°C before being added to the column. (c) reduced amounts of this buffer (50–100 ml) were used to capture DNA.

Chapter 4

Taxonomy and geographic distribution of the Pitheciidae

José de Sousa e Silva Júnior, Wilsea M.B. Figueiredo-Ready & Stephen F. Ferrari

Family Pitheciidae

The monophyly of the group formed by the titis (*Callicebus*), sakis (*Chiropotes* and *Pithecia*) and uacaris (*Cacajao*) is well established (Rosenberger *et al.* 1990). While some analyses have indicated that it may be a subfamily of the Atelidae, the current consensus (Schneider & Rosenberger 1996; Goodman *et al.* 1998; Schneider 2000; Groves 2005) is that Pitheciidae is a valid family, in which *Callicebus* is the most basal genus. Whereas Schneider (2000) recognized a single subfamily, with two tribes (Callicebini and Pitheciini), the latter divided into the subtribes Pitheciina (*Pithecia*) and Chiropotina (*Chiropotes* and *Cacajao*), Groves (2005) opted for two subfamilies, Callicebinae and Pitheciinae, the classification followed here.

Subfamily Callicebinae

This subfamily has a single, relatively complex genus, *Callicebus*, which has been organized differently by a variety of authors over the past five decades (Hershkovitz 1963, 1988, 1990; Kobayashi 1995; Groves 2001, 2005; van Roosmalen *et al.* 2002). All of these authors have organized the genus in four or five species groups. *Callicebus* is the most widespread pitheciid genus, and the only one found outside the Amazon–Orinoco basins, with an isolated group of species in the Brazilian Atlantic Forest.

Genus *Callicebus*

Taxonomy

While recognizing the considerable diversity of the genus, early classifications, such as those of Hill (1960) and Hershkovitz (1963), identified relatively few species, but a large number of subspecies. In fact, Hill identified a total of 34 subspecific forms, slightly more than the number of taxa recognized currently (Table 4.1).

Given its considerable diversity, especially in comparison with other pitheciids, an important first level of organization of the Callicebinae is the species groups, which are informal arrangements based on morphological, cytogenetic and zoogeographic similarities. Kobayashi (1995) and van Roosmalen *et al.* (2002) recognized five species groups: *Callicebus cupreus*, *Callicebus donacophilus*, *Callicebus moloch*, *Callicebus personatus* and *Callicebus torquatus*. This contrasts with Hershkovitz (1988, 1990), who did not recognize the *C. cupreus* or *C. personatus* groups (both were included in the *C. moloch* group), but placed *Callicebus modestus* in a group of its own.

The other basic difference between these recent classifications is in the interpretation of species-level diversity. Kobayashi (1995), for example, did not recognize either *Callicebus oenanthe* or *Callicebus dubius*, based on an analysis of morphometric parameters, whereas both forms have been considered true species by all other authors. In other cases, e.g. *pallescens*, *baptista*, and the *C. personatus* group, a given form has been identified alternately as a species or a subspecies. Ultimately, van Roosmalen *et al.* (2002) considered all previously recognized forms as true species, a viewpoint subsequently upheld by Groves (2005), despite his initial support for some subspecies (Groves 2001). An additional species – *Callicebus aureipalatii* – was described subsequently (Wallace *et al.* 2006).

Groves (2005) also considered the *C. torquatus* group to be so distinct from the others that he allocated it to the subgenus *Torquatus*, with the remaining species in the subgenus *Callicebus*. The classification adopted here (Table 4.1) combines this approach with the arrangement of van Roosmalen *et al.* (2002). It seems likely that other morphological forms have yet to be described, considering the lack of samples from wide areas within the range of the genus, especially in the Amazon basin.

Geographic distribution

The relatively ample and disjunct geographic distribution of *Callicebus* is unique among platyrrhine genera. It is also unusual in the occurrence of intrageneric sympatry (otherwise seen only in *Cebus* and *Saguinus*), which involves species of the *C. cupreus* and *C. torquatus* groups. While the distribution of

Evolutionary Biology and Conservation of Titis, Sakis and Uacaris, eds. Liza M. Veiga, Adrian A. Barnett, Stephen F. Ferrari and Marilyn A. Norconk. Published by Cambridge University Press.

Table 4.1 Taxonomy of the titis, Callicebinae, according to van Roosmalen *et al.* (2002), Groves (2005) and Wallace *et al.* (2006).

Subgenus	Species group	Species	Common name	Distribution (countries)
Callicebus	*Callicebus donacophilus*	*Callicebus donacophilus*	Bolivian gray titi	Bolivia [Brazil][1]
		Callicebus modestus	Rio Beni titi	Bolivia
		Callicebus olallae	Olalla Brothers' titi	Bolivia
		Callicebus oenanthe	Rio Mayo titi	Colombia
		Callicebus pallescens	White-coated titi	Brazil, Paraguay, [Argentina][1]
	Callicebus cupreus	*Callicebus cupreus*	Coppery titi	Brazil, Peru
		Callicebus aureipalatii	Golden Palace titi	Bolivia
		Callicebus caligatus	Booted titi	Brazil
		Callicebus discolor	Discolored titi	Brazil, Colombia, Ecuador, Peru
		Callicebus dubius	Doubtful titi	Brazil [Peru][1]
		Callicebus ornatus	Ornate titi	Colombia
		Callicebus stephennashi	Stephen Nash's titi	Brazil
	Callicebus moloch	*Callicebus moloch*	Dusky titi	Brazil
		Callicebus baptista	Baptista Lake titi	Brazil
		Callicebus bernhardi	Prince Bernhard's titi	Brazil
		Callicebus brunneus	Brown titi	Brazil
		Callicebus cinerascens	Ashy black titi	Brazil
		Callicebus hoffmannsi	Hoffmann's titi	Brazil
	Callicebus personatus	*Callicebus personatus*	Masked titi	Brazil
		Callicebus barbarabrownae	Barbara Brown's titi	Brazil
		Callicebus coimbrai	Coimbra Filho's titi	Brazil
		Callicebus melanochir	Southern Bahian masked titi	Brazil
		Callicebus nigrifrons	Black-fronted titi	Brazil
Torquatus	*Callicebus torquatus*	*Callicebus torquatus*	Collared titi	Brazil, Colombia
		Callicebus lucifer	Widow monkey	Brazil, Colombia, Ecuador, Peru
		Callicebus lugens	Collared titi	Brazil, Colombia, Venezuela
		Callicebus medemi	Colombian black-handed titi	Colombia, Ecuador
		Callicebus purinus	Widow monkey	Brazil
		Callicebus regulus	Widow monkey	Brazil, Colombia

[1] Unconfirmed or poorly documented.

the *C. personatus* group – restricted primarily to the Atlantic Forest – is obviously allopatric, the *C. donacophilus*, *C. cupreus* and *C. moloch* groups are parapatric in the southern and western Amazon basin (Figure 4.1).

As for most other platyrrhines, especially those of the Amazon basin, rivers play a fundamental role in the zoogeography of the titis. Ecological factors are also important, however. This is demonstrated most clearly by the absence of these monkeys from most of the savanna-like *Cerrado* and *Caatinga* ecosystems which separate the Amazon basin from the Atlantic Forest. Despite ongoing fieldwork in many areas, the overall scarcity of data is the main obstacle to a more systematic understanding of the zoogeography of the genus.

Subgenus *Callicebus*

Callicebus donacophilus species group

The species of this group are distributed south of the Madre de Díos and Mamoré rivers in Bolivia, Paraguay and Brazil (Figure 4.1), with the exception of *C. oenanthe*, which has an isolated range on the upper Rio Mayo, Peru. *Callicebus donacophilus* and *C. pallescens* range further south than any of the

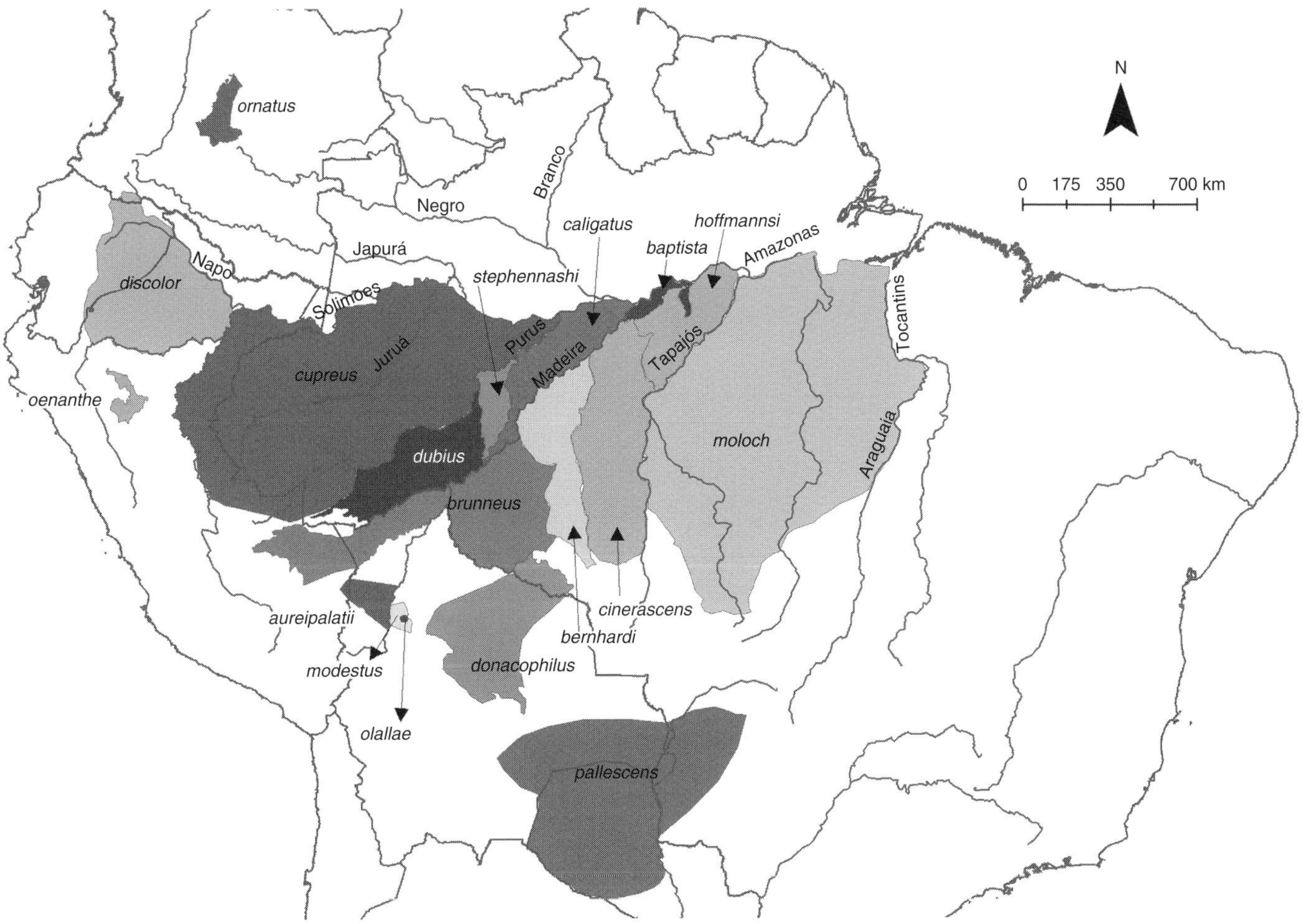

Figure 4.1 Geographic distribution of the *C. donacophilus*, *C. cupreus* and *C. moloch* species groups, based on van Roosmalen *et al.* (2002), Mark (2003), Rowe & Martinez (2003), Carrillo-Bilbao *et al.* (2005), Felton *et al.* (2006), and Wallace *et al.* (2006).

Amazonian species, in the *Cerrado*, and the Chaco scrublands and Pantanal swamps of Paraguay and Brazil, respectively.

Callicebus cupreus species group

This group is found mainly in the western Amazon basin, west of the Madeira River and south of the Solimões/Amazonas (Figure 4.1), although *C. discolor*, and especially *C. caquetensis* and *C. ornatus*, are found further north. The distribution of *C. ornatus*, in particular, is poorly known, and the lack of data from the upper Rio Caquetá proscribes a more systematic analysis of the distribution of titis in this region of southern Colombia. The Purus and the Ucayali–Solimões rivers are the main zoogeographic barriers, delimiting the ranges of all species except *C. ornatus*. Sympatry with *Torquatus* group species is widespread in the northern half of this group's range.

Callicebus moloch species group

The range of this group is restricted to the southern Amazon basin in Brazil, south of the Amazon, and between the Madeira-Mamoré to the west, and the Tocantins-Araguaia to the east (Figure 4.1). Within this area, the main barriers are the Tapajós, Roosevelt-Aripuanã and Jiparaná rivers, which influence the distribution of all species except *C. baptista*, found primarily on Tupinambaranas Island. The geographic configuration of the group in the extreme south of its range, where the Amazonian *Hylea* gives way to the *Cerrado* is poorly known, however.

Callicebus personatus species group

Endemic to eastern Brazil, this group is found in the Atlantic Forest between the Tietê–Paraná–Paraíba rivers in the south and west, and the São Francisco in the north and northwest (Figure 4.2). While these rivers form well-defined boundaries, the exact southern and western limits of the group's range are still poorly known, not least because *C. nigrifrons* occurs on both banks of the upper São Francisco, and south of the Tietê. Other rivers, such as the Paraguaçu and Jequitinhonha, may influence species distributions on a local scale. Ecological factors may be at least as important as physical barriers in some cases, most notably *C. barbarabrownae*, which may be restricted to the *caatinga* biome (see Printes *et al.*, Chapter 5).

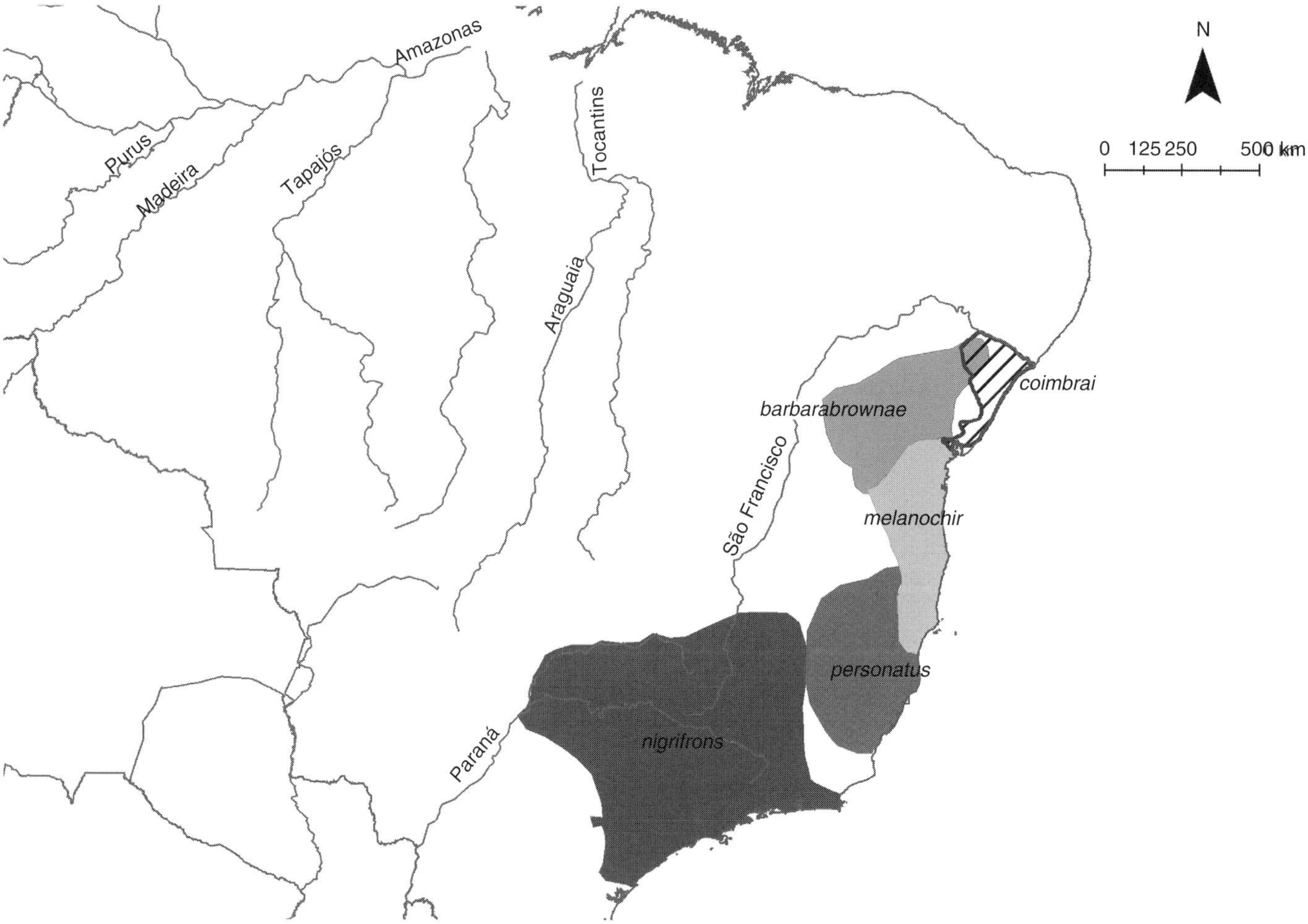

Figure 4.2 Geographic distribution of the *C. personatus* species group, following Vasconcelos & Hirsch (2000), van Roosmalen *et al.* (2002), Oliveira *et al.* (2003), and Printes *et al.* (Chapter 5).

Subgenus *Torquatus*

Callicebus torquatus species group

This is the northernmost group, which ranges from south of the Amazon, west of the Purus, as far north as the Orinoco river in Venezuela (Figure 4.3). Species ranges are delimited primarily by major rivers, including the Solimões–Amazonas, Juruá, Purus, Japurá–Caquetá, and Negro. The group is sympatric with *C. cupreus* group species in the southern half of this range.

Subfamily *Pitheciinae*

Taxonomy

Collectively, the pitheciines are the platyrrhines most highly specialized for the predation of seeds (Kay *et al.*, Chapter 1). There is a general consensus on the monophyly of the group, the close relationship of *Cacajao* and *Chiropotes*, and the relatively external position of *Pithecia*, which is also the least-specialized for seed predation (Ford 1986; Kay 1990; Schneider & Rosenberger 1996; Figueiredo 2006). Barnett and Brandon-Jones (1997) have argued that *Cacajao* and *Chiropotes* are similar enough morphologically to be members of the same genus.

Of the three genera, only *Chiropotes* has been reviewed systematically within the past few years (Silva Jr. & Figueiredo 2002; Bonvicino *et al.* 2003), and even then, a number of questions remain unresolved, due primarily to the paucity of samples, and the lack of consensus on subspecific diversity. Until more reliable analyses become available, the classifications of Hershkovitz (1987a, 1987b) are the current standards for *Pithecia* and *Cacajao*, although Rylands *et al.* (2000) and Groves (2005) have provided alternative viewpoints on some specific points.

Geographic distribution

All three pitheciine genera are essentially Amazonian. While *Pithecia* is sympatric with both *Cacajao* and *Chiropotes* throughout most of its range, the latter two genera are almost totally allopatric (see Ayres & Prance, Chapter 12), although Boubli (2002) recorded peripatry between *Chiropotes chiropotes* and *Cacajao melanocephalus* on the upper Rio Negro. Within each genus, species are normally parapatric, with the

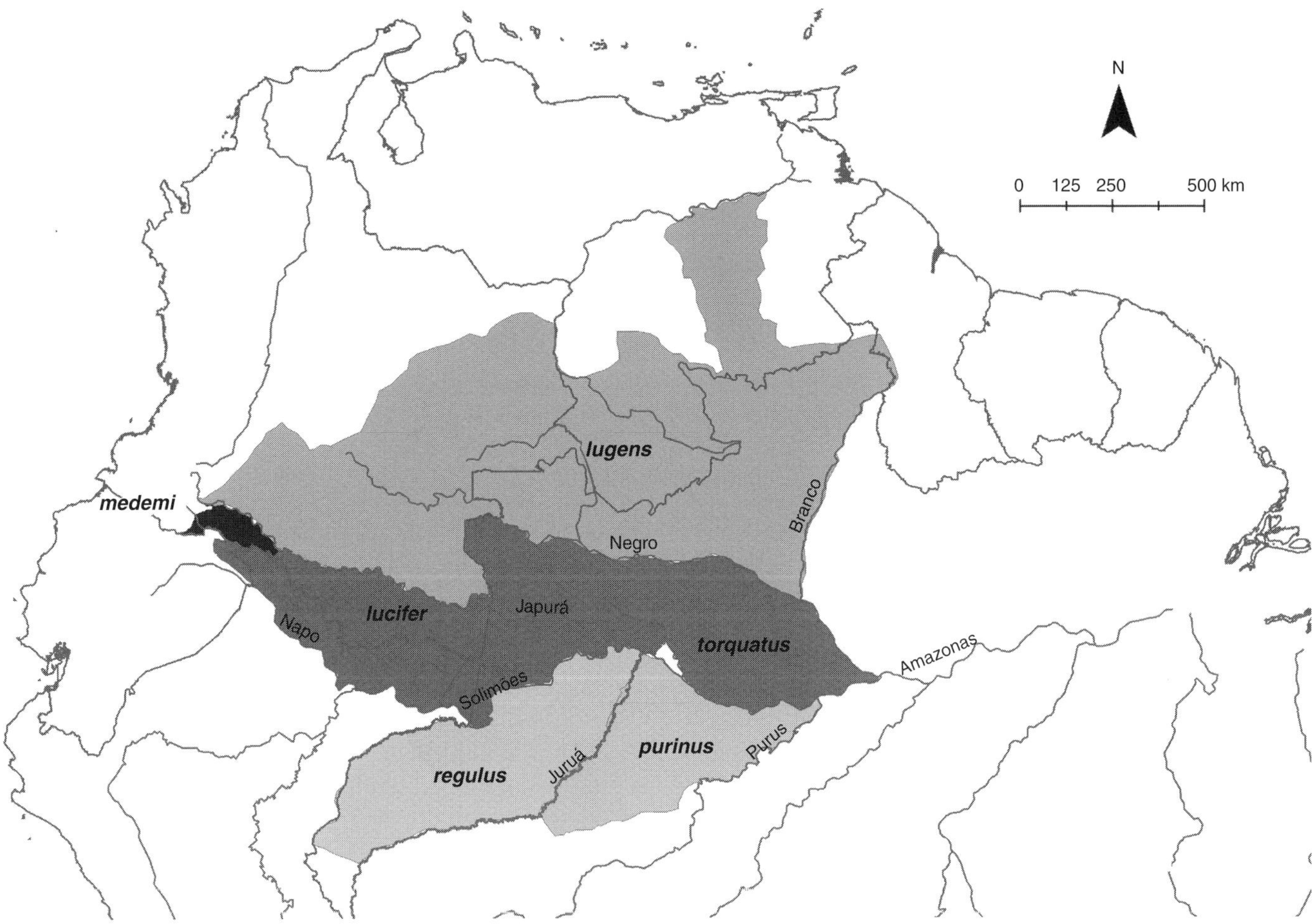

Figure 4.3 Geographic distribution of the *Torquatus* subgenus (adapted from van Roosmalen *et al.* 2002).

exception of some uacaris, which are either allopatric or have widely disjunct ranges, which may be at least partly related to their specialization for flooded forest habitats.

Genus *Pithecia*

Taxonomy

The sakis are one of only two platyrrhine genera (the other is *Alouatta*) that present marked sexual dichromatism in some species, which has interfered with the identification of taxa in the past (Hershkovitz 1979, 1987a). Hill (1960) recognized only two species, with a total of six subspecific forms, whereas Hershkovitz (1979) identified four monotypic species – *Pithecia pithecia*, *Pithecia hirsuta*, *Pithecia monachus* and *Pithecia albicans*. In his subsequent revision, Hershkovitz (1987a) established two species groups (*P. pithecia* and *P. monachus*), with a total of eight taxa (Table 4.2).

Rylands *et al.* (2000) followed this arrangement, except for recognizing an additional subspecies, *Pithecia monachus napensis*, and a possible fourth form (*hirsuta*) of this species. By contrast, Groves (2005) did not recognize species groups, and considered all subspecies to be junior synonyms of their respective species. This does appear to underestimate the true diversity of the genus, however (see Marsh 2004, 2006).

Geographic distribution

Pithecia is amply distributed in the Amazon–Orinoco basin west and north of the Juruena–Tapajós–Amazon river system (Figure 4.4), except for parts of southern Colombia and northern Bolivia. The two species groups are parapatric along the Amazon between the Tapajós and the Negro.

Pithecia pithecia species group

The range of this group coincides with the Guyanan Shield between the Orinoco and the Amazon, and the Negro, to the southwest (Figure 4.4). Hershkovitz (1987a) placed *P. p. pithecia* throughout the northern half of this region in Venezuela and the Guyanas, and as far south as the Rio Amapari in the Brazilian state of Amapá. However, Silva Júnior and Cerqueira (1997) recorded this subspecies throughout Amapá, and westwards as far as the Rio Jari, which restricts the distribution of *P. p. chrysocephala* to the area between the Branco-Negro river system and the Rio Nhamundá. Specimens from the region between the Nhamundá and the Jari present an intermediate morphotype, which suggests intergradation, and supports the subspecific classification, which is further upheld by Figueiredo's (2006) analysis of the genetic structure of these populations. There may also be an enclave of

Table 4.2 Taxonomy of the sakis, genus *Pithecia*, according to Hershkovitz (1987a).

Species group	Species	Subspecies	Common name	Distribution (countries)
Pithecia pithecia	*P. pithecia*	*P. p. pithecia*	White-faced saki	Brazil, French Guiana, Guyana, Suriname, Venezuela
		P. p. chrysocephala	Gold-faced saki	Brazil
Pithecia monachus	*P. monachus*	*P. m. monachus*	Geoffroy's monk saki	Brazil, Ecuador, Peru
		P. m. milleri	Miller's monk saki	Colombia
	P. aequatorialis		Equatorial saki	Ecuador, Peru
	P. albicans		Buffy saki	Brazil
	P. irrorata	*P. i. irrorata*	Gray's bald-faced saki	Bolivia, Brazil, Peru
		P. i. vanzolinii	Vanzolini's bald-faced saki	Brazil

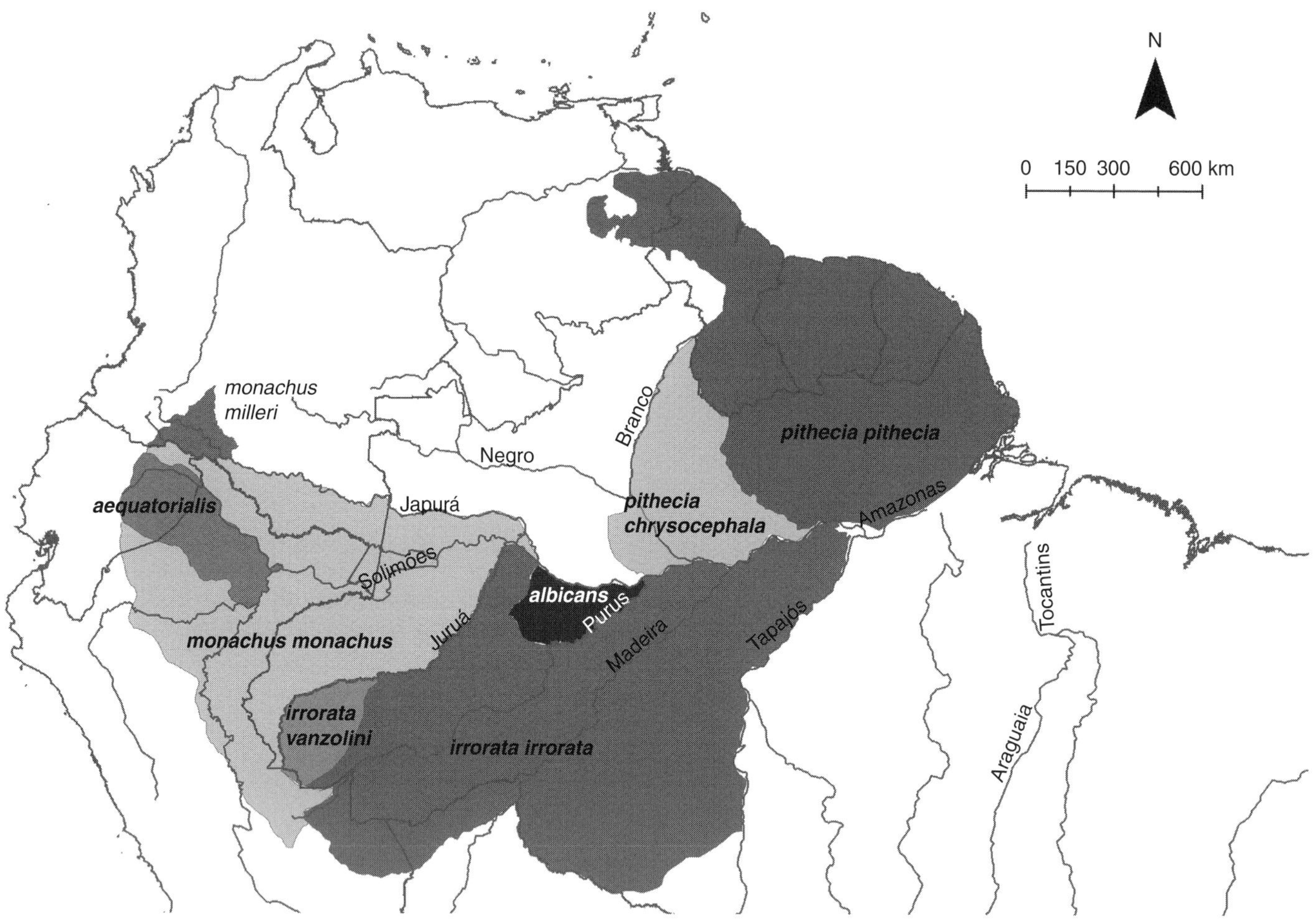

Figure 4.4 Geographic distribution of *Pithecia*, based on Hershkovitz (1987a) and Silva Jr. & Cerqueira (1997).

P. p. chrysocephala west of the Negro, in the region of Manacapuru (Hershkovitz 1987a).

Pithecia monachus species group

This group is distributed throughout most of the southern Amazon basin, west of the Tapajós River (Figure 4.4). The internal arrangement of this group is less well-defined, given that, in most cases, range limits are only partially defined by river boundaries. In fact, only *P. m. milleri*, which is isolated north of the Rio Caquetá, can be considered unproblematic in this sense. In particular, while the range of *P. albicans* is well-defined to the west, north and east by the Tefé, Solimões and Purus rivers, respectively, there is no reliable information on its southern limit, and possible contact with the range of *P. i. irrorata*, which occurs further south in the Juruá–Purus interfluvium. The distribution of the group is especially poorly understood in the western extreme, where a number of forms – including at least one new taxon, according to Marsh (2004, 2006) – coexist.

Genus *Cacajao*

Taxonomy

The uacaris are easily distinguished from all other platyrrhines by their short tails, which are no more than a third of body length. Hill (1960) recognized four species, although Hershkovitz (1985) verified that this author's *Cacajao roosevelti* from the southern Amazon basin did in fact refer to *Chiropotes albinasus*. Hershkovitz (1987b) subsequently reduced the number of species to two – *Cacajao calvus* and *Cacajao melanocephalus* – easily distinguished as the bald and black-faced uacaris, respectively, with a total of six subspecies. This arrangement is widely accepted (see Rylands *et al.* 2000). Figueiredo's (2006; Figueiredo *et al.*, Chapter 3) molecular analysis supports the species status of the two black-faced forms, but is inconclusive on that of the bald uacaris. Boubli *et al.* (2008) have recently identified a third black-faced form, with a restricted distribution, which was described as a new species, although Ferrari *et al.* (2010) considered this form to be a subspecies of *Cacajao melanocephalus*. Given these differences, the classification of Hershkovitz (1987b) is adopted here for the bald uacaris (Table 4.3), whereas that of the black-faced forms is adapted from Rylands *et al.* (2000) and Boubli *et al.* (2008).

Table 4.3 Taxonomy of the uacaris, genus *Cacajao*, adapted from Hershkovitz (1987b), Figueiredo (2006) and Boubli *et al.* (2008).

Group	Taxon	Common name	Distribution (countries)
Bald uacaris	*C. c. calvus*	White bald-headed uacari	Brazil
(*C. calvus*)	*C. c. novaesi*	Novaes's bald-headed uacari	Brazil
	C. c. rubicundus	Red bald-headed uacari	Brazil [Colombia][1]
	C. c. ucayalii	Ucayali bald-headed uacari	Brazil, Peru
Black uacaris	*C. ayresi*	Ayres's bald-headed uacari	Brazil
	C. melanocephalus	Humboldt's black-headed uacari	Brazil, Venezuela
	C. ouakary	Spix's black-headed uacari	Brazil, Colombia

[1] Unconfirmed.

Geographic distribution

While they are found over a wide area of the western Amazon basin between the Negro–Branco and Ucayali–Solimões–Juruá river systems, data on the distribution of the uacaris are sketchy at best. In some cases, such as Fernandes (1990) and Silva Jr. and Martins (1999), the presence of bald uacaris was confirmed in isolated areas of southwest Amazonas and Acre, but the taxon was not identified reliably. The problem of widely dispersed and allopatric ranges is exacerbated by the paucity of data from most areas.

Bald uacaris

In contrast with most other Amazonian platyrrhines, bald uacaris tend to be distributed *along* rather than between river systems (Figure 4.5), a pattern associated with their ecological specializations for flooded forest habitats (Ayres 1989; Ayres & Prance, Chapter 12). With the possible exception of *C. c. ucayalii*, the subspecies have relatively restricted and poorly defined ranges.

Cacajao calvus calvus

The distribution of this subspecies is centred on the confluence of the Japurá, Juruá and Solimões rivers, although Silva Jr and Martins (1999) identified an isolated population on the Jurupari, in the upper Juruá basin, south of the range of *C. c. novaesi*. Despite the morphological similarities of the Jurupari form with *C. c. calvus*, the molecular analysis of Figueiredo (2006; Figueiredo *et al.*, Chapter 3) indicates that this population is closely related genetically to *C. c. novaesi*, whereas *C. c. calvus* from the Japurá–Solimões aligns with the other subspecies.

Cacajao calvus novaesi

This subspecies is restricted to the middle–upper Juruá, upriver from the local population of the *C. c. calvus* form. As the headwaters of the Juruá system come to within a few kilometers of the Uacayali basin on the border of Brazil and Peru, there may be some form of contact with the range of *C. c. ucayalii* in this region (Fernandes 1990), although there is no evidence of parapatry at the present time.

Cacajao calvus rubicundus

Like *C. c. calvus*, this subspecies has a relatively complex and apparently disjunct distribution, with one population in the Japurá–Solimões interfluvium, west of *C. c. calvus*, and a second population on the Solimões upriver from the Içá (Putumayo). It is unclear whether the gap of approximately 100 km between populations is real, or a sampling effect, although Defler (2004) reports that this uacari has

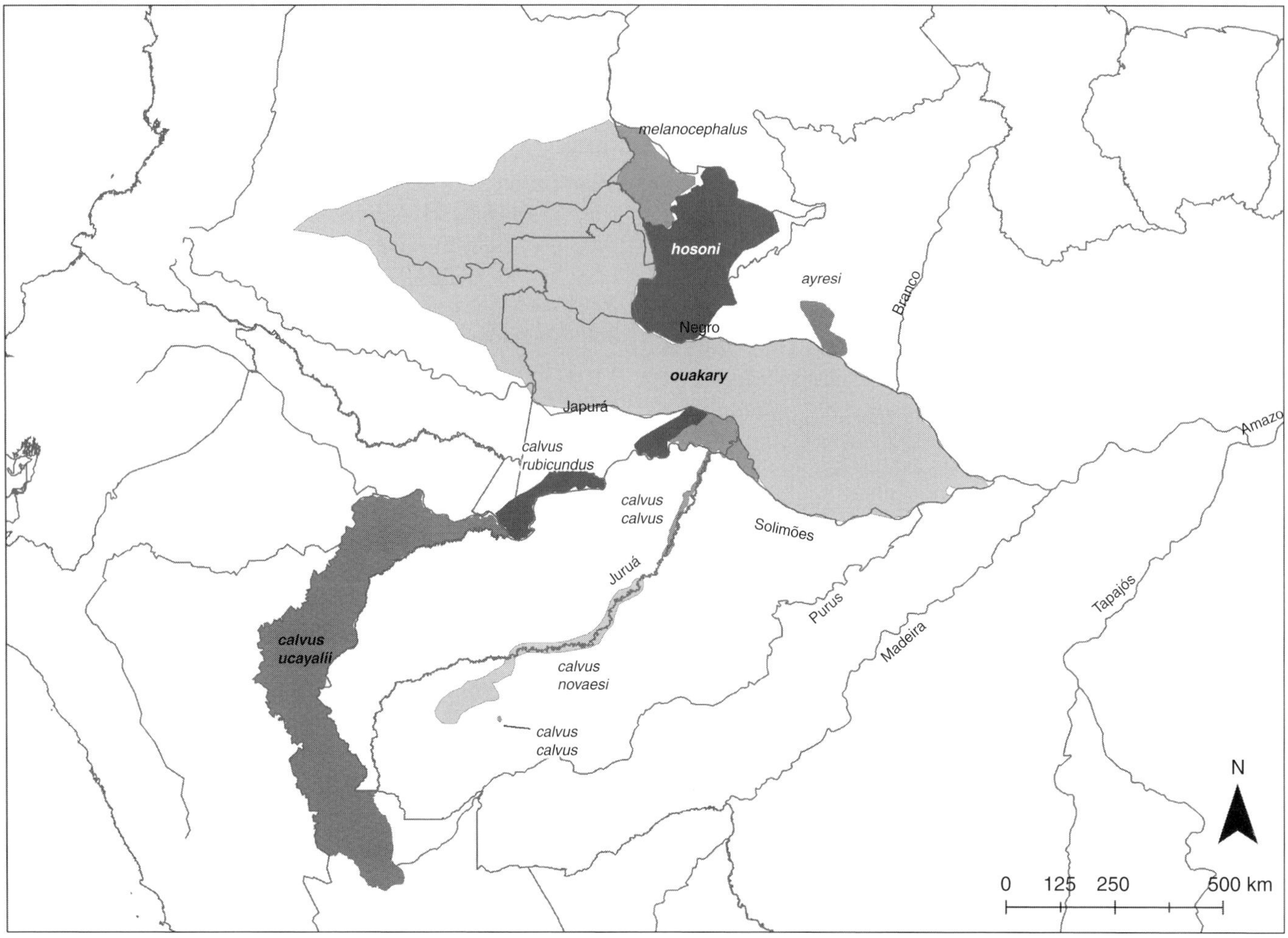

Figure 4.5 Geographic distribution of *Cacajao*, adapted from Hershkovitz (1987b), Silva Jr. & Martins (1999), and Boubli *et al.* (2008).

apparently disappeared from the Putumayo in southern Colombia in historic times, possibly as a result of hunting pressure.

Cacajao calvus ucayalii

This subspecies is the most widely distributed of the bald-headed forms, and is found all along the eastern Ucayali valley in eastern Peru as far east as the Rio Javari on the Brazilian border. Its distribution is relatively well-documented in comparison with the other subspecies (Heymann 1990; Aquino 1998; Ward & Chism 2003). It is known from only one site in Brazil, but may range further east (Fernandes 1990).

Black-faced uacaris

The black-faced uacaris are amply distributed between the Solimões–Japurá river system in the south, and the Negro–Branco in the east, and as far north as the upper Orinoco. These uacaris may also inhabit extensive areas of *terra firme* forest (Barnett & Brandon-Jones 1997; Boubli & Tokuda 2008).

Cacajao melanocephalus

This species has a relatively restricted distribution in an area characterized primarily by relatively high elevations and poor soils (Boubli 1994, 1997). Whereas Hershkovitz (1987b) restricted the taxon to the upper Orinoco in Venezuela, Boubli (1997) and Boubli *et al.* (2008) extended its range south into Brazil as far as the Rio Negro, in an area originally thought to be occupied by *C. ouakary*. The species may be limited to the east and north by the presence of bearded sakis (*Chiropotes chiropotes*), with which it may be sympatric in some areas (Boubli 2002).

Cacajao ouakary

By far the most widespread of the uacaris, *C. ouakary* ranges from the confluence of the Solimões and Negro rivers in the east to the foothills of the Andes in southern Colombia. A number of recent studies (Cunha & Barnett 1990; Defler 2004; Valsecchi 2005; Valsecchi *et al.* 2005) have provided important information on the limits of the range of this species.

Cacajao ayresi

The description of this species (Boubli *et al.* 2008) was based on specimens from two sites in the Demeni river basin, on the left bank of the Rio Negro, separated from *C. melanocephalus* by a distance of little more than 50 km. On the basis of the available evidence, the authors restrict the distribution of the new species to a small area of little more than 5000 km^2, which would be one of the smallest ranges of any platyrrhine.

Genus *Chiropotes*

Taxonomy

Once again, Hill (1960) and Hershkovitz (1985) were responsible for the standards of bearded saki taxonomy in the second half of the twentieth century. Hill recognized three species, *Chiropotes satanas*, *Chiropotes chiropotes* and *Chiropotes albinasus*, which Hershkovitz modified by considering *C. chiropotes* a subspecies of *C. satanas*, and adding a fourth taxon, *Chiropotes satanas utahicki*. In their preliminary analysis, Silva Jr. and Figueiredo (2002) considered all four of these taxa to be species, and divided *C. chiropotes* into two species, separated by the Rio Branco, with *C. chiropotes* to the west of this river, and *Chiropotes sagulatus* to the east. Bonvicino *et al.* (2003) also supported this arrangement, but interpreted the nomenclature of the two northern species differently, considering the species to the east of the Rio Branco to be *C. chiropotes*, and western species to be *Chiropotes israelita*.

More recently, Figueiredo (2006) reconfirmed this species arrangement through molecular (cytochrome *b*) analyses. As *C. israelita* is a junior synonym of *C. chiropotes*, the original classification of Silva Jr. and Figueiredo (2002) is upheld here (Table 4.4), based on the priority rule of the International Commission of Zoological Nomenclature (1999). These authors also corrected the *utahicki* epithet to *utahickae*, considering that the name honours a female scientist, Uta Hick.

Table 4.4 Taxonomy of the bearded sakis, genus *Chiropotes*, according to Silva Jr. & Figueiredo (2002).

Species	Common name	Distribution (countries)
C. albinasus	White-nosed bearded saki	Brazil
C. chiropotes	Tawny-olive bearded saki	Brazil, Venezuela
C. sagulatus	Reddish-brown bearded saki	Brazil, Guyanas, French Guiana, Suriname
C. satanas	Black bearded saki	Brazil
C. utahickae	Uta Hick's bearded saki	Brazil

Geographic distribution

Bearded saki species are all parapatric, with ranges delimited by major rivers, including the Amazon, Madeira, Negro–Branco, Tocantins and Xingu (Figure 4.6). The western limits are well defined by the Solimões–Negro, Orinoco and Madeira, but the southern and eastern extremes are less clear, being related to the Amazon–*Cerrado* ecotone (Silva Jr. 1991). While records are scant for some areas, there may be major lacunas in the distribution of the genus, such as the Xingu–Iriri and Araguaia–Tocantins interfluvia, and the lower Madeira (Hershkovitz 1985; Silva Jr. & Noronha 2000). The genus may also be absent from most of the Marajó archipelago.

Chiropotes albinasus

This species is found south of the Amazon, between the Xingu and Madeira rivers, southwards to the *Hylea–Cerrado* ecotone (Pimenta & Silva Jr. 2005). It is still unclear whether and to what extent the species occurs east of the Rio Iriri (Martins *et al.* 1988). Ferrari *et al.* (1999) found that the species was absent from most of Rondônia, west of the Rio Jiparaná, although Wallace *et al.* (1996) reported the species much further south, on the right bank of the Rio Guaporé.

Chiropotes chiropotes

The distribution of this species is delimited by the Branco, Negro and Orinoco rivers to the east, southwest and northwest, respectively. Norconk (2011) suggests that the species dispersed into Venezuela along the east bank of the Orinoco, reaching the Rio Caroni (Bodini & Pérez-Hernández 1987; Kinzey *et al.* 1988), but not areas further east, which might account for the absence of the genus from western Guyana.

Chiropotes sagulatus

Found north of the lower Amazon as far west as the Negro–Branco river system, this species ranges north into the Guyanas, where it appears to be patchily distributed (Thoisy *et al.* 2000) and absent west of the Essequibo River (Norconk 2011). The species may also be absent from neighboring areas of northern Brazil.

Chiropotes satanas

This is the easternmost species, and has by far the smallest geographic range of any bearded saki. It is found south of the lower Amazon and east of the Tocantins, eastwards into the state of Maranhão, to the eastern limit of the Amazonian Hylea (Silva Jr. 1991).

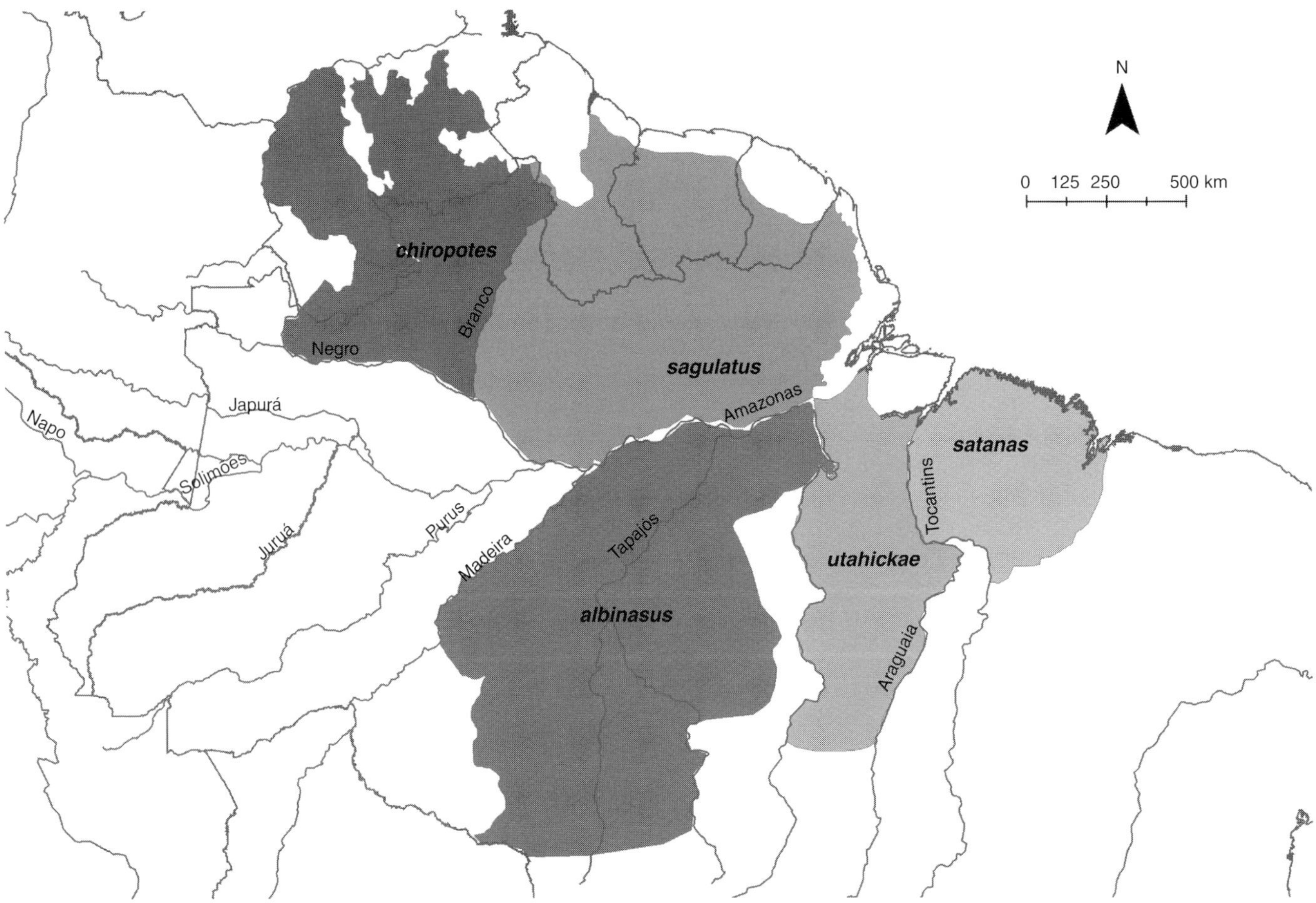

Figure 4.6 Geographic distribution of *Chiropotes*, based on Hershkovitz (1985), Silva Jr. (1991), Ferrari *et al.* (1999), Wallace *et al.* (1996) and Silva Jr. & Figueiredo (2002).

Chiropotes utahickae

Also found south of the Amazon, this species ranges between the Xingu and the Tocantins. Hershkovitz (1985) defined the Rio Itacaiunas in the Serra dos Carajás as the southern limit of this species, although it has been recorded further south (National Museum specimen MN-28799), on the lower Rio Tapirapé, in northern Mato Grosso.

Overview

As for other platyrrhine families, the Pitheciidae is characterized by a considerable diversity of taxa, but is also plagued by a dearth of adequate data on many genetic, morphological, ecological and zoogeographic parameters. This is especially the case throughout much of the Amazon and Orinoco basins, where the scale of the region limits the potential for the collection of data. Despite considerable recent advances in our understanding of all four genera, it seems likely that their true diversity is still underestimated considerably.

On the one hand, while they have been reviewed extensively over recent years, the systematics of both *Callicebus* (Kobayashi 1995; Kobayashi & Langguth 1999; van Roosmalen *et al.* 2002) and *Chiropotes* (Silva Jr. & Figueiredo 2002; Bonvicino *et al.* 2003; Figueiredo 2006) are still open to controversy, and future studies may yet reveal new taxa. On the other hand, *Cacajao* and in particular *Pithecia* have received little attention in the more than 20 years since the reviews of Hershkovitz (1987a, 1987b), although this situation is also now shifting (Figueiredo 2006; Marsh 2006; Boubli *et al.* 2008).

As for many other Neotropical vertebrate groups, the pitheciids suffer not only from a scarcity of specimens, but also a lack of multidisciplinary studies, and the inadequate application of the subspecific category (Vivo 2007). Nevertheless, it is hoped that ongoing research will result in an ever-more adequate pool of samples, and that the integration of morphological studies with complementary approaches, in particular molecular genetics, will provide increasingly reliable estimates of the true diversity of these primates.

Acknowledgments

We are grateful to Liza Veiga and Luís Barbosa for producing the maps.

References

Aquino, R. (1998). Some observations on the ecology of *Cacajao calvus ucayalii* in the Peruvian Amazon. *Primate Conservation*, **18**, 21–24.

Ayres, J.M. (1989). Comparative feeding ecology of the uakari and bearded saki, *Cacajao* and *Chiropotes*. *Journal of Human Evolution*, **18**, 697–716.

Barnett, A.A. & Brandon-Jones, D. (1997). The ecology, biogeography and conservation of the uacaris, *Cacajao* (Pitheciinae). *Folia Primatologica*, **68**, 223–235.

Bodini, R. & Pérez-Hernández, R. (1987). Distribution of the species and subspecies of cebids in Venezuela. *Fieldiana Zoology, New Series*, **39**, 231–244.

Bonvicino, C.R., Boubli, J.P., Otazú, I.B., *et al.* (2003). Morphologic, karyotypic, and molecular evidence of a new form of *Chiropotes* (Primates, Pitheciinae). *American Journal of Primatology*, **61**, 123–133.

Boubli, J.P. (1994). The black uakari monkey in the Pico da Neblina National Park. *Neotropical Primates*, **2**, 11–12.

Boubli, J.P. (1997). Feeding of black-headed uacaris (*Cacajao melanocephalus melanocephalus*) in Pico da Neblina National Park, Brazil. *International Journal of Primatology*, **20**, 719–749.

Boubli, J.P. (2002). Western extension of the range of bearded sakis: a possible new taxon of *Chiropotes* sympatric with *Cacajao* in the Pico da Neblina National Park, Brazil. *Neotropical Primates*, **10**, 1–4.

Boubli, J.P. & Tokuda, M. (2008). Socioecology of black uakari monkeys, *Cacajao hosomi*, in Pico da Neblina National Park, Brazil. The role of the peculiar spatial-temporal distribution of resources in the Neblina forests. *Primate Report*, **75**, 3–10.

Boubli, J.P., da Silva, M.N.F., Amado, M.V., *et al.* (2008). A taxonomic reassessment of *Cacajao melanocephalus* Humboldt (1811), with the description of two new species. *International Journal of Primatology*, **29**, 723–741.

Carrillo-Bilbao, G., Di Fiore, A. & Fernández-Duque, E. (2005). Dieta, forrajeo e presupuesto de tiempo en cotoncillos (*Callicebus discolor*) del Parque Nacional Yasuní em la Amazonia ecuatoriana. *Neotropical Primates*, **13**, 7–11.

Cunha, A.C. & Barnett, A.A. (1990). Sightings of the golden-backed uacari, *Cacajao melanocephalus ouakary*, on the upper Rio Negro, Amazonas, Brazil. *Primate Conservation*, **11**, 8–11.

Defler, T.R. (2004). *Primates of Colombia*. Bogotá: Conservation International.

Felton, A., Felton, A.M., Wallace, R.B., *et al.* (2006). Identification, behavioral observations, and notes on the distribution of the titi monkeys *Callicebus modestus* Lönnberg, 1939 and *Callicebus olallae*, Lonnberg, 1939. *Primate Conservation*, **20**, 41–46.

Fernandes, M.C.A.G. (1990). Distribuição de primatas não-humanos no Estado do Acre e vizinhanças: Um estudo preliminar. Unpublished undergraduate dissertation, Universidade Federal do Acre.

Ferrari, S.F., Guedes, P.G., Figueiredo, W.B., *et al.* (2010). Re-evaluation of the nomenclature of the black-faced uacaris (*Cacajao melanocephalus* group, *sensu* Hershkovitz, 1987). Abstracts of the XXIII International Primatology Congress, 2010, Kyoto.

Ferrari, S.F., Iwanaga, S., Coutinho, P.E.G., *et al.* (1999). Zoogeography of *Chiropotes albinasus* (Pltyrrhini, Atelidae) in southwestern Amazônia. *International Journal of Primatology*, **20**, 995–1004.

Figueiredo, W.M.B. (2006). Estimativas de tempos de divergência em platirrinos e filogenia molecular e filogeografia dos uacaris, parauacus e cuxiús. Unpublished PhD thesis, Universidade Federal do Pará.

Ford, S.M. (1986). Systematics of the New World monkeys. In *Comparative Primates Biology, Vol. 1: Systematics, Evolution and Anatomy*, ed. D.R. Swindler & J. Erwin. New York, NY: Alan R. Liss, pp. 73–135.

Goodman, M., Porter, C.A., Czelusniak, J., *et al.* (1998). Toward a phylogenetic classification of primates based on DNA evidence complemented by fossil evidence. *Molecular Phylogenetics and Evolution*, **9**, 585–598.

Groves, C.P. (2001). *Primate Taxonomy*. Washington, DC: Smithsonian Institution Press.

Groves, C.P. (2005). Order Primates. In *Mammal Species of the World: A Taxonomic and Geographic Reference, 3rd edition*, ed. D.E. Wilson & D.M. Reeder. Baltimore, MD: The Johns Hopkins University Press, pp. 111–184.

Hershkovitz, P. (1963). A systematic and zoogeographic account of the monkeys of the genus *Callicebus* (Cebidae) of the Amazonas and Orinoco River basins. *Mammalia*, **27**, 1–80.

Hershkovitz, P. (1979). The species of sakis, genus *Pithecia* (Cebidae, Primates), with notes on sexual dichromatism. *Folia Primatologica*, **31**, 1–22.

Hershkovitz, P. (1985). A preliminary taxonomic review of the South American bearded saki monkeys genus *Chiropotes* (Cebidae, Platyrrhini), with description of a new subspecies. *Fieldiana Zoology*, **27**, 1–45.

Hershkovitz, P. (1987a). The taxonomy of South American sakis, genus *Pithecia* (Cebidae, Platyrrhini): a preliminary report and critical review with the description of a new species and a new subspecies. *American Journal of Primatology*, **12**, 386–468.

Hershkovitz, P. (1987b). Uakaris, New World monkeys of the genus *Cacajao* (Cebidae, Platyrrhini): a preliminary review with the description of a new subspecies. *American Journal of Primatology*, **12**, 1–57.

Hershkovitz, P. (1988). Origin, speciation, dispersal of South American titi monkeys, genus *Callicebus* (Cebidae, Platyrrhini). *Proceedings of the Academy of Natural Sciences of Philadelphia*, **140**, 240–272.

Hershkovitz, P. (1990). Titis, New World monkeys of the genus *Callicebus* (Cebidae, Platyrrhini): a preliminary taxonomic review. *Fieldiana Zoology, New Series*, **55**, 1–109.

Heymann, E.W. (1990). Further field notes on red uacaris, *Cacajao calvus ucayalii*, from Quebrada Blanco, Amazonian Peru. *Primate Conservation*, **11**, 7–8.

Hill, W.C.O. (1960). *Primates. Comparative Anatomy and Taxonomy. IV Cebidae. Part A*. Edinburgh: Edinburgh University Press.

International Commission on Zoological Nomenclature (1999). *International Code of Zoological Nomenclature*. London: British Museum of Natural History. http://www.iczn.org/iczn/index.jsp

Kay, R.F. (1990). The phyletic relationships of extant and fossil Pitheciinae (Platyrrhini, Anthropoidea). *Journal of Human Evolution*, **19**, 175–208.

Kinzey, W.G., Norconk, M.A. & Alvarez-Cordero, E. (1988). Primate survey of eastern Bolívar, Venezuela. *Primate Conservation*, **9**, 66–70.

Kobayashi, S. (1995). A phylogenetic study of titi monkeys, genus *Callicebus*, based on cranial measurements: I. Phyletic groups of *Callicebus*. *Primates*, **36**, 101–120.

Kobayashi, S. & Langguth, A. (1999). A new species of titi monkey, *Callicebus* Thomas, from north-eastern Brazil (Primates, Cebidae). *Revista Brasileira de Zoologia*, **16**, 531–551.

Mark, M.M. (2003). Some observations on *Callicebus oenanthe* in the upper Río Mayo valley, Peru. *Neotropical Primates*, **11**, 183–187.

Marsh, L.K. (2004). Primate species at the Tiputini Biodiversity Station, Ecuador. *Neotropical Primates*, **12**, 75–78.

Marsh, L.K. (2006). Identification and conservation of a new species of *Pithecia* in Amazonian Ecuador. *International Journal of Primatology*, **27**, 509.

Martins, E.S., Ayres, J.M. & Valle, M.B.R. (1988). On the status of *Ateles belzebuth marginatus* with notes on other primates of the Iriri River Basin. *Primate Conservation*, **9**, 87–90.

Norconk, M.A. (2011). Sakis, uakaris, and titi monkeys: behavioral diversity in a radiation of primate seed predators. In *Primates in Perspective* (2nd edn), ed. C.J. Campbell, A. Fuentes, K.C. MacKinnon, S.K. Bearder & R.M. Stumpf. New York, NY: Oxford University Press, pp. 122–139.

Oliveira, C.R., Coelho, A.S. & Melo, F.R. (2003). Estimativa de densidade e tamanho populacional de sauá (*Callicebus nigrifrons*) em um fragmento de mata em regeneração, Viçosa, Minas Gerais, Brasil. *Neotropical Primates*, **11**, 91–94.

Pimenta, F.E. & Silva Jr., J.S. (2005). An update on the distribution of primates of the Tapajós–Xingu interfluvium, Central Amazonia. *Neotropical Primates*, **13**, 23–28.

Rosenberger, A.L., Setoguchi, T. & Shigehara, N. (1990). The fossil record of callitrichine primates. *Journal of Human Evolution*, **19**, 209–236.

Rowe, N. & Martinez, W. (2003). *Callicebus* sightings in Bolivia, Peru and Ecuador. *Neotropical Primates*, **11**, 32–35.

Rylands, A.B., Schneider, H., Langguth, A., *et al.* (2000). An assessment of the diversity of New World primates. *Neotropical Primates*, **8**, 61–93.

Schneider, H. (2000). The current status of the New World monkey phylogeny. *Anais da Academia Brasileira de Ciências*, **72**, 165–172.

Schneider, H. & Rosenberger, A.L. (1996). Molecules, morphology, and platyrrhine systematics. In: *Adaptative Radiations of Neotropical Primates*, ed. M.A. Norconk, A.L. Rosenberger & P.A. Garber. New York, NY: Plenum Press, pp. 3–19.

Silva Jr., J.S. (1991). Distribuição geográfica do cuxiú-preto (*Chiropotes satanas satanas* Hoffmannsegg, 1807) na Amazônia Maranhense (Cebidae, Primates). In *A Primatologia no Brasil – 3*, ed. A.B. Rylands & A.T. Bernardes. Belo Horizonte: Sociedade Brasileira de Primatologia, pp. 275–284.

Silva Jr., J.S. & Cerqueira, R. (1997). Variação geográfica de *Pithecia pithecia* Linnaeus, 1766 (Primates, Cebidae). *Resumos do VIII Congresso Brasileiro de Primatologia*, **65**.

Silva Jr., J.S. & Figueiredo, W.M.B. (2002). Revisão sistemática dos cuxiús, gênero *Chiropotes* Lesson, 1840 (Primates, Pitheciidae). *Livro de resumos do X Congresso Brasileiro de Primatologia*, **21**.

Silva Jr., J.S. & Martins, E.S. (1999). On a new white bald uakari population in southwestern Brazilian Amazonia. *Neotropical Primates*, **7**, 119–121.

Silva Jr., J.S. & Noronha, M.A. (2000). Resultados de uma pequena expedição primatológica à Amazônia Central (Primates, Platyrrhini). In *A Primatologia no Brasil – 7*, ed. C. Alonso & A. Langguth. João Pessoa: Sociedade Brasileira de Primatologia, pp. 291–304.

Thoisy, B., Massemin, D. & Dewynter, M. (2000). Hunting impact on Neotropical primates: a preliminary case study in French Guiana. *Neotropical Primates*, **8**, 141–144.

Valsecchi, J. (2005). Diversidade de mamíferos e uso da fauna nas Reservas de Desenvolvimento Sustentável Mamirauá e Amanã. Unpublished Masters dissertation, Museu Paraense Emílio Goeldi.

Valsecchi, J., Muniz, I.C.M., Avelar, A.A., *et al.* (2005). Primatas da Reserva de Desenvolvimento Sustentável Amanã, Amazonas. *Livro de Resumos do XI Congresso Brasileiro de Primatologia*, **70**.

van Roosmalen, M.G.M., van Roosmalen, T. & Mittermeier, R.A. (2002). A taxonomic review of the titi monkeys, genus *Callicebus* Thomas, 1903, with the description of two new species, *Callicebus bernhardi* and *Callicebus stephennashi*, from Brazilian Amazonia. *Neotropical Primates*, **10** (Suppl.), 1–52.

Vasconcelos, M.F. & Hirsch, A. (2000). A new locality for the masked titi monkey, *Callicebus personatus nigrifrons*, in a protected área in Minas Gerais, Brazil. *Neotropical Primates*, **8**, 153–154.

Vivo, M. de (2007). Problemas da Mastozoologia brasileira. *Boletim da Sociedade Brasileira de Mastozoologia*, **48**, 1–4.

Wallace, R.B., Gómez, H., Felton, A., *et al.* (2006). On a new species of titi monkey, genus *Callicebus* Thomas (Primates, Pitheciidae), from western Bolivia with preliminary notes on distribution and abundance. *Primate Conservation*, **20**, 29–39.

Wallace, R.B., Painter, R.L.E., Taber, A.B., *et al.* (1996). Notes on distributional river boundary and southern range extension for two species of Amazonian primates. *Neotropical Primates*, **4**, 149–151.

Ward, N.S. & Chism, J. (2003). A report on a new geographic location of red uacaris (*Cacajao calvus ucayalii*) on the Quebrada Tahuaillo in northeastern Peru. *Neotropical Primates*, **11**, 19–22.

Chapter 5

Zoogeography, genetic variation and conservation of the *Callicebus personatus* group

Rodrigo C. Printes, Leandro Jerusalinsky, Marcelo C. Sousa, Luis Reginaldo R. Rodrigues & André Hirsch

Introduction

The genus *Callicebus* occurs in two disjunct areas in South America: the tropical forests of the Amazon and Orinoco basins (extending southwards to the Paraguay and Paraná rivers), and the Brazilian Atlantic Forest. These two areas are separated by over 500 km of drier vegetation types (Hershkovitz 1988). The geographical origin of the contemporary forms of *Callicebus* has caused considerable controversy and is still unclear (Kinzey 1982; Hershkovitz 1988).

Twenty-nine species of *Callicebus* are currently recognized (van Roosmalen *et al.* 2002; Wallace *et al.* 2006), making this one of the most diverse of Neotropical primate genera, comparable to *Saguinus* (32 taxa) and *Cebus*, with 33 taxa (Rylands *et al.* 1996–1997). Kobayashi & Langguth (1999) and van Roosmalen *et al.* (2002) recognize five species groups – *cupreus*, *donacophilus*, *moloch*, *personatus* and *torquatus* – of which only the *personatus* group has an exclusively extra-Amazonian distribution. The *personatus* group currently has five recognized species: *Callicebus barbarabrownae*, *Callicebus coimbrai*, *Callicebus melanochir*, *Callicebus nigrifrons* and *Callicebus personatus*.

Although distributed predominantly in the Atlantic Forest, some members of the *personatus* group also occur in other ecosystems, including arboreal *Caatinga* and riparian forests within the *Cerrado* (Hershkovitz 1988). Inhabiting the most developed and populous region of Brazil, all five species have suffered extensive habitat loss and fragmentation over the past five centuries. Anthropogenic pressures of all kinds, combined with a lack of basic biological information about the species, have resulted in the entire *personatus* group being considered threatened with extinction, with two species, *C. barbarabrownae* and *C. coimbrai*, classified as critically endangered and endangered, respectively (Machado *et al.* 2005; Veiga *et al.*, 2008a, 2008b), an extremely worrying situation. *Callicebus melanochir* and *C. personatus* are considered vulnerable, although *C. nigrifrons* is currently classified only as "near threatened".

Zoogeography of the genus *Callicebus*

As for all Neotropical primates, the origin of *Callicebus* is hard to elucidate, due primarily to the lack of an adequate fossil record. *Miocallicebus villaviejai* Takai *et al.* 2001, from the middle Miocene of Colombia, appears to be the oldest fossil with a resemblance to modern *Callicebus* (see Kay *et al.*, Chapter 1), although the paucity of samples prohibits a more systematic analysis.

Hershkovitz (1963, 1977, 1988, 1990) believed that *Callicebus* originated in the upper Amazon basin, subsequently dispersing to lowland areas along river systems during the climatic changes of the Quaternary. According to this author's centripetal dispersal theory, populations – including those of what is now the *personatus* group – became isolated during further river formation, which disrupted dispersal routes, resulting in patterns of sympatry and gradual speciation.

By contrast, Kinzey (1982) emphasized the central role of Pleistocene refugia in *Callicebus* speciation. These refugia were relict tracts of forest isolated during periods of drier climate (Ab'Saber 1977; Defler 2004). Kinzey (1997) affirmed that *Callicebus* fossils found by Lund in 1939, at Lagoa Santa (Minas Gerais), represent the western-most extreme of the distribution of *Callicebus melanochir*, dispersing from a forest refuge in central Bahia.

While the debate is ongoing, most authors prefer to view the formation of the Pleistocene refugia as the fundamental process underlying platyrrhine speciation (Futuyma 1996; Defler 2004). Wherever *Callicebus* originated, it persists in both the Atlantic Forest and Amazonia, and the understanding of the historical factors that permitted its dispersal between the two biomes will be important not only for the definition of zoogeographic processes, but also the systematic assessment of the ecological and geographic determinants of present-day ranges.

Present-day diversity and distribution of the *Callicebus personatus* group

The current distribution of the *Callicebus personatus* species group extends across the Brazilian states of Bahia, Sergipe, Minas Gerais, Espírito Santo, Rio de Janeiro and São Paulo, predominantly in the Atlantic Forest biome, but also in neighboring *Caatinga* and *Cerrado* ecosystems (van Roosmalen *et al.* 2002). In central Brazil, the distribution of *Callicebus* is interrupted by

Evolutionary Biology and Conservation of Titis, Sakis and Uacaris, eds. Liza M. Veiga, Adrian A. Barnett, Stephen F. Ferrari and Marilyn A. Norconk. Published by Cambridge University Press.

the *Cerrado*, a drier more open habitat which extends from the Araguaia–Tocantins and Paraguai–Taquari Novo river systems, in the west, to the São Francisco and Paraná–Paraíba in the east (see Hershkovitz 1988). Hershkovitz believed that, in the past, riparian forest along *Cerrado* river systems would have provided suitable habitats for *Callicebus*, facilitating dispersal between the Amazonian and Atlantic Forest biomes. He proposed that these *Cerrado*-dwelling titis disappeared during the climatic changes of the Quaternary, which caused the biome to become even drier, restricting riparian forests to small, isolated valleys. With this, the *Cerrado* became an ecological barrier for the titis. Current distributions reflect this, with members of the *personatus* group being restricted to the right bank of São Francisco River, in northeastern Brazil, and to the right bank of Tietê River, in the southeast.

Distribution of the *personatus* group species

Callicebus personatus (Geoffroy, 1812)

This species occurs in the state of Espírito Santo, northwestern Minas Gerais and northern Rio de Janeiro (van Roosmalen *et al.* 2002). Oliver and Santos (1991) indicated that the region of the Itaúnas and Mucuri rivers might be a zone of intergradation between *C. personatus* and *C. melanochir*. *Callicebus personatus* occurs on the right bank of the Jequitinhonha River, but it remains unclear whether this species or another titi occurs to the northwest of this river (van Roosmalen *et al.* 2002). The range of this species extends westwards along the Rio Doce valley into Minas Gerais as far as the Mantiqueira Mountains. It is possible that the distribution of *C. personatus* is more restricted than previously supposed, as *C. nigrifrons* and not *C. personatus* was recorded at the frontier between Minas Gerais and Espírito Santo, an area formerly considered to be within the range of *C. personatus*.

Callicebus nigrifrons (Spix, 1823)

Callicebus nigrifrons is restricted to the states of São Paulo and Minas Gerais. It has the most ample distribution of any *personatus* group species (van Roosmalen *et al.* 2002), and occurs in both mature rainforest and disturbed fragments. It is found north of the Tietê and east of the Paraná and Paranaíba rivers, and on both margins of the São Francisco. The species also occurs east of the Mantiqueira and Espinhaço ranges, where it meets *C. personatus* (see above).

Callicebus melanochir (Wied-Neuwied, 1820)

Hershkovitz (1990) placed the geographic distribution of *C. melanochir* between the Mucuri River, in Espírito Santo and the Paraguaçu, in Bahia. However, van Roosmalen *et al.* (2002) considered the possibility of an overlap zone with *C. personatus* in the valleys of the Itaúnas and Mucuri rivers, in the north of Espírito Santo.

As noted above, the *Cerrado* proper acts as a barrier to the distribution of modern-day titi taxa, and *C. melanochir* probably ranges no further west into Minas Gerais than Montes Claros (Printes 2007). To the north, the main obstacle to the dispersal of *C. melanochir* appears to be the Paraguaçu River. It is likely that *C. melanochir* once occurred in the Pardo–Jequitinhonha interfluvium, but the forests of this region have now been replaced by cattle pasture and our surveys found no *Callicebus* there. *Callicebus melanochir* does occur in the Pau Brasil National Park, south of the Jequitinhonha. More surveys are needed, but it is possible that the species may reach the Rio Doce, some 300 km further south. To the south of the Jequitinhonha, *C. melanochir* is restricted to remnant coastal forests. To the west, it is replaced by *C. personatus*.

Callicebus barbarabrownae (Hershkovitz, 1990)

Until the survey of Printes (2007), this species was known from only four localities, but *C. barbarabrownae* has now been confirmed at no less than 37 sites. The species appears to be endemic to the state of Bahia, where it inhabits upland areas between 241 and 908 m above sea level (Printes & Rylands 2008; Printes *et al.* 2011). Using the minimum polygon convex approach (IUCN 1994), its total range is believed to cover an area of some 252,546 km^2, although actual occupancy is estimated at 2636 km^2 (Figure 5.1).

No titis were recorded in the region west of Araci and Nova Soure (Figure 5.1), which is dominated by the *Cerrado*. However, *C. barbarabrownae* was recorded in the *Caatinga* of the moister uplands northeast of Araci (Mandacaru, Mirandela and Serra Branca). The species was also recorded in *Caatinga* habitat further north as far as the Salitre River, 170 km from Juazeiro, but titis were not recorded west of the Chapada Diamantina, which supports Hershkovitz's (1988, 1990) conclusion: that topographic relief, or the shifts in vegetation associated with it, limit the dispersal of titis in this region.

In eastern Bahia, the *Agreste* forests of the coastal range probably supported large populations of *C. barbarabrownae* in the past, although this type of vegetation is rare in the present day. We nevertheless recorded populations in five municipalities (Cícero Dantas, Antas, Sítio do Quinto, Jeremoabo and Pedro Alexandre).

Callicebus coimbrai (Kobayashi and Langguth, 1999)

Kobayashi and Langguth (1999) originally restricted the distribution of *C. coimbrai* to the coastal rainforests between the mouths of the São Francisco River in Sergipe and the Itapicuru in Bahia. Nevertheless, Printes (2007) recorded the species at Lamarão do Passé (Figure 5.1), extending its range some 200 km further south. Kobayashi and Langguth (1999) also believed that *C. coimbrai* was restricted to primary rainforest, although more recent surveys have shown that the species also occurs in fragments of seasonal forest and forests with a *Caatinga* influence, along the border between the states of Sergipe and Bahia (Sousa 2003; Jerusalinsky *et al.* 2006). However, further west,

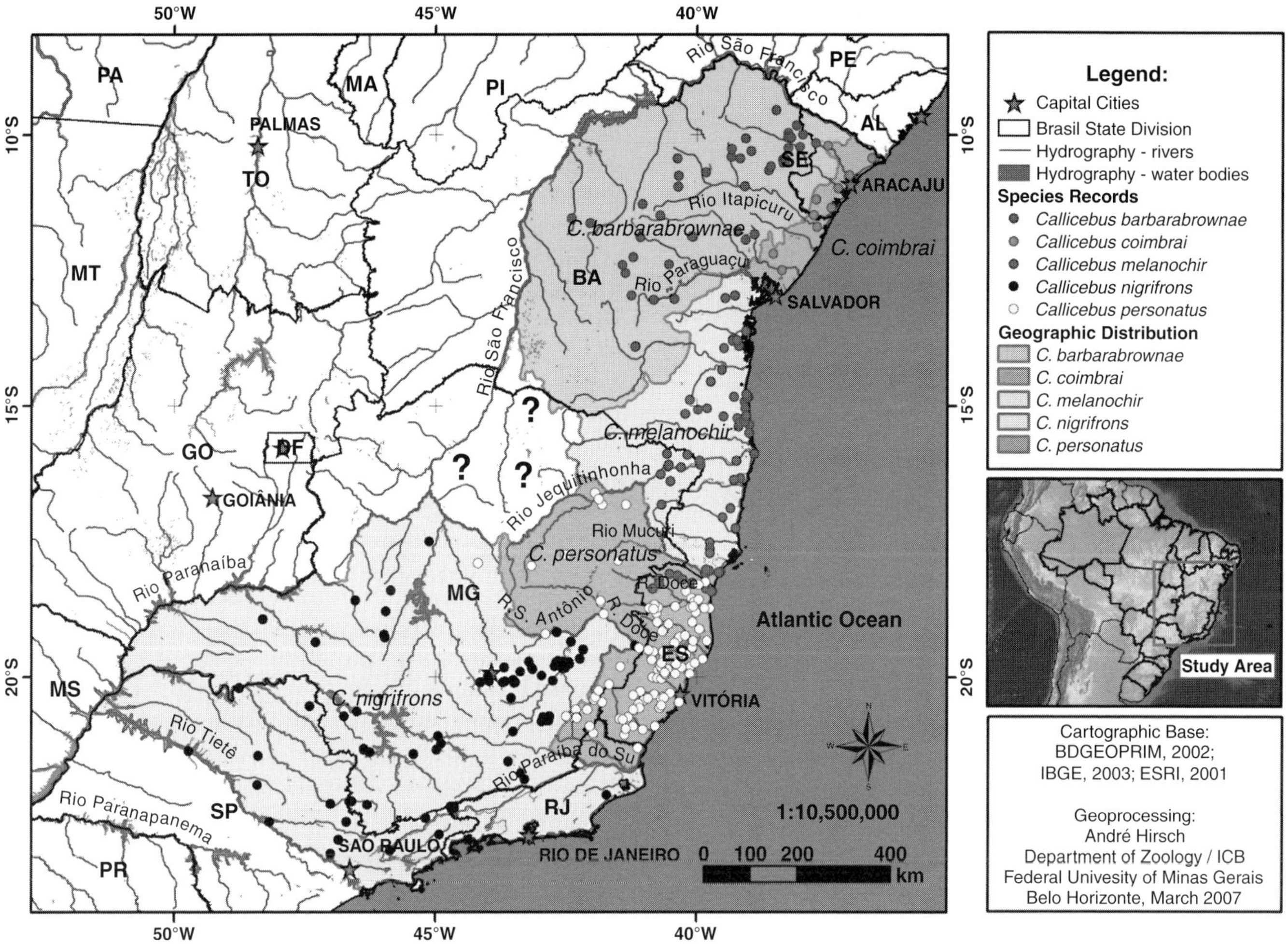

Figure 5.1 Geographic distribution of *personatus* group of *Callicebus*.

the species is replaced by *C. barbarabrownae* in more typical *Caatinga* ecosystems (Printes *et al.* 2011). The species distribution covers little less than 30,000 km^2 (Jerusalinsky 2008), and our surveys confirmed its presence in at least 35 forest remnants (Sousa 2000, 2003; Jerusalinsky *et al.* 2006; Printes *et al.* 2011). The largest of these fragments is a 3000 ha tract in São Francisco do Paraguaçu, Bahia, whereas the smallest is a 3 ha fragment in Arauá, Sergipe (Jerusalinsky *et al.* 2006).

Distributional limits of *Callicebus* in the Brazilian northeast

The recent surveys of *C. barbarabrownae* and *C. coimbrai* have provided a number of insights into the factors limiting the distribution of *Callicebus* species in northeastern Brazil. For example, *C. melanochir* replaces *C. barbarabrownae* in the region known as the "Recôncavo Baiano" (south of the Paraguaçu River in the municipality of Igrapiuna). The individuals observed in this area presented the unmistakable *C. melanochir* color pattern of uniform gray trunk with chestnut-brown tail.

The lower Paraguaçu forms the limit between the geographic ranges of *C. coimbrai*, restricted to the left or north bank, and *C. melanochir*, found on the right bank. Further west, in the region of Feira de Santana, *C. barbarabrownae* occurs on both banks of the Paraguaçu. This suggests to us that *C. barbarabrownae* dispersed prior to the formation of the Paraguaçu river system, and that the process was influenced by regional phytogeography, rather than physical barriers. There does not appear to be a physical barrier between the eastern limit of the range of *C. barbarabrownae* and the western limit of *C. coimbrai*, although there may be an association with altitude and related humidity gradients, with the latter species typically occurring at lower, and relatively more humid altitudes.

We also recorded *C. coimbrai* in areas of *Caatinga* with a strong rainforest influence (e.g. Nossa Senhora da Glória, Sergipe). Coimbra-Filho and Câmara (1996) suggested that anthropogenic disturbance over the past 450 years has favored the expansion of the more xeric *Caatinga* vegetation into areas once covered by Atlantic Forest. If this is true, *C. barbarabrownae* may now occupy areas that were inhabited by *C. coimbrai* in recent times.

Genetic diversity

Genetic studies of *Callicebus* have mainly involved cytogenetic analyses, with a few contributions from molecular biology (Schneider *et al.* 1993; Bonvicino *et al.* 2003). The results have revealed a high karyotypic diversity in the genus, with diploid numbers ranging from 2n = 16 in *Callicebus lugens* (the lowest number found in any primate) to 2n = 50 in *Callicebus donacophilus* (De Boer 1974; Minezawa & Borda 1984), *Callicebus pallescens* (Stanyon *et al.* 2000; Dumas *et al.* 2005) and *Callicebus hoffmannsi* (Rodrigues *et al.* 2001). This diversity is the result of genomic reorganization involving centric fusion and fissions, based primarily on tandem fusions and pericentric inversions.

Before the current study, three species from the *personatus* group had been karyotyped: *C. nigrifrons*, 2n = 42 (Nagamachi *et al.* 2003), and *C. personatus* and *C. coimbrai*, both with 2n = 44 (Rodrigues *et al.* 2004, 2006). These karyotypes are completely homologous in G banding patterns and in the number and position of NOR. The difference in diploid numbers is due to a centric fusion/fission rearrangement that transformed two acrocentric pairs of *C. personatus* and *C. coimbrai* chromosomes into a metacentric one in *C. nigrifrons* (Rodrigues *et al.* 2004). The C banding revealed a slight reduction in the amount of constitutive heterochromatin in the *C. coimbrai* karyotype. Comparative genomic mapping (ZOO-FISH) of human and *C. personatus* chromosomes showed that the *personatus* group represents a distinct lineage supported by five chromosomal synapomorphies (Rodrigues 2006).

The role of chromosomal rearrangements in the speciation process is controversial. Many neighboring pairs of species are known to have different karyotypes, e.g. *Callicebus moloch* (2n = 48) and *Callicebus hoffmannsi* (2n = 50); *Callicebus donacophilus* (2n = 50) and *Callicebus brunneus* (2n = 48); *C. nigrifrons* (2n = 42) and *C. personatus* (2n = 44). In all these cases, divergence is due to a centric fusion/fission rearrangement, which could result in unviable or unfertile hybrids because of problems in meiotic division, i.e. the formation of misbalanced gametes. Such rearrangements may represent a post-zygotic reproductive barrier, which would effectively isolate species (King 1995).

The karyotypes of *C. melanochir* and *C. barbarabrownae* are unknown. Additional genetic studies are necessary to understand the evolutionary relationships of *Callicebus* species, and to clarify the role of chromosomal rearrangements in the speciation process. Molecular studies may also contribute to our understanding of the phylogenetic relationships between the species of *personatus* group, and of this group with the rest of the genus.

Land-use patterns and conservation issues

The *personatus* species group occurs in the Brazilian region with the longest history of European colonization, which has resulted in widespread deforestation and habitat fragmentation. The impact of land-use patterns on *Callicebus* populations needs to be analyzed as a global phenomenon with regional characteristics. Although strategies for the conservation of biodiversity are usually defined and implemented at national and international levels (e.g. ECO 92 and COP 8), the success of these strategies depends ultimately on the practices of local communities, and the understanding of their exploitation of natural resources (Silvano & Begossi 2005).

The reduction of available habitat for the eastern Brazilian titis is ongoing, mainly due to demands for land for cattle ranching and agriculture. These economic activities, together with urbanization, are driven by the developmental paradigms of local governments. In Bahia, for example, 115 new municipalities were created between 1970 and 2000, while the urban population increased from 41% to 67% of the total (IBGE 2005). Even the most remote localities in the Bahian *Caatinga*, considered to be demographic deserts a century ago (Cunha 1901), are now undergoing development, facilitated by an extensive network of highways. The progressive expansion of both minor and major urban centres, such as Porto Alegre, São Paulo, Salvador and Belo Horizonte, has occurred at the cost of natural habitats. It has also generated secondary impacts on primate populations, with potential dangers ranging from roads and powerlines to domestic pets as potential predators (Printes *et al.* 2010).

The resulting small, isolated populations of titis are exposed to synergistic genetic and demographic risks (Lacy 1992), although hunting pressure is probably negligible to moderate in most cases, given their small body size (Jerusalinsky 2008). Titis are also rarely kept as pets, in comparison with the larger-bodied capuchins (*Cebus* spp.) and the much smaller marmosets (*Callithrix* spp.) from the same region. It remains unclear to what extent this reflects the relative abundance of the different species – marmosets, in particular, are very common in many urban areas – or the difficulty of keeping titis in captivity. Certainly, these monkeys are as rare in zoos as they are among pet keepers. During our extensive, long-term surveys of *Callicebus coimbrai*, for example, we have only observed three individuals being kept as pets (in contrast with dozens of *Callithrix jacchus*), and all three had been found wandering in open areas, dispersing between fragments (Jerusalinsky *et al.* 2006; Jerusalinsky 2008). Similarly, we observed only two individuals of *C. barbarabrownae* being kept as pets.

In the specific case of *C. melanochir*, populations in southern Bahia had been afforded a degree of protection by the traditional *cabruca* agroforestry system based on the cultivation of cocoa trees in the shade of remnants of native forest, which provided habitable cover for the titis. However, following widespread infestation by the fungal parasite *Moniliophthora* (*Crinipellis*) *perniciosa* at the end of the 1980s, cocoa cultivation declined abruptly, and the *cabrucas* have largely been replaced by pastures and eucalyptus plantations, neither of which are titi-friendly.

An additional current problem is the Agrarian Reform Program, which has stimulated the growth of political organizations such as the *Movimento dos Sem Terra* (Landless Agricultural Workers' Movement), which are able to exert considerable pressure on governments for the occupation of large rural properties throughout the Atlantic Forest. Some settlements have resulted in effective and well-integrated programs of sustainable development, but more often than not, the primary objective of these groups is the occupation of forested areas. This is partly because these areas are considered to have more productive soils, but also because landowners are more likely to relinquish forested land than that already under cultivation. Ultimately, government agencies will also often prioritize the areas because they are considered to be "unproductive" and as such, less valuable for the assessment of compensation due to the landowner on confiscation.

Despite increasing attempts in recent decades to integrate conservation efforts with economic development, the protection of natural populations in conservation units, such as parks and reserves, still seems to be the most effective approach. In this light, the five species of *personatus* group face widely differing scenarios. Considering only federal- and state-protected areas, the three least-threatened species of the group are represented by populations in relatively extensive areas of forest: *C. nigrifrons* in seven reserves, totalling more than 100,000 ha, *C. personatus* at nine localities, with a combined area of more than 100,000 ha, and *C. melanochir* in four areas with more than 50,000 ha of forest.

At the opposite extreme, neither *C. barbarabrownae* nor *C. coimbrai* – the two most endangered taxa of this group – are known to occur in any kind of official reserve. Jerusalinsky *et al.* (2006), for example, estimated that the total remnant population of *C. coimbrai* comprises some 500–1000 individuals, occupying a total area of 100–150 km^2, pulverized into numerous unprotected forest fragments, most of which cover less than 100 ha. Surveys of the 100,000 ha Raso da Catarina Ecological Station have failed to confirm the presence of *C. barbarabrownae* in this potentially important federal reserve (Printes 2007). These and other findings have led the Brazilian Federal Environment Institute's Centre for the Protection of Brazilian Primates (ICMBIO/CPB) to create an interinstitutional working group for the conservation of *C. barbarabrownae* and *C. coimbrai*, and to coordinate proposals for the creation of federal reserves aimed at the protection of these species. These species, together with *C. personatus* and *C. melanochir* (and *Cebus xanthosternos* and *Cebus robustus*) have been included in the portfolio of the International Committee for the Conservation and Management of Primates of the Northern Atlantic Forest and Caatinga, established in 2006 with the objective of assisting the Federal Institute in strategic decision-making for *in situ* and *ex situ* conservation (MMA 2006).

In the specific case of *C. coimbrai*, privately owned reserves – known as *Reservas Particulares de Patrimônio Natural* or RPPNs – may play an especially important role in the conservation of the species. While they may protect only relatively small areas of forest, and proportionately small populations of titis, such areas may nevertheless represent a substantial contribution to the conservation of the species, especially for the development of metapopulation management strategies, in combination with an eventual network of public reserves. A number of potential sites are currently under study in Sergipe, and may also contribute to the conservation of other endangered endemics, such as the yellow-breasted capuchin, *Cebus xanthosternos*, and the maned sloth, *Bradypus torquatus*.

Acknowledgments

We would like to thank Drs. Anthony Rylands, Stephen Ferrari and Júlio César Bicca-Marques. Roberto Groehs, Luisa Lokschin and André Alonso assisted in the surveys of *C. barbarabrownae*. Fieldwork was supported by Conservation International, Critical Ecosystem Partnership Fund, Fundação Biodiversitas, Margot Marsh Biodiversity Foundation, Instituto de Estudos Sócio-Ambientais do Sul da Bahia, CODEVASF, Universidade Federal de Sergipe, and Programa Macacos Urbanos (UFRGS–INGA). We are also grateful to André Hirsch for Figure 5.1.

References

Ab'Saber, A.N. (1977). Os domínios morfoclimáticos da América do Sul – primeira aproximação. *Geomorfologia*, **52**, 1–23.

Bonvicino, C.R., Penna-Firme, V., Nascimento, F.F., *et al.* (2003). The lowest diploid number (2n = 16) yet found in any primate: *Callicebus lugens* (Humboldt, 1811). *Folia Primatologica*, **74**, 141–149.

Coimbra-Filho, A.F. & Câmara, I.G. (1996). *Os Limites Originais do Bioma Mata Atlântica na Região Nordeste do Brasil.* Rio de Janeiro: Fundação Brasileira para a Conservação da Natureza.

Cunha, E. (1901). *Os Sertões*. Rio de Janeiro: Otto Pierre Editores (1979 edition).

De Boer, L.E.M. (1974). Cytotaxonomy of the Platyrrhini (Primates). *Genen en Phaenen*, **17**, 1–115.

Defler, T.R. (2004). *Primates de Colômbia.* Bogotá: Conservación Internacional.

Dumas, F., Bigoni, F., Stone, G., *et al.* (2005). Mapping genomic rearrangements in titi monkeys by chromosome flow sorting and multidirectional *in-situ* hybridization. *Chromosome Research*, **13**, 85–96.

Futuyma, D.J. (1996). *Biologia Evolutiva.* São Paulo: Sociedade Brasileira de Genética/CNPq.

Hershkovitz, P. (1963). A systematic and zoogeographic account of the monkeys of the genus *Callicebus* (Cebidae) of the Amazonas and Orinoco river basins. *Mammalia*, **27**, 1–80.

Hershkovitz, P. (1977). *Living New World Monkeys (Platyrrhini) with an Introduction to Primates.* Vol. **1**. Chicago, IL: University of Chicago Press.

Hershkovitz, P. (1988). Origin, speciation, and distribution of South American titi monkeys, genus *Callicebus* (Family Cebidae, Plathyrrhini). *Proceedings of The Academy of Natural Sciences of Philadelphia*, **140**, 240–272.

Hershkovitz, P. (1990). Titis, New World monkeys of the genus *Callicebus* (Cebidae, Platyrrhini): a preliminary taxonomic review. *Fieldiana Zoology*, **55**, 1–109.

IBGE (2005). *Censo Agropecuário*. Sistema IBGE da Agricultura (SIDRA). www.ibge.gov.br

IUCN (1994). Species Survival Commission. 1994. *IUCN Red List Categories*. The World Conservation Union, Gland, Switzerland, 30 November 1994.

IUCN (2008). *2008 IUCN Red List of threatened Species*. Gland: The World Conservation Union. www.redlist.org

Jerusalinsky, L. (2008). *Callicebus coimbrai* Kobayashi & Langguth 1999. In: *Livro vermelho das espécies da fauna brasileira ameaçadas de extinção*, ed. A.B.M. Machado, G.M. Drummond & A.P. Paglia. Belo Horizonte: Fundação Biodiversitas, **V.II**, pp. 769–771.

Jerusalinsky, L., Oliveira, M.M., Pereira, R.F., *et al.* (2006). Preliminary evaluation of the conservation status of *Callicebus coimbrai* Kobayashi & Langguth, 1999 in the Brazilian state of Sergipe. *Primate Conservation*, **21**, 25–32.

King, M. (1995). *Species Evolution: The Role of Chromosome Change*. Cambridge: Cambridge University Press.

Kinzey, W.G. (1982). Distribution of primates and forest refuges. In *Biological Diversification in the Tropics*, ed. G.T. Prance. New York, NY: Columbia University Press, pp. 455–482.

Kinzey, W.G. (1997). *Callicebus*. In *New World Primates – Ecology, Evolution and Behavior*, ed. W.G. Kinzey. New York, NY: Aldine de Gruyter, pp. 213–221.

Kobayashi, S. & Langguth, A. (1999). A new species of titi monkeys, *Callicebus* Thomas, from north-eastern Brazil (Primates, Cebidade). *Revista Brasileira de Zoologia*, **16**, 531–551.

Lacy, R.C. (1992). The effects of inbreeding on isolated populations: are minimum viable population sizes predictable? In *Conservation Biology – The Theory and Practice of Nature Conservation*, ed. P.L. Fiedler & S.K. Jain. London: Chapman & Hall, pp. 277–296.

Machado, A.B.M., Martins, C.S. & Drummond, G.M. (2005). *Lista da Fauna brasileira ameaçada de Extinção incluindo as Listas das Espécies quase ameaçadas e Deficientes em Dados*. Brasília: Fundação Biodiversitas & IBAMA/MMA.

Minezawa, M. & Borda, J.V. (1984). Cytogenetic study of the Bolivian Titi and revision of its cytotaxonomic state. *Kyoto University Overseas Research Reports on New World Monkeys*, **4**, 39–45.

MMA (2006). Portaria N° 26, de 09 de março de 2006. *Diário Oficial da União – Seção 1*. www.diario-oficial.com.br/

Nagamachi, C.Y., Rodrigues, L.R.R., Galetti Jr., P.M., *et al.* (2003). Cytogenetic studies in *Callicebus personatus nigrifrons* (Platyrrhini, Primates). *Caryologia*, **56**, 47–52.

Oliver, W.L.R. & Santos, I.B. (1991). Threatened endemic mammals of the Atlantic Forest region of south-east Brazil. *Wildlife Preservation Trust, Special Scientific Report*, **4**, 1–26.

Printes, R.C. (2007). *Avaliação taxonômica, distribuição e status do guigó-da-caatinga.* Unpublished PhD thesis, Universidade Federal de Minas Gerais.

Printes, R.C. & Rylands, A.B. (2008) *Callicebus barbarabrownae* Hershkovitz 1990. In *Livro vermelho das espécies da fauna brasileira ameaçadas de extinção*, ed. A.B.M. Machado, G.M. Drummond & A.P. Paglia. Belo Horizonte: Fundação Biodiversitas, **V.II**, pp. 766–768.

Printes, R.C., Buss, G., Jardim, M.M., *et al.* (2010). The urban monkey program: a survey of *Alouatta clamitans* in the south of Porto Alegre and its influence on land use policy between 1997 and 2007. *Primate Conservation*, **25**, 11–19.

Printes, R.C., Rylands, A.B. & Bicca-Marques, J.C. (2011). Distribution and status of the Critically Endangered blond titi monkeys of north-east of Brazil. *Oryx*, **45**, 439–443.

Rodrigues, L.R.R. (2006). Estudos citogenéticos comparativos por bandeamentos e pintura cromossômica (ZOO-FISH) em *Callicebus* (Platyrrhini – Primates). Unpublished PhD thesis, Universidade Federal do Pará.

Rodrigues, L.R.R., Barros, R.M.S., Pissinati, A., *et al.* (2001). Cytogenetic study of *Callicebus hoffmannsi* (Cebidae, Primates) and comparison with *C. m. moloch*. *Cytobios*, **105**, 137–145.

Rodrigues, L.R.R., Barros, R.M.S., Pissinati, A., *et al.* (2004). A new karyotype of an endangered primate species (*Callicebus personatus*) from the Brazilian Atlantic Forests. *Hereditas*, **140**, 87–91.

Rodrigues, L.R.R., Sousa, M.C., Pieczarka, J.C., *et al.* (2006). Karyotypic study of *Callicebus coimbrai*: a rare and threatened primate species from Brazil. *Caryologia*, **59**, 248–252.

Rylands, A.B., Rodríguez-Luna, E. & Cortés-Ortiz, L. (1996–1997). Neotropical primate conservation – the species and the IUCN/SSC primate specialist group network. *Primate Conservation*, **17**, 46–49.

Schneider, H., Schneider, M.P., Sampaio, M.I., *et al.* (1993). Divergence between biochemical and cytogenetic differences in three species of the *Callicebus moloch* group. *American Journal of Physical Anthropology*, **90**, 345–350.

Silvano, R.A.M. & Begossi, A. (2005). Local knowledge on a cosmopolitan fish: ethnoecology of *Pomatomus saltatrix* (Pamatomidae) in Brazil and Australia. *Fisheries Research*, **71**, 43–59.

Sousa, M.C. (2000). New localities for Coimbra-Filho's titi monkey, *Callicebus coimbrai*, in north-east Brazil. *Neotropical Primates*, **8**, 151.

Sousa, M.C. (2003). Distribuição do guigó (*Callicebus coimbrai*) no Estado de Sergipe. *Neotropical Primates*, **11**, 89–91.

Stanyon, R., Consigliere, S., Müller, S., *et al.* (2000). Fluorescence *in situ* hybridization (FISH) maps chromosomal homologies between the dusky titi and squirrel monkey. *American Journal of Primatology*, **50**, 95–107.

Takai, M., Anaya, F., Suzuki, H., *et al.* (2001). A new platyrrhine from the middle Miocene of La Venta, Colombia, and the phyletic position of Callicebinae. *Anthropological Science (Japan)*, **109**, 289–307.

van Roosmalen, M.G.M., van Roosmalen, T. & Mittermeier, R.A. (2002). A taxonomic review of the titi monkeys, genus *Callicebus* Thomas, 1903, with the description of two new species, *Callicebus bernhardi* and *Callicebus stephennashi*, from Brazilian Amazonia. *Neotropical Primates*, **10**, 1–52.

Veiga, L.M., Printes, R.C., Rylands, A.B., *et al.* (2008a). *Callicebus barbarabrownae.* In *2008 IUCN Red List of Threatened Species*, ed. IUCN. Gland: IUCN. http://www.iucnredlist.org/apps/redlist/details/39929/0.

Veiga, L.M., Sousa, M.C., Jerusalinsky, L., *et al.* (2008b). *Callicebus coimbrai.* In *2008 IUCN Red List of Threatened Species*, ed. IUCN. Gland: IUCN. http://www.iucnredlist.org/apps/redlist/details/39954/0.

Wallace, R.B., Gómez, H., Felton, A., *et al.* (2006). On a new species of titi monkey, genus *Callicebus* Thomas (Primates, Pitheciidae), from western Bolivia with preliminary notes on distribution and abundance. *Primate Conservation*, **20**, 29–39.

Part II

Comparative Pitheciid Ecology

Marilyn A. Norconk

Authors of the chapters in this section seek to understand pitheciid adaptations through comparative methods and by developing or applying models. Their questions deal with the place pitheciids occupy in the ecologies of South America and the guild of arboreal frugivores.

As seed predators, pitheciines may avoid competition with other primates for ripe fruit, but do they trade one liability for another? Do sakis and uacaris compete with other vertebrate seed predators – parrots and macaws – for unripe seeds (Palminteri *et al.*, Chapter 11)? Most tropical plants depend on the services of seed-dispersing vertebrates. As seed predators, do pitheciines do more harm than good? Ayres and Prance (Chapter 12) ask if sakis and uacaris are likely to provide any benefits to plants, particularly species in the family Lecythidaceae. And, has canopy-level seed predation selected for specific locomotor adaptations in this group? Davis and Walker-Pacheco (Chapter 8) provide the first comparative study to investigate similarities and differences in the postcranial anatomy of all four pitheciine genera. Norconk *et al.* (Chapter 6) note that primate seed predators are found in all major primate radiations. Is seed predation a

Evolutionary Biology and Conservation of Titis, Sakis and Uacaris, eds. Liza M. Veiga, Adrian A. Barnett, Stephen F. Ferrari and Marilyn A. Norconk. Published by Cambridge University Press.

single strategy that requires specific dental, masticatory or gut adaptations, or is there more than one way to crack a nut? Have phylogenetic constraints placed their stamp of distinctiveness on these species?

Body mass of pitheciines ranges from about 1 to 3.5 kg. Lehman (Chapter 10) and Garber and Kowalewski (Chapter 9) examine how body mass and group size may have affected energy expenditure and the evolution of male–male cooperation, respectively. And why do the larger-bodied pitheciines seem to require more space than any other platyrrhine? Setz *et al.* (Chapter 7) review use of time and space for sakis and uacaris.

If the diverse topics found in this section can be distilled to one concept, it is "relationships" – relationships among males, among sympatric pitheciines, among sakis and other vertebrates or plant species. The long-term study of Suzanne Palminteri and colleagues on bald-faced sakis and three species of macaws showed that, despite initial impressions (Ayres 1986; Norconk *et al.* 1997; Barnett *et al.* 2005), the overlap between pitheciines and macaws is probably low and direct competition may be negligible. Seeds represented > 85% of the diets of both vertebrate groups, but monthly overlap ranged from only 3% to 25% (Palminteri *et al.*, Chapter 11). Interestingly, Poulsen *et al.* (2002) found very similar (low) levels of overlap between primates and hornbills in Cameroon. Relative to macaws that fly > 10 km a day (Munn, 1992), sakis lack the ability to track widely dispersed resources over a short period of time. Limited mobility, compared with large frugivorous birds, may force primates to use a more diverse set of resources (Poulsen *et al.* 2002; Palminteri *et al.*, Chapter 11).

Primate seed predators found in the Neotropics, Africa, Asia, and Madagascar exploit more than 300 genera of seeds found in 81 families (Norconk *et al.*, Chapter 6). They divided these seed predators into four functional groups: folivores that also ingest seeds, extractive foragers, primates practicing durophagy, and sclerocarpic foragers (pitheciines) and propose that only the latter are primarily adapted to seed predation.

The late Marcio Ayres and Sir Ghillean Prance (Chapter 12) approach perhaps the most intriguing problem in the relationship of pitheciines to their food resources. Sakis are clearly adapted to destroying seeds during ingestion and some studies suggest that seed predators can do a very thorough job of destroying an entire food crop (Peres, 1991). Two-thirds of the species of the brazil nut family (Lecythidaceae) worldwide (212 species: Ayres and Prance, Chapter 12) are found in the New World. Genera such as *Eschweilera*, *Couratari* and *Lecythis* often rank high in the annual and seasonal diets of pitheciines (Norconk 2011). Using a biogeographical approach, Ayres and Prance (Chapter 12) found a significant overlap between bearded sakis or uacaris and the percentage of Lecythidaceae species. A positive relationship between primates and a specific food sources is rare, because primates are quite eclectic feeders and certainly suggests, contrary to expectations, that seed predators may have a beneficial effect on their major resources. Ayres and Prance propose that bearded sakis and uacaris may act as beneficial pruners, taking some flowers and fruit and in effect channeling the tree's resources into the remaining fruit. They predict that seeds that survived would be larger, better prepared for germination and establishment.

We do not know very much yet about the relationships among sympatric pitheciines, but *Pithecia* spp. overlap geographically with both *Chiropotes* and *Cacajao*: *P. pithecia* with *Chiropotes sagulatus* and *P. monachus* with *Cacajao calvus*. Furthermore, there are areas in the western Amazon basin where titis can be added to the list. Has sympatry among closely related species influenced habitat preferences or locomotor anatomy? Perhaps it has. With one exception, Lesa Davis and Suzanne Walker-Pacheco (Chapter 8) were struck by the short list of shared anatomical characteristics in their study of six species of sakis and uacaris, despite the fact that all of them are arboreal quadrupeds and leapers. White-faced sakis were most divergent and specialized in terms of their locomotor anatomy. They also share *terra firme* habitats in the Guianas and Brazil with bearded sakis, thus one cannot rule out the possibility that competition between these two seed predators drove differences in locomotion and habitat preferences. *Pithecia monachus* and *Cacajao calvus* are also sympatric in some areas in the western Amazon Basin, but they may have few opportunities to interact. *Pithecia monachus* does not range into flooded forests that are seasonally occupied by uacaris. Shawn Lehman's findings (Chapter 10) on the relationship of body mass and population densities to habitat use lend support to the anatomical findings of Davis and Walker for sympatric *P. pithecia* and *C. sagulatus*.

Using the energetic equivalence model, which predicts that body mass and population densities vary inversely, Lehman (Chapter 10) suggests that smaller-bodied pitheciines may exploit patches more efficiently and occur in higher densities than larger-bodied pitheciines. The latter are predicted to invest more heavily in search efforts and occur in lower densities. He further predicts that the body size of the larger-bodied pitheciines (about 3 kg) is the result of stabilizing selection, where bearded sakis and uacaris are sufficiently large to maintain adequate fat reserves in the face of variable resource availability and energy expended in long daily travel. Threat of predation or food/energy relationships could set the upper limit of body mass for these species.

While Lehman's predictions are derived from mathematical models, Eleonore Setz *et al.* (Chapter 7) provide data on what is known about habitat use in sakis and uacaris. Food availability appears to drive troop movements that are fast and widespread in bearded sakis and uacaris. Instead of shifting to less-preferred resources seasonally, these species may search out new resources and then exploit them thoroughly. For uacaris, this means exploiting different habitat types seasonally, moving between flooded and non-flooded forests (Setz *et al.* Chapter 7). Population pressure on resources may vary as group sizes change to conform to food abundance.

Body size differences in pitheciines has not only had an effect on habitat use and foraging strategies, but also on social organization and mating systems. Bearded sakis and uacaris live in large groups of multiple adult males and females; titis and sakis live in smaller groups with one or two breeding males and females. Paul Garber and Martin Kowalewski (Chapter 9) review similarities between atelines and pitheciines and expand on recent field data on pitheciine social behavior. Male–male cooperation could be a phylogenetically conserved feature of larger social groups of both subfamilies of the Atelidae, facilitated perhaps by female transfer and male relatedness. While a growing number of studies support the view that male bearded sakis and uacaris are affiliative and males demonstrate a preference for other males, we do not yet know what underlies the behavior – is it kinship, cooperative mating strategy, cooperative foraging strategy? How does male–male cooperation facilitate female reproductive success? Do males provide some ecological benefit for females, e.g. predator avoidance or defense, or provide better access to food than females could do alone or with just one male?

As evident in the next seven chapters, pitheciine research is still at the point of engendering many questions. Still, the groundwork in the form of basic field research has been laid for many species and now is a good time to formulate research questions using broad theoretical or phylogenetic frameworks.

References

Ayres, M. (1986). Uakaries and Amazonian flooded forests. Unpublished PhD dissertation, Cambridge University.

Barnett, A.A., Volkmar de Castilho, C., Shapley, R.L., *et al.* (2005). Diet, habitat selection and natural history of Cacajao melanocephalus ouakary in Jau National Park, Brazil. *International Journal of Primatology*, **26**, 949–969.

Munn, C.A. (1992). Macaw biology and ecotourism, or "When a bird in the bush is worth two in the hand." In *New World Parrots in Crisis: Solutions from Conservation Biology*, ed. S.R. Beissinger & N.F.R. Snyder. Washington, DC: Smithsonian Institution Press, pp. 47–72.

Norconk, M.A. (2011). Sakis, uakaris, and titi monkeys: behavioral diversity in a radiation of primate seed predators . In *Primates in Perspective* (2nd edn), ed. C.J. Campbell, A. Fuentes, K.C. MacKinnon, S.K. Bearder R.M. Stumpf. Oxford: Oxford University Press, pp. 122–139.

Norconk, M.A., Wertis, C. & Kinzey, W.G. (1997). Seed predation by monkeys and macaws in eastern Venezuela: preliminary findings. *Primates*, **38**, 177–184.

Peres, C.A. (1991). Seed predation of *Cariniana micrantha* (Lecythidaceae) by brown capuchin monkeys in central Amazonia. *Biotropica*, **23**, 262–270.

Poulsen, J.R. Clark, C.J., Conner, E.F., *et al.* (2002). Differential resource use by primates and hornbills: implications for seed dispersal. *Ecology*, **83**, 228–240.

Chapter 6

Morphological and ecological adaptations to seed predation – a primate-wide perspective

Marilyn A. Norconk, Brian W. Grafton & W. Scott McGraw

Introduction

As an order of mammals, Primates are morphologically and behaviorally adapted to eating fruit (Kay 1984; Fleagle & McGraw 1999). While almost all primates will eat fruit when it is available, it is an ephemeral resource (Chapman *et al.* 1999; ter Steege & Persaud 1999) that provides variable amounts of important nutrients (e.g. protein and lipids: Oftedal 1991; Conklin-Brittain *et al.* 1998; Milton 1998; Norconk & Conklin-Brittain 2004; Norconk *et al.* 2009). Many frugivorous primates add leaves or insects to their diets and these food items contribute significantly to balancing nutritional intake as well as filling gaps in food availability. In addition to the major dietary categories of fruit, leaves and insects, there is a long list of relatively incidental items, including seeds, bark, fungus, lichen, nectar, pith, gum and flowers that may provide important nutrients (Norconk *et al.* 2009) or serve as fallback resources (Conklin-Brittain *et al.* 1998; Lambert 1998; Lambert *et al.* 2004; Sayers & Norconk 2008; Grueter *et al.* 2009; Marshall *et al.* 2009). For 31 species of primates dispersed among three major primate radiations (Table 6.1), seeds are important resources used seasonally or routinely and selected at least as frequently as, and sometimes in lieu of, fruit pulp.

We define seed predation as the act of ingesting and masticating seeds or whole fruits that include seeds. Seeds that are eaten and destroyed may or may not be covered with attractive flesh like arils or pulp, and as seeds are often destroyed during ingestion, seed predators make poor seed dispersers for some plant species (Norconk *et al.* 1998). Given that seeds are often well protected from predation (see below), seed predators exhibit a variety of adaptations to extract seeds. We have five goals in this chapter: (1) to suggest a revision in terminology related to seed predation in primates; (2) to review the strategies that plants use to protect seeds; (3) to review primate adaptations that appear to be linked to seed predation; (4) to review what is known about the ecology of seed predation; and (5) to construct an explanatory model that suggests that primates have used four independent evolutionary pathways to seed predation.

We set an annual intake of 20% seeds to identify primate seed predators for this review. While this criterion may seem low, some species exhibit significant seasonal variation in their use of seeds (e.g. *Cebus* spp., Table 6.1). Primate seed predators were divided into four taxonomic/geographic groups: (1) the pitheciids and capuchins from South and Central America (Figure 6.1); (2) colobines (Africa and Asia); (3) papionins (mangabeys, mandrills, baboons, geladas, and macaques) from Africa and Asia; and (4) two lemuroids from Madagascar, sifakas and aye-ayes.

Sclerocarpy and durophagy: terms used to describe the ingestion and mastication of seeds

Embedded seeds may be ingested whole or removed from the pericarp or seed coat before ingestion. What is ingested and how it is prepared will influence feeding behavior (e.g. handling time) and food choice (e.g. avoidance of secondary compounds). Primate seed predators use two strategies: (a) *sclerocarpic foraging* defined as the extraction of seeds using anterior dentition primarily (incisors, canines and/or the first premolar in the tooth row) and hands, followed by mastication of seeds by the molars (Kinzey & Norconk 1990) and (b) *durophagy*, the ingestion of fruit whole with minimal processing followed by mastication or crushing by the posterior dentition (e.g. Daegling & McGraw 2007).

The term "sclerocarpy" or "sclerocarpic foraging" was first used to describe the multistage feeding process used by sakis to extract embedded seeds (Kinzey & Norconk 1990; Kinzey 1992). Handling time may exceed 10 min for some fruit ingested by white-faced sakis (e.g. *Gustavia augusta*), during which time the pericarp is planed by procumbent lower incisors (figure 2 in Kinzey 1992) to gain access to seeds and arils (video at http://www.personal.kent.edu/~mnorconk/saki-feeding-ecology.html). See also Photo 6.1.

The term durophagy was originally applied to the processing of hard-shelled crustaceans by marine vertebrates using molariform teeth (Grubich 2003; Ramsey & Wilga 2007). In

Evolutionary Biology and Conservation of Titis, Sakis and Uacaris, eds. Liza M. Veiga, Adrian A. Barnett, Stephen F. Ferrari and Marilyn A. Norconk. Published by Cambridge University Press.

Table 6.1 Primate seed predators, proportion of seeds in their diet, seasonal use of seeds, whether seeds are ingested pre- or post-dispersal, and principal plant family food sources.

Species	% Seeds in diet	% Seeds in top-ranked month	Top-ranked month/ season eating seeds	Pre- or post-dispersal seed predation?	Principal seed source families	Source
STREPSIRRHINES						
Daubentonia madagascariensis		> 80.0 (hot, dry season)	hot, dry	pre-	Burseraceae, Combretaceae	Sterling 1994; Sterling & McCreless 2006
Propithecus diadema	39.2			pre-	Clusiaceae, Sapotaceae	Powzyk & Mowry 2003
P. diadema edwardsi	35.4				Sapotaceae	Hemingway 1998
PLATYRRHINES						
Cebus albifrons		56.0	May–June (early dry)	pre-	Arecaceae	Terborgh 1983
Cebus apella	12.5	64.0	May–June (early dry)	pre-	Arecaceae	Terborgh 1983
Cebus apella	16.0	24.0	dry	pre-	Fabaceae (caesal)[a], Lecythidaceae, Euphorbiaceae	Galetti & Pedroni 1994
Cebus apella	25.0	39.0		pre-	Lecythidaceae	Peres 1991
Callicebus personatus	26.4			pre-		Heiduck 1998
Callicebus torquatus lugens	26.9	55.0		pre-		Palacios *et al.* 1997
Callicebus torquatus torquatus	37.0			pre-		Kinzey 1977
Pithecia albicans	46.2			pre-		Peres 1993
Pithecia monachus	40.0	74.0		pre-		Soini 1986
Pithecia p. pithecia	53.3	86.2	Nov–Mar (dry)	pre-		Homburg 1998
Pithecia p. chrysocephala				pre-		Setz 1993
Chiropotes albinasus	35.9	81.8	dry	pre-		Ayres 1989
Chiropotes chiropotes	74.8			pre-		Kinzey & Norconk 1993
Chiropotes chiropotes	50.7	80.8	Nov–Jan (dry)	pre-		Peetz 2001
Chiropotes sagulatus	63.2			pre-	Lecythidaceae, Sapotaceae, Moraceae, Fabaceae (mim)	Ayres 1989
Chiropotes sagulatus	66.2			pre-	Lecythidaceae, Sapotaceae, Chrysobalanaceae, Fabaceae (caesal), Capparadaceae	van Roosmalen *et al.* 1988

Table 6.1 *(cont.)*

Species	% Seeds in diet	% Seeds in top-ranked month	Top-ranked month/ season eating seeds	Pre- or post-dispersal seed predation?	Principal seed source families	Source
Chiropotes satanas	54.3	>80%	Mar (wet)	pre-	Sapotaceae, Lecythidaceae, Burseraceae, Simaroubaceae, Arecaceae, Fabaceae (caesal), Annonaceae,	Veiga 2006
Cacajao c. calvus	66.9	97.0	Sept & Jan (no marked dry season)	pre-	Hippocrateaceae, Lecythidaceae, Annonaceae, Sapotaceae	Ayres 1986, 1989
Cacajao c. ucayalii	46.0			pre-		Aquino 1998
Cacajao melanocephalus	71.2	83[1], 59.5[2]	wet[1], dry[2]	pre-	Euphorbiaceae, Fabaceae (caesal), Lecythidaceae, Sapotaceae	Boubli 1999
COLOBINES						
Colobus angolensis	50.0			pre-		Maisels *et al.* 1994
Colobus polykomos	32.0			pre-		Dasilva 1994
Colobus satanas	58.0	80+	Jan, June (Oct to Aug)	pre-	Euphorbiaceae, Fabaceae (caesal), Olacaceae, Loganiaceae, Connaraceae	McKey 1978
Colobus satanas	60.0	82.0	little variation	pre-		Harrison 1986
Colobus satanas	40.6	c. 50.0	Jan-Mar (minor dry – minor wet)	pre	Fabaceae (caesal); Olacaceae	Brugiere *et al.* 2002
Nasalis larvatus	37.0		Jan–May (no dry season)	pre-	Celastraceae, Myrtaceae, Euphorbiaceae	Yeager 1989
Presbytis melalophos	39.5^b			pre-	Fabaceae (caesal), Sterculiaceae, Celastraceae, Myristicaceae	Davies *et al.* 1988
Presbytis rubicunda	30.1			pre-	Sapindaceae, Meliaceae, Fabaceae (caesal), Lauraceae, Verbenaceae	Davies *et al.* 1988
Procolobus badius	25			pre-	Apocynaceae, Fabaceae (caesal)	Davies *et al.* 1999
Trachypithecus auratus	42.1			pre-	Moraceae, Meliaceae	Kool 1993
Trachypithecus phayrei	22			pre-		Suarez 2006
PAPIONINS						
Cercopithecus wolfi	27.3	60 legumes[1]; 15% non legumes[2]	Aug–Nov[1] – wet Dec[2] – late wet	pre-	Fabaceae (mim?)	Gautier-Hion *et al.* 1993
Cercopithecus pogonias	49.8	c. 70.0	Apr–Aug (minor wet to major dry)	pre-	Fabaceae (caesal); Burseraceae	Brugiere *et al.* 2002

Table 6.1 *(cont.)*

Species	% Seeds in diet	% Seeds in top-ranked month	Top-ranked month/ season eating seeds	Pre- or post-dispersal seed predation?	Principal seed source families	Source
Cercopithecus nictitans	50.2	c. 55.0	May–July (minor wet to major dry)	pre-	Fabaceae (caesal); Burseraceae	Brugiere *et al.* 2002
Cercocebus atys	270	78	April–June	post-	Humeriaceae	McGraw *et al.* 2011
Cercocebus galeritus	80[c]			pre- and post-		Homewood 1978
Cercocebus torquatus	25.6			pre-	Fabaceae (caesal), Annonaceae, Sterculiaceae	Mitani 1989
Cercocebus torquatus	26			post-	Humeriaceae	Cooke *et al.*, 2009
Lophocebus albigena	29	58	Aug (Oct–Dec) wet	pre-		Poulsen *et al.* 2001
Lophocebus albigena	57.1	c. 65%	Apr–Jul (minor wet to major dry)	pre-	Fabaceae (caesal); Burseraceae	Brugiere *et al.* 2002
Lophocebus albigena[d]					Ebenaceae	Waser 1977
Macaca sylvanus	26.7; 32.1		summer & autumn	post- (?)	Fagaceae (*Quercus* spp.)	Ménard & Vallet 1996
Macaca fuscata	29.5; 43.6	60–80	Feb (winter)	post- (and pre-)	Fagaceae (*Lithocarpus*)	Agetsuma & Nakagawa 1998
Mandrillus sphinx					Burseraceae, Fabaceae (caesal + papil), Sterculiaceae, Guttiferae, Arecaceae	Jouventin 1975
Mandrillus sphinx	39		Apr–Jul (minor wet season)	pre- and post- (68%)	Annonaceae, Fabaceae (caesal), Connaraceae, Euphorbiaceae	Hoshino 1985
Mandrillus sphinx					Olacaceae, Annonaceae, Sapotaceae	Lahm 1986
Mandrillus sphinx					Annonaceae, Dilleniaceae, Fabaceae (mim)	Rogers *et al.* 1996
Papio anubis	54.4				Fabaceae (mim)	Kunz & Linsenmair 2007
Papio cynocephalus[1], spp.[2]	20.2[1]		large portion of year, incl dry season[1]		Fabaceae (mim)[1,2], Poaceae[1]	[1]Altmann 1998; [2]Barton *et al.* 1993
Theropithecus gelada		69.7	Nov (early dry)		Poaceae	Dunbar 1977

[a] Subfamilies of Fabaceae (caesal = Caesalpinioideae; mim = Mimosoideae; papil = Papilionoideae) (Seigler 2004); [b]includes 14.2% classified as fruits and/or seeds; [c] fruit and seeds; [d]seeds from diverse species are eaten, but only seeds from *Dyospyros abyssinica* were well represented in the diet.

Figure 6.1 Diversity of extraction strategies: young male white-faced saki (left) using procumbent lower incisors to extract mesocarp from fruit. Brownsberg Nature Park, Suriname. Photo: Nick Robi; young female red uacari (right) using canines to puncture a Brazil nut's hard husk. Eco-Park, Manaus, Brazil. Photo: Bruna Bezerra.

primates, the term has been used to denote the incision and mastication of mechanically resistant foods by *Cebus apella* (Daegling 1992) and mangabeys (Daegling & McGraw 2007; McGraw *et al.* 2011). We use it to describe the feeding behavior of some papionins.

We prefer either sclerocarpy or durophagy to describe seed predation over "hard object feeding" (e.g. Kay 1981; Strait 1997; Martin *et al.* 2003; Lambert *et al.* 2004; Lucas 2004; Lucas *et al.* 2008). Despite its widespread use, "hard object feeding" is a better description of the food itself than how the animals process it.

Photo 6.1 Young male white-faced saki using procumbent lower incisors to extract mesocarp from fruit. Brownsberg Nature Park, Suriname. Photo: Nick Robl. (See color plate section.)

Point-counterpoint: plant defenses and primate seed predation

The major challenges of seed-eating appear to be related to mechanical and chemical investments made by plants to protect seeds during development. There may also be ecological challenges related to the dispersion of fruiting individual plants which could account for the large home ranges and nomadic habits of some seed predators (e.g. bearded sakis and uacaris; Ayres 1986, 1989; Bowler & Bodmer 2009; Norconk 2011). Renton (2001) found that parrots that depended on seeds for > 80% of their diet displayed high mobility and dietary flexibility in coping with fluctuating seed crops.

Mechanical properties of fruit and seeds

Plants have evolved various mechanical and chemical means to deter herbivory and seed predation (Wink 1988; Bennet & Wallsgrove 1994; Coley & Barone 1996; Lucas *et al.* 2000). Both leaves and seeds present mechanical defenses in the form of tissues that combine hardness (resistance to crack initiation) and toughness, i.e. resistance to crack propagation (Maas & Dumont 1999; Teaford *et al.*, 2006). Defensive structures include fibrous or hard pericarps and seed coats, spines, thorns and hairs, and are often further strengthened by the inclusion of minerals such as silica (Lucas *et al.* 2000). See Photo 6.2.

Fruit pericarp

Primate seed predators short-circuit the relationship between plant and seed disperser, and are able to breech woody (e.g. Lecythidaceae), bony (Chrysobalanaceae), fibrous (Fabaceae),

Photo 6.2 While most seeds eaten by golden-backed uacaris are unripe, the fruits containing them differ greatly in size and shape. Uacaris also eat flowers, insects and pith. Photo: Adrian Barnett. (See color plate section.)

thick (Sapotaceae), and/or sticky (Sapotaceae, Apocynaceae) pericarps to gain access to embedded seeds before the fruit is ripe. Pericarps, the outer layers of a fruit, function mechanically to protect the seed from damage and desiccation during development. The external presentation of fruit to a potential disperser or predator differs widely across plant taxa in terms of fruit size, color, texture and thickness (Janson 1983; van Roosmalen 1985; Willson & Whelan 1990; Stevenson *et al.* 2000). Pericarps may change in color or texture during development as a visual cue of fruit maturity. They may also dehisce when ripe, providing a mechanism for the dispersal of mature seeds. In some species, mature fruit remains attached to the plant and opens to reveal colorful or nutritionally rich arils or pulp surrounding the seeds, e.g. *Inga* spp. (Fabaceae), *Virola* spp. (Myristicaceae), and *Tetragastris* spp. (Burseraceae).

Seed coats

Seed coats serve a number of functions, including protection and the promotion of dormancy (Bewley & Black 1994). They consist of layers of specialized maternal cells providing an important boundary between the embryo and the external environment during development, dispersal, and germination (Haughn & Chaudhury 2005). Development of the seed coat from maternal integuments ultimately results in a multilayered structure comprised of dead tissue (Boesewinkel & Bouman 1995; Haughn & Chaudhury 2005). The hardness of seed coats is related primarily to the density of the cell walls of the sclerenchyma (Lucas *et al.* 2000), which are composite materials comprised of proteins, cellulose, lignins and silica (Boesewinkel & Bouman 1995; Lucas *et al.* 1995).

Chemical defenses

Many tropical plants produce chemical deterrents in the form of secondary metabolites (Wink 1988; Bennet & Wallsgrove 1994; Pichersky & Gang 2000). While this subject has been reviewed extensively in these and other papers (e.g. Rosenthal & Berenbaum 1992; Lambert 1998), we should make a few general points here. Secondary metabolites include such substances as cyanogenic glycosides, alkaloids, flavonoids, terpenoids, and tannins (Harbourne 1991; Bennet & Wallsgrove 1994). These substances are produced in a wide variety of edible tissues, including fruit flesh (Cipollini & Levey 1997), seeds and seed coats (Bell 1978), and leaves (Glander 1978; Chapman & Chapman 2002). While it is likely that these substances evolved primarily in response to insect predation (Fraenkel 1959), they may also affect primate feeding behavior (Glander 1982; Waterman & Kool 1994). See Photo 6.3.

Photo 6.3 Sometimes diet item exploitation is very subtle. For *Swartzia laevicarpa* (Fabaceae, Papilionoidae), the tough husk is opened at the suture, but only the sweet aril is eaten. Although around 85% of the non-husk weight, the seed is rich in toxic coumarins and saponins, and is avoided. Photo: Adrian Barnett.

Adaptations of the masticatory apparatus

Our review of cranio-dental adaptations indicates that primates have responded to the requirements of processing seeds in a variety of ways, and that this, together with the lack of a standardized terminology in the literature, makes it difficult to identify morphological adaptations "universally" associated with seed predation in primates.

Enamel

Tooth crowns are covered by enamel, which serves at least two functions: resisting abrasion on the occlusal surface and inhibiting fracture of the tooth crown. Enamel thickness varies considerably across the order Primates (Shellis *et al.* 1998; Maas & Dumont 1999; Martin *et al.* 2003) and it seems reasonable to link the variation in enamel thickness to the potential of different foods to abrade or fracture teeth. Primates that feed on mechanically resistant resources, including some seeds, seed pods and unripe fruits, tend to have relatively thick molar enamel, which would appear to be an adaptation that slows wear and ultimately prolongs the life of the tooth.

Attrition may also result from greater cusp-on-cusp contact during the more powerful chewing used to break down hard seeds (Kay 1977, but see Wall *et al.* 2005).

Among catarrhine primates, the thickest enamel is found in *Pongo*, and the mangabeys, *Cercocebus* and *Lophocebus* (Kay 1981), all well-known seed predators. *Lophocebus* and *Cercocebus* have especially thick enamel and, while food-processing modes appear to be genus-specific (Shah 2003; McGraw *et al.*, 2011) their diets have converged on very hard seeds (Chalmers 1968; Homewood 1978; Fleagle & McGraw 1999, 2002; Wieczkowski 2003, 2009; Lambert *et al.* 2004; Lucas *et al.* 2008; Cooke *et al.* 2009).

Cebus apella, a medium-sized platyrrhine, has the thickest molar enamel of all primates (Dumont 1995; Shellis *et al.* 1998; Martin *et al.* 2003, but see Wright 2005, table 7) and frequently feeds on very hard, brittle seeds that other sympatric monkeys – even congeners (Terborgh 1983) – cannot open (Kinzey 1974; Peres 1991; Janson & Boinski 1992). Not all primates that ingest seeds have thickened molar enamel (Cuozzo & Sauther 2006; Martin *et al.* 2003), but given that those primates with the thickest enamel consume at least a moderate amount of seeds, it is probably fair to say that there is some association between enamel thickness and seed predation (Lucas *et al.* 2008). Variation in enamel volume between cusps and across teeth further confounds attempts at generating species-specific thickness profiles (Shellis *et al.* 1998; Grine *et al.* 2005). Given this, and other intra-taxon variation (Teaford 2007), attention is beginning to focus on other characteristics of the enamel, such as its microstructure, which may also evolve in response to the challenges of processing mechanically resistant foods (Lucas *et al.* 2008; Chai *et al.* 2009).

Decussation (interweaving) of enamel prisms into visible lines called Hunter-Shreger bands is observed in the enamel of many primates known to process hard objects dentally, e.g. *Pongo*, *Cercocebus*, *Lophocebus*, some *Lemur*, *Macaca* and *Papio* species, the pitheciines, and *Cebus* (Maas 1991, 1994; Delgado & van Schaik 2000; Martin *et al.* 2003; Cuozzo & Sauther 2006). These prisms are also convergent on similar characteristics found in protective seed coverings (Lucas *et al.* 2008). However, decussated enamel is also present in primates that do not typically feed on hard-cased fruit, such as *Callithrix* spp., *Hylobates*, *Pan* and modern *Homo* (Maas & Dumont 1999). This indicates that additional factors such as allometry, jaw mechanics and phylogeny likely contribute to the variation in enamel microstructure seen in primates.

Tooth size

Tooth enlargement is a second potential response to the requirements of processing mechanically resistant foods. Grine (1986) argued that the broadened premolars and molars of robust australopithecines were adaptations to a diet of small, fracture-resistant foods obtained under xeric conditions. Similarly, Fleagle and McGraw (1999, 2002) proposed that the molarized premolars of *Cercocebus* and *Mandrillus* are a synapomorphy related to their reliance on very large, hard seeds and fruits that resist decomposition for months on the forest floor. Recent field data indicate that the expanded post-canine teeth of at least one *Cercocebus* species is associated with powerful crushing behaviors used to open seeds that are the most frequently consumed and hardest items in the sooty mangabey diet (McGraw *et al.*, in press). Capuchins have relatively large premolars and block-like molars used to crush mechanically resistant foods, such as palm nuts (Fragaszy *et al.* 2004) and Wright (2005) found that second molar tooth size was significantly larger in *Cebus apella* compared with *C. olivaceus*. As with enamel thickness, however, the correlation between cheek tooth size and the ingestion of well-protected fruit is inconclusive. Compared to the cheek teeth of mangabeys and mandrills, for example, the terrestrial geladas (*Theropithecus*) have diminutive premolars (Fleagle & McGraw 1999, 2002). On the other hand, pitheciines masticate relatively soft seeds yet possess molarized upper and lower fourth premolars (Kinzey 1992; Martin *et al.* 2003). Clearly, then, expanded cheek teeth are associated with seed predation, but molarized premolars are not consistently associated with this dietary strategy (Kay 1990; Daegling *et al.* 2011).

Modifications of the anterior dentition have also been linked to seed-eating. Hylander (1975) showed that primates that feed on large, hard foods tend to have wider incisors than those that feed on small, soft foods which require less incisal preparation. Using its enlarged incisors, for example, the arboreal mangabey *Lophocebus albigena* is able to open hard seeds that sympatric monkeys cannot exploit (Chalmers 1968; Poulsen *et al.* 2001; Shah 2003; Lambert *et al.* 2004). This may account for the reduction in cheek tooth size in this species in comparison with terrestrial mangabeys (*Cercocebus*) and mandrills (Daegling & McGraw 2007). Observations of *Lophocebus* and *Cercocebus* feeding behavior highlight the contrast between seed incision and mastication (Shah 2003).

Anterior tooth modification for the purpose of exploiting seeds is well-illustrated by the pitheciines. *Pithecia*, *Chiropotes* and *Cacajao*, all sclerocarpic foragers, have enlarged, laterally splayed canines and narrowly compressed procumbent incisors that facilitate pericarp processing and seed extraction (Kinzey & Norconk 1990; Kinzey 1992; see Rosenberger and Tejedor, Chapter 2). These adaptations have defined the clade since at least the middle Miocene (Meldrum & Kay 1997). An even more dramatic example is found in *Daubentonia madagascariensis*. Using a probe-like third digit and remarkable, chisel-like incisors, the aye-aye is able to both access insect larvae from trees in woodpecker fashion (Cartmill 1974; Tattersall 1982) and process the hard seeds of *Canarium* fruits (Sterling 1994; Kitko *et al.* 1996).

Finally, we would be negligent if we failed to note that among primates incisor *reduction* is also associated with seed predation in the very specialized case of gelada baboons who feed largely on tiny grass seeds and grass blades on the Ethiopian plateau (Jolly 1970; Iwamoto, 1993).

Crown morphology

Molar crests also vary according to dietary patterns (Kay 1975; 1984; Strait 1997; Lucas 2004; Cuozzo & Yamashita 2006). Folivores typically have high molar shearing crests (for shredding leaves) while more frugivorous species have lower crests. Seed predators tend to have the lowest "shearing quotient" of all primates. This relationship is most apparent in the *Cercocebus* mangabeys, which have particularly flat molars (Fleagle & McGraw 1999, 2002), also the "featureless molars" of *Daubentonia* (Cuozzo & Yamashita 2006, p. 86) and flat, crenulated molars of the pitheciines (Kinzey 1992). The low occlusal relief on the cheek teeth of pitheciines appears to be an adaptation to chewing relatively soft, tough seeds, but not hard ones (Martin *et al.* 2003).

High shearing crests do not necessarily preclude adaptations for seed eating. Lucas and Teaford (1994) argued that, from an engineering perspective, the bilophodont molars of colobines, with their high, sharp shearing crests and deep wedges, are well designed for both folivory and seed eating. This is supported by other non-cercopithecoid primates, including *Propithecus diadema*, which feed on seeds and possess pseudo-bilophodont molars (Kay & Hylander 1978; Cuozzo & Yamashita 2006). Once again, the correspondence between occlusal morphology and food items is not one-to-one.

Crenulated molar enamel may help grip seeds during mastication or promote more efficient crushing and grinding. The most extensive crenulation is found on the cheek teeth of the grass-eating geladas (Iwamoto 1993), but they are also present in primates that feed on harder and larger seeds. For example, orangutans have crenulations on their molar occlusal surfaces that are believed to prevent seeds from slipping during mastication (Delgado & van Schaik 2000). The molars of *Chiropotes* and *Cacajao* have crenate occlusal surfaces which, in combination with low molar relief, may "facilitate secondary breakdown of seed particles during grinding" (Kinzey 1992, p. 503).

Jaw form

Mandibular geometry should also reflect adaptations to different occlusal loads. A diet of mechanically resistant foods presumably requires greater recruitment of jaw adductors for the application of greater bite force. Primates that eat more obdurate foods should therefore have stronger jaws with features reflecting greater occlusal forces. While there is little doubt that masticating hard vs. soft foods results in different in vivo stresses, understanding how different forces are reflected in mandibular shape has proven difficult (Hylander 1979; Taylor 2006). Thus attempts to relate jaw morphology to specific dietary items have met with mixed success.

For example, in addition to expanded cheek teeth and thickened enamel, the mandible of *Cebus apella* appears well adapted to the demands of a diet rich in mechanically resistant seeds (Terborgh 1983; Galetti & Pedroni 1994; Fragaszy *et al.* 2004; Wright 2005). The mandibular corpus is robustly built and CT scans indicate that the distribution of cortical bone is designed to offset high masticatory stresses (Daegling 1992). Similar buttressing has not been observed in other seed predators, however. Sympatric West African colobus monkeys illustrate this point well. The diet of *Colobus polykomos* in the Ivory Coast's Taï Forest contains significant quantities of seeds and the heavily lignified seed pods of large *Pentaclethera macrophylla* fruits (Korstjens *et al.* 2007). These fruits are rarely, if ever, consumed by *Procolobus badius*, yet the jaws of *C. polykomos* are not significantly more robust (Daegling & McGraw 2001). Results from a recent comparison of mangabey jaws have proven equally inconclusive (Daegling & McGraw 2007).

Thus, while some primate seed predators have jaws that conform to the biomechanical predictions of increased load resistance, others do not. Clearly, further studies should consider allometric effects, functional trade-offs, the material properties of foods and bone, seed size, and, most importantly, how differing loads should be reflected in jaw geometry (Hylander 1979; Strait 1997; Daegling 2002; Dechow & Hylander 2000; Vinyard & Ryan 2006). Fortunately, this is an active area of research and we are confident that improved experimental studies combined with more detailed field data on feeding behavior will provide greater resolution (Cuozzo & Sauther 2006). In the meantime, we can only conclude that the association between jaw form and seed predation is complex, by no means consistent, and that there is no "universal" way in which primate seed predators reinforce their jaws.

Ecological characteristics of seed predation

Primate seed predators exploit at least 331 genera of plants, distributed in 79 families. Eighteen (22.7%) of these families are common to the Neotropics, Madagascar, and Africa-Asia (Table 6.2). Platyrrhine seed predators use the highest diversity of plant genera (190 genera) compared with 111 genera in cercopithecoids and 49 in prosimians. These plant taxa tend to follow two seed dispersal strategies: (a) non-dehiscent drupes with seeds dispersed by endozoochory (e.g. Anacardiaceae, Annonaceae, Chrysobalanaceae, Melastomataceae, Rubiaceae) and (b) dehiscent fruit (Euphorbiaceae, Fabaceae, Rutaceae). In each case, but particularly the latter, seed predators must breach immature pericarp to ingest seeds.

The proportion of seeds in the diets of platyrrhines (particularly pitheciines) is significantly higher than that of other taxa (Kruskal–Wallis $H = 12.75$, $p < 0.002$). This may be due to a data-collecting tradition of differentiating seed eating from other forms of feeding strategies in the pitheciines, but some primate species clearly use fewer seed sources than others. For example, *Daubentonia* (Sterling & McCreless 2006) exploits only 1–3 species of seeds, whereas bearded sakis use more than 100 plant species in an annual cycle (Norconk 1996; Peetz 2001; Veiga 2006).

Table 6.2 Primate seed predators ingest seeds from a total of 79 plant families of which 18 (22.7%) families are common to all and a total of 331 genera. The 18 plant families in common are listed below with the total number of genera used by all, the specific plant genera shared between groups, and the subtotal of plant genera used by each group of primates (in parentheses).

Plant family	# Genera	Madagascar (*Propithecus, Daubentonia*)	Asia & Africa (cercopithecoids)	Neotropics (pitheciines, *Cebus* spp.)
Anacardiaceae	5	*Mangifera* (3)	*Mangifera* (2)	(1)
Annonaceae	5	*Xylopia* (1)	*Xylopia* (1)	*Xylopia* (5)
Apocynaceae	12	(2)	(3)	(7)
Celastraceae	6	(1)	(2)	(3)
Chryso-balanaceae	5	(1)	(1)	(3)
Clusiaceae	5	*Garcinia, Calophyllum* (4)	*Garcinia, Calophyllum* (2)	(1)
Combretaceae	2	*Terminalia* (1)	*Terminalia, Combretum* (2)	*Combretum* (1)
Euphorbiaceae	21	(2)	*Drypetes, Sapium* (13)	*Drypetes, Sapium* (8)
Fabaceae				
Mimosoideae	16	*Albizia* (2)	*Albizia* (7)	(8)
Papilionoideae	18	(1)	(5)	(12)
Lauraceae	5	(1)	(2)	(2)
Melastom-ataceae	4	(2)	(1)	(1)
Meliaceae	10	(1)	*Swietenia* (6)	*Swietenia* (4)
Moraceae	14	*Ficus* (2)	*Ficus* (3)	*Ficus* (11)
Myrtaceae	1	*Eugenia*	*Eugenia*	*Eugenia*
Rubiaceae	14	(3)	(4)	(7)
Rutaceae	5	(1)	*Zanthoxylum* (2)	*Zanthoxylum* (3)
Sapindaceae	12	(3)	(3)	(6)
Sterculiaceae	6	(1)	*Sterculia* (2)	*Sterculia* (4)

Like primates adapted to other feeding strategies, seed predators are selective feeders. Hoshino (1985) observed mandrills discarding the seeds of *Staudtia stipitata* and *Coelocaryon preussi* (Myristicaceae) which contain toxins used by people to kill fish, and Norconk and Conklin-Brittain (2004) found that sakis spat out *Strychnos* spp. seeds unlike langurs that ate seeds containing strychnine (Glander 1982). Some primates (*Papio* spp.) consume legume seeds (Fabaceae) while others (e.g. pitheciines) ingest only their pulp. Euphorbiaceae species are among the most common taxa exploited by primate seed predators, even though their latex, leaves, and stems (and seeds, in some cases) are toxic (Smith *et al.* 2004). There is much yet to be discovered about the relationship between seed chemistry and primate seed selection and predation.

The seed predators on our list represent species that ingest seeds seasonally at relatively high frequencies (*Cebus apella* and *Cebus albifrons*: Terborgh 1983, Galetti & Pedroni 1994; *Lophocebus albigena:* Poulsen *et al.* 2001; *Nasalis larvatus*: Waterman 1984, Yeager 1989) and those primates that ingest seeds at relatively high frequencies year-round, but with seasonal fluctuation (e.g. pitheciines: Ayres 1986; Norconk 1996; Veiga 2006; Gregory 2011; Norconk & Veres, 2011; some colobines: Harrison 1986 and *Cercocebus* spp.: McGraw *et al.* 2011). Aside from nutritional value, the major attraction of seeds may be their relative durability – long period of availability of seeds compared with ripe fruit and leaf flush (McKey 1978; Harrison 1986; Mitani 1989; Gautier-Hion *et al.* 1993; Norconk 1996; Boubli 1999; Veiga 2006; Bowler & Bodmer 2009).

Finally, the actual impact of primate seed predation on plant reproductive success is unknown for any primate taxon. Even though studies of seed crop losses up to 100% have been documented for insects (Janzen 1971; Crawley 1992), rodents (e.g. Jansen *et al.* 2004; Xiao *et al.* 2005), and birds (e.g. Dirzo & Domínguez 1986; Francisco *et al.* 2002), only Peres (1991) has documented the substantial removal (70%) of seeds from an individual tree by *Cebus apella*. Pre-disperser seed predators (see Table 6.1) should have the largest effect on plant

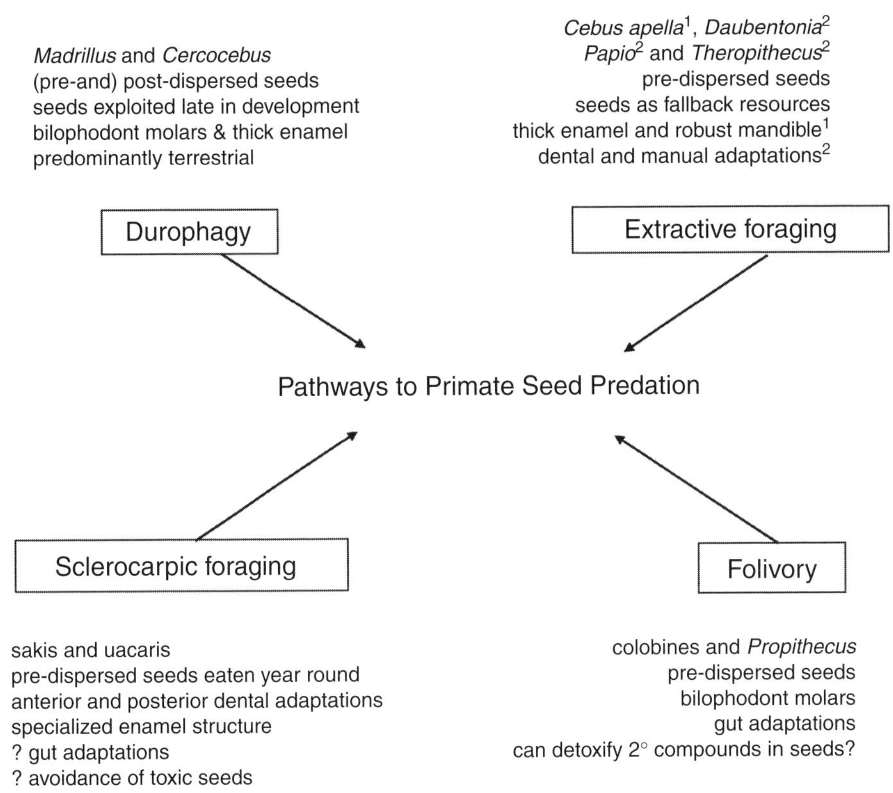

Figure 6.2 Four independent pathways to seed predation in primates. Strategies include morphological, physiological, and behavioral adaptations that contribute to seed predation as a primary (sclerocarpic foragers and durophagic feeders) or secondary (folivores and extractive foragers) feeding adaptation.

reproductive success by early seed destruction, but if these seed predators inhabit large home ranges and are semi-nomadic in their ranging patterns (e.g. bearded sakis and uacaris) then return rates to specific trees might be irregular or infrequent. Many fruit could be left to mature and therefore available for seed dispersal by typical frugivores or scatter-hoarders. Post-disperser seed predators extract seeds from mature fruit that has fallen to the ground presumably after the best window for seed dispersal has closed. Their effect on plant fitness may be minimal, but specific examples of the impact of primate seed predation are few and far between.

Evolution of seed predation strategies in primates

Taxonomically diverse groups of primates have converged on seed predation from four directions (Figure 6.2). Extant colobines and *Propithecus* ingest seeds, but are primarily folivores. *Cebus apella*, *Daubentonia*, *Theropithecus* and *Papio* are extractive foragers using manual and dental adaptations to open foods that include seeds. Durophagy and sclerocarpic foraging appear to be strategies specifically related to seed predation by mangabeys and pitheciines, respectively. We characterized these species as adapted primarily for seed eating, and folivores and extractive foragers as secondary seed predators. We also identified what appear to be opportunistic seed eaters in other lineages of Old World monkeys.

Folivory pathway

This pathway characterizes primates for which leaves constitute both a dietary staple and a fallback strategy. Few of these species met our 20% seed ingestion criterion in their annual diets, but we may have underestimated seed intake as many seeds are probably ingested and masticated as a portion of (whole) fruit (*Trachypithecus*, *Presbytis*, *Semnopithecus* and *Procolobus*: Oates *et al.* 1980; Davies *et al.* 1988; Sayers & Norconk 2008). Colobine dental, masticatory and gut adaptations, which evolved in the context of folivory, may also facilitate the ingestion and digestion of fibrous seeds in some lineages. We also placed *Propithecus* in this group, as an obligate folivore (Yamashita 2003) that satisfies our 20% cut-off for seed eating.

Extractive foraging pathway

Cebus apella and *Daubentonia* are extractive foragers *sensu stricto* (cf. Gibson 1986); both have a high intake of hidden invertebrates (Rosenberger 1992). Their ability to detect hidden resources, in combination with physical strength and/or learning capabilities, appear to have given them broad adaptive capabilities that includes seed predation. Wright's (2005) comparative study of food mechanical properties and cranial–dental morphology of wild *Cebus apella* and *C. olivaceus* suggests that these species have a generalized strategy of opening tough foods using both sclerocarpic foraging techniques (where the emphasis is on anterior dentition processing) and durophagy (posterior dentition processing). *Cebus apella* was capable of opening food items that were up to four times the maximum toughness of foods opened by *C. olivaceus* (Wright 2005).

Durophagy pathway

Mandrillus and *Cercocebus* exhibit adaptations in the mandible and molars that appear to converge on durophagous fish, but behaviorally they are convergent on the forest-floor gleaning peccaries (Bodmer 1989). Interestingly, the seed-eating, papionins *Papio* and *Theropithecus*, probably fit the "extractive foraging" pathway better than the "durophagic" one. Species in these genera are well known for the manipulative skill and manual strength in extracting underground storage organs as well as small grass seeds (Dunbar 1977; Altmann 1998; Kunz & Linsenmair 2007).

Sclerocarpic foraging pathway

The pitheciines share dental adaptations that are specifically designed to open well-protected fruit and extract and masticate seeds (Kinzey 1992). These adaptations provide them with early and sustained access to seeds year-round. Seed consumption may decline in some months, but seeds never drop out of monthly diets (Norconk 1996; Peetz 2001; Veiga 2006). The morphological specializations of pitheciines put them in a class of their own, but other species (e.g. *Cebus apella*) also use anterior dentition extensively when opening mechanically protected fruit (Wright 2005).

Opportunistic seed predators

Some populations of macaques and guenons have seasonally high intake of seeds or live in habitats where seeds are relatively abundant. Two species of macaques living in cool temperate habitats or high latitudes (Thierry 2011): *Macaca fuscata* (Agetsuma & Nakagawa 1998) and *Macaca sylvanus* (Ménard & Vallet 1996) met our 20% annual intake criterion for seed predation (Table 6.1). Seeds also reached a high of 27% in the seasonal (but not annual) diet of Tibetan macaques, *Macaca thibetana* (Zhao 1996). Finally, Brugiere *et al.* (2002) found a high proportion of seeds in the annual diets of two cercopithecines (*Cercopithecus nictitans* and *C. pogonias*) in a Caesalpiniaceae-dominated forest in Gabon, but the authors noted that ripe fruit seeds were not preferred food items for guenons. Orangutans may also fit into this category. Taylor (2006, p. 378) reviewed dietary data as part of a study on jaw form and commented that "little is known about how orangutans actually process various seed species in the wild."

Conclusions and suggestions for future research

We suggest that committed primate seed predators exhibiting a variety of craniodental, ecological and perhaps physiological adaptations have arrived at seed-eating from four independent starting points. Seeds can be an excellent source of nutrition and energy, rich in protein and with lipid values equivalent to insect prey in some species (Norconk & Conklin-Brittain 2004), but are also often protected mechanically or chemically. The window of resource availability is also wider for seeds than for ripe fruit or leaf flush. This is especially true if seed predators ingest seeds from both young and mature fruits (Terborgh 1983; Norconk 1996; Boubli 1999). Interestingly, the semi-nomadic ranging patterns of some seed predators (e.g. bearded sakis and uacaris) suggest either that they have preferences for seeds of specific plant species or fruiting sources are widely dispersed. Finally, the ability to use resources such as winged (e.g. Bignoniaceae) seeds that are not attractive to ripe fruit frugivores may limit resource overlap and competition with ripe fruit frugivores.

This is the first attempt to synthesize data on primate seed predators, to set criteria to recognize the taxa, and to identify the ecological opportunities and constraints that may have led to using seeds as frequent resources. While there are now substantial data on the craniodental anatomy of seed predators, data on gut physiology and seed detoxification, specific characteristics of seed choice, the ecology of temporal and spatial distribution of suitable seeds, and the impact of primate seed predators on preferred plants are far from well understood. Even though we know more about the feeding ecology of pitheciines than any other aspect of their behavior, there is still plenty of room for detailed, hypothesis-driven studies of food choice and habitat use. Continued research into pitheciine feeding strategies will contribute to a better understanding of the seed predation feeding strategy in primates as a whole.

Acknowledgments

We are very grateful to Delanie Hurst, Stephen Ferrari, Tremie Gregory, Richard Kay, Orin Neal, Andrew Ritchie, Alfred Rosenberger, Kenneth Sayers, Cynthia Thompson, Christopher Vinyard and Ari Vreedzaam for comments on earlier versions of the manuscript. WSM acknowledges the financial support of National Science Foundation through grants BCS 0840110, BCS 0921770 and BCS 0922429. Logistical help during fieldwork in Ivory Coast provided by the Centre Suisse de Recherches Scientifique in Abidjan.

References

Agetsuma, N. & Nakagawa, N. (1998). Effects of habitat differences on feeding behaviors of Japanese monkeys: comparison between Yakushima and Kinkazan. *Primates*, **39**, 275–289.

Altmann, S. (1998). *Foraging for Survival: Yearling Baboons in Africa*. Chicago: Chicago University Press.

Aquino, R. (1998). Some observations on the ecology of *Cacajao calvus ucayalii* in the Peruvian Amazon. *Primate Conservation*, **18**, 21–24.

Ayres, J.M. (1986). The white uakaris and the Amazonian flooded forests. Unpublished Ph.D thesis, Cambridge University.

Ayres, J.M. (1989). Comparative feeding ecology of the uakari and bearded saki, *Cacajao* and *Chiropotes*. *Journal of Human Evolution*, **18**, 697–716.

Barton, R.A., Whiten, A., Byrne, R.W., *et al.* (1993). Chemical composition of baboon plant foods: implications for the interpretation of intra- and interspecific differences in diet. *Folia Primatologica*, **61**, 1–20.

Bell, E.A. (1978). Toxins in seeds. In *Biochemical Aspects of Plant and Animal Coevolution*, ed. J.B. Harbourne. New York, NY: Academic Press, pp. 143–161.

Bennet, R.N. & Wallsgrove, R.M. (1994). Secondary metabolites in plant defense mechanisms. *New Phytologist*, **127**, 617–633.

Bewley J.D. & Black, M. (1994). *Seeds: Physiology of Development and Germination*. New York, NY: Plenum Press.

Bodmer, R.E. (1989). Frugivory in Amazonian artiodactyla: evidence for the evolution of the ruminant stomach. *Journal of Zoology, London*, **219**, 457–467.

Boesewinkel, F.D. & Bouman, F. (1995). The seed: structure and function. In *Seed Development and Germination*, ed. J. Kigel and G. Galil. New York, NY: Marcel Dekker, Inc., pp. 1–24.

Boubli, J.P. (1999). Feeding ecology of black-headed uacaris (*Cacajao melanocephalus melanocephalus*) in Pico da Neblina National Park, Brazil. *International Journal of Primatology*, **20**, 719–749.

Bowler, M. & Bodmer, R. (2009). Social behaviour in fission–fusion groups of red uakari monkeys (*Cacajao calvus ucayalii*). *American Journal of Primatology* **71**, 976–987.

Brugiere, D., Gautier, J-P., Moungazi, A., *et al.* (2002). Primate diet and biomass in relation to vegetation composition and fruiting phenology in a rain forest in Gabon. *International Journal of Primatology*, **23**, 999–1024.

Cartmill, M. (1974). *Daubentonia*, *Dactylopsila*, woodpeckers and kinorhychy. In *Prosimian Biology*, ed. R.D. Martin, G.A. Doyle and A.C. Walker. London: Duckworth, pp. 655–672.

Chai, H., Lee, J.W., Constantino, P.J., *et al.* (2009). Remarkable resilience of teeth. *Proceedings of the National Academy of Science*, **106**, 7289–7293.

Chalmers, N.R. (1968). Group composition, ecology and daily activities of free living mangabeys in Uganda. *Folia Primatologica*, **8**, 247–262.

Chapman, C.A. & Chapman, L.J. (2002). Foraging challenges of red colobus monkeys: influence of nutrients and secondary compounds. *Comparative Biochemistry and Physiology A*, **133**, 861–875.

Chapman, C.A., Wrangham, R.W., Chapman, L.J., *et al.* (1999). Fruit and flowering phenology at two sites in Kibale National Park, Uganda. *Journal of Tropical Ecology*, **5**, 189–211.

Cipollini, M.L. & Levey, D.J. (1997). Secondary metabolites of fleshy vertebrate-dispersed fruits: adaptive hypotheses and implications for seed dispersal. *American Naturalist*, **150**, 346–372.

Coley, P.D. & Barone, J.A. (1996). Herbivory and plant defenses in tropical forests. *Annual Review of Ecology and Systematics*, **27**, 305–335.

Conklin-Brittain, N.L., Wrangham, R.W. & Hunt, K.D. (1998). Dietary response of chimpanzees and cercopithecines to seasonal variation in fruit abundance. *International Journal of Primatology*, **19**, 971–998.

Cooke, C., Moussopo, I.R. & McGraw, W.S. (2009). New information on the feeding and grouping behavior of *Cercocebus torquatus*, the red capped mangabey, from southwestern Gabon. *American Journal of Physical Anthropology*, **Suppl 2009**, 108–109.

Crawley, M.J. (1992). Seed predators and plant population dynamics. In *Seeds: The Ecology of Regeneration in Plant Communities*, ed. M. Fenner. Wallingford, Oxon: C.A.B. International, pp. 157–191.

Cuozzo, F.P. & Sauther, M.L. (2006). Severe wear and tooth loss in wild ring-tailed lemurs (*Lemur catta*): a function of feeding ecology, dental structure, and individual life history. *Journal of Human Evolution*, **51**, 490–505.

Cuozzo, F.P. & Yamashita, N. (2006). Impact of ecology on the teeth of extant lemurs: a review of dental adaptations, function, and life history. In *Lemurs - Ecology and Adaptation*, ed. L. Gould and M.L. Sauther. New York, NY: Springer Science, pp. 67–96.

Daegling, D.J. (1992). Morphology and diet in the genus *Cebus*. *International Journal of Primatology*, **13**, 545–570.

Daegling, D.J. (2002). Bone geometry in cercopithecoid mandibles. *Archives of Oral Biology*, **47**, 315–325.

Daegling, D.J. & McGraw, W.S. (2001). Feeding, diet and jaw form in West African *Colobus* and *Procolobus*. *International Journal of Primatology*, **22**, 1033–1055.

Daegling, D.J. & McGraw, W.S. (2007). Functional morphology of the mangabey mandibular corpus: relationship to dental specializations and feeding behavior. *American Journal of Physical Anthropology*, **134**, 50–62.

Daegling, D., McGraw, W.S., Ungar, P., *et al.* (2011). Hard-object feeding in sooty mangabeys (*Cercocebus atys*) and interpretation of early hominin feeding ecology. *PLoS One*, **6**, 1–12.

Dasilva, G.L. (1994). Diet of *Colobus polykomos* on Tiwai Island: selection of food in relation to its seasonal abundance and nutritional quality. *International Journal of Primatology*, **15**, 655–680.

Davies, A.G., Bennett, E.L. & Waterman, P.G. (1988). Food selection by two south-east Asian colobine monkeys (*Presbytis rubicunda* and *Presbytis melalophos*) in relation to plant chemistry. *Biological Journal of the Linnaean Society*, **34**, 33–56.

Davies, A.G., Oates, J.F. & Dasilva, G.L. (1999). Patterns of frugivory in three west African colobine monkeys. *International Journal of Primatology*, **20**, 327–358.

Dechow, P.C. & Hylander, W.L. (2000). Elastic properties and masticatory bone stress in the macaque mandible. *American Journal of Physical Anthropology*, **112**, 553–574.

Delgado R.A. & van Schaik, C.P. (2000). The behavioral ecology and conservation of the orangutan (*Pongo pygmaeus*): a tale of two islands. *Evolutionary Anthropology*, **9**, 201–218.

Dirzo, R. & Domínguez, C.A. (1986). Seed shadows, seed predation and the advantages of dispersal. In *Frugivores and seed dispersal*, ed. A. Estrada and T.H. Fleming. Dordrecht: Junk Publishers, pp. 237–249.

Dumont, E.R. (1995). Enamel thickness and dietary adaptation among extant primates and chiropterans. *Journal of Mammalogy*, **76**, 1127–1136.

Dunbar, R.I.M. (1977). Feeding ecology of gelada baboons: a preliminary report. In *Primate Ecology*, ed. T.H. Clutton Brock. London: Academic Press, pp. 251–273.

Fleagle, J.G. & McGraw, W.S. (1999). Skeletal and dental morphology supports diphyletic origin of baboons and mandrills. *Proceedings of the National Academy of Sciences*, **96**, 1157–1161.

Fleagle, J.G. & McGraw, W.S. (2002). Skeletal and dental morphology of African papionins: unmasking a cryptic clade. *Journal of Human Evolution*, **42**, 267–292.

Fraenkel, G.S. (1959). The raison d'être of secondary plant substances. *Science*, **129**, 1466–1470.

Fragaszy, D.M., Visalberghi, E. & Fedigan, L.M. (2004). *The Complete Capuchin: The Biology of the Genus Cebus*. Cambridge: Cambridge University Press.

Francisco, M.R., Lunardi V. de Oliveira, & Galetti, M. (2002). Massive seed predation of *Pseudobombax grandiflorum* (Bombacaceae) by parakeets *Brotogeris versicolurus* (Psittacidae) in a forest fragment in Brazil. *Biotropica*, **34**, 613–615.

Galetti, M. & Pedroni, F. (1994). Seasonal diet of capuchin monkeys (*Cebus apella*) in a semideciduous forest of south-east Brazil. *Journal of Tropical Ecology*, **10**, 27–39.

Gautier-Hion, A., Gautier, J.P. & Maisels, F. (1993). Seed dispersal versus predation: an inter-site comparison of two related African monkeys. *Vegetatio*, **107/108**, 237–244.

Gibson, K. (1986). Cognition, brain size and the extraction of embedded food resources. In *Primate Ontogeny, Cognition and Social Behavior*, ed. J.G. Else and P.C. Lee. Cambridge: Cambridge University Press, pp. 93–103.

Glander, K.E. (1978). Howling monkey feeding behavior and plant secondary compounds: a study of strategies. In *The Ecology of Arboreal Folivores*, ed. G.G. Montgomery. Washington, DC: Smithsonian Institution Press, pp. 561–574.

Glander, K.E. (1982). The impact of plant secondary compounds on primate feeding behavior. *Yearbook of Physical Anthropology*, **25**, 1–18.

Gregory, L.T. (2011). Socioecology of the Guianan bearded saki, *Chiropotes sagulatus*. Unpublished doctoral dissertation, Kent State University.

Grine, F.E. (1986). Dental evidence for dietary differences in *Australopithecus* and *Paranthropus*: a quantitative analysis of permanent molar microwear. *Journal of Human Evolution*, **15**, 783–822.

Grine, F.E., Spencer, M.A., Demes, B., *et al.* (2005). Molar enamel thickness in the Chacma baboon, *Papio ursinus* (Kerr 1792). *American Journal of Physical Anthropology*, **128**, 812–822.

Grubich, J. (2003). Morphological convergence of pharyngeal jaw structure in duriphagous perciform fish. *Biological Journal of the Linnaean Society*, **80**, 147–165.

Grueter, C.C., Li, D., Ren, B., *et al.* (2009). Fallback foods of temperate-living primates: a case study on snub-nosed monkeys. *American Journal of Physical Anthropology*, **140**, 700–715.

Harbourne, J.B. (1991). The chemical basis of plant defense. In *Plant Defenses against Mammalian Herbivory*, ed. R.T. Palo and C.T. Robbins. Boston: CRC Press, Inc., pp. 45–60.

Harrison, M. (1986). Feeding ecology of black colobus, *Colobus satanas*, in central Gabon. In *Primate Ecology and Conservation*, ed. J.G. Else and P.C. Lee. Cambridge: Cambridge University Press, pp. 31–37.

Haughn, G. & Chaudhury, A. (2005). Genetic analysis of seed coat development in Arabidopsis. *Trends in Plant Science*, **10**, 472–477.

Heiduck, S. (1998). Feeding ecology of the masked titi, *Callicebus personatus*. *Neotropical Primates*, **6**, 92–93.

Hemingway, C.A. (1998). Selectivity and variability in the diet of Milne-Edwards' sifakas (*Propithecus diadema edwardsi*): implications for folivory and seed-eating. *International Journal of Primatology*, **19**, 355–377.

Homburg, I. (1998) Ökologie und Sozialverhalten von Weißgesicht-sakis. Unpublished PhD thesis, Universität Bielefeld, Germany.

Homewood, K.M. (1978). Feeding strategy of the Tana mangabey (*Cercocebus galeritus galeritus*) (Mammalia: Primates). *Journal of Zoology, London*, **186**, 375–391.

Hoshino, J. (1985). Feeding ecology of mandrills (*Mandrillus sphinx*) in Campo Animal Reserve, Cameroon. *Primates*, **26**, 248–273.

Hylander, W.L. (1975). Incisor size and diet in anthropoids with special reference to Cercopithecoidea. *Science*, **189**, 1095–1098.

Hylander, W.L. (1979). The functional significance of primate mandibular form. *Journal of Morphology*, **160**, 223–240.

Iwamoto, T. (1993). The ecology of *Theropithecus gelada*. In *Theropithecus: The Rise and Fall of a Primate Genus*, ed. N.G. Jablonski. Cambridge: Cambridge University Press, pp. 441–452.

Jansen, P.A., Bongers, F. & Hemerik, L. (2004). Seed mass and mast seeding enhance dispersal by a neotropical scatter-hoarding rodent. *Ecological Monographs*, **74**, 569–589.

Janson, C.H. (1983). Adaptation of fruit morphology to dispersal agents in a Neotropical forest. *Science*, **219**, 187–189.

Janson, C.H. & Boinski, S. (1992). Morphological and behavioral adaptations for foraging in generalist primates: the case of the cebines. *American Journal of Physical Anthropology*, **88**, 483–498.

Janzen, D.H. (1971). Seed predation by animals. *Annual Review of Ecology and Systematics*, **2**, 465–492.

Jolly, C.J. (1970). The seed eaters: a new model of hominid differentiation based on a baboon analogy. *Man*, **5**, 5–26.

Jouventin, P. (1975). Observations sur la socio-ecologie du mandrill. *Terre et la Vie*, **29**, 493–532.

Kay, R.F. (1975). The functional adaptations of primate molar teeth. *American Journal of Physical Anthropology*, **43**, 195–215.

Kay, R.F. (1977). The evolution of molar occlusion in the Cercopithecidae and early catarrhines. *American Journal of Physical Anthropology*, **46**, 327–352.

Kay, R.F. (1981). The nut-crackers – a new theory on the adaptations of the Ramapithecinae. *American Journal of Physical Anthropology*, **55**, 141–151.

Kay, R.F. (1984). On the use of anatomical features to infer foraging behavior in extinct primates. In *Adaptations for Foraging in Nonhuman Primates*, ed. P.S. Rodman and J.G.H. Cant. New York, NY: Columbia University Press, pp. 21–53.

Kay, R.F. (1990). The phyletic relationships of extant and fossil Pitheciinae (Platyrrhine, Anthropoidea). *Journal of Human Evolution*, **19**, 175–208.

Kay, R.F. & Hylander, W.L. (1978). The dental structure of mammalian folivores with special reference to primates and phalangeroids (Marsupialia). In *The Ecology of Arboreal Folivores*, ed. G.G. Montgomery. Washington, DC: Smithsonian Institution Press, pp. 173–192.

Kinzey, W.G. (1974). Ceboid models for the evolution of hominoid dentition. *Journal of Human Evolution*, **3**, 193–203.

Kinzey, W.G. (1977). Diet and feeding behaviour of *Callicebus torquatus*. In *Primate Ecology: Studies of Feeding and Ranging Behaviour in Lemurs, Monkeys, and Apes*, ed. T.H. Clutton-Brock. New York, NY: Academic Press, pp. 127–151.

Kinzey, W.G. (1992). Dietary and dental adaptations in the Pitheciinae. *American Journal of Physical Anthropology*, **88**, 499–514.

Kinzey, W.G. & Norconk, M.A. (1990). Hardness as a basis of fruit choice in two sympatric primates. *American Journal of Physical Anthropology*, **81**, 5–15.

Kinzey, W.G. & Norconk, M.A. (1993). Physical and chemical properties of fruit and seeds eaten by *Pithecia* and *Chiropotes* in Surinam and Venezuela. *International Journal of Primatology*, **14**, 207–228.

Kitko, R.E., Strait, S.G. & Overdorff, D.J. (1996). Physical properties of *Canarium* seeds and food processing strategies of the Aye-aye in Ranomafana, Madagascar. *American Journal of Physical Anthropology*, **22**(Suppl), 139.

Kool, K.M. (1993). The diet and feeding behavior of silver leaf monkey (*Trachypithecus auratus sondaicus*) in Indonesia. *International Journal of Primatology*, **14**, 667–700.

Korstjens, A.H., Schippers, E.P., Nijssen, E.C., *et al.* (2007). The influence of food on the social organization of the three colobine species. In *Monkeys of the Taï Forest: An African Primate Community*, ed. W.S. McGraw, K. Zuberbuhler and R. Noe. Cambridge: Cambridge University Press, pp. 72–108.

Kunz, B.K. & Linsenmair, K.E. (2007). Changes in baboon feeding behavior: maturity-dependent fruit and seed size selection within a food plant species. *International Journal of Primatology*, **28**, 819–835.

Lahm, S.A. (1986). Diet and habitat preference of *Mandrillus sphinx* in Gabon: implications of foraging strategy. *American Journal of Primatology*, **11**, 9–26.

Lambert, J.E. (1998). Primate digestion: interactions among anatomy, physiology, and feeding ecology. *Evolutionary Anthropology*, **7**, 8–20.

Lambert, J.E., Chapman, C.A., Wrangham, R.W., *et al.* (2004). Hardness of cercopithecine foods: implications for the critical function of enamel thickness in exploiting fallback foods. *American Journal of Physical Anthropology*, **125**, 363–368.

Lucas, P. (2004). *Dental Functional Anatomy: How Teeth Work*. Cambridge: Cambridge University Press.

Lucas, P. & Teaford, M. (1994). Functional morphology of colobine teeth. In *Colobine Monkeys: Their Ecology, Behavior and Evolution*, ed. A.G. Davies and J.F. Oates. Cambridge: Cambridge University Press, pp. 173–203.

Lucas, P., Constantino P., Wood, B., *et al.* (2008). Dental enamel as a dietary indicator in mammals. *BioEssays*, **30**, 374–385.

Lucas, P.W., Darvell, B.W., Lee, P.K.D., *et al.* (1995). The toughness of plant cell walls. *Philosophical Transactions of the Royal Society of London, Series B: Biological Sciences*, **348**, 363–372.

Lucas, P.W., Turner, I.M., Dominy, N.J., *et al.* (2000). Mechanical defences to herbivory. *Annals of Botany*, **86**, 913–920.

Maas, M.C. (1991). Enamel structure and microwear: an experimental study of the response of enamel to shearing force. *American Journal of Physical Anthropology*, **85**, 31–50.

Maas, M.C. (1994). Enamel microstructure in Lemuridae (Mammalia, Primates): assessment of variability. *American Journal of Physical Anthropology*, **95**, 221–241.

Maas, M.C. & Dumont, E.R. (1999). Built to last: the structure, function, and evolution of primate dental enamel. *Evolutionary Anthropology*, **8**, 133–152.

Maisels, F., Gautier-Hion, A. & Gautier, J.P. (1994). Diets of two sympatric colobines in Zaire: more evidence on seed-eating in forests on poor soils. *International Journal of Primatology*, **15**, 681–701.

Marshall, A.J., Boyko, C.M., Feilen, K.L., *et al.* (2009). Defining fallback foods and assessing their importance in primate ecology and evolution. *American Journal of Physical Anthropology*, **140**, 603–614.

Martin, L.B., Olejniczak, A.J. & Maas, M.C. (2003). Enamel thickness and microstructure in pitheciin primates, with comments on dietary adaptations of the middle Miocene hominoid *Kenyapithicus*. *Journal of Human Evolution*, **45**, 351–367.

McGraw, W.S., Pampush, J. & Daegling, D. (in press). Enamel thickness and hard-object feeding in mangabeys. *American Journal of Physical Anthropology*.

McGraw, W.S., Vick, A. & Daegling, D.J. (2011). Sex and age differences in the diet and ingestive behaviors of sooty mangabeys (*Cercocebus atys*) in the Tai Forest, Ivory Coast. *American Journal of Physical Anthropology*, **114**, 140–153.

McKey, D. (1978). Soils, vegetation, and seed-eating by black colobus monkeys. In *The Ecology of Arboreal Folivores*, ed. G.G. Montgomery. Washington, DC: Smithsonian Institution Press, pp. 423–437.

Meldrum, D.J. & Kay, R.F. (1997). *Nuciruptor rubricae*, a new Pitheciin seed predator from the Miocene of Colombia. *American Journal of Physical Anthropology*, **102**, 407–427.

Ménard, N. & Vallet, D. (1996). Demography and ecology of Barbary macaques (*Macaca sylvanus*) in two different habitats. In *Evolution and Ecology of Macaque Societies*, ed. J.E. Fa and D.G. Lindburg. Cambridge: Cambridge University Press, pp. 106–131.

Milton, K. (1998). Physiological ecology of howlers (*Alouatta*): energetic and digestive considerations and comparison with the Colobinae. *International Journal of Primatology*, **19**, 513–548.

Mitani, M. (1989). *Cercocebus torquatus*: adaptive feeding and ranging behaviours related to seasonal fluctuations of food resources in the tropical rain forest of south-western Cameroon. *Primates*, **30**, 307–323.

Norconk, M.A. (1996). Seasonal variation in the diets of white-faced and bearded sakis (*Pithecia pithecia* and *Chiropotes satanas*) in Guri Lake, Venezuela. In *Adaptive Radiations of Neotropical Primates*, ed. M.A. Norconk, A.L. Rosenberger and P.A. Garber. New York, NY: Plenum, pp. 403–423.

Norconk, M.A. (2011). Sakis, uakaris, and titi monkeys: behavioral diversity in a radiation of primate seed predators. In *Primates in Perspective*, ed. C.J. Campbell, A. Fuentes, K.C. MacKinnon, S.K. Bearder and R.M. Stumpf. Cambridge: Cambridge University Press, pp. 122–139.

Norconk, M.A. & Conklin-Brittain, N.L. (2004). Variation on frugivory: the diet of Venezuelan white-faced sakis. *International Journal of Primatology*, **25**, 1–26.

Norconk, M.A. & Veres, M. (2011). Physical properties of fruit and seeds ingested by primate seed predators with emphasis on sakis and bearded sakis. *Anatomical Record*, **294**, 2092–2111.

Norconk, M.A., Grafton, B.W. & Conklin-Brittain, N.L. (1998). Seed dispersal by Neotropical seed predators. *American Journal of Primatology*, **45**, 103–126.

Norconk, M.A., Wright, B.W., Conklin-Brittain, N.L., *et al.* (2009). Mechanical and nutritional properties of foods as factors in platyrrhine dietary adaptations. In *South American Primates: Testing New Theories in the Study of Primate Behaviour, Ecology, and Conservation*, ed. P.A. Garber, A. Estrada, C. Bicca-Marques, E. Heymann and K. Strier. New York, NY: Springer Science, pp. 279–319.

Oates, J.F., Waterman, P.G. & Choo, G.J. (1980). Food selection by the south Indian leaf-monkey, *Presbytis johnii*, in relation to leaf chemistry. *Oecologia*, **45**, 45–56.

Oftedal, O.T. (1991). The nutritional consequences of foraging in primates: the relationship of nutrient intakes to nutrient requirements. *Philosophical Transactions of the Royal Society of London, Series B*, **334**, 161–170.

Palacios, E., Rodriguez, A. & Defler, T.R. (1997). Diet of a group of *Callicebus torquatus lugens* (Humboldt, 1812) during the annual resource bottleneck in Amazonian Colombia. *International Journal of Primatology*, **18**, 503–522.

Peetz, A. (2001). Ecology and social organization of the bearded saki (*Chiropotes satanas chiropotes*) (Primates: Pitheciinae) in Venezuela. *Ecotropical Monographs*, **1**, 1–170.

Peres, C.A. (1991). Seed predation of *Cariniana micranthra* (Lecythidaceae) by brown capuchin monkeys in central Amazonia. *Biotropica*, **23**, 262–270.

Peres, C.A. (1993). Notes on the ecology of buffy saki monkeys (*Pithecia albicans*, Gray 1860): a canopy seed-predator. *American Journal of Primatology*, **31**, 129–140.

Pichersky, E. & Gang, D.R. (2000). Genetics and biochemistry of secondary metabolites in plants: an evolutionary perspective. *Trends in Plant Science*, **5**, 439–445.

Poulsen, J.R., Clark, C.J. & Smith, T.B. (2001). Seasonal variation in the feeding ecology of the grey-cheeked mangabey (*Lophocebus albigena*) in Cameroon. *American Journal of Primatology*, **54**, 91–201.

Powzyk, J.A. & Mowry, C.B. (2003). Dietary and feeding differences between sympatric *Propithecus diadema diadema* and *Indri indri*. *International Journal of Primatology*, **24**, 1143–1162.

Ramsey, J.B. & Wilga, C.D. (2007). Morphology and mechanics of teeth and jaws of white-spotted bamboo sharks (*Chiloscyllium plagiosum*). *Journal of Morphology*, **268**, 664–682.

Renton, K. (2001) Lilac-crowned parrot diet and food resource availability: resource tracking by a parrot seed predator. *The Condor*, **103**, 62–69.

Rogers, M.E., Abernethy, K.A., Fontaine, B., *et al.* (1996). Ten days in the life of a mandrill horde in the Lope Reserve, Gabon. *American Journal of Primatology*, **40**, 297–313.

Rosenberger, A.L. (1992). Evolution of feeding niches in New World monkeys. *American Journal of Physical Anthropology*, **88**, 525–562.

Rosenthal G.A. & Berenbaum, M.R. (1992). *Herbivores, Their Interactions with Secondary Plant Metabolites. Vol 2, Ecological and Evolutionary Processes*. San Diego, CA: Academic Press.

Sayers, K.A. & Norconk, M.A. (2008). Himalayan *Semnopithecus entellus* at Langtang National Park, Nepal: diet, activity patterns, and resources. *International Journal of Primatology*, **29**, 509–530.

Seigler, D.S. (2004). Fabaceae (pea or bean family). In *Flowering Plants of the Neotropics*, ed. N. Smith, S.A. Mori, A. Henderson, D.W. Stevenson and S.V. Heald. Princeton, NJ: Princeton University Press, pp. 151–156.

Setz, E.Z.F. (1993). Ecologia alimentar de um grupo de parauacus (*Pithecia pithecia chrysocephala*) em um fragmento florestal na Amazônia central. Unpublished PhD thesis, Universidade Estadual de Campinas.

Shah, N. (2003) Foraging strategies in two sympatric mangabey species (*Cercocebus agilis and Lophocebus albigena*). PhD thesis, Stony Brook University.

Shellis, R.P., Beynon, A.D., Reid, D.J., *et al.* (1998). Variations in molar enamel thickness among primates. *Journal of Human Evolution*, **35**, 507–522.

Smith, N., Mori, S.A., Henderson, A., *et al.* (2004). *Flowering Plants of the Neotropics*. Princeton, NJ: Princeton University Press.

Soini, P. (1986). A synecological study of a primate community in Pacaya-Samiria National Reserve, Peru. *Primate Conservation*, **7**, 63–71.

Sterling, E.J. (1994). Aye-ayes: specialists on structurally defended resources. *Folia Primatologica*, **62**, 142–154.

Sterling, E.J. & McCreless, E.E. (2006). Adaptations in the aye-aye: a review. In *Lemurs: Ecology and Adaptation*, ed. L. Gould and M.L. Sauther. New York, NY: Springer Scientific, pp. 159–184.

Stevenson, P.R., Quiñones, M.J. & Castellanos, M.C. (2000). *Guía de frutos de los bosques del río Duda La Macarena, Colombia*. Netherlands Committee for IUCN: Asociacion para la Defensa de la Reserva de la Macarena, Tropical Rain Forest Programme.

Strait, S.G. (1997). Tooth use and the physical properties of food. *Evolutionary Anthropology*, **5**, 199–211.

Suarez, S.A. (2006). Phayre's leaf monkeys (*Trachypithecus phayrei*) as seed predators in the Phu Khieo Wildlife Sanctuary, Thailand. *American Journal of Physical Anthropology*, **129**(Suppl 42), 173.

Tattersall, I. (1982). *The Primates of Madagascar*. New York, NY: Columbia University Press.

Taylor, A. (2006). Diet and mandibular morphology in African apes. *International Journal of Primatology*, **27**, 181–201.

Teaford, M.F. (2007). What we do know and not know about diet and enamel structure. In *Evolution of the Human Diet: the Known, the Unknown and the Unknowable*, ed. P. Ungar. Oxford: Oxford University Press, pp. 56–76.

Teaford, M.F., Lucas, P.W., Ungar, P.S., *et al.* (2006). Mechanical defenses in leaves eaten by Costa Rican howling monkeys (*Alouatta palliata*). *American Journal of Physical Anthropology*, **129**, 99–104.

Terborgh, J. (1983). *Five New World Primates*. Princeton, NJ: Princeton University Press.

ter Steege, H. & Persaud, C.A. (1999). The phenology of Guyanese timber species: a compilation of a century of observations. *Vegetatio*, **95**, 177–198.

Thierry, B. (2011). The macaques: a double-layered social organization. In *Primates in Perspective*, ed. C.

Campbell, A. Fuentes, K.C. MacKinnon, S.K. Bearder and R.M. Stumpf. Oxford: Oxford University Press, pp. 229–241.

van Roosmalen, M.G.M. (1985). *Fruits of the Guianan Flora*. Utrecht: Utrecht University.

van Roosmalen, M.G.M., Mittermeier, R.A. & Fleagle, J.G. (1988). Diet of the northern bearded saki (*Chiropotes satanas chiropotes*): a neotropical seed predator. *American Journal of Primatology*, **14**, 11–35.

Veiga, L.M. (2006). Ecologia e comportamento do cuxiú-preto (*Chiropotes satanas*) na paisagem fragmentada da Amazônia oriental. Unpublished PhD thesis, Universidade Federal do Pará.

Vinyard, C.J. & Ryan, T.M. (2006). Cross-sectional bone distribution in the mandibles of gouging and non-gouging Platyrrhini. *International Journal of Primatology*, **27**, 1461–1490.

Wall, C.E., Vinyard, C.J., Johnson, K.R., *et al.* (2005). Phase II jaw movements and masseter muscle activity during chewing in *Papio anubis*. *American Journal of Physical Anthropology*, **129**, 215–224.

Waser, P. (1977). Feeding, ranging and group size in the mangabey *Cercocebus albigena*. In *Primate Ecology: Studies of Feeding and Ranging Behaviour in Lemurs, Monkeys and Apes*, ed. T.H. Clutton-Brock. New York, NY: Academic Press, pp. 183–222.

Waterman, P.G. (1984). Food acquisition and processing as a function of plant chemistry. In *Food Acquisition and Processing in Primates*, ed. D.J. Chivers, B.A. Wood and A. Bilsborough. New York, NY: Plenum, pp. 177–211.

Waterman, P.G. & Kool, K. (1994). Colobine food selection and plant chemistry. In *Colobine Monkeys: Their Ecology, Behaviour, and Evolution*, ed. A.G. Davies and J.F. Oates. Cambridge: Cambridge University Press, pp. 251–284.

Wieczkowski, J. (2003). Aspects of the ecological flexibility of the Tana mangabey (*Cercocebus galeritus*) in its fragmented habitat, Tana River, Kenya. PhD dissertation, University of Georgia.

Wieczkowski, J. (2009). Brief communication: Puncture and crushing resistance scores of the Tana River mangabey (*Cercocebus galeritus*) diet items. *American Journal of Physical Anthropology*, **140**, 572–577.

Willson, M.F. & Whelan, C.J. (1990). The evolution of fruit color in fleshy-fruited plants. *The American Naturalist*, **136**, 790–809.

Wink, M. (1988). Plant breeding: importance of plant secondary metabolites for protection against pathogens and herbivores. *Theoretical and Applied Genetics*, **75**, 225–233.

Wright, B.W. (2005), Craniodental biomechanics and dietary toughness in the genus *Cebus*. *Journal of Human Evolution*, **48**, 473–492.

Xiao, Z., Zhang, Z. & Wang, Y. (2005). The effects of seed abundance on seed predation and dispersal by rodents in *Castanopis fargesii* (Fagaceae). *Plant Ecology*, **177**, 249–257.

Yamashita, N. (2003). Food procurement and tooth use in two sympatric lemur species. *American Journal of Physical Anthropology*, **121**, 125–133.

Yeager, C.P. (1989). Feeding ecology of the proboscis monkey (*Nasalis larvatus*). *International Journal of Primatology*, **10**, 497–530.

Zhao, Q.-K. (1996). Etho-ecology of Tibetan macaques at Mount Emei, China. In *Evolution and Ecology of Macaque Societies*, ed. J.E. Fa and D.G. Lindburg. Cambridge: Cambridge University Press, pp. 263–289.

Chapter 7

Pitheciins: use of time and space

Eleonore Z.F. Setz, Liliam P. Pinto, Mark Bowler, Adrian A. Barnett, Jean-Christophe Vié, Jean P. Boubli & Marilyn A. Norconk

Introduction

Spatio-temporal variation in the abundance and distribution of edible plant parts plays a major role in population size and dispersion, social structure, and range use in highly mobile vertebrates such as primates (e.g. Crook & Gartlan 1966; Terborgh & van Schaik 1987; Milton 2000). Phenology, or the production of plant parts, is highly seasonal in the Neotropics (van Schaik *et al.* 1993; Chapman *et al.* 1999) and has important consequences for the behavior of primates (e.g. Terborgh 1983). Despite wide variation in monthly rainfall volumes (Sombroek 2001), each year Amazonian forests typically have one rainy season and one dry season (monthly rainfall < 100 mm; Nimer 1977). Rainfall in Guyana Shield forests is influenced by the Intertropical Convergence Zone (ITCZ) and there are annual "long" wet and "long" dry seasons, each lasting about 4 months, and shorter, more variable wet and dry periods between the longer periods (ter Steege & Persaud 1991).

Most Neotropical primates rely heavily on available fruits to meet their daily energetic and nutritional needs (e.g. Milton, 1998; Oftedal, 1991; Norconk *et al.* 2009). Pitheciins are specialists among frugivores, distinguished by dental and behavior adaptations to exploit large-seeded fruits that are well protected by hard husks and often unavailable to other primates (Ayres 1986; Kinzey 1992; Rosenberger 1992; also see Norconk *et al.* 1997; Palminteri *et al.*, Chapter 11, for dietary overlap of sakis with psittacids). Sakis and uacaris consume unripe seeds, benefiting from their long availability compared with fleshy fruits (Norconk *et al.* 1998; Boubli & Tokuda 2008). Nevertheless, seeking available resources in seasonal environments poses challenges even for pitheciins, which adapt by shifting time budgets and ranging behavior to accommodate seasonal changes in the availability of seeds (e.g. Barnett 2010; Boubli & Tokuda 2008). Knowledge of variable schedules of fruiting and flowering is critical to understanding resource availability for frugivores, including pitheciins, and how this affects ranging patterns and social behavior. In this chapter, we investigate how sakis and uacaris respond to seasonality behaviorally and ecologically.

Use of time – preferred plant families and phenological patterns

Long-term phenological studies in Guyana have shown that peak flowering corresponds to peak periods of sunshine in the dry season (although the trigger for flowering may be declining rainfall) and fruiting peaks in the wet season (ter Steege & Persaud 1991). Dry fruits and wind-dispersed seeds generally mature earlier in the dry season or in the transition from dry to rainy seasons (Opler *et al.* 1980; Foster 1982; Terborgh 1983; Boubli & Couto-Santos 2007). In seasonally flooded várzea and igapó forests, inundation level (rather than rainfall) is the main driver of plant community phenology (Ferreira & Almeida 2005; Ferreira *et al.* 2005; Parolin 2000) and fruiting occurs predominantly during flooded periods (Schöngart *et al.* 2002). At Jaú, Boubli & Tokuda (2008) found that fruiting was more labile in terra firme forests (peaking in December–January) than flooded (igapó) forests (i.e. fewer species produced fruit in flooded forest, but the number of productive species dropped gradually over 6 months). The extended period of unripe fruit availability compared with ripe fruit availability (e.g. Pinto 2008) benefits consumers, like pitheciins, that ingest both mature and young fruits and their seeds.

Fruit and seeds of Sapotaceae are often high-ranked in pitheciin feeding records, particularly for *Chiropotes* (Table 7.1). Sapotaceae is the highest ranking family in five of eight studies and in the top three most frequently used families for all eight studies of bearded sakis. Lecythidaceae is also high-ranking in *Chiropotes* diets and ranked either first or second in five studies. Both Sapotaceae and Lecythidaceae were in the top five most-often used plant families for all of the *Cacajao* studies (Table 7.1). The very large plant family, Fabaceae (particularly species of the genus *Inga*), is also frequently cited in the top five families. Arecaceae (palms) appears to be more important for some *Cacajao* compared with the other two pitheciins – for example, Sapotaceae ranked a distant second behind Arecaceae in Peruvian uacaris (Bowler & Bodmer 2011). This dominance may be because populations of uacaris far more commonly live

Evolutionary Biology and Conservation of Titis, Sakis and Uacaris, eds. Liza M. Veiga, Adrian A. Barnett, Stephen F. Ferrari and Marilyn A. Norconk. Published by Cambridge University Press.

Table 7.1 The five top-ranked plant families in pitheciin diets. Sources of data are documented below and are also from Barnett (2010: table V-21). The top two ranked plant families overall are indicated by bold-faced type (#1) and italicized (#2).

Species	First	Second	Third	Fourth	Fifth
Pithecia p. chrysocephala[1]	**Sapotaceae**	Moraceae	Chrysobalanaceae	Annonaceae	Burseraceae
P. p. pithecia[2]	Connaraceae	Erythroxylaceae	Fabaceae*	Chrysobalanaceae	Rubiaceae
P. p. pithecia[3]	Connaraceae	*Lecythidaceae*	Loganiaceae	Fabaceae	Erythroxylaceae
P. p. pithecia[4]	*Lecythidaceae*	Fabaceae	Euphorbiaceae	Menispermaceae	Simaroubaceae
Ch. albinasus[5]	Arecaceae	**Sapotaceae**	Fabaceae	Caryocaraceae	Moraceae
Ch. albinasus[6]	**Sapotaceae**	*Lecythidaceae*	Moraceae	Polygalaceae	Celastraceae
Ch. chiropotes[7]	**Sapotaceae**	Loranthaceae	Moraceae	Fabaceae	Bignoniaceae
Ch. chiropotes[5]	Moraceae	Fabaceae	**Sapotaceae**	*Lecythidaceae*	Melastomataceae
Ch. sagulatus[8]	**Sapotaceae**	*Lecythidaceae*	Moraceae	Vochiciaceae	Chrysobalanaceae
Ch. sagulatus[9]	**Sapotaceae**	*Lecythidaceae*	Chrysobalanaceae	Euphorbiaceae	Fabaceae
Ch. sagulatus[10]	**Sapotaceae**	*Lecythidaceae*	Fabaceae	Burseraceae	Moraceae
Ch. satanas[11]	*Lecythidaceae*	**Sapotaceae**	Fabaceae	Burseraceae	Chrysobalanaceae
Ca. c. calvus[12]	*Lecythidaceae*	Moraceae	Celastraceae	**Sapotaceae**	Annonaceae
Ca. c. ucayalii[13]	Arecaceae	**Sapotaceae**	Fabaceae	*Lecythidaceae*	Chrysobalanaceae
Ca. c. ucayalii[14]	**Sapotaceae**	Apocynaceae	Humiriaceae	*Lecythidaceae*	Arecaceae
Ca. melanocephalus[15]	Euphorbiaceae	Fabaceae	**Sapotaceae**	*Lecythidaceae*	Arecaceae
Ca. ouakary[16]	**Sapotaceae**	*Lecythidaceae*	Fabaceae	Combretaceae	Euphorbiaceae

[1] Setz (1993); [2]Homburg (1997); [3]Norconk (1996); [4]Norconk (unpublished); [5]Ayres (1981); [6]Pinto (2008); [7]Peetz (2001); [8]Gregory (2011); [9]Boyle (2008); [10]van Roosmalen *et al.* (1988); [11]Veiga (2006); [12]Ayres (1986); [13]Bowler and Bodmer (2011); [14]Aquino & Encarnación (1999); [15]Boubli (1997); [16]Barnett (2010).

* Three major groups of legumes are grouped into Fabaceae as suggested by recent molecular analysis (Wojciechowski *et al.* 2004). The Fabaceae is comprised of three subfamilies, Mimosoideae, Caesalpinoideae, and Faboideae.

in, or frequently use, flooded forests, and palm swamps (Heymann & Aquino 2010), especially the stand-forming *Mauritia flexuosa*. However, palms may also be important for other pitheciins: top two feeding sources for wet season *Ch. albinasus* were palms and legumes (Ayres 1981). Palms are relatively rare in upland forest sites occupied by many populations of *Chiropotes*, particularly those in Guyana Shield habitats: Arecaceae does not appear in the top food sources (genera) for *Ch. sagulatus* and *Ch. chiropotes*.

Pithecia feeding sources are much more diverse in terms of plant families. However, the sample of long-term studies (Table 7.1) is not as great as that from *Chiropotes* and *Cacajao*. For example, the data from both Homburg (1997) and Norconk (1996) are from the same site in a tropical seasonally dry forest in Venezuela, although there are temporal differences in sampling, and Norconk (unpublished) is only a 4-month sample from a Guyana Shield study. Setz (1993) is perhaps the most representative of the genus and her plant family rankings bear a closer resemblance to *Chiropotes* and *Cacajao* (see Table 7.1). More studies are needed on *Pithecia* before dietary generalizations can be made at the level of plant family, but higher diversity of feeding sources (following the higher diversity of habitat use) may be the norm for this genus.

Ter Steege and Persaud (1991) extracted fruiting and flowering data from unpublished records of phenological samples from Guyana. Data were continuous from 1930 to 1950 with sporadic data collected before (from 1887) and after (to 1989). Fruiting is strongly associated with rainfall and Sapotaceae are particularly abundant and durable producers (some of the 17 species fruited for months). Some *Licania* spp. (Chrysobalanaceae) fruited for 10–12 months (ter Steege & Persaud 1991, Appendix 1). *Lecythidaceae* spp. in *terra firme* are more synchronous and primarily flower in the dry season and fruit in the short dry and early long wet season, but there are no mast fruiting trees in South American forests that are comparable with Asian dipterocarp forests. The long daily paths of *Chiropotes* and *Cacajao* may enable them to take advantage of slightly asynchronized, long-producing species or families of plants (Norconk 1996; Boubli & Tokuda 2008).

Use of time – daily time budgets

As most *Pithecia* studies have been conducted in forest fragments (or land-bridge islands; Norconk 1996; Homburg 1997; Vié *et al.* 2001; Setz 1993), it is not yet possible to sort the variation in activity patterns by habitat type for sakis. Two studies have wet/dry season data (Norconk 1996; Homburg 1997), showing mild to moderate increases in feeding in the dry season and Setz (1993) suggested that duration of feeding bouts varied seasonally and with size of tree crowns. She found that feeding bouts were longer in the dry season when smaller food patches were exploited and large food patches were unavailable.

More time budget data exist for *Chiropotes* and although there is considerable variation in percentage of activities spent feeding (one-sample t-test = 8.47, df = 8, $p < 0.01$, two-tailed), the differences are not aligned by habitat type (forest fragment vs. continuous forest) (Mann–Whitney U = 4.0, ns). Between-site comparisons among studies are not as reliable as a controlled study at a single site, i.e. Boyle *et al.* (2009a) found that bearded sakis travelled more than expected in smaller fragments (10 and 100 ha), covered a larger area (km/ha) in small fragments, and revisited feeding trees more often when in forest fragments. They also found that seasonal variation in the extent of travel occurred only in continuous forest. About 1/3 of daily activities were travel-related in free-ranging *Ch. sagulatus* in Suriname (Gregory 2011). The proportion of feeding time doubled in the short dry season (compared with long dry and short wet seasons) at the expense of resting, but travel time stayed constant. For *Ch. albinasus* and *Ch. satanas*, traveling time correlated positively with daily average group size (Veiga 2006; Pinto 2008). In months when immature seeds were the most important dietary item, feeding occupied more of the daily activity budget; in those months when ripe fruits were most consumed, *Ch. chiropotes* spent more time traveling (Peetz 2001). As there were no differences in seed and ripe fruit densities between these times, Peetz (2001) suggested that ripe fruit sources may have depleted faster than immature seeds, and longer travel time was required to move between more distant feeding trees.

Ca. c. ucayalii at Lago Preto, Peru (Bowler 2007) and *Ca. c. calvus* at Tefé, Brazil (Ayres 1986) exhibited similar proportions of activity patterns (Table 7.2); however, Ayres found that the activity patterns varied seasonally, with the uacaris spending more time moving and less time resting during the high-water season. Time spent feeding did not vary significantly between high- and low-water periods for *Ca. c. calvus*. Bowler (2007) found that feeding on *Mauritia flexuosa* (Arecaceae) fruits corresponded to especially long resting bouts in *Ca. c. ucayalii*. These palms occur in large patches and produce great quantities of fruit. Boubli and Tokuda (2008) found that, when in *terra firme* where distance between food trees was greatest, *Ca. ouakary* spent more time traveling than in igapó, where food trees were less widely spaced. However, in igapó, when little fruit was available and the diet was dominated by leaves and small fruits from minor trees, *Ca. ouakary* both travelled more and exhibited more on-the-move feeding than at other times. At this time, the monkeys also moved in smaller groups (Boubli & Tokuda 2008).

Diurnal variation in activity patterns suggest that black uacaris were more likely to be seen traveling in the early and late hours of the day when the monkeys were moving away from their sleeping site or moving towards a new sleeping site (Boubli 1997). Traveling was also observed throughout the day at times when the monkeys were moving from one feeding patch to another. Feeding remained relatively constant throughout the day, with a small peak in the mid to late afternoon. Homburg (1997) documented very little variation in white-faced saki activities except for the early and late periods of the day when resting decreased and traveling increased. Feeding activities were relatively stable throughout the day.

Studies from both forest fragments and *terra firme* indicate that *Pithecia* is active for 8–10 h/day, rising between 0600 and 0700 and retreating to sleeping sites long before sunset, often by mid-afternoon (Setz 1993, 1999; Homburg 1997; Vié *et al.* 2001). In comparison, *Chiropotes* and *Cacajao* are often active for 12 h or longer – beginning their movements at dawn and moving and foraging until dusk (Boubli 1997; Bowler 2007; Barnett and Bezerra, unpublished data).

Use of space – general considerations

Members of the genus *Pithecia* occupy diverse habitats ranging from *terra firme* to sand-ridge forests and coastal swamp forests to upper lowland forests (Husson 1957; Buchanan 1978). They are also found in tropical dry forests surrounded by savanna (Buchanan 1978; Izawa 1975; Norconk 1996), in *Cecropia*-dominated secondary growth (Buchanan 1978; Barnett, pers obs), flooded forest (Heymann *et al.* 2002; Barnett, pers. obs), and montane forest (Norconk *et al.* 2003).

In contrast, the two larger genera demonstrate more restricted habitat use. *Chiropotes* species have been most frequently recorded from *terra firme* forests (e.g. Ayres 1981; van Roosmalen *et al.* 1981; Kinzey & Norconk 1990; Ferrari *et al.* 1999; Gregory & Norconk, Chapter 28), but they have also been found in other habitats including secondary forest, savanna-like transition forests (Ayres 1981; Wallace *et al.* 1996; Ferrari *et al.* 1999, 2003; Peetz 2001) and igapó forest of the Demeni river (Boubli *et al.* 2008). Populations of *Cacajao* are most frequently recorded from flooded habitats, although some may spend a high proportion of the year in non-flooded habitats including mountains at altitudes of 1500 m a.s.l. (Boubli 1999; see Barnett *et al.*, Chapter 16).

The spatial distribution of flooded habitats strongly influences the ranging patterns of *Cacajao*, in particular. For example, the narrow floodplain of the black water rivers of the Rio Negro and its tributaries (Goulding *et al.*, 2003) means that most igapó occurs in narrow ribbons of forest some only 50–200 m wide. In contrast, várzea floodplains on the neighboring white water rivers of the Solimões–Japurá may be several kilometers wide (Ayres 1986). However, the complex system of parallel leveés and sloughs produces a habitat mosaic that also limits the movement of the monkeys. Restinga forest, on the leveés, is arranged in long strips averaging 224 m wide bordered by open water and areas of scrubby slough-inhabiting chavascal (Ayres 1986). Consequently, the *Ca. c. calvus* group studied by Ayres (1986) were

Table 7.2 Activity budgets of pitheciins by habitat type. Dry season data are in bold-faced type; wet season data are italicized; unspecified seasonal data are in normal font. Habitat types = forest fragment (fragment surrounded by terrestrial matrix); land-bridge island (fragment surrounded by water).

Species	Habitat type	Activity budget (%)				
		Feed	Forage	Travel	Rest	Other
Pithecia p. chrysocephala[1]	Forest fragment	**27.1** *22.4*		**46.8** *38.4*	**26.1** *39.2*	
P. p. pithecia[2]	Land-bridge island	**49.2** *35.4*		**15.5**	**38.3**	**2.8**
P. p. pithecia[3]	Continuous forest	20.9		49.7	25.7	3.7
P. p. pithecia[4]	Land-bridge island	35.0		15.0	43.0	7.0
Ch. albinasus[5]	Continuous forest	23.8	0.5	36.2	27.5	8.8
Ch. chiropotes[6]	Land-bridge island	37.0	10.1	18.7	21.4	11.4
Ch. sagulatus[7]	Continuous and fragmented forests	24.9		21.3	46.6	7.2
Ch. sagulatus[8]	Continuous forest	20.3		31.1	48.5	
Ch. satanas[9,10]	Land-bridge island	19.8		58.5	13.8	7.9
Ch. satanas[11]	Continuous forest	21.7	3.6	55.8	16.1	2.8
Ch. satanas[11]	Land-bridge island	22.4	1.4	45.8	27.0	3.4
Ch. satanas[12]	Forest fragment	26.0	3.0	36.0	25.0	9.0
Ch. satanas[12]	Land-bridge island	30.0	4.0	27.0	23.0	15.0
Ch. utahickae[13]	Land-bridge island	58.8		30.8	9.5	0.9
Ch. utahickae[14]	Land-bridge island	31.9	5.4	50.6	10.6	1.2
Ca. calvus calvus[15]	Várzea	36.0		35.0	29.0	
Ca. calvus ucayalii[16]	Várzea, *terra firme* and palm swamp	34.0		36.0	24.0	5.0
Ca. melanocephalus[17]	*Caatinga* forest	20.0	31.0	27.0	22.0	
Ca. ouakary[18]	Igapó	36.0	10.0	47.0	4.0	3.0

[1] Setz (1993), [2]Homburg (1997), [3]Vié (1998), [4]Vié *et al.* (2001), [5]Pinto (2008), [6]Peetz (2001), [7]Boyle (2008), [8]Gregory (2011), [9]Port-Carvalho (2002), [10]Port-Carvalho & Ferrari (2004), [11]Silva (2003), [12]Veiga (2006), [13]Santos (2002), [14]Vieira (2005), [15]Ayres (1986), [16]Bowler (2007), [17]Boubli (1997), [18]Barnett (2010).

observed to spread out and move continuously in one direction through the restingas, forming dispersed foraging groups where the forested strips were wider. Modeling distances travelled between food trees in this long and narrow, but continuous habitat, Ayres (1986) concluded that the very long distances travelled daily by *Ca. c. calvus* during high-water season were largely a function of habitat topography.

Movement of *Ca. ouakary* through the largely ribbon-like igapó is similar to *Ca. c. calvus*, with dispersed groups moving in a broad front through the vegetation. Groups appear to spend 2 or 3 days in an area, moving back-and-forth until current crops are exhausted. They then move to another area, returning to the former area in 4–7 days (A. Barnett, unpublished data). However, for those parts of the year when the flooded forests are dry and lack both fruits and new leaves (key dietary resources), *Ca. ouakary* move to the immediately adjacent *terra firme* forests where fruit, and their immature seeds, are available (Boubli & Tokuda 2008).

Use of space – home range size, body mass and group size

Pithecia spp. are the smallest pitheciins (body mass range: 1.35–3.1 kg). They form smaller groups (2–12 individuals), and have smaller home ranges, than bearded sakis and uacaris (Norconk 2011). *Chiropotes* spp. (body mass range: 2.5–3.6 kg) and *Cacajao* spp. (body mass range: 2.8–3.5 kg) are medium-sized platyrrhines. They form groups that often exceed 50 individuals (sometimes reaching 200 in *Cacajao* spp.), have daily travel distances of up to 8 km, and home ranges that exceed 1000 ha (Ayres 1986; Boubli 1997; Pinto 2008; Boyle *et al.* 2009b; Gregory 2011; Table 7.3). *Chiropotes* spp. have also been studied on land-bridge islands (the result of flooding from hydroelectric dams; e.g. Kinzey & Norconk 1993; Peetz 2001) and in forest fragments as small as 14 ha (Silva 2003; Veiga 2006; Boyle *et al.* 2009a, 2009b). Bezerra *et al.* (Chapter 30) report on a group of *Ca. ouakary* (8–15 animals) inhabiting a natural 74 ha igapó island.

Table 7.3 Group size and use of space by pitheciins.

Species	Group size	Home range size (ha)	Habitat size[1] (ha)	Day-range (m)	Study site	Source
Pithecia albicans		170	CF		Lake Teiú, Mamirauá, Brazil	Peres 1993
P. hirsuta	2–3		CF		Loreto, Peru	Happel 1982
P. p. chrysocephala	2–3	8–9	20	300–500	Central Amazon, Brazil	Oliveira *et al.* 1985
P. p. chrysocephala	4–7	10	10	721–1076	Central Amazon, Brazil	Setz 1993
P. p. pithecia	8	12.75	12.75	1104	Lago Guri, Venezuela	Homburg 1997
P. p. pithecia	7	12.75	12.75	1772 ± 274	Lago Guri, Venezuela	Cunningham 2003
P. p. pithecia	1–5	287	CF	1880 ± 520 [1100–2700]	Petit Saut dam, French Guiana	Vié *et al.* 2001
P. p. pithecia	2.3	17.6	28	776±182		Vié 1998
Cacajao calvus calvus	48	493	CF	2710 [1200–6200]	Lake Teiú Mamirauá, Brazil	Ayres 1986
Ca. calvus ucayalii	150	>1200	CF		Lago Preto, Peru	Bowler 2007
Ca. c. ucayalii		3000	CF	7300	Loreto, Peru	Leonard & Bennett 1996
Ca. c. ucayalii			CF	2000	Loreto, Peru	Swanson-Ward & Chism 2003
Ca. c. ucayalii		15 000	CF		Reserva Comunal Tamshiyacu-Tahuayo	Aquino 1995
Ca. ouakary	20	500	CF	3000 [50–5000]	Apaporis River, Colombia	Defler 2004
Ca. melanocephalus	70	1053	CF	2300 ± 1197 [1200–4400]	Pico da Neblina	Boubli 1997
Ch. albinasus	–	200–350[2]	CF	2500–5000[1]	Aripuanã, Brazil	Ayres 1981, 1989
Ch. albinasus	56+	>1000[1]	CF	3667 [1840–7809]	Tapajós, Brazil	Pinto 2008
Ch. chiropotes	17	122.25	180	1600 [500–2700]	Lago Guri, Venezuela	Peetz 2001
Ch. sagulatus	9; 13[3]	–	CF	3200	Raleighvallen-Voltzberg Nature Reserve, Suriname	Norconk & Kinzey 1994
Ch. sagulatus	–	10	10	1300	Central Amazon, Brazil	Ayres 1981, 1989
Ch. sagulatus	–	200–250[1]	CF	2500[1]	Raleighvallen-Voltzberg Nature Reserve, Suriname	van Roosmalen *et al.* 1981, 1988
Ch. sagulatus	30	–	CF	1096 [240–6500]	Central Amazon, Brazil	Frazão 1992
Ch. sagulatus	3.79 (±0.21)	13.7	13.7	1720 ± 90	#1202, Central Amazon, Brazil	Boyle *et al.* 2009a
Ch. sagulatus	3.79 (±0.21)	12.1	14.0	1720 ± 90)	#2206, Central Amazon, Brazil	Boyle *et al.* 2009a
Ch. sagulatus	22.9 (±5.09)	559	CF	2990 ± 20	Km 41, Central Amazon, Brazil	Boyle *et al.* 2009a
Ch. sagulatus	22.9 (±5.09)	300	CF	2990 ± 20	Cabo Frio, Central Amazon, Brazil	Boyle *et al.* 2009a

Table 7.3 *(cont.)*

Species	Group size	Home range size (ha)	Habitat size[1] (ha)	Day-range (m)	Study site	Source
Ch. sagulatus	3–45	382–742	CF	2362 ± 821	Brownsberg Nature Park, Suriname	Gregory 2011
Ch. satanas	27	57	1300	–	Tucuruí, Brazil	Santos 2002
Ch. satanas	34	68.9	1300	–	Tucuruí, Brazil	Silva 2003
Ch. satanas	7	16.3	16.3	–	Tucuruí, Brazil	Silva 2003
Ch. satanas	39	75	1300	4025 (± 994)	Tucuruí, Brazil	Veiga 2006
Ch. satanas	8	19	19.4	2807 (± 289)	Tucuruí, Brazil	Veiga 2006
Ch. utahickae	24	100[2]	129		Tucuruí, Brazil	Santos 2002
Ch. utahickae	23	57.5	129	2530 [1940–4080]	Tucuruí, Brazil	Vieira 2005

[1] CF = continuous forest; [2]estimated; [3]two groups.

Group sizes as large as 12 *Pithecia pithecia* individuals were observed during repeated censuses in Guiana (Lehman *et al.* 2001), although the average group size was 4.8 individuals. Habituated *P. pithecia* groups at Brownsberg Nature Park, Suriname (free-ranging) and Lago Guri, Venezuela (land-bridge islands) were "cohesive" and ranged from 2 to 9 individuals (Cunningham 2003; Thompson & Norconk, Chapter 27; Gregory & Norconk, Chapter 28; Thompson 2011). *Pithecia albicans* (but not *P. pithecia*) groups were observed to fragment temporarily into feeding parties (Peres 1993).

Chiropotes spp. groups tend to be larger in continuous forest than in forest fragments or on islands (Table 7.3). Boyle (2008) and Boyle *et al.* (2009a) detected significant differences in average group size of *Ch. sagulatus* in continuous forest and in various sized forest fragments in Central Amazonia: as forest size class increased, group size was progressively larger. Some studies show that *Chiropotes* groups travel cohesively, fissioning only on arrival at a feeding place whereupon they spread out over distances of up to 300 m (Ayres 1981, 1989; van Roosmalen *et al.* 1981, 1988; Frazão 1992; Norconk & Kinzey 1994). However, other studies have found *Chiropotes* foraging in subgroups that remain separated for periods from a few hours to several days (Veiga *et al.* 2006; Pinto 2008; Gregory 2011; Shaffer 2012). See Photo 7.1.

Cacajao spp. can form very large groups, occasionally between 100 and 200 individuals (Boubli 1997; Defler 1999; Barnett *et al.* 2005; Bowler 2007). Small groups (5–15 animals) are common in *Ca. ouakary* where larger groups may fission into subunits that remain apart for days (B. Bezerra, unpublished data). At Lago Preto, Peru, the mean foraging group size of *Ca. c. ucayalii* was 43.5 (± 24.1). Typically, a large group of 100+ animals would come together to sleep in the evenings, and separate into smaller foraging groups during the day (Bowler 2007). Similarly, in Brazil, *Ca. ouakary* subgroups united to sleep, only to split again the next day (Barnett *et al.* 2005, 2012). In Pico da Neblina National Park, *Ca. melanocephalus* travelled in a dispersed group of about 70 individuals that did not split into smaller groups (Boubli 1997). Boubli and Tokuda (2008) posit that this is due to the unusual nature of annual fruit availability at this *campinarana* (also known locally as *caatinga*) study site (see below). Fruit production was found to be strongly equitable across months (but not across years); productive trees were very patchy and highly dispersed; and patches were sufficient to accommodate entire groups.

In the highly seasonal forests of the Colombian igapó, Defler (1999, 2004) reported that small groups of *Ca. ouakary* were most often observed during the season with fewest available fruits. This view is supported by observations in a similar habitat in Jaú, where *Ca. ouakary* foraged in groups of 2–5 when fruit availability was low and foods were in small patches (young leaves, fruit from understory trees), but in groups of 12–15 (subgroups of a 30- to 40-member group) when fruit was more abundant and occurred in canopy trees. The largest groups (50–70 and sometimes 100+) occurred in *terra firme* during the seasonal igapó fruit dearth (but when immature fruit availability in *terra firme* was high: see Boubli & Tokuda 2008).

Cacajao subgroups may also be widely dispersed: Ayres (1989) reported that foraging subgroups of *Ca. c. calvus* could be 1–2 km apart. *Cacajao* day ranges are large, reaching a maximum of 7500 m (Leonard & Bennett 1996), but averaging about 2750 m (Ayres 1986; Boubli 1997; Defler 2004; Table 7.3).

Use of space – seasonal variation in travel patterns

In a study of *Pithecia p. pithecia* foraging behavior on an island in Lago Guri, Venezuela, Cunningham and Janson (2007) found that white-faced sakis made feeding choices based on both tree productivity and relative distance to the next

Photo 7.1 *Chiropotes sagulatus* male resting on large branch, Brownsberg Nature Park, Suriname. Photo: Tremaine Gregory.

feeding tree: (1) the monkeys passed up resources in order to feed at preferred feeding trees; (2) relative productivity of feeding trees predicted direction of travel; (3) the monkeys employed memory to return to highly productive feeding trees from a variety of starting points; and (4) they expressed a preference for near trees only if near and far trees were equally productive. As the productivity of trees declined in the dry season, they included more trees of low productivity in their daily ranges (Cunningham & Janson 2007). During the transition to the wet season, a period of pronounced fruit scarcity in the tropical dry forests of eastern Venezuela, sakis revisited trees less often and traveled shorter distances between feeding trees. Setz (1993) found that *P. p. chrysocephala* travelled farther and faster in the rainy season when fruits were abundant. The study groups occupied small forest fragments in both studies.

On average, travel paths are longer for *Chiropotes* than for *Pithecia* (Table 7.3), but there is no consensus about the direction of seasonal differences in the length of distances travelled daily. Ayres (1981) found that *Ch. albinasus* increased day-range in the dry season (to 620 m/h from 330 m/h in the wet season) when availability and consumption of ripe fruits decreased, but Frazão (1992) observed the converse for a *Ch. sagulatus* group. The average day-range was longer in the rainy season (1334 m/day) than the dry season (835 m/day) (Frazão 1992). Gregory (2011) found no seasonal variation in path length except in the long dry season. At that time, both foraging group size and foraging distances were significantly correlated. At the Lago Guri site, Peetz (2001) observed that *Ch. chiropotes* traveled more widely during the transition between dry and rainy seasons. At this time, diet breadth declined and feeding was focused on the fruits of just two plant species (*Oryctanthus alveolatus*, Loranthaceae and *Lepidocordia punctata*, Boraginaceae). Each had small crop volumes which, depleting quickly, drove an increase in day-range. For *Ch. satanas* at Tucuruí, travel time, average day-range and area used each month increased in the rainy season, and these variations were positively correlated to group size (Veiga 2006). The studies above suggest that the interaction between travel distance and seasonality is complex and may be affected by group size, flexibility in troop fission, availability of specific resources, and (for *Cacajao*) flooding regimes.

Some populations appear to be strongly influenced by single, locally abundant resources. For example, Bobadilla (1998) found that areas favored by *Ch. utahickae* at Arataú were positively correlated with the abundance of energy-rich fruits of the babaçu palm (*Attalea speciosa*, Arecaceae: Cavalcante 1988). Norconk *et al.* (1998) reported a similar situation for *Ch. chiropotes* in Lago Guri where fruit and seeds of *Pradosia caracasana* (Sapotaceae) were eaten at various stages of maturity for 11 months of the year.

Ca. ouakary is largely found in seasonally flooded igapó, and most populations show evidence of seasonal use of *terra firme* forests during the dry season (Barnett *et al.* 2005; Barnett & da Cunha 1991; da Cunha & Barnett 1989; Defler 1989, 2004; Hernández-Camacho & Defler 1989; Palacios & Peres 2005). Seasonal movements range from as little as 200–300 m into the *terra firme* on the lower Rio Negro (Barnett *et al.* 2005) to 1500 m or more on the Apaporis River, Colombia (Defler 1989, 2004; see Chapter 16). These shifts in travel patterns appear to be related to seasonal dearth of fruit production in igapó. Other very specific incursions into *terra firme* forests are presumably associated with exploitation of fruiting stands of *Mauritia flexuosa* palms along forest streams inside the *terra firme* forest (Barnett *et al.* 2005).

Some populations of *Ca. melanocephalus* occur in igapó (Lehman & Robertson 1994a, 1994b), but igapó is absent in the foothills of the Pico de Neblina Mountains in northwest Amazonia and there the black-backed uacaris inhabit non-flooded forests types (*campinarana* = *caatinga* and *terra firme*). Boubli (1997, 1999) observed no seasonal migrations, although *Ca. melanocephalus* groups covered very large distances daily (2300 ± 1970 m). Boubli and Tokuda (2008) suspected that uacaris tracked resources with supra-annual fruiting cycles by ranging very widely. They found that uacaris shifted their ranging patterns in their core area and migrated beyond the large trail system (> 1000 ha) for the months of January and September 1995 (Boubli 1997). They were likely using a different part of the Pico de Neblina forest during these months, shifting their core area to track resources seasonally on a very large scale.

Ayres (1986) found that group size and travel patterns of *Ca. c. calvus* were a function of fruit quality and its dispersion over the year. During the high-water period, high-quality foods (those rich in proteins and lipids) were rare and scattered, while only fruits rich in glycids or carbohydrates (but poor in proteins and lipids) were common. He concluded that this forced *Ca. c. calvus* to travel large distances for higher-quality foods. During the low-water season, travel distances were shorter, and group sizes were smaller. During this period *Ca. c. calvus* exhibited two foraging strategies: between August and October when few tree species had fruits, *Ca. c. calvus* foraged on the ground or ate the fruits of lianas. Later, between November and February (during an abundance of large seeded, fat-rich, fruits) *Ca. c. calvus* returned to the canopy to exploit these abundant and widespread resources and travelled shorter distances to meet daily energy requirements (Ayres 1986).

Populations of *Ca. c. ucayalii* at Lago Preto, Peru, move between seasonally flooded várzea, *terra firme* and *Mauritia* palm swamp habitats (Bowler 2007), as relative fruit availability dictates. Várzea forests were most used in the high-water season, while *terra firme* forests were most used in the low-water season. *Mauritia* palm swamp forests were used extensively when ripe *M. flexuosa* fruit pulp was available. At this time, *C. c. ucayalii* groups traveled directly between the palm swamp patches (Bowler 2007). The mosaic nature of contiguous habitat patches facilitates the flexible response of the uacaris to tracking resource availability.

Use of space – vertical strata

Guianan sakis (*P. pithecia*) are recorded in the middle to lower strata of the forest more frequently than Amazonian sakis (*P. albicans*, *P. irrorata*, *P. monachus*: Happel 1982; Walker 1996; S. Palminteri, personal communication; Davis & Walker-Pacheco, Chapter 8). Use of the lower half of the forest by Guyanan sakis correlates with the frequency of using vertical stems for locomotion, relatively long hind limbs, and relatively frequent vertical clinging and leaping (Walker 1996; see Davis & Walker-Pacheco, Chapter 8). Sympatric *Chiropotes* and *Pithecia* species use different forest strata, although they also have different home range sizes, travel path distances (Table 7.2) and somewhat different diets (see Table 7.1). Thus the hypothesis that vertical stratification in pitheciins might be driven by interspecific competition has not been fully explored (Gregory & Norconk, Chapter 28). Both *P. hirsuta* and *P. p. chrysocephala* travel and feed in lower forest strata, although they feed seasonally in the upper strata (Happel 1982; Setz 1993). *P. p. pithecia* studied by Vié (1998) in continuous forest and fragments in French Guiana spent 65% of the time below 20 m, and 35% in the understory. Resting and social activities occurred only in the middle to higher strata.

Cacajao generally use the upper and middle canopy strata (Ayres 1986; Boubli 1997; Bowler 2007; A. Barnett, unpublished data). The mean height used by *Ca. c. calvus* was 16.2 m (Walker 1996). At Lago Preto, Bowler (2007) observed that *Ca. c. ucayalii* foraged at lower levels in várzea when the forest was inundated than when the forest was dry. *Ca. c. ucayalii* were often feeding on low shrubs with ripe fruit pulp during the high-water season, so it is possible that the change in foraging level may reflect the relative abundance of fruits in the lower strata at this time. Furthermore, big cats may enter inundated várzea less frequently (Bowler 2007), thus reducing predation risk for uacaris on the ground. *Ca. ouakary* use the understory more often when the canopy has few fruit; they increase consumption of fruits from understory trees (Boubli & Tokuda 2008) and extract fruits floating on the surface of the water (Barnett, unpubl. data).

Predator avoidance may also influence choice of sleeping site. On the Apaporis River, Colombia, sleeping heights for *Ca. ouakary* were 10–15 m in flooded igapó, and 25–30 m when it was not flooded (Defler 2004). Similarly, *Ca. c. ucayalii* at Lago Preto slept at low levels in flooded várzea, but when the várzea dried out the animals moved their sleeping site to *terra firme* forests, where they slept in taller trees. Barnett *et al.* (2012) found *Ca. ouakary* selected trees with uncluttered interiors and large horizontal limbs arranged in broad open canopies. The two most commonly used trees (*Hydrochorea marginata* and *Ormosia paraensis*, both Fabaceae), were uncommon. Individual sleeping trees were larger and taller than average, lacked touching canopies, lianas and wasp nests and were often close to open water.

Cacajao c. ucayalii at Lago Preto were never recorded foraging or locomoting on the ground, but *Ca. c. calvus* at Lake Teiú forage terrestrially in times of canopy food shortage (Ayres 1986). Terrestrial feeding in *Ca. ouakary* has been reported by Defler (2004) in Colombia and observed by Barnett in Jaú (Barnett *et al.*, Chapter 14). Big cats were rare at Lake Teiú (although they did enter the area during the low-water season), but jaguar, ocelot and puma prints and scats were regularly encountered at Lago Preto. However, this explanation does not hold at Jaú, where the jaguar population is dense and big cats used unflooded igapó contemporaneously with foraging uacaris. Uacaris at Jaú slept singly or in small groups on the outer third of branches. Barnett *et al.* (2012) considered this a compromise to maximize detection of arboreal predators (such as cats) and aerial predators (such as large owls).

Conclusions

Sakis and uacaris are able to capitalize on inherently long periods of fruit production in some plant families (e.g. Sapotaceae and Arecaceae) through their long-distance travel/search patterns and their ability to gain access dentally to both young and mature fruit and seeds. While these qualities may dampen the dramatic seasonal effects on fruit choice, ranging patterns and social behavior seen in some ripe fruit specialists, it is clear that seasonality (driven by both rainfall and flooding patterns) affects travel distances, foraging group size and diet breadth in sakis. Thus, pitheciins display flexibility of behavior that might not have been expected from their well-known preference for seeds. In *Pithecia*, the least specialized of the pitheciins, this flexibility extends to successful use of small fragments and secondary forest. *Chiropotes* adjusts group size and home-range characteristics to deal with seasonal variation in food availability (e.g. Peetz 2001; Veiga 2006; Pinto 2008; Boyle *et al.* 2009a, 2009b; Gregory 2011; Shaffer 2012). With the availability of key dietary resources and absence of hunting, they also can survive in small fragments (e.g. Bobadilla 1998; Bobadilla & Ferrari 2000). Superimposed on seasonal variations will be Amazon basin-wide geographical gradients of climate and soil quality. These substantial environmental factors may have dramatic effects on the nature of resources available to primates (see Schwartzkopf & Rylands 1989). In turn, this may well explain some of the intersite variation observed for widespread species, such as *P. pithecia*.

While there are many aspects of pitheciin foraging behavior that are not fully understood, it is clear that sakis and uacaris alter ranging and foraging patterns during an annual cycle and that this is related to availability (and size) of food sources. Ecological decision-making is evident in shifting day-ranges seasonally, in decisions about which trees to feed in on a daily basis, and in changes in group cohesion and foraging group sizes. At the social level, fission–fusion grouping permits fine-tuned responses to seasonal variation in fruit distribution and between-site variations in patch size and resource density. With more data we will be able to test socioecological models (e.g. Wrangham 1980; Isbell & Young 2002) with more rigor, but it is clear that sakis and uacaris make behavioral decisions in feeding and foraging that are based on generalized and predictable ecological conditions.

References

Aquino, R. (1995). Conservation of *Cacajao calvus ucayalii* in Amazonian peru. *Neotropical Primates*, **3**, 40–42.

Aquino, R. & Encarnacion, F. (1999). Preliminary observations on the diet of *Cacajao calvus ucayalii* in northeastern Peru. *Neotropical Primates*, **7**, 1–5.

Ayres, J.M. (1981). Observações sobre a ecologia e o comportamento dos cuxiús (*Chiropotes albinasus* e *Chiropotes satanas*, Cebidae: Primates). Unpublished Master thesis, Instituto Nacional de Pesquisas da Amazônia (INPA), Manaus.

Ayres, J.M. (1986). The white uakaris and the Amazonian flooded forests. Unpublished PhD thesis, Cambridge University.

Ayres, J.M. (1989). Comparative feeding ecology of the uakari and bearded saki, *Cacajao* and *Chiropotes*. *Journal of Human Evolution*, **18**, 697–716.

Barnett, A.A. (2010). Diet, habitat use and conservation ecology of the golden-backed uacari, *Cacajao melanocephalus ouakary*, in Jaú National Park, Amazonian Brazil. PhD Dissertation, Roehampton University, London. http://roehampton.openrepository.com/roehampton/.

Barnett, A. & da Cunha A.C. (1991). The golden-backed uacari on the upper Rio Negro, Brazil. *Oryx*, **25**, 80–88.

Barnett, A.A., Shaw, P., Spironello, W.R., *et al.* (2012). Sleeping site selection by golden-backed uacaris, *Cacajao melanocephalus ouakary* (Pitheciidae), in Amazonian flooded forests. *Primates*, DOI 10.1007/s 10329-012-0296-4.

Barnett, A.A., Volkmar de Castilho, C., Shapley, R.L., *et al.* (2005). Diet, habitat selection and natural history of *Cacajao melanocephalus ouakary* in Jaú National Park, Brazil. *International Journal of Primatology*, **26**, 949–969.

Bobadilla, U.L. (1998). Abundância, tamanho de agrupamento e uso do habitat por cuxiús de Utahick, *Chiropotes satanas utahicki* Hershkovitz, 1985 em dois sítios na Amazônia Oriental: implicações para a conservação. Unpublished Master thesis, Universidade Federal do Pará, Belém.

Bobadilla, U.L. & Ferrari, S.F. (2000). Habitat use by *Chiropotes satanas utahicki* and syntopic platyrrhines in eastern Amazonia. *American Journal of Primatology*, **50**, 215–224.

Boubli, J.P. (1997). A study of the black uakari, *Cacajao melanocephalus melanocephalus*, in the Pico da Neblina National Park, Brazil. *Neotropical Primates*, **5**, 113–115.

Boubli, J.P. (1999). Feeding ecology of black-headed uacaris (*Cacajao melanocephalus melanocephalus*) in Pico da Neblina National Park, Brazil. *International Journal of Primatology*, **20**, 719–749.

Boubli, J.P. (2002). Lowland floristic assessment of Pico de Neblina National Park, Brazil. *Vegetatio*, **160**, 149–167.

Boubli, J.P. & Couto-Santos. (2007). Phenology of canopy trees in the ever-wet lowland forest mosaic of Pico da Neblina national Park, Amazonas, Brazil. *Ecotropica*, **13**, 17–26.

Boubli, J.P. & Tokuda, M. (2008). Socioecology of black uakari monkeys, *Cacajao hosomi*, in Pico da Neblina National Park, Brazil: the role of the peculiar spatial–temporal distribution of resources in the Neblina forests. *Primate Report*, **75**, 3–10.

Boubli, J.P., da Silva, M.N.F., Amado, M.V., *et al.* (2008). A taxonomic reassessment of *Cacajao melanocephalus* Humboldt (1811), with the description of two new species. *International Journal of Primatology*, **29**(3), 723–741.

Bowler, M. (2007). The ecology and conservation of the red uakari monkey on the Yavarí River, Peru. Unpublished PhD thesis, University of Kent, Canterbury.

Bowler, M. & Bodmer, R.E. (2011). Diet and food choice in Peruvian red uakaris (*Cacajao calvus ucayalii*): selective or opportunistic seed predation? *International Journal of Primatology*, **32**, 1109–1122.

Boyle, S.A. (2008). The effects of forest fragmentation on primates in the Brazilian Amazon. Unpublished doctoral dissertation, Arizona State University, Tempe.

Boyle, S.A., Lourenco, W.C., da Silva, L.R., *et al.* (2009a). Travel and spatial patterns change when *Chiropotes satanas chiropotes* inhabit forest fragments. *International Journal of Primatology*, **30**, 515–531.

Boyle, S.A., Lourenco, W.C., da Silva, L.R., *et al.* (2009b). Home range estimates vary with sample size and methods. *Folia Primatologica*, **80**, 33–42.

Buchanan, D.B. (1978). Communication and ecology of pithecine monkeys, with special reference to *Pithecia pithecia*. Unpublished PhD thesis, Wayne State University, Detroit.

Cavalcante, P.B. (1988). *Frutos comestíveis da Amazônia, 4ª ed.* Belém: Museu Paraense Emílio Goeldi/Souza Cruz.

Chapman, C.A., Wrangham, R.W., Chapman, L.J., *et al.* (1999). Fruit and flower phenology at two sites in Kibale National Park, Uganda. *Journal of Tropical Ecology*, **15**, 189–211.

Crook, J.H. & Gartlan, J.S. (1966). Evolution of primate societies. *Nature*, **210** 1200–1203.

da Cunha, A.C. & Barnett, A. (1989). *Project uakari: First report – preliminary survey. Part one – Zoology.* Rio de Janeiro: Pronatura. (Portuguese summary), p. 108.

Cunningham, E.P. (2003). The use of memory in *Pithecia pithecia's* foraging strategy. Unpublished PhD thesis, City University of New York.

Cunningham, E. & Janson, C. (2007). Integrating information about location and value of resources by white-faced saki monkeys (*Pithecia pithecia*). *Animal Cognition*, **10**, 293–304.

Defler, T.R. (1989). The status and some ecology of primates in the Colombian Amazon. *Primate Conservation*, **10**, 51–56.

Defler, T.R. (1999). Fission–fusion in the black-headed uacari (*Cacajao melanocephalus*) in eastern Colombia. *Neotropical Primates*, **7**, 5–8.

Defler, T.R. (2004). *Primates de Colombia.* Bogota: Conservacion Internacional.

Ferrari, S.F., Iwanaga, S., Coutinho, P.E.G., *et al.* (1999). Zoogeography of *Chiropotes albinasus* (Platyrrhini, Atelidae) in southwestern Amazonia. *International Journal of Primatology*, **20**, 995–1004.

Ferrari, S.F., Iwanaga, S., Ravetta, A.L., *et al.* (2003). Dynamics of primate communities along the Santarém–Cuiabá Highway in south-Central Brazilian Amazonia. In *Primates in Fragments: Ecology and Conservation*, ed. L. Marsh. New York, NY: Kluwer Academic/ Plenum Publishers, pp. 123–144.

Ferreira, L.V. & Almeida, S.S. (2005). Relação entre a altura de inundação, riqueza específica de plantas e o tamanho de clareiras naturais em uma floresta inundável de igapó, na Amazônia Central. *Revista Árvore*, **29**, 445–453.

Ferreira, L.V., Almeida, S.S., Amaral, D.D., *et al.* (2005). Riqueza e composição de espécies da floresta de igapó e várzea da Estação Científica Ferreira Penna: Subsídios para o plano de Manejo da Floresta Nacional de Caxiuanã. *Pesquisas, Botânica*, **56**, 103–116.

Foster, R.B. (1982). The seasonal rhythm of fruitfall on Barro Colorado Island. In *The Ecology of a Tropical Forest, Seasonal Rhythms and Long-term Changes*, ed. E.G. Leigh, A.S. Rand & D.M. Windsor. Washington, DC: Smithsonian Institution, pp.151–172.

Frazão, E.R. (1992). Dieta e estratégia de forragear de *Chiropotes satanas chiropotes* (Cebidae: Primates) na Amazônia Central Brasileira. Unpublished Master thesis, Instituto Nacional de Pesquisas da Amazônia (INPA), Manaus.

Gregory, T. (2011). Socioecology of the Guianan bearded saki, *Chiropotes sagulatus*. Unpublished doctoral dissertation, Kent State University, Kent, OH.

Goulding, M., Barthem, R. & Ferreira, E. (2003). *The Smithsonian Atlas of the Amazon*. Washington, DC: Smithsonian Institution Press.

Happel, R.E. (1982). Ecology of *Pithecia hirsuta* in Peru. *Journal of Human Evolution*, **11**, 581–590.

Hernandez-Camacho, J. & Defler, T.R. (1989) Algunos aspectos de la conservación de primates no-humanos en Colômbia. In *La Primatologia en Latinoamerica*, ed. C.J., Saavedra, R.A. Mittermeier & I.B. Santos. Bairro Cincao, Brazil: Ed. Littera Maciel, pp. 67–100.

Heymann, E.W. & Aquino, R. (2010). Peruvian red uakaris (*Cacajao calvus ucayalii*) are not flooded-forest specialists. *International Journal of Primatology*, **31**, 751–758.

Heymann, E.W., Canaquin Y.J.E. & Encarnacion, C.F. (2002). Primates of the Río Curaray, northern Peruvian Amazon. *International Journal of Primatology*, **23**, 191–201.

Homburg, I. (1997). Ökologie und Sozialverhalten von Weissgesicht-sakis, ein Freilandstudie in Venezuela. Doctoral thesis, Universität Bielefeld. Göttingen: Cuvillier Verlag.

Husson, A.M. (1957). Studies on the fauna of Suriname and other Guyanas: II. Notes on the primates of Suriname. *Natuurwetenschappelijke Studiekring voor Suriname en de Nederlandse Antillen*, **16**, 13–40.

Isbell, L.A. & Young, T.P. (2002). Ecological models of female social relationships in primates: similarities, disparities, and some directions for future clarity. *Behaviour*, **139**, 177–202.

Izawa, K. (1975). Foods and feeding behavior of monkeys in the Upper Amazon Basin. *Primates*, **16**, 295–316.

Kinzey, W.G. (1992). Dietary and dental adaptations in the Pitheciinae. *American Journal of Physical Anthropology*, **88**, 499–514.

Kinzey, W.G. & Norconk, M.A. (1990). Hardness as a basis for fruit choice in two sympatric primates. *American Journal of Physical Anthropology*, **81**, 5–15.

Kinzey, W.G. & Norconk, M.A. (1993). Physical and chemical properties of fruit and seeds eaten by *Pithecia* and *Chiropotes* in Surinam and Venezuela. *International Journal of Primatology*, **14**(2), 207–227.

Lehman, S.M. & Robertson K.L. (1994a). Preliminary survey of *Cacajao melanocephalus melanocephalus* in Southern Venezuela. *International Journal of Primatology*, **15**, 927–934.

Lehman, S.M. & Robertson, K.L. (1994b). Survey of Humboldt's black head uakari (*Cacajao melanocephalus melanocephalus*) in southern Amazonas,

Venezuela. *American Journal of Primatology*, **33**, 223.

Lehman, S.M., Prince, W. & Mayor, M. (2001). Variations in group size in white-faced sakis (*Pithecia pithecia*): evidence for monogamy or seasonal congregations? *Neotropical Primates*, **9**, 96–101.

Leonard, S. & Bennett, C. (1996). Associative behavior of *Cacajao calvus ucayalii* with other primate species in Amazonia Peru. *Primates*, **37**, 227–230.

Milton, K. (1998). Physiological ecology of howlers (*Alouatta*): energetic and digestive considerations and comparison with the Colobinae. *International Journal of Primatology*, **19**, 513–548.

Milton, K. (2000). Quo vadis? Tactics of food search and group movement in primates and other animals. In *On the Move: How and Why Animals Travel in Groups*, ed. S. Boinski & P.A. Garber. Chicago, IL: University of Chicago Press, pp. 375–417.

Nimer, E. (1977). Clima. In *Geografia do Brasil*, vol. **1**, ed. M.V. Galvão. Rio de Janeiro: Fundação Instituto Brasileiro de Geografia e Estatística, pp. 39–58.

Norconk, M.A. (1996). Seasonal variation in the diets of white-faced and bearded sakis (*Pithecia pithecia* and *Chiropotes satanas*) in Guri Lake, Venezuela. In *Adaptive Radiations of Neotropical Primates*, ed. M.A. Norconk, A.L. Rosenberger & P.A. Garber. New York, NY: Plenum, pp. 403–423.

Norconk, M.A. (2011). Saki, uakaris, and titi monkeys: behavioral diversity in a radiation of primate seed predators. In: *Primates in Perspective*, 2nd edition, ed. C.J. Campbell, A. Fuentes, K.C. MacKinnon, M. Panger & S.K. Bearder. New York, NY: Oxford University Press, pp. 123–138.

Norconk, M.A. & Kinzey, W.G. (1994). Challenge of Neotropical frugivory: travel patterns of spider monkeys and bearded sakis. *American Journal of Primatology*, **34**, 171–183.

Norconk, M.A., Grafton, B.W. & Conklin-Brittain, N.L. (1998). Seed dispersal by neotropical seed predators. *American Journal of Primatology*, **45**, 103–126.

Norconk, M.A. Raghanti, M.A., Martin, S.K., *et al.* (2003) Primates of Brownsberg Natuurpark, Suriname, with particular attention to the Pitheciins. *Neotropical Primates*, **11**, 94–100.

Norconk, M.A., Wertis, C.A. & Kinzey, W.G. (1997). Short communication: seed predation by monkeys and macaws in eastern Venezuela: preliminary findings. *Primates*, **38**, 177–184.

Norconk, M.A., Wright, B.W., Conklin-Brittain, N.L., *et al.* (2009). Mechanical and nutritional properties of food as factors in platyrrhine dietary adaptations. In Garber, P.A., Estrada, A., Bicca-Marques, C., Heymann, E., & Strier, K. (Eds.). *South American Primates: Testing New Theories in the Study of Primate Behavior, Ecology, and Conservation.* New York: Springer Science, 279–319.

Oftedal, O. (1991). The nutritional consequences of foraging in primates: the relationships of nutrient intakes to nutrient requirements. *Philosophical Transactions, Royal Society London, B*, **234**, 161–170.

Oliveira, J.M.S., Lima, M.G., Bonvicino, C., *et al.* (1985). Preliminary notes on the ecology and behavior of the guianan saki *Pithecia pithecia*, Linnaeus 1766: Cebidae, Primates). *Acta Amazonica*, **15**, 249–263.

Opler, P.A., Frankie, G.W. & Baker, H.G. (1980). Comparative phenological studies of treelet and shrub species in tropical wet and dry forests in the lowlands of Costa Rica. *Journal of Ecology*, **68**, 167–189.

Palacios, E. & Peres, C.A. (2005). Primate population densities in three nutrient-poor Amazonian terra firme forests of south-eastern Colombia. *Folia Primatologica*, **76**, 135–145.

Parolin, P. (2000). Phenology and CO_2-assimilation of trees in Central Amazonian floodplains. *Journal of Tropical Ecology*, **16**, 465–473.

Peetz, A. (2001). *Ecology and Social Organization of the Bearded Saki Chiropotes satanas chiropotes (Primates: Pitheciinae) in Venezuela.* Bonn: Society of Tropical Ecology.

Peres, C.A. (1993). Notes on the ecology of buffy saki monkeys (*Pithecia albicans*, Gray 1860): a canopy seed-predator. *American Journal of Primatology*, **31**, 129–140.

Pinto, L.P. (2008). Ecologia alimentar do cuxiú-de-nariz-vermelho *Chiropotes albinasus* (Primates: Pitheciidae) na Floresta Nacional do Tapajós, Pará. Unpublished Doctoral Thesis, Universidade Estadual de Campinas, Campinas.

Port-Carvalho, M. (2002). Dieta, comportamento e densidade populacional do cuxiú-preto *Chiropotes satanas satanas* (Primates: Pitheciinae) na paisagem fragmentada do oeste do Maranhão. Unpublished Master thesis, Universidade Federal do Pará, Belém.

Port-Carvalho, M. & Ferrari, S.F. (2004). Occurrence and diet of the black bearded saki (*Chiropotes satanas satanas*) in the fragmented landscape of western Maranhao, Brazil. *Neotropical Primates*, **12**, 17–21.

Rosenberger, A.L. (1992). Evolution of feeding niches in New World monkeys. *American Journal of Physical Anthropology*, **88**, 525–562.

Santos, R.R. (2002). Ecologia de cuxiús (*Chiropotes satanas*) na Amazônia Oriental: perspectivas para a conservação de populações fragmentadas. Unpublished Master thesis, Museu Paraense Emílio Goeldi e Universidade Federal do Pará, Belém.

Schöngart, J., Piedade, M.T.F., Ludwigshausen, S., *et al.* (2002). Phenology and stem-growth periodicity of tree species in Amazonian floodplain forests. *Journal of Tropical Ecology*, **18**, 581–597.

Schwarzkopf, L. & Rylands, A.B. (1989). Primate species richness in relation to habitat structure in Amazonian rainforest fragments. *Biological Conservation*, **48**, 1–12.

Setz, E.Z.F. (1993). Ecologia alimentar de um grupo de parauacus *Pithecia pithecia chrysocephala* em um fragmento florestal na Amazônia Central. Unpublished Doctoral thesis, Universidade Estadual de Campinas, Campinas.

Setz, E.Z.F. (1999). Sleeping habits of golden-faced sakis in a forest fragment in the Central Amazon. *American Journal of Primatology*, **49**, 100.

Shaffer, C.A. (2012). Ranging behavior, group cohesiveness, and patch use in northern bearded sakis (*Chiropotes sagulatus*) in Guyana. Unpublished doctoral dissertation, Washington University.

Silva, S.S.B. (2003). Comportamento alimentar do cuxiú-preto (*Chiropotes satanas*) na área de influência do reservatório da usina hidrelétrica de Tucuruí–Pará. Unpublished Masters thesis, Museu Paraense Emílio Goeldi e Universidade Federal do Pará, Belém.

Sombroek, W. (2001). Spatial and temporal patterns of Amazon rainfall. *Ambio*, **30**, 388–396.

Swanson Ward, N. & Chism, J. (2003). A report on a new geographic location of red uakaris (*Cacajao calvus ucayalii*) on the Quebrada Tahuaillo in northeastern Peru. *Neotropical Primates*, **11**, 19–22.

ter Steege, H. & Persaud, C.A. (1991). The phenology of Guyanese timber species: a compilation of a century of observations. *Plant Ecology*, **95**, 177–198.

Terborgh, J. (1983). *Five New World Primates, A Study in Comparative Ecology*. Princeton, NJ: Princeton University Press.

Terborgh, J. & van Schaik, C.P. (1987). Convergence vs. nonconvergence in primate communities. In *Organization of Communities Past and Present*, ed. J.H.R. Gee & P.S. Giller. Oxford: Blackwell Scientific Publications, pp. 205–226.

Thompson, C.L. (2011). Sex, aggression and affiliation: the social system of white-faced saki monkeys (*Pithecia pithecia*). PhD Dissertation, Kent State University, OH.

van Roosmalen, M.G.M., Mittermeier, R.A. & Fleagle, J.G. (1988). Diet of the northern bearded saki (*Chiropotes satanas chiropotes*): a Neotropical seed predator. *American Journal of Primatology*, **14**, 11–35.

van Roosmalen, M.G.M., Mittermeier, R.A. & Milton, K. (1981). The bearded sakis, genus *Chiropotes*. In *Ecology and Behavior of Neotropical Primates*, vol. **1**, ed. A.F. Coimbra-Filho & R.A. Mittermeier. Rio de Janeiro: Academia Brasileira de Ciências, pp. 419–441.

van Schaik, C.P., Terborgh, J.W. & Wright, S.J. (1993). The phenology of tropical forests: adaptive significance and consequences for primary consumers. *Annual Review of Ecology and Systematics*, **24**, 353–377.

Veiga, L.M. (2006). Ecologia e comportamento do cuxiú-preto (*Chiropotes satanas) na paisagem fragmentada da Amazônia Oriental.* Unpublished Doctoral thesis, Universidade Federal do Pará, Belém.

Veiga, L.M., Pinto, L.P. & Ferrari, S.F. (2006). *Fission–fusion sociality in bearded sakis (Chiropotes albinasus and Chiropotes satanas) in Brazilian Amazonia.* Proceedings of the XXIst Congress of the International Primatological Society, Uganda.

Vié, J.-C. (1998). Les effects d´une perturbation majeure de l'habitat sur deux espèces de primates en Guyane Française: translocations de singes hurleurs roux (*Alouatta seniculus*) et translocation et insularisation de sakis à face pale (*Pithecia pithecia*). Unpublished Doctoral thesis. Université de Montpellier II, Montpellier.

Vié, J.-C., Richard-Hansen, C. & Fournier-Chambrillon, C. (2001). Abundance, use of space, and activity patterns of white-faced sakis (*Pithecia pithecia*) in French Guiana. *American Journal of Primatology*, **55**, 203–221.

Vieira, T. (2005). Aspectos da ecologia do cuxiú de Uta Hick, *Chiropotes utahickae* (Hershkovitz, 1985), com ênfase na exploração alimentar de espécies arbóreas da ilha de Germoplasma, Tucuruí–PA. Unpublished Master thesis, Museu Paraense Emílio Goeldi and Universidade Federal do Pará, Belém.

Walker, S.E. (1996). The evolution of positional behavior in the saki–uakaris (*Pithecia, Chiropotes*, and *Cacajao*). In *Adaptive Radiations of Neotropical Primates*, ed. M.A. Norconk, A.L. Rosenberger & P.A. Garber. New York, NY: Plenum Press, pp. 335–367.

Wallace, R.B., Painter, R.L.E., Taber, A.B., *et al.* (1996). Notes on a distributional river boundary and southern range extension for two species of Amazonian primates. *Neotropical Primates*, **4**, 149–151.

Wojciechowski, M.F., Lavin, M. & Sanderson, M.J. (2004). A phylogeny of legumes (Leguminosae) based on analysis of the plastid *MATK* gene resolves many well-supported subclades within the family. *American Journal of Botany*, **91**, 1846–1862.

Wrangham, R.W. (1980). An ecological model of female-bonded groups. *Behaviour*, **75**, 262–300.

Chapter 8

Functional morphology and positional behavior in the Pitheciini

Lesa C. Davis & Suzanne E. Walker-Pacheco

Introduction

The pitheciins are the last major platyrrhine group to undergo long-term observations in their natural habitat. Recent studies, including many in this volume, have detailed similarities and distinctions in the ecology and positional behavior of pitheciins *Cacajao*, *Chiropotes* and *Pithecia*, and have enhanced our understanding of this monophyletic lineage and its rich evolutionary history. Pitheciin postcranial anatomical studies have primarily focused on *Pithecia pithecia* and *Chiropotes satanas*, with the most comprehensive treatment being that of Fleagle and Meldrum (1988) 24 years ago. While *Cacajao* skeletal anatomy has received little attention to date (but see Walker & Davis, 2007), numerous similarities to *Chiropotes* in body size, diet, canopy use, locomotion and posture have been documented. Field data on *Pithecia* species other than *P. pithecia* suggest numerous similarities with the *Chiropotes*/*Cacajao* taxa rather than with the distinctive *Pithecia pithecia*. These behavioral and ecological data along with anatomical studies to date suggest a lineage that includes a distinct leaper and vertical clinger, *P. pithecia*, on the one hand, and a more pronograde quadrupedal contingent of taxa on the other.

New morphological data collected on rare *Cacajao* specimens as well as on additional *Pithecia* and *Chiropotes* species provide an opportunity to update and broaden our understanding of pitheciin functional anatomy and further explore positional behavior adaptations in the lineage. Here we report on the comparative postcranial anatomy and positional behavior adaptations of seven pitheciin species: *Cacajao calvus*, *C. melanocephalus*, *Chiropotes satanas* (now referred to by some authors as *Ch. chiropotes*, a new designation; Silva Jr & Figueiredo 2002), *Ch. albinasus*, *Pithecia hirsuta*, *P. monachus* and *P. pithecia*. *Callicebus* is used as an out-group for comparative purposes.

The Pitheciini

The Pitheciini comprise the genera *Pithecia*, *Chiropotes* and *Cacajao*. For the taxonomy of the group, and a recent taxonomic revision, see chapters in Part I. It is generally agreed that *Chiropotes* and *Cacajao* are more closely related to each other than either is to *Pithecia*, the former two sharing a number of derived features, particularly in dental and mandibular morphology (Ford 1986; Rosenberger 1988; Kinzey 1992). The pitheciins range in body size from 1.7 to at least 3.2 kg: *C. melanocephalus*, 2.8 kg; *C. calvus*, 3.2 kg; *Ch. albinasus*, 2.7 kg; *Ch. satanas*, 2.8 kg; *P. hirsuta*, 2.4 kg; *P. monachus*, 2.3 kg; *P. pithecia*, 1.7 kg (Ford & Davis 1992; Smith & Jungers 1997).

All pitheciins are found in northern South America, with sympatry between *Pithecia* and each of the other genera relatively common, while the more similar genera *Chiropotes* and *Cacajao* are allopatric primarily. More specific geographical distribution information is found in Silva Jr *et al.* (Chapter 4).

Most researchers agree that *Callicebus* represents the sister group to the pitheciins. *Callicebus* also fits the general ancestral pattern of a basal pitheciine (cf. Ford 1988; Kay 1990). Here, *Callicebus* will be used for comparative purposes to help frame possible anatomical adaptations of the pitheciins.

Previous studies of pitheciin positional behavior

Few long-term studies have focused on pitheciin positional behavior, although data have been collected on pitheciins as part of larger multi-species studies or on a short-term basis. Early studies included Fleagle and Mittermeier (1980), whose short-term observations of *Pithecia pithecia* and *Chiropotes satanas* set the stage for more in-depth study, and whose early results were supported by later work. Buchanan *et al.* (1981), Happel (1982) and Peres (1993) also contributed positional behavior and/or habitat use data on *Pithecia*. Early data for *Chiropotes* were also presented by van Roosmalen *et al.* (1981), and for captive *Cacajao*, Fontaine (1981). Ayres (1986) collected some locomotor data as part of his extensive studies of *Cacajao* ecology. A short-term study of *Cacajao* positional behavior was reported by Walker and Ayres (1996) and a long-term study of both *Pithecia* and *Chiropotes* was conducted by Walker (1996). Long-awaited and important comparative data on the positional behavior of one of the

Evolutionary Biology and Conservation of Titis, Sakis and Uacaris, eds. Liza M. Veiga, Adrian A. Barnett, Stephen F. Ferrari and Marilyn A. Norconk. Published by Cambridge University Press.

Table 8.1 Overall positional behavior of *Pithecia pithecia*, *Chiropotes satanas*, and *Cacajao calvus*.

Positional behavior	*P. pithecia*		*Ch. satanas*		*C. calvus*	
	Freq.	%	Freq.	%	Freq.	%
Pronograde clamber	823	9.4	672	11.0	110	16.3
Quadrupedalism	628	7.2	845	13.8	121	18.0
Leap/hop	1253	14.3	498	8.1	60	8.9
Drop	32	0.4	138	2.3	29	4.3
Climb	62	0.7	38	0.6	6	0.9
Bridge	0	0	36	0.6	2	0.3
Sit	3798	43.4	2334	38.1	213	31.6
Stand	442	5.0	493	8.1	39	5.8
Vertical cling	737	8.4	0	0	5	0.7
Perch	299	3.4	231	3.8	12	1.8
Bipedal stand	55	0.6	98	1.6	4	0.6
Hindlimb suspension	0	0	33	0.5	16	2.4
Lie	289	3.3	543	8.9	45	6.7
Other	235	2.7	171	2.8	11	1.6
Total observations	8653	98.8	6130	100.2	673	99.9

members of the *P. monachus*-group (all *Pithecia* species other than *P. pithecia*) are now available (Youlatos 1999).

Positional behavior and habitat use of the study species are briefly reviewed before we present our data on the relationship between morphology and behavior below. Direct, cross-study comparisons of the quantitative data were made difficult by the various methodologies, categories used, and contexts for data collection (e.g. documentation of positional behavior during travel, rest, and feed/foraging, vs. positional behavior throughout all contexts, etc.). In addition, for some pitheciin field studies, data were collected on both locomotion and posture; in others, only on locomotion. Some categories from the literature were pooled. A more complete positional behavior data set for one member of each genus, *P. pithecia*, *Ch. satanas* and *C. calvus*, adapted from Walker (1996), is presented in Table 8.1.

Callicebus

Callicebus moloch, observed in southeastern Ecuador, is an understory primate, using quadrupedal walk (45.2%) on horizontal supports as their primary locomotor mode (Youlatos 1999). It also uses considerable clambering (17.5%) and leaping (17.7%); significantly, 12% of leaps were from a vertical support.

Kinzey (1977) conducted an early locomotor study of C. torquatus and his results compare well with those of Lawler *et al.* (2006). Lawler *et al.* report on both *Callicebus brunneus* and *C. torquatus*. These two taxa were observed allopatrically in Peru, and found to be divergent in their habitat use and locomotor behavior. *Callicebus brunneus* uses the understory more than does *C. torquatus*, which more frequently uses the main canopy. In terms of locomotor behavior (in all contexts), *Callicebus brunneus* exhibited more leaping (38.3%) and running (16.8%), while *Callicebus torquatus* engaged in more quadrupedal walking (62.4%) (Lawler *et al.* 2006). Interestingly, *C. brunneus* used vertical cling as a frequent posture during foraging (11.1% compared to *C. torquatus* at 3.6%) (Lawler *et al.* 2006).

Pithecia

Pithecia pithecia was observed in Guri Lake, Venezuela (Walker 1996). Its main form of locomotion is leaping (39.7% during travel); all quadrupedal behaviors taken together are used somewhat less (quadrupedal walk, quadrupedal run, and pronograde clamber make up 35.3% of travel samples). It is often found in the lower part of the main canopy or in the understory, most often using oblique supports on or near the trunk. *Pithecia pithecia*'s body orientation during much of its positional repertoire is orthograde, particularly during its vertical clinging (which occurs in 10.5% of feeding samples and 11.3% of rest samples) and leaping. Almost one-half of the leap take-offs were from a vertical position. See Photo 8.1.

Youlatos' (1999) data on the positional behavior of *P. monachus* support the preliminary information of Peres (1993) on *P. albicans* and Happel (1982) on *P. hirsuta*, indicating that these *P. monachus*-group members are main-canopy primates, using mostly quadrupedal walk (38.4%), but with significant leaping (28.4%) (Youlatos 1999). Approximately 5% of leaps are from a vertical take-off position. Horizontal supports are most commonly used, followed by frequent use of oblique supports as well.

Chiropotes

Chiropotes satanas (see Photo 8.2) were observed in Guri Lake, Venezuela (Walker 1996). It uses a variety of quadrupedal behaviors for travel, with quadrupedal walk the most frequent (22.9%), followed by pronograde clamber (16.7%) and quadrupedal run (12.3%). The frequency of leaping is actually slightly greater than quadrupedal walk alone (24.5%), although conducted in a different manner than *P. pithecia*'s leaping. It is most often found in the upper canopy, upper mid-canopy, or emergent trees. Frequently used supports are most often the deformable terminal branches, but when the inclination of branches was distinguishable, they were most often horizontal. A notable behavior for *Chiropotes* is pedal suspension, observed during feeding (3.5%) and (even more frequently) in play (17.1%). There are no positional behavior data for *Chiropotes albinasus*.

Photo 8.1 Female white-faced saki (*P. pithecia*) crossing a gap between trees at Brownsberg Nature Park, Suriname. Photo: Christopher Little.

Photo 8.2 Black-bearded saki (*Cho Satanas*) running. Photo: Liza Veiga.

Cacajao

Cacajao calvus calvus was observed in the seasonally flooded "varzea" forest at Lake Teiú, Brazil (Walker & Ayres 1996). Like *Chiropotes*, a variety of quadrupedal behaviors makes up the bulk of *Cacajao*'s positional repertoire (54.4% for quadrupedal walk and run, and pronograde clamber taken together), but leaping (21.3%) is almost as frequent as quadrupedal walk alone (22.2%). This is true for *C. calvus*, for *C. melanocephalus* (Boubli, personal communication), and is similar to *Chiropotes satanas* reviewed above. *Cacajao* is an upper-canopy primate, also often found in emergent trees. Angled supports are most often used, followed by the indistinguishable inclination of deformable terminal branch ends, with substantial use of horizontal supports as well. Also like *Chiropotes*, *Cacajao* uses pedal suspension (Walker & Ayres 1996; Fontaine 1981; Boubli, personal communication).

Summary of behavioral data

As has been noted previously, some primary features of pitheciin locomotion are the overall similarity of habitat use and positional behavior for *Chiropotes* and *Cacajao* and the distinctiveness of *Pithecia pithecia*, particularly in terms of its behavioral specializations for clinging and leaping (Fleagle & Mittermeier 1980; Walker 1996, 2005; Davis *et al.* 2006). It is also differentiated from other *Pithecia* species in this way; interestingly, these more orthograde positional behaviors link it behaviorally to *Callicebus*, particularly *C. moloch* and *C. brunneus*.

We can make several predictions of the anatomy based on these behavioral profiles. While some skeletal features will likely follow taxonomic distinctions between the three pitheciin genera, we expect to see leaping and vertical clinging adaptations distinguish *P. pithecia* from the more quadrupedal taxa. We expect to find skeletal features consistent with pronograde quadrupedalism in the remaining taxa. We may also find anatomical correlates of pedal suspension in the lower leg and foot of the largest taxa, *Ch. satanas* and *C. calvus*.

Methods

Quantitative data were collected on seven pitheciine species and a *Callicebus* out-group (Table 8.2). A total of 58 measurements including four indices were examined in this study (Table 8.3). Joint surface morphology and other anatomical features were examined on the scapula, humerus, radius, ulna, carpus, os coxa, femur, tibia, calcaneus, astragalus and distal foot. Only adult, non-pathological specimens were included.

Primate postcranial remains are generally sparse in museum collections and some taxa, including *Cacajao*, several *Callicebus* species, and some *Pithecia* species, are extremely rare. We made efforts to collect data on as many of the pitheciin taxa as possible, and the full complement of 58 skeletal traits were examined in *P. pithecia*, *P. monachus*, *Ch. satanas*, and the *Callicebus* out-group. It was possible to examine 23 of the 58 traits for the entire pitheciine sample (*P. pithecia*, *P. monachus*, *Ch. satanas*, *Ch. albinasus*, *Cacajao calvus*, *C. melanocephalus* and *Callicebus*). These traits are indicated in Table 8.3.

Data were analyzed using multivariate and univariate techniques. Principal Components Analysis (PCA) was performed on 23 traits that were available for all taxa to look for patterns that might discriminate between taxa. *Cacajao calvus* was dropped from this analysis due to missing values for some of the 23 traits. Variables with large eigenvector values were noted and further explored in the univariate analysis.

For the univariate tests, data were size-corrected using ratios to a single measure, humeral head height (HHH), shown to have a strong (high correlation coefficient) and nearly isometric relationship to body weight (see Davis 2002). This method was chosen over ratio to a geometric mean because specimens varied significantly in which elements were present (particularly for hand and foot elements). Indices were computed on raw data. Univariate analyses were performed on all size-corrected variables and indices, using analysis of variance (ANOVA) and the post-hoc Tukey–Kramer Honestly

Table 8.2 Skeletal sample.

Taxon	*N*	Institutional origin[1]	Specimen origin		Sex			Measured by
			Captive	Wild	Male	Female	Unknown	
Pithecia	**17**							
pithecia	9	AMNH; FMNH; USNM	2	7	5	3	1	LCD
monachus	5	FMNH	0	4	3	1	0	LCD
	1	AMNH						SEW-P
hirsuta	2	MZUSP	0	2	2	0	0	SEW-P
Chiropotes	**12**							
albinasus	4	MZUSP	0	4	3	1	0	SEW-P
satanas	4	FMNH	0	4	3	1	0	LCD
	4	MZUSP; AMNH; MPEG	0	4	2	0	2	SEW-P
Cacajao	**10**							
calvus	6	MNRJ; MZUSP	0	6	2	3	1	SEW-P
melanocephalus	3	MNRJ; MZUSP	0	3	0	3	0	SEW-P
Callicebus	**13**							
cupreus	7	FMNH; USNM	3	4	5	2	0	LCD
donacophilus	2	FMNH	2	0	1	1	0	LCD
moloch	1	FMNH	1	0	0	0	1	LCD

[1] AMNH, American Museum of Natural History (New York); FMNH, Field Museum of Natural History (Chicago); USNM, United States National Museum of Natural History – Smithsonian Institution (Washington, DC); MZUSP, Museo de Zoología – São Paulo (São Paulo, Brazil); MNRJ, Museo Nacional (Rio de Janeiro, Brazil); MPEG, Museo Paraense Emilio Goeldi (Belém, Brazil).

Table 8.3 Quantitative traits.[1]

Indices
- **IMI**[2] – intermembral index (H3PL+RL)/(FL+TL)
- **BI**[2] – brachial index (RL/H3PL)
- **CI**[2] – crural index (TL/FL)
- **ARML** – arm length ((H3PL + RL)/HHH)

Measurements

Scapula
- **SGFH**[2] – glenoid fossa height
- **SGFW**[2] – glenoid fossa width
- **SISH**[2] – height of infraspinous fossa
- **SSSH**[2] – height of supraspinous fossa
- **SVH** – ventral height
- **SVLB**[2] – axial border length

Humerus
- **H3PL**[2] – maximum humerus length
- **HHH** – humeral head height (body size character)
- **HDTL** – deltoid tuberosity length
- **HMSD** – midshaft transverse diameter
- **HBEW**[2] – biepicondylar width
- **HAT1**[2] – superior anterior trochlear width
- **HACW**[2] – anterior capitulum width
- **HACH**[2] – anterior capitulum height
- **HPTW** – posterior trochlear width
- **HMEW** – medial epicondyle width

Radius
- **RL**[2] – maximum radius length
- **RNTD** – radial neck transverse diameter
- **RCFW** – mediolateral radiocarpal facet width
- **RDST** –distal shaft transverse diameter
- **RDSS** – distal shaft sagittal diameter

Ulna
- **UL**[2] – ulnar length
- **UOPL**[2] – olecranon process length
- **UMST** – midshaft transverse diameter
- **UMSS** – midshaft sagittal diameter
- **UCD** – anteroposterior projection of coracoid process

Carpus
- **MC3L** – third metacarpal length

Os Coxa
- **IDPL** – dorsal pelvis length
- **ISHL**[2] – ischium length

Femur
- **FL**[2] – femoral medial length
- **FH/NL** – head and neck length
- **FNW**[2] – femoral neck width
- **FBCW**[2] – bicondylar width
- **FBCD**[2] – bicondylar depth
- **FSPG** – superior patellar groove width
- **FPGL** – patellar groove length
- **FMCH** – medial condyle height
- **FMCW** – medial condyle width
- **FLCH** – lateral condyle height
- **FLCW** – lateral condyle width

Tibia
- **TL**[2] – tibial length
- **TPW** – tibial plateau width
- **TMST** – midshaft transverse diameter
- **TMSS** – midshaft sagittal diameter
- **TDFH** – distal fibular facet height

Astragalus
- **AML** – astragalar maximum length
- **AMBH** – astragalar medial body height
- **ATRL** – trochlear length
- **ATW2** – anterior trochlear width at midpoint

Calcaneus
- **CML**[2] – calcaneal medial length
- **CAL** – anterior calcaneal length
- **CTH** – calcaneal tuberosity height
- **CTL** – calcaneal tuberosity length

Distal tarsus
- **XT3L** – 3rd metatarsal length

[1] Complete measurement definitions may be found in Davis (2002).
[2] Variables collected on the entire pitheciin and *Callicebus* sample.

Significantly Difference (HSD) test, which is designed to accommodate unequal sample sizes, to test for generic differences (Sokal & Rohlf 1981; Tukey, unpubl. ms., cited in SAS Institute, Inc. 2001). The Tukey–Kramer HSD test is more conservative than Student's *t*-test and decreases the likelihood of Type 1 experiment-wise errors. Finally, in order to control for the number of comparisons, Bonferroni correction of the significance value of 0.05 was made (Leigh & Jungers 1994), and the level of significance was set at 0.006 (0.05/8). All analyses were done using JMP v.4.0.4 (SAS Institute, Inc. 2001).

Results and discussion: comparative pitheciin functional anatomy

The above review of pitheciine locomotion highlights the differences between *Pithecia pithecia*, a trunk leaper and vertical clinger (see Photo 8.3), and the far more quadrupedal pitheciins, who use leaping to varying degrees but comparatively little, if any, trunk leaping. In addition to *Chiropotes* and *Cacajao* species, this latter group includes *P. pithecia*'s congeners *P. monachus* and apparently *P. hirsuta* as well. Below we

Photo 8.3 Adult white-faced saki male clinging vertically. Brownsberg Nature Park, Suriname. Photo: Nick Robl.

discuss the nature of these positional behavior differences as reflected in bone.

Multivariate analysis

Results from the PCA reveal that PC1 accounted for the largest amount of variation (77%), and is assumed to represent size. Principal component 2 (5% of the variation) is plotted against PC1 in Figure 8.1. *Chiropotes satanas* and *Ch. albinasus* generally cluster together; their body size differences are evident in the cluster. *Pithecia monachus* and the only *P. hirsuta* specimen in the multivariate analysis cluster below the *Chiropotes* species. The two *Cacajao melanocephalus* individuals in the plot are widely separated on both axes, possibly reflecting their small sample size. *Pithecia pithecia* appears as a tight grouping, distinct from the other taxa. Anatomical variables with the highest eigenvector values for PC2 include (eigenvector value in parentheses): CI (0.49554); SSSH (0.46911); IMI (0.25576); SGFW (0.24843); HAT1

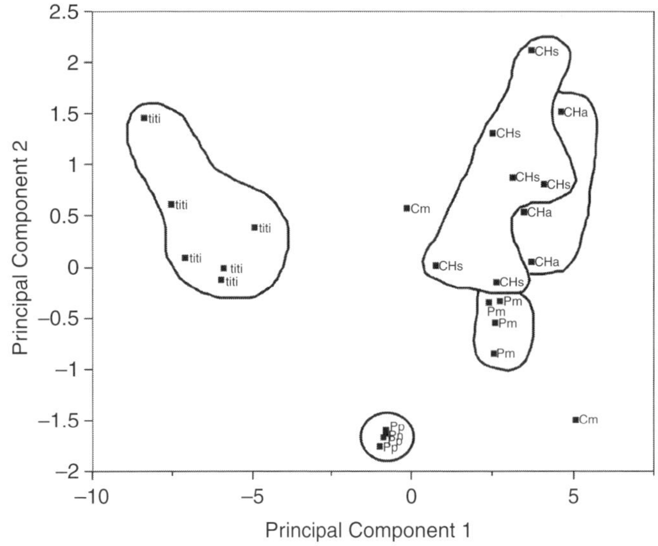

Figure 8.1 Principal components analysis of anatomical data, plot of PC2 to PC1. Species abbreviations: Cm, *Cacajao melanocephalus*; CHa, *Chiropotes albinasus*; CHs, *Chiropotes satanas*; Ph, *Pithecia hirsuta*; Pm, *P. monachus*; Pp, *P. pithecia*; titi, *Callicebus cupreus*, *C. donacophilus*, and *C. moloch*.

(0.20311). These and additional variables are explored further below.

Univariate analysis

The size-adjusted anatomical data are presented in Table 8.4. Of the 58 traits, 27 (47%) were found not to be statistically significantly different among the study taxa. We believe these traits attest to the monophyly of Pitheciini and its close alliance with *Callicebus*. The remaining traits showed significant differences between species, variously including patterns within and across taxonomic boundaries. These are discussed more fully below.

Long bone lengths and indices

Lengths of the humerus (H3PL), radius (RL) and ulna (UL) are significantly greater in *Pithecia pithecia* compared to nearly all other taxa, and *Cacajao calvus* is least different from *P. pithecia* in these measurements (Table 8.4). Interestingly, these two taxa not only have distinctly different behavioral repertoires, but they also represent the size extremes of the pitheciins: *P. pithecia* as the lightest species and *C. calvus* as the heaviest. Figure 8.2. shows relative arm length among the species. In their discussion of *P. pithecia*, Fleagle and Meldrum (1988) suggest that a relatively elongated forearm with a shortened humerus is favorable for clinging, but our findings show that each of the three forelimb long bones is elongated in this taxon.

Long forelimbs in other platyrrhine species have been associated with extractive foraging (Garber 1992; Hershkovitz 1977; Peres 1986, Stafford *et al.* 1996), landing forelimb-first after a leap (Garber 1991; Garber & Leigh 2001) and, of course, forelimb suspensory positional activities (Dykyj 1982; Fleagle

Table 8.4 Species means for size-adjusted data.[1]

Trait	*Chiropotes*						*Cacajao*						*Pithecia*									*Callicebus*			
	albinasus			*satanas*			*calvus*			*melanocephalus*			*hirsuta*			*monachus*			*pithecia*			*cupreus, moloch*			
	n	mean	SD	*n*	mean	SD	*n*	mean	SD	*n*	mean	SD	*n*	mean	SD	*n*	mean	SD	*n*	mean	SD	*n*	mean	SD	*p*
IMI	4	80.78	0.71	7	81.24	1.61	2	84.43	1.25	2	81.93	2.00	1	76.18	.	5	77.18	1.31	4	75.75	1.27	9	74.93	1.80	<.00001
BI	4	89.38	1.58	7	87.40	3.34	2	87.53	2.21	2	87.08	1.22	1	88.27	.	5	87.17	3.51	6	91.62	2.09	10	84.00	1.95	.0004
CI	4	92.31	1.57	7	90.86	1.66	3	89.27	2.60	2	92.22	3.97	2	91.13	2.90	5	90.53	2.96	6	94.98	3.58	9	93.13	4.22	NS
ARML	4	15.22	0.44	7	15.45	0.53	2	16.35	0.23	2	15.56	0.17	1	14.45	2.13	5	16.02	0.45	6	16.95	0.69	10	15.41	0.90	0.0043
SGFH	4	0.77	0.03	7	0.87	0.06	2	0.81	0.06	2	0.77	0.03	2	0.83	0.08	5	0.86	0.04	8	0.89	0.07	10	0.87	0.05	NS
SGFW	4	0.51	0.04	7	0.55	0.04	2	0.48	0.02	2	0.42	0.05	2	0.50	0.08	5	0.54	0.03	8	0.58	0.04	10	0.55	0.02	.0001
SISH	4	2.63	0.16	7	2.40	0.19	2	2.11	0.00	2	2.16	0.18	2	2.40	0.02	5	2.32	0.20	8	2.50	0.26	10	2.03	0.16	<.00001
SSSH	4	0.95	0.09	7	0.93	0.12	2	0.66	0.03	2	0.63	0.02	2	0.65	0.02	5	0.76	0.09	8	0.73	0.07	10	0.84	0.10	<.00001
SVH	0	.	.	3	3.28	0.18	0	.	.	0	.	.	1	3.48	.	4	3.11	0.28	7	3.57	0.33	10	3.17	0.35	NS
SVLB	4	4.04	0.12	7	4.08	0.14	2	3.39	0.04	2	3.49	0.31	2	4.06	0.22	5	4.20	0.15	7	4.42	0.32	10	3.89	0.42	.0009
H3PL	4	8.04	0.27	7	8.25	0.32	2	8.72	0.23	2	8.32	0.04	1	8.36	.	5	8.56	0.14	7	8.91	0.38	10	8.37	0.46	.0054
HMSD	0	.	.	5	0.52	0.04	0	.	.	0	.	.	0	.	.	5	0.54	0.03	7	0.55	0.06	6	0.56	0.04	NS
HDTL	0	.	.	5	3.07	0.17	0	.	.	0	.	.	0	.	.	4	3.01	0.30	8	2.42	0.36	6	3.10	0.30	.0012
HPTW	0	.	.	5	0.63	0.02	0	.	.	0	.	.	0	.	.	5	0.57	0.03	8	0.55	0.02	6	0.59	0.02	<.00001
HMEW	0		.	7	0.47	0.08	0	.	.	0	.	.	0	.	.	5	0.53	0.08	8	0.51	0.02	10	0.49	0.04	NS
HBEW	4	1.49	0.09	7	1.55	0.10	2	1.47	0.08	2	1.37		1	1.40	.	5	1.54	0.10	8	1.51	0.07	10	1.50	0.06	NS
HAT1	3	0.46	0.01	7	0.43	0.03	1	0.45	.	2	0.40	0.05	1	0.39	.	5	0.42	0.04	8	0.41	0.04	10	0.46	0.08	NS
HACW	3	0.56	0.05	7	0.50	0.07	1	0.67	.	2	0.55	0.01	1	0.56	.	5	0.47	0.03	8	0.64	0.05	10	0.53	0.05	.0001
HACH	3	0.53	0.08	7	0.53	0.03	1	0.60	.	2	0.54	0.04	1	0.52	.	5	0.53	0.02	8	0.49	0.02	10	0.52	0.02	NS
RL	4	7.18	0.19	7	7.20	0.29	2	7.63	0.01	2	7.25	0.13	1	7.59	.	5	7.46	0.36	7	8.19	0.39	10	7.04	0.45	<.00001
RNTD	0	.	.	3	0.39	0.03	0	.	.	0	.	.	0	.	.	4	0.39	0.04	8	0.40	0.03	10	0.38	0.03	NS
RDST	0	.	.	3	0.54	0.04	0	.	.	0	.	.	0	.	.	4	0.59	0.05	8	0.53	0.04	10	0.53	0.08	NS
RDSS	0	.	.	3	0.40	0.02	0	.	.	0	.	.	0	.	.	4	0.38	0.03	8	0.38	0.04	10	0.38	0.05	NS
RCFW	0	.	.	4	0.70	0.03	0	.	.	0	.	.	0	.	.	5	0.70	0.04	6	0.70	0.03	5	0.67	0.04	NS
UL	4	8.01	0.25	7	8.17	0.28	2	8.44	0.06	2	8.05	0.16	1	7.54	.	5	8.47	0.26	8	9.27	0.41	10	7.93	0.47	<.00001
UOPL	4	0.62	0.04	7	0.58	0.03	2	0.53	0.10	2	0.51	0.04	1	0.54	.	5	0.62	0.03	8	0.60	0.04	9	0.60	0.06	NS
UMST	0	.	.	5	0.30	0.05	0	.	.	0	.	.	0	.	.	5	0.30	0.03	8	0.27	0.04	6	0.25	0.03	NS
UMSS	0	.	.	5	0.54	0.03	0	.	.	0	.	.	0	.	.	5	0.54	0.05	8	0.47	0.04	6	0.45	0.04	0.0007

UCD	0	.	.	5	0.26	0.03	0	.	.	0	.	.	0	.	.	4	0.23	0.01	8	0.37	0.05	6	0.21	0.03	<.00001
MC4L	0	.	.	3	1.71	0.22	0	.	.	0	.	.	0	.	.	4	1.72	0.03	7	2.08	0.28	5	1.71	0.07	NS
ISHL	4	1.83	0.26	7	1.92	0.12	2	1.90	0.03	2	1.69	0.26	2	1.51	0.18	5	2.00	0.07	6	2.18	0.15	10	2.29	0.18	<.00001
IDPL	0	.	.	5	6.34	0.21	0	.	.	0	.	.	1	6.77	.	5	6.19	0.12	6	6.62	0.44	6	6.43	0.35	NS
FL	4	9.80	0.27	7	9.97	0.40	2	10.18	0.15	2	9.89	0.07	2	9.91	1.17	5	10.89	0.21	6	11.81	0.45	9	10.70	0.79	<.00001
FNW	4	0.51	0.01	7	0.54	0.06	2	0.52	0.07	2	0.54	0.06	1	0.57	.	5	0.61	0.04	8	0.66	0.03	10	0.62	0.05	<.00001
FHNL	0	.	.	5	0.90	0.07	0	.	.	0	.	.	0	.	.	5	0.95	0.07	8	0.96	0.06	6	0.95	0.06	NS
FPGL	0	.	.	3	0.83	0.02	0	.	.	0	.	.	0	.	.	4	0.83	0.03	7	0.89	0.06	10	0.93	0.10	NS
FSPG	0	.	.	3	0.69	0.10	0	.	.	0	.	.	0	.	.	4	0.75	0.06	7	0.80	0.05	10	0.64	0.04	<.00001
FBCW	4	1.28	0.04	7	1.38	0.07	2	1.37	0.09	2	1.32	0.06	2	1.36	0.04	5	1.51	0.05	7	1.58	0.07	10	1.44	0.07	<.00001
FBCD	4	0.96	0.04	7	1.01	0.05	2	0.96	0.06	2	0.89	0.01	1	1.02		5	1.03	0.03	7	1.13	0.05	10	1.15	0.03	<.00001
FMCH	0	.	.	7	0.71	0.03	2	.	.	2	.	.	0	.	.	5	0.71	0.02	7	0.77	0.05	10	0.74	0.05	NS
FMCW	0	.	.	5	0.47	0.02	0	.	.	0	.	.	0	.	.	5	0.52	0.04	7	0.54	0.03	6	0.50	0.03	.0043
FLCH	0	.	.	5	0.70	0.04	0	.	.	0	.	.	0	.	.	5	0.73	0.04	7	0.78	0.03	6	0.80	0.03	.0010
FLCW	0	.	.	5	0.44	0.03	0	.	.	0	.	.	0	.	.	5	0.46	0.02	7	0.50	0.03	6	0.48	0.04	.0054
TL	4	9.05	0.36	7	9.05	0.29	2	9.19	0.41	2	9.12	0.33	1	9.57	.	5	9.86	0.32	8	11.17	0.60	10	9.93	0.55	<.00001
TPW	0	.	.	3	1.35	0.06	0	.	.	0	.	.	0	.	.	4	1.40	0.02	8	1.50	0.06	10	1.47	0.06	.0020
TMST	0	.	.	3	0.45	0.02	0	.	.	0	.	.	0	.	.	4	0.50	0.11	8	0.47	0.04	10	0.43	0.03	NS
TMSS	0	.	.	3	0.61	0.03	0	.	.	0	.	.	0	.	.	4	0.60	0.06	8	0.64	0.04	10	0.62	0.06	NS
TDFH	0	.	.	3	0.16	0.01	0	.	.	0	.	.	0	.	.	4	0.17	0.03	8	0.27	0.13	9	0.14	0.03	NS
AML	0	.	.	3	1.28	0.08	0	.	.	0	.	.	0	.	.	4	1.27	0.02	7	1.36	0.06	8	1.38	0.07	NS
AMBH	0	.	.	3	0.43	0.01	0	.	.	0	.	.	0	.	.	4	0.48	0.05	7	0.48	0.06	8	0.52	0.02	NS
ATRL	0	.	.	3	0.69	0.01	0	.	.	0	.	.	0	.	.	4	0.70	0.03	8	0.77	0.05	8	0.74	0.04	NS
ATW2	0	.	.	3	0.52	0.04	0	.	.	0	.	.	0	.	.	4	0.54	0.02	8	0.54	0.05	8	0.55	0.04	NS
CML	4	1.88	0.08	6	1.84	0.08	1	1.77	.	2	1.90	0.04	1	1.89	.	5	1.85	0.05	6	1.99	0.04	8	2.00	0.10	.0025
CAL	0	.	.	3	0.72	0.07	0	.	.	0	.	.	0	.	.	4	0.70	0.08	8	0.71	0.05	8	0.80	0.04	.0010
CTH	0	.	.	3	0.73	0.03	0	.	.	0	.	.	0	.	.	4	0.77	0.05	7	0.77	0.08	8	0.86	0.06	NS
CTL	0	.	.	3	0.46	0.02	0	.	.	0	.	.	0	.	.	4	0.49	0.04	7	0.53	0.06	8	0.49	0.05	.0008
XT4L	0	.	.	2	2.84	0.08	0	.	.	0	.	.	0	.	.	4	3.12	0.06	7	3.41	0.19	5	3.14	0.11	.0008

[1] Complete trait names are provided in Table 8.3. Size-adjusted data are not provided for the HHH variable as it is the body size variable. NS = correlation not significant at the $>p = .006$ level.

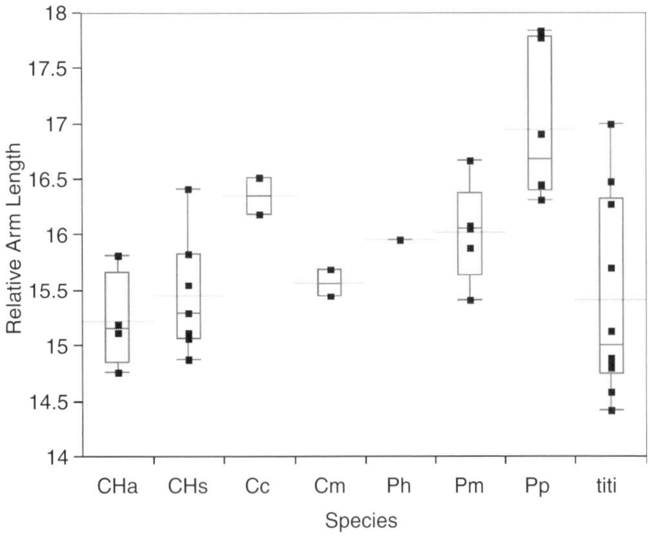

Figure. 8.2 Box plot of relative arm length adjusted for size (ARML). Species abbreviations as in Figure 8.1, with the addition of Cc, *Cacajao calvus.*

1999; Preuschoft & Demes 1985; Rosenberger & Strier 1989). In addition, Garber (1991; Garber & Leigh 2001) finds that forelimb elongation in some callitrichid species aids in repositioning the body mid-air during a leap between vertical trunks, a maneuver commonly used by *P. pithecia* as well (Walker 2005). This adaptation may be even more significant for sakis/uacaris as additional force is needed to rotate the body during leaps between vertical supports given their larger body size. It is possible that long forelimb elements of the heavier *Cacajao calvus* function to increase its reach during pedal suspension in its foraging activities, although the other pitheciin pedal suspensor, *Chiropotes satanas*, does not share this adaptation. The brachial index is relatively uninformative for this group although *Callicebus* has a significantly lower brachial index than *P. pithecia* and *Ch. albinasus*. See Photo 8.4.

The femur of *P. pithecia* is longer (FL) than all other species except *P. monachus*, and the *P. pithecia* tibia (TL) is longer than all others. The long hindlimb of *P. pithecia* (and less so of *P. monachus*) are consistent with their shared leaping proclivities. While the two *Pithecia* species utilize different types of leaping (leaping between vertical supports in *P. pithecia* and between non-vertical supports in *P. monachus*), long hindlimbs would be advantageous for both. There were no significant differences in the crural index for the sample.

The intermembral index (IMI) is significantly lower in *Callicebus* and the *Pithecia* species, and is significantly higher in the *Chiropotes* species and especially *Cacajao* species. Interestingly, the IMI of *P. pithecia* is low despite its elongated forelimb. The index appears to sort by genus and by body size (lower indices for the smaller taxa, etc.), rather than by degree of leaping and quadrupedalism, although *P. pithecia* does have the second lowest IMI value.

Photo 8.4 Black-bearded saki pedal suspension. Photo: Liza Veiga.

Forelimb features

In the scapula, *Cacajao* species and especially *C. melanocephalus* have a significantly narrowed glenoid fossa (SGFW) (Table 8.4). Glenoid fossa height (SGFH) is also shortest in the *Cacajao* genus, although not significantly so. Fleagle and Meldrum (1988) found the glenoid fossa of *P. pithecia* to be less elongate than that of *Chiropotes satanas*. Results here show the opposite to be the case, although differences are not significant. While *Cacajao* emerges as having the shortest, most narrow glenoid fossa, it is difficult to see how it might be advantageous for a relatively large, primarily quadrupedal platyrrhine. This feature would seem to allow a greater range of movement in the glenohumeral joint, and it may be that it is favorable for negotiating deformable terminal branches. It would be helpful to know if this milieu is regularly used by *C. melanocephalus* and if *Cacajao* species in general use this habitat more than other pitheciins, including *Ch. satanas*.

Chiropotes albinasus and *Ch. satanas* are characterized by a significantly deep (or tall) scapula in its craniocaudal aspect that is formed at least in part by tall infraspinous (SISH) and supraspinous fossae (SSSH). Fleagle and Meldrum (1988) noted a mediolaterally expanded scapula for *Ch. satanas* relative to *P. pithecia* that they related to the more quadrupedal activities of the former. Axial border length (SVLB), measured from the glenoid fossa to the inferior angle, also serves as an indicator of scapular length, and is found here to

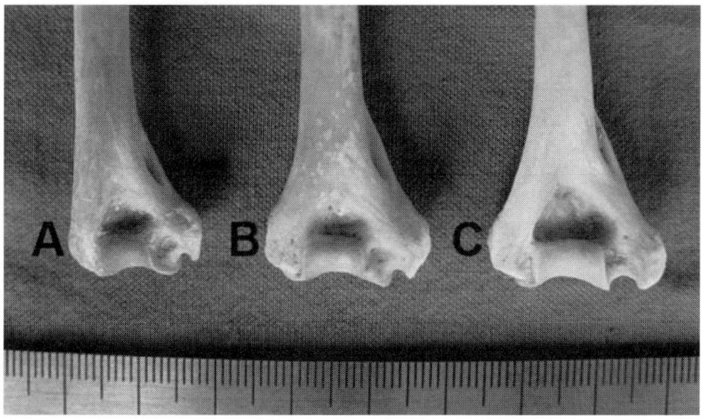

Figure 8.3 Photograph showing humeral trochlear width in posterior view (HPTW); A, *Pithecia pithecia*; B, *P. monachus*; C, *Chiropotes satanas*.

be significantly longer in the *Pithecia* species, compared to the short condition found in *Cacajao* species. *Chiropotes* species fall in between the two groups. While this measure is also influenced by scapular height, it does provide an indication of scapular size when used with other measures. Direct form/function relationships are difficult to detail in the scapula because both dorsal and ventral surfaces support a large number of muscles. Results here suggest that locomotor behavior correlations in the pitheciin scapula are less clear than previously thought and certainly warrant further study.

In the elbow, *Pithecia pithecia* has a wide capitulum (HACW), narrow posterior humeral trochlea (HPTW; Figure 8.3), and a relatively deep ulnar coracoid process (UCD). The two *Cacajao* species also have a wide capitulum, but the nature of the other two elbow traits is not known in this genus. A narrowed posterior humeral trochlea also characterizes strepsirhine vertical clingers and leapers (Rose 1993) and the New World vertically clinging callitrichids, *Callimico* and *Cebuella* (Davis 2002). The wider capitulum and deep coracoid of *P. pithecia* may serve to enhance radiohumeral stability and humeroulnar stability, respectively, especially in a flexed arm such as that seen in vertical clinging. These three features together may constitute a suite of vertical clinging adaptations in *P. pithecia*, although the degree to which they are present in the *Cacajao* elbow needs to be investigated.

Measures of shaft robusticity for the forelimb long bones were examined in the reduced sample (*P. pithecia*, *P. monachus*, *Ch. satanas* and *Callicebus*), and no significant differences were noted in the humeral midshaft, distal radius shaft, or the transverse aspect of the ulnar midshaft. However, *Ch. satanas* and *P. monachus* each have a significantly deepened ulnar midshaft in its sagittal aspect compared to that of *P. pithecia*. This is in accordance with Fleagle and Meldrum's (1988) findings for *P. pithecia* and *Ch. satanas*. Their elucidation of the functional import of a buttressed ulnar midshaft (see also Preuschoft 1974) as serving as an adaptation for withstanding the bending forces incurred during quadrupedal walking and running, is supported here.

Hindlimb features

The functional significance of length of the ischium in the os coxa has been explored in several studies (cf. Anemone 1993; Fleagle 1977; Fleagle & Meldrum 1988; Fleagle & Simons 1995; Napier & Walker 1967; Walker 1974). Anemone (1993) found that prosimian leapers and slow climbers both have short ischia and that ischium length is increased in more quadrupedal species. Fleagle and Simons (1995) argue that a long ischium characterizes primates that leap from quadrupedal, pronograde positions, whereas leapers using orthograde postures (e.g. vertical clingers and leapers) benefit from a relatively short ischium that is expanded dorsally. They include *P. pithecia* in the former group and this species, along with *Callicebus*, does have the longest ischium (ISHL) of the present sample (Table 8.4). However, *P. pithecia*'s primary leaping mode between vertical trunks employs an orthograde posture, and the more quadrupedal *Cacajao melanocephalus* and likely *Chiropotes albinasus*, have significantly shorter ischia. Dorsal elongation of the ischium has not been examined for the pitheciins, and this information would help clarify the role of ischial length in primate locomotor activities.

In the proximal femur, head/neck length (FHNL) was found not to sort species, but *P. pithecia* exhibits a wider neck than all taxa except *P. monachus* and *C. melanocephalus*. A wide or thick femoral neck has been noted for some strepsirhine vertical clingers and leapers (Anemone 1990), and Fleagle and Meldrum (1988) previously noted the buttressed neck in *P. pithecia*. We agree here that it likely serves to accommodate the significant bending forces incurred in leaping and especially during landing.

Pithecia pithecia has a distinctive knee region. The distal femur of *P. pithecia* is significantly deeper in its anteroposterior aspect (FBCD) than all other pitheciines, except *P. hirsuta* ($n = 1$) and *Callicebus*. Many of the remaining traits in the knee, leg and foot were only examined in the partial sample of *P. pithecia*, *P. monachus*, *Chiropotes satanas*, and *Callicebus* and unless otherwise noted, these taxa comprise the sample discussed from here on.

P. pithecia has a distinctively wide patellar groove (FPGW), wide tibial plateau (TPW), and has the highest value for femoral bicondylar width (FBCW), although FBCW in the full pitheciine sample is not statistically significantly greater than that of any other taxon. The patellar groove is slightly less wide in *P. monachus*, but is significantly narrower in *Ch. satanas* and *Callicebus*. Relative femoral condyle articular surface width (FMCW and FLCW) is symmetrical in all four taxa, although *P. pithecia* has a significantly wider medial condyle articular surface, and wider and taller lateral condyle articular surface (FLCW and FLCH) than most of the other taxa. Each of these features in the *P. pithecia* knee serve to increase the articular surface area of this weight-bearing joint or increase the mediolateral surface area for the patella's excursion in the *m. quadriceps* tendon. These features would be advantageous for stabilizing the knee during vertical clinging and for accommodating the landing forces generated by this relatively large-bodied leaper. The presence of a relatively wide patellar groove in *P. monachus* may indicate this trait is derived for the genus *Pithecia*, although patellar groove data are not yet available for other *Pithecia* species.

In the foot, *P. pithecia* and *Callicebus* share an elongated calcaneus (CML) compared to *Chiropotes satanas* and *P. monachus*. The *P. pithecia* foot shows further elongation in the third metatarsal (XT3L), suggesting an overall longer foot that functions as an effective lever for trunk leaping. Features of the talus could only be examined in *Ch. satanas*, *Pithecia monachus*, *P. pithecia* and *Callicebus* and no significant differences were found (Table 8.4). However, other studies have noted differences in talar features among extant and extinct pitheciins (Fleagle & Meldrum 1988; Kay *et al.* 1998; Meldrum 1990).

Conclusions

Recent published reports of pitheciin species positional behavior, coupled with our skeletal data on seven pitheciin species, presented the opportunity to more fully explore the anatomical correlates of locomotion and posture in this lineage. *Pithecia pithecia* remains the sole known vertically oriented leaper in this tribe of pronograde quadrupeds who, when they do leap, tend to do so more from horizontal substrates or deformable terminal branches.

The uniqueness of *Pithecia pithecia*'s positional behavior within this group is clearly evident in its skeleton. Trait clusters in different anatomical regions appear to relate to its trunk leaping and vertical clinging habits. These include limbs elongated in every element, low IMI, three features in the elbow, seven in the knee region, and two in the foot. As for exclusive features of the other two genera, the lists are much shorter. Two traits in the scapula are unique to the *Cacajao* species, and *Chiropotes* is distinguished by a narrow patellar groove.

In stark contrast to *P. pithecia*, the primarily quadrupedal pitheciins were not found to share any exclusive suite of traits. *Cacajao* and *Chiropotes* are generally characterized by a short ischium, a high intermembral index, a narrow and short scapular glenoid fossa, and a tall scapula. These traits are variously related to quadrupedal movements, but are notably absent or not found to the same degree in *P. monachus*. *Pithecia monachus* and *Ch. satanas* do share a deep ulnar midshaft and a short calcaneus, both of which are consistent with a reliance on quadrupedalism. However, various combinations of the *Cacajao* and *Chiropotes* species with *P. pithecia* – exclusive of *P. monachus* – were found to share more traits in common than expected given their significant positional behavior differences. These include long forelimbs, a wide humeral capitulum and a wide femoral neck. No anatomical correlates to pedal suspension were identified for *Ch. satanas* and *C. calvus*, although very few ankle and foot features were able to be examined in the study.

It is interesting that few traits were shared exclusively between all three or even two of the *Pithecia* species. The few traits shared include a low IMI and a wide patellar groove. The extremely small sample size for *P. hirsuta* may have confounded the identification of more commonalities within the genus.

The lack of a large number of traits shared among all of the more quadrupedal taxa is at least partially explained by the fact that "quadrupedalism" is not just one behavior, nor is there only one way to be an arboreal quadruped. Body size, support inclination and stability, canopy level, and possible phylogenetic constraints are some of the variables that could affect the quadrupedal habits of a primate, and all may play a role here. The question of qualitative differences within behavioral categories and their possible influence on anatomical correlates is addressed in Walker (1998), and calls for data collection on finer subcategories of behavior, including orientation of the body to the substrate.

This study has featured new data and analysis of the skeletal diversity and functional morphology found among seven pitheciin species that have not been previously compared. It covers distinct form/function relationships that cross taxonomic lines, and highlights the distinctive anatomical and locomotor behavior of *Pithecia pithecia* among known pitheciins. While beyond the scope of the present chapter, these data provide a wider lens through which the growing pitheciin fossil record may be explored and understood.

Acknowledgments

We thank the editors of this volume, Drs. Liza Veiga, Adrian Barnett, Stephen Ferrari and Marilyn Norconk for the invitation to contribute to this important body of work. We are also grateful to two anonymous reviewers and Marilyn Norconk for their comments on this work, to Dr. Jean-Phillipe Boubli for helpful information on *Cacajao melanocephalus*, and to Drs. Susan Ford and David Byers for useful discussion. We appreciate the curatorial staff for their assistance during data collection at AMNH, FMNH, USNM, MPEG, MNRJ, and MZUSP. The contributions of the late José Márcio Ayres, whose pioneering work on *Cacajao calvus* and later establishment of the species' habitat comprising the Mamirauá Sustainable Development Reserve, will continue to be appreciated by all who study this rare and fascinating species.

References

Anemone, R.L. (1990). The VCL hypothesis revisited: patterns of femoral morphology among quadrupedal and saltatorial prosimian primates. *American Journal of Physical Anthropology*, **83**, 373–393.

Anemone, R.L. (1993). The functional anatomy of the hip and thigh in primates. In *Postcranial Adaptations in Nonhuman Primates*, ed. D.L. Gebo. DeKalb, IL: Northern University Press, pp. 150–174.

Ayres, J.M. (1986). Uakaris and Amazonian flooded forest. Unpublished PhD dissertation, University of Cambridge.

Buchanan, D.B., Mittermeier, R.A. & Van Roosmalen, M.G.M. (1981). The saki monkeys, genus *Pithecia*. In *Ecology and Behavior of Neotropical Primates*, ed. R.A. Mittermeier & A.F. Coimbra-Filho. Rio de Janeiro: Academia Brasileira de Ciencias, pp. 391–418.

Davis, L.C. (2002). Functional morphology of forelimb and long bones in the Callitrichidae (Platyrrhini, Primates). Unpublished PhD dissertation, Southern Illinois University.

Davis, L.C., Walker, S.E. & Ford, S.M. (2006). Locomotion and skeletal differentiation within the Pitheciini. *American Journal of Physical Anthropology*, Suppl. 42.

Dykyj, D. (1982). Allometry of trunk and limbs in New World primates. Unpublished Doctoral dissertation, City University of New York, USA.

Fleagle, J.G. (1977). Locomotor behavior and skeletal anatomy of sympatric Malaysian leaf-monkeys (*Presbytis obscura* and *Presbytis melalophos*). *Yearbook of Physical Anthropology*, **20**, 440–453.

Fleagle, J.G. (1999). *Primate Adaptation and Evolution*, 2nd edn. New York, NY: Academic Press.

Fleagle, J.G. & Meldrum, J.D. (1988). Locomotor behavior and skeletal morphology of two sympatric pitheciine monkeys, *Pithecia pithecia* and *Chiropotes satanas*. *American Journal of Primatology*, **16**, 227–249.

Fleagle, J.G. & Mittermeier, R.A. (1980). Locomotor behavior, body size and comparative ecology of seven Surinam monkeys. *American Journal of Physical Anthropology*, **52**, 301–314.

Fleagle, J.G. & Simons, E.L. (1995). Limb skeleton and locomotor adaptations of *Apidium phiomense*, an Oligocene anthropoid from Egypt. *American Journal of Physical Anthropology*, **97**, 235–289.

Fontaine, R. (1981). The uakari, genus *Cacajao*. In *Ecology and Behavior of Neotropical Primates*, ed. R.A. Mittermeier & A.F. Coimbra-Filho. Rio de Janeiro: Academia Brasileira de Ciencias, pp. 443–494.

Ford, S.M. (1986). Systematics of the new world monkeys. In *Comparative Primate Biology, Vol. 1: Systematics, Evolution, and Anatomy*, ed. D.R. Swindler & J. Erwin. New York, NY: Alan R. Liss, pp. 73–135.

Ford, S.M. (1988). Postcranial adaptations of the earliest platyrrhine. *Journal of Human Evolution*, **17**, 155–192.

Ford, S.M. & Davis, L.C. (1992). Systematics and body size: implications for feeding adaptations in New World monkeys. *American Journal of Physical Anthropology*, **88**, 415–468.

Garber, P.A. (1991). A comparative study of positional behavior in three species of tamarin monkeys. *Primates*, **32**, 219–230.

Garber, P.A. (1992). Vertical clinging, small body size, and the evolution of feeding adaptations in the Callitrichinae. *American Journal of Physical Anthropology*, **88**, 469–482.

Garber, P.A. & Leigh, S.L. (2001). Patterns of positional behavior in mixed species troops of *Callimico goeldii*, *Saguinus labiatus*, and *Saguinus fuscicollis* in Northwestern Brazil. *American Journal of Primatology*, **54**, 17–31.

Happel, R. (1982). Ecology of *Pithecia hirsuta* in Peru. *Journal of Human Evolution*, **11**, 581–590.

Hershkovitz, P. (1977). *Living New World Monkeys (Platyrrhini), with an Introduction to Primates, Vol. 1*. Chicago, IL: University of Chicago Press.

Kay, R.F. (1990). The phyletic relationships of extant and fossil Pitheciinae (Platyrrhini, Anthropoidea). *Journal of Human Evolution*, **19**, 175–208.

Kay, R.F., Johnson, D. & Meldrum, D.J. (1998). A new pitheciin primate from the middle Miocene of Argentina. *American Journal of Physical Anthropology*, **45**, 317–336.

Kinzey, W.G. (1977). Positional behavior and ecology in *Callicebus torquatus*. *Yearbook of Physical Anthropology*, **20**, 468–480.

Kinzey, W.G. (1992). Dietary and dental adaptations in the Pitheciinae. *American Journal of Physical Anthropology*, **88**, 499–514.

Lawler, R.R., Ford, S.M., Wright, P.C., *et al.* (2006). The locomotor behavior of *Callicebus brunneus* and *Callicebus torquatus*. *Folia Primatologica*, **77**, 228–239.

Leigh, S.R. & Jungers, W.L. (1994). A re-evaluation of subspecific variation and canine dimorphism in woolly spider monkeys (*Brachyteles arachnoides*). *American Journal of Physical Anthropology*, **95**, 435–442.

Meldrum, D.J. (1990). New fossil platyrrhine tali from the early Miocene of Argentina. *American Journal of Physical Anthropology*, **82**, 403–418.

Napier, J.R. & Walker, A.C. (1967). Vertical clinging and leaping – a newly recognized category of locomotor behaviour of primates. *Folia Primatologica*, **6**, 204–219.

Peres, C.A. (1986). Golden lion tamarin project. II. Ranging patterns and habitat selection in golden lion tamarins *Leontopithecus rosalia* (Linnaeus, 1766) (Callitrichidae, Primates). In *A Primatologia no Brasil – 2*, ed. M. Thiago de Mello. Brasilia: Sociedade Brasileira de Primatologia, pp. 223–233.

Peres, C.A. (1993). Notes on the ecology of buffy saki monkeys (*Pithecia albicans*, Gray 1860): a canopy seed-predator. *American Journal of Primatology*, **31**, 129–140.

Preuschoft, H. (1974). Body posture and mode of locomotion in fossil primates. Methods and example: *Aegyptopithecus zeuxis*. *Proceedings of the Symposium of the 5th Congress of the International Primatological Society*, pp. 345–359.

Preuschoft, H. & Demes, B. (1985). Influence of size and proportions on the biomechanics of brachiation. In *Size and Scaling in Primate Biology*, ed. W.L. Jungers. New York, NY: Plenum Press, pp. 383–399.

Rose, M.D. (1993). Functional anatomy of the elbow and forearm in primates. In *Postcranial Adaptations in Nonhuman Primates*, ed. D.L. Gebo. DeKalb, IL: Northern Illinois University Press, pp. 70–95.

Rosenberger, A.G. (1988). Pitheciinae. In *Encyclopedia of Human Evolution and Prehistory*, ed. I. Tattersall, E. Delson & J. van Couvering. New York, NY: Garland Press, pp. 454–455.

Rosenberger, A.L. & Strier, K.B. (1989). Adaptive radiation of the ateline primates. *Journal of Human Evolution*, **18**, 717–750.

SAS Institute Inc. (2001). SAS Campus Drive, Cary, North Carolina 27513, USA.

Silva Jr., J.S. & Figueiredo, W.M.B. (2002). Revisão sistemática dos cuxiús, gênero *Chiropotes* Lesson, 1840 (Primates Pithecidae). *Livro de Resumos do XO. Congresso da Sociedade Brasileira de Primatologia, Amazônia – A Última Fronteira*, Belém do Pará, p. 21.

Smith, R.J. & Jungers, W.L. (1997). Body mass in comparative primatology. *Journal of Human Evolution*, **32**, 523–559.

Sokal, R.R. & Rohlf, F.J. (1981). *Biometry: The Principles and Practice of Statistics in Biological Research (2nd Edition)*. New York, NY: W.H. Freeman and Co.

Stafford, B.J., Rosenberger, A.L., Baker, A.J., *et al.* (1996). Locomotion of golden lion tamarins (*Leontopithecus rosalia*): the effects of foraging adaptations and substrate characteristics on locomotor behavior. In *Adaptive Radiations of Neotropical Primates*, ed. M.A. Norconk, A.L. Rosenberger & P.A. Garber. New York, NY: Plenum Press, pp. 111–132.

van Roosmalen, M.G.M., Mittermeier, R.A. & Milton, K. (1981). The bearded sakis, genus *Chiropotes*. In *Ecology and Behavior of Neotropical Primates*, ed. R.A. Mittermeier & A.F. Coimbra-Filho. Rio de Janeiro: Academia Brasileira de Ciencias, pp. 419–442.

Walker, A. (1974). Locomotor adaptations in past and present prosimian primates. In *Primate Locomotion*, ed. F.A. Jenkins, Jr. New York, NY: Academic Press, pp. 349–381.

Walker, S.E. (1996). Evolution of positional behavior in the saki–uakaris (*Pithecia*, *Chiropotes*, and *Cacajao*). In *Adaptive Radiations of Neotropical Primates*, ed. M.A. Norconk, A.L. Rosenberger & P.A. Garber. New York, NY: Plenum Press, pp. 335–366.

Walker, S.E. (1998). Fine-grained differences within positional categories: a case study of *Pithecia* and *Chiropotes*. In *Primate Locomotion*, ed. E. Strasser, J.G. Fleagle, H. McHenry & A.L. Rosenberger. New York, NY: Plenum Press, pp. 31–43.

Walker, S.E. (2005). Leaping behavior of *Pithecia pithecia* and *Chiropotes satanas* in Eastern Venezuela. *American Journal of Primatology*, **66**, 369–387.

Walker, S.E. & Ayres, J.M. (1996). Positional behavior of the white uakari (*Cacajao calvus calvus*). *American Journal of Physical Anthropology*, **101**, 161–172.

Walker, S.E. & Davis, L.C. (2007). Postcranial features of *Cacajao* with comparisons to *Chiropotes* and *Pithecia*. *American Journal of Physical Anthropology*, Suppl. **44**.

Youlatos, D. (1999). Comparative locomotion of six sympatric primates in Ecuador. *Annales des Sciences Naturelles*, **20**, 161–168.

Chapter 9

Male cooperation in Pitheciines: the reproductive costs and benefits to individuals of forming large multimale/multifemale groups

Paul A. Garber & Martin M. Kowalewski

Primates differ from many other animal lineages in that they show rather good evidence of cooperation, especially in long-term relationships. (van Schaik & Kappeler 2006, p. 13)

Introduction

Traditional approaches to the study of primate sociality have highlighted predator protection as a primary benefit of group living, and argue that factors such as aggression, dominance and feeding competition constrain or determine group size, group composition and the nature of social interactions among individual group members (van Schaik 1983, 1989; Janson & Goldsmith 1995; Sterck *et al.* 1997; Chapman & Chapman 2000; Janson 2000; Kappeler & van Schaik 2002). In particular, both the socioecological model and the ecological constraints model describe a set of resource conditions (clumped and monopolizable resources of intermediate size relative to group size) and competitive regimes under which within-group contest competition for food, especially among females and unrelated individuals, is expected to be severe, agonistic and frequent (van Schaik 1989; Sterck *et al.* 1997).

Moreover, it is argued that the costs of feeding competition increase with group size, as individuals are required to travel greater distances to locate more productive feeding sites or experience elevated predation risk by foraging in small, scattered parties in order to encounter sufficient resources (Janson & Goldsmith 1995; Sterck *et al.* 1997; Koenig *et al.* 1998; Chapman & Chapman 2000; Janson 2000). Even under conditions in which food patches are small, dispersed, and nonmonopolizable, these models suggest that group members compete through a process of avoidance and scramble competition (Sterck *et al.* 1997; Chapman & Chapman 2000; Steenbeek & van Schaik 2001; Snaith & Chapman 2005). The major limitations of these approaches are their focus on competition and aggression as the main factors driving within-group social relationships and primate social organization, and their lack of attention to the joint reproductive benefits both males and females receive as members of a functioning, integrated, cooperative and stable social unit (Sussman & Garber 2007, 2011).

Over the past few years, a series of papers have been published (Dugatkin 1997; Nunn 2000; Chapais 2006; van Schaik & Kappeler 2006; Sussman & Garber 2007, 2011) that provide strong empirical and theoretical support for models of primate sociality based on patterns of affiliation, cooperation, reciprocity and mutualism. These studies build on the work of Hamilton (1964), Trivers (1971) and Axelrod and Hamilton (1981) in defining and expanding the set of social and ecological conditions under which cooperative behaviors are expected to evolve and be maintained as an evolutionary stable strategy.

Here we define cooperation as coordinated "actions or traits that benefit other individuals" (van Schaik & Kappeler 2006, p. 3). Cooperation can take several forms, including altruism, reciprocity or mutualism. Altruism is a unilateral interaction with high benefits for the recipient and high costs for the donor, and no expectation of reciprocation, which is usually explained through kin selection (Hamilton 1964). In reciprocity, benefits to the recipient are variable, and immediate costs to the donor are low to moderate. In this case, the donor receives eventual benefits through future reciprocal actions of the recipient. Where the immediate costs to the donor are low [Trivers's (1971) tit-for-tat strategy] or the long-term benefits are high (e.g. friendships, cooperative hunting, alliance partners: Boesch *et al.* 2006; Smuts 1997), reciprocally cooperative relationships are expected to be based more on partner competency than kinship (Goldberg & Wrangham 1997; Noë 2001; Chapais 2006; Mitani 2006). Mutualism brings either benefits to both actors, or benefits the recipient at no cost to the donor (Dugatkin 1997). As returns are immediate, such relationships do not require book-keeping of cost–benefit ratios or sanctions against defectors. We argue that mutualisms and reciprocity, based principally on partner competency, represent a critical factor driving affiliative social interactions and social organization in many primate species (Chapais 2006; Sussman & Garber 2007).

Evolutionary Biology and Conservation of Titis, Sakis and Uacaris, eds. Liza M. Veiga, Adrian A. Barnett, Stephen F. Ferrari and Marilyn A. Norconk. Published by Cambridge University Press.

Pitheciines

In this chapter we examine the published literature on patterns of social interactions and cooperative behavior among adult male pitheciines (*Pithecia*, *Chiropotes* and *Cacajao*). Compared to most other New World primate taxa, bearded sakis and uacaris live in relatively large social units that contain as many as 10–16 adult males (Table 9.1). Although it is often assumed that resident adult male primates compete aggressively for access to reproductive partners, in several species of New World monkeys there is evidence of high levels of within-group tolerance among males (Table 9.1). We explore several theories of social cooperation in order to evaluate the set of factors that promote adult male affiliation in pitheciines.

Pithecia

Pithecia appears to represent a basal member of the pitheciine clade (Bonvicino *et al.* 2003; Schneider & Rosenberger 1996). Although once considered to mate monogamously and live in a pair-bonded social system (Buchanan *et al.* 1981), there is growing evidence, at least in *Pithecia pithecia* (Lehman *et al.* 2001; Norconk 2011), *Pithecia hirsuta* (Soini 1986) and *Pithecia albicans* (Peres 1993), that stable groups may contain as many as 8–12 individuals including 1–4 adult males and 1–3 adult females (Table 9.1).

Pithecia are characterized by a well-developed set of scent glands (Buchanan *et al.* 1981; Norconk 2011) and, in the case of white-faced sakis (*P. pithecia*), males are reported to scent mark more than females. Norconk (2011, p. 131), citing a personal communication from T. Gleason, writes "... adult males huddled, rubbed their chests against each other, and possibly exchanged scent just prior to some inter-troop encounters". In addition, monk sakis (*Pithecia monachus*) have been observed to engage in frequent allogrooming involving most group members (Moynihan 1976). Scent marking and allogrooming appear to represent important tactile and olfactory mechanisms used by male sakis to promote affiliative bonds, coordinate social interactions and maintain group cohesion.

Table 9.1 Demographic, social and reproductive characteristics of pitheciine and atelid genera. Based on a compilation of data from Masterson and Hartwig (1998), Ford (1994), Di Fiore and Campbell (2007), Kowalewski (2007), and Norconk (2006, 2011).

Variable	Pitheciinae			Atelidae			
	Pithecia	*Chiropotes*	*Cacajao*	*Alouatta*	*Lagothrix*	*Ateles*	*Brachyteles*
Group size	1–12	19–44	1–>108 (200?)	6–40	10–49	12–42	7–63
# of males per group	1–4	1–8	16	1–6	2–11	3–6	4–21
# of females per group	1–3	8–10	16	1–15	3–15	5–18	3–28
Within-group male affiliation and social tolerance (grooming, proximity, lining up, tail wagging)	Low/medium	High	High	Low/medium	Medium	High	High
Within-group contest feeding competition (aggressive)	Low	Low	Low	Low	Low	Low	Low
Dispersal	Bisexual	Female-biased	?	Bisexual	Bisexual (?)	Female-biased	Female-biased
Sexual dimorphism – body mass[1]	Medium/high 1.02–1.47	Low 1.08–1.26	Low 1.08–1.26	High 1.31–1.48	Low–medium 1.04–1.45	Low 0.85–1.2	Low 1.283
Sexual dimorphism – canine size[2]	1.17	1.06	1.19	1.25–1.33	1.07–1.22	1.10–1.13	1.15
Sexual dichromatism	Yes (some species)	No	Very low to some (1 species)	Yes (some species)	No	No	No
Testis size/color difference	?	Yes (color)	?	Yes (color)	?	No	Yes (size)
Testis size/body mass[3]	5.8×10^{-4}	?	1.7×10^{-3}	3.2×10^{-3}	2.1×10^{-3}	1.7×10^{-3}	?
Intergroup encounter (collective male defense)	Yes	Yes	?	Yes	Yes	Yes	Yes
Females receptive across year	?	Yes	?	Yes	Yes	Yes	Yes

Table 9.1 *(cont.)*

Variable	Pitheciinae			Atelidae			
	Pithecia	*Chiropotes*	*Cacajao*	*Alouatta*	*Lagothrix*	*Ateles*	*Brachyteles*
Female promiscuity	?	?	Some evidence	Yes	Yes	Yes	Yes
Within-group male intrasexual tolerance during copulations	?	?	Yes (limited data)	In some species	Yes	No (?)	Yes
Groups with within-group adult sex ratios approaching 1 : 1 (female : male)	Yes (0.67–3)	Yes (1.12–5)	Yes (1)	Yes (0.7–4.2)	No (1.3–2.8)	Yes (0.8–4.4)	Yes (0.9–2.4)
Interbirth interval (months)	15–36	24	?	15.9–22.5	36.7±4.7	31.9–50	36.4±4.3
Female mate choice	Weak	Strong	Strong	Weak in some species and strong in others	Strong	Strong	Strong
Seasonality of births	Dry season (also late dry to early wet)	Wet season (also late wet to dry)	Oct–Nov (dry season)	Depends on the site (with and without)	Late wet/ early dry season	Some (it varies)	Dry season
Expected forms of male social cooperation	Mutualism	Mutualism/ reciprocity	Mutualism/ reciprocity	Mutualism	Mutualism/ reciprocity	Reciprocity/ mutualism	Reciprocity/ mutualism

[1] Kay *et al.* (1988), Bowler (pers. commun.); [2]canine dimorphism index (CDI) from Kay *et al.* (1988). If this index is 1.00, there is no canine dimorphism and it would be 2.00 if male canines are twice as large as those of females; [3]Harcourt *et al.* (1995).

Chiropotes

Species of the genera *Chiropotes* live in multimale/multifemale groups containing up to 44 individuals, although groups of approximately 20 individuals are more common (Table 9.1). van Roosmalen *et al.* (1981) and Norconk (2011) indicate that *Chiropotes* social groups travel as a cohesive unit, although they may divide into smaller subgroups to exploit nearby feeding trees. This has been described as a type of fission–fusion social system (Peetz 2001; Veiga *et al.* 2006). In *Chiropotes albinasus* and *C. satanas*, stable groups composed of as many as 15 adult males and 13 adult females have been reported (Norconk 2011; Veiga & Ferrari 2006). Regardless of group size, the ratio of adult males to adult females is often close to 1 : 1 (Table 9.1; see also table 8.1 in Norconk 2011). Although only a few long-term studies of bearded saki social organization exist, observations of lone females in *C. chiropotes* (Peetz 2001) and *C. satanas* (Veiga & Silva 2005) suggest the probability of female dispersal. Male dispersal is less clear, but it is possible that as in atelines (Strier 2000; Di Fiore & Campbell 2007), male bearded sakis either are philopatric or disperse less frequently than females (Veiga, pers. commun.).

Recent studies indicate high levels of affiliative behavior among male *Chiropotes*. In a 16-month study of a group containing 12–15 adult male and 9–13 adult female *C. satanas* in Brazil, Veiga *et al.* (2006) found that males were consistently more affiliative to each other than they were to females. In addition, adult males were each other's nearest neighbors 63.7% of the time compared to 15.9% for adult females. In addition to social proximity, over 90% of adult male social interactions were affiliative and included social resting, allogrooming, alliance formation and "lining-up" behavior. Lining-up is a highly ritualized behavioral pattern of body contact between adults which appears to include the exchange of tactile and olfactory information through scent-marking and grooming. In a second study, Peetz (2001) also observed lining-up behavior in male *C. chiropotes* in Venezuela. In this study, 17.9% of the activity budget involved social interactions, and 83.7% of which were affiliative.

A second ritualized behavioral pattern described for *Chiropotes* (and *Cacajao*) is tail wagging, which is "accompanied by vocalizations, occurs in a variety of contexts from mild to severe agitation … [and functions during the] … reunion of group members, and group reorganization after rest or feeding periods" (Norconk 2011, p. 129). Both behavioral patterns function to reinforce social bonds, reunite group members, and promote affiliative and cooperative behaviors.

Finally, recent observations by Veiga (pers. commun.) of *C. satanas* at Tucuruí in Brazil indicate that females are sexually receptive throughout the year, including periods in which they are transporting small infants and lactating. Female receptivity during non-fertile periods is likely to have an important effect on reproductive strategies, male–female social and reproductive bonds, and the degree to which any single male can monopolize access to females.

Cacajao

Uacaris are the largest-bodied pitheciines, and live in extremely large multimale/multifemale groups (Ayres 1989; Defler 1999; Barnett 2005). In *Cacajao calvus*, groups can be as large as 70–150 individuals, including possibly 30 adult males (Bowler 2007). Defler (2001) indicates that, although most groups of *Cacajao oukary* contain 20–30 animals, he observed a temporary group that may have approached 200 animals. Bowler (2007) reported an average group size of 43.5 at Largo Preto, Peru. Despite their large size, uacari troops appear to be relatively stable, traveling and foraging together "in a coordinated fashion for several days" (Defler 1999, p. 6).

Uacari social interactions and spatial organization have been described as a type of fission–fusion social system, possibly in response to changes in the distribution, availability and productivity of feeding sites (Barnett 2005). In comparison to chimpanzees and spider monkeys (*Ateles*), however, uacari (*Cacajao melanocephalus*) subgroups are relatively large, with counts of 21–30 animals the most common, and subgroups containing 31–100 animals were as common as those of 11–20 (Defler 1999).

In *C. calvus*, larger subgroups or fission–fusion parties occur during periods of mature fruit abundance and smaller and more scattered foraging parties are reported in the dry season (Ayres 1989). Subgroups may disperse over an area of 1–2 km, but coordinate their activities using an extensive repertoire of contact vocalizations.

Despite large group size and the presence of many resident adult males, Boubli (1999) observed virtually no aggressive interactions during a 16-month study of *C. melanocephalus* in Brazil. Similarly, Cunha and Barrett (1989, cited in Barnett & Brandon-Jones 1997) indicate high levels of male–male tolerance in their study groups of *C. ouakary*. Male–male tolerance and cooperation may play a role in predator protection and intergroup encounters. In *C. ouakary* and *C. melanocephaus*, Barnett (2005) report that males may remain behind after the rest of the group has left to distract or confront a potential threat. In captivity, same-sex adult uacaris (*C. c. calvus*) also are reported to be highly tolerant of one another (Barnett & Brandon-Jones 1997).

In a recent study of *C. calvus ucayalii*, Bowler (2007) observed all male subgroups of up to 10 individuals on the periphery of large bisexual groups. This author suggests that similar-aged males develop strong social bonds as juveniles which continue when they form bachelor groups. While at least some of these males are almost certainly related (as sibs, half sibs or cousins), the absence of genetic studies impedes a more systematic understanding of the role of kinship in the cooperative relationships of male uacaris.

Discussion

The pitheciines are characterized by social groups containing multiple adult males and females. Direct evidence of within-group contest competition at feeding sites is extremely limited, whereas affiliative and cooperative social interactions among same-sex adult group members, especially males, appears to be common (Peetz 2001; Veiga & Silva 2005). Given the paucity of long-term studies of individually marked pitheciines, the discussion that follows is best considered a starting point from which to examine the social and reproductive costs and benefits of forming large multimale–multifemale ("multi") groups. The presence of stable multi groups across a wide range of primate taxa highlights the reproductive advantages for both males and females of residing in a large troop (Table 9.2).

Why form multimale–multifemale groups?

Due to the relatively high costs and slow rate of reproduction, related primate females may tend to aggregate in an attempt to increase their access to feeding sites, reduce predation risk and enhance offspring survival (van Schaik 1983, 1989). Although this may help to explain the spatial distribution of many matrilineal species, it is less clear how it applies to primates with male philopatry (Strier 1999; Kowalewski 2007). Females may benefit from the presence of several adult males if this guarantees them access to high-quality feeding sites, protection from predators, group stability (reduction in the invasion of potentially infanticidal males) and increased mate choice (van Schaik *et al.* 1999; Crockett & Janson 2000; Pereira *et al.* 2000; Kowalewski 2007; Kowalewski & Garber 2010).

Males, in contrast, may respond to the distribution of females by developing strategies for increasing their access to mating partners (Altmann 1990; Nunn 2000). This can occur in several ways, including (a) forming a pair bond or harem (monopolization), (b) forming small multi groups in which access to fertile females is constrained by a rigid male dominance hierarchy or mate guarding, and (c) forming larger multi groups in which male social interactions are cooperative, non-hierarchical, and all or most resident males have access to receptive females (Table 9.2) and act collectively to exclude extragroup males (Nunn 2000; Pereira *et al.* 2000; Strier 2000; Kowalewski 2007; Kowalewski & Garber 2010).

Nunn (2000) describes a set of alternative models for examining within-group cooperative behaviors in male primates that promote what he terms the "collective good". Under conditions in which multimale groups are characterized by a rigid linear male hierarchy, the dominant male is the primary breeder and females are receptive only during ovulation, subordinate resident males obtain few direct benefits (possibly long-term expectation of inheriting the dominant breeding position) while incurring potential costs (e.g. injuries) when aiding the dominant male to repel intruders. Under

Table 9.2 Expected effects of within-group male intrasexual social tolerance on reproductive strategies and biology in species characterized by different dispersal patterns. We assume that the majority of new groups form by the fissioning of established groups and therefore individuals may receive benefits from maintaining long-term social bonds with natal group members. In species in which males have reduced intrasexual tolerance, the number of co-resident males will be relatively small.

Dispersal pattern:	**Bisexual dispersal**		**Female dispersal/ male philopatry**
Sociosexual parameter — Level of male tolerance within group:	**Intolerant or lower male intrasexual tolerance**	**Higher male intrasexual tolerance**	**Higher male intrasexual tolerance**
Per capita cost of female defense	Higher	Lower	Lower
Risk of infanticide	Higher	Low	Lowest
Per capita cost of predator detection	Higher	Lower	Lower
Male reproductive tenure	Lower	High	Highest
Ability of males to monopolize matings	Highest	Lower	Lower
Access to a larger number of females	Lower	Higher	Higher
Mate guarding	Higher	Lower	Lower
Cost to male of entering a new group	High	Lower	Highest
Body/canine size dimorphism	Higher	Lower	Low
Testis size/sperm competition	Lower	Higher	Higher
Opportunities for long-term social bonds among males	Low	High	Highest
Opportunities for male–male cooperation	Low	High	Highest
Opportunities for female mate choice	Low	Highest	High
Increased genetic diversity of a female's successive offspring	Lower–higher depending on male tenure in group	Higher	High
Effects of female receptivity during non-mating periods in strengthening male–female social bonds	Lower	Higher	High
Opportunities for females to mate promiscuously	Lower	Higher	Higher

these conditions, male cooperation will have little effect on reproductive skew, as the dominant male is the principal breeder. In contrast, in species characterized by less-rigid hierarchies in which all or most males have at least some access to receptive females, all resident males benefit from the collective defense of receptive females. Assuming that collective action is more effective than individual defense, each male increases his probability of siring offspring in the group (Table 9.2), while the per capita costs of collective action, whether fighting, chasing, or vocal battles, decrease.

A model of benefits and costs of male cooperation in pitheciines

We begin with the assumption that group size, social organization and behavioral interactions in many primate lineages are determined more by reproductive strategies and demographic parameters than foraging constraints. Bisexual dispersal and female-biased dispersal are common in platyrrhines (Strier 1999, 2000; Steenbeek & van Schaik 2001; Jack 2007). Whereas in the former both males and females must make dispersal decisions, in species with male philopatry, the selection of target groups for dispersal is the responsibility of females.

Within this framework, comparisons with the atelids (Table 9.1) may provide important insights into male cooperation and social strategies in pitheciines, especially considering that this platyrrhine group probably represents the sister taxon of the Pitheciidae (Schneider & Rosenberger 1996). In *Brachyteles* and *Ateles*, dispersal is female-biased, established groups are largely composed of male patrilines with unrelated adult females, and within-group social bonds among adult males are highly affiliative and cooperative (Strier 1992; Di Fiore & Campbell 2007). In these taxa, females are promiscuous, actively solicit copulations from males, appear to exercise mate choice and are receptive outside of their fertile periods (Table 9.1). Female receptivity during non-fertile periods may reinforce male cooperative mate defense, as well as strengthen sociosexual bonds between males and females. In the case of the northern muriquis (*Brachyteles hypoxanthus*), Strier (2000, p. 72) reports that "male kin display tolerance toward one another's sexual activities in exchange for cooperation in competitive contests with other male kin groups over access to females".

In *Ateles*, adult males are each other's closest social partners (Symington 1990). Resident females mate with multiple males, and males jointly patrol the borders of their range and act collectively to repel intruders (Table 9.1). Although kinship may play an important role in understanding the benefits of male mating tolerance in spider monkeys and muriquis, many cooperative behaviors such as tolerance at feeding sites, minimal within-group intrasexual aggression, greeting behaviors, and cooperative group or mate defense may be best classified as mutualisms – relatively low-cost behaviors from which several individuals receive immediate benefits.

In contrast with *Ateles* and *Brachyteles*, dispersal in woolly monkeys (*Lagothrix*) appears to be bisexual, and affiliative bonds between adult males are reported to be weaker (Table 9.1). Grooming relationships and spatial proximity, however, are strong between adult and subadult males, and adult males appear to cooperate during intergroup encounters (Di Fiore & Campbell 2007). Female woolly monkeys solicit copulations, copulate outside their ovulatory period, and adult males are highly tolerant of within-group copulations by other resident males (Peres 1996).

Alouatta is characterized by bisexual dispersal, and therefore adult group members are generally unrelated (Table 9.1). Although in red howlers (*Alouatta seniculus*) related males may disperse together into the same group and females are more tolerant of adult daughters than unrelated females (Pope 2000), this does not appear to be the case for other species. In *Alouatta palliata*, the overwhelming majority of females and males disperse from their natal group (Clarke & Glander 2001), and stable bisexual groups with as many as 8–12 adult males have been reported (Chapman 1990). Wang and Milton (2003) identified an alpha male in an *A. palliata* group with 6 males and 12 females in Panama, but could not discern any hierarchy among the other resident males of the group. In addition, these authors observed several males copulating with the same female during her estrous period in full view of other males, including the alpha male, without eliciting an aggressive response.

Although *A. palliata* is unusual in living in groups that can exceed 20 individuals, other howler species are known to form smaller bisexual groups composed of one or a small number of adult males (Brockett *et al.* 2001; Crockett & Janson 2000; Kowalewski & Zunino 2004). In *Alouatta caraya*, groups may contain up to four adult males (Kowalewski 2007), within-group agonism is rare, and adult males spend as much as 5.2% of their activity budget engaged in affiliative behavior including grooming and embracing. In addition, females may mate with most or all resident male group members during both fertile and non-fertile periods, and in full view of other male group members (Table 9.1). Unrelated resident adult male *A. caraya* act cooperatively to defend the group from incursions by extra-group males throughout the year (Table 9.2).

Overall, we feel that atelids offer an instructive model for understanding the basis of male cooperation in the pitheciines. Depending on patterns of natal and secondary migration, adult sex ratio and group stability, we expect partner reliability and competence, kinship and the mutual benefits of cooperative mate defense to be primary factors in the evolution of male social tolerance (Table 9.2). Given long-term residence in the same social group, opportunities for cooperative reciprocity among males are expected to be greater in male philopatric species than in species in which males migrate (unless male relatives migrate in pairs: Garber *et al.* 1993; Jack 2007). In addition, mutualistic behaviors among either related or unrelated males are likely to be very common given their immediate return and low cost to all participants (Dugatkin 1997). Moreover, if the presence of several adult males offers advantages to females in terms of enhanced infant survivorship or greater genetic diversity for their offspring (Garber 1997; Altmann 1990; Mitani *et al.* 1996; Nunn 2000), then females may prefer groups composed of cooperative adult males. Female promiscuity and sexual receptivity during non-fertile periods would serve to promote male–female bonds and reduce the potential for infanticide.

In conclusion, present evidence indicates high levels of social tolerance among resident male pitheciines. The basis of male–male cooperation in the pitheciines requires additional long-term behavioral and genetic investigation. We view intrasexual social cooperation as an important component of a male reproductive strategy, and have tried to identify factors that are likely to influence the costs and benefits of male cooperation. Based on comparisons with the atelids, low-cost forms of mutualism appear to best explain male cooperation in *Pithecia*. In *Chiropotes* and *Cacajao*, we anticipate that mutual and immediate benefits of cooperation are common, but where males reside in the same group over long periods of time, low-cost forms of reciprocity also are expected (Table 9.2). Male–male tolerance, affiliation and cooperation are found in many platyrrhine taxa (e.g. Boinski 1994; Di Fiore & Campbell 2007; Digby *et al.* 2007). It is likely that long-term familiarity, partner competence and the mutual and reciprocal benefits of male collective action represent the primary adaptive basis of male–male cooperation in many primate species (Chapais 2006; Dugatkin 1997).

Acknowledgments

PAG wishes to acknowledge the nonhuman primates of Central and South America for allowing me to observe the intimate details of their lives for over 30 years. PAG thanks Sara and Jenni for teaching me about the virtues of tolerance and cooperation, and Chrissie McKenney for her love and kindness. MK thanks all his friends and especially dedicates this chapter to Polo – Ramon Romero – for years of cooperation, friendship, assistance and collaboration. We thank Mark Bowler for allowing us to cite his unpublished data on *Cacajao calvus*, and for his sharing his insights on pitheciine social behavior.

References

Altmann, J. (1990). Primate males go where the females are. *Animal Behaviour*, **39**, 193–195.

Axelrod, R. & Hamilton, W.D. (1981). The evolution of cooperation. *Science*, **211**, 1390–1396.

Ayres, J.M. (1989). Comparative feeding ecology of the uakari and bearded saki, *Cacajao* and *Chiropotes*. *Journal of Human Evolution*, **18**, 697–716.

Barnett, A.A. (2005). *Cacajao melanocephalus*. *Mammalian Species*, **776**, 1–6.

Barnett, A.A. & Brandon-Jones, D. (1997). The ecology, biogeography and conservation of the uakaris, *Cacajao* (Pitheciinae). *Folia Primatologica*, **68**, 223–235.

Boesch, C., Boesch, H. & Vigilant, L. (2006). Cooperative hunting in chimpanzees: kinship or mutualism In *Cooperation in Primates and Humans: Mechanisms and Evolution*, ed. P.M. Kappeler & C.P. van Schaik. Berlin: Springer Verlag, pp. 139–150.

Boinski, S. (1994). Affiliation patterns among male Costa Rican squirrel monkeys. *Behaviour*, **130**, 191–209.

Bonvicino, C.R., Boubli, J.P., Otazu, I.B., *et al.* (2003). Morphologic, karyotypic, and molecular evidence of a new form of *Chiropotes* (Primates, Pitheciinae). *American Journal of Primatology*, **61**, 123–133.

Boubli, J.P. (1999). Feeding ecology of black-headed uacaris (*Cacajao melanocephalus melanocephalus*) in Pico do Neblina National Park, Brazil. *International Journal of Primatology*, **20**, 719–749.

Bowler, M. (2007). The ecology and conservation of the red uakari monkey on the Yavarí River, Peru. Unpublished PhD thesis, University of Kent.

Brockett, R.C., Horwich, R.H. & Jones, C.B. (2001). Reproductive seasonality in the Belizean black howling monkey (*Alouatta pigra*). *Neotropical Primates*, **8**, 136–138.

Buchanan, D.B., Mittermeier, R.A. & van Roosmalen, M.G.M. (1981). The saki monkeys, genus *Pithecia*. In *Ecology and Behavior of Neotropical Primates*, Vol. 1., ed. A.F. Coimbra-Filho & R.A. Mittermeier. Rio de Janeiro: Academia Brasileira de Ciências, pp. 391–417.

Chapais, B. (2006). Kinship, competence and cooperation in primates. In *Cooperation in Primates and Humans: Mechanisms and Evolution*, ed. P.M. Kappeler & C.P. van Schaik, Berlin: Springer Verlag, pp. 47–64.

Chapman, C.A. (1990). Ecological constraints on group size in three species of Neotropical primates. *Folia Primatologica*, **55**, 1–9.

Chapman, C.A. & Chapman, L.J. (2000). Determinants of group size in primates: the importance of travel costs. In *On the Move: How and Why Animals Travel in Groups*, ed. S. Boinski & P.A. Garber. Chicago, IL: University of Chicago Press, pp. 24–42.

Clarke, M.R. & Glander, K.E. (2001). Presence of mother and juvenile dispersal in free-ranging howling monkeys (*Alouatta palliata*) in the tropical dry forest of Costa Rica. *American Journal of Physical Anthropology*, Suppl **32**, 50.

Crockett, C.M. & Janson, C. (2000). Infanticide in red howlers: female group size, male membership, and a possible link to folivory. In *Infanticide by Males and its Implications*, ed. C.P. van Schaik & C. Janson. Cambridge: Cambridge University Press, pp. 75–98.

Defler, T.R. (1999). Fission–fusion in the black-headed uacari (*Cacajao melanocephalus*) in eastern Colombia. *Neotropical Primates*, **7**, 5–8.

Defler, T.R. (2001). *Cacajao melanocephalus ouakary* densities on the lower Apaporis River, Colombian Amazon. *Neotropical Primates*, **9**, 31–36.

Di Fiore, A. & Campbell, C.J. (2007). The atelines: variation in ecology, behavior, and social organization. In *Primates in Perspective*, ed. C.J. Campbell, A. Fuentes, K.C. MacKinnon, M. Panger & S.K. Bearder. New York, NY: Oxford University Press, pp. 155–185.

Digby, L.J., Ferrari, S.F. & Saltzman, W. (2007). Callitrichines: the role of competition in coopertatively breeding species. In *Primates in Perspective*, ed. C.J. Campbell, A. Fuentes, K.C. MacKinnon, M. Panger & S.K. Bearder. New York, NY: Oxford University Press, pp. 85–106.

Dugatkin, L.A. (1997). *Cooperation among Animals: An Evolutionary Perspective.* New York, NY: Oxford University Press.

Ford, S.M. (1994). Evolution of sexual dimorphism in body weight in platyrrhines. *American Journal of Primatology*, **34**, 221–244.

Garber, P.A. (1997). One for all and breeding for one: cooperation and competition as a tamarin reproductive strategy. *Evolutionary Anthropology*, **5**, 187–199.

Garber, P.A., Encarnación, FF., Moya, L., *et al.* (1993). Demographic and reproductive patterns of mating in moustached tamarin monkeys (*Saguinus mystax*): implications for reconstructing platyrrhine mating systems. *American Journal of Primatology*, **29**, 235–254.

Goldberg, T.L. & Wrangham, R.W. (1997). Genetic correlates of social behaviour in wild chimpanzees: evidence from mitochondrial DNA. *Animal Behaviour*, **54**, 559–570.

Hamilton, W.D. (1964). The genetical evolution of social behavior. I and II. *Journal of Theoretical Biology*, **7**, 17–52.

Harcourt, A.H., Purvis, A. & Liles, L. (1995). Sperm competition: mating system, not breeding season, affects testes size of primates. *Functional Ecology*, **9**, 468–476.

Jack, K.M. (2007). The cebines: toward an explanation of variable social structure. In *Primates in Perspective*, ed. C.J. Campbell, A. Fuentes, K.C. MacKinnon, M. Panger & S.K. Bearder. New York, NY: Oxford Univesity Press, pp. 107–122.

Janson, C. (2000). Primate socio-ecology: the end of a golden age. *Evolutionary Anthropology*, **9**, 73–86.

Janson, C. & Goldsmith, M.L. (1995). Predicting group size in primates: foraging costs and predation risks. *Behavioral Ecology and Sociobiology*, **6**, 326–336.

Kappeler, P.M. & van Schaik, C.P. (2002). Evolution of primate social systems. *International Journal of Primatology*, **23**, 707–740.

Kay, R.F., Plavcan, J.M., Glander, K.E., *et al.* (1988). Sexual selection and canine dimorphism in New World monkeys. *American Journal of Primatology*, **77**, 385–397.

Koenig, A., Beise, J., Chalise, M.K., *et al.* (1998). When females should contest for food: testing hypotheses about resource density, distribution, size, and quality with Hanuman langurs (*Presbytis entellus*). *Behavioral Ecology and Sociobiology*, **42**, 225–237.

Kowalewski, M.M. (2007). Patterns of affiliation and co-operation in howler monkeys: an alternative model to explain social organization in non-human primates. PhD thesis, University of Illinois.

Kowalewski, M.M. & Garber, P.A. (2010). Mating promiscuity and reproductive tactics in female black and gold howler monkeys (*Alouatta caraya*) inhabiting an island on the Parana River, Argentina. *American Journal of Primatology*, **72**, 734–748.

Kowalewski, M.M. & Zunino, G.E. (2004). Birth seasonality in *Alouatta caraya* in Northern Argentina. *International Journal of Primatology*, **25**, 383–400.

Lehman, S.M., Prince, W. & Mayor, M. (2001). Variations in group size in white-faced sakis (*Pithecia pithecia*): evidence for monogamy or seasonal congregations? *Neotropical Primates*, **9**, 96–101.

Masterson, T.J. & Hartwig, W.C. (1998). Degrees of sexual dimorphism in *Cebus* and other New World monkeys. *American Journal of Physical Anthropology*, **107**, 243–256.

Mitani, J.C. (2006). Reciprocal exchange in chimpanzees and other primates. In *Cooperation in Primates and Humans: Mechanisms and Evolution*, ed. P.M. Kappeler & C.P. van Schaik. Berlin: Springer Verlag, pp. 107–120.

Mitani, J.C., Gross-Louis, J. & Manson, J.H. (1996). Number of males in primate groups: comparative tests of competing hypotheses. *American Journal of Primatology*, **38**, 315–332.

Moynihan, M. (1976). *The New World Primates*. Princeton, NJ: Princeton University Press.

Noe, R. (2001). Biological markets: partner choice as the driving force behind the evolution of mutualisms. In *Economics in Nature: Social Dilemmas, Mate Choice and Biological Markets*, ed. R. Noe, J.A.R. A.M. van Hoof & P. Hammerstein. Cambridge: Cambridge University Press, pp. 93–118.

Norconk, M.A. (2006). Long-term study of group dynamics and female reproduction in Venezuelan *Pithecia pithecia*. *International Journal of Primatology*, **27**, 653–674.

Norconk, M.A. (2011). Sakis, uakaris, and titi monkeys: behavioral diversity in a radiation of primate seed predators In *Primates in Perspective* (2nd end), ed. C.J. Campbell, A. Fuentes, K.C. MacKinnon, S.K. Bearder & R.M. Stumpf. New York, NY: Oxford University Press, pp. 122–139.

Norconk, M.A. & Conklin-Brittain, N.L. (2004). Variation on frugivory: the diet of Venezuelan white-faced sakis. *International Journal of Primatology*, **25**, 1–26.

Nunn, C.L. (2000). Collective benefits, free-riders, and male extra-group conflict. In *Primate Males: Causes and Consequence of Variation in Group Composition*, ed. P.M. Kappeler. Cambridge: Cambridge University Press, pp. 192–204.

Peetz, A. (2001). Ecology and social organisation of the bearded saki, *Chiropotes satanas chiropotes* (Primates: Pitheciinae) in Venezuela. *Ecotropical Monographs*, **1**, 1–170.

Pereira, M.E., Clutton-Brock, T.H. & Kappeler, P.M. (2000). Understanding primate males. In: *Primate Males: Causes and Consequence of Variation in Group Composition*, ed. P.M. Kappeler. Cambridge: Cambridge University Press, pp. 271–277.

Peres, C.A. (1993). Notes on the ecology of buffy saki monkeys (*Pithecia albicans*, Gray 1860): a canopy seed predator. *American Journal of Primatology*, **31**, 129–140.

Peres, C.A. (1996). Use of space, spatial group structure, and foraging groups size of gray woolly monkeys (*Lagothrix lagotricha cana*) at Urucu, Brazil: a review of the Atelinae. In *Adaptive Radiations of Neotropical Primates*, ed. M.A. Norconk, A.L. Rosenberger & P.A. Garber. New York, NY: Plenum Press, pp. 467–488.

Pope, T.R. (2000). Reproductive success increases with degree of kinship in cooperative coalitions of female red howler monkeys (*Alouatta seniculus*). *Behavioral Ecology and Sociobiology*, **48**, 253–267.

Schneider, H. & Rosenberger, A.L. (1996). Molecules, morphology, and platyrrhine systematics. In *Adaptive Radiation of Neotropical Primates*, ed. M.A. Norconk, A.L. Rosenberger & P.A. Garber. New York, NY: Plenum Press, pp. 3–20.

Smuts, B.B. (1997). Social relationships and life histories of primates. In *The Evolving Female: A Life-History Perspective*, ed. M.E. Morbeck, A. Galloway & A.L. Zihlman. Princeton, NJ: Princeton University Press, pp. 60–68.

Snaith, T.V. & Chapman, C.A. (2005). Towards an ecological solution to the folivore paradox: patch depletion as an indicator of within-group scramble competition in red colobus monkeys (*Piliocolobus tephrosceles*). *Behavioral Ecology and Sociobiology*, **59**, 185–190.

Soini, P. (1986). A synecological study of a primate community in Pacaya–Samiria National Resere, Peru. *Primate Conservation*, **7**, 63–71.

Steenbeek, R. & van Schaik, C.P. (2001). Competition and group size in Thomas's langurs (*Presbytis thomasi*): the folivore paradox revisited. *Behavioral Ecology and Sociobiology*, **49**, 100–110.

Sterck, E.H.M., Watts, D.P. & van Schaik, C.P. (1997). The evolution of female social relationships in nonhuman primates. *Behavioral Ecology and Sociobiology*, **41**, 291–309.

Strier, K.B. (1992). Causes and consequences of nonaggression in the woolly spider monkey, or muriqui (*Brachyteles arachnoides*). In *Aggression and Peacefulness in Humans and Other Primates*, ed. J. Silverberg & J.P. Gray. New York, NY: Oxford University Press, pp. 100–116.

Strier, K.B. (1999). Why is female kin bonding so rare? Comparative sociality of neotropical primates. In *Comparative Primate Socioecology*, ed. P.C. Lee. Cambridge: Cambridge University Press, pp. 300–319.

Strier, K.B. (2000). From binding brotherhoods to short-term sovereignty: the dilemma of male Cebidae. In *Primate Males: Causes and Consequence of Variation in Group Composition*, ed. P.M. Kappeler. Cambridge: Cambridge University Press, pp. 72–83.

Sussman, R.W. & Garber, P.A. (2007). Cooperation and competition in primate social interactions. In *Primates in Perspective*, ed. C.J. Campbell, A. Fuentes, K.C. MacKinnon, M. Panger & S.K. Bearder. New York, NY: Oxford University Press, pp. 636–651.

Sussman, R.W. & Garber, P.A. (2011). Cooperation, collective action, and competition in primate social interaction. In *Primates in Perspective*, Vol. 2, ed. C.J. Campbell, A. Fuentes, K.C. MacKinnon, S.K. Bearder & R.M. Stumpf. New York, NY: Oxford University Press, pp. 587–598.

Symington, M.M. (1990). Fission–fusion social organization in *Ateles* and *Pan*. *International Journal of Primatology*, **11**, 47–61.

Trivers, R. (1971). The evolution of reciprocal altruism. *Quarterly Review of Biology*, **46**, 25–57.

van Roosmalen, M.G.M., Mittermeier, R.A. & Milton, K. (1981). The bearded sakis, genus *Chiropotes*. In *Ecology and*

Behavior of Neotropical Primates, Volume 1, ed. A.F. Coimbra-Filho & R.A. Mittermeier. Rio de Janeiro: Academia Brasileira de Ciências, pp. 419–442.

van Schaik, C.P. (1983). Why are diurnal primates living in groups? *Behaviour*, **87**, 120–144.

van Schaik, C.P. (1989). The ecology of social relationships amongst female primates. In *Comparative Socioecology. The Behavioural Ecology of Humans and Other Mammals*, ed. V. Standen & R.A. Foley. Oxford: Blackwell Scientific, pp. 195–218.

van Schaik, C.P. & Kappeler, P.M. (2006). Cooperation in primates and humans: closing the gap. In: *Cooperation in Primates and Humans: Mechanisms and Evolution*, ed. P.M. Kappeler & C.P. van Schaik. Berlin: Springer Verlag, pp. 3–21.

van Schaik, C.P., van Noordwijk, M.A. & Nunn, C.L. (1999). Sex and social evolution in primates. In *Comparative Primate Socioecology*, ed. P.C. Lee. Cambridge: Cambridge University Press, pp. 204–231.

Veiga, L.M. & Ferrari, S.F. (2006). Predation of arthropods by southern bearded sakis (*Chiropotes satanas*) in Eastern Brazilian Amazonia. *American Journal of Primatology*, **68**, 209–215.

Veiga, L.M. & Silva, S.S.B. (2005). Relatives or just good friends? Affiliative relationships among male southern bearded sakis (*Chiropotes satanas*). *Livro de Resumos, XI Congresso Brasileiro de Primatologia*, Porto Alegre: Sociedade Brasileira de Primatologia, p. 174.

Veiga, L.M., Pinto, L.P. & Ferrari, S.F. (2006). Fission–fusion sociality in bearded sakis (*Chiropotes albinasus* and *Chiropotes satanas*) in Brazilian Amazonia. *International Journal of Primatology*, **27**(Suppl 1), 224.

Wang, E. & Milton, K. (2003). Intragroup social relationships of male *Alouatta palliata* on Barro Colorado Island, Republic of Panama. *International Journal of Primatology*, **24**, 1227–1244.

Chapter 10

Evolutionary ecology of the pitheciinae: evidence for energetic equivalence or phylogenetically structured environmental variation?

Shawn M. Lehman

Introduction

A fundamental issue in evolutionary ecology is understanding interspecific variations in body size. Peters (1986) noted that this issue is important because body size correlates with a variety of important traits (e.g. abundance, lifespan and metabolism). Consequently, the relationship between body size and abundance has been widely recognized as an important area of research (Damuth 2007; White *et al.* 2007). For example, many researchers have noted a pattern of high densities for small-bodied organisms compared to low densities for large-bodied organisms in temperate and tropical habitats (Damuth 1981). This inverse relationship between population density and body size is referred to as the energetic equivalent rule (EER; Damuth 1981). The EER predicts that density (D) and body mass (M) are negatively correlated and will exhibit a $D : M^{-0.75}$ relationship. The EER ultimately expresses the relationship between body size and metabolic rates, which scales to the +0.75 power (Kleiber 1962). Large-bodied taxa tend to have higher energy requirements than small-bodied taxa (Peters 1986). As such, it is assumed that equal amounts of energy are available to each species in a community, which results in species' abundance being limited by its energetic requirements (Damuth 1981). Moreover, associated slope variances can provide insights into energy use and community dynamics. Griffiths (1986) suggested that if the slope of the $D : M$ relationship is greater than −0.75, then this pattern may indicate a disproportionately greater amount of energy controlled by larger individuals. Conversely, if the slope is less than −0.75, then a disproportionately greater amount of energy is controlled by smaller individuals. Despite considerable debate regarding the applicability of the EER and −0.75 slope to mammals (Blackburn *et al.* 1993), the model has been applied successfully to many primate taxa (Clutton-Brock & Harvey 1977; Ganzhorn 1999; Harcourt *et al.* 2005; Lehman 2007), including *Aotus* sp. and the Pitheciinae (Robinson *et al.* 1986). However, there is cause for re-evaluation of the EER in the Pitheciinae (*Callicebus*, *Pithecia*, *Chiropotes* and *Cacajao*) given recent advances in analytical techniques as well as new data on density estimates, body size, habitat use, and phylogenetic relationships in this taxon (e.g. Smith & Cheverud 2002; Garland *et al.* 2005; Opazo *et al.* 2006).

Previous investigations of $D : M$ relationships in primates have tended to use mean density estimates for species in a variety of habitats (Robinson *et al.* 1986), a method that has been termed a species average. Although this method may be acceptable for studies at a continental scale, there are issues with the use of species averages in investigations at smaller biogeographic scales (Silva & Downing 1995). For example, many Pitheciinae range into *terra firme* (non-flooded) and flooded forests (Barnett & Brandon-Jones 1997; Peres 1997; Haugaasen & Peres 2005a,b). There are considerable differences in forest structure, nutrient levels and fruit phenology between these habitats (Terborgh & Andresen 1998). Concomitantly, primate densities and biomass tend to be higher in flooded versus non-flooded forests in Amazonia (Peres 1997; Haugaasen & Peres 2005a,b). Pitheciini (*Pithecia*, *Chiropotes* and *Cacajao*) exploit seeds from unripe fruit during the high-water season in flooded forests, which may result in reduced competition with other frugivores seeking ripe fruit in this habitat (Ayres 1989). Conversely, Pitheciini compete with other animals for scarce, ripe fruits in *terra firme* forests. As a result, some Pitheciini supplement their frugivorous diet with insects and leaves when ranging into *terra firme* forests (Oliveira *et al.* 1985; Ayres 1986; Peres 1993; Norconk 1997). These spatial variations in foraging patterns and diet may influence energy use and density patterns in Pitheciini.

Understanding the effects of phylogeny on body size is critical in any study of interspecific allometry. As noted by Felsenstein (1985), statistical tests often require independence between data points, a requirement that can be violated when comparing taxa with close phylogenetic relationships. Phylogenetically independent contrasts (IC) is the standard method to control for phylogeny in comparative studies (e.g. Felsenstein 1985; Garland *et al.* 1992, 1993). These IC data are derived by calculating pairwise comparisons at all levels of a phylogeny, and using each branch only once. Resulting analyses of the transformed data can then be considered to be free of the confounding effects of phylogenetic

Evolutionary Biology and Conservation of Titis, Sakis and Uacaris, eds. Liza M. Veiga, Adrian A. Barnett, Stephen F. Ferrari and Marilyn A. Norconk. Published by Cambridge University Press.

history among taxa. Many earlier studies of $D : M$ relationships were unable – due to lack of analytical tools – to consider the confounding effects of phylogenetic relationships among the taxa being studied (Damuth 1981; Griffiths 1986). For example, Nunn and Barton (2000) tested Kleiber's Law by investigating metabolic needs and ranging patterns in primates. Although Nunn and Barton's (2000) interspecific allometry data produced slopes similar to the predicted +0.75 slope value, use of IC data revealed a higher than predicted slope value for the relationship between home range size and metabolic needs. The authors concluded that phylogenetic relatedness and interspecific variations in diet were important covariates influencing the results.

There is growing evidence that interspecific body size variations should not be discussed exclusively in terms of ecology or phylogenetic inertia (Lord *et al.* 1995). Desdevises *et al.* (2003) noted the value of testing for phylogenetically structured environmental variation (PSEV) in studies of mammalian phylogeny and ecology. PSEV can be defined as the shared attributes that related taxa have acquired because they tend to have occupied similar niches during their evolutionary history (Grafen 1989). Thus, IC data take into consideration the confounding effects of phylogeny whereas PSEV considers a portion of the phylogenetic variation of a trait that may be related to ecology (i.e. the common influence of phylogeny and ecology). For example, Wiens (2004) noted that some plants and animals tend to maintain their ancestral ecological niche (i.e. PSEV) and as a result, do not adapt to new environments. Consequently, PSEV can result in vicariance and the adaptive radiation of multiple taxa in spatially and temporally fluctuating habitats (Kozak *et al.* 2006). Although PSEV has received attention in biogeographic research on temperate plants and animals (Wiens & Donoghue 2004), the model has rarely been applied to studies of extant primates living in tropical ecosystems (Lehman 2007). Thus, the question arises as to the phylogenetic basis of Pitheciinae $D : M$ relationships in *terra firme* and flooded forests.

Because my chapter is among the first to investigate the evolutionary ecology of extant Pitheciinae in both *terra firme* and flooded forests, a-priori hypothesis testing was not conducted. Instead, I sought to answer the following questions: (1) do Pitheciinae $D : M$ relationships support the EER in *terra firme* and flooded forests, and (2) is there support for phylogenetic niche conservatism as an explanatory model for Pitheciinae evolutionary ecology?

Methods

Species-specific data on phylogenetic relationships, density (number of individuals/km^2) and body mass (kg) of Pitheciinae were collected from the literature (Figure 10.1, Table 10.1). I did not create IC data for species for which I could not obtain molecular phylogenetic data (i.e. Callicebiini; Kobayashi 1995) because of issues associated with phylogenies based only on morphological characters of platyrrhines (Corruccini 2001).

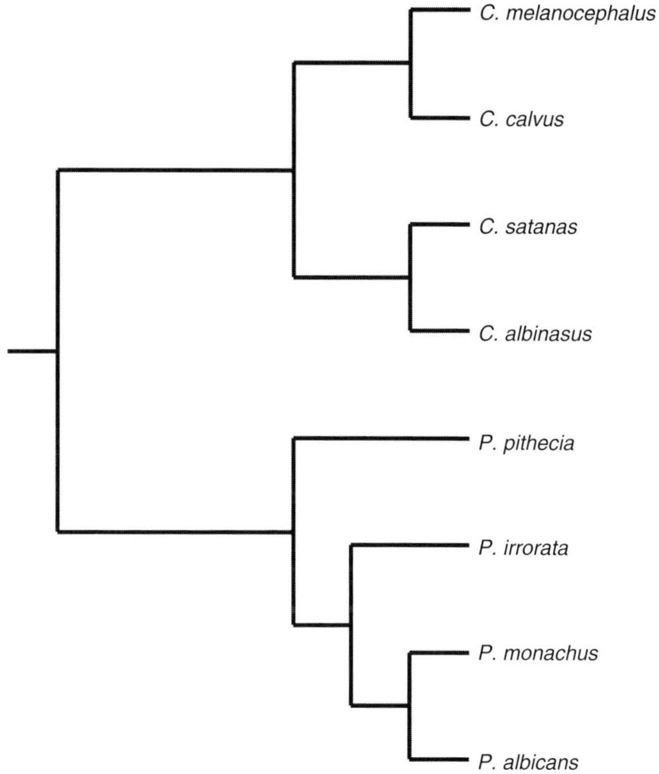

Figure 10.1 Phylogeny of the Pitheciini (*Chiropotes*, *Cacajao* and *Pithecia*) used in this study. Based on Bonvicino *et al.* (2003), Canavez *et al.* (1999) and Groves (2001).

Density estimates were selected only from comprehensive, longitudinal surveys rather than short-term, rapid assessments in *terra firme* and flooded forests. Thus, the density estimates used here represent population values rather than species averages. All variables were subject to natural logarithmic transformations to linearize relationships. I first used least-squares regression models of interspecific allometry data to investigate $D : M$ relationships at various taxonomic levels in the Pitheciinae. I then used IC data to determine $D : M$ relationships in the Pitheciini after taking phylogenetic relatedness into account. Pitheciini body size and ecological data were transformed to IC data using Mesquite software (Maddison & Maddison 2006). A default procedure was used in which all branch lengths were assigned the same length (Purvis *et al.* 1994). For all IC regressions, the slope of the regression equation was forced through the origin (Garland *et al.* 1992). These analyses provide information on Pitheciini $D : M$ relationships once phylogeny is taken into account.

Following Desdevises *et al.* (2003), I used a partitioning method to determine if Pitheciini body size variations reflect PSEV. This method partitions variation among the following four components: (a) a part strictly due to ecology, (b) PSEV, (c) a part strictly due to phylogeny, and (d) an unexplained component. For example, component (a) will represent the

Table 10.1 Body size and density estimates for 15 taxa of Pitheciinae in *terra firme* and flooded forests.

Taxa[1]	Mean body size (kg)	Density (no. individuals/km^2)	
		Terra firme forests	Flooded forests
Callicebus aureipalatii	0.950	6.20	–
Callicebus brunneus	0.830	4.84	–
Callicebus cupreus	1.070	6.00	14.62
Callicebus moloch	0.988	10.00	–
Callicebus nigrifons	1.350	3.50	–
Callicebus personatus	1.325	10.23	–
Callicebus torquatus	1.245	8.00	7.50
Cacajao calvus	3.165	17.00	26.00
Cacajao melanocephalus	2.935	12.00	12.00
Chiropotes albinasus	2.820	8.95	4.97
Chiropotes satanas	3.030	7.53	–
Pithecia albicans	3.000	9.40	4.10
Pithecia irrorata	2.160	9.20	13.30
Pithecia monachus	2.360	9.07	17.21
Pithecia pithecia	1.760	3.60	15.00

[1] Data from Branch 1983; Peres 1997; Smith & Jungers 1997; Wallace *et al.* 1998; Lehman 1999; Reed 1999; Chiarello 2000; Bennett *et al.* 2001; Defler 2001; Bernardo & Galetti 2004; Porter 2006.

relationship between ln(primate body mass) and ln(primate population density) after considering the effects of phylogeny. A complete description and theoretical explanation of variation partitioning can be found in Borcard *et al.* (1992) and Borcard and Legendre (1994). First, a least-squares regression model of $D : M$ in each habitat (*terra firme* and flooded) was conducted, retaining the r^2 coefficient for the components a + b. Second, a patristic distance matrix was derived for the phylogenetic tree using PDDIST software (Garland *et al.* 2002). A principal coordinate (PC) analysis was performed on this matrix using The R Package version 4.0 (Casgrain & Legendre 2000). A forward regression selection procedure was used to select the PC(s) that significantly contributed to the explanation of body size in each habitat. A multivariate regression conducted on the relationship between body size in each habitat and the PC(s) yielded an r^2 coefficient that represents the fraction b + c. A final multivariate regression of $D : M$ in each habitat and the PC(s) provided an r^2 coefficient equal to the fraction of a + b + c. Fractions a, b, c and d were obtained by subtraction: a = (a + b + c) – (b + c); b = (a + b) – (a); c = (b + c) – (b); and d = 1 – (a + b + c). Regressions were conducted using SPSS 14.0.

Results

Interspecific allometry for the Pitheciinae (*Callicebus*, *Cacajao*, *Chiropotes* and *Pithecia*) indicate that there is no significant correlation between density (dependent variable) and body size (independent variable) in either *terra firme* or flooded forests (Table 10.2). The slopes for the regression analyses are positive for *terra firme* and flooded forests. For the Callicebiini (*Callicebus*), there is no significant correlation between density and body size in *terra firme* forests. However, the slope of the best-fit line in the Callicebiini is greater than but not significantly different from the predicted value of –0.75 ($t = 0.132$, $P > 0.10$). Although body size is a significant determinant of Pitheciini (*Cacajao*, *Chiropotes* and *Pithecia*) density in *terra firme* forests, there is no significant correlation for the $D : M$ relationship in flooded forests. The negative slope for Pitheciini in flooded forests is less than but not significantly different from the value predicted by the EER ($t = 0.050$, $P > 0.10$).

For the IC data on Pitheciini, density remains a significant covariate of body size in *terra firme* forests. The slope of the $D : M$ regression is greater than and significantly different from the predicted value of –0.75 in *terra firme* forests ($t = 3.617$, $P < 0.01$). Although IC data for Pitheciini $D : M$ relationships in flooded forests is not statistically significant, the slope of the best-fit line is negative and is similar to the predicted slope value ($t = 0.635$, $P > 0.10$).

I then tested for PSEV in Pitheciini $D : M$ variations in *terra firme* and flooded forests (Figure 10.2). In *terra firme* forests, the r^2 value (0.592) from the interspecific allometry regression model of $D : M$ was retained as the component a + b. Only PC1 and PC5 from the principal component analysis of the patristic data matrix were retained by the forward regression model ($r^2 = 0.819$, $P < 0.05$). A final multivariate regression of density, PC1, and PC5 on Pitheciini body size yields the component a + b + c ($r^2 = 0.849$, $P < 0.05$). The resulting model indicates that the PSEV component (56.8%) comprises a larger proportion of Pitheciini body size variations than either density (2.4%) or phylogeny (25.1%) in *terra firme* forests.

For Pitheciini in flooded forests, the r^2 value (0.068) from the interspecific allometry regression model of density and body size was retained as the component a + b. Two principal components (PC1 and PC5) were again retained for the component b + c ($r^2 = 0.819$, $P < 0.05$). The final regression of PCs and density on body size resulted in the component a + b + c ($r^2 = 0.833$, $P > 0.05$). After solving for each component, the final partitioning model reveals that phylogeny (76.5%) makes up a larger proportion of body size variations than either density (1.4%) or PSEV (5.4%) in flooded forests.

Discusssion

My first question was to determine if Pitheciinae $D : M$ relationships support the EER in *terra firme* and flooded forests. $D : M$ relationships in the Callicebiini provide support for the EER in *terra firme* forests and for Pitheciini in flooded forests.

Table 10.2 Least-squares regression models of the relationship between Pitheciinae ln density (dependent variable) and ln body size (independent variable) in *terra firme* and flooded forests.

Taxon	Habitat[1]	Analysis[2]	r^2	*b*	a^3	*n*	*P*
Pitheciinae	T	IA	0.200	0.417 (0.241)	1.823 (0.181)	13	0.109
	F	IA	0.001	0.046 (0.557)	2.466 (0.458)	8	0.628
Callicebiini	T	IA	0.049	−0.581 (1.284)	2.007 (0.248)	5	0.674
	F	IA	–	–	–	–	–
Pitheciini	T	IA	0.592	1.637 (0.555)	0.615 (0.542)	8	0.026
	F	IA	0.068	−0.817 (1.350)	3.178 (1.292)	7	0.571
	T	IC	0.500	1.583 (0.645)	0.000	7	0.049
	F	IC	0.193	−1.785 (1.630)	0.000	6	0.323

[1] T, *terra firme* forests; F, flooded forests; IA, interspecific allometry; IC, independent contrasts; r^2, correlation coefficient; *b*, slope (standard error); *a*, intercept (standard error); *n*, sample size; *P*, significance value.
[2] Incomplete *Callicebus* phylogeny precludes IC analyses for Pitheciinae and Callicebiini, and insufficient sample size for regression analysis of Callicebiini in flooded forests.
[3] Intercept is always 0.000 for IC data.

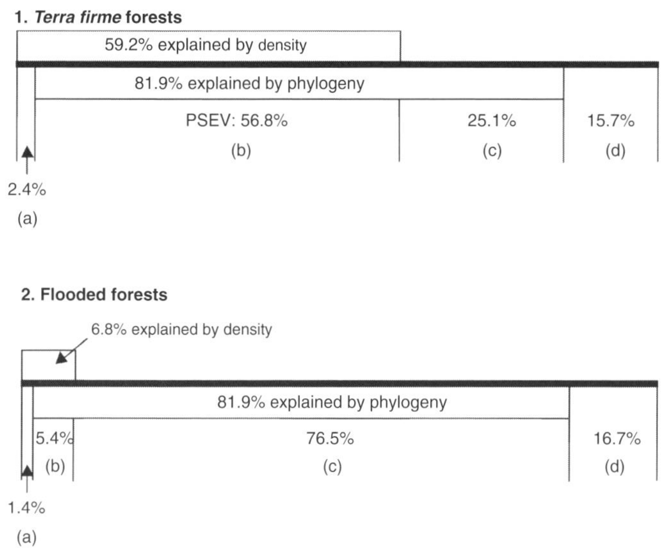

Figure 10.2 Pitheciini (*Chiropotes*, *Cacajao* and *Pithecia*) body size variations (thick horizontal line) partitioned among density and phylogeny in *terra firme* forests (1) and flooded forests (2). Individual components are as follows: (a) a part strictly due to ecology, (b) phylogenetically structured environmental variation (PSEV), (c) a part strictly due to phylogeny, and (d) unexplained.

However, slope orientations and values were similar to those predicted by the EER only for Callicebiini in *terra firme* forests and for Pitheciini in flooded forests. Griffiths (1986) suggested that a slope greater than the predicted value of −0.75 indicates a disproportionately greater amount of energy controlled by larger individuals. The general patterns that emerge from the least squares models are that large-bodied *Callicebus* control more energy than relatively small-bodied congenera in *terra firme* forests. For Pitheciini in flooded forests, large-bodied species are controlling more energy than small-bodied species. Although there are few field studies of the effects of body size on interspecific competition in *Callicebus* and the Pitheciini, Basset and Angelis (2007) used a spatially explicit simulation model to explain body size mediated coexistence and competition by consumers competing for resources in a habitat. The model predicts that a small consumer will exploit patches more efficiently, but search for profitable patches less effectively than a larger competitor. Body size related constraints should, therefore, result in niche partitioning. Consequently, a large animal should control resource dynamics in its habitat, which results in competitive coexistence with other, smaller-boded taxa. Although the model is individually based, it does have intriguing possibilities for hypothesis testing in future studies of the socioecology of Pitheciinae.

The question arises then as to the reasons for slope variances in Pitheciini $D : M$ relationships between *terra firme* and flooded forests (i.e. positive slopes in *terra firme* forests and negative slopes in flooded forests). These variances are due to the considerable differences in feeding ecology for Pitheciini between *terra firme* and flooded forests. Pitheciini tend to exploit insects, leaves and flowers during the low-water period in *terra firme* forests, and then shift their diet to fruit seeds during the high-water period in flooded forests (Ayres 1989; Barnett & Brandon-Jones 1997; Boubli 1997). Fruit resources are particularly relevant to the EER because they contain more easily extractable energy, in the form of sugars, than either leaves or insects (Dasilva 1992; Milton 1999; Wasserman & Chapman 2003). There are also concomitant changes in Pitheciini day ranges, group cohesion and activity patterns between these habitats, which will influence detection probabilities and density estimates. Ayres (1989) found that *Chiropotes albinasus* forages in cohesive multimale/multifemale groups in *terra firme* forests, whereas groups of *Cacajao calvus* form flexible subunits as they forage for fruit from young seeds in flooded forests. Furthermore, the overall positive slope values for Pitheciini in *terra firme* and flooded forests has been observed in other mammalian trophic groups

(Silva & Downing 1995; Marquet & Taper 1998). For example, Silva and Downing (1995) obtained positive slopes for $D : M$ relationships in medium-sized (1–10 kg range), seed-eating mammals. This size range and diet category applies to the Pitheciini, which are also medium-sized seed predators. Although there are few data on energy availability and use in Pitheciinae, studies of other mammalian communities have revealed that resources are not always equally available across body sizes (Silva *et al.* 1997; Morgan Ernest 2005). Therefore, the slope variations for Pitheciini in *terra firme* and flooded forests may be a direct result of different energy use and expenditures for animals between these habitats.

My second question was to determine if there is support for PSEV as an explanatory model for Pitheciinae evolutionary ecology. Due to limited sample sizes, it is only possible to suggest that *terra firme* forests represent an ancestral ecological niche for the Pitheciini (*Cacajao*, *Chiropotes* and *Pithecia*). In other words, selective pressures operating on Pitheciini in *terra firme* forests are likely influenced by phylogeny, the environment, and by competition from other taxa present in this habitat. If closely related taxa within the Pitheciini had completed the invasion of *terra firme* forests, and these taxa and their descendants had remained in similar environments with similar selective pressures, then stabilizing selection might have selected for densities and body sizes in extant taxa that approximated that of the ancestral condition (Harvey & Pagel 1991). Phylogenetic evidence for stabilizing selection can be found when basal branch lengths are longer than terminal (species) branch lengths because phenotypic changes are small as a result of constraints (Diniz-Filho 2004). Although there are few data on species-specific branch lengths in the Pitheciinae, phylogenetic studies provide some support for stabilizing selection in this taxon. Opazo *et al.* (2006) analyzed Pitheciinae nuclear genes and found that divergence times were longer for basal branches compared to terminal (species) branches. Stabilizing selection has also been noted as a strong force operating on other Neotropical mammals (Peterson *et al.* 1999; Smith *et al.* 2004; Lemos *et al.* 2005). For example, Smith *et al.* (2004) found that mammalian body size ranges were particularly constrained in South America. They concluded that the observed body size patterns were best explained by strong stabilizing selection. There is also paleontological evidence of stable body size for Pitheciinae compared to other fossil platyrrhines. Specifically, there is no paleontological evidence for fossil Pitheciinae with body sizes larger than extant taxa in this subfamily (Kay *et al.* 1998) although giant (*ca.* 25 kg) primate taxa closely related to spider monkeys existed in the Pleistocene (Hartwig & Cartelle 1996).

If stabilizing selection has selected for intermediate body sizes in the Pitheciini, then it is necessary to identify the constraints at each end of the body size spectrum. Individuals at a smaller than average body size tend to have low fat reserves and are more at risk of starvation (Cuthill & Houston 1997; Altmann & Alberts 2003a,b), particularly in environments like flooded forests that experience seasonal variations in water levels and fruit availability. For example, the large day ranges in *Cacajao melanocephalus* (maximum of 4400 m/day) results in considerable energy expenditure (Boubli 1997), which places strong selective pressures on females to maintain adequate body size because low maternal body size is associated with reduced fertility (Altmann & Alberts 2003b). Although individuals with larger than average body size are better able to deal with prolonged resource fluctuations (McNamara & Houston 1990), these individuals incur higher predation pressures (reviewed in Ekman 2004). Hart (2000) conducted a meta-analysis of predation pressures on primates and found high rates of avian predation on primates in the Neotropics. Furthermore, large-bodied individuals would require more energy, which would result in increased travel distances for taxa living in large groups (e.g. *Cacajao*, *Chiropotes*).

In flooded forests, the evolutionary ecology of Pitheciini body size is better explained as the result of phylogeny rather than density or PSEV (synergistic ecological and phylogenetic effects). One possible explanation for lack of support for PSEV and density as covariates to Pitheciini body size in flooded forests is reduced interspecific competition with mammals for seeds from fruit in this habitat. As noted by Ayres (1986), large macaws are the only competitors faced by *Cacajao* for seeds from young fruits in flooded forests. There is low species diversity of primates and non-primate mammals in flooded forests (Haugaasen & Peres 2005a,b), further reducing competitive interactions for fruit in this habitat. Moreover, the addition of data on habitat related variations in diet for each primate species would enable use of multivariate rather than univariate models (Carbone *et al.* 2007; Lehman 2007).

It is important to note methodological issues with investigating the evolutionary ecology of body size in the Pitheciinae. For the models used here, small sample sizes combined with lack of data on density estimates and body size for many species of Pitheciinae indicate that analyses specific to this taxon should be viewed with caution. Furthermore, resolution of the Callicebiini phylogeny and accurate information on branch lengths within this taxon are needed to ensure that any analyses of ecological relationships are free of the confounding effects of phylogeny. I used a phylogenetic model in which all branch lengths were considered to be equal, which implies a speciational model of evolution. However, there are few genetic data supporting a speciational model of evolution (Smith & Cheverud 2002). These methodological issues may be mitigated to some degree by the fact that stabilizing selection is better understood via interspecific allometry methods rather than IC methods (Harvey & Pagel 1991; Smith & Cheverud 2002). With the addition of accurate data on branch lengths and other ecological variables, different evolutionary models can be evaluated for Pitheciine $D : M$ relationships as well as statistical methods specific to testing for stabilizing selection (Hansen 1997).

Acknowledgments

I thank Adrian Barnett, Liza Veiga, Stephen Ferrari and Marilyn Norconk for their kind invitation to participate in this edited volume. I am grateful to Robert Sussman and Linda Taylor for supporting my initial research interests in pitheciines, Vicki Funk and Carol of the Biological Diversity of the Guiana Shield Program at the Smithsonian Institution for supporting my research in Guyana, Richard Smith for his advice on independent contrasts, and Marilyn Norconk and the reviewer for excellent comments on a previous draft of the manuscript. My research is supported by The University of Toronto and a Discovery Grant from the Natural Sciences and Engineering Research Council of Canada.

References

Altmann, J. & Alberts, S.C. (2003a). Intraspecific variability in fertility and offspring survival in a non-human primate: behavioral control of ecological and social sources. In *Offspring: Human Fertility Behavior in a Biodemographic Perspective*, ed. K.W. Wachter & R.A. Bulatao. Washington, DC: National Academy Press, pp. 140–169.

Altmann, J. & Alberts, S.C. (2003b). Variability in reproductive success viewed from a life-history perspective in baboons. *American Journal of Human Biology*, **15**(3), 401–409.

Ayres, J.M. (1986). Uakaries and Amazonian flooded forests. Unpublished PhD dissertation, Cambridge University.

Ayres, J.M. (1989). Comparative feeding ecology of the uakari and bearded saki, *Cacajao* and *Chiropotes*. *Journal of Human Evolution*, **18**(7), 697–716.

Barnett, A.A. & Brandon-Jones, D. (1997). The ecology, biogeography, and conservation of the uakaris *Cacajao* (Pitheciinae). *Folia Primatologica*, **68**, 223–235.

Basset, A. & Angelis, D.L. (2007). Body size mediated coexistence of consumers competing for resources in space. *Oikos*, **16**(8), 1363–1377.

Bennett, C.L., Leonard, S. & Carter, S. (2001). Abundance, diversity, and patterns of distribution of primates on the Tapiche River in Amazonian Peru. *American Journal of Primatology*, **54**(2), 119–126.

Bernardo, C.S. & Galetti, M. (2004). Density and population size of primates in a forest fragment in southeastern Brazil. *Revista Brasileira de Zoologia*, **21**(4), 827–832.

Blackburn, T.M., Brown, V.K., Doube, B.M., *et al.* (1993). The relationship between abundance and body size in natural animal assemblages. *The Journal of Animal Ecology*, **62**(3), 519–528.

Bonvicino, C.R., Boubli, J.P., Otazú, I.B., *et al.* (2003). Morphologic, karyotypic, and molecular evidence of a new form of *Chiropotes* (primates, pitheciinae). *American Journal of Primatology*, **61**(3), 123–133.

Borcard, D. & Legendre, P. (1994). Environmental control and spatial structure in ecological communities: an example using oribatid mites (Acari, Oribatei). *Environmental and Ecological Statistics*, **1**(1), 37–61.

Borcard, D., Legendre, P. & Drapeau, P. (1992). Partialling out the spatial component of ecological variation. *Ecology*, **73**(3), 1045–1055.

Boubli, J.P. (1997). Ecology of the black uakari monkey, *Cacajao melanocephalus melanocephalus*, in Pico da Neblina National Park, Brazil. Unpublished PhD dissertation, University of California, Berkeley.

Branch, L.C. (1983). Seasonal and habitat differences in the abundance of primates in the Amazon (Tapajos) National Park, Brazil. *Primates*, **24**(3), 424–431.

Canavez, F.C., Moreira, M.A.M., Ladasky, J.J., *et al.* (1999). Molecular phylogeny of New World primates (Platyrrhini) based on beta 2-microglobulin DNA sequences. *Molecular Phylogenetics and Evolution*, **12**(1), 74–82.

Carbone, C., Rowcliffe, J.M., Cowlishaw, G., *et al.* (2007). The scaling of abundance in consumers and their resources: implications for the energy equivalence rule. *American Naturalist*, **170**(3), 479–484.

Casgrain, P. & Legendre, P. (2000). *The R package for multivariate and spatial analysis*. Montreal: University of Montreal.

Chiarello, A. (2000). Conservation value of a native forest fragment in a region of extensive agriculture. *Revista Brasileira de Biologia*, **60**(2), 237–247.

Clutton-Brock, T.H. & Harvey, P.H. (1977). Primate ecology and social organization. *Journal of Zoology (London)*, **183**, 1–39.

Corruccini, R.S. (2001). Confidence intervals for morphology-based cladistic trees: a platyrrhine phylogeny test case. *International Journal of Primatology*, **22**(6), 1007–1019.

Cuthill, I.C. & Houston, A. (1997). Managing time and energy. In *Behavioural Ecology: An Evolutionary Approach*, ed. J.R. Krebs & N.B. Davies. Oxford: Blackwell Scientific, pp. 97–120.

Damuth, J. (1981). Population density and body size in mammals. *Nature*, **290**, 699–700.

Damuth, J. (2007). A macroevolutionary explanation for energy equivalence in the scaling of body size and population density. *American Naturalist*, **169**(5), 621–631.

Dasilva, G.L. (1992). The western black-and-white colobus as a low-energy strategist: activity budget, energy expenditure and energy intake. *Journal of Animal Ecology*, **61**, 79–91.

Defler, T.R. (2001). *Cacajao melanocephalus ouakary* densities on the lower Apaporis River, Colombian Amazon. *Primate Report*, **61**, 31–36.

Desdevises, Y., Legendre, P., Azouzi, L., *et al.* (2003). Quantifying phylogenetically structured environmental variation. *Evolution*, **57**(11), 2647–2652.

Diniz-Filho, A.J.F. (2004). Phylogenetic diversity and conservation priorities under distinct models of phenotypic evolution. *Conservation Biology*, **18**(3), 698–704.

Ekman, J. (2004). Mass-dependence in the predation risk of unequal competitors. *Oikos*, **105**(1), 109–116.

Felsenstein, J. (1985). Phylogenies and the comparative method. *American Naturalist*, **125**, 1–15.

Ganzhorn, J.U. (1999). Body mass, competition and the structure of primate communities. In *Primate Communities*, ed. J.G. Fleagle, C.H. Janson & K. Reed. Cambridge: Cambridge University Press, pp. 141–157.

Garland, T., Jr., Bennett, A.F. & Rezende, E.L. (2005). Phylogenetic approaches in comparative physiology. *Journal of Experimental Biology*, **208**, 3015–3035.

Garland, T., Jr., Dickerman, A.W., Janis, C.M., *et al.* (1993). Phylogenetic analysis of covariance by computer simulation. *Systematic Biology*, **42**, 265–292.

Garland, T., Jr., Harvey, P.H. & Ives, A.R. (1992). Procedures for the analysis of comparative data using phylogenetically independent contrasts. *Systematic Biology*, **41**, 18–32.

Garland, T., Jr., Midford, P.E. & Jones, J.A. (2002). PDAP: Phenotypic Diversity Analysis Programs. Riverside, Department of Biology, University of California-Riverside.

Grafen, A. (1989). The phylogenetic regression. *Philosophical Transactions of the Royal Society London (Series B)*, **326**, 119–157.

Griffiths, D. (1986). Size–abundance relations in communities. *American Naturalist*, **127**(2), 140–166.

Groves, C.P. (2001). *Primate Taxonomy.* Washington, DC: Smithsonian Institution Press.

Hansen, T.F. (1997). Stabilizing selection and the comparative analysis of adaptation. *Evolution*, **51**(5), 1341–1351.

Harcourt, A.H., Coppeto, S.A. & Parks, S.A. (2005). The distribution–abundance (density) relationship: its form and causes in a tropical mammal order, Primates. *Journal of Biogeography*, **32**(4), 565–579.

Hart, D. (2000). Primates as prey: ecological, morphological and behavioral relationships between primate species and their predators. Unpublished PhD dissertation, Washington University.

Hartwig, W.C. & Cartelle, C. (1996). A complete skeleton of the giant South American primate *Protopithecus*. *Nature*, **381**(6580), 307–311.

Harvey, P.H. & Pagel, M.D. (1991). *The Comparative Method in Evolutionary Biology*. London: Oxford University Press.

Haugaasen, T. & Peres, C.A. (2005a). Mammal assemblage structure in Amazonian flooded and unflooded forests. *Journal of Tropical Ecology*, **21**, 133–145.

Haugaasen, T. & Peres, C.A. (2005b). Primate assemblage structure in Amazonian flooded and unflooded forests. *American Journal of Primatology*, **67**(2), 243–258.

Kay, R.F., Johnson, D. & Meldrum, D.J. (1998). A new pitheciin primate from the middle miocene of Argentina. *American Journal of Primatology*, **45**(4), 317–336.

Kleiber, N. (1962). *The Fire of Life: An Introduction to Animal Energetics*. New York, NY: Wiley.

Kobayashi, S. (1995). A phylogenetic study of titi monkeys, genus *Callicebus*, based on cranial measurements: I. Phyletic groups of *Callicebus*. *Primates*, **36**(1), 101–120.

Kozak, K.H., Weisrock, D.W. & Larson, A. (2006). Rapid lineage accumulation in a non-adaptive radiation: phylogenetic analysis of diversification rates in eastern North American woodland salamanders (Plethodontidae: Plethodon). *Proceedings: Biological Sciences*, **273**(1586), 539–546.

Lehman, S.M. (1999). Biogeography of the primates of Guyana. Unpublished PhD dissertation, Washington University.

Lehman, S.M. (2007). Ecological and phylogenetic correlates to body size in the Indriidae. *International Journal of Primatology*, **28**(1), 183–210.

Lemos, B., Meiklejohn, C.D., Caceres, M. & Hartl, D.L. (2005). Rates of divergence in gene expression profiles of primates, mice, and flies: stabilizing selection and variability among functional categories. *Evolution*, **59**, 126–137.

Lord, J.M., Westoby, M. & Leishman, M.R. (1995). Seed size and phylogeny in six temperate floras: constraints, niche conservatism and adaptation. *American Naturalist*, **146**, 349–364.

Maddison, W.P. & Maddison, D.R. (2006). Mesquite: a modular system for evolutionary analysis. Version 2.75. http://mesquiteproject.org

Marquet, P.A. & Taper, M.L. (1998). On size and area: patterns of mammalian body size extremes across landmasses. *Evolutionary Ecology*, **12**, 127–139.

McNamara, J.M. & Houston, A.I. (1990). The value of fat reserves and the trade-off between starvation and predation. *Acta Biotheoretica*, **38**, 37–61.

Milton, K. (1999). Nutritional characteristics of wild primate foods: do the diets of our closest living relatives have lessons for us? *Nutrition*, **15**(6), 488–498.

Morgan Ernest, S.K. (2005). Body size, energy use, and community structure of small mammals. *Ecology*, **86**, 1407–1413.

Norconk, M.A. (1997). Seasonal variations in the diets of white-faced and bearded sakis (*Pithecia pithecia* and *Chiropotes satanas*) in Guri Lake, Venezuela. In *Adaptive Radiations of Neotropical Primates*, ed. M.A. Norconk, A.L. Rosenberger & P.A. Garber. New York, NY: Plenum Press, pp. 403–426.

Nunn, C.L. & Barton, R.A. (2000). Allometric slopes and independent contrasts: a comparative test of Kleiber's Law in primate ranging patterns. *American Naturalist*, **156**(5), 519–533.

Oliveira, J.M.S., Lima, J.G., Bonvicino, C., *et al.* (1985). Preliminary notes of the ecology and behavior of the Guianan saki (*Pithecia pithecia*, Linnaeus 1766; Cebidae, Primates). *Acta Amazonica*, **15** (1–2), 249–263.

Opazo, J.C., Wildman, D.E., Prychitko, T., *et al.* (2006). Phylogenetic relationships and divergence times among New World monkeys (Platyrrhini, Primates). *Molecular Phylogenetics and Evolution*, **40**(1), 274–280.

Peres, C.A. (1993). Notes on the ecology of buffy saki monkeys (*Pithecia albicans*, Gray 1860): a canopy seed-predator. *American Journal of Primatology*, **31**, 129–140.

Peres, C.A. (1997). Primate community structure at twenty western Amazonian flooded and unflooded forests. *Journal of Tropical Ecology*, **13**, 381–405.

Peters, R.H. (1986). *The Ecological Implications of Body Size*. Cambridge: Cambridge University Press.

Peterson, A.T., Soberon, J. & Sanchez-Cordero, V. (1999). Conservatism of ecological niches in evolutionary time. *Science*, **285**, 1265–1267.

Porter, L.M. (2006). Distribution and density of *Callimico goeldii* in the department of Pando, Bolivia. *American Journal of Primatology*, **68**(3), 235–243.

Purvis, A., Gittleman, J.L. & Luh, H.-K. (1994). Truth or consequences: effects of phylogenetic accuracy on 2 comparative methods. *Journal of Theoretical Biology*, **167**(3), 293–300.

Reed, K.E. (1999). Population density of primates in communities: differences in community structure. In *Primate Communities*, ed. J.G. Fleagle, C.H. Janson & K.E. Reed. Cambridge: Cambridge University Press, pp. 116–140.

Robinson, J.G., Wright, P.C. & Kinzey, W.G. (1986). Monogamous cebids and their relatives: intergroup calls and spacing. In *Primate Societies*, ed. B.B. Smuts, D.L.

Cheney, R.M. Seyfarth, R.W. Wrangham & T.T. Struhsaker. Chicago, IL: University of Chicago Press, pp. 44–53.

Silva, M. & Downing, J.A. (1995). The allometric scaling of density and body mass: a nonlinear relationship for terrestrial mammals. *American Naturalist*, **145**, 704–727.

Silva, M., Brown, J.H. & Downing, J.A. (1997). Differences in population density and energy use between birds and mammals: a macroecological perspective. *The Journal of Animal Ecology*, **66**(3), 327–340.

Smith, F.A., Brown, J.H., Haskell, J.P., *et al.* (2004). Similarity of mammalian body size across the taxonomic hierarchy and across space and time. *American Naturalist*, **163**(5), 672–691.

Smith, R.J. & Cheverud, J.M. (2002). Scaling of sexual dimorphism in body mass: a phylogenetic analysis of Rensch's Rule in primates. *International Journal of Primatology*, **23**(5), 1195–1135.

Smith, R.J. & Jungers, W.L. (1997). Body mass in comparative primatology. *Journal of Human Evolution*, **32**, 523–559.

Terborgh, J. & Andresen, E. (1998). The composition of Amazonian forests: patterns at local and regional scales. *Journal of Tropical Ecology*, **14**, 645–664.

Wallace, R.B., Painter, R.L.E. & Taber, A.B. (1998). Primate diversity, habitat preferences, and population density estimates in Noel Kempff Mercado National Park, Santa Cruz Department, Bolivia. *American Journal of Primatology*, **46**(3), 197–211.

Wasserman, M.D. & Chapman, C.A. (2003). Determinants of colobus monkey abundance: the importance of food energy, protein, and fiber content. *Journal of Animal Ecology*, **72**, 650–659.

White, E.P., Ernest, S.K., Kerkhoff, A.J., *et al.* (2007). Relationships between body size and abundance in ecology. *Trends in Ecology & Evolution*, **22**(6), 323–330.

Wiens, J.J. (2004). Speciation and ecology revisited: phylogenetic niche conservatism and the origin of species. *Evolution*, **58**(1), 193–197.

Wiens, J.J. & Donoghue, M.J. (2004). Historical biogeography, ecology and species richness. *Trends in Ecology & Evolution*, **19**(12), 639–644.

Chapter 11

Competition between pitheciines and large *Ara* macaws, two specialist seed-eaters

Suzanne Palminteri, George Powell, Krista Adamek & Raul Tupayachi

Introduction

Interspecific competition may arise from individuals of different species either exhibiting aggressive or other damaging behavior (interference competition; see Schoener 1983) or sharing the same limited resources (exploitation competition; Connell 1983; Schoener 1983). Interspecific competition among sympatric primates appears to be variable but in general relatively minor, due in part to changes in preferred foods during times of fruit scarcity (Peres 1994; Wahungu 1998; Stevenson *et al.* 2000) or use of different canopy heights (Terborgh 1983; Peres 1991).

Gautier-Hion *et al.* (1985) observed considerable overlap in the fruit diets of taxonomically distinct vertebrate groups in a Gabonese forest community, while Poulsen *et al.* (2002) found that hornbills and primates in Cameroon each showed greater dietary overlap within their respective taxonomic groups than between hornbill–primate pairings.

Pitheciines' specialization on hard, well-protected and often immature seeds minimizes dietary competition with other primates (Soini 1987; Ayres 1989; Kinzey & Norconk 1993; Peres 1993; Aquino & Encarnación 1999; Barnett *et al.* 2005). However, the importance of competition with other sympatric seed predators remains largely unexplored. Based on data collected in Venezuela, Norconk *et al.* (1997) suggested that pitheciines may compete with large macaws (*Ara* spp.), another group that specializes on hard, often unripe, seeds (Gilardi *et al.* 1999; Powell *et al.* 1999; Berg *et al.* 2007). More recently, Barnett *et al.* (2005) documented feeding by scarlet macaws (*A. macao*) in Brazil on fruits that were also eaten by golden-backed uakaris, *Cacajao melanocephalus ouakary*.

Up to three species of large macaws – *Ara chloropterus*, *A. ararauna* and *A. macao* – are sympatric with pitheciines across much of the Amazon biome (Patterson *et al.* 2005; Ridgely *et al.* 2007). We examined the extent to which the diets of *Pithecia* and large *Ara* macaws overlap and whether the overlap varies with respect to availability of preferred foods, to indicate the degree of competition between these bird and primate taxa.

Methods

Study site

Data on diet, behavior and habitat use of *Pithecia* and *Ara* were collected concurrently in the Los Amigos Conservation Concession, 140,000 ha of largely intact evergreen seasonal forest (Osher & Buol 1998) at 250 m a.m.s.l. in southeastern Peru (Figure 11.1). Most of the annual 2200 mm of rain falls between November and May (Pitman 2005).

Foraging data collection

Between June 2004 and December 2005, we recorded feeding patterns of 5 habituated groups of bald-faced sakis (*Pithecia irrorata*) during 5-min instantaneous scans, monitoring each group for approximately 5 days each month. A separate field team collected macaw feeding data between January 2004 and December 2005 from observations of radio-tagged birds and birds seen on foraging walks each month. Each visit by a group or individual of either species to a food tree, regardless of the number of individuals or length of the observation, was considered one feeding bout. For each feeding bout, we recorded: plant species, plant part consumed (seed, mesocarp, whole fruit, flower, leaf), and maturity of fruits or seeds consumed. Non-plant feeding bouts were ignored.

Diet analysis

Feeding bouts representing 18 months of observations served as the unit of analysis. For this analysis, we grouped the three *Ara* species, considering all as potential competitors to the seed-eating sakis. We compared the richness and diversity of saki and macaw diets using standard indices. We calculated the generic richness of each taxon's diet using the Margalef (1958) index:

$$R' = (G - 1)/\ln(n)$$

where G is the total number of plant genera consumed by each consumer taxon, and n is the number of feeding records for

Evolutionary Biology and Conservation of Titis, Sakis and Uacaris, eds. Liza M. Veiga, Adrian A. Barnett, Stephen F. Ferrari and Marilyn A. Norconk. Published by Cambridge University Press.

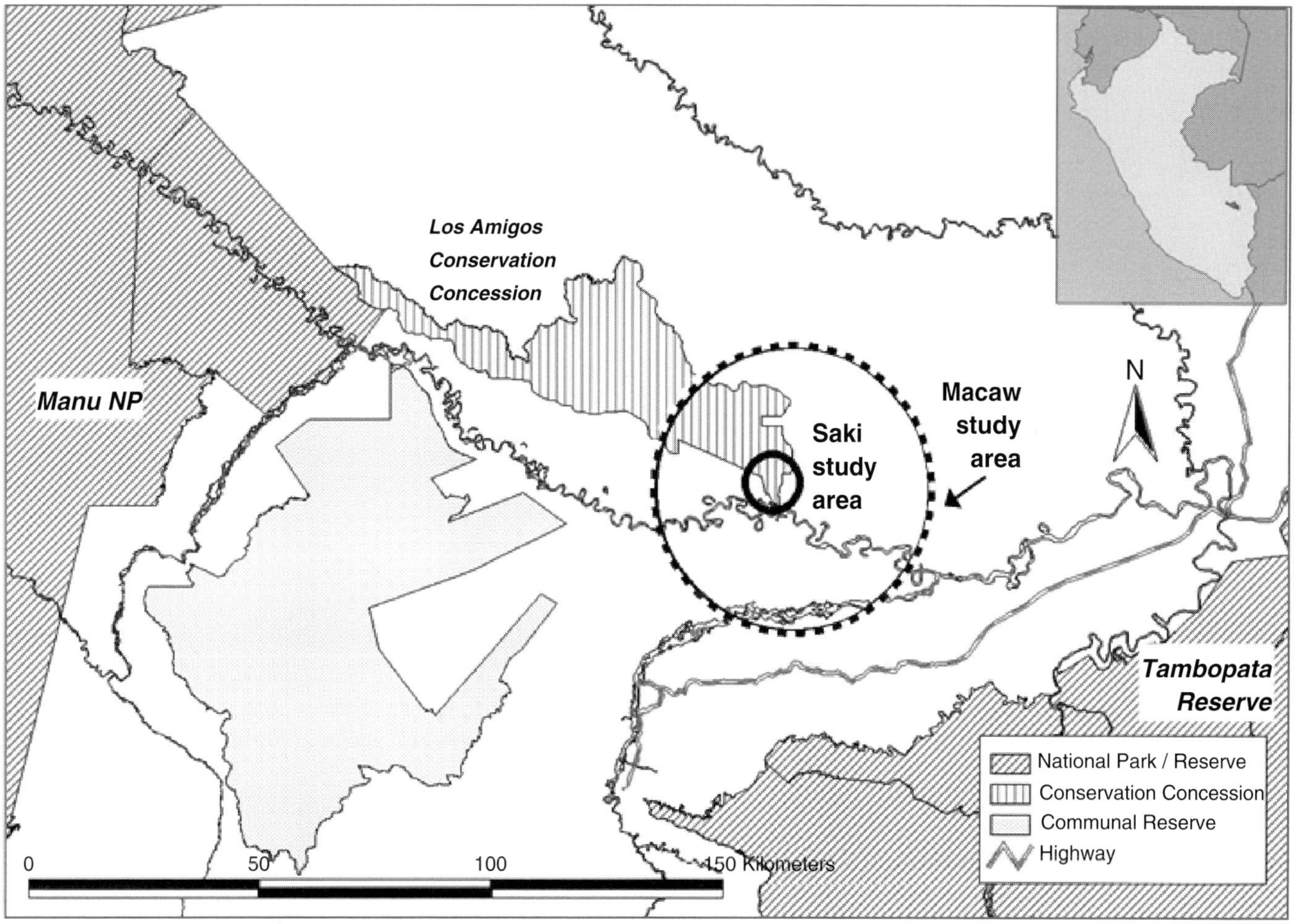

Figure 11.1 Study area for saki and macaw studies in Madre de Dios, southeastern Peru.

each genus. We calculated the diversity of food plant genera in each taxon's diet with the Shannon index (Shannon 1948):

$$H' = -\sum(p_i)(lnp_i)$$

where H' is the diversity index and p is proportion of individuals of the sample belonging to the ith genus.

To analyze seasonal variation in diet, while correcting for variation in monthly sampling effort, we compared the number of genera consumed per month as a function of the number of feeding observations that month.

Fruit abundance and availability

To determine whether the relative availability of food resources throughout the year influenced seasonal dietary overlap, we recorded the phenology of 839 marked trees, representing 116 plant genera, from February 2005 to December 2006. We averaged each month's data to produce a single 12-month cycle. Each month, we estimated the percentage of maximum production of flowers, immature fruits and mature fruits in up to five individuals of each plant species (Fournier 1974). Based on these percentage values, we assigned each tree a monthly value of 0–5 for each plant reproductive part. We used t-tests to compare monthly fruit production values of immature and mature fruits for all 116 plant genera and to compare immature fruit production among preferred saki and macaw foods. See Photo 11.1.

Dietary overlap

We generated monthly and overall indices of dietary overlap between sakis and macaws using Schoener's (1974) resource overlap index:

$$R_o = 1 - \tfrac{1}{2}\sum_{i=1}^{n}\left|p_{ij} - p_{ik}\right|$$

where R_o is resource overlap, and p_{ij} and p_{ik} are the proportions of the observations in which species j and k consume resource i. The resulting value ranges from 0 to 1, with 0 representing no overlap and 1 representing complete overlap (Poulsen *et al.* 2002). We also analyzed separately the 20 plant genera eaten most frequently by each consumer group to determine whether the species' preferred food items overlapped differently from their overall diets. We analyzed

Photo 11.1 Psittacines (macaws and parrots) and pitheciines are all predators of immature seeds, frequently of hard-husked fruit. Bald-faced sakis and other pitheciines use their: enlarged canine teeth, while blue and yellow macaws (*Ara ararauna*) and other large *Ara* species use their strong bills, to open and eat the seeds of hard-husked fruit. Photo: Krista Adamek.

monthly dietary overlap values for the plant genera with the greatest overall *Pithecia–Ara* overlap.

To compare our results to those of Stevenson *et al.* (2000) of dietary overlap among four Colombian primates, we also calculated Morisita's overlap index (Morisita 1959, cited in Horn 1966). In Morisita's index:

$$O = 2\sum(x_i y_i) \Big/ \left(\sum x_i^2 + \sum y_i^2\right)$$

x_i is the proportion of food genus i in the diet of animal species x, y_i is the proportion of food genus i in the diet of animal species y, and the sum includes all fruit genera that are consumed by both x and y. This value also varies between 0 and 1, with higher values indicating greater overlap. We compared monthly dietary overlap with immature and mature fruit production values using Pearson's correlations and linear regression.

Results

Dietary characteristics

We analyzed 1171 feeding bouts involving consumption of plant parts by bald-faced sakis ($n = 585$) and macaws ($n = 586$). Together, fruits and seeds comprised 98% (575) and 88% (516) of the feeding bouts of sakis and macaws, respectively (Figure 11.2). Seeds alone made up 83% of the feeding bouts of sakis and 68% of the feeding bouts of macaws. While seeds played an important role in saki and macaw diets throughout the year, pulp was taken in over 20% of saki feeding bouts in February and March, the time when it was least consumed by macaws. Overall, flowers comprised 5% of macaws' diet but just 0.5% of the sakis' diet. Macaws consumed both flowers and leaves in the dry season (primarily April–September); with flowers contributing nearly 30% to their diet in June and leaves comprising 23% of their diet in July.

During the study, we observed sakis and macaws consuming parts of plants of 109 genera in 43 families (Appendix). Sakis' diet included more plant families (37), genera (66) and morphospecies (at least 118) than macaws (27 families, 64 genera and at least 104 morphospecies), and scores for overall dietary richness (Margalef 1958), diversity and evenness (Shannon 1948) were all higher for sakis (Table 11.1).

Fruit abundance and availability

Sakis and macaws typically ate the seeds of immature fruits and the pulp of mature fruits. Immature fruit was most abundant from November to January, the initial months of rainy season, and least abundant in April, at the end of the rainy season (Figure 11.3). Mature fruit production remained below 4% of maximum potential throughout the year, except for a peak in production in the late rainy season of February and March. Immature fruit abundance was significantly higher than mature fruit abundance in all months ($t = -11.106$, df $= 22$, $p < 0.0001$). Several important food species of the Euphorbiaceae (*Hevea*), Sapotaceae (*Pouteria*) and Fabaceae (*Inga*) families reach their highest immature fruit production in early rainy season (November–January), while production of several species of Moraceae (*Pseudolmedia, Castilla*) peaked in September and October. Three palm genera (*Mauritia, Iriartea, Socratea*) showed a constant fruit production throughout the year.

The 20 preferred food plant genera of each consumer (32 total genera) accounted for 79% and 76% of all feeding bouts recorded for *Pithecia* and *Ara*, respectively (see Appendix 11.1). Only

Table 11.1 Dietary diversity of sakis (*Pithecia irrorata*) and macaws (*Ara* spp.) in Madre de Dios, Peru, 2004–2005.

	Families (unique to taxon)	Genera (unique to taxon)	Generic richness[1]	Generic diversity[2]	Generic evenness[2]	No. genera in 50% of obs	No. genera in 75% of obs
Sakis	37 (16)	66[b] (45)	10.36	3.36	0.80	6	15
Macaws	27 (6)	64[b] (43)	10.04	3.11	0.75	5	16
Total	43[a]	109[b]					

[a]Known families; [b]109 genera with known families + 1 additional category called "family unknown". [1]Margalef (1958). [2]Shannon (1948).

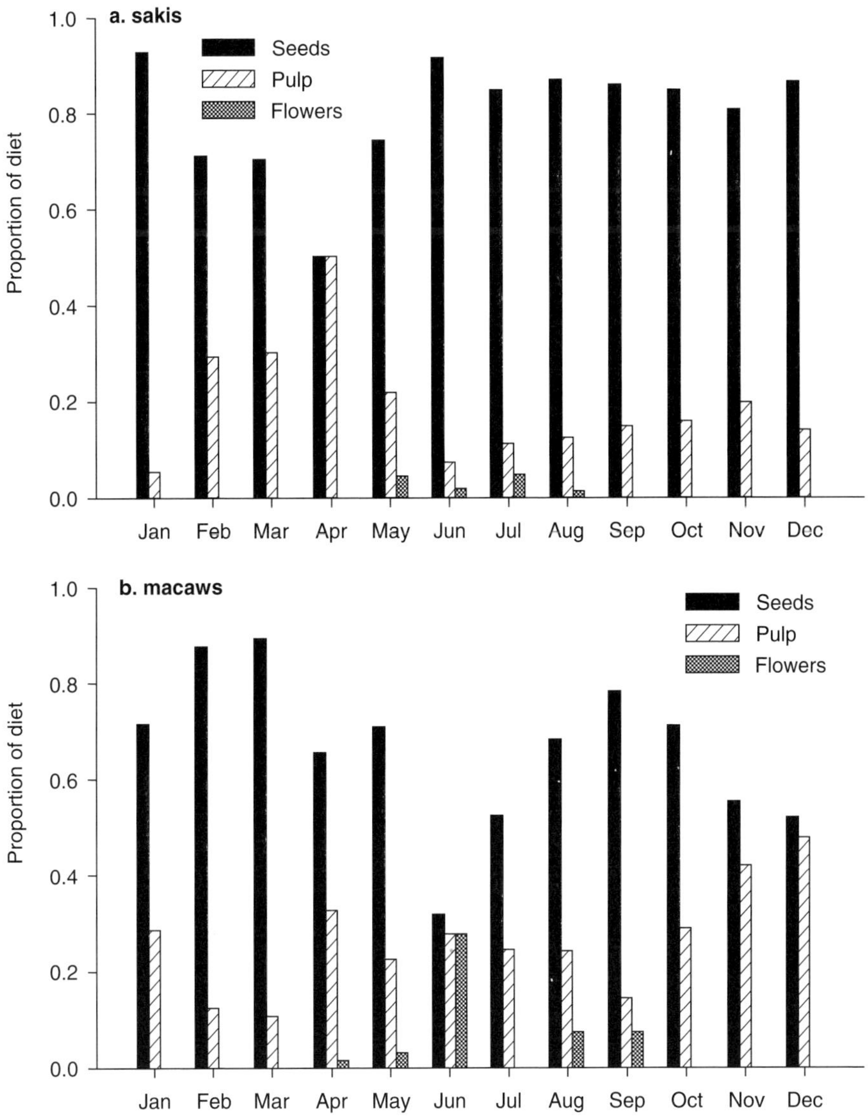

Figure 11.2 Proportion of seeds, pulp (including whole fruits) and flowers in the monthly diets of *P. irrorata* (a) and *Ara* macaws (b), 2004–2005. Leaves comprised an additional 5% of macaws' overall diet (not shown).

six (18%) of these genera were preferred foods of both taxa. Mean levels of immature fruit production among preferred saki food plant genera were similar to those among preferred macaw food plant genera ($t = 1.750$, df $= 38$, $p = 0.088$).

Dietary overlap

At the generic level, the diet of *Pithecia* showed little resemblance to that of *Ara* spp. (see Appendix 11.1). Of the 109 genera consumed by these two consumers, only 21 (19%), each

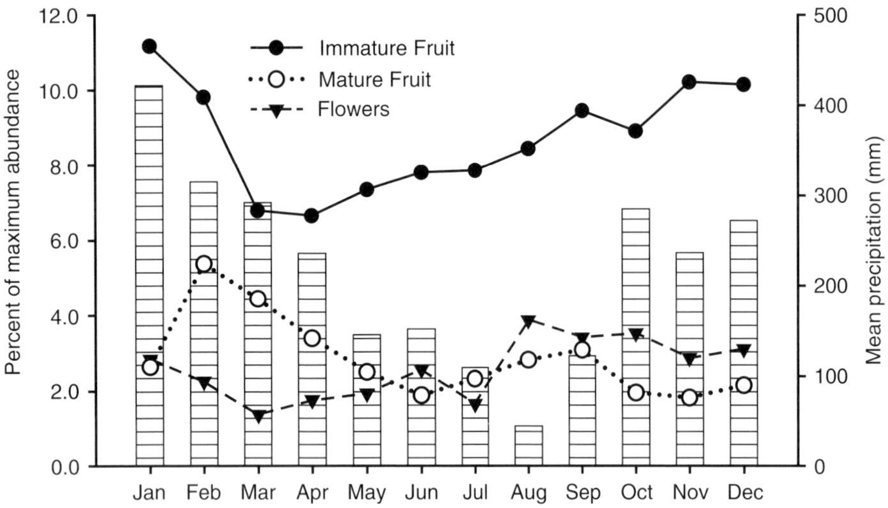

Figure 11.3 Monthly precipitation and mean phenological cycle for flowers and fruits from 839 marked trees from 116 plant genera in Madre de Dios, southeastern Peru, 2005–2006.

representing from 1 to 12 species, were consumed by both (Table 11.2). Schoener's (1974) resource overlap values of 0.244, calculated for the complete diets, and 0.238, calculated for the 20 genera most frequently consumed by each study animal, indicate a less than 25% dietary overlap between sakis and macaws during this period.

Individual monthly overlap values ranged from a low of 0.032 in May to a high of 0.253 in September (Figure 11.4). With the exception of September, all individual monthly overlap values (Schoener 1974), as well as their mean (0.147 ± 0.07 SD), were lower than the annual overlap value of 0.244. Dietary overlap values were lowest during the changeover from wet to dry season (April–June) and highest during the changeover from dry to wet season (September–January). Monthly overlap values correlated significantly with mean production values of immature fruit ($r = 0.778$, $p = 0.003$, $n = 12$) but not of mature fruit ($r = -0.200$, $p = 0.534$, $n = 12$). The pattern of monthly overlap values obtained using Morisita's (1959) index was similar to that obtained using Schoener's index (Figure 11.4). The overall Morisita index value was 0.195, and the mean monthly dietary overlap measure was 0.152 ± 0.11 SD.

Sakis consumed foods of more plant genera in April and May (Figure 11.5), when dietary overlap with macaws was lowest, than at the end of the dry season (September–November), when overlap was highest. The higher dietary diversity in April and May explains part of the decrease in dietary overlap in that period ($r^2 = 0.692$, $p = 0.0008$, $\mathrm{df} = 11$). Macaws' dietary diversity did not correlate with overlap value ($r = -0.283$, $p = 0.372$, $n = 12$). Use of particular genera may have contributed to lower overlap values in certain months. Species of four genera– *Mauritia*, *Sloanea*, *Matisia* and Sapotaceae sp. – made up over 60% of sakis' diet in May. These same genera were absent from macaws' diet in May, but present in their diet during other months. Conversely, *Bertholletia* and *Euterpe* contributed 58% and 12%, respectively, to macaws' diet in May, while neither genus contained foods eaten by sakis. Even in September, the month with the highest overlap, 30% of the macaws' observations were of a genus of Fabaceae (*Parkia*) not eaten by sakis.

The six genera that showed the greatest average overlap values (Schoener 1974; Figure 11.6) were consumed at different times of the year and in different amounts by *Pithecia* and *Ara*. For example, in September, *Pseudolmedia* represented over 36% of the diet of *Pithecia*, but just 7% of that of *Ara* (Figure 11.6a). Levels of *Inga* consumption by macaws and sakis were similar from June to September but diverged over the rest of the year (Figure 11.6b) as production of immature and mature *Inga* fruit increased. Consumption of three genera – *Mauritia*, *Pouteria* and *Eschweilera* – was concentrated in key months for one consumer and spread across the year by the other (Figures 11.6c–11.6e). Finally, *Pithecia* and *Ara* ate *Sloanea* in different months, with no temporal overlap (Figure 11.6f).

Interspecific interactions

During over 1800 h of following sakis, we never recorded their feeding in the same tree with macaws. On two occasions, we watched macaws either leave a feeding tree or wait nearby until a saki group finished feeding and left the tree, implying avoidance of interference competition (*sensu* Case & Gilpin 1974; Schoener 1983). During this same period, the team studying macaws recorded one incidence of *Pithecia* and *A. ararauna* feeding interspersed in the same tree. Thus, in almost 2000 combined hours of observation, we have observed only three potential incidences of interference competition between *Pithecia* and *Ara*.

Discussion

Dietary characteristics

The contribution of seeds (83% of feeding bouts) to the diet of *P. irrorata* in southeastern Peru was higher than that of *P. pithecia* (61%) and *Chiropotes satanas* (75%) in Venezuela (Kinzey & Norconk 1993), *P. albicans* (45%) in Brazil (Peres 1993) and *P. monachus* in northern Peru (3%: Happel 1982; and 40%: Soini 1987).

Our results in southeastern Peru indicate that while macaws generally ate more different parts of plants, sakis maintained a taxonomically more varied diet than the macaws,

Table 11.2 Twenty-one food plant genera shared by *Pithecia* and *Ara*, with number of species and months consumed by each animal taxon.

Family	Genus	No. species		Key months consumed	
		Sakis	Macaws	Sakis	Macaws
Fabaceae	*Acacia*	2	1	Jun–Aug, Oct	Sept
Combretaceae	*Buchenavia*	1	1	Aug	Jul
Sterculiaceae	*Byttneria*	2	2	May, Jul–Sep	Jul–Aug
Cucurbitaceae	*Cayaponia*	2	3	Jun, Aug–Sep	Jul–Aug
Chrysobalanaceae	*Couepia*	2	1	Sep–Dec	Oct
Lecythidaceae	*Eschweilera*	3	5	Aug, Oct–Nov, Jan	Jan–Feb, Apr, Jun–Nov
Euphorbiaceae	*Hevea*	1	1	Mar, Jul–Aug	May–Aug, Nov–Mar
Euphorbiaceae	*Hura*	1	1	Jun	Mar, May–Jun
Fabaceae	*Inga*	12	5	Jun–Apr	Mar–Jan
Arecaceae	*Iriartea*	1	1	May	Mar–Aug, Nov
Malvaceae	*Matisia*	1	1	May–Jun	Jul, Oct–Nov
Arecaceae	*Mauritia*	1	1	Oct–Aug	Feb, Apr, Jun–Aug, Oct–Nov
Cecropiaceae	*Pourouma*	6	1	Jun–Sep, Dec	Nov
Sapotaceae	*Pouteria*	9	7	Feb, Jun–Dec	Apr, Jul, Nov–Dec
Moraceae	*Pseudolmedia*	3	3	May–Dec	Sep, Nov
Celastraceae	*Salacia*	1	1	Jul–Aug	Nov
Elaeocarpaceae	*Sloanea*	1	2	May–Jun	Aug, Nov–Dec
Arecaceae	*Socratea*	1	1	Aug	May
Bignoniaceae	Sp	5*	1	Jun–Oct, Dec	Sep
Fabaceae	Sp	1*	4	Oct–Nov	May, Jul–Sep
Vochysiaceae	*Vochysia*	1	1	Sep	Jul, Sep

*Identification to genus level is ongoing in these families.

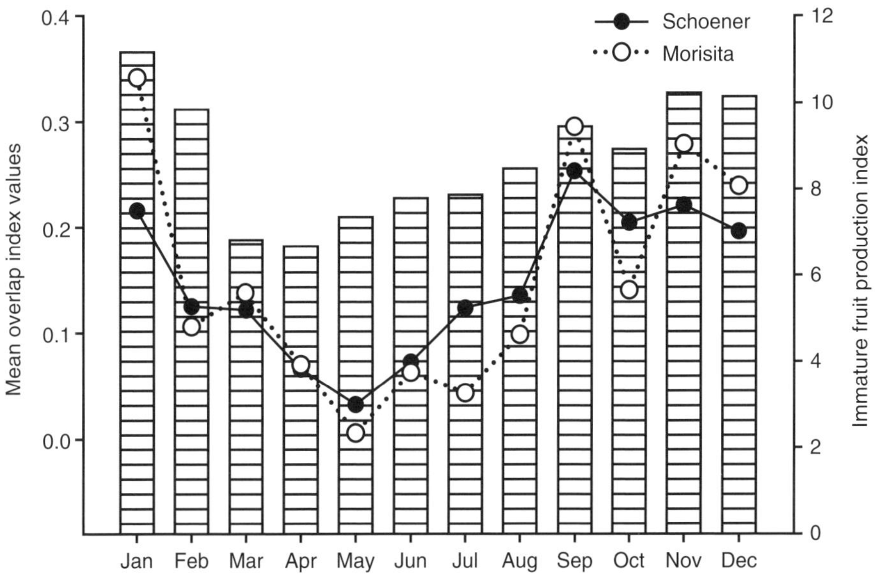

Figure 11.4 Mean monthly dietary overlap values (Schoener 1974; Morisita, cited in Horn, 1966) between sakis and macaws, 2004–2005. 0.0, no overlap; 1.0, complete overlap for both indices. Immature fruit production index is shown for comparison.

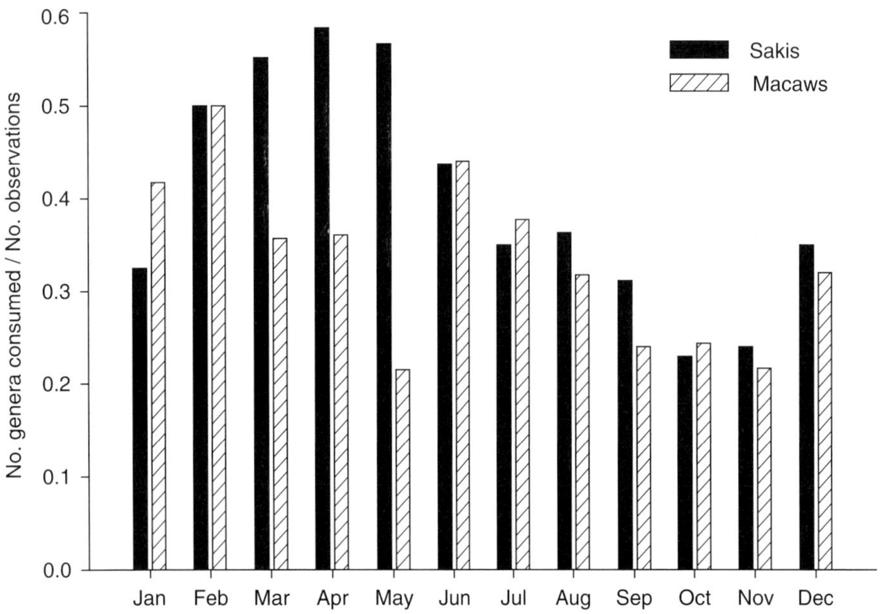

Figure 11.5 Number of food plant genera consumed monthly by sakis and macaws in 2004–2005, standardized by number of feeding observations per month.

particularly during the period of lowest immature fruit production. They fed on more plant families, genera and species than did the macaws, and their scores for richness, diversity and evenness were all higher than corresponding indices for macaws.

Dietary overlap: interference competition

Our observations suggest that interference competition is virtually non-existent between the two study taxa, with three interactions seen in over 2000 h of observation.

Dietary overlap: exploitation competition in southeastern Peru

While months of higher dietary overlap tended to be those with greater production of immature fruit, months with lower overlap tended to be those with high production of fruit of a few preferred plant genera. Quantities of fruit production among the preferred saki food plant genera and the preferred macaw food plant genera were similar, suggesting that differences in diet between sakis and macaws were not a function of relative availability of immature fruit.

The use of food plant genera, rather than species, in the analysis may overestimate potential competition. In some cases, it is possible that the two consumers were eating different species of the same genus in a given time period. Monthly variation in dietary overlap and the variable relationship between overlap and fruit production further complicate evaluation of the level of exploitation competition between these two taxa.

Dietary overlap: regional exploitation competition

In their 3-month study of dietary overlap between a Pitheciine, *Chiropotes satanas*, and an *Ara*, *A. chloropterus*, Norconk *et al.* (1997) provide qualitative data from a forest remnant in Venezuela's recently flooded Guri Lake. While six of the seven food plant species recorded in the diet of *A. chloropterus* were also eaten by *C. satanas*, over half of the 17 species of plants consumed by *C. satanas* were not consumed by *A. chloropterus*. While sharing of food species in a short period suggests competition, the extent and seasonality of dietary overlap were not quantified, due the limited number of observations of *Ara*.

Despite the substantial shift in plant species composition across the Amazon basin (ter Steege *et al.* 2006) and the resulting divergence in food plants consumed, Pitheciine monkeys and *Ara* macaws in Venezuela and Peru both focus their feeding on hard, unripe seeds. Calculation and comparison of dietary overlap for *C. satanas* (or *P. pithecia*) and *A. chloropterus* in Venezuela at geographically opposite ends of the Amazon basin would provide an interesting look at how a foraging guild adapts to variations in food resources.

Bearing in mind possible overestimation from generic-level plant identification, our overlap value was intermediate when compared with those of sympatric primates in Colombia. Our overall value for Morisita's overlap index (0.195) was greater than those for pairings of *Alouatta seniculus* with *Cebus apella*, *Lagothrix lagotricha*, or *Ateles belzebuth* (all less than 0.15, Stevenson *et al.* 2000) but less than those between pairings of species with more frugivorous diets: *C. apella*, *L. lagotricha*, and *A. belzebuth* (each overlap > 0.20). Our overlap values for three individual months (January, September and November) fell within the average range of overlap among even these more frugivorous primates, though during certain biweekly periods, Stevenson *et al.* (2000) reported primate–primate overlap values almost three times greater than the highest saki/macaw values we observed. Interspecific competition among sympatric primates may be both more variable and potentially much greater than that across vertebrate orders.

Monthly overlap values (after Schoener 1974) between primates and hornbills in Cameroon ranged from 0.025 and 0.350 (Poulsen *et al.* 2002). The mean monthly primate–hornbill overlap value of 0.159 is not significantly different

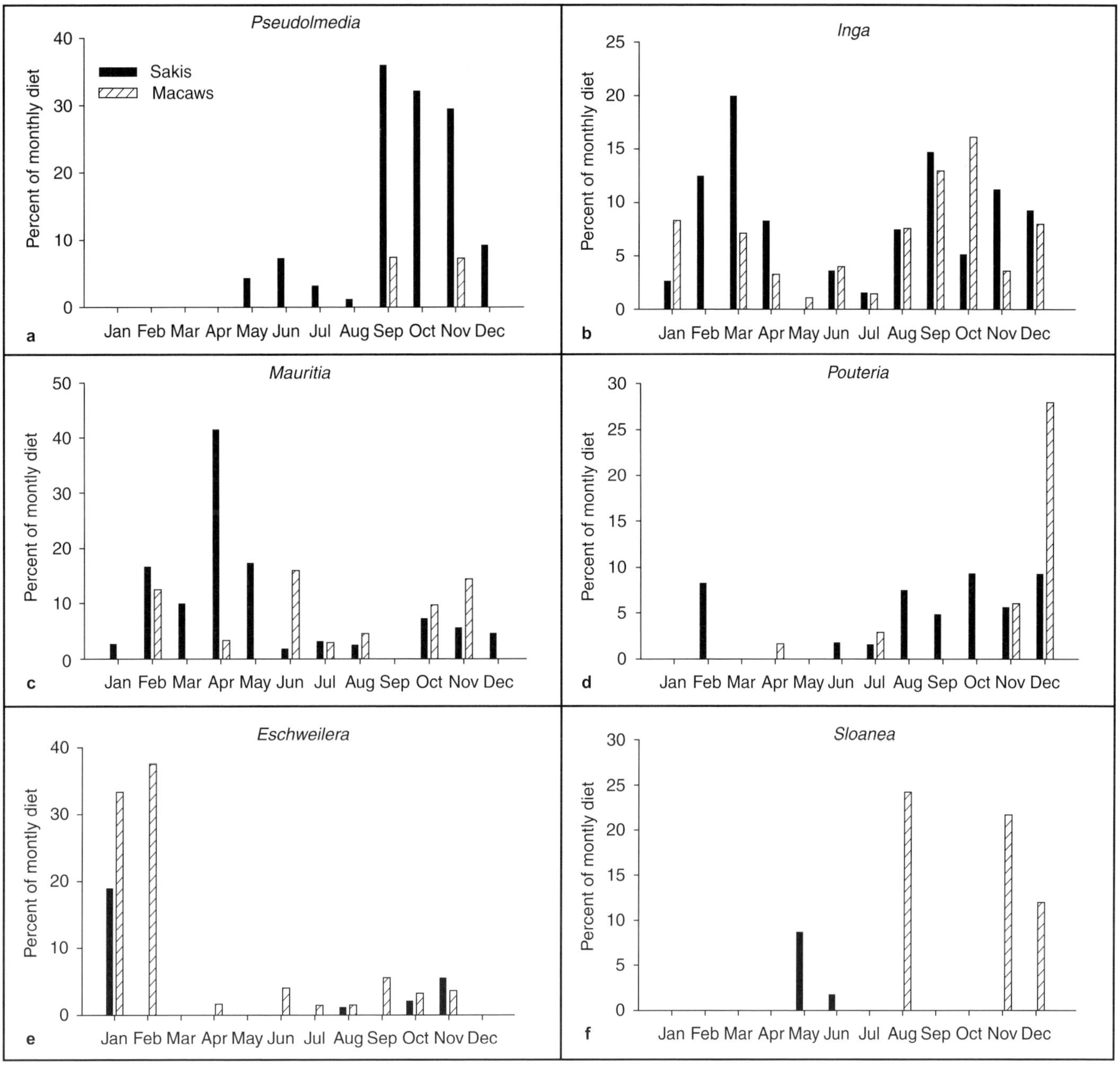

Figure 11.6 Percent contribution to monthly diets of sakis and macaws for the six genera (a–f) of greatest dietary overlap, 2004–2005.

from the 0.147 mean monthly overlap value obtained with the same index for sakis and macaws in our study ($t = -0.297$, $p = 0.769$, df $= 221$). Poulsen *et al.* (2002) concluded that, despite some 36 plant species consumed by both hornbills and primates in Cameroon, actual dietary overlap between these groups was low due to differences in fruit characteristics, proportion of food plants in each consumer's diet, and canopy height frequented by each group. While we could not obtain precise foraging height data for the macaws, we observed that they tended to feed on the outside of the canopy at the ends of branches (G. Powell, pers. obs.), while *Pithecia* tended to feed within the canopy (S. Palminteri, pers. obs.).

Poulsen *et al.* (2002) suggested that primates are more limited in their mobility than frugivorous birds and thus may be under greater pressure to diversify their diet. In southeastern Peru, sakis did maintain a taxonomically more diverse diet than macaws. Radio-tagged macaws had far larger home ranges, traveling up to 50 km from their nest site, including a region devoid of sakis (G. Powell *et al.*, unpubl. data), demonstrating their ability to track the location and production of fruit over far greater distances than *Pithecia*.

We have found that these two seed predators, *P. irrorata* and *Ara* macaws, live sympatrically with limited dietary overlap. Evidence that their respective diets are influenced by competition is equivocal. They rarely ate fruit and seeds from trees of the same genera at the same time, in the same proportions. At the generic level, the availability of fruits and seeds was not the main source of this variation in consumption patterns. Plant identification to species level will improve this analysis; additional research on the density and/or population size of sakis and macaws in this region, and their seasonal changes due to macaw migrations, will also help to determine the extent to which consumption of food resources by one taxon affects their availability to the other.

Acknowledgments

This study was supported by The Gordon and Betty Moore Foundation, through a grant to World Wildlife Fund-US. Donald Brightsmith and Marilyn Norconk provided helpful comments on the manuscript. We thank Mario Albites, Ulises Aleman, Rufo Bustamante, Julio Ccoycosi, Edgard Collado, Juan Alberto Escudero, Segundo Imunda, Pedro Maceda, Paola Martinez, Ivan Sandoval, and Silver Sandoval and a host of volunteers for dedicated field assistance and plant identification, and Cintya Verastegui and Arnaldo Marquez for facilitating logistics.

References

Aquino, R. & Encarnación, F. (1999). Observaciones preliminares sobre la dieta de *Cacajao calvus ucayali* en el Nor-oriente Peruano. *Neotropical Primates*, 7 (1), 1–5.

Ayres, J.M. (1989). Comparative feeding ecology of the uakari and bearded saki, *Cacajao* and *Chiropotes*. *Journal of Human Evolution*, **18**, 697–716.

Barnett, A., Volkmar de Castilho, C., Shapley, R., *et al.* (2005). Diet, habitat selection and natural history of *Cacajao melanocephalus ouakary* in Jau National Park, Brazil. *International Journal of Primatology*, **26**(4), 949–969.

Berg, K., Socola, J. & Angel, R. (2007). Great green macaws and the annual cycle of their food plants in Ecuador. *Journal of Field Ornithology*, **78**(1), 1–10.

Case, T. & Gilpin, M. (1974). Interference competition and niche theory. *Proceedings of the National Academy of Sciences*, **71**, 3073–3077.

Connell, J. (1983). On the prevalence and relative importance of interspecific competition: evidence from field experiments. *The American Naturalist*, **122**, 661–696.

Fournier, L.A. (1974). Un método cuantitativo para la medición de características fenológicas en árboles. *Turrialba*, **24**, 422–423.

Gautier-Hion, A., Duplantier, J.-M., Quris, R., *et al.* (1985). Fruit characters as a basis of fruit choice and seed dispersal in a tropical forest vertebrate community. *Oecologia (Berlin)*, **65**, 324–337.

Gilardi, J., Duffey, S., Munn, C., *et al.* (1999). Biochemical functions of geophagy in parrots: detoxification of dietary toxins and cytoprotective effects. *Journal of Chemical Ecology*, **25**, 897–922.

Happel, R. (1982). Ecology of *Pithecia hirsuta* in Peru. *Journal of Human Evolution*, **11**, 581–590.

Horn, H. (1966). Measurement of "overlap" in comparative ecological studies. *The American Naturalist*, **100**(914), 419–424.

Kinzey, W. & Norconk, M. (1993). Physical and chemical properties of fruit and seeds eaten by *Pithecia* and *Chiropotes* in Surinam and Venezuela. *International Journal of Primatology*, **14**(2), 207–227.

Margalef, R. (1958). Information theory in ecology. *General Systems*, **3**, 36–71.

Norconk, M., Wertis, C., & Kinzey, W. (1997). Seed predation by monkeys and macaws in eastern Venezuela: preliminary findings. *Primates*, **38**(2), 177–184.

Osher, L.J. & Buol, S.W.(1998). Relationship of soil properties to parent material and landscape position in eastern Madre de Dios, Peru. *Geoderma*, **83**, 143–166.

Patterson, B.D., Ceballos, G., Sechrest, W., *et al.* (2005). Digital Distribution maps of the mammals of the Western Hemisphere, version 2.0. Arlington, VA: NatureServe,

Peres, C. (1991). Diet and feeding ecology of saddle-back (*Saguinus fuscicollis*) and moustached (*S. mystax*) tamarins in an Amazonian terra firme forest. *Journal of Zoological Society of London*, **230**, 567–592.

Peres, C. (1993). Notes on the ecology of buffy saki monkeys (*Pithecia albicans*, Gray 1860): a canopy seed-predator. *American Journal of Primatology*, **31**, 129–140.

Peres, C. (1994). Primate responses to phenological changes in an Amazonian terra firme forest. *Biotropica*, **26**(1), 98–112.

Pitman, N. (2005). An overview of the Los Amigos watershed, Madre de Dios, southeastern Peru. Unpublished report. June 2005.

Poulsen, J., Clark, C., Connor, E., *et al.* (2002). Differential resource use by primates and hornbills: implications for seed dispersal. *Ecology*, **83**(1), 228–240.

Powell, G., Wright, P., Aleman, U., *et al.* (1999). Research findings and conservation recommendations for the great green macaw in Costa Rica. Report to Tropical Science Center, Costa Rica. 39 pages.

Ridgely, R.S., Allnutt, T.F., Brooks, T., *et al.* (2007). Digital distribution maps of the birds of the Western Hemisphere, version 3.0. Arlington, VA: NatureServe.

Schoener, T. (1974). Resource partitioning in ecological communities. *Science*, **185** (4145), 27–39.

Schoener, T. (1983). Field experiments on interspecific competition. *The American Naturalist*, **122**(2), 240–285.

Shannon, C. (1948). A mathematical theory of communication. *Bell System Technical Journal*, **27**, 379–423.

Soini, P. (1987). *La dieta del mono huapo (Pithecia monachus). Reporte Pacaya-Samiria, Informe No. 25.* Universidad Nacional de la Amazonia Peruana.

Stevenson, P., Quiñones, M. & Ahumana, J. (2000). Influence of fruit availability on ecological overlap among four Neotropical primates at Tinigua National Park, Colombia. *Biotropica*, **32**(3), 533–544.

Terborgh, J. (1983). *Five New World Primates: A Study in Comparative Ecology*. Princeton, NJ: Princeton University Press.

Ter Steege, H., Pitman, N., Phillips, O., *et al.* (2006). Continental-scale patterns of canopy tree composition and function across Amazonia. *Nature*, **443**, 444–447.

Wahungu, G. (1998). Diet and habitat overlap in two sympatric primate species, the Tana crested mangabey *Cercopithecus galeritus* and yellow baboon *Papio cynocephalus*. *African Journal of Ecology*, **36**, 159–173.

Appendix 11.1 Number of feeding bout records in 109 plant genera (plus unidentified samples) in the diets of *Pithecia irrorata* and *Ara* spp. in SE Peru in 2004–2005

Plant family	Plant genus		No. of species used in genus		Item consumed		Number of feeding bout records	
			P. irrorata	*Ara* spp.	*P. irrorata*	*Ara* spp.	*P. irrorata*	*Ara* spp.
Annonaceae	***Guatteria***	*P*	**1**		**S**		**7**	
Annonaceae	Sp.			1		S		1
Araceae	*Heteropsis*		1		S		3	
Araceae	*Philodendron*		1		S		1	
Arecaceae	*Attalea*			1		P		3
Arecaceae	***Euterpe***	*P*		**1**		**F**		**16**
Arecaceae	***Iriartea***	*P*	1	**1**	P	**L,S**	1	**33**
Arecaceae	***Mauritia***	*P*	**1**	**1**	**P**	**P**	**34**	**30**
Arecaceae	*Oenocarpus*			1		P		1
Arecaceae	*Socratea*		1	1	F,P	?	1	1
Arecaceae	Sp.			1		F		3
Aristolochiaceae	*Aristolochia*		1		S		1	
Bignoniaceae	***Jacaranda***	*P*		**1**		**S**		**9**
Bignoniaceae	*Adenocalymma*		1		S		3	
Bignoniaceae	**Sp.**		**5**	1	**S**	S	**19**	1
Bixaceae	*Bixa*			1		S		1
Boraginaceae	*Cordia*			1		S		2
Boraginaceae	Sp.			1		F		1
Caryocaraceae	*Anthodiscus*			1		S		1
Caryocaraceae	*Caryocar*			1		S		2
Cecropiaceae	***Cecropia***	*P*		**2**		**F**		**15**
Cecropiaceae	***Pourouma***	*P*	**6**	1	**S**	S	**28**	1
Celastraceae	Sp. *Pith11*		1		S		1	
Celastraceae	*Salacia*		1	1	P,S	S	4	1
Chrysobalanaceae	**Sp. *Pith60***		**1**		**S**		**16**	
Chrysobalanaceae	Sp. *Pith95*		1		S		2	
Chrysobalanaceae	***Couepia***		**2**	1	**S**	S	**16**	3
Clusiaceae	*Clusia*	*P*	3		S		4	

Plant family	Plant genus		No. of species used in genus		Item consumed		Number of feeding bout records	
			P. irrorata	*Ara* spp.	*P. irrorata*	*Ara* spp.	*P. irrorata*	*Ara* spp.
Clusiaceae	*Caraipa*		1		S		3	
Clusiaceae	*Symphonia*	*P*		1		FL		6
Combretaceae	*Buchenavia*		1	1	S	S	1	1
Combretaceae	*Combretum*		1		S		3	
Connaraceae	*Connarus*		1		S		1	
Cucurbitaceae	***Cayaponia***	***P***	**2**	**3**	**S**	**S**	**5**	**8**
Cucurbitaceae	*Gurania*	*P*		1		F		5
Cucurbitaceae	*Helmontia*		1		S		1	
Cucurbitaceae	Sp.		1		S		5	
Elaeocarpaceae	***Sloanea***	***P***	1	**2**	S	**S**	3	**37**
Euphorbiaceae	***Alchornea***	***P***	**1**		**S**		**7**	
Euphorbiaceae	*Dysidendrum*			1		F		1
Euphorbiaceae	*Nealchornea cf*		1		S		1	
Euphorbiaceae	***Hevea***	***P***	1	**1**	S	**S**	4	**28**
Euphorbiaceae	*Hura*	*P*	1	1	S	S	1	5
Euphorbiaceae	*Maniot*	*P*		1		S		4
Euphorbiaceae	*Pausandra*		1		S		2	
Euphorbiaceae	*Plukenetia*		1		S		2	
Euphorbiaceae	*Sapium*	*P*		1		F		4
Fabaceae	***Acacia***	***P***	**1**	1	**S**	S	**6**	1
Fabaceae	*Apuleia*			1		S		2
Fabaceae	*Cedrelinga*			1		F		3
Fabaceae	*Ducia*			1		S		1
Fabaceae	***Dipteryx***	***P***		**1**		**S**		**7**
Fabaceae	*Enterolobium*	*P*		1		S		4
Fabaceae	***Erythrina***	***P***		**1**		**?**		**9**
Fabaceae	***Inga***	***P***	**12**	**5**	**P,S**	**S**	**44**	**24**
Fabaceae	*Lecointea*		1		P			
Fabaceae	*Parkia*	*P*		2		S		6
Fabaceae	*Phyllocarpus*			1		FL		3
Fabaceae	*Pterocarpus*		1		S		1	
Fabaceae	**Sp. *Ara39***			**1**		**S**		**27**
Fabaceae	Sp.		1	3	P	S	4	5
Lecythidaceae	***Bertholletia***	***P***		**1**		**S**		**135**
Lecythidaceae	*Cariniana*	*P*		1		S		6
Lecythidaceae	***Couratari***	***P***		**2**		**S**		**7**
Lecythidaceae	***Eschweilera***	***P***	**3**	**5**	**S**	**S**	**14**	**19**
Lecythidaceae	Sp.			1		S		3

Plant family	Plant genus		No. of species used in genus		Item consumed		Number of feeding bout records	
			P. irrorata	*Ara* spp.	*P. irrorata*	*Ara* spp.	*P. irrorata*	*Ara* spp.
Loganiaceae	*Strychnos*		2		P		5	
Malpighiaceae	*Byrsonima cf*			1		F		1
Malpighiaceae	Sp.		2		S		3	
Malvaceae	*Apeiba*	*P*		1		A,S		6
Malvaceae	*Ceiba*	*P*		2		F,FL		5
Malvaceae	*Huberodendron*			2		S		3
Malvaceae	***Matisia***	***P***	1	**1**	S	**S**	3	**8**
Melastomaceae	*Bellucia*		1		S		1	
Meliaceae	*Cedrela*			1		S		1
Meliaceae	*Trichilia*		1		S		1	
Memecylaceae	*Mouriri*		1		S		2	
Menispermaceae	*Anomospermun*			1		F		1
Moraceae	***Brosimum***	***P***	**5**		**S**		**34**	
Moraceae	***Castilla***	***P***		**1**		**P,S**		**7**
Moraceae	*Clarisia*	*P*	1		S,F		6	
Moraceae	*Ficus*		1		S		2	
Moraceae	*Naucleopsis*	*P*	1		S		5	
Moraceae	***Pseudolmedia***	***P***	**3**	**3**	**S,P**	**P,S**	**86**	**10**
Myristicaceae	***Iryanthera***	***P***	**2**		**S**		**22**	
Myristicaceae	***Otoba***	***P***	**1**		**S**		**10**	
Myrtaceae	*Calycolpus*		1		S		1	
Olacaceae	***Minquartia***	***P***	**1**		**S**		**10**	
Passifloriaceae	*Dilkea*		1		S		1	
Phytolacaceae	*Gallesia*			1		FL		1
Quiinaceae	*Quiina*		1		S		3	
Rhizophoraceae	*Cassipourea*		1		S		1	
Rubiaceae	*Cinchona*			1		S		1
Rubiaceae	Sp.		2		S		3	
Sabiaceae	*Meliosma*			1		S		2
Sapindaceae	*Paullinia*		2		S		3	
Sapotaceae	*Micropholis*		1		S		1	
Sapotaceae	***Pouteria***	***P***	**9**	**7**	**S**	**P,S**	**30**	**15**
Sapotaceae	**Sp.**		**5**		**S**		**64**	
Simaroubaceae	*Simarouba*	*P*	1		S		5	
Siparunaceae	*Siparuna*		1		S		1	
SP	*Euphonia?*			1		F		2
SP	*Gabaretia?*			1		?		1
SP	Sp. *Pith152*		1		S		4	

Plant family	Plant genus		No. of species used in genus		Item consumed		Number of feeding bout records	
			P. irrorata	*Ara* spp.	*P. irrorata*	*Ara* spp.	*P. irrorata*	*Ara* spp.
SP	Sp. *Pith67*		1		S		1	
SP	*Ochroma*			1		FL		3
Sterculiaceae	***Byttneria***	***P***	**2**	2	**S**	S	**11**	5
Ulmaceae	*Celtis*		1		S		3	
Vochysiaceae	*Vochysia*		1	1	S	S	1	3
Unidentified	Sp.		2	14			12	25

A, Aril; FL, Flower; F, Fruit; L, Leaf; P, Pulp; S, Seed; ?, unknown.
Thirty-two items in **bold** represent the 20 food plant genera most frequently consumed each by sakis and macaws.
Forty-one of the 109 genera, denoted with a "P", were represented in the phenology study. Plant species collection and identification are ongoing.

Chapter 12

On the distribution of Pitheciine monkeys and Lecythidaceae trees in Amazonia

J. Márcio Ayres[+] & Ghillean T. Prance (with endnotes by Adrian A. Barnett Liza M. Veiga and Jean-Phillipe Boubli)

Preface (by Ghillean Prance)

This chapter was prepared by Márcio Ayres with my help in 1990. It was submitted as a paper to a journal and after a long period of review returned for revision. By that time, because Marcio was too involved with conservation work and also fighting ill health, he never had time to revise the paper. After the untimely death of Marcio in 2003 I filed the paper as unpublished work. In a casual conversation I mentioned the existence of this paper to Adrian Barnett and he asked for it to be updated and included in this book. I have updated the information and the nomenclature of the Lecythidaceae following Mori & Prance (1990). I have also made other minor revisions to the botany, but left the details about primates as they were in the original manuscript. Some footnotes by the editors and J.P. Boubli have provided updates on the primates. I am happy that through this paper Márcio Ayres will have a contribution to this book about the primates that he spent so much time studying.

Introduction

Primate distributions and densities in the Amazon forest have been related to a number of historical factors, soils and/or vegetation types (Hershkovitz 1977; Kinzey & Gentry 1979; Emmons 1984).[1]

All phenological studies conducted so far in tropical rainforests have demonstrated that there is community seasonality in flowering, fruiting and vegetative growth. The most common overall pattern in fruit production in the Neotropics is the existence of two distinct seasons (Foster 1982; Terborgh 1983; Ayres 1986). At one time of the year, usually coinciding with the early rainy season, there is a peak of fruit maturation (see Mori & Prance 1987). During the drier part of the year only very mature fruits are available. Thus, arboreal frugivores experience a period of scarcity of potential foods then and frugivorous primates will complement their diets with a number of other items such as foliage, nectar, insects and seeds from immature fruits to cope with the lack of other foods such as pulp and arils of mature fruits. The relatively few plant items that feed animals during the period of dearth are known as "keystone" plant resources (*sensu* Terborgh 1986a, 1986b).

Because "keystone" plant resources play an important role in the maintenance of populations of frugivores during the scarcity period, it is possible that the distribution and availability of these key plant resources influence the ranging patterns of animal taxa that depend on them (see Dew & Boubli 2005). Terborgh and Stern (1987) have shown that the size of tamarin (*Saguinus*) territories is dependent on the availability of a sufficient number of *Combretum* sp. (Combretaceae) vines during the dry season to supply nectar as an energy source after the competing needs of other monkeys and birds have been met. During the last three decades, many field studies have been carried out on Amazonian primates and it is now possible to detect some dietary specializations on particular plant taxa during the period of fruit paucity. Botanical inventories also have increased comparably in the last few years (Gentry 1982; Campbell *et al.* 1986; Campbell & Hammond 1989; ter Steege *et al.* 2003). Thus, it is possible to correlate the distribution of "keystone" plant resources with those of primates in the region.

In this chapter, the distributions of the two largest pithecines, *Cacajao* and *Chiropotes*, are discussed in relation to densities of Lecythidaceae trees in Amazonia, their "keystone" plant resource.

Background

The subfamily Pitheciinae comprises three genera and eight[2] species of diurnal and arboreal monkeys, all endemic to the Amazon region. *Pithecia* is the smallest of the three genera, individuals weighing between 1.5 and 2 kg, living in groups of 2–8 and found in a great variety of vegetation types from primary *terra firme* forests to old secondary forests and even seasonally flooded forests (Buchanan *et al.* 1981; Oliveira *et al.* 1985; Haugaasen & Peres 2005). *Pithecia* has a wide geographical distribution extending through a large part of the Amazon basin. The only place where *Pithecia* species are not found in Amazonia is between the right bank of Rio Tapajós and the Atlantic coast in eastern Amazonia, south of the Amazon river (Hershkovitz 1987a).[3] *Pithecia* forage in the

Evolutionary Biology and Conservation of Titis, Sakis and Uacaris, eds. Liza M. Veiga, Adrian A. Barnett, Stephen F. Ferrari and Marilyn A. Norconk. Published by Cambridge University Press.

lower strata of the forest where resources are probably distributed in smaller clumps; they are mainly frugivores, although their diets include seeds and some foliage.[4] In several places within Amazonia, *Pithecia* is syntopic with the two larger genera of pitheciines, *Cacajao* and *Chiropotes* (Ayres 1986).

In contrast to *Pithecia*, *Chiropotes* and *Cacajao* usually inhabit the upper part of the canopy. *Chiropotes* and *Cacajao* are heavier, weighing on average between 2.5 and 4 kg,[5] living in larger troops ranging from 15 up to 70+ individuals (Ayres 1986).[6] *Cacajao* and *Chiropotes* have never been reported inhabiting the same site.[7] Daily distances travelled by the larger monkeys range from 1.5 to 6 km and the home ranges are on the order of 300–600 ha (Ayres 1981, 1986).[8] *Chiropotes* occurs mainly in the eastern part of Amazonia. North of the Amazon, *Chiropotes satanas chiropotes* is found throughout the area between the east part of lower Rio Negro and the south of Venezuelan savannas.[9] *Chiropotes satanas satanas* and *Chiropotes satanas utahickae* occur south of the Amazon between the right bank of Rio Xingú and the limits of the Amazon forest in the east. *Chiropotes albinasus* is found south of the Amazon between the left bank of Rio Xingú and the right bank of Rio Madeira in the west (Ayres 1981; Hershkovitz 1985; see Figure 12.1). The preferred habitat of *Chiropotes* is tall primary *terra firme* forest. The genus is occasionally found in flooded habitats of clear water rivers (*igapó*) (Ayres 1981). There are no reports on the occurrence of *Chiropotes* in the extensive flood plains along the lower part of the Amazon, such as in the eastern part of Marajó Island (Ayres *et al.* 1989).

Both bald uacaris (*Cacajao calvus*) and the black-headed uacaris (*Cacajao melanocephalus*) are found in western Amazonia.[10] There is only a small area of sympatry between *Chiropotes* and *Cacajao*, on the northern bank of the Rio Negro, but there is no indication of real syntopy between these two genera (Hershkovitz 1985, 1987b).[11] The reason for this probable competitive exclusion is that the two genera are similar in several aspects of their ecology. Both are mainly frugivores but complement their diets with other plant materials and insect larvae. During the season of fruit maturation, like other sympatric monkeys, *Cacajao* and *Chiropotes* feed mainly on the sugar- and lipid-rich mesocarps of ripe fruits. During the period of scarcity of mature fruits, both *Cacajao* and *Chiropotes* feed on seeds from immature fruits. No other living primate has been reported feeding on such high percentages of seeds in their diets (above 60% per year and up to 93% during the period of scarcity; Ayres 1981, 1986; van Roosmalen *et al.* 1981, 1988).

Other primates that include seeds in their diets are the colobines, but they select fruits from different plant taxa (mainly Leguminosae), and have a different morphology than those selected by the two pitheciines (compare Struhsaker 1975; McKey *et al.* 1981; Davies *et al.* 1988; with Ayres 1986).[12] Pitheciines, like other New World monkeys, lack the sacculated stomach which is characteristic of colobines. The forestomach of colobines serves as a chamber for bacterial fermentation by which secondary compounds and cellulose are broken down. Two strategies have been reported for the protection of tropical fruits against seed predation: soft husks combined with seeds high in secondary compounds or mechanical protection by a hard husk and seeds that are

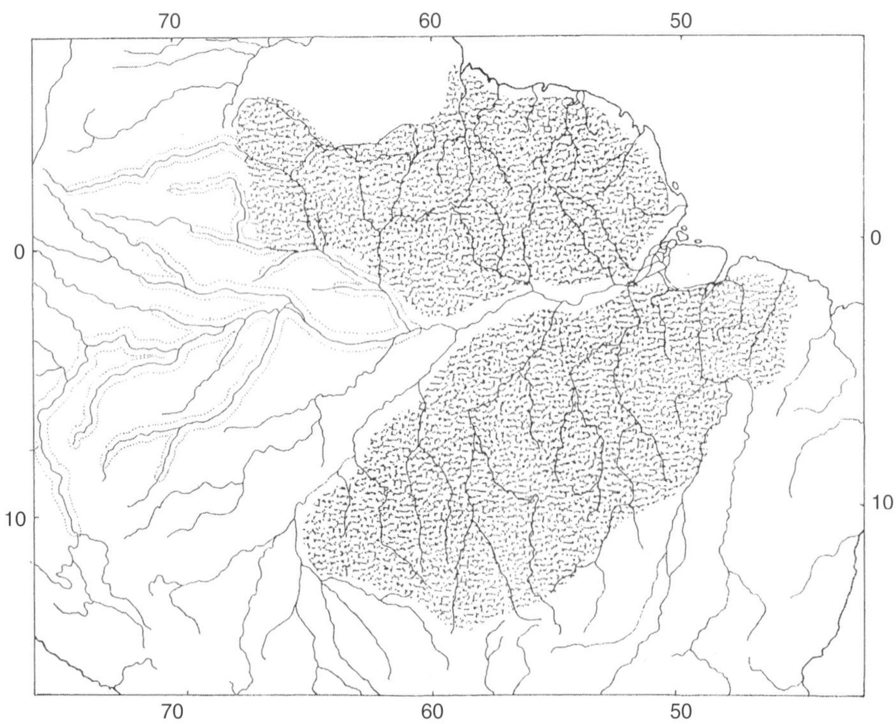

Figure 12.1 Distribution of *Cacajao calvus* and *Cacajao melanocephalus* (dotted lines): *Chiropotes albinasus* and *Chiropotes satanas* (stippled area).

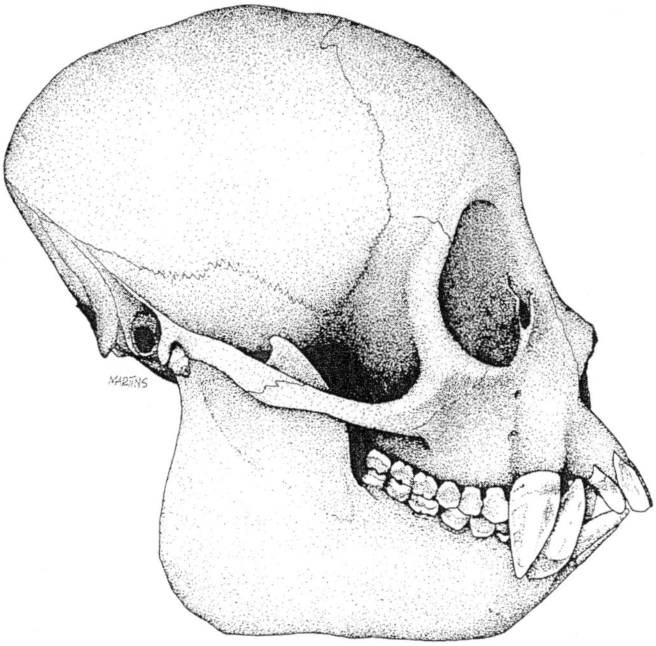

Figure 12.2 Skull and dentition of *Chiropotes satanas*. Illustrations: Antonio Martins.

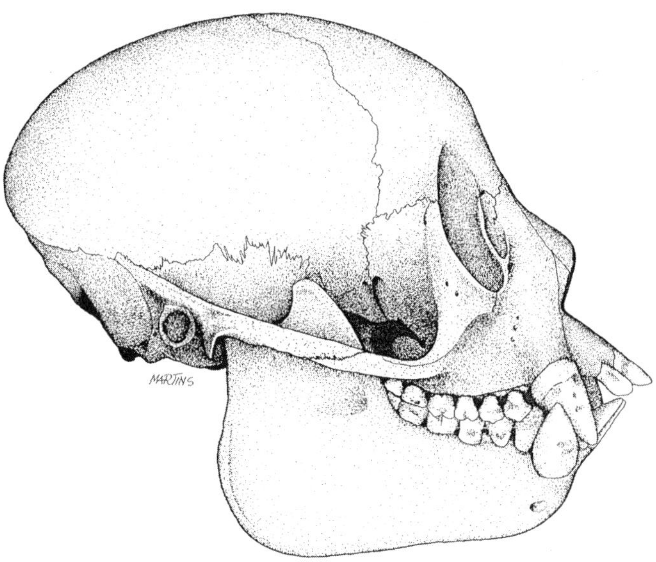

Figure 12.3 Skull and dentition of *Cacajao calvus*. Dental alignment rearranged to show lower canine.

low in secondary compounds (Janzen 1985; Ayres 1986). The former tend to be animal-dispersed and the latter dispersed by other means (e.g. wind, water, explosive dehiscence).[13] Because of their lack of a sacculated stomach, Pitheciines feed mainly on those seeds which are low in secondary compounds, and as a consequence have hard husks. Unlike other Neotropical primates, Pitheciines have a peculiar dentition associated with strong jaw musculature.[14] Their incisors are projected forward, with the upper overbiting the lower ones, and the canines are extremely stout, not recurved but strongly divergent laterally (Hershkovitz 1985, 1987b; see

Photo 12.1 Flat-winged seeds of *Couratari stellata* (Lecythidaceae) exposed by stripping away the thin, leathery pericarp. Seeds ingested by bearded sakis. Fruit dimensions: 7.3–11.5 cm long; 3.6–5.8 cm wide ($n = 7$). Brownsberg Nature Park, Suriname. Photo: Marilyn Norconk.

Figures 12.2 and 12.3). The unusual dentition of pitheciines is associated with a strong jaw musculature which enables them to break the hard husks, and access the seeds of fruits which are inaccessible to most other primates.

Uacaris and bearded sakis feeding on Lecythidaceae

The Lecythidaceae is an important plant family in the lowland areas of the Neotropics. Lecythidaceae is quite abundant in some areas within Amazonia but quite scarce in other areas of the Neotropical region. The family is represented by 20 genera and about 300 species throughout the world, of which 11 genera and about 212 species are endemic to the New World (Prance & Mori 1979; Mori *et al.* 2007; Tsou 1994). To this plant family belong the Brazil nut (*Bertholletia excelsa*) and many other trees whose seeds are protected by a hard shell (including *Eschweilera*, *Lecythis*, *Couratari*, *Cariniana*, *Corythophora*). The fruits are often large and well protected (Prance & Mori 1979; Mori & Prance 1990). See Photo 12.1.

Table 12.1 Lecythidaceae eaten by *Cacajao* and *Chiropotes*

Plant species	Primate species	Rank in diet	Parts eaten	Source
Couratari oblongifolia	*Chiropotes satanas*	10	YS	1
Couratari sp l.	*Chiropotes albinasus*	?	YS	3
Couratari stellata	*Chiropotes satanas*	*	YS, BC	1
Eschweilera albiflora	*Cacajao calvus*	1	YS, Fl	2
Eschweilera amazonica	*Cacajao calvus*	*	YS	2
Eschweilera congestiflora	*Chiropotes satanas*	*	YS	1
Eschweilera simiorum	*Chiropotes satanas*	*	YS	1
Eschweilera sp. 1	*Chiropotes albinasus*	?	YS	3
Eschweilera sp. 2	*Chiropotes satanas*	?	YS, Fl	3
Gustavia hexapetala	*Chiropotes satanas Cacajao calvus*	* *	YS, MS YS, MS, Fl, Rp	1 2
Lecythis chartacea	*Chiropotes satanas*	*	YS	1
Lecythis corrugata	*Chiropotes satanas*	2	YS	1
Lecythis poiteaui	*Chiropotes satanas*	*	YS	1
Lecythis zabucaja	*Chiropotes satanas*	*	YS, MS	1

YS, young seeds; MS, mature seeds; Fl, flower or nectar; BC, base of columella; Rp, ripe pulp.
Rank = * = below 10th annual diet.
Sources: 1, van Roosmalen *et al.* (1988); 2, Ayres (1986); 3, Ayres (1981).

Several species of Lecythidaceae from at least four genera (*Couratari*, *Eschweilera*, *Lecythis* and *Gustavia*) provide young seeds and some species provide nectar for *Chiropotes* and *Cacajao* (Table 12.1). Van Roosmalen *et al.* (1988) reported nine species as sources of food for *Chiropotes satanas chiropotes*, in the Raleighvallen-Voltzberg Nature Reserve in Suriname. *Lecythis corrugata* Poiteau (*Eschweilera corrugata* (Poit.) Miers in van Roosmalen *et al.* 1988) was the second most important plant species in the annual diet of *Chiropotes* and the most eaten of the nine species of Lecythidaceae. *Lecythis* and *Eschweilera* accounted for 12% of the annual diet of *C. satanas chiropotes* at the Suriname site. Ayres (1981) also observed *C. s. chiropotes* in forest fragments north of Manaus, Brazil where flowers (nectar) and young seeds of *Eschweilera* sp. accounted for at least 11% of the *Chiropotes satanas* diet. During a study at Aripuanã, Ayres (1981) twice observed the red-nosed saki (*Chiropotes albinasus*) feeding on immature seeds of *Eschweilera* sp. In this study it was not possible to determine the relative importance of Lecythidaceae during fruit scarcity, because most of the data were gathered during the rainy season when more ripe pulp is available. See Photos 12.2 and 12.3.

The only long-term study of *Cacajao calvus* to date was carried out at the *várzea* forest of Lake Mamiraua-Teiú, at the mouth of the Rio Japurá in Brazilian Amazonia (Ayres 1986) and two species of *Eschweilera* were sources of nectar and young seeds for *Cacajao*. *Eschweilera albiflora* (A.P.DC.) Miers accounted for 15.4% of the annual diet of the uacaris and was the most important plant in their diet; nectar accounted for 5.9% and the rest was seeds from immature fruits. *E. albiflora* was eaten from early October until the end of May. The flowering season is clearly synchronized each year, starting in the last week of September and ending in the second half of October. At this time of year (October), nectar from *E. albiflora* accounts for over 55% of the uacari's diet. Uacaris usually feed during the first part of the day on the nectar of *E. albiflora* and destroy the flowers. A month later, when the fruits of *E. albiflora* are large enough, uacaris feed intensively on the seeds, which account for over 40% of their diet in December and January (Figure 12.4a and 12.4b). Quantitative observations were carried out in 1986 and 1988, confirming the importance of the Lecythidaceae as a "keystone" food resource for uacaris.[15]

Pithecia also feeds on Lecythidaceae and two *Eschweilera* species recorded in Suriname were among the top five plant species in the diet of *P. pithecia* (Buchanan *et al.* 1981).[16] Soini (1986) also recorded *Eschweilera* among the top three sources of seeds in the diet of *Pithecia irrorata* in the Pacaya-Saimiria reserve in Peru. Three other short-term studies, however, have failed to record Lecythidaceae in the diet of *Pithecia* (Happel 1982; Oliveira *et al.* 1985; Johns 1986). The fact that *Pithecia* supplements its diet with larger

Photo 12.2 Examples of Lecythidaceae fruit ingested by *Chiropotes*. Fibrous pyxidium of *Lecythis idatimon* (Lecythidaceae) fruit opened by *Chiropotes sagulatus* to remove seeds. Brownsberg Nature Park, Suriname. Photo: Marilyn Norconk.

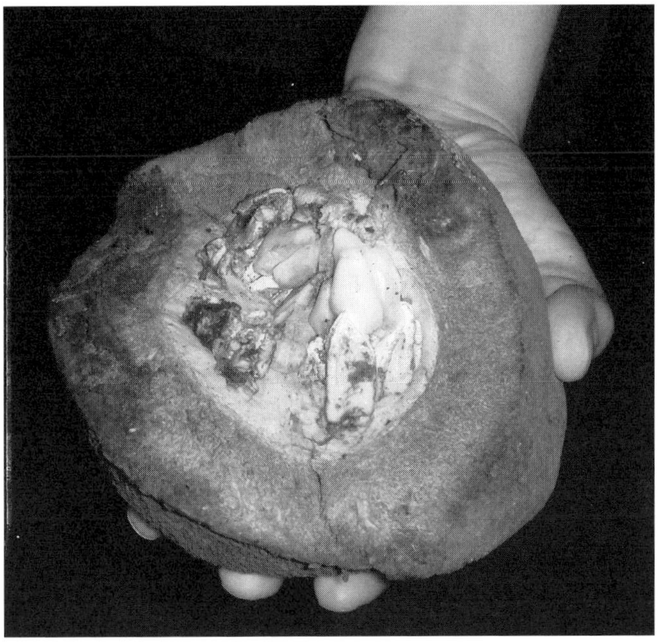

Photo 12.3 *Lecythis zabucajo* is one of the largest of the Lecythidaceae fruit in the New World. Fruit dimensions: approximately 12.0 cm wide and 11.5 cm long and fruit weight is more than 570 g (wet weight). Brownsberg Nature Park, Suriname. Photo: Marilyn Norconk. See Mori *et al.* (2010, http://sweetgum.nybg.org/lp/index.php). (See color plate section.)

quantities of leaves (Happel 1982; Johns 1986) could be a reason why Lecythidaceae seeds are not an important resource for some of its populations. There are occasional reports on other Neotropical primates feeding on Lecythidaceae fruits, but it is usually not an important item on their annual diets (e.g. J.G. Robinson, pers. commun., 1988).[17]

Eschweilera and *Lecythis* distribution and densities in Amazonia

In the previous section we have seen how important these two genera of the Lecythidaceae are to the diet of *Cacajao* and *Chiropotes* during the period of food scarcity. Several botanical inventories have been conducted in Amazonia which give a picture of the densities of *Eschweilera*.

Prance and Mori (1979) and Mori and Prance (1990) listed the density of Lecythidaceae species in several Amazonian sites in eastern Amazonia which was complemented with data from other botanical plots published subsequently (Table 12.2). Although there is considerable variation between authors in the minimum diameter used for botanical sampling, it is possible to verify the differences in Lecythidaceae and *Eschweilera* densities in the plots because the genus is well represented in all classes of diameter (the important species of *Lecythis* as food for pitheciines were formerly included in *Eschweilera* and these species were transferred to *Lecythis* by Mori & Prance, 1990). We gathered information on the densities of *Eschweilera* for 19 sites in northern South America, most of which are within Amazonia. Densities per hectare in Amazonian plots of *Eschweilera* ranged from 121 per hectare in Capitão Poço (in Pará) to none in Cocha Cashu in Peru (Terborgh, pers. commun., 1986). See Photo 12.4.

Of the 15 sites located within Amazonia east of the Rios Negro and Madeira within the geographical range of *Chiropotes*, only two, in Marajó and the *várzea* of the Rio Xingú, had no *Eschweilera*. The Marajó site, located in Rio Rabelo, is a tidal *várzea* where Ayres *et al.* (1989) found only two

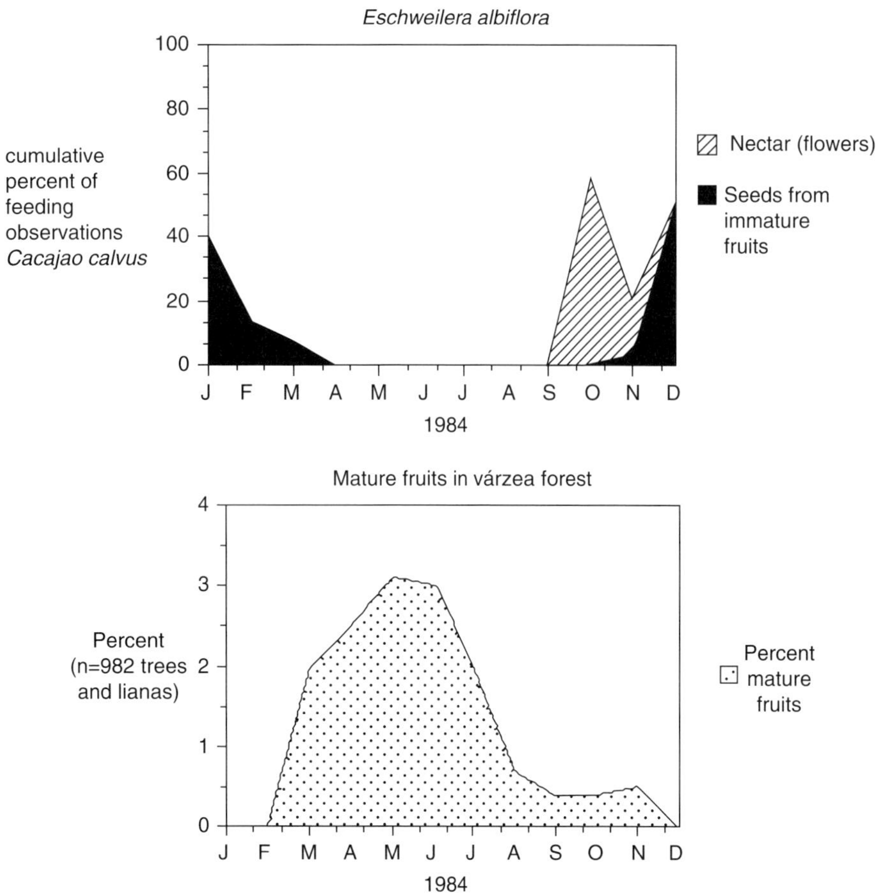

Figure 12.4 Phenology of feeding behavior of *Cacajao calvus* on *Eschweilera albiflora* (top); and availability of mature fruits in a *várzea* forest (bottom). Data from Ayres (1986).

Photo 12.4 *Lecythis lurida* (Lecythidaceae) pyxidium and embedded seeds (right) and operculum that forms the base of the fruit (left). Tapajós National Forest, Pará, Brazil. Photo: Liliam Pinto.

Table 12.2 Densities of trees from family Lecythidaceae and genus *Eschweilera* in several sites in the Neotropical region relative to the distribution of *Chiropotes* and *Cacajao*.

Site	Forest type	DBH minimum (cm)	Lecythidaceae per hectare	*Eschweilera* per hectare	Pitheciinae found (excluding *Pithecia*)	Reference
Lake Teiú (Brazil)	Upper Amazon *várzea*	10	32	21.5	*Cacajao calvus*	Ayres (1986)
R. Aripuanã (Brazil)	*terra firme*	15	15	14	*Chiropotes albinasus*	Ayres (1986)
R. Xingú (Brazil)	*terra firme*	10	25	1.7	*Chiropotes satanas*	Campbell *et al.* (1986)
R. Xingú (Brazil)	Lower Amazon *várzea*	10	16	0	*Chiropotes satanas* (possibly only seasonal appearances)	Campbell *et al.* (1986)
Reserva Ducke, Manaus (Brazil)	*terra firme*	15	67	48	*Chiropotes satanas*	Prance *et al.* (1976)
Mocambo, Belém, (Brazil)	*terra firme*	15	117.7	120.9	*Chiropotes satanas**	Prance & Mori (1979)
Catú and Aurá Belém (Brazil)	Tidal *várzea*	15	36.2	24	*Chiropotes satanas**	Prance & Mori (1979)
Marajó Island (Brazil)	Tidal *várzea*	10	0	0	None	Ayres *et al.* (1989)
3 de Outubro, Castanhal (Brazil)	*terra firme*	10	78	?	*Chiropotes satanas*	Pires *et al.* (1953)
Capitão Poço (Brazil)	*terra firme*	9.55	128	121	*Chiropotes satanas*	Dantas *et al.* (1980)
Tranzamazon highway (Brazil)	*terra firme*	9.55	23	5.3	*Chiropotes albinasus*	Dantas & Muller (1979)
Serra do Navio, Amapá, (Brazil)	*terra firme*	15	23	14	*Chiropotes satanas*	Rodrigues (1963)
Rio Gurupi (Brazil)	*terra firme*	10	83	68	*Chiropotes satanas*	Balée (1987)
Guyana (five different sites)	*terra firme*	10	48.5	47.5	*Chiropotes satanas*	Davis & Richards (1934) in Prance & Mori (1979)
Alto Ivon (Bolivia)	*terra firme*	10	1	1	None	Boom (1985)
Cocha Cashu, Manu (Peru)	Floodplain	10	?	0	None	Terborgh (pers. commun., 1986)
Macuira, Santa Ana and Margarita (Colombia, Venezuela)	Cloud forest	2 m height	0	0	None	Sugden (1983)
Masaguaral (Venezuela)	Gallery forest and savannah	4 m height	0.1	0	None	Robinson (1986)

* Within the geographical range of the species, but there is no indication of the occurrence of the primate in that particular site. In some sites near Belém it is possible that the species has disappeared already from the spots due to human disturbance. Skins at Museu Goeldi and other museums confirm the existence of the species some 30 years ago in some of these areas around Belém.

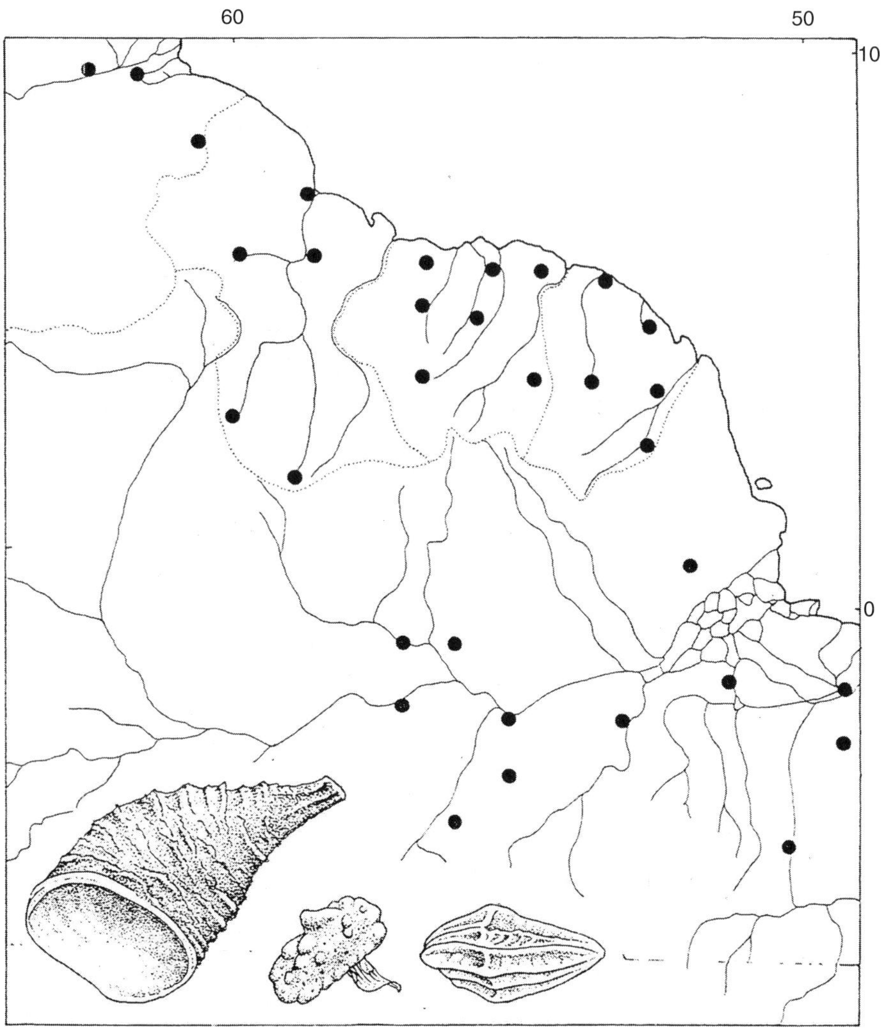

Figure 12.5 Distribution and fruits of *Lecythis corrugata* subsp. *corrugata* (fruit from C.W. Anderson 490).

species of primates, neither belonging to the genus *Chiropotes*. The 0.5 ha site on the *várzea* of Rio Xingu (Campbell *et al.* 1986) is also within the range of *Chiropotes satanas*, but it is not known whether it occurs there. However, a survey done along the Xingú and Iriri rivers (Martins *et al.* 1988) failed to record *Chiropotes* in their floodplains. These *várzeas* on the tributaries of the Amazon are usually not extensive enough to include an entire home range of *Chiropotes*, and the monkeys go there only during the period of fruit maturation. All other sites in eastern Amazonia usually contained more than ten individuals of *Eschweilera* per hectare. The distribution of *Lecythis corrugata* (= *Eschweilera corrugata*) coincides with that of *Chiropotes satanas*. It is a common tree of poorly drained clay soils, sandy soils of dryland forests of eastern Amazonia (Mori & Prance 1990); see Figure 12.5.

Inventory sites in the western part of Amazonia differ strikingly from those of the east. The 1-ha plot in the Alto Ivon, Bolivia sampled by Boom (1985) contained no *Eschweilera* in an area where there are no reports of *Chiropotes* and *Cacajao*. Cocha Cashu in Peru has the highest number of primate species ever found in a single Amazonian site (13 species); however, *Cacajao* does not occur there, and it is surely significant that *Eschweilera* trees are also absent from the region (Terborgh, pers. commun., 1986). The only western site which contained high densities of *Eschweilera* (21.5 trees/ per hectare, $n = 2$) was located in the extensive Holocene *várzea* of the lower Rio Japurá, where *Cacajao calvus* occurs. *E. albiflora*, the number one species of tree in the diet of *C. calvus*, is distributed in the periodically flooded forests of central and western Amazonia, and matches to a large extent the distribution of *Cacajao calvus* and part of the range of *Cacajao melanocephalus* (Mori & Prance 1990; see Figure 12.6).

The distribution of the two larger pitheciine genera in Amazonia is influenced by the density as well as the distribution of Lecythidaceae. High densities of Lecythidaceae trees, the seeds of which are eaten by pitheciines, are found throughout the *terra firme* of eastern Amazonia and the *várzea* of western Amazonia, which coincides, respectively, with the distributions of *Chiropotes* and *Cacajao*.[18] If one groups all those sites (Table 12.2) according to the occurrence or not of *Chiropotes* or *Cacajao* in each area,

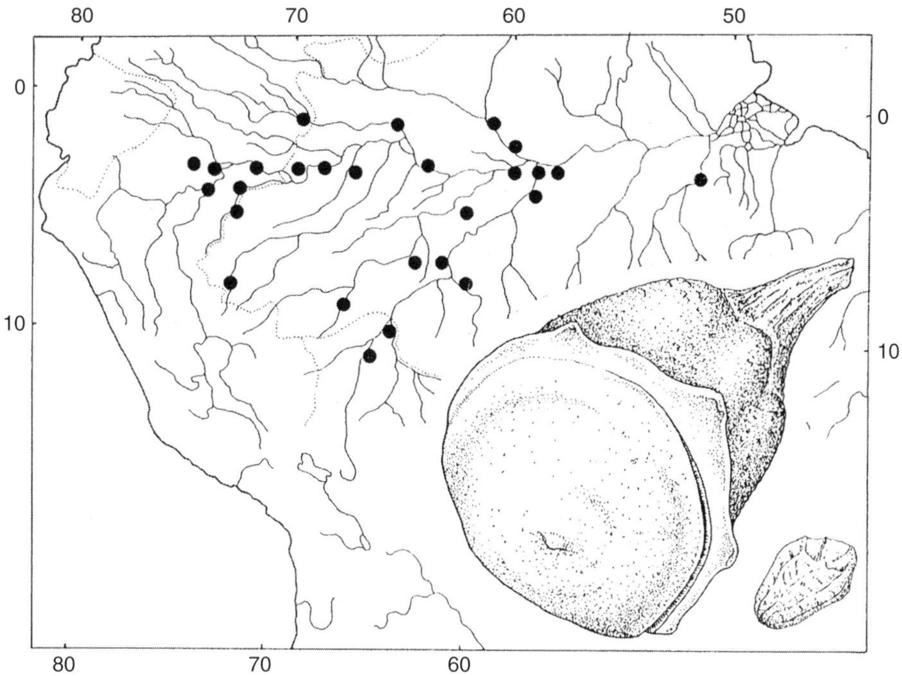

Figure 12.6 Distribution and fruits of *Eschweilera albiflora* (fruit from Prance *et al.*, 9199, seed from W.R. Anderson 11914).

there is a significant difference in the mean percentage of Lecythidaceae between these sites (Mann–Whitney U test statistic = 58; df = 1; $p < 0.01$). If we group the sites similarly in relation to the number of *Eschweilera* trees per hectare, even higher differences will be found (Mann–Whitney U test statistic = 60; df = 1; $p < 0.01$).

Another factor which might influence the distribution of *Chiropotes* and *Cacajao* is the phenology of Lecythidaceae species. Although there are few quantitative botanical inventories in areas south of the Amazon, between the Rio Juruá and Rio Purus, Silva *et al.* (1992) showed Lecythidaceae to be common in the Rio Juruá basin near Carauari. They listed 14 species including 3 *Lecythis* (*L. corrugata* was one) and 6 *Eschweilera*. *Eschweilera alba* (= *E. grandiflora*) was the most abundant species in the four hectares and *E. odora* (= *E. coriacea*) ranked third. Lecythidaceae also seems to be abundant in some areas such as the Rio Tefé basin (A.D. Johns & C. Peres, pers. commun., 1987) where the only pitheciine to be found is *Pithecia albicans*. In the *várzea* at Lake Mamirauá, during four years (1983, 1984, 1986 and 1988) one of us (J.M.A.) observed a closely synchronized pattern of flowering in *Eschweilera albiflora*. In the last week of September a large number of *E. albiflora* trees produce flower buds. By early October the floor of this *várzea* forest is conspicuously covered by the white and yellow flowers of *E. albiflora*. During his one-year study on the phenology of trees and primates, Johns (1986) did not observe a similar pattern of flowering in *terra firme* at Lake Tefé. This indicates that besides densities, there are also striking habitat differences in the phenology of flower and fruit production of Lecythidaceae species. This factor might be another important component in the proposed association between this plant family and pitheciine monkeys, but more phenological studies on Lecythidaceae are needed to investigate the extent of these differences. In addition to these factors, the lack of *Cacajao* or *Chiropotes* in the Rio Tefé basin could be due to the absence of the preferred species of Lecythidaceae or presence of *Pithecia albicans*. The only body weight data available for a wild animal of this species (3.0 kg) indicates that this is by far the largest of the monkeys from genus *Pithecia* (Johns 1986).[19] Thus, *Pithecia albicans* seems to be as big as adult males of *Cacajao* and *Chiropotes*. Adult individuals of other *Pithecia* species weigh less than 2.0 kg.

Discussion and conclusions

The distribution of "keystone" plant resources can be an important component to understand both local and large scale range patterns of Amazonian primates. It would be easier to understand how this plant–animal interaction had evolved if seed dispersal was involved. Uacaris (*Cacajao*) and bearded sakis (*Chiropotes*) are the largest of the Neotropical arboreal primate seed predators. Ayres (1986) estimated that an average troop of *Cacajao calvus*, containing between 45 and 50 individuals, reduces the flower and fruit crop of *E. albiflora* by at least 60% each year. Nectar feeding by *Cacajao calvus* destroys the flowers and makes no contribution to pollination. Despite this massive reduction of seeds, *Eschweilera albiflora* is still the second most abundant tree in the Holocene *várzea* forests of the middle–upper section of the Amazon (Ayres 1986).

We postulate that a possible mechanism of interaction between Lecythidaceae and pitheciines could be as follows. The reduction of the flower crop and fruits at an early stage of their development could cause the plant to divert more resources to the remaining fruits, thus increasing seed size. Consequently, larger seeds will have a higher probability of establishing themselves under the unpredictable conditions of the forest. It is well known in some domestic seed crops that removal of flowers increases the individual seed weight and total seed mass by up to 50% (Harper *et al.* 1970). It also increases their germination rates (Stephenson 1984). Another possible mechanism of interaction between pitheciines and Lecythidaceae would be similar to the whole of large animal seed predators. By decreasing the seed pool of a well-adapted species, pitheciines could be influencing forest dynamics by giving a chance for other plant species to establish.

Pitheciines are unique among primates. While there is considerable dietary overlap among other sympatric frugivorous primates, *Cacajao* and *Chiropotes* are specialized feeders on seeds from fruits with hard husks. Unlike other frugivorous primates, a few fruits account for a large part of the annual diet of *Cacajao* and *Chiropotes*. Most important are those available during the period of fruit shortage (dry season) when immature fruits of Lecythidaceae are available, and are essential for their survival. The immature fruits are softer than mature fruits making them more readily accessible to primate predators.

Although information for many other sites is still needed to confirm the hypothesis proposed in this paper, especially in western Amazonia, it seems that the distribution of *Cacajao* and *Chiropotes* is related in part to that of some species of *Eschweilera* and *Lecythis*. Present published information on quantitative botanical inventories indicate that *Eschweilera* has higher densities on *terra firme* of eastern Amazonia, where the soils are on the Crystalline basement and pre-Tertiary sediments, in contrast to those areas in the upper part of the Amazon basin with a predominance of late Tertiary soils. The extensive floodplains of western Amazonia, of Pleistocene and Holocene, seem to have higher production of *Eschweilera* fruits than its surrounding *terra firme* forests.

Acknowledgments

Special thanks to Drs D.J. Chivers, E.L. Bennett, R.A. Mittermeier and Scott Mori for their comments on earlier drafts of this manuscript. J.M. Ayres was supported for this work in Cambridge by a post-doctoral scholarship from CNPq/ Museu Goeldi and an IBM-Brazil grant, and fieldwork in Amazonia by G.T. Prance was supported by the National Science Foundation. We are grateful to Christine Grey-Wilson and Antonio Martins for preparing the plates. GTP is grateful to Adrian Barnett and Liza Veiga for suggesting that this paper be included in this book and for the extensive updating work and also to J.P. Boubli for updating the primate data.

Endnotes

1. See also Peres (1990).
2. This figure has since been revised, see Silva Junior *et al.*, Chapter 4. Recent fieldwork on *Chiropotes* suggests that further revision may be needed.
3. The genus is also absent from the basins of the Negro and Orinoco, west of the Rio Branco, although *P. pithecia* is recorded in Guyana (Barnett *et al.* 2000; Lehman 2004).
4. See Norconk & Setz, Chapter 25, for an up-to-date summary of *Pithecia* ecology.
5. See Barnett *et al.*, Chapter 14 and Veiga & Ferrari, Chapter 22, for *Cacajao* and *Chiropotes*, respectively.
6. Recent studies have shown this number may exceed 100, sometimes reaching 200 animals (Defler 1989, 2004).
7. While this is still generally true, Boubli (2002) reported sympatric occurrence of the two genera in the Pico de Neblina region of northwestern Amazonia.
8. Or more, see, Boubli (1997).
9. Recent fieldwork has revealed that *Chiropotes* also occupies the area between the Rio Branco's west bank and the Rio Negro's east bank in the Pico de Neblina region (see Bonvicino *et al.* 2003).
10. Effectively the two species occupy different drainage systems, with *melanocephalus* uacaris restricted to the Negro/Orinoco basins and *calvus* white uacaris restricted to the Amazon and its tributaries, west of the Madeira river (Ayres 1986; Boubli 1993, 1997; Barnett *et al.* 2005).
11. This overlap has recently been shown to be more geographically extensive than previously supposed (see Boubli, 2002), but the degree of ecological overlap may still be minimal (see Barnett *et al.* 2005).
12. And with Boubli (1997); Aquino (1998) and Aquino and Encarnación (1999).
13. See also van Roosmalen (1985).
14. See Kinzey (1992) for details.
15. Although fruits from Lecythidaceae trees ranked fourth only in the annual diet of black-backed uacaris at Pico da Neblina National Park, Brazil, they included the seeds of 11 species in 3 genera (*Lecythis*, *Couratari* and *Eschweilera*) of this family (Boubli 1997). The lower importance of Lecythidaceae in this case might result from the peculiar floristic composition at a site with a high dominance of plants of Euphorbiaceae and Leguminosae (Boubli 1999, 2001, 2005). Elsewhere within the range of the genus Lecythidaceae have been found to be high-ranking diet components for *C. c. uacayalii* (Aquino 1998; Aquino & Encarnación 1999), and also appear to be dietary mainstays for golden-backed uacaris in Colombia (Defler 2004) and Brazil (Barnett *et al.* 2005, in press). Thus Marcio Ayres' observations allowed him to presciently recognize a widespread pattern in the genus.
16. Peres (1993) recorded nine species of Lecythidaceae in the diet of the buffy saki, *Pithecia albicans*: Lecythidaceae was one of the two most important families in the diet.
17. See also Peres (1991).
18. They are also common at several central Amazonian sites, e.g. Mori and Lepsch-Cunha (1995) and also occur in the Pantepui, in the foothills of Neblina, Imeri and Tapirapeco – see Boubli (2005).
19. See also Ayres and Clutton-Brock (1992); Peres (1993); Lindenfors (2002).

References

Aquino, R. (1998). Some observations on the ecology of *Cacajao calvus ucayalii* in the Peruvian Amazon. *Primate Conservation*, **18**, 21–24.

Aquino, R. & Encarnación, F. (1999). Preliminary observations on the diet of *Cacajao calvus ucayalii* in northeast Peru. *Neotropical Primates*, **7**, 1–5.

Ayres, J.M. (1981). Observações sobre a ecologia e comportamento dos cuxius (*Chiropotes satanas* and *Chiropotes albinasus*, Cebidae, Primates). Belém: Grafisa.

Ayres, J.M. (1986). Uakaris and the Amazonian flooded forest. Unpublished PhD thesis, University of Cambridge.

Ayres, J.M. & Clutton-Brock, T.H. (1992). River boundaries and species range size in Amazonian primates. *The American Naturalist*, **140**, 531–539.

Ayres, J.M., Bonsiepe, J.I. & Townshend, T.J. (1989). Notes on monkeys and habitat in the northeast of Marajo Island, Brazil. *Primate Conservation*, **10**, 21–22.

Balée, W.A. (1987). Etnobotánica quantitative dos Indios Tembé (Rio Gurupi, Pará). *Boletim do Museu Paraense Emílio Goeldi, série Botanica*, **3**, 29–50.

Barnett, A.A., de Castilho, C.V., Shapley, R.L., *et al.* (2005). Diet, habitat selection and natural history of *Cacajao melanocephalus ouakary* in Jaú National Park, Brazil. *International Journal of Primatology*, **26**, 949–969.

Barnett, A.A., MacLarnon, A., Almeida, T., *et al.* (in press). Seasonal changes in diet and habitat use by golden-backed uacari, *Cacajao melanocephalus ouakary* (Spix, 1823) in Jaú National Park, Amazonian Brazil. *American Journal of Primatology*.

Barnett, A.A., Shapley, B., Lehman, S., *et al.* (2000). Primate records from the Potaro Plateau, western Guyana, including the first for *Cebus albifrons* east of the Rio Branco, Brazil. *Neotropical Primates*, **8**, 35–40.

Bonvicino, C.R., Boubli, J.P., Otazu, I.B., *et al.* (2003). Morphologic, karyotypic and molecular evidence of a new form of *Chiropotes* (Primates, Pitheciinae). *American Journal of Primatology*, **61**, 123–133.

Boom, B. (1985). A forest inventory in Amazonian Bolivia. *Biotropica*, **18**, 287–294.

Boubli, J.P. (1993). Southern expansion of the geographical distribution of the black uakari monkey *Cacajao melanocephalus melanocephalus*. *International Journal of Primatology*, **14**, 933–937.

Boubli, J.P. (1997). The black uakari monkey, *Cacajao melanocephalus*, in the Pico da Neblina National Park, Brazil. PhD Dissertation, University of California, Berkeley.

Boubli, J.P. (1999). The feeding ecology of the black-headed uacari *Cacajao melanocephalus*, in Pico da Neblina National Park, Brazil. *International Journal of Primatology*, **20**, 719–749.

Boubli, J.P. (2001). Floristic assessment of the lowlands of Pico da Neblina National Park, Brazil. *Plant Ecology*, **160**, 149–167.

Boubli, J.P. (2002). Western extension of the range of bearded sakis: a possible new taxon of *Chiropotes* sympatric with *Cacajao* in Pico da Neblina National Park, Brazil. *Neotropical Primates*, **10**, 1–4.

Boubli, J.P. (2005). Floristics, primary productivity and primate diversity in Amazonia: contrasting a eutrophic varzea forest and an oligotrophic caatinga forest in Brazil. In *Tropical Fruits and Frugivores: The Search for Strong Interactors*, ed. J.L. Dew and J.P. Boubli. Dordrecht: Springer, pp. 59–73.

Buchanan, D.B., Mittermeier, R.A. & van Roosmalen, M.G.M. (1981). The saki monkeys, genus *Pithecia*. In *Ecology and Behavior of Neotropical Primates*, ed. A.F. Coimbra-Filho and R.A. Mittermeier. Rio de Janeiro: Academia Brasileira de Ciencias.

Campbell, D.G. & Hammond, H.D. (eds.). (1989). *Floristic Inventory of Tropical Countries*. New York, NY: The New York Botanical Garden.

Campbell, D.G., Daly, D.C., Prance, G.T., *et al.* (1986). Quantitative ecological inventory of *terra firme* and várzea tropical forest on the rio Xingú, Brazilian Amazon. *Brittonia*, **38**, 369–393.

Dantas, M. & Muller, N.R.M. (1979). Estudos fito-ecológicos do trópico úmido brasileiro l-aspectos fitosociológicos de mata sobre terra roxa na região de Altamira. *Anais da Sociedade Botânica do Brasil*, **30**, 205–208.

Dantas, M., Rodrigues, I.M.A. & Muller, N.R.M. (1980). Estudos fito-ecológicos do trópico úmido brasileiro: aspectos fitosociológicos de uma mata sobre latosolo amarelo em Capitão Poço. *Boletim de Pesquisa*, **9**, Embrapa, Belém.

Davies, A.G., Bennett, E.L. & Waterman, P.G. (1988). Food selection by two South-east Asian colobine monkeys (*Presbytis rubicunda* and *Presbytis melalophos*) in relation to plant chemistry. *Biological Journal of the Linnaean Society*, **34**, 33–56.

Defler, T.R. (1989). The status and some ecology of primates in the Colombian Amazon. *Primate Conservation*, **10**, 51–56.

Defler, T.R. (2004). *Primates of Colombia*. Bogotá: Conservation International. 550 pages.

Dew, J.L. & Boubli, J.P. (2005). *Tropical Fruits and Frugivores: the search for strong interactions*. Dordrecht: Springer. 260 pages.

Emmons, L. (1984). Geographic variation in densities and diversities of non-flying mammals in Amazonia. *Biotropica*, **16**, 210–222.

Foster, R.B. (1982). A seasonal rhythm of fruit fall on Barro Colorado Island. In *The Ecology of a Tropical Forest: Seasonal Rhythms and Long-term Changes*, ed. E.G. Leigh, Jr., A.S. Rand and D.M. Windsor. Washington, D.C.: Smithsonian Institution Press.

Gentry, A.H. (1982). Patterns in neotropical plant species diversity. *Evolutionary Biology*, **15**, 1–80.

Happel, R.E. (1982). Ecology of *Pithecia hirsuta* in Peru. *Journal of Human Evolution*, **11**, 581–590.

Harper, J.L., Lovell, P.H. & Moore, K.G. (1970). The shapes and sizes of seeds. *Annual Review of Ecology & Systematics*, **1**, 327–355.

Haugaasen, T. & Peres, C.A. (2005). Mammal assemblage structure in Amazonian flooded and unflooded forests. *Journal of Tropical Ecology*, **21**, 133–145.

Hershkovitz, P. (1977). *Living New World Monkeys (Platyrrhini) with an Introduction to Primates*. Vol. 1. Chicago, IL: The University of Chicago Press, pp. xiv, 1117.

Hershkovitz, P. (1985). A preliminary taxonomic review of the South American bearded saki monkeys, genus *Chiropotes* (Cebidae, Platyrrhini), with the description of a new subspecies. *Fieldana, Zoology*, **27**, 1–46.

Hershkovitz, P. (1987a). The taxonomy of South American bearded saki genus *Pithecia* (Cebidae, Platyrrhini): a preliminary report and critical review with the description of a new species and a new subspecies. *American Journal of Primatology*, **12**, 387–468.

Hershkovitz, P. (1987b). Uakaries, New World monkeys of the genus *Cacajao* (Cebidae, Platyrrhini), a preliminary review. *American Journal of Primatology*, **12**, 1–53.

Janzen, D.H. (1985). Plant defences against animals in the Amazonian rainforest. In *Amazonia*, ed. G.T. Prance and T.E. Lovejoy. Oxford: Pergamon Press.

Johns, A.D. (1986). Notes on the ecology and current status of the buffy saki, *Pithecia albicans*, Gray 1860. *Primate Conservation*, 7, 26–29.

Kinzey, W.G. (1992). Dietary and dental adaptations in the Pitheciinae. *American Journal of Physical Anthropology*, **88**, 499–514.

Kinzey, W.G. & Gentry, A.H. (1979). Habitat utilization in two species of *Callicebus*. In *Primate Ecology: Problem-oriented Field Studies*, ed. R.W. Sussman. New York, NY: John Wiley & Sons.

Lehman, S.M. (2004). Distribution and diversity of primates in Guyana: species–area relationships and riverine barriers. *International Journal of Primatology*, **25**, 73–95.

Lindenfors, P. (2002). Sexually antagonistic selection on primate size. *Journal of Evolutionary Biology*, **15**, 595–607.

Martins, E.S., Ayres, J.M. & Ribeiro Do Valle, M.B. (1988). On the status of *Ateles belzebuth marginatus* and notes on other monkeys of the Iriri River basin. *Primate Conservation*, **9**, 87–91.

McKey, D., Gartlan, J.S., Waterman, P.G., *et al.* (1981). Food selection by black colobus monkeys (*Colobus satanas*) in relation to plant chemistry. *Biological Journal of the Linnean Society*, **16**, 115–146.

Mori, S.A. & Lepsch-Cunha, N. (1995). Lecythidaceae of a central Amazonian moist forest. *Memoirs of The New York Botanical Garden*, **75**, 1–55.

Mori, S.A. & Prance, G.T. (1987). Phenology. Chapter 11 in S.A. Mori, *et al.*, The Lecythidaceae of a lowland forest: La Fumée Mountain. *Memoirs of The New York Botanical Garden*, **44**, 124–136.

Mori, S.A. & Prance, G.T. (1990). *Monograph of Zygomorphic-flowered Lecythidaceae*. Flora Neotropica. Monograph **21** (2). The New York Botanical Garden, New York.

Mori, S.A., Tsou, C.-H., Cronholm, B., *et al.* (2007). Evolution of Lecythidaceae with an emphasis on the circumscription of Neotropical genera: information from combined NDHF and TRNL-F sequence data. *American Journal of Botany*, **94**, 289–301.

Oliveira, J.M.S., Guerreiro De Lima, M., Bonvincino, C., *et al.* (1985). Preliminary notes on the ecology and behaviour of the brown-faced saki (*Pithecia pithecia*, Linnaeus, 1766; Cebidae, Primates). *Acta Amazonica*, **15**, 249–263.

Peres, C.A. (1990). Effects on hunting of western Amazonian primate communities. *Biological Conservation*, **54**, 47–59.

Peres, C.A. (1991). Seed predation of *Cariniana micrantha* (Lecythidaceae) by brown capuchin monkeys in Central Amazonia. *Biotropica*, **23**, 262–270.

Peres, C.A. (1993a) Structure and spatial organization of an Amazonian terra firme forest primate community. *Journal of Tropical Ecology*, **9**, 259–276.

Peres, C.A. (1993b). Notes on the ecology of Buffy Saki Monkeys (*Pithecia albicans*, Gray 1860): a canopy seed-predator. *American Journal of Primatology*, **31**, 129–140.

Peres, C.A. (1997). Primate community structure at twenty western Amazonian flooded and unflooded forests. *Journal of Tropical Ecology*, **13**, 381–405.

Pires, J.M., Dobzhansky, T. & Black, G.A. (1953). An estimate of the number of species of trees in an Amazonian forest community. *Botanical Gazette*, **14**, 467–477.

Prance, G.T. & Mori, S.A. (1979). *Lecythidaceae – Part I. Flora Neotropica Monograph* **21**, 1–270. The New York Botanical Garden.

Prance, G.T., Rodrigues, W.A. & Silva, M.F. (1976). Inventário florestal de um hectare de mata de terra firme km 30 da Estrada Manaus-Itacoatiara. *Acta Amazonica*, **6**, 9–35.

Robinson, J.G. (1986). Seasonal variation in time and space by the wedge-capped capuchin monkey, *Cebus olivaceus*: implications for foraging theory. *Smithsonian Contributions to Zoology*, **431**, 149–179.

Rodrigues, W.A. (1963). Estudo de 2.6 hectares de mata de terra firme da Serra do Navio, Território do Amapá. *Boletim do Museu Paraense Emílio Goeldi, Botanica*, **19**, 1–43.

Rodrigues, W.A. (1967). Inventário florestal piloto ao longo da Estrada Manaus-Itacoatiara, Estado do Amazonas: Dados preliminaries. *Atas do Simpósio sobre a Biota Amazonica*, 7, 25 (*Conservação da Natureza e Recursos Naturais*).

Silva, A.S. da, Lisboa, P.L.B. & Maciel, U.N. (1992). Diversidade florística e estrrutura em floresta densa da Bacia do Rio Juruá-Am. *Boletim do Museu Paraense Emílio Goeldi, Série Botánica*, **8**, 203–258.

Soini, P. (1986). A synecological study of primate community in the Pacaya-Saimiria National Reserve, Peru. *Primate Conservation*, **7**, 63–71.

Stephenson, A.G. (1984). The cost of over-initiating fruit. *American Midland Naturalist*, **112**, 379–386.

Struhsaker, T.T. (1975). *The Red Colobus Monkey*. Chicago, IL: University of Chicago Press.

Sugden, A.M. (1983). Determinant of species composition in some isolated neotropical cloud forests. In *Tropical Rain Forest: Ecology and Management*, ed. S.L. Sutton, T.C. Whitmore and A.C. Chadwick. Oxford: Blackwell Scientific.

Terborgh, J. (1983). *Five New World Primates: A Study in Comparative Ecology*. Princeton, NJ: Princeton University Press.

Terborgh, J. (1986a). Community aspects of frugivory in Neotropical Primates. In *Frugivores and Seed Dispersal*, ed. A. Estrada and T.H. Fleming. Dordrecht: Dr W. Junk Publishers.

Terborgh, J. (1986b). Keystone plant resources in the tropical forest. In *Conservation Biology*, ed. M.E. Soulé. Sunderland, MA: Sinauer Associates.

Terborgh, J. & Stern, M. (1987). The surreptitious life of the saddle-back tamarind. *American Scientist*, 75, 260–269.

Ter Steege, H., Pitman, N., Sabatier, D., *et al.* (2003). A spatial model of tree diversity and tree density for the Amazon. *Biodiversity and Conservation*, **12**, 2255–2277.

Tsou, C.-H. (1994). The embryology, reproductive morphology and systematics of Lecythidaceae. *Memoirs of the New York Botanical Garden*, **71**, 1–110.

van Roosmalen, M.G.M. (1985). *Fruits of the Guianan Flora.* Institute of Systematic Botany, University of Utrecht, 483 pages.

van Roosmalen, M.G.M., Mittermeier, R.A. & Fleagle, J. (1988). Diet of the northern bearded saki (*Chiropotes satanas chiropotes*): a Neotropical seed predator. *American Journal of Primatology*, **14**, 11–35.

van Roosmalen, M.G.M., Mittermeier, R.A. & Milton, K. (1981). The bearded sakis, genus *Chiropotes*. In *Ecology and Behavior of Neotropical Primates*, ***Vol. 1***, ed. A.F. Coimbra-Filho and R.A. Mittermeier. Rio de Janeiro: Academia Brasileira de Ciências.

Part III

Genus Reviews and Case Studies

Stephen F. Ferrari

While they share many morphological and ecological traits, each of the four pitheciid genera is also distinguished by its own unique characteristics, ranging from body size, pelage and even tail length, through dietary adaptations and habitat preferences to social organization. This diversity is fundamental to the understanding of evolutionary processes and the phylogenetic relationships among genera, as well the interpretation of ecological and zoogeographical patterns. Knowledge of this diversity is equally essential to the development of effective conservation programs, an unfortunate necessity for a growing number of pitheciid species. Behavioral field

Evolutionary Biology and Conservation of Titis, Sakis and Uacaris, eds. Liza M. Veiga, Adrian A. Barnett, Stephen F. Ferrari and Marilyn A. Norconk. Published by Cambridge University Press.

studies constitute the primary source of information for the understanding of ecological processes and patterns, whether at the level of population, species or subfamily.

As Russ Mittermeier has pointed out in the foreword, we knew virtually nothing of the natural history of the pitheciids prior to the late 1970s. In fact, pitheciines, in particular, were rarely kept in captivity, and few researchers were lucky enough to have even seen one of these elusive primates, let alone studied them in the wild. The titis were slightly better understood (Mason 1974; Kinzey *et al.* 1977), although this knowledge was still relatively limited, especially considering the diversity of the genus. While it included chapters covering all four pitheciid genera, then, Mittermeier and Coimbra-Filho's (1981) ground-breaking compendium is also testimony to the overall lack of information available at the time.

There has been considerable progress over the past three decades, however, and if they were rewritten on the basis of present-day knowledge, all four chapters would be far more substantial, although they would still be relatively sparse in comparison with those of some other platyrrhine genera, such as *Alouatta* and *Saguinus*. Much of this progress has been due to the pioneering work of the late Márcio Ayres (1981, 1986) and Warren Kinzey (1977, 1992), which not only provided a great deal of original and enduring data, but also inspired subsequent generations of researchers. At the present time, research is ongoing at a dozen or more sites distributed throughout the geographic range of the Pitheciidae, although this is still a relatively small number, considering that there are at least 48 recognized taxa, most of which have ample ranges, encompassing an often considerable variety of ecological conditions.

In any case, the overall number of studies may not be quite as important as the species involved, and the type of research being carried out. It would seem obvious to prioritize the species most threatened with extinction, for example, although it may be equally important to sample the full amplitude of the ecological variation experienced by a genus or species group in order to more fully understand its adaptations and limitations. In this sense, the geographical spread of the chapters presented in this section is in itself an encouraging sign, although there is still room for improvement and, hopefully, the next generation of pitheciid researchers will explore ever-expanding frontiers, in ecological, geographic and taxonomic terms. For the time being, suffice to say that the data now available on the Pitheciidae represent a quantum leap over the state of our knowledge only a generation ago (i.e. Mittermeier & Coimbra-Filho, 1981).

Monitoring the behavior of any free-ranging primate is a challenging task, but pitheciids are notoriously difficult to study in the wild (see Pinto *et al.*, Chapter 13), albeit for slightly different reasons, depending on the genus. Bearded sakis, in particular, are known for both their agility and their preference for the highest forest strata, which reduces visibility and hampers habituation. In many parts of the Amazon, sakis are known as *macacos voadores* or *monos voladores* – "flying monkeys" – because of their typical rapid leaping style of movement through the canopy, but perhaps also for their almost ethereal presence in the forest. While uacaris may not be quite as agile, many populations present the fieldworker with a unique challenge in their preference for flooded forest habitats, which may be submerged in up to 4 or 5 m of water during half the year.

By contrast, titis are almost pathologically shy, and seem to spend most of their time moving quietly in dense vegetation. However, they will normally respond to playbacks of the genus' characteristic duet vocalization (Melo & Mendes 2000; Santos Junior 2007), and will even approach to within a few meters of the source, which at least facilitates contact, if not necessarily monitoring. To further assist observation, some researchers (e.g. Müller 1996; Neri 1997) have successfully used radio telemetry, although this does require trapping, which may not always be feasible under all field conditions.

In fact, considering the difficulties of observing pitheciids in the wild, on the one hand, and the availability of the technology, on the other, it is perhaps surprising that more studies have not made use of radio telemetry. The only pitheciine research to have involved telemetry is that of Vié *et al.* (2001) on *Pithecia pithecia*. Perhaps significantly, both this study, and that of Neri (1997) involved animals rescued during the flooding of hydroelectric reservoirs. Otherwise, most researchers appear to have assumed that the potential benefits, in terms of the efficiency of monitoring, may be outweighed considerably by the challenges – and especially the risks – of capturing these elusive monkeys. The more systematic use of telemetry may nevertheless constitute an important next step in the advancement of our knowledge of their ecology, although this may depend on significant advances in capture techniques.

This section is concerned with the identification and evaluation of the ecological differences and similarities that constitute the essence of a unique primate radiation. While the titis are allocated to a distinct subfamily, a more useful subdivision, from a purely ecological perspective, would appear to be that between the specialized seed predators (*Cacajao* and *Chiropotes*) on the one hand, and the smaller-bodied and much less-specialized sakis and titis on the other.

Sakis and titis not only have relatively similar, more generalized diets, for example, but they also typically form relatively small, territorial family groups in the wild, in contrast with the much larger aggregations and characteristic fission–fusion organization of *Cacajao* and *Chiropotes*, a pattern typical of specialized frugivores such as *Ateles* and *Pan*. The virtual absence of sympatry between *Cacajao* and *Chiropotes* points to a limited potential for the exploitation of such a specialized niche, which also appears to be related to the utilization of specific types of forest (Ayres & Prance, Chapter 12). By contrast, both *Callicebus* and *Pithecia* appear to be less selective of their habitat, overall, and may even be relatively successful in marginal and secondary habitats. However, a number of recent studies (see e.g. Boyle *et al.*, Chapter 24 and Santos *et al.*, Chapter 23) have shown that bearded sakis, at least, can be relatively tolerant of habitat fragmentation.

From a simplistic evolutionary perspective, then, the sakis and titis appear to represent a more primitive pitheciid condition of smaller body size, relatively unspecialized feeding behavior and more catholic habitat preferences. However, there are also major differences between the two genera. *Callicebus* is the only pitheciid found outside the Amazon–Orinoco basins, for example, and, in stark contrast with the pitheciines, interspecific sympatry can be found in some areas of the western Amazon basin, where species of the yellow-handed titi (*Callicebus torquatus*) group coexist with those of the *Callicebus cupreus* and *Callicebus moloch* groups (Defler 1994; van Roosmalen *et al.* 2002; Bicca-Marques & Heymann, Chapter 17).

Intrageneric sympatry is rare in platyrrhines – apart from *Callicebus*, it is found only in *Cebus* and *Saguinus* – and where it does occur, the two species typically form systematic, long-term associations, mediated by subtle distinctions in foraging strategies (Terborgh, 1983). In stark contrast, but in parallel with the *Cacajao–Chiropotes* discontinuum, the coexistence of titi species appears to depend on the occupation of distinct habitats, with yellow-handed titis preferring pristine *terra firme* forest, and its local congener, more marginal habitats, including swampy floodplain forests and anthropogenic habitats (Peres 1988; Defler 1994). Even so, there are precious few data on the ecological characteristics of such populations, and while the ecological differences among the four pitheciid genera are now relatively well understood, the relationships among sympatric populations are still poorly known (but see Gregory & Norconk, Chapter 28).

While the chapters in this section explore many of the patterns – both unifying and differentiating – that characterize the ecology of this diverse group of primates, they also emphasize the underlying paucity of our knowledge, even though some (Bezerra *et al.*, Chapter 30; Cunningham *et al.*, Chapter 26; Thompson & Norconk, Chapter 27; and Wright, Chapter 21) focus on surprisingly specific aspects of pitheciid behavior. This contrasts considerably with the fact that few, if any detailed ecological data are available for many species. Unfortunately, "classic" studies of behavioral ecology, such as that of Ayres (1986), in which a social group is monitored continuously over at least an annual cycle, are still few and far between, and even where some level of monitoring has extended over longer periods (see Santos *et al.*, Chapter 23), virtually nothing is known of generation-scale population dynamics or longitudinal ecological processes in most species. In the case of *Cacajao* and *Chiropotes*, in particular, our knowledge of social organization, mating systems and dispersal patterns remain incipient, at best.

In presenting the state of the art of our knowledge of the ecology of the pitheciids, then, the 18 chapters of this section also highlight many of its deficiencies. While we now know far more about what the species of each genus eat, where they live and how they behave than we did a generation ago, the same cannot be said for each and every species, or many of the more specific but nevertheless important questions. In addition, as our knowledge has grown, so has our understanding of the considerable variability that characterizes any ecological trait in such a diverse group of animals. Even such an apparently simple task as cataloguing all the plant species exploited by a single pitheciid might demand another generation, or even two.

Clearly, many challenges still lie ahead, and perhaps one of the greatest will be the identification of priorities for future studies. Conservation is the most obvious of these, and in general, studies of endangered species should probably be prioritized over those of less-threatened ones, as far as possible. However, the understanding of broader ecological processes and evolutionary patterns may be at least important for the development of effective management strategies. Ultimately, the long-term survival of pitheciid populations will depend on the protection of the ecosystems they inhabit, and all the cases studies presented in this section will surely contribute to this process.

References

Ayres, J.M. (1981). Observações sobre a ecologia e o comportamento dos cuxius (*Chiropotes albinasus* e *Chiropotes satanas,* Cebidae: Primates). Unpublished Masters' dissertation, Universidade do Amazonas.

Ayres, J.M.C. (1986). Uakaris and Amazonian flooded forests. Unpublished PhD thesis, Cambridge University.

Defler, T.R. (1994). *Callicebus torquatus* is not a white-sand specialist. *American Journal of Primatology*, **33**, 149–154.

Kinzey, W.G. (1977). Diet and feeding behaviour of *Callicebus torquatus*. In *Primate Ecology: Studies of Feeding and Ranging Behaviour in Lemurs, Monkeys and Apes*, ed. T.H. Clutton-Brock. London: Academic Press, pp. 127–151.

Kinzey, W.G. (1992). Dietary and dental adaptations in the Pitheciinae. *American Journal of Physical Anthropology*, **88**, 499–514.

Kinzey, W.G., Rosenberger, A. L., Heisler, P.S., *et al.* (1977). A preliminary field investigation of the yellow handed titi monkey, *Callicebus torquatus torquatus*, in northern Peru. *Primates*, **18**, 159–181.

Mason, W.A. (1974). Comparative studies of social behavior in *Callicebus* and *Saimiri*: behavior of male–female pairs. *Folia Primatologica*, **22**, 1–8.

Melo, F.R. & Mendes, S.L. (2000). Emissão de gritos longos por grupos de *Callicebus nigrifrons* e suas reações a playbacks. In *A Primatologia no Brasil*, vol. **7**, ed. C. Alonso & A. Langguth. João Pessoa: SBPr & Editora Universitária, pp. 215–222.

Mittermeier, R.A. & Coimbra-Filho, A.F. (1981). *Ecology and Behavior of Neotropical Primates*, vol. **1**. Rio de Janeiro: Academia Brasileira de Ciências.

Müller, K.-H. (1996). Diet and feeding ecology of masked titis (*Callicebus personatus*). In *Adaptive Radiations of Neotropical Primates*, ed. M.A. Norconk, A.L. Rosenberger & P.A. Garber. New York, NY: Plenum Press, pp. 383–401.

Neri, F.M. (1997). Manejo de *Callicebus personatus*, Geoffroy 1812, resgatados: uma tentativa de reintrodução e estudos ecológicos de um grupo silvestre na Reserva do Patrimônio Natural Galheiro – Minas Gerais. Unpublished Master's dissertation, Universidade Federal de Minas Gerais.

Peres, C.A. (1988). Primate community structure in western Brazilian Amazônia. *Primate Conservation*, **9**, 83–87.

Santos Junior, E.M. (2007). Observações preliminares sobre a ecologia comportamental do *Callicebus comibrai* na Mata Atlântica de Sergipe. Unpublished undergraduate dissertation, Universidade Federal de Sergipe.

Terborgh, J. (1983). *Five New World Monkeys: A Study in Comparative Ecology*. Princeton, NJ: Princeton University Press.

van Roosmalen, M.G.M., van Roosmalen, T. & Mittermeier, R.A. (2002). A taxonomic review of the titi monkeys, genus *Callicebus* Thomas, 1903, with the description of two new species, *Callicebus bernhardi* and *Callicebus stephennashi*, from Brazilian Amazonia. *Neotropical Primates*, **10** (Suppl.), 1–52.

Vié, J.C., Richard-Hansen, C. & Fournier-Chambrillon, C. (2001). Abundance, use of space, and activity patterns of white-faced sakis (*Pithecia pithecia*) in French Guiana. *American Journal of Primatology*, **55**, 203–221.

Chapter 13

Why we know so little: the challenges of fieldwork on the Pitheciids

Liliam Patricia Pinto, Adrian A. Barnett, Bruna Martins Bezerra, Jean Philippe Boubli, Mark Bowler, Nayara de Alcântara Cardoso, Christini Barbosa Caselli, Maria Juliana Ospina Rodríguez, Ricardo Rodrigues Santos, Eleonore Zulnara Freire Setz & Liza Maria Veiga

Introduction

Possessing a suite of unusual and interesting features, Pitheciids are at the extremes of many of primatology's ecological and sociological continua (see Norconk 2011). Pitheciids should provide acute tests of many primatological models; however, this is frequently thwarted by the lack of even the most basic quantitative information concerning ecology, behavior and social organization. Such gaps are due not only to the small number of studies, but also to difficulties in obtaining data. This chapter considers why, given that these primates possess such aesthetic and intellectual appeal and high conservation value, they have been so little studied.

The peculiarities of pitheciids, combined with their native habitats' inherent challenges, have often undermined potentially successful fieldwork. Some researchers who began working with species of *Callicebus* or *Chiropotes*, for example, simply gave up because of problems habituating the animals or because the study was so difficult that the quality and quantity of gained data would not be worth the effort. Several researchers who persisted were either unable to obtain the expected volume of data or were incapable of answering many of their initial research questions, sometimes both.

Using field experiences from a variety of research projects across a range of habitat types and pitheciid taxa, this chapter reports on the problems that arose during these studies and presents suggestions to minimize similar difficulties in the future.

Photo 13.1 Small wooden canoes – the work horses for flooded forest primate research. (See color plate section).

Evolutionary Biology and Conservation of Titis, Sakis and Uacaris, eds. Liza M. Veiga, Adrian A. Barnett, Stephen F. Ferrari and Marilyn A. Norconk. Published by Cambridge University Press.

Photo 13.2 A phenology quadrat in flooded forest. Plastic belts are easier to move as the water rises and falls than classic metal tags (which anyway, get bitten by river turtles and capuchin monkeys). Photo: Adrian Barnett.

Problems 1: Study area features, low species densities and home-range size

Pitheciids are often rare and present at low densities. In most areas where uacaris occur, for example, they are encountered infrequently, are only briefly present before moving elsewhere and may not be found again for many months (R. Bodmer, pers. commun.; E. Heymann, pers. commun.). Such factors, combined with large daily and annual ranges, have resulted in a failure to successfully establish research sites where uacaris can be followed consistently.

Successful sites are often of a special nature. Ayres' (1986) study on *Cacajao calvus calvus* at Mamirauá, Brazil, focused on a group whose home range was bordered on all sides by *chavascal* (*várzea baixa*), a habitat they rarely entered. Within the favored range, called *restinga* (*várzea alta*), habitats occurred in narrow strips, facilitating group encounters and monitoring. *Cacajao calvus ucayalii* has an extensive geographical range but is rarely seen, consistently or in high numbers. However, in 1990, R. Bodmer and P. Puertas discovered Lago Preto on the Yavarí River, Peru, where groups could be found relatively easily. Although probably less constrained by unfavorable habitats than Mamirauá's white uacaris, topography may nevertheless be a factor in their constant presence and the ease with which groups here are found (Bowler 2007; Bowler & Bodmer 2010).

Naturally low densities may also present profound problems when locating study groups of *Chiropotes* and *Pithecia*, contributing to the scarcity of field studies in areas of continuous forest (Mittermeier & van Roosmalen 1981; Vié *et al.* 2001; Pinto 2008). *Pithecia* tolerate disturbance and may be found in forest fragments near urban settlements, but bearded saki densities may be substantially lower close to human communities, due a combination of anthropic factors, including hunting (Ferrari *et al.* 1999, 2003).

Cacajao ouakary primarily inhabits seasonally inundated black-water forests, requiring flooded-season observations to be made almost entirely from wooden canoes. While studying these animals in Jaú National Park, A. Barnett found that, although small and unstable, such canoes are almost silent, and a stealthy approach can greatly facilitate observation (Barnett 2010), while B. Bezerra found they serve as a good platform for vocalization studies (see Figure 13.1).

Cacajao calvus ucayalii at Lago Preto uses both flooded and non-flooded forests (Bowler 2007; Bowler & Bodmer 2010), again requiring a canoe. Collecting discarded fruit parts that do not float can be difficult, but canoes positioned below feeding monkeys can catch fruit on occasion. Daily uacari movement between forest types was a major problem at Lago Preto. Because canoes were not always available when monkeys moved from *terra firme* forests into flooded areas, animals were frequently lost during the high-water season. Lago Preto uacaris often used palm swamps, where following them is not enhanced by knee-deep (or more) mud and thick vegetation. Groups of *C. ouakary* were lost to canoe-based researchers when they moved fast through dense flooded *igapó* or entered into thickets of the razor-edged scandent sedge *Scleria* sp. (Cyperaceae). Additionally, fallen fruit and feces were often eaten by fish before they could be collected (Barnett 2010).

Uacaris have some of the largest known daily ranges of any primate. Because of this, M. Bowler could not cut trails in all areas at Lago Preto and most *C. c. ucayalii* following was off-trail (Bowler & Bodmer 2010). A. Barnett and B. Bezerra's studies of *C. ouakary* and L. Pinto's study of *Chiropotes albinasus* suffered similarly. While radiotracking offers a possible solution, their agility, large groups and preference for higher forest strata may make capturing uacaris and bearded sakis with anesthetizing darts especially difficult. Avoiding anesthetized animals drowning in flooded habitats is an additional challenge.

Figure 13.1 Work on Pitheciines sometimes requires innovative approaches. Bruna Bezerra and Eduardo de Souza study uacari responses to call playback, Jaú National Park, Brazil. Photo: Bruna Bezerra.

In addition, seasonal habitat choices may further complicate pitheciine fieldwork. For example, *C. ouakary* move to *terra firme* during the dry season, making it very difficult to find them, as walking through the forest causes excessive noise from dried-out fallen leaves. Working with *Pithecia pithecia chrysocephala* in the north of Manaus, Brazil, E. Setz solved this by sweeping up fallen leaves on the trails. However, such care is viable only in studies of taxa like *Pithecia*, with small home ranges.

For islands or forest fragments, such areas' physical circumscription facilitates fieldwork, while fragments' frequent proximity to towns or roads can ease logistics. However, at another scale, working in such habitats is also problematic, because the target primate's ecology and behavior probably will itself be influenced by isolation effects and habitat impoverishment.

Problems 2: Pitheciid behavioral characteristics

Pitheciids have particular behavioral characteristics that complicate their in-field localization and monitoring. This includes habituation, which for pitheciids usually lasts longer than for other Neotropical primates; while a *Callithrix* group may get used to observers within a few weeks (Bezerra & Souto 2008), the process can take many months with pitheciids. Ayres' (1986) *C. c. calvus* focal group never became completely habituated even after 700 h of observation, and rarely tolerated observers nearer than 30 m. Closer observations were always made from hiding. *Cacajao ouakary* are also extremely timid, which greatly complicated the initial fieldwork of A. Barnett and B. Bezerra. To begin with, the animals disappeared very quickly into dense vegetation as soon as they perceived human presence. Even after more than 6 months of habituation, uacaris remained very shy compared to Jaú National Park's other primate species (Barnett 2010). Although *Pithecia* habituation can be a protracted process, observation distance generally shortens progressively. When fully habituated, *Pithecia* will pass by at arms' reach (Setz 1991).

Timidity is a challenge, but is not uniform across species. Site characteristics also influence habituation speed. Working around 10 days a month, it took C. Caselli 604 h to habituate a *Callicebus nigrifrons* pair in continuous forest at Serra do Japi (Caselli 2008). In contrast, M. Rodríguez required only 75 h to habituate *Callicebus cupreus ornatus* in gallery forest on the eastern Colombian *llanos*. Because of their cryptic behavior and agility *C. nigrifrons* were seen only a few times and contact was often very quickly lost, even after the successful habituation. Meanwhile, previous primate research at the more open *llanos* site had probably pre-accustomed the titis to human presence.

M. Rodríguez habituated *C. c. ornatus* in 37 contact hours across 2 months by using a small area and concentrating search effort on known sleeping sites. She discovered the titis were very sensitive to fast movements beneath them, but usually remained calm if she walked placidly. In a private reserve in the Serra do Teimoso, Brazil, N. Cardoso required a similarly short period to habituate *Callicebus melanochir*, as sleeping localities were already known and the study area was small (Cardoso 2003). Prior to habituation, however, the animals could move very quickly, and completely conceal themselves, even when the entire study area was searched intensively. Later, a good trail system facilitated movement and assisted localization of these titis.

Natural timidity may partly explain such extreme habituation times, but social structure also plays a part: *Cacajao* and

Chiropotes often spend long periods split into subgroups (Ayres 1986; Barnett & Brandon-Jones 1997; Boubli 1999; Silva 2003; Veiga 2006; Bowler 2007; Pinto 2008; Barnett 2010; Bowler & Bodmer 2010). Consequently, observers may sequentially expend time in contact with only one subpart of a social group, and so require longer overall to habituate all members. Characteristically large group sizes (see Norconk 2011) do not help.

A. Barnett observed that once found, uacaris are often quite obvious, with loud calls and crashing leaps. However, some pitheciids can be difficult to detect. *Pithecia* live in small families and are very quiet, but as they are mostly trunk leapers they can be detected by the thumps accompanying their movement. Bearded sakis mostly employ quick quadrupedal locomotion, mainly on the thicker more horizontal branches of large trees (Ayres 1981). They are therefore quieter when moving through vegetation and so less detectable at medium-range distances than many similar-sized primates.

Bearded sakis are also timid, and use escape strategies as soon as they perceive humans close by. During habituation of *Chiropotes satanas* by L. Veiga in Tucuruí, Brazil and *C. albinasus* by L. Pinto in Flona Tapajós, the animals displayed two main researcher-avoidance tactics. The first was to escape by moving away swiftly, rarely stopping to eat, often grabbing food items on the move. Observers spent all day chasing them. The second tactic was to hide, usually in dense lianas; in this case, researchers spent hours below them, waiting patiently for them to emerge. L. Veiga and her assistants found a way of dealing with this second tactic: pretending to go away, making considerable noise as they "left", talking and crunching noisy leaves. They would then sneak back very quietly and hide. This artifice was very effective; almost without exception, the monkeys would appear.

Most problems encountered by R. Santos while habituating *Chiropotes utahickae* and *C. satanas* at Tucuruí were caused by females with infants, which were more cryptic than the rest of the group. In the studies of R. Santos, L. Veiga and L. Pinto, juveniles would emit alarm calls more frequently than adults, mainly when researchers were close (< 15 m). In these situations, monkeys became calmer when observers avoided eye contact with them, and crouched down or hid behind trees. Initially, many of the observations made by A. Barnett and B. Bezerra on flooded-forest-dwelling *C. ouakary* were likewise made with canoe and observers positioned behind (or within) canopies of small trees.

At Flona Tapajós, temporary associations of *C. albinasus* with *Cebus apella* and *Saimiri sciureus* increased the likelihood that the study group would move more quickly and so be easily lost (Pinto 2008). When the study began, these non-pitheciids would emit constant alarm calls while being followed, intensifying bearded saki nervousness. M. Bowler observed a similar situation with Peruvian red uacaris and *C. albifrons*, and (to a lesser extent) with *Lagothrix lagotricha*. For this reason, collateral habituation of these associated species at the study sites was needed to reduce interference with the target species.

As with other taxa, there are aspects of pitheciid ecology that are only revealed with experience. For example, in an exception to standard patterns of primate activity, *C. albinasus* at Flona Tapajós sometimes began their activities before dawn (Pinto 2008). This became especially pronounced when they were eating significant volumes of flowers whose peak nectar production was at night (e.g. *Caryocar glabrum* and *Caryocar pallidum*, Caryocaceae). Activity after dusk was also observed for *C. albinasus*, as well as for *C. c. calvus* by Ayres (1986) at Mamirauá. In both situations low light impedes the observer's ability to observe subject behaviors.

In *Cacajao* and *Chiropotes*, researches involving social issues are hindered by difficulties of distinguishing individual group members (e.g. Ayres 1981, 1986; Frazão 1992; Santos 2002; Veiga 2006). In both genera, large groups (40–100 animals in *Cacajao*), great mobility, visibility problems and low aggression levels complicate the search for parameters, such as natural marks and scars, which might otherwise permit individual characterization.

Despite having smaller home ranges than *Cacajao* and *Chiropotes*, *Pithecia* and *Callicebus* are also far from easy to study. In addition to being timid, these primates are small, cryptic and often prefer dense vegetation, traits which together make habituation singularly difficult (Müller *et al.* 1993; Müller & Schildger 1994; Neri 1997; Vié *et al.* 2001; Norconk 2006). With *Pithecia pithecia* of Guri Lake, Venezuela, for example, despite long-term ecological studies, investigation of social and sexual behavior has still not proved possible (Norconk 2006). Contrastingly, although habituation of a *Pithecia pithecia chrysocephala* group in continuous forest in Central Brazilian Amazonia by Calouro and Setz (unpublished data) was not itself difficult, finding them each month was a real challenge.

Pithecia and some *Callicebus* have a short activity period, entering their sleeping trees by early afternoon. Consequently, if the animals are not located before noon, there is little point continuing, because by then they are already calming down and going (quietly) to their sleeping sites (E. Setz, unpubl. data). *Pithecia*'s habit of halting and staying utterly immobile for more than 30 min is discouraging for many researchers. However, they continue to make characteristic bird-like vocalizations and realizing this can encourage a researcher to wait and persist (Setz 1991, 1993).

While most searches have been visual, sound-based techniques also show promise. Playback trials with selected *Callicebus* species having shown good results (Müller & Schildger 1994; Neri 1997; Melo & Mendes 2000), with titis usually vocally responding to playback (so revealing their presence and position) and even moving towards the sound source. For C. Caselli and A. Nascimento (pers commun.), this tool was decisive in locating the study group on a regular basis. It has also been used successfully with *Pithecia* (S. Palminteri, pers. commun.). Work by B. Bezerra on *C. ouakary* in Jaú National Park has shown they respond well to playback (Bezerra *et al.* 2010), and Bezerra *et al.* (2008) recommended that playback be used when biogeographic field surveys are attempting to assess the presence of uacaris. Playback does not help actually follow the monkeys, for which Müller and Schildger (1994) eventually used radio-tracking. However, radio-tracking has problems; expensive and invasive, its utilization may increase study animal stress and mortality (Müller & Schildger 1994; Neri 1997).

Final considerations

Because there are few long-term pitheciid studies, most investigations are pioneering. During a field study of wild pitheciids, the methodological and logistic problems may be so many and so extensive that researchers must devote considerable field time just to solving them. Given the difficulties inherent in studying pitheciids, initial fieldwork planning should include the strong possibility of needing to stay beyond the originally programmed period, requiring a more generous budget than anticipated.

Because pitheciids are demanding in terms of time, money and physical effort, we also suggest that those planning a field project attempt it using an integrated team of researchers, with each one focusing on one aspect of the life of the animals. Besides increasing fieldwork efficiency, such joint effort reduces costs and observer stress, while broadening research outcomes.

Pitheciids are hard. Overcoming the challenges they present requires perseverance, patience, dedication and, most importantly, a great deal of enthusiasm. Such qualities, allied with straightforward hard work, should improve the chances of project success and, hopefully, yield information that enhances our knowledge of, and with it our capacity to conserve, the members of this most unusual group of Neotropical primates.

Acknowledgments

The authors would like to thank the editors of this volume for inviting them to contribute this chapter. The authors would collectively like to thank all their field assistants and the respective organizations and funding agencies for making possible the fieldwork that led to this chapter: Akodon Ecological Consulting, Amazon Ecopark Lodge, American Society of Primatologists, CAPES, CNPq, Colombus Zoo, Conservation International, Eletronorte, Fundação de Amparo à Pesquisa do Estado de São Paulo, Fundação O Boticário de Proteção à Natureza, Idea Wild, Kapok Foundation, Living Rain Forest Foundation, LSB Leakey Foundation, Leakey Fund (UK), Mammal Society (UK), Percy Sladen Memorial Fund (Linnean Society), Pontificia Universidad Javeriana, Primate Action Fund, Primate Conservation Inc., Primate Society of Great Britain, Programme Alban, Sophie Danforth Conservation Fund, The LA Zoo, Roehampton University, The Rufford Foundation, University of Bristol, Wildlife Conservation Society. Thanks to R. Bodmer, E. Heymann, A. Nascimento and S. Palminteri for sharing unpublished observations.

References

Ayres, J.M. (1981). Observações sobre a ecologia e o comportamento dos cuxiús (*Chiropotes albinasus e Chiropotes satanas, Cebidae: Primates*). Unpublished Master's thesis, Instituto Nacional de Pesquisas da Amazônia, Manaus.

Ayres, J.M. (1986). Uakaris and the Amazonian flooded forests. Unpublished PhD thesis, Cambridge University.

Barnett, A.A. (2010). Diet, habitat use and conservation ecology of the golden-backed uacari, *Cacajao melanocephalus ouakary*, in Jaú National Park, Amazonian Brazil. PhD dissertation, Roehampton University, London. http://roehampton.openrepository.com/roehampton/.

Barnett, A.A. & Brandon-Jones, D. (1997). The ecology, biogeography and conservation of the uacaris, *Cacajao* (Pitheciinae). *Folia Primatologica*, **68**, 223–235.

Bezerra, B.M. & Souto, A.S. (2008). Structure and usage of the vocal repertoire of *Callithrix jacchus*. *International Journal of Primatology*, **29**, 671–701.

Bezerra, B., Barnett, A., Silva Junior, S.J., *et al.* (2008). Sounding out species: holes in our distributional knowledge of Amazonian primates, and the potential for call playback as a survey technique. *Presentation # 695, Abstracts, XXII.* Congress, International Primatological Society, Edinburgh, August 2–8, 2008: 249–250.

Bezerra, B., Souto, A. & Jones, G. (2010). Responses of golden-backed uacaris, *Cacajao melanocephalus*, to call playback: implications for surveys in the flooded Igapó forest. *Primates*, **51**, 327–336.

Boubli, J.P. (1999). Feeding ecology of black-headed uacaris (*Cacajao melanocephalus melanocephalus*) in Pico da Neblina National Park, Brazil. *International Journal of Primatology*, **20**, 719–749.

Bowler, M. (2007). The ecology and conservation of the red uakari monkey on the Yavarí River, Peru. Unpublished PhD thesis, University of Kent, Canterbury.

Bowler, M. & Bodmer, R. (2010). Social behavior in fission–fusion groups of red uakari monkeys (*Cacajao calvus ucayalii*). *American Journal of Primatology*, **71**, 976–987.

Cardoso, N. (2003). Uso de recursos alimentares por *Callicebus melanochir* (Wied-Neuwied, 1820) (Primates: Pithecidae) em uma área perturbada na Reserva Natural Serra do Teimoso, Jussari-BA. Monografia de graduação (Ciências Biológicas), Universidade Estadual de Santa Cruz, Bahia.

Caselli, C.B. (2008). Ecologia alimentar, padrão de atividade e uso de espaço por *Callicebus nigrifrons* (Primates: Pitheciidae). Master's thesis, Unicamp, SP.

Ferrari, S., Emidio-Silva, C., Lopes, M., *et al.* (1999). Bearded sakis in south-eastern Amazonia – back from the brink? *Oryx*, **33**, 346–351.

Ferrari, S.F., Iwanaga, S., Ravetta, A.L., *et al.* (2003). Dynamics of primate communities along the Santarém-Cuiabá Highway in south-Central Brazilian Amazonia. In *Primates in Fragments: Ecology and Conservation*, ed. L. Marsh. New York, NY: Kluwer Academic/ Plenum Publishers, pp. 123–144.

Frazão, E.R. (1992). Dieta e estratégia de forragear de *Chiropotes satanas chiropotes* (Cebidae: Primates) na Amazônia Central Brasileira. Unpublished Master's thesis, Instituto Nacional de Pesquisas da Amazônia, Manaus.

Melo, F.R. & Mendes, S.L. (2000). Emission of long calls by groups of *Callicebus nigrifrons* and their reactions to playbacks. *A Primatologia no Brasil*, 7, 215–222.

Mittermeier, R.A. & van Roosmalen, M.G.M. (1981). Preliminary observations on habitat utilization and diet in eight Surinam monkeys. *Folia Primatolica*, **36**, 1–39.

Müller, K.H. & Schildger, J. (1994). Capture and radio-telemetry of masked titi monkeys, *Callicebus personatus melanochir*. *Neotropical Primates*, **2**, 7–8.

Müller, K.H., Heiduck, S. & Shultze, S. (1993). Field investigations on masked titi monkeys (*Callicebus personatus melanochir*) in Una, Bahia, Brazil. *Neotropical Primates*, **1**, 21–22.

Neri, F.M. (1997). Manejo de *Callicebus personatus* (Geoffroy 1812) resgatados: uma tentativa de translocação e estudos ecológicos de um grupo silvestre na Reserva Particular do Patrimônio Natural Galheiro – Minas Gerais. Unpublished Master's thesis, Universidade Federal de Minas Gerais, Belo Horizonte.

Norconk, M.A. (2006). Long-term study of group dynamics and female reproduction in Venezuelan *Pithecia pithecia*. *International Journal of Primatology*, **27**, 653–674.

Norconk, M.A. (2011). Sakis, uakaris, and titi monkeys: behavioral diversity in a radiation of primate seed predators. In: *Primates in perspective*, 2nd edition, ed. C.J. Campbell, A. Fuentes, K.C. MacKinnon, M. Panger, S.K. Bearder. New York, NY: Oxford University Press, pp. 123–138.

Pinto, L.P. (2008). Ecologia alimentar do cuxiú-de-nariz-vermelho *Chiropotes albinasus* (Primates: Pitheciidae) na Floresta Nacional do Tapajós, Pará. Unpublished Doctoral thesis, Universidade Estadual de Campinas, Campinas.

Santos, R.R. (2002). Ecologia de cuxiús (*Chiropotes satanas*) na Amazônia Oriental: Perspectivas para a conservação de populações fragmentadas. Unpublished Master's thesis, Museu Paraense Emílio Goeldi and Universidade Federal do Pará, Belém.

Setz, E.Z.F. (1991). Métodos de quantificação de comportamento de primatas em estudos de campo. In *A primatologia no Brasil – 3*, ed. A.B. Rylands and A.T. Bernardes. Belo Horizonte: Fundação Biodiversitas, pp. 411–435.

Setz, E.Z.F. (1993). Ecologia alimentar de um grupo de parauacus *Pithecia pithecia chrysocephala* em um fragmento florestal na Amazônia Central. Unpublished Doctoral thesis, Universidade Estadual de Campinas, Campinas.

Silva, S.S.B. (2003). Comportamento alimentar do cuxiú-preto (*Chiropotes satanas*) na área de influência do reservatório da usina hidrelétrica de Tucuruí–Pará. Unpublished Master's thesis, Museu Paraense Emílio Goeldi and Universidade Federal do Pará, Belém.

Veiga, L.M. (2006). Ecologia e comportamento do cuxiú-preto (*Chiropotes satanas*) na paisagem fragmentada da Amazônia Oriental. Unpublished Doctoral thesis, Universidade Federal do Pará, Belém.

Vié, J.-C., Richard-Hansen, C. & Fournier-Chambrillon, C. (2001). Abundance, use of space and activity patterns of white-faced sakis (*Pithecia pithecia*) in French Guiana. *American Journal of Primatology*, **55**, 203–221.

Chapter 14

Ecology and behavior of uacaris (genus *Cacajao*)

Adrian A. Barnett, Mark Bowler, Bruna M. Bezerra & Thomas R. Defler

Introduction

Modern uacaris are representatives of a lineage that diverged from its sister clade, *Chiropotes*, some 9 my bp (Bonvicino *et al.* 2003; and Kay *et al.*, Chapter 1). *Cacajao* is the only pitheciine genus (indeed, the only cebine genus) endemic to the Amazon basin. Its distribution is the most restricted of the four pitheciine genera, occurring only in the northwestern sector of the Amazon basin (see Figure 4.5, Chapter 4). It has been posited that the evolution of the genus was prompted by the development of substantial flooded forests along the riverine margins in the upper Amazon basin (Ayres 1986, 1989; Ayres & Prance, Chapter 12), although Boubli and Ditchfield (2000) point out that the time of divergence precedes the formation of the Holocene *várzeas*. Hershkovitz's (1987) proposal that speciation in the genus was related to isolation of populations in forest refugia during the Pleistocene has been refuted by analysis of mitochondrial DNA (Boubli & Ditchfield 2000), although the model itself (if not the timing) may be sound.

Until recently, Hershkovitz (1987) provided the most comprehensive revision of *Cacajao* taxonomy. This has latterly been supplemented by the work of Figueiredo (2006), Boubli *et al.* (2008), Silva Junior *et al.* (Chapter 4) and Figueiredo *et al.* (Chapter 3). From this emerges the consensus that there are two uacari lineages: the *calvus*-group (the bald or red-faced uacaris) and the *melanocephalus*-group (the tonsured or black-faced uacaris). The former includes (*sensu* Hershkovitz, 1987) the white uacari (*C. calvus calvus*), Brazilian red uacari (*C. c. rubicundus*), Noveas' uacari (*C. c. novaesi*), and the Peruvian red uacari (*C. c. ucayalii*). The status of the very pale white uacari population on the Jurua River (Silva Junior & Martins 1999) is currently uncertain (Silva Junior *et al.*, Chapter 4, and Figueiredo *et al.*, Chapter 3). While the taxonomy of Hershkovitz (1987) remains unchanged for the red-faced uacaris, the same is not true for the black-faced clade, and the two sub-species of *C. melanocephalus* recognized by Hershkovitz (*C. m. melanocephalus*, the black-backed uacari, and the golden-backed uacari, *C. m. ouakary*) are now considered full species (Figueiredo 2006; Boubli *et al.* 2008). Although Boubli *et al.* (2008) offer an alternative view, here we follow Ferrari *et al.* (2009, submitted) and Silva Junior *et al.* (Chapter 4) in calling them *C. ouakary* and *C. melanocelphalus*, respectively. A third taxon, *C. ayresi*, was described by Boubli *et al.* (2008). Figure 4.5 of Chapter 4 shows the distribution of the currently recognized taxa.

Although the subject of some factual field-based natural history reporting in the first 75 years of the twentieth century (e.g. Lönnberg 1938; Olalla 1955), field studies of uacaris began late, compared with other Neotropical primates, with the first PhD study occurring only in the mid 1980s. Several taxa (e.g. *C. c. novaesi*, *C. c. rubicundus*) have received little attention beyond basic surveys. Others have been the subjects of longer-term studies (e.g. *C. c. calvus*: Ayres 1986; *C. c. ucayalii*: Bowler 2007; Bowler & Bodmer 2009, 2010; *C. melanocephalus*: Boubli, 1997, 1999; *C. ouakary*: Barnett *et al.* 2005, 2008, 2011, 2012; Bezerra *et al.* 2007, 2008, 2010a, 2010b; Bezerra, Barnett *et al.* 2011; Bezerra, Souto *et al.* 2011). Due in no small part to the difficulties of working on the genus (Pinto *et al.*, Chapter 13), much basic information is still lacking and the ecology of the genus as a whole is very much less known than that of both other pitheciines (particularly *Pithecia*) and other cebids (such as *Ateles*, *Cebus* and *Saimiri*). To add to the difficulties, all red-furred uacaris were once included under the rubric "*C. rubicundus*", or conflated as a single subspecies *C. c. rubicundus*. This has led to information on red-furred uacaris being confounded and conflated, and many of the reports referring to "rubicundus" uacaris (e.g. Davis 1931; Fontaine 1981; Fontaine & Hench 1982; Ayres 1986) do, in fact, refer to *C. c. ucayalii* (Barnett 2010a, submitted).

Although substantially more has become known since the last major review of the biology of the genus *Cacajao* (Barnett & Brandon-Jones 1997), there are still enormous gaps in our knowledge. It is the aim of this chapter to summarize current knowledge of the ecology, natural history, social behavior and conservation ecology of the uacaris, and to suggest areas where future research might be most effectively directed to both increase our knowledge of these animals and to maximize the chances of their long-term future survival.

Evolutionary Biology and Conservation of Titis, Sakis and Uacaris, eds. Liza M. Veiga, Adrian A. Barnett, Stephen F. Ferrari and Marilyn A. Norconk. Published by Cambridge University Press.

Photo 14.1 Juvenile *Cacajao calvus novaesi* kept as a pet by members of the Morro Alto community. Uacari Sustainable Development Reserve on the Rio Jurua, Brazil. Photo: Waldener Endo. (See color plate section).

Data on *Cacajao* in captivity is given in Becker *et al.* (Chapter 34) and Bezerra, Barnett *et al.* (2011). Information on the taxonomic and evolutionary histories, anatomy (corporal and dental), cytology, etymology, morphology, myology, paleontology, parasitology and physiology of the genus may be found in Barnett (submitted), while Barnett (2005) provides an overview for *C. ouakary* and *C. melanocephalus*. Considerations of the evolutionary scenarios that resulted in the naked red heads of the *C. calvus* group occur in Barnett & Ross (submitted) and Bowler (2007). The origins of the genus have been considered by Ayres (1986), Hershkovitz (1987) and Boubli and Ditchfield (2000).

Habitat preferences

Cacajao spp. are found in the seasonally flooded forests of the Negro and Japurá-Solimões river systems and their affluents, as well as in *terra firme*, white-sand campinas and palm swamps forests (Brazil, Colombia, Peru: Boubli 1997; Barnett 2005, 2010a; Bowler 2007). In the Pico de Neblina region of Venezuela and Brazil they also inhabit higher elevation forests (Boubli *et al.* 2008).

Amazonian flooded forests occur in two forms (Prance 1979): *igapó* (on blackwater rivers, e.g. Rio Negro) and *várzea* (on white water ones, e.g. Rio Japurá-Solimões). One subspecies (*C. c. calvus*) is entirely restricted to *várzea* (Ayres 1986), others (e.g. *C. ouakary*) are known to seasonally migrate between seasonally flooded forest and *terra firme* (never-flooded) forest (Barnett 2005; Barnett *et al.*, Chapter 16). The timing of such movements appears to be determined by seasonal availability of food items (Defler 2004; Barnett *et al.* 2005; Bowler 2007; Barnett 2010a; Barnett *et al.*, Chapter 16).

Still other uacari taxa have some populations that live in flooded forest most months of the year (e.g. *C. melanocephalus*, *igapó*: Lehman & Robertson 1994a, 1994b; *C. c. ucayalii*, *igapó* and buritizal: Aquino 1995, 1998; Aquino & Encarnación 1999; *C. ayresi*, buritizal and *terra firme*: Boubli & de Lima 2009), while some live permanently in never-flooded forest habitats (e.g. *C. melanocephalus*, white sand campina: Boubli 1997, 1999; *C. c. ucayalii*, Heymann & Aquino 2010). *Cacajao calvus ucayalii* at Lago Preto move between *várzea*, *terra firme* and swamps of the palm *Mauritia flexuosa*. *Terra firme* is most used, overall, but preferences are seasonal (Bowler 2007).

For *C. c. calvus*, Ayres (1986) recorded movements between "*restinga*" and "*chavascal*" *várzea* subtypes, and reported they related to the availability of hard-husked unripe fruits. Bowler (2007) noted some populations making seasonal resource-related movements between *terra firme* and *várzea*, and between two types of "*buritizal*", specifically in search of *Mauritia flexuosa* pulp. *Cacajao ouakary* has recorded seasonal resource-tracking movements between *igapó* and *terra firme* forests at three sites in the Amazon (see Barnett *et al.*, Chapter 16).

For *C. ouakary*, migrations into *terra firme* rarely appear to last more than 3 months (Almeida *et al.* 2008; Barnett 2005, 2010a; Barnett *et al.* 2008; Silva *et al.* 2008). In some cases, population movements can be quite limited, with animals reported to be never more than 500 m from *igapó* (Barnett *et al.* 2005). Elsewhere, *C. ouakary* has been reported to range several kilometers into *terra firme* (Defler 2004). Possible reasons for such interpopulation differences are discussed by Barnett *et al.* (Chapter 16).

In the populations of *C. melanocephalus* studied by Jean-Philippe Boubli (e.g. Boubli 1997, 1999; Boubli & Tokuda 2008; Boubli & de Lima 2009), the study animals were in an area of highly unusual botanical composition (Boubli 2002a), where extended fruiting season, large crop size and patchiness of the most tree species playing the largest role in the uacari diet means that considerable within-habitat migration is sufficient to sustain the *C. melanocephalus* troops (Boubli & de Lima 2009). Other *C. melanocephalus* populations live in *igapó* (Lehman & Robertson 1994a, 1994b) and their biology may be more comparable to those of *C. ouakary* (e.g. Defler 1999b; Barnett *et al.* 2005; Bezerra, Barnett *et al.* 2011).

Uacaris have long been regarded as flooded-forest specialists (e.g. Hershkovitz 1987; Barnett & Brandon-Jones 1997). However, recent research has revealed a more finessed situation. For, while such forest types are clearly important, many populations make near-permanent use of lowland *terra firme*, *caatinga*, palm swamp and other forest types, while altitude ranges from 100–1500 m (*C. melanocephalus*) to some 50–100 m (almost all other taxa). The forest types used by uacaris vary greatly in composition and structure (compare, for example, the plant lists and structural descriptions of Ayres 1986 for *várzea*; Boubli 2002a for *caatinga*; Boubli *et al.* 2008 for montane forests; Ferreira 1997, 1999 for *igapó*; and Ferreira & Prance 1998 for *terra firme* forests at Jaú; and Bowler 2007 for four separate habitat types on the Yavari River, Peru). However, in the Solimões, Negro and Ucayali basins the habitats used by uacaris all share the characteristic of being nutrient-restricted and often richer than average in hard-shelled large-seeded families (ter Steege *et al.* 2006; Bowler 2007). The consequences of this are explored elsewhere (see below, plus Setz *et al.*, Chapter 7; Ayres & Prance, Chapter 12).

The division of the genus *Cacajao* (*vide* Hershkovitz 1987) into a clade that occurs on black-waters (*melanocephalus* – *sensu* Hershkovitz 1987), and another (*calvus*) that occurs on white-water, is an attractive scenario, but insufficiently nuanced. Both *C. c. rubicundus* (Mittermeier & Coimbra-Filho 1977) and *C. c. ucayalii* (Swanson-Ward & Chism 2003) are known to have populations in forests bordering small black-water rivers. Similarly, *C. ouakary* at the headwaters of the Rio Jaú in the Amanã Extractive Reserve occur in areas that are, at least seasonally, white water (de Souza *et al.* 2004; Amaral *et al.* 2005). Ayres (1986) also recorded *C. ouakary* in predominantly white-water areas, but noted that all these records were from vegetation surrounding black-water lakes. In addition, while some populations of *C. melanocephalus* (Boubli 1997, 1999; Boubli & de Lima 2009) live in areas where inundated forest does not occur in any relevant quantity, the newly described *C. ayresi* lives in a clear-water system of the Curuduri/Madixi rivers which is dominated by huge patches of *Mauritia flexuosa* palm (Boubli *et al.* 2008). The area is a huge flooded forest, but the monkeys also utilize the upland *terra firme* when the water (and or *Mauritia* crop) is low (see Boubli *et al.* 2008). Additionally, some populations of *C. c. ucayalii* are also almost exclusively palm-swamp dwelling (Heymann & Aquino 2010). Clearly the situation with respect to habitat choice and river hydrology is both varied and complex (see Bowler 2007).

Unlike the situation recorded for *Pithecia* (Norconk 2011) and *Chiropotes* (Boyle *et al.* 2006), there are no reliable records of any *Cacajao* taxon inhabiting secondary, fragmented or degraded habitats for any extended period. Jean-Philippe Boubli, for example, reports (pers. commun.) having seen *C. ouakary* passing through a large area of *capoeira* (forest regenerating from land used by subsistence agriculture), but that the animals did not remain there. Such preferences probably result from feeding specializations and the requirement for a continuous supply of fruits (see below, and Setz *et al.*, Chapter 7), a community phenology likely to be absent in a plant community that has been disrupted by human action (see Johns & Skorupa 1987; Saunders *et al.* 1991).

Annual rainfall patterns in habitats used by *Cacajao* range from highly seasonal (e.g. Jaú and Amanã), to permanently pluvial (Caparú, Lago Preto and Neblina). Annual rainfall ranges from 2900 mm for Amanã (Barnett *et al.*, Chapter 16), to as much as 3836 mm (Caparú, Colombia: Defler & Defler 1996). Annual ambient temperatures average 26°C for *C. ayresii* and *C. c. ucayalii* (Boubli *et al.* 2008 and Bowler 2007, respectively), and down to 15°C for *C. melanocephalus*, in those populations that live at higher altitudes (Boubli, pers. commun.).

Population ecology

Frequencies, densities and ranging patterns

When in large noisy groups, uacaris are often the easiest primates to see in a region. Locally abundant, they are often the most frequently encountered primate in flooded forests (Aquino 2005; da Cunha & Barnett 1989; Barnett *et al.* 2005; Bennett *et al.* 2001; Bezerra, Barnett *et al.* 2011) and *terra firme* areas (Bowler 2007; Bowler *et al.*, Chapter 15). Densities are given in Table 14.1. The extreme variation may be due in part to such natural factors as soil fertility and variation in forest productivity between habitats, as well as anthropic impacts such as the intensities of past and present hunting practices (Aquino 1995; Palacios & Peres 2005; Bowler 2003, 2007). Further surveys are needed with comparable methodology and a thorough analysis of variables before true patterns can be discerned.

Uacaris have some of the largest home ranges and daily travel distances of any Neotropical primate. Data for uacaris and comparative data for other Neotropical taxa of comparative weight and bigger are given in Table 14.2. The extreme values given for *Cacajao* doubtless reflect the fact that, as originally pointed out by Norconk and Kinzey (1994), when foraging for unripe hard fruits uacaris are exploiting a narrow window of resource availability, one which leads to resources that are highly spatially dispersed and often very quickly depleted. Long maturation time of the fruits of many flooded forest trees (Parolin *et al.* 2002) does not necessarily ameliorate this situation, as only part of the fruit crop may be at the appropriate stage at any one time (Barnett 2005, 2010a; Barnett *et al.* 2008).

Group size

Reported group sizes in *Cacajao* differ widely (Table 14.3). This is due to the fission–fusion behavior (*sensu* Aureli *et al.* 2008) of the genus (Defler 1999a; Di Fiore & Campbell 2007; Bowler & Bodmer 2009), with groups often splitting into smaller bands that travel

Table 14.1 Densities of uacari taxa at sites throughout the Amazon.

Locality	Habitat	Species	Density (animals/km^2)	Author
Lago Teiú, Amazonas, Brazil,	*Várzea*	*C. c. calvus*	20	Ayres (1989)
193.5 km survey across three river systems in northeastern Amazonian Peru	Various	*C. c. ucayalii*	0.08[1]	Aquino (1988)
Yavarí and Yavarí-Miri, northeastern Amazonian Peru	Various	*C. c. ucayalii*	4.94	Salovaara *et al.* (2003)
Lago Preto Conservation Concession, Yavari River, Peru	Various	*C. c. ucayalii*	6.4	Bowler (2007)
Pico de Neblina, Amazonas, Brazil	*Caatinga*	*C. melanocephalus*	14	Boubli (2002a)
Taraira Lake, lower Apaporis River, Colombia	*Igapó*	*C. ouakary*	12	Defler (2004)
Caño Pintadillo, tributary of Río Apaporis, Colombia	*Igapó*	*C. ouakary*	1.44	Palacios & Peres (2005)

Table 14.2 Uacari group sizes across the Amazon.

Locality	Habitat type	Species	Group size	Author
Lago Teiú, Brazil	*Várzea*	*C. c. calvus*	20–50	Ayres (1989)
North-eastern Amazonian Peru	Flooded forest	*C. c. ucayalii.*	20–50	Bartecki & Heymann (1987)
Lago Preto Conservation Concession, Yavari River, Peru	Various	*C. c. ucayalii*	Up to 200, mean 43	Bowler (2007)
Pico de Neblina, northwest Amazonia, Brazil	*Caatinga*	*C. melanocephalus*	70	Boubli (1997)
Jaú National Park, central Amazonas, Brazil	*Igapó*	*C. ouakary*	100	Barnett *et al.* (2005)
Rios Curicuriari and Uapes, western Amazonas, Brazil	*Igapó*	*C. ouakary*	4–22	da Cunha & Barnett (1989)
Taraira Lake, lower Apaporis River, Colombia	*Igapó*	*C. ouakary*	200	Defler (2004)
Central Amazonia, Brazil	*Igapó*	*C. ouakary*	15–50	Mittermeier & Coimbra-Filho (1977)

Table 14.3 Day-ranges and home-ranges of uacari taxa at sites throughout the Amazon.

Locality	Habitat	Species	Day ranges (km)	Home range (km^2)	Author
Lago Teiú, central Amazonian Brazil	*Várzea*	*C. c. calvus*	Up to 5 km	500–550 ha	Ayres (1989)
Lago Preto Conservation Concession, Yavari River, Peru	Various	*C. c. ucayalii*	Regularly exceeded 6 km	1200 ha	Bowler (2007)
North-eastern Amazonian Peru	Various	*C. c. ucayalii*	7.3 km	–	Leonard & Bennett (1995)
Lowland forests, Pico de Nebina foothills, northeast Brazil	*Caatinga*	*C. melanocephalus*	2.3 ± 1.2 km	1053 ha	Boubli (1997, 1999)
Taraira Lake, lower Apaporis River, Colombia	*Igapó*	*C. ouakary*	100 m – 5 km	–	Defler (2004)

1–2 km apart. Similar events have been recorded from a number of other genera – *Ateles*, *Chiropotes*, *Macaca*, *Pan*, *Pongo* and *Varecia* (Fukuda 1989; van Schaik 1999; Anderson *et al.* 2002; Vasey 2006; Veiga *et al.* 2006; Asensio *et al.* 2008; Coles *et al.* 2008) – in some species and populations of *Cebus* (Overdorff *et al.* 2003) and in *Eulemur* (Izar & Nakai 2006; Toborowsky 2008). The strategy is associated with avoidance of scramble competition while exploiting highly depletable resources that show high spatial and temporal dispersion (MacFarland 1986; Lehmann *et al.* 2007).

At Pico de Neblina, Boubli (1997, 1999) did not record fission–fusion in his *C. melanocephalus* study group, which maintained a consistent composition of some 70 individuals. This is unlike Caparú, Jaú, or any other study site where uacaris have been studied in detail (e.g. Ayres 1986; Barnett 2010a; Bezerra, Barnett *et al.* 2011; Defler 1999b; Bowler 2007; Bowler & Bodmer 2009, 2010). The habitat at Pico de Neblina is *caatinga*, a species-poor community on exceptionally nutrient impoverished non-inundated soils (see Boubli 2002a, 2005 for details). The lack of clumping in the tree community, the relatively homogeneous within-habitat distribution of tree species, and the extended fruiting periods of some of the numerically dominant tree species may explain why *C. melanocephalus* at Boubli's study site did not exhibit this otherwise widespread phenomenon (Boubli & de Lima 2009; Boubli & Tokuda 2008). There are indications from short-term studies of *C. melanocephalus* by Lehman and Robertson (1994a, 1994b) that elsewhere within its range, in *igapó*, fission–fusion does occur.

Bowler (2007) recorded regular fission–fusion in a study population of over 200 individuals of *C. c. ucayalii* that sometimes foraged as a single unit. Mean subgroup size was larger in *terra firme* forest and *buritizal* palm swamp than in *várzea* forests. Bowler (2007) and Bowler and Bodmer (2010) related this to the distribution of resources in the different habitats, but as habitat use changed seasonally, the effect may also have been associated with seasonal variation in the distribution of resources, particularly for large seasonal patches of *Mauritia flexuosa* palms.

Mating and reproduction

Copulation in *C. ouakary* has been observed by Boubli (1997), Barnett *et al.* (2005) and Bezerra, Barnett *et al.* (2011). It is brief, lasting no more than 30 s. In one instance, two males sequentially mated with the same female (Barnett 2005; Barnett *et al.* 2005). Fontaine (1981) provides a detailed description of the mating process in semi-captive red uacaris. The sequence involves genital rubbing of various substrates by the female, sniffing of the female's ano-genital region by the male, and a specific quadrupedal presenting position. Copulation was said to last for more than 5 min and involve multiple intromissions. Bowler (2007) recorded copulation six times in wild *C. c. ucayalii*, with mating lasting between 30 and 57 s.

Hayssen *et al.* (1993) listed an estrus cycle of 14–48 days for captive *C. calvus* and a lactation period of some 21 months. Ayres (1989) reported *C. c. calvus* mating between March and April, and newborn young between September and November (the late dry season at his Lake Teiú study site) – a gestation time of 6 months. This agrees with Eisenberg's (1977) general calculations for large Neotropical primates, showing that the genus has a gestation time that is as would be predicted for its size. For *C. c. ucayalii*, Aquino and Encarnación (1994) report wild births between December and March, while Swanson-Ward and Chism (2003) observed young infants clinging dorsally to females in late June. Bowler (2007) noted that the even smaller newborn infants cling laterally around the mothers' waist. Bowler (2007) observed mating in April to June (at the end of the high-water season in the *várzea*) and births from the end of August to December (during the low-water season in the *várzea* and just after the fruiting season for *Mauritia flexuosa*, a key food source for this uacari population).

For *C. melanocephalus*, Boubli (1997), in Pico de Neblina National Park, Brazil, observed adult females with very small infants in March, the beginning of the rainy season and the period of peak fruit availability. For *C. ouakary*, small young uacaris were reported in March, April and May in Brazil (da Cunha & Barnett 1989; Rylands 1992; Barnett *et al.* 2005; Barnett 2010a; Bezerra, Barnett *et al.* 2011) and Defler (2004) reports that birthing generally occurs in January–March in Colombia. In both locations this is just after the periods of greatest fruit production by *igapó* trees (see data in Goulding *et al.* 1988; Parolin *et al.* 2002). However, out-of-season births (November) have been observed at both sites (Barnett 2010a; Barnett *et al.*, Chapter 16).

Group composition

The variation in size of the groups and the fission–fusion nature of uacari society means that, to date, no studies have been carried out on group composition, its nature and continuity (Bowler & Bodmer 2009). Encounters with small foraging parties of 3–5 individuals are often mentioned in the literature, but it is currently unclear if these (or indeed any of the other larger groups) have any form of consistent composition, except for mother–infant diads. Bowler and Bodmer (2009) observed a high degree of affiliation between adult male red uacaris that suggests that small groups of males do consistently forage together.

Groups of around 40 appear to consistently occupy areas and, in golden-backed uacaris, individuals exhibiting the same unusual features (a lot of yellow on the back, a black tip to the tail) seem to be seen in the same groups repeatedly (Barnett *et al.*, unpubl.). However, nothing is known about long-term group composition. Similarly, the age at which dispersal occurs and which sex is involved also remains to be investigated. Although the concealed nature of development of genitalia in males and the delay in maturity until 4 years old (Defler, pers. obs.) may well be relevant here.

The sex ratio of adults appears to be roughly 1 : 1, but further data are needed.

Social ecology

Activity budget

As with titi monkeys (see Bicca-Marques & Heymann, Chapter 17), there is little information on the time budgets of wild uacaris. What is known is presented in Table 14.4. Known behaviors, derived largely from Fontaine (1981) and Bezerra, Barnett *et al.* (2011), are presented in Table 14.5. There is, as yet, no way to know if uacaris engage in such subtleties of foraging as eating harder-to-digest materials in

Table 14.4 Time budgets of uacari taxa at sites throughout the Amazon.

Habitat type	Species	Moving (%)	Foraging (%)	Feeding (%)	Resting (%)	Social + other (%)	Author
Várzea	*C. c. calvus*	35	36	*	29	–	Ayres (1989)
Various	*C. c. ucayalii*	38.2	3.5	28.2	24.4	5.8	Bowler (2007)
Caatinga	*C. melanocephalus*	27	31	20	22	–	Boubli (1997, 1999)
Igapó and *terra firme*	*C. ouakary*	42.7	48.2	*	5.1	4.0	Barnett (2010a)
Igapó	*C. ouakary*	52.53	39.91	*	6.65	0.89	Bezerra *et al.* (2010c)

* foraging and feeding combined.

Table 14.5 Proportionality of uacari diet items at sites throughout Amazonia.

Species	Author	Immature fruits and seeds (%)	Mature fruits, pulp, seeds (%)	Flowers and nectar (%)	Leaves and buds (%)	Arthropods (%)	Other[1] (%)
C. c. calvus	Ayres (1986)[2]	57.1	28.2	6.2	3.3	5.2	0.0
C. c. ucayalii	Bowler (2007)	50.6	38.4	3.7	1.7	2.8	2.8
C. melanocephalus	Boubli (1997)	63.8	27.2	4.0	3.0	2.0	0.0
C. ouakary	Barnett (2010a)	68.0	11.0	8.0	9.0	3.0	1.0

[1] includes arils, bark, gums, pith, stems, [2]only data from fruits collected.

the afternoon so that it may be digested during sleeping hours (a situation reported by Chapman & Chapman 1991 for *Ateles* and by Heymann & Smith 1999 for *Saguinus*).

Social behavior

The best-documented studies of uacari social behavior were made in semi-captivity by Roy Fontaine and co-workers, studying red uacaris at the Florida Monkey Jungle. The behavioral results of these studies are summarized in Table 14.5. Vocalizations are considered by Bezerra *et al.* (Chapter 30). Given how little we know about this genus, it is perhaps unsurprising to learn that wild-based studies have tended to focus on such fundamentals as diet and habitat use. With the exception of Bowler and Bodmer (2009), observations on wild uacari social behavior have, to date, tended to be qualitative rather than quantitative (Bartecki & Heymann 1987; Heymann 1990; Aquino 1995, 1998; Leonard & Bennett 1995, 1996; Boubli 1999; Bennett *et al.* 2001; Bowler 2003; Swanson Ward & Chism 2003). The following are, however, noteworthy.

- In Colombia, *Cacajao* are highly playful and adult males readily play with juveniles and younger. The cryptic sex of juvenile males may subvert aggressive tendencies that adult males certainly have.
- Juveniles are very playful. At Jaú, regular associations have been observed between adolescent animals that have lasted over months. Chasing and wrestling on branches and lianas occupy a significant amount of the day of such animals. Still younger animals (newly locomoting juveniles) may also join in on a regular basis.
- Although aggressive interactions have been reported for *C. ouakary* (Bezerra, Barnett *et al.* 2011), aggression seems overall to be a rare event in both *C. melanocephalus* and *C. ouakary*, both in comparison with other primates (e.g. Fairbanks 2009) and with other uacaris (Bowler 2007; Bowler & Bodmer 2009). Over 12 months of observation at Jaú no aggressive interactions (such as chasing or biting) were observed between adult *C. ouakary*, and the only times fear-squeals were heard, they seemed to be associated with playing juveniles (Barnett *et al.*, unpubl. obs.). Boubli (1997, 1999) did not report any aggressive interactions during a 16-month study of *C. melanocephalus*. However, in *C. c. ucayalii*, Bowler (2007) recorded aggression to be a common component of male–male interactions.
- Allo-grooming appears to be a rare event in uacaris. For example, Bezerra, Barnett *et al.* (2011) report social grooming constituted 0.89% of the time budget of *C. ouakary* at Jaú. This is especially notable given the long shaggy nature of the *Cacajao* pelage. Nor does it seem common in *C. c. ucayalii* (Bowler 2007) or *C. melanocephalus* (Boubli 1997).

These observations are an indication that uacaris may be limiting aggression between adults by simple avoidance and the use of distance: while juveniles are tolerated, the delayed-onset of sexual characters in juvenile male uacaris may be a female-mimicking strategy designed to reduce aggression from adult males. Such strategies have been widely recorded in bird species where reproductive competition is intense due to brief reproductive seasons and inter-male aggression is high (see Senar 1999 for a topic review). It has also been reported in other mammals, such as hyaenas (Muller & Wrangham 2002). While this factor is unlikely to be predominant in the evolution of such a strategy in uacaris, the presence of effective female-mimicry may be related to the potentially lethal combination of large canines and powerful jaw muscles possessed by all uacaris (see Lorenz 1966).

Mother–infant interactions are also little recorded to date. However, observations of interest include the following.

- Food-sharing – for black-faced uacaris this appears uncommon, but it has been observed both between adult diads and between mothers and weaning infants (Boubli 1997; Barnett 2010a; Bezerra, Barnett *et al*., 2011). Females have been observed to offer small fragments of fruits to their young, who have also been observed to steal food from their mother's mouth and hands (Boubli 1997). Food transfers have also been observed between older and younger *C. c. ucayalii*. Reporting this in *C. melanocephalus*, Boubli (1997, 1999) posited that this may allow less-powerful young to access seeds from fruit they are physically incapable of breaking. This may also be true for *C. c. ucayalii* (Bowler & Bodmer 2010).
- Very young golden-backed uacaris are often concealed and parked in trees while their mothers forage. Because mothers were always close to the infant, it is unlikely that this is undertaken to reduce the mother's energetic expenditure while foraging, as suggested by Ross (2001). Instead, maintaining infant safety is considered a more likely cause, as immediately after the majority of observed parking events, mothers proceeded to feed by pedal suspension on pendant infructescences at the ends of branches. Under perceived threat infants are quickly retrieved and transferred to the mother's back for transport. Uacaris, then, seem to be pursuing a mixed strategy, combining features of both "park" and "ride" syndromes distinguished by Ross (2001).

Communication

Vocal communication is almost entirely uninvestigated in the genus *Cacajao*. What little is known is summarized by Bezerra *et al.* (Chapter 30). It is a clear and open field for useful and interesting research. This also applies to the field of chemical communication in this genus about which nothing is known save from a few cursory observations by Fontaine (1981) that ano-genital sniffing accompanies mating in semi-captive *C. c. ucayalii*, and the observations of Bezerra, Barnett *et al.* (2011) of ano-genital rubbing on branches and of the sniffing of genitals prior to copulations. This also occurs in *C. c. ucayalii* (Bowler 2007).

Postural communication in uacaris has likewise been little explored and is still at the qualitative and descriptive stages. Postures of note include the following.

- Tail wagging – this has been observed in all taxa for which a detailed study has been conducted (e.g. Walker & Ayres 1996). Slow lateral movements (around one per second) are often seen when animals are foraging. More rapid movements, with the tail held down and accompanied by "chucks", may indicate distress (in the Brazilian population – although not in the Colombian ones, see Barnett *et al.*, Chapter 16).
- Tail held down and to one side – in Colombian populations of golden-back uacaris, this appears to be associated with individuals feeling insecure or sensing danger. Under such circumstances no wagging occurs. In Brazilian golden-backed uacaris, in accordance with the observations of Fernandes (1993) that this action in pitheciines functions as a tension-relief mechanism, increasing tension is associated with increasingly frequent movements of a depressed tail. Bowler (2007) did not find this in *C. c. ucayaii*, noting that stressed or intimidated animals ceased all movement of the tail, possibly, he speculated, to minimize the possibility of drawing attention to themselves. It may, in fact, be that tail-wagging is a graduated signal with, for example, young animals wagging and calling to the female parent when nervous, adults making no movement when in proximity to perceived danger (whether field biologist or large bird), then call and wag at different frequency when danger is perceived to be less (when, for example, the "eagle" turns out to be a toucan).
- Piloerection, tail – in *C. c. ucayalii* a stiff piloerected tail often accompanies aggressive displays by males (Bowler 2007; Bowler & Bodmer 2009). Perhaps associated with this display function, the tail of adult male *C. c. ucayalii* is bushier than that of females.
- Piloerection, body – when an animal is excited, especially in situations of potential danger (such as the presence of predators or unknown humans) the body hair on the back and flanks is puffed up, increasing apparent size. Seen in *C. ouakary* (da Cunha & Barnett 1990; Barnett & da Cunha 1991; Defler 2004; Bezerra, Barnett *et al.* (2011), and in *C. c. ucayalii* (Bowler 2007; Bowler & Bodmer 2009).
- Penile display – in golden-backs an adult male rests on a branch with one leg hanging, displaying an upright penis. The distal part of the penis is pink (Bruna Bezerra; Adrian Barnett, pers. obs.) and contrasts strongly with the black belly. *Cacajao c. ucayalii* has an entirely pink penis, which is also used in display (Fontaine 1981; Bowler 2007). Bowler (2007; Bowler & Bodmer 2009) also recorded the penis display in combination with other displays including hanging upside down while swinging backwards and forwards. This display also occurs in *Saimiri* (see Wooley *et al.* 1978 and illustrations in Defler 2004), and by analogy with *Saimiri* are probably a dominance display, although

there is not yet sufficient information to be certain. Similar displays have yet to be reported for other members of the genus.

- Posterior display – flat footed, and with the tail erect, the posterior is presented to the observer. Possibly an appeasement gesture or indication of friendly intent. Seen in *C. ouakary* (Bezerra, Barnett *et al.* 2011) and *C. c. ucayalii* (Bowler 2007), but not yet observed in other uacari taxa.

Displays related to olfaction have also been observed.

- Urine washing – urine is deposited on the feet and hands and then transferred to the chest, belly and back. Seen in *C. ouakary* (Defler 2004) and *C. c. ucayalii* (Bowler 2007; Bowler & Bodmer 2009), urine washing serves a wide variety of functions in different species (ranging from dominance, and submission to group identity and parasite repulsion – see Campos *et al.* 2007). Under the current state of knowledge, it is impossible to say which of these is of greatest importance in the genus *Cacajao*. Fontaine (1981) identified a display component to urine washing in captive *C. c. ucayalii*, and wild *C. c. ucayalii* engaged in urine washing between bouts of other display behaviors.
- scent-marking – both ano-genital and sternal rubbing have been observed, in adult *C. ouakary* (Bezerra, Barnett *et al.* 2011). Similar behavior was observed by Bowler (2007) in *C. c. ucayalii*. As with urine washing, current knowledge is insufficient to inform us about the likely functions of these behaviors.

Foraging ecology

Uacaris are sclerocarpic foragers (Kinzey 1992; Rosenberger & Kinzey 1976; Rosenberger 1992), hard-fruit specialists that use their enlarged canines to puncture and rip the husks of a variety of fruits to gain access to the seeds inside. Procumbent incisors are use to pick out seeds or their fragments. Seeds are triturated by the low-crowned molars (Kinzey 1992). Although the teeth are strong and the accompanying musculature powerful, uacaris also use finesse in processing their food: in *C. ouakary*, animals choose lines of weakness (natural sutures) in fruit as points at which pressure is applied via the canines. Significantly, up to one-third less force is needed to puncture the husk at such points (Barnett, Bezerra *et al.* 2007; Barnett, Veiga *et al.* 2007).

The procumbent incisors appear to be used as pincers, allowing extraction of seeds as small as 0.25 cm across from their matrix. This allows very fine management of fruit resources as opposed to the rather messy way in which similar fruits are exploited by *Lagothrix lagotricha* (T. Defler, pers. obs., and see Photo 14.2).

Seeds dominate the diet of all studied uacari taxa and the seeds are mostly unripe (Table 14.6). As might be expected with an already specialist diet, uacaris appear to eat whatever of the items met by their selection criteria are most abundant in the habitat at the time (see Ayres 1986; Boubli 1997, 1999; Bowler 2007; Barnett *et al.* 2008; Silva *et al.* 2008). Plant phenologies mean that there are significant parts of the year when unripe fruits are either not available (Silva *et al.* 2008; Boubli & de Lima 2009), or less available (Ayres 1986; Bowler 2007). At these times daily energy needs are met by ripe fruit, pulp and non-fruit plant items. Buds, new leaves and flowers are eaten occasionally, but each individual item is only included in the diet when it is abundant (Ayres 1986; Barnett, Bezerra *et al.* 2007; Bowler 2007; Barnett *et al.* 2008; Silva *et al.* 2008). Deliberate insectivory has been reported for *C. calvus* by Ayres (1986: caterpillars) and by Bowler (2007:

Photo 14.2 *Cacajao calvus calvus* feeding. Photo: Rachel Acosta. (See color plate section).

Table 14.6 Interactions of various uacari taxa with other primates.

Habitat type	Species	Other diurnal primate species present	Observed interactions	Author
Várzea	*C. c. calvus*	*Alouatta seniculus* *Cebus apella* *Saimiri vanzolini*	None reported	Ayres (1986)
Various	*C. c. ucayalii*	(full site list not provided)	*Lagothrix lagotricha* 51% of the time, *S. sciureus* 37% of the time, occasionally with *Cebus albifrons* and *C. apella*	Aquino (1995)
Various	*C. c. ucayalii*	*Ateles chamek* *Alouatta seniculus* *Callicebus moloch* *Cebuella pygmaea* *Cebus apella* *Cebus albifrons* *Lagothrix lagothricha* *Pithecia monarchus* *Saimiri sciureus* *Saguinus fusicollis* *Saguinus mystax*	21.3% of time spent in association with *Lagothrix lagotricha*, *Saimiri sciureus*, *Cebus apella* and *Cebus albifrons*	Bowler (2007)
Caatinga	*C. melanocephalus*	*Ateles belzebuth* *Aotus* sp. *Alouatta seniculus* *Callicebus torquatus* *Cebus albifrons* *Chiropotes* sp. *Saimiri* sp.	None reported	Boubli (1997, 1999)
Terra firme	*C. melanocephalus*	*Chiropotes chiropotes*	*C. chiropotes* and *C. melanocephalus* reported traveling together in fully integrated troops, at least during some parts of the year	Boubli (2002b)
Igapó	*C. ouakary*	*Cebus albifrons* *Lagothrix lagotricha* *Saimiri sciureus*	Often associates with *Lagothrix lagothricha* and *Saimiri sciureus* in *terra firme* forest, and with *Cebus albifrons*, *L. lagotricha* and *S. sciureus* in *igapó*	Defler (2004)
Igapó	*C. ouakary*	*Alouatta seniculus* *Aotus* sp. *Callicebus torquatus* *Cebus albifrons* *Cebus apella* *Saimiri sciureus* *Pithecia pithecia*	Interactions very rare. Occasional associations with *Saimiri*, but neither fully integrated nor long-lasting. Aggressive reactions observed when *C. albifrons* in close proximity	Barnett *et al.* (2005), Barnett, Defler *et al.* (Chapter 16), Bezerra *et al.* (2010c)

ants, caterpillars, termites, mantids and katydids), while Barnett *et al.* (2008; Barnett 2010a; Barnett, Ronchi-Teles *et al.*, submitted) and Bezerra, Barnett *et al.* (2011) report it for *C. ouakary* (ants, caterpillars, mayflies, spiders, termites, whip scorpion). Bowler (2007) also records a single incidence of predation on a small lizard. However, as originally mooted by Ayres (1986), a significant proportion of the animal protein in some *Cacajao* populations may well come from the insect larvae inside fruits. For some species in the diet of *C. ouakary* up to 70% of collected fruits contained adult and/or larval insects, constituting up to 17% of the wet weight of the fruit, and up to 30% the seed(s) (Barnett 2010a). The late Nick Gordon photographed golden-backed uacaris raiding river turtle nests and eating the eggs, but to date this is the only record in pitheciines of a behavior otherwise quite well known for such opportunistic species as *Cebus albifrons* which frequently forage on the ground (Barnett *et al.*, Chapter 16; Defler 1979). At Jaú, turtle nests are raided by *C. albifrons*, but uacaris have yet to be seen to do this.

Photo 14.3 (1) Leaves are seasonally important for golden-backed uacaris, especially in the dry season when there is little fruit available. (2) The variety of size, shape and hardness of uacari diet fruits is impressively varied. (3) Samples of fruits eaten, rejected or ignored by golden-backed uacaris were studied to determine foraging rules underlying their preferences. Flowers were most prevalent in the diet during those periods of the year when they were numerically dominant over fruits and young leaves. The flowers of *Eschweilera tenuifolia* (Lecythidaceae) dominated the diet on some days. Flowers were destructively consumed both for anthers (pollen) and nectar. (See color plate section.) Photo by Adrian A. Barnett

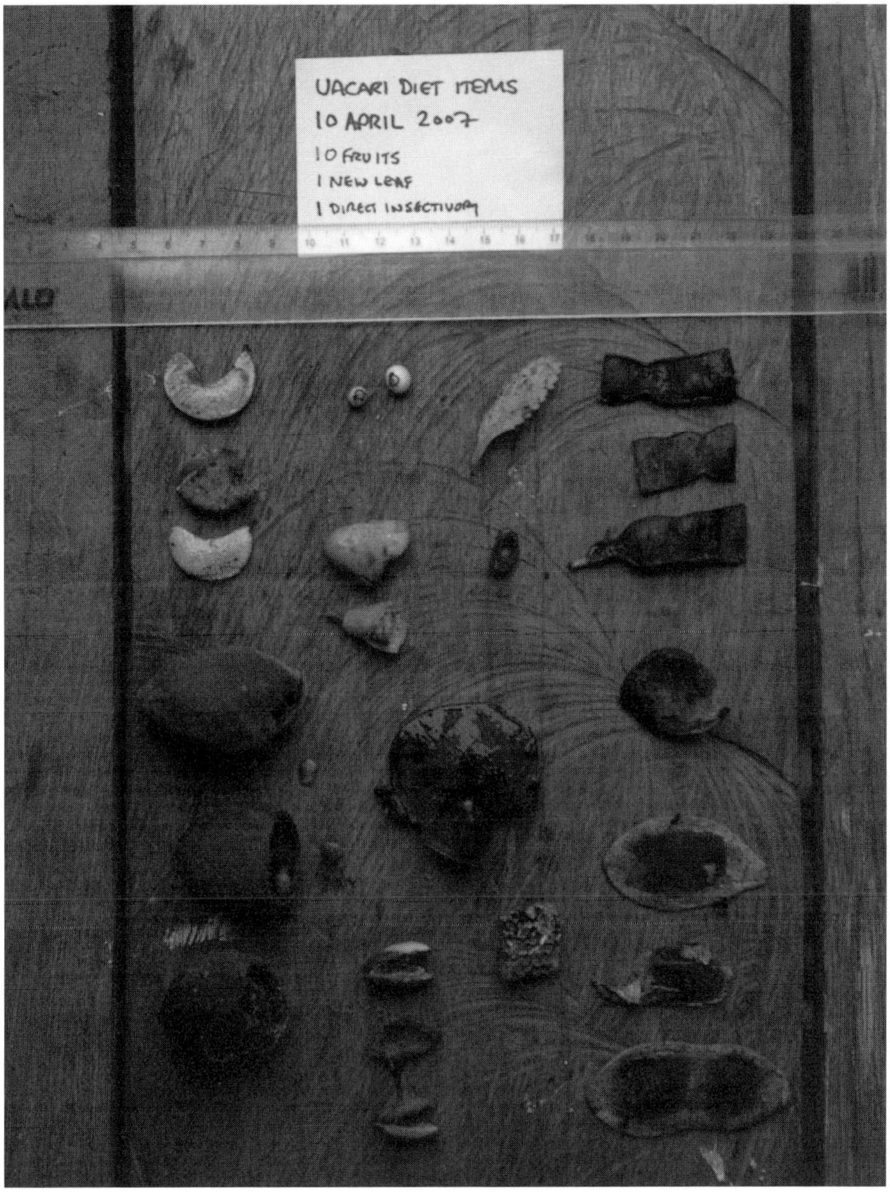

Photo 14.3 *(cont.)*

Photo 14.3 (cont.)

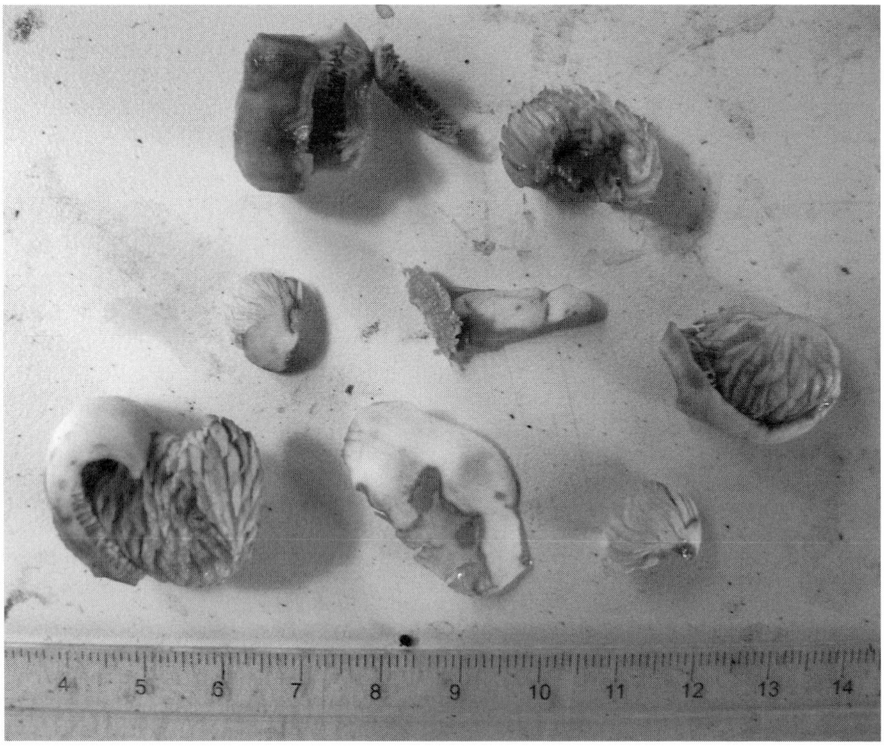

Use of space

The seasonal migration of some uacari populations has already been referred to above. Use of space within the forest habitat, has been the subject of studies that have focused on posture (Walker & Ayres 1996), and the architecture of sleeping and feeding trees (Barnett 2010a, 2010b; Barnett *et al.* 2012). There are also general reports of which part of the forest canopy are preferred by uacaris (Norconk 2011).

Aquino (1995) noted that sleeping trees always exceeded 20 m in height and Aquino (1998) reported that sleeping sites were mostly in the crowns of trees from 17 to 32 m tall. Such trees were primarily of *Pouteria* and *Eschweilera* spp. Defler (2004) reported that, around Lake Taraira, on the lower Río Apaporis, groups of sleeping golden-backed uacaris in flooded *igapó* occurred on the ends of large strong branches in leafless canopy, up to 15 m above the floodwaters, while in non-flooded *terra firme* forest they chose trees with a much higher (25–30 m) and less-exposed canopy. They tend to nap for 1–2 h at mid-day. Using data from 43 sleeping trees and 100 randomly selected trees, Barnett *et al.* (2012) found that most trees were emergents close to the water's edge, had non-touching canopies and a lack of lianas and wasp nests. The leguminaceous tree *Hydrochorea marginata* (Fab.: Mim.) appeared to be especially favored. Like other sleeping trees it has an uncluttered canopy interior with clean, broad, near-horizontal branches. Adult uacaris at Jaú generally slept alone midway between trunk and canopy edge.

Competitors and predators

Competitors

Examples of sympatry are rare between members of the genus *Cacajao*, primarily because of the largely disjunct ranges of its component taxa (see Figure 4.5, Chapter 4). There have been unverified reports of mixed bands of *C. c. calvus* and *rubicundus* on the Solimões-Japurá (A. Rylands, pers. commun.; H. Queiroz, pers. commun.), and *C. c. rubicundus* and *C. ouakary* have both been reported from the Amanã Sustainable Development Reserve (Amaral *et al.* 2005). However, the latter contains a mixture of white- and black-water areas and it is still unclear if the two species do or do not remain stratified by *várzea*/*igapó* habitat preferences. In the Araça River region, the studies of Jean-Philippe Boubli (Boubli *et al.* 2008) have revealed that *C. ayresi* uses *Mauritia* palm swamp and *terra firme* forest in what Boubli believes to be a lowland enclave within the distribution of *Chiropotes*.

The three pitheciine genera *Pithecia*, *Chiropotes* and *Cacajao* show an increasing trend for specialization in foraging on hard-husked (generally immature) fruits (Kinzey 1992; Norconk 2011; and see Rosenberger & Tejedor, Chapter 2). While patterns of forest use and a predominance of smaller softer fruits in the diet serve to minimize overlap between *Pithecia* and its two larger congeners (Norconk 2011; Norconk *et al.*, Chapter 6), similarities in dentition, size and canopy use mean that *Cacajao* and *Chiropotes* are potential competitors in the hard-fruit niche (Ayres 1989). It was long thought that the distribution of the two genera was mutually exclusive and this was taken as an example of competitive exclusion (see Ayres 1986, 1989). Observations by Boubli (2002b) have show that the two genera do overlap in Pico de Neblina region. However, as pointed out by Barnett *et al.* (2005), this may not indicate that competition occurs on a regular basis. Data from a variety of mammal and other vertebrate groups (see references in Barnett *et al.* 2005) show conclusively that competition for food resources may be a rare occurrence that happens only during infrequent incidents of resource dearth. Indeed, the very fact that Boubli (2002b) reported the two species as traveling in the same mixed species group may argue against the existence of intense competition at the time of observation. Boubli now believes that this is an occasional event that occurs because *Chiropotes* sometimes venture in *Cacajao* range and may end up pairing up in the group; likewise the odd *Cacajao* may venture into *Chiropotes* range (J.-P. Boubli, pers. commun.).

Predators

Uacaris frequently forage on the tops and outer surfaces of the vegetation (whether it is that of tall, often emergent, trees over land or the those of often very much smaller trees and even bushes in flooded *igapó*). Many of the foraging surfaces used are open to aerial attack and this may make them vulnerable to attacks by raptors (and see below). As a result, it is perhaps not surprising that a number of fieldworkers have noted that uacaris are notably nervous in the presence of large birds, giving alarm calls to overflying vultures, toucans and raptors too small to harm them (Boubli 1997; Bowler, 2007; Barnett, 2010a, 2010b; T. Defler, pers. obs.; R. Mittermeier, pers. commun.; A. Rylands, pers. commun.). This is the same alarm call given when large raptors are present. They also give a different kind of alarm call when tayras (*Eira barbara*) are present (Bowler 2007).

Observations of the architecture of uacari sleeping trees and the positioning of chosen sleeping spots therein led Barnett *et al.* (2012) to propose that this constituted anti-predator behavior, a strategy to avoid both aerial attacks and those of night-hunting felids and mustelids. These authors also reported the highest levels of open canopy activity in the last hours of the afternoon, a period when diurnal raptor activity is at a nadir.

Despite (or perhaps because of) this, reports of actual predation on uacaris are rare, with only one published record; that of Barnett *et al.* (2011) who observed a harpy eagle (*Harpia harpyja*) pluck an adult uacari from a canopy-top branch. They also reported that uacaris were frequently nervous in the presence of large birds, dropping from the canopy to cling, *Pithecia*-like, to tree trunks if a large bird flew near a group or if the call of a raptor was heard close by. Harpy eagles are known to take *Chiropotes* (Martins *et al.* 2005), and *Pithecia* (Rettig 1978; Aguiar 2007), and both Boubli (1997) and Barnett (2010b) reported that uacaris reacted nervously to large birds, including vultures and toucans. Other pitheciines have been reported as being preyed upon by large snakes (Ferrari *et al.* 2004), and ocelot (Bianchi & Mendes 2007). Jaguars have been reported to take howlers (Garla *et al.* 2001), and muriqui (Imos 1994), primates of a larger size than an adult uacari. Although these predators have yet to be recorded for the genus *Cacajao*, Thomas Defler (unpublished data) not only sighted an ocelot near a group of *C. ouakary* foraging in unflooded *igapó*, but also received a report of a jaguar following this group in the same habitat.

Predation of uacaris by humans occurs sporadically across their range. Some populations of *calvus* uacaris appear to be protected by local taboos, because the naked head makes them look "too human to hunt", while other populations are hunted quite extensively (e.g. *C. c. ucayalii* in Peru: Aquino 1988; Puertas & Bodmer 1993; Swanson Ward & Chism 2003; Bowler 2007; *C. melanocephalus* in the Pico de Neblina region of Brazil: Boubli 1997, 1999; *C. ouakary* in the Amanã Sustainable Development Reserve: Barnett *et al.*, Chapter 16). Mittermeier and Coimbra-Filho (1977) mentioned that uacaris were sometimes caught and used as meat for trapping large felids, a practice that has not been reported since (possibly due to changes in laws regarding the export of pelts). Kinzey (1997) also mentioned that the tails of *Chiropotes* are sometimes sold for use as "feather dusters". Uacari tails prepared like this have also been seen (Figure 14.5), but not as a result of commercial activity.

The (very) selective logging on the Yavarí in Peru is unlikely to impact uacari populations much at current rates, except by associated hunting (Bowler 2007).

Interactions with other species

Primates

Like many aspects of uacari biology (see above), the extent and nature of interactions of uacaris with other primates seems to show extensive regional variation (see Table 14.6). Low densities may explain the lack of associations in the "*caatinga*" site studied by Boubli (1997, 1999); Boubli (2002a) reported *Alouatta seniculus* densities there were 2 animals/km^2, while *Cebus albifrons* and *Callicebus torquatus* densities were, respectively, 1.4 and 1.6 animals/km^2. However, low densities do not explain the between-site differences in the tendencies of uacaris to associate with other primates observed between Jaú and Caparú for *C. ouakary* (Table 14.6), or between the surveys of Swanson-Ward and Chism (2003) and Aquino (1995) for *C. c. ucayalii* in essentially the same area and season. Clearly there is a need for detailed fieldwork on local resource availability and (perhaps) the transmission of regional traditions. Without this, the reasons for such variations are likely to remain obscure.

Aquatic vertebrates

Flooded forests are generally mammal-poor (e.g. Ayres 1993), limiting the opportunities for cross-species mammalian interactions. The literature appears to contain few reports of how (or, indeed, if) uacaris interact with other mammals (other than predators, see below). Bezerra, Barnett *et al.* (2011) report that *C. ouakary* appeared to ignore the tamanduas and sloths they encountered in the *igapó* of Jaú National Park. Defler (2004) reported that pink river dolphins (*Inia geoffrensis*: Platanistidae) followed and ate the fish (including pacu, *Colossoma macropomum*: Characidae, Serrasalmidae and matrixã, *Brycon matrinchao*: both Characidae, Bryconidae: see Goulding *et al.* 1988) that were attracted to the fruits knocked down by feeding *C. ouakary*. Similar events have been noted in Peru with *C. c. ucayalii* (M. Bowler, unpubl. data). Although *Inia* have been observed in *igapó* forests at other uacari-containing sites (e.g. Jaú), they have yet to be seen to behave in this way, nor has this association been reported for other uacaris, even though Bezerra, Barnett *et al.* (2011) report that fish species were attracted to fruit fall below trees where uacaris were feeding. The fish also ate the uacari feces that fell into the water. These associations may be incidental, and not an adaptive analogue of a multi-species bird flock (Mark Bowler, pers. obs.).

Birds

Small raptors are also known to follow primate bands. For example, Zhang and Wang (2000) reported on the association between *Cebus apella* and the White Hawk *Leucopternis albicollis*, while Ferrari (1990) noted that *Callithrix flaviceps* was followed by Plumbaceous *Ictinea plumbea* and Gray-headed *Leptodon cayanensis* kites. In inland forest at Caparú, Defler (2004) reported that White Hawk *Leucopternis albicollis* and Double-toothed Kite *Harpagus bidentatus* both commonly associate with *C. ouakary*. These birds were also observed in association with *Cebus*, *Saimiri* and *Lagothrix* (Defler 2004). At the Capurú site, *H. bidentatus* is almost always found in proximity to a foraging group of large primates (T. Defler, unpubl.), while Mark Bowler (unpubl.) notes that uacaris at Lago Preto also almost always seem to be followed by this species. Such associations have yet to be reported from other sites and for other uacari taxa.

Long-term associations between Neotropical primates and small insectivorous birds (jacamars, woodcreepers, antshrikes, antwrens and flycatchers) have been reported for several primate species and from a number of locations (Boinski & Scott 1988: *Cebus* and *Saimiri*). However, to date the only reports of associations between uacaris and birds appear to be fortuitous. Both Defler (2004) and Barnett *et al.* (2005) reported associations between *C. ouakary* and Greater Ani (*Crotophagus major*: Cuculidae), and Bowler (unpubl.) noted this for *C. c. ucayalii*. Barnett (2010a) noted occasional following of small uacari bands (3–5 animals) in *igapó* by black-fronted nunbirds (*Monasa nigrifrons*: Bucconidae), woodcreepers (*Dendrocolaptes* and *Xiphorhynchus* spp.: Dendrocolaptidae), *Thamnophilus* antshrikes and *Myrmotherula* antwrens (both Thamnophilidae). Associations appeared temporary. As many *Xiphorhynchus* and *Thamnophilus* are territorial (e.g. Rivera *et al.* 2003; Duca *et al.* 2006), with territory sizes often less than 2 ha in the case of the latter, this impermanence may have been because such associations only endure while the uacaris are in transit across the territories of the birds concerned (Munn & Terbough 1979). However, Barnett (unpubl.) has shown that foraging rates increased for two bird species (*Monasa nigrifrons* and *Galbula leucogastra*, both sit-and-wait predators) when uacaris passed through their territories, compared to when uacaris were absent. However, uacari presence did not influence foraging success by antwrens and antshrikes

(Thamnophidae), birds that actively seek prey. This may be attributed to nunbird and jacamar preference for large insects (such as Orthoptera) that would be disturbed by primate passage, while the passage of uacaris would be unlikely to induce the smaller arthropods eaten by antwrens to leave concealment. In *terra firme* at Jaú 20-strong bands of uacaris were observed for over an hour to be moving in loose association with Yellow-ridged *Ramphastos culminates* and Cuvier's *R. cuvieri* toucans (A. Barnett, pers. obs.), after which the uacaris moved on, but the birds stayed. No obvious interactions were observed.

This lack of regular association may be because of three aspects of uacari biology: first, as high-canopy specialists, uacaris leap frequently rather than run; second, because their foraging rarely involves the investigative rummaging so characteristic of *Saimiri* and *Cebus*. Together these two aspects may mean that the passage of a band of uacaris disturbs fewer insects than that of a smaller band of squirrel or capuchin monkeys. Finally, uacaris rarely pause to feed extensively in one area and are often moving very quickly in comparison to *Saimiri* and *Cebus*, so that birds associating with them may simply not have time to forage. In this context it is notable that most transient uacari–bird associations are recorded when the monkeys are moving back and forth in a limited area.

Competition with other species

Although it has yet to be quantified, competition for food resources between uacaris and other species may well be quite limited. Within the pitheciines themselves, animal size and diet preferences mitigate against competition between *Callicebus* and *Cacajao*. For the larger genera, competition between uacaris and the two genera of sakis appears to be limited by distribution (*Cacajao* and *Chiropotes*) and by habitat preferences (*Cacajao* and *Pithecia*). In the latter case, pitheciines seem to regularly enter flooded forest only in areas outside the geographical range of *Cacajao* (Haugaasen & Peres 2005, 2009 in eastern Amazonia for *Pithecia* and Wallace *et al.* 1996 for *Chiropotes* in the extreme south of Amazonia). This may be a possible example of competitive release, although fieldwork would be needed to confirm this. Competition between *Pithecia* and *Cacajao* is potentially possible during the period of flooded forest fruit dearth when uacaris migrate to *terra firme*. However, although the diet of uacaris at this time is currently totally unstudied (except for the briefest observations by Barnett *et al.* 2005 and Barnett 2010a; Barnett, MacLarnon *et al.*, submitted), it is likely that the known preferences of *Pithecia* for smaller softer fruits, and the higher incidence of leaves in their diet (see Setz *et al.*, Chapter 7), should minimize competition at this time. In addition, while uacaris seem to be canopy specialists in whatever habitat they are in (Barnett & Brandon-Jones 1997; Barnett *et al.* 2005; Bowler 2007; Boubli & de Lima 2009; Barnett 2010a; Barnett, MacLarnon *et al.*, submitted), the preference of *Pithecia* for the mid-canopy (Norconk 2011) should further avoid competition between the two (although this contention has yet to be tested in the field). The genera *Cacajao* and *Chiropotes*, of similar sizes and with very similar dental specializations for sclerocarpic foraging (Kinzey 1992), have long been regarded as an example of competitive exclusion, as their ranges appeared to be mutually exclusive (see Hershkovitz 1985, 1987). However, observations by Boubli (2002b) suggest that, in the Pico de Neblina region of Brazil at least, representatives of the two genera do sometimes occur together, and even form mixed species bands. The foraging ecology of this has yet to be studied but, as Barnett *et al.* (2005) have pointed out, such incidences may well be occurring at times of fruit superabundance, when resources are not limiting and competition is therefore effectively non-existent, or *Cacajao* and *Chiropotes* may exist in a state of fluctuating competition (see Schoener 1983; Chesson & Huntley 1997 for concept overviews).

At Caparú, *Cacajao* and *Lagothrix* compete directly for immature seeds of *Micrandra spruceana* (Euphorbiaceae). Both eat the fruits of *Brosimum utile* (Moraceae), *Minquartia guianensi* and *Minquartia punctata* (Olacaceae). However, while for all three *Lagothrix* mostly eat the flesh, the uacaris eat some flesh but primarily the seeds (T. Defler, unpubl. data).

Competition with other primates has yet to be quantified. Diet overlap certainly occurs between the various *igapó*-dwelling primate species (see Barnett *et al.* 2005 for a list of species), and van Roosmalen and Garcia (2000) listed several instances where uacaris and other cebids have fed from the same individual Sapotaceae trees. However, as fruit crops of *igapó* trees tend to be large and available for extended periods (Parolin 2000, 2001; Parolin *et al.* 2002), the presence of diet overlap may not mean that competition is also occurring (see Powell & Zielinski 1983; Wissinger 1992; and Mysterud 2000 for general topic discussions; Holt & Kotler 1987 for review). In addition, the primate species that are sympatric with uacaris may use fruits of different ages to those eaten by uacaris themselves. For example, Barnett (2010a) found that, while both *Cacajao* and *Cebus* will eat fruits of the same Sapotaceaous trees, the former eats only the unripe fruits and the latter the ripe ones. This could lead to asymmetric competition (*sensu* Law *et al.* 1997: see Torok & Toth 1999; Rodriguez-Girones *et al.* 2003 for examples) because while the uacaris would be diminishing the resources for the primates eating ripe fruit, the reverse could not occur. However, in other cases, the shared resource (small ripe fruit) is so abundant (e.g. the fruits of several genera of Myrtaceous treelets and large shrubs collectively known as *araça*) that it is difficult to imagine the resource being limiting enough for competition to occur.

Resource competition has yet to be quantified for uacaris, or indeed studied in any detail. The potential competitors are likely to be other pitheciines, large parrots and macaws (for large and thick-husked fruits) and other primates (other fruit types). As noted above, *Chiropotes* and *Cacajao* are largely allopatric and their diets have been neither studied nor compared in the small areas of geographical overlap that occur (Pico de Neblina region – see Boubli 2002b and Boubli & de

Photo 14.4 *Hevea spruceana* (Euphorbiaceae): toothmarks of golden-backed uacari are concentrated on a suture, the area where the husk is weakest. Photo: Adrian Barnett.

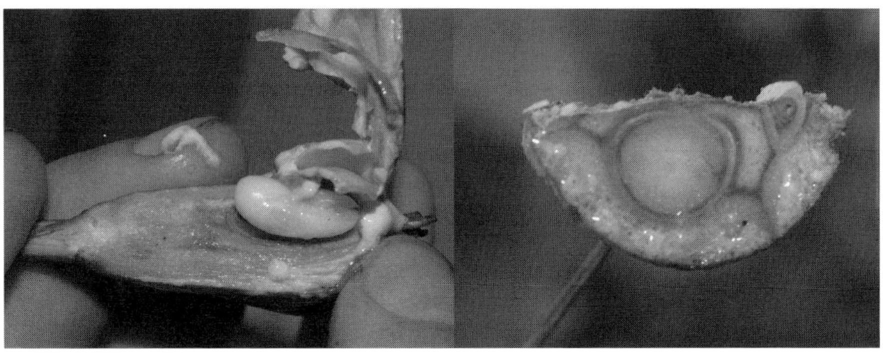

Photo 14.5 The pericarp of *Lorostemon negrense* (Clusiaceae) fruits are rich in canals. From these, a very sticky latex rushes abundantly when the fruit is broken. Taking hours to harden, it appears insoluble in spittle, alcohol or water. Despite this, golden-backed uacaris fed intensively on *L. negrense* fruit during their brief period of availability. Photo: Adrian Barnett. (See color plate section).

Lima 2009). Norconk *et al.* (1997) reported on *Pithecia–Ara* (macaw) competition and Palminteri *et al.* (Chapter 11) provide similar data from Bolivia. Barnett *et al.* (2005) reported that macaws at Jaú ate two fruits (*Parkia discolour* Fab.: Mim and *Eschweilera tenuifolia*, Lecythidaceae) that were important uacari food items. Preliminary data from Jaú National Park indicates that *C. ouakary* may share 12 diet species, including *Hevea*, *Mabea* (Euphorbiaceae), *Macrolobium* (Fab.: Caes.) and *Tabebuia* (Bignoniaceae), with the region's larger parrots (Almeida *et al.* 2008; T. Almeida, unpubl. data). *Maprounea* (Euphorbiaceae) and *Tabernaemontana* (Apocynaceae) were used by both uacaris and small parrots of the genus *Brotogeris* (Almeida *et al.* 2008; T. Almeida, unpubl. data). Barnett *et al.* (2005) also reported that *Amazona* parrots also ate *Eschweilera tenuifolia* seeds, although they were less successful at opening the pyxidia than either *Ara* or *Cacajao*. However, many of the tree species concerned have both large crop volumes and extended fruiting periods (Barnett, unpubl.) and so scramble competition for limiting resources is unlikely.

Barnett *et al.* (2005) noted that *C. ouakary* and squirrels shared diet items, but quantified studies have yet to be undertaken. Bowler (2007) reported that *C. c. uacayalii* seemed to depress the density of squirrels wherever they were common.

Conservation and future research directions

Currently, it is not possible to give even a rough estimate of the populations of any *Cacajao* taxon, and for many the distributional limits are uncertain. Nor is it known how uacari density varies across various vegetation types. All this makes impossible even the most cursory of vegetation-map-based population estimates. For this reason the announcement that IBAMA's Amazon Primates Committee has teamed up with PAG (the Pithecine Action Group) to plan a series of field surveys for the *calvus* uacaris is very welcome, although this still leaves the *melanocephalus* uacaris in an uncertain position.

Given our state of ignorance of even the most basic aspects of the biology, baseline ecological studies are clearly still essential for most uacaris. This includes the known taxa of *C. c. rubicundus*, *C. c. novaesi.* Such studies should include preliminary analyses of diet and identification of those species of key importance to their survival. For all taxa (with the probable exception of *C. c. calvus* for which conservation plans are in force: see Chapter 33), it is also important to identify regional threats from hunting and habitat destruction and other forms of resource exploitation. To allow preventative conservation schemes to be implemented, trends in such socioeconomic phenomena should also be identified. In addition, to ensure maximum effectiveness of conservation initiatives, the genetic structuring of populations should be investigated, especially when the same taxon appears to occur on two sides of a major river or other potential barrier to gene flow. At the more local level, it would be useful to know which sex is the major disperser as this would permit an analysis of likely pathways of genetic differentiation and the likely consequences of inbreeding (for a review of the effects of male- and female-based migration on genetic integrity, see Prugnolle & de Meeus 2002, while Cheyney & Seyfarth 1983 provide a classic field-based example). Furthermore, an understanding of the patterns of annual dispersal is of key importance if appropriate sized areas of habitat are to be protected, each with the appropriate mix of habitats that permit the full range of seasonally adjusted resource exploitation. Finally, a comprehensive inventory of uacari vocalizations and field signs would greatly assist surveys and allow rapid assessments of the presence and density of *Cacajao* in a selected area via playback experiments.

Acknowledgments

The authors would like to thank the anonymous reviewers whose suggestions greatly improved the manuscript, Dominic Lewis, Martin Griffiths, Lynette Talbot and the production team at Cambridge University Press and all those who gave freely of their time and expertise to make this chapter a reality, including: Liza Veiga, Jean Philippe Boubli, Marilyn Norconk, Ann MacLarnon and Caroline Ross. AB thanks Eliana Andrade for her unflagging support, and American Society of Primatologists, Columbus Zoo Conservation Fund, Sophie Danforth Conservation Fund, LSB Leakey Foundation (US), Leakey Fund (UK), Margot Marsh Foundation, Oregon Zoo Conservation Fund, Percy Sladen Memorial Fund, Pittsburgh Zoo and Aquarium Conservation Fund, Primate Action Fund, Primate Conservation Inc., Roehampton University, Wildlife Conservation Society. MB would like to thank Richard Bodmer, Pablo Puertas and all the biologists, students and field assistants who have worked at Lago Preto. BMB is supported by Programme Alban; ORS award; University of Bristol; a Rufford Small Conservation Grant; an IDEA WILD grant, Amazon Ecopark Lodge and Living Rain Forest foundation. TRD would like particularly to acknowledge Angel and Ovidio Yacuna for their dedicated field assistance.

References

Aguiar, H. (2007). Dieta do gavião-real Harpia harpyja (Aves: Accipitridae) em florestas de terra firme de Parintins, Amazonas, Brasil. Master's thesis, INPA/UFAM, Manaus.

Almeida, T., de Almeida, R. & Barnett, A.A. (2008). Parrots of Jaú National Park: diet composition and competition with seed-predating primates. *Abstracts of the Association for Tropical Biology and Conservation 2008 meeting* (June 9–13, Paramaribo, Suriname), 14.

Amaral, J.V., Muniz, I.C.M., Avelar, A.A., *et al.* (2005). Primatas de Reserva de Desenvolvimento Sustenivel Amana. In *Programa e Livro de resumos, XI Congresso Brasiliero de Primatologia*, ed. J.C. Bicca-Marques. Porto Alegre: Sociedade Brasileira de Primatologia, p. 70.

Anderson, D.P., Nordheim, E.V., Boesch, C., *et al.* (2002). Factors influencing fission–fusion grouping in chimpanzees in the Tai National Park, Cote d'Ivoire. In: *Behavioural Diversity in Chimpanzees and Bonobos*, ed. C. Boesch, G. Hohmann & L.F. Marchant. New York, NY: Cambridge University Press, pp. 90–101.

Aquino, R. (1988). Preliminary survey on the population densities of *Cacajao calvus ucayalii. Primate Conservation*, **9**, 24–26.

Aquino, R. (1995). Conservacion de *Cacajao calvus ucayalii* en el Amazon Peruana. *Neotropical Primates*, **3**, 40–42.

Aquino, R. (1998). Some observations on the ecology of *Cacajao calvus ucayalii* in the Peruvian Amazon. *Primate Conservation*, **18**, 21–24.

Aquino, R. (2005). Alimentación de mamíferos de caza en los aguajales de la Reserva Nacional de Pacaya-Samiria (Iquitos, Perú). *Revista Peruana de Biología*, **12**, 417–425.

Aquino, R. & Encarnación, F. (1994). Primates of Peru. *Primate Report*, **40**, 1–127.

Aquino, R. & Encarnación, F. (1999). Observaciones preliminares sorbe la dieta de *Cacajao calvus ucayalii* en el Nor-Oriente Peruano. *Neotropical Primates*, **7**, 1–5.

Asensio, N., Korstjens, A.H., Schaffner, C.M., *et al.* (2008). Intragroup aggression, fission–fusion dynamics and feeding competition in spider monkeys. *Behaviour*, **145**, 983–1001.

Aureli, F., Schaffner, C.M., Boesch, C., *et al.* (2008). Fission–fusion dynamics: new research frameworks. *Current Anthropology*, **49**, 627–646.

Ayres, J.M. (1986). The white uakaris and the Amazon flooded forests. Unpublished PhD thesis, University of Cambridge.

Ayres, J.M. (1989). Comparative feeding ecology of the uakari and bearded saki, *Cacajao* and *Chiropotes*. *Journal of Human Evolution*, **18**, 697–716.

Ayres, J.M. (1993). *As Matas de Várzea do Mamirauá*. Tefé: Sociedade Civil Mamirauá.

Barnett, A.A. (2005). *Cacajao melanocephalus*. *Mammalian Species*, **739**, 1–7.

Barnett, A.A. (2010a). Diet, habitat use and conservation ecology of the golden-backed uacari, *Cacajao ouakary*, in Jaú National Park, Amazonian Brazil. PhD thesis, Roehampton University, London. http://roehampton.openrepository.com/roehampton/

Barnett A.A. (2010b). Preliminary analysis of potential predator recognition and distinction of imminent threat by golden-backed uacari, *Cacajao ouakary* (Primates), in Brazilian Amazonia. Association for the Study of Animal Behaviour Annual Conference, December 3, London.

Barnett, A.A. (Submitted). The genus *Cacajao* (Pitheciidae, Primates), a review of literature from 1812 to 2012: part 1 – taxonomy, etymology, anatomy, morphology, karyology and evolutionary history. *Mammal Review*.

Barnett, A.A. & Brandon-Jones, D. (1997). The ecology, biogeography and conservation of the uacaris, *Cacajao* (Pitheciinae). *Folia Primatologica*, **68**, 223–235.

Barnett, A. & da Cunha, A.C. (1991). The golden-backed uacari on the upper Rio Negro, Brazil. *Oryx*, **25**, 80–88.

Barnett, A.A. & Ross, C. (Submitted). The naked red head of *Cacajao calvus:* facts, fantasies and filaria. *Quarterly Review of Biology*.

Barnett, A.A., Bezerra, B., Ross, C., *et al.* (2008). Hard fruits and black waters: the conservation ecology of the Golden-backed uacari, *Cacajao melanocephalus ouakary*, an extreme diet- and habitat-specialist. Presentation # 692, Abstracts, XXII Congress, International Primatological Society, Edinburgh, August 2–8, 2008, pp. 248–250.

Barnett, A.A., Bezerra B.M. & Spironello, W.R. (2007). Diet of the Golden-backed Uacari, *Cacajao melanocephalus ouakary* (Primates): seasonal strategies and manipulational subtleties. *Abstracts of the Association for Tropical Biology and Conservation 2007 meeting* (July 15–19, Morelia, Mexico). Washington, DC: Association for Tropical Biology and Conservation, p. 7.

Barnett, A.A., de Castilho C.V., Shapley, R.L., *et al.* (2005). The golden-backed uacari, *Cacajao melanocephalus ouakary* in Jaú National Park, Brazil: preliminary observations on wet and dry season diet and habitat selection. *International Journal of Primatology*, **26**, 961–981.

Barnett, A.A., MacLarnon, A., Almeida, T., *et al.* (Submitted). Seasonal changes in diet and habitat use by the golden-backed uacari, *Cacajao melanocephalus ouakary* (Spix, 1823), in Jaú National Park, Amazonian Brazil. *American Journal of Primatology*.

Barnett, A.A., Ronchi-Teles, B., Almeida, T., *et al.* (Submitted). Covert carnivory: insects, infested fruits and the golden-backed uacari, a seed-predating primate. *International Journal of Primatology*.

Barnett, A.A., Schiel, V., Deveny, A., *et al.* (2011). Predation on *Cacajao ouakary* and *Cebus albifrons* (Primates: Platyrrhini) by harpy eagles. *Mammalia*, **75**, 169–172.

Barnett, A.A., Shaw, P., Spironello, W.R., *et al.* (2012). Sleeping site selection by golden-backed uacaris, *Cacajao melanocephalus ouakary* (Pitheciidae), in Amazonian flooded forests. *Primates*, **53**, 273–285.

Barnett, A.A., Veiga, L.M. & Spironello, W. R. (2007). Combining force with finesse: food selection and processing in *Cacajao* and *Chiropotes*. In *Resumos XII Congresso Brasileiro de Primatologia, Belo Horizonte, MG*, ed. F.R. Melo, A. Hirsch, C.G. Costa, L.G. Dias, I.M.C. Mourthe, F.P. Tabacow & L.M. Scoss, p. 7.

Bartecki, U. & Heymann, E.W. (1987). Sightings of red uakaris, *Cacajao calvus rubicundus*, at the Rio Blanco, Peruvian Amazonia. *Primate Conservation*, **8**, 34–36.

Bennett, C.L., Leonard, S. & Carter, S. (2001). Abundance, diversity and patterns of distribution of primates on the Tapiche river in Amazonian Peru. *American Journal of Primatology*, **54**, 119–126.

Bezerra, B.M., Barnett, A., da Silva Junior, J., *et al.* (2008). Sounding out species: holes in our distributional knowledge of Amazonian primates, and the potential for call playback as a survey technique. In *XXII International Primatological Society Congress*, August 3–8, 2008, Primate Eye, **96** (Sp CD-ROM iss – IPS 2008), Abstract 695.

Bezerra, B.M., Barnett, A.A., Souto, A.S., *et al.* (2007). Preliminary recordings on the vocalisations of golden-backed uacari (*Cacajao melanocephalus ouakary*), in Jaú National Park, Amazon, Brazil. In *XII Brazilian Primatological Conference, Belo Horizonte, Sociedade Brasileira de Primatologia, CD-ROM*, ed. F.R. Melo, A. Hirsch, C.G. Costa, L.G. Dias, I.M.C. Mourthe, F.P. Tabacow & L.M. Scoss, p. 9.

Bezerra, B.M., Barnett, A.A., Souto, A., *et al.* (2010c). Ethogram and natural history of golden-backed uakaris (*Cacajao melanocephalus*). *International Journal of Primatology*, **32**, 46–68.

Bezerra, B.M., Barnett, A., Souto, A.S., *et al.* (2011). Ethogram and natural history of the golden-backed uakari, *Cacajao ouakary*. *International Journal of Primatology*, **32**, 46–68.

Bezerra, B.M., Souto, A.S. & Jones, G. (2010a). Vocal repertoire of golden-backed uacaris (*Cacajao melanocephalus*): call structure and context. *International Journal of Primatology*, **31**, 759–778.

Bezerra, B.M., Souto, A.S. & Jones, G. (2010b). Responses of golden-backed uacaris, *Cacajao melanocephalus*, to call playback: implications for surveys in the flooded Igapó forest. *Primates*, **51**, 327–336.

Bezerra, B.M., Souto, A.S., Radford, A.N., *et al.* (2011). Brevity is not always a virtue in primate communication. *Biology Letters*, **7**, 23–25.

Bianchi, R.C. & Mendes, S.L. (2007). Ocelot (*Leopardus pardalis*) predation on primates in Caratinga Biological Station, Southeast Brazil. *American Journal of Primatology*, **69**, 1173–1178.

Boinski, S. & Scott, P.E. (1988). Association of birds with monkeys in Costa Rica. *Biotropica*, **20**, 136–143.

Bonvicino, C.R., Boubli, J.P., Otazu, I.B., *et al.* (2003). Morphologic, karyotypic, and molecular evidence of a new form of *Chiropotes* (Primates, Pitheciinae). *American Journal of Primatology*, **61**, 123–133.

Boubli, J.P. (1997). Ecology of black uakari monkeys in Pico da Neblina National Park, Brazil. PhD dissertation, University of California, Berkeley.

Boubli, J.P. (1999). Feeding ecology of black-headed uacaris (*Cacajao melanocephalus melanocephalus*) in Pico da Neblina National Park, Brazil. *International Journal of Primatology*, **20**, 719–749.

Boubli, J.P. (2002a). Lowland floristic assessment of Pico da Neblina National Park, Brazil. *Plant Ecology*, **160**, 149–167.

Boubli, J.P. (2002b). Western extension of the range of bearded sakis: a possible new taxon of *Chiropotes* sympatric with *Cacajao* in the Pico da Neblina National Park, Brazil. *Neotropical Primates*, **10**, 1–4.

Boubli, J.P. (2005). Floristics, primary productivity and primate diversity in Amazonia: contrasting a eutrophic varzea forest and an oligotrophic caatinga forest in Brazil. In *Fruits and Frugivores: in Search of Strong Interactors*, ed. J.L. Dew & J.P. Boubli. Dordrecht: Kluwer Academic/Plenum Publishers, pp. 59–73.

Boubli, J.P. & de Lima, M.G. (2009). Modeling the geographical distribution and fundamental niches of *Cacajao* spp. and *Chiropotes israelita* in northwestern Amazonia via a maximum entropy algorithm. *International Journal of Primatology*, **30**, 217–228.

Boubli, J.P. & Ditchfield, A.D. (2000). The time of divergence between the two species of uacari monkeys: *Cacajao calvus* and *Cacajao melanocephalus*. *Folia Primatologia*, **71**, 387–391.

Boubli, J.P. & Tokuda, M. (2008). Socioecology of black uakari monkeys, *Cacajao hosomi*, in Pico da Neblina National Park, Brazil: the role of the peculiar spatial–temporal distribution of resources in the Neblina forests. *Primate Report*, **75**, 3–10.

Boubli, J.P., da Silva, M.N.F., Amado, M.V., *et al.* (2008). A taxonomic reassessment of *Cacajao melanocephalus* Humboldt (1811), with the description of two new species. *International Journal of Primatology*, **29**, 723–741.

Bowler, M. (2003). The ecology and conservation of the red uakari monkey *Cacajao calvus ucayalii* in the Yavari Valley of Peru. *Primate Eye*, **79**, 3–4.

Bowler, M. (2007). The ecology and conservation of the red uakari monkey on the Yavarí River, Peru. PhD thesis, University of Kent, Canterbury.

Bowler, M. & Bodmer, R. (2009). Social behavior in fission–fusion groups of red uakari monkeys (*Cacajao calvus ucayalii*). *American Journal of Primatology*, **71**, 1–12.

Bowler, M. & Bodmer, R. (2010). Diet and food choice in Peruvian red uakaris (*Cacajao calvus ucayalii*): selective or opportunistic seed predation? *International Journal of Primatology*, **32**, 1109–1122.

Boyle, S.A., Spironello, W.R. & Smith, A.T. (2006). The effects of forest fragmentation on the bearded saki monkey (*Chiropotes sagulatus*) in the Brazilian Amazon. *American Journal of Primatology*, **68**, 87.

Campos, F., Manson, J.H. & Perry, S. (2007). Urine washing and sniffing in wild white-faced capuchins (*Cebus capucinus*): testing functional hypotheses. *International Journal of Primatology*, **28**, 55–72.

Chapman, C.A. & Chapman, L.J. (1991). The foraging itinerary of spider monkeys: when to eat leaves? *Folia Primatologia*, **56**, 162–166.

Chesson, P. & Huntley, N. (1997). The roles of harsh and fluctuating conditions in the dynamic of ecological communities. *American Naturalist*, **150**, 519–553.

Cheyney, D.L. & Seyfarth R.M. (1983). Nonrandom dispersal in free-ranging vervet monkeys: social and genetic consequences. *American Naturalist*, **122**, 392–412.

Coles, R.C., Talebi, M.G. & Lee, P.C. (2008). Fission–fusion sociality in southern muriquis (*Brachyteles arachnoides*) in the continuous Atlantic forest of Brazil. In *XXII International Primatological Society Congress*, August 3–8, Primate Eye, **96** (Sp CD-ROM iss – IPS 2008), Abstract 652.

da Cunha, A.C. & Barnett, A.A. (1989). *Project Uakari: First Report – Preliminary Survey. Part One – Zoology*. Rio de Janeiro: Pronatura.

da Cunha, A.C. & Barnett, A.A. (1990). Sightings of the golden-backed uacari, *Cacajao melanocephalus ouakary*, on the Upper Rio Negro, Amazonas, Brazil. *Primate Conservation*, **11**, 8–11.

Davis, N.C. 1931. The transmission of yellow fever further experiments with monkeys of the new world. *American Journal of Tropical Medicine*, **11**, 113–125.

Defler, T.R. (1979). On the ecology and behavior of *Cebus albifrons* in eastern Colombia: I. Ecology. *Primates*, **20**, 475–490.

Defler, T.R. (1991). Preliminary observations of *Cacajao melanocephalus* Humboldt, 1811 (Primates, Cebidae) in Colombia. *Trianea*, **4**, 557–558.

Defler, T.R. (1999a). Fission–fusion in the black-headed uacari (*Cacajao melanocephalus*) in eastern Colombia. *Neotropical Primates*, **7**, 5–8.

Defler, T.R. (1999b). *Cacajao melanocephalus ouakary* densities on the lower Apaporis River, Colombian Amazon. *Primate Report*, **61**, 31–36.

Defler, T.R. (2004). *Primates of Colombia*. Bogota: Conservacion Internacional.

Defler, T.R. & Defler, S.B. (1996). The diet of a group of woolly monkeys (*Lagothrix lagothricha lagothricha*) in southeastern Colombia. *International Journal of Primatology*, **17**, 161–190.

de Souza, L.L., Queiroz, H.L. & Ayres, J.M. (2004). The mottled-face tamarin, *Saguinus inustus*, in the Amanã Sustainable Development Reserve, Amazonas, Brazil. *Neotropical Primates*, **12**, 121–122.

Di Fiore, A. & Campbell, C.J. (2007). The Atelines: variations in ecology, behavior, and social organization. In: *Primates in Perspective*, ed. C.J. Campbell, A. Fuentes, K.C. MacKinnon, M. Panger & S.K. Bearder. New York, NY: Oxford University Press, pp. 155–185.

Duca, C., Guerra, T.J. & Marini, M.Â. (2006). Territory size of three Antbirds (Aves, Passeriformes) in an Atlantic Forest fragment in southeastern Brazil. *Revista Brasiliera da Zoolologia*, **23**, 692–698.

Eisenberg, J.F. (1977). Comparative ecology and reproduction of New World monkeys. In *The Biology and Conservation of the Callitrichidae*, ed. D.G. Kleinman. Washington, DC: Smithsonian Institution Press.

Fairbanks, L.A. (2009). Hormonal and neurochemical influences on aggression in group-living monkeys. In: *Endocrinology of Social Relationships*, ed. P.T. Ellison & P.B. Gray. Cambridge, MA: Harvard University Press, pp. 159–195.

Fernandes, M.E.B. (1993). Tail-wagging as a tension relief mechanism in pithecines. *Folia Primatologia*, **61**, 52–56.

Ferrari, S.F. (1990). A foraging association between two kite species (*Ictinea plumbea* and *Leptodon cayanensis*) and buffy-headed marmosets (*Callithrix flaviceps*) in southeastern Brazil. *Condor*, **93**, 781–783.

Ferrari, S.F., Figueiredo, W.M.B., Guedes, P., *et al.* (Submitted). Reconsidering the nomenclature of the black-faced uacaris (*Cacajao melanocephalus* group, *sensu* Hershkovitz, 1987) from the northern Amazon basin. *American Journal of Primatology*.

Ferrari, S.F., Guedes, P., Figueiredo, W.M.B., *et al.* (2009). A nomenclatura dos uacaris de cara-preta (grupo *Cacajao melanocephalus*, *sensu* Hershkovitz, 1987): observações preliminares. In

Programa e Livro de resumos, XIV Congresso Brasiliero de Primatologia. Blumenau: Sociedade Brasileira de Primatologia.

Ferrari, S.F., Pereira, W.L.A., Santos, R.R., *et al.* (2004). Fatal attack of a boa constrictor on a bearded saki (*Chiropotes satanas utahicki*). *Folia Primatologica*, **75**, 111–113.

Ferreira, L.V. (1997). Effects of the duration of flooding on species richness and floristic composition in three hectares in the Jaú National Park in floodplain forests in central Amazonia. *Biodiversity and Conservation*, **6**, 1353–1363.

Ferreira, L.V. (1999). Effects of river level fluctuations on plant species richness, diversity, and distribution in a floodplain forest in Central Amazonia. *Oecologia*, **120**, 582–587.

Ferreira, L.V. & Prance, G.T. (1998). Species richness and floristic composition in four hectares in the Jaú National Park in upland forests in Central Amazônia. *Biological Conservation*, **7**, 1349–1364.

Figueiredo, W.M.B. (2006). Estimativas de tempos de divergência em platirrinos e filogenia molecular e filogeografia dos uacaris, parauacus e cuxius. Doctoral dissertation, Universidade Federal do Pará, Belém, Brazil.

Fontaine, R. (1981). The Uakaris, genus *Cacajao*. In: *Ecology and Behaviour of Neotropical Primates*, ed. A.F. Coimbra-Filho & R.A. Mittermeier. Rio de Janeiro: Academia Brasiliera da Ciencias, pp. 443–493.

Fontaine, R. & Hench, M. (1982). Breeding New World monkeys at Miami's Monkey Jungle. *International Zoo Yearbook*, **22**, 77–84.

Fukuda, F. (1989). Habitual fission–fusion and social organization of the Hakone troop T of Japanese macaques in Kanagawa Prefecture, Japan. *International Journal of Primatology*, **10**, 419–439.

Garla, R.C., Setz, E.Z.F. & Gobbi, N. (2001). Jaguar (*Panthera onca*) food habits in Atlantic Coastal Forest, southeastern Brazil. *Biotropica*, **33**, 691–696.

Goulding, M., Carvalho, M.L. & Ferreira, E.J.G. (1988). *Rio Negro: Rich Life in Poor Water, Amazonian Diversity and Foodchain Ecology as Seen Through Fish Communities.* The Hague: SPB Academic Publishing bv.

Haugaasen, T. & Peres, C.A. (2005). Primate assemblage structure in Amazonian flooded and unflooded forests. *American Journal of Primatology*, **67**, 243–258.

Haugaasen, T. & Peres, C.A. (2009). Interspecific primate associations in Amazonian flooded and unflooded forests. *Primates*, **50**, 239–251.

Hayssen, V., van Tienhoven, A. & van Tienhoven, A. (1993). *Asdell's Patterns of Mammalian Reproduction: A Compendium of Species-specific Data.* Ithaca, NY and London: Cornell University Press.

Hershkovitz, P. (1985). A preliminary taxonomic review of the South American bearded saki monkeys genus *Chiropotes* (Cebidae, Platyrrhini), with the description of a new subspecies. *Fieldiana, Zoology*, **27**, 1–46.

Hershkovitz, P. (1987). Uacaries, New World monkeys of the genus *Cacajao* (Cebidae, Platyrrhini): a preliminary taxonomic review with the description of a new subspecies. *American Journal of Primatology*, **12**, 1–53.

Heymann, E.W. (1990). Further field notes on red uacaris, *Cacajao calvus ucayalii*, from the Quebrada Blanco, Amazonian Peru. *Primate Conservation*, **11**, 7–8.

Heymann, E.W. & Aquino, R. (2010). Peruvian red uakaris, *Cacajao calvus ucayalii*, are not flooded-forest specialists. *International Journal of Primatology*, **31**, 751–758.

Heymann, E.W. & Smith, A.C. (1999). When to feed on gums: temporal patterns of gummivory in wild tamarins, *Saguinus mystax* and *Saguinus fuscicollis* (Callitrichinae). *Zoo Biology*, **18**, 459–471.

Holt, R.D. & Kotler, B.P. (1987). Short-term apparent competition. *American Naturalist*, **130**, 412–430.

Imos, F. (1994). Jaguar predation on muruiqui, *Brachyteles arachnoids. Neotropical Primates*, **2**, 16.

Izar, P. & Nakai, E.S. (2006). Fission–fusion in tufted capuchin monkeys (*Cebus apella nigritus*) in Brazilian Atlantic Forest. *International Journal of Primatology*, **27** (Suppl. 1), Abstract 226.

Johns, A.G. & Skorupa, J.P. (1987). Responses of rain-forest primates to habitat disturbance: a review. *International Journal of Primatology*, **8**, 157–191.

Kinzey, W. (1992). Dietary and dental adaptations in the Pitheciinae. *American Journal of Physical Anthropology*, **88**, 499–514.

Kinzey, W.G. (1997). Synopsis of New World primates: *Chiropotes*. In: *New World Primates: Ecology, Evolution, and Behavior*, ed. W.G. Kinzey. New York, NY: Aldine de Gruyter, pp. 258–263.

Law, R., Marrow, P. & Diekmann, U. (1997). Evolution under asymmetric competition. *Evolutionary Ecology*, **11**, 485–501.

Lehman, S. & Robertson, K.L. (1994a). Preliminary survey of *Cacajao melanocephalus melanocephalus* in Southern Venezuela. *International Journal of Primatology*, **15**, 927–934.

Lehman, S.M. & Robertson, K.L. (1994b). Survey of Humboldt's black head uakari (*Cacajao melanocephalus melanocephalus*) in southern Amazonas, Venezuela. *American Journal of Primatology*, **33**, 223.

Lehmann, J., Korstjens, A.H. & Dunbar, R.I.M. (2007). Fission–fusion social systems as a strategy for coping with ecological constraints: a primate case. *Evolutionary Ecology*, **21**, 613–634.

Leonard, S. & Bennett, C. (1995). Behavioral ecology study of red uakari, *Cacajao calvus ucayalii*, in northeastern Peru. *Neotropical Primates*, **3**, 84.

Leonard, S. & Bennett, C. (1996). Associative behavior of *Cacajao calvus ucayalii* with other primate species in Amazonia Peru. *Primates*, **37**, 227–230.

Lönnberg, E. (1938). Remarks on some members of the genera *Pithecia* and *Cacajao* from Brazil. *Arkiv für Zoologi*, **30**A, 1–25.

Lorenz, K. (1966). *On Aggression*. London: Methuen.

MacFarland, M.J. (1986). Ecological determinants of fission–fusion sociality in *Ateles* and *Pan*. In: *Primate Ecology and Conservation*, ed. J.G. Else & P.C. Lee. New York, NY: Cambridge University Press, pp. 181–190.

Martins, S.S., Lima, E.M. & Silva, J.S. Jr. (2005). Predation of bearded saki (*Chiropotes utahicki*) by a harpy eagle (*Harpia harpyja*). *Neotropical Primates*, **13**, 7–10.

Mittermeier, R.A. & Coimbra-Filho, A.F. (1977). Primate conservation in Brazilian

Amazonia. In: *Primate Conservation*, ed. Prince Rainier III & G.H. Bourne. New York, NY: Academic Press, pp. 117–166.

Muller, M.N. & Wrangham, R. (2002). Sexual mimicry in hyaenas. *Quarterly Review of Biology*, **77**, 3–16.

Munn, C.A. & Terbough, J.W. (1979). Multi-species territoriality in Neotropical foraging flocks. *Condor*, **81**, 338–347.

Mysterud, A. (2000). Diet overlap among ruminants in Fennoscania. *Oecologia*, **124**, 130–137.

Norconk, M.A. (2011). Sakis, Uakaris, and Titi monkeys: behavioral diversity in a radiation of primate seed predators. In: *Primates in Perspective*, 2nd edition, ed. C.J. Campbell, J. Fuentes, K. MacKinnon, M. Panger & S.K. Bearder. Oxford: Oxford University Press, pp. 123–138.

Norconk, M.A. & Kinzey, W.G. (1994). Challenge of Neotropical frugivory: travel patterns of spider monkeys and bearded sakis. *American Journal of Primatology*, **34**, 171–183.

Norconk, M.A., Wertis, C.K. & Kinzey, W.G. (1997). Seed predation by monkeys and macaws in eastern Venezuela: preliminary findings. *Primates*, **38**, 177–184.

Olalla, A.M. (1955). Notas biologicas sobre los uacaries. *Geografica, Sao Paulo*, **5**(5), 17–23.

Overdorff, D.J., Erhart, E.M. & Mutschler, T. (2003). Fission–fusion in *Eulemur fulvus rufus* in southeastern Madagascar. *American Journal of Primatology*, **60** (Suppl. 1), 42–43.

Palacios, E. & Peres, C.A. (2005). Primate population densities in three nutrient-poor Amazonian terra firme forests of south-eastern Colombia. *Folia Primatologica*, **76**, 135–145.

Parolin, P. (2000). Seed mass in Amazonian floodplain forests with contrasting nutrient supplies. *Journal of Tropical Ecology*, **16**, 417–428.

Parolin, P. (2001). Seed germination and early establishment in 12 tree species from nutrient-rich and nutrient-poor Central Amazonian floodplains. *Aquatic Botany*, **70**, 89–103.

Parolin, P., Armbruester, N., Wittmann, F., *et al.* (2002). A review of tree phenology in Central Amazonian floodplains. *Pesquisas Bôtanica*, **52**, 195–222.

Powell, R.A. & Zielinski, W.J. (1983). Competition and coexistence in mustelid communities. *Acta Zoologica Fennica*, **174**, 223–227.

Prance, G.T. (1979). Notes on the vegetation types of Amazonia III: the terminology of Amazonian forest types subject to inundation. *Brittonia*, **31**, 43–61.

Prugnolle, F. & de Meeus, T. (2002). Inferring sex-biased dispersal from population genetic tools: a review. *Heredity*, **88**, 161–165.

Puertas, P. & Bodmer, R.E. (1993). Conservation of a high diversity primate assemblage. *Biodiversity and Conservation*, **2**, 586–593.

Rettig, N. (1978). Breeding behavior of the harpy eagle (*Harpia harpyja*). *The Auk*, **95**, 629–643.

Rivera, J.H.V., Ayala, D. & Haas, C.A. (2003). Home-range size, habitat use, and reproduction of the Ivory-billed Woodcreeper (*Xiphorhynchus flavigaster*) in dry forest of western Mexico. *Journal of Field Ornithology*, **74**, 141–151.

Rodriguez-Girones, M.A. Sandsten, H. & Santamaria, L. (2003). Asymmetric competition and the evolution of propagule size. *Journal of Ecology*, **91**, 554–562.

Rosenberger, A. (1992). Evolution of feeding niches in New World monkeys. *American Journal of Physical Anthropology*, **88**, 525–562.

Rosenberger, A. & Kinzey, W. (1976). Functional patterns of molar occlusion in platyrrhine primates. *American Journal of Physical Anthropology*, **45**, 281–297.

Ross, C. (2001). Park or ride? Evolution of infant carrying in primates. *International Journal of Primatology*, **22**, 749–771.

Rylands, A. (1992). Relatório: Viagem ao Parque Nacional do Jaú. Unpublished report to Fundação Vitória Amazônica, Manaus.

Salovaara, K., Bodmer, R., Recharte, M., *et al.* (2003). Diversity and abundance of mammals. In *Perú: Yavarí (Rapid Biological Inventories. 11)*, ed. N. Pitman, *et al.* Chicago, IL: Field Museum, pp. 156–164

Saunders, D.A., Hobbs, R.J. & Margules, C.R. (1991). Biological consequences of ecosystem fragmentation: a review. *Conservation Biology*, **5**, 18–32.

Schoener, T.W. (1983). Field experiments on interspecific competition. *American Naturalist*, **122**, 240–285.

Senar, J.C. (1999). Plumage coloration as a signal of social status. In: *Proceedings of the 22nd. International Ornithological Congress*, ed. N.J. & R.H. Slotow. Johannesburg: Birdlife-South Africa, pp. 1669–1686.

Silva, W., Almeida, T. & Barnett, A.A. (2008). Resource availability and seasonal use of contiguous forest types by the golden-backed uacari (*Cacajao melanocephalus ouakary*: Primates). In *Abstracts of the Association for Tropical Biology and Conservation 2008 meeting (June 9–13, Paramaribo, Suriname)*, p. 37.

Silva Junior, J.S. & Martins, E.S. (1999). On a new white bald uakari population in Southwestern Brazilian Amazonia. *Neotropical Primates*, **7**, 119–121.

Swanson Ward, N. & Chism, J. (2003). A report on a new geographic location of red uakaris (*Cacajao calvus ucayalii*) on the Quebrada Tahuaillo in northeastern Peru. *Neotropical Primates*, **11**, 19–22.

ter Steege, H., Pitman, N., Phillips, O.L., *et al.* (2006). Continental-scale patterns of canopy tree composition and function across Amazonia. *Nature*, **443**, 28.

Toborowsky, C. (2008). Preliminary evidence of fission–fusion sub-grouping behavior in white-fronted brown lemurs (*Eulemur fulvus albifrons*) in Betampona Natural Reserve, Madagascar. *American Journal of Physical Anthropology*, Suppl 46, p. 208.

Torok, J. & Toth L. (1999). Asymmetric competition between two Tit species: a reciprocal removal experiment. *Journal of Animal Ecology*, **68**, 338–345.

van Roosmalen, M.G.M. & Garcia, O.M.C.G. (2000). Fruits of the Amazonian forest. Part II: Sapotaceae. *Acta Amazonica*, **30**, 187–290.

van Schaik, C.P. (1999). The socioecology of fission–fusion sociality in orangutans. *Primates*, **40**, 69–86.

Vasey, N. (2006). Impact of seasonality and reproduction on social structure, ranging patterns, and fission–fusion social organization in red ruffed lemurs. In: *Lemurs: Ecology and Adaptation*, ed. L. Gould, & M.L. Sauther. New York, NY: Springer, pp. 275–304.

Veiga, L.M., Pinto, L.P. & Ferrari, S.F. (2006). Fission–fusion sociality in bearded sakis (*Chiropotes albinasus* and *Chiropotes satanas*) in Brazilian

Amazonia. *International Journal of Primatology*, **27**(Suppl. 1), abstract 224.

Walker, S.E. & Ayres, J.M. (1996). Positional behavior of the white uakari (*Cacajao calvus calvus*). *American Journal of Physical Anthropology*, **101**, 161–172.

Wallace, R.B., Painter, R.L.E., Taber, A.B., *et al.* (1996). Notes on a distributional river boundary and southern range extension for two species of Amazonian primates. *Neotropical Primates*, **4**, 149–151.

Wissinger, S.A. (1992). Niche overlap and the potential for competition and intraguild predation between size-structured populations. *Ecology*, **73**, 1431–1444.

Wooley, M.J., Clark, D.L. & Stevens, J.J. (1978). Squirrel monkey dominance and social behavior as related to group size and group structure. *Primates*, **19**, 169–177.

Zhang, S. & Wang, L. (2000). Following of brown capuchin monkeys by white hawks in French Guiana. *Condor*, **102**, 198–201.

Chapter 15

Annual variation in breeding success and changes in population density of *Cacajao calvus ucayalii* in the Lago Preto Conservation Concession, Peru

M. Bowler, C. Barton, S. McCann-Wood, P. Puertas & R. Bodmer

Introduction

Cacajao calvus has a patchy distribution and generally occurs at low densities (Aquino 1988; Ayres 1986; Puertas & Bodmer 1993; Aquino *et al.* 2005). Salovaara *et al.* (2003) demonstrated a patchy distribution for *Cacajao calvus ucayalii* between the Yavarí and Ucayali Rivers and where the species occurred, its abundance varied from 6.07 to 47.33 individuals/100 km. Barnett *et al.* (Chapter 16) also found widely variable densities for *Cacajao* in various studies, and although differing methodologies may account for some of the variation, it is clear that *Cacajao* can both be locally abundant or locally rare. The Lago Preto Conservation Concession on the Yavarí River contains large numbers of red uacari monkeys, and is the only known site where regular sightings can be guaranteed.

Ayres and Prance (Chapter 12) and Bowler (2007) considered the effects of the distribution of key food species on the distribution and population density of *Cacajao calvus* and Ayres (1986) recorded seasonal variation in the availability of key resources for *Cacajao calvus calvus*. However, nothing is known about how populations of *Cacajao calvus* react to interannual variation in food availability. Human influences also affect the distribution and abundance of *Cacajao calvus*. Densities may change in response to timber extraction in some areas (Ayres & Johns 1987; Bowler 2007), or through extraction of non-timber resources that are important diet components, such as *Mauritia flexuosa* fruits (Aquino 1999; Bowler 2007). The effect of hunting on all large-bodied primate densities, including *Cacajao calvus ucayalii*, was demonstrated by Puertas and Bodmer (1993), Bodmer *et al.* (1997, 2003) and Peres (2000).

Notwithstanding the above studies, fundamental demographic variables, such as birth rates and the age–sex composition of populations, are almost unknown for *Cacajao*, and changes in population size and density through time have not been monitored. Ayres (1986) recorded seasonality of births in *Cacajao calvus calvus*, but until now, no studies on wild *Cacajao calvus ucayalii* have monitored births and mating year-round. Similarly, interbirth intervals, and the proportion of the population made up by breeding females, have never been estimated.

Photo 15.1 Red uacari male (*Cacajao calvus ucayalii*), Yavari River, Peru. Photo: Mark Bowler. (See color plate section.)

Low levels of hunting have long occurred at Lago Preto. Hunting diminished still further after field research by

Evolutionary Biology and Conservation of Titis, Sakis and Uacaris, eds. Liza M. Veiga, Adrian A. Barnett, Stephen F. Ferrari and Marilyn A. Norconk. Published by Cambridge University Press.

R. Bodmer and co-workers began there in 1990 (Bodmer *et al.* 2003). Since 2003, the population at Lago Preto has received unofficial protection from the almost constant presence of researchers (Bowler 2007), and subsequently "concession guards" when in 2006 the area became a "conservation concession" managed by Wildlife Conservation Society (WCS) and The Durrell Institute of Conservation and Ecology (DICE), in partnership with the Peruvian National Institute for Natural Resources (INRENA). As a result, practically all hunting in the immediate area has ceased. This chapter reports on group composition and breeding seasonality in the red uacari population in the Lago Preto Conservation Concession, and changes in population density following the declaration of the concession.

Study site

The 9926.19 ha Lago Preto Conservation Concession (LPCC: 04°27.5′S, 071°45.9′W) lies 175 km southeast of Iquitos on the Yavarí River. It was established because of an exceptionally high density of red uacaris, *Cacajao calvus ucayalii*. The concession is bounded by the Yavarí River to the south, the Yavarí–Mirín River and proposed Greater Yavarí Reserve to the west and extensive logging concessions to the north and east (Figure 15.1). There are three main forest habitats: *terra firme* (non-flooding), *várzea* (seasonally flooding) and *aguajal* palm swamp. The *várzea* is generally flooded between November and May, but flooding periods vary depending on rainfall both locally and in the Andes.

The behavioral and ecological data for this chapter were collected within an area of around 2200 ha on the west side of the concession. Extending several kilometers outside the northernmost reserve boundary, this contains 1400 ha of *terra firme*, 500 ha of *várzea* and 300 ha of *aguajal*. Census data were collected from a wider area around the study area, on trails though all three major habitat types.

Methods

Group composition and breeding seasonality

A total of 945:10 h of behavioral data were collected between April 2003 and January 2004, March 2004 and April 2005 and June and July 2005. Uacaris were located by walking the trail system and were followed until they settled at dusk or the group was lost. When the *várzea* forest was flooded, the monkeys were often followed by canoe.

Group size was recorded at 10-min intervals and, where possible, age and sex classes were counted, using Fontaine's (1981) definitions; *Infant-one* (0–3 months: small young carried on their mother's waists or bellies with gray-brown or partially gray-brown faces), *Infant-two* (3–12 months), *Juvenile* (12–36 months), *Subadult* (where the sex could not

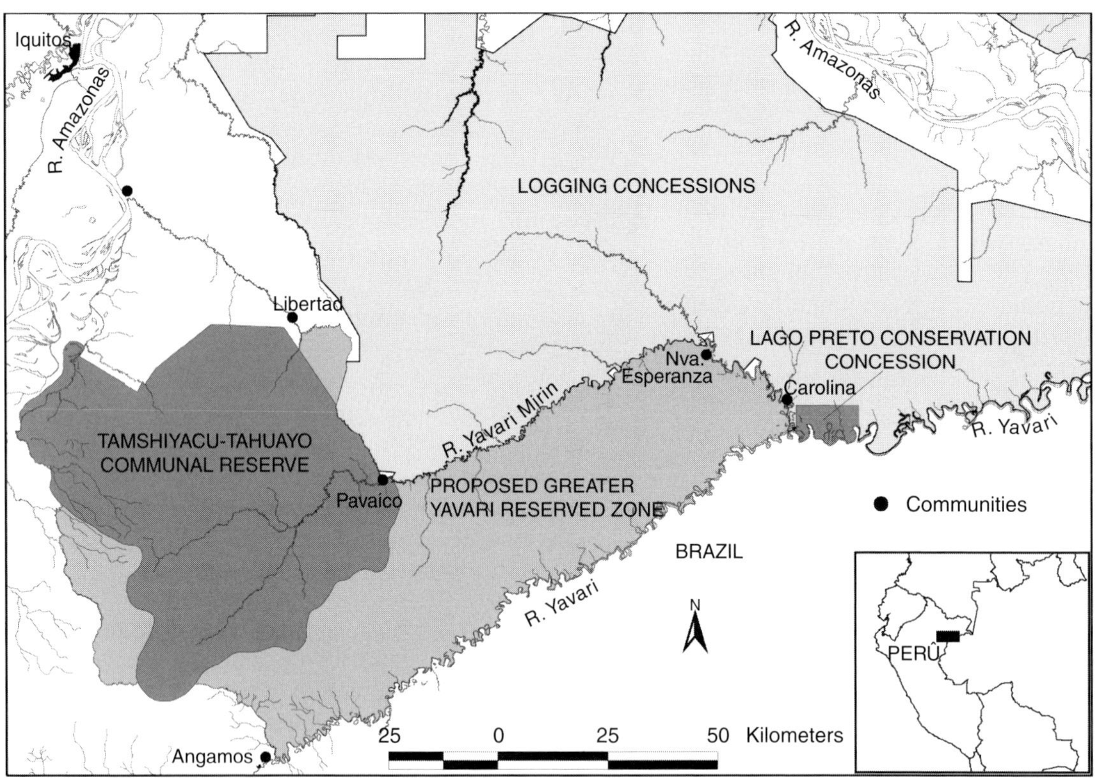

Figure 15.1 Map of the Lago Preto Conservation Concession and logging concessions on the Yavarí and Yavarí–Mirín Rivers with other key areas for the conservation of *Cacajao calvus ucayalii* (figure adapted from INRENA).

be determined; juvenile and subadult uacaris are sexually cryptic), *Subadult* male (full size, but not showing the typical features of an adult male; principally the large cranial muscles), *Adult Male* and *Adult Female*. By recording the presence of infant-one uacaris, the seasonality of births at Lago Preto could be deduced. Counts of infants were also made during mammal censuses in April 2006, and an expedition was made in November 2007 to record the proportion of females carrying infant-one young, when attempts were made to inspect all adult female uacaris for infants.

Population density

Between 2001 and 2007, uacaris were censused in the LPCC using the DISTANCE sampling method (Buckland *et al.* 1993). A total of 1248.4 km was walked, at a rate of approximately 1 km/h; 217.2 km in 2001, 382.6 km in 2005, 305 km in 2006 and 343.6 km in 2007. When uacaris were encountered, the group size and perpendicular distance from the trail of the first animal sighted were recorded. Densities were calculated using the DISTANCE program. Census effort was concentrated between May and July each year.

Results

Group size and composition

Group sizes ranged from 1 to 150+ individuals (mean 43.5 ± 24.1), sometimes foraging as a single unit, but at other times fissioning and fusing during the day. On several occasions, group size estimates for several different uacari groups were simultaneously obtained by different observers. Using the lower estimates for each group, we obtain a conservative minimum estimate of 160 uacaris simultaneously using the study area.

Age and sex classes were not evenly distributed through uacari groups. All-male parties of 1–10 adult and subadult males were often observed within the groups. For this reason, the group composition of small groups could not be used to determine the average composition and age/sex ratios. Furthermore, it was never possible to record the complete composition once groups exceeded 31 animals because individuals were then generally dispersed over several hectares. Partial counts allow the crude estimation that the Lago Preto population comprises approximately 33.3% adult males, 33.3% adult females and 33.3% juveniles and infants.

Mating

Mating was observed on six occasions during the study. On 11 June 2003 mating was observed three times within 47 min. It was not possible to keep track of individuals between observations, but at least two different males were involved. The number of females involved was unclear. Mating was also observed on 12 June 2003, 30 April 2004 and 16 May 2004.

Births

Infant-one uacaris were recorded on multiple occasions in August, September, October and November 2003; minimum and maximum counts were not obtained. One infant-one was recorded in April 2004, indicating a birth between February and March 2004. In December 2004 a single infant-one was recorded. In April 2006, a minimum count of 5 young in the mid to later part of the infant-two stage were observed in a group of 50–80 individuals, indicating births between August and October 2005. In November 2007, all adult females that

Photo 15.2 Adult male uacari (*Cacajao calvus ucayalii*) in aguajal palm (*Mauritia flexuosa*), Yavari River, Peru. Photo: Mark Bowler. (See color plate section.)

were closely inspected carried infant-one young. Individual females could not be recognized, so while a minimum of 6 adult females were inspected, many more were probably checked.

Breeding in *Cacajao calvus ucayalii* at Lago Preto was seasonal; only one birth was detected outside the period from August to November. In 2003, 2005 and 2007 a large proportion of adult females were carrying infant uacaris that would have been born between these months. 2004 was a poor year for uacari births at Lago Preto, with one birth during the normal birth season and one apparently outside it. Counts were not made in 2006, but there is a possibility that births at Lago Preto alternated between good and poor years. Annual birth-rate estimates depend on whether the high proportion of females observed with infant-one young in 2003, 2005 and 2007 is typical and the poor year in 2004 was a relatively infrequent occurrence, or if *Cacajao calvus ucayalii* generally gives birth every 2 years, and births have become synchronized. Given that estimates suggest that one-third of the population at Lago Preto are adult females, the maximum annual birth rate could be as high as 33% of the population if females give birth every year, or 17% of the population if they give birth every 2 years.

Population density

The density of *Cacajao calvus ucayalii* recorded at Lago Preto increased in each year that censuses were conducted (Table 15.1). Population density estimates ranged from 19.0 Ind./km^2 in 2001 to 116.8 Ind./km^2 in 2007. From this we can calculate that uacari individuals in the 2200 ha study area increased from 418 in 2001 to 2570 in 2007. These estimates are much larger than minimum estimates obtained by simultaneous group counts in 2005.

Between 2001 and 2007, estimated uacari population density grew at an average annual rate of 35.3%. Within the limits of confidence of the density calculations, this growth rate is consistent with upper estimates for birth rate at Lago Preto, assuming low levels of mortality, but is not consistent with the apparently bad year of 2004.

Table 15.1 Densities of *Cacajao calvus ucayalii* recorded in different years in the Lago Preto Conservation Concession.

Year	Distance censused (km)	Density of *Cacajao calvus ucayalii* (Ind./km^2)	C.V.	Estimated population size for the study area of 2200 ha (22 km^2)
2001	217.2	19.0	55.5	418.0
2005	382.6	34.7	25.56	763.4
2006	305.0	112.1	21.15	2466.2
2007	343.6	116.8	34.74	2569.6

Discussion

Breeding in *Cacajao calvus ucayalii* at Lago Preto was seasonal. The six observations of mating occurred between April and mid June, and almost all births between August and November. Typically, Neotropical primate species conform to a birthing pattern concentrated shortly before a food availability peak, allowing most lactation (small-sized species) or weaning itself (capuchins) to occur before the lean season commences (Di Bitetti & Janson 2000). The uacari birthing period falls after the seasonal abundance of *Mauritia flexuosa* fruits, a key resource for *Cacajao calvus ucayalii* between April and August, and at a time when relatively few other fruits are available. Fruit availability for uacaris at Lago Preto (including both ripe and unripe fruits) peaked in February (Bowler 2007), around 4 months after the middle of the birthing season. This birthing pattern differs from that observed in *Pithecia*, *Chiropotes* and *Cacajao calvus calvus* where parturition occurs shortly before peak fruit availability (van Roosmalen *et al.* 1981; Ayres 1986; Soini 1986), although births of *Cacajao calvus calvus* at Lago Tefé occurred at approximately the same time of year as *Cacajao calvus ucayalii* at Lago Preto. The *C. c. ucayalii* pattern more closely resembles that of *Ateles*, *Lagothrix* and *Brachyteles*, which also give birth during the lean season (Di Bitetti & Janson 2000). No data are available on the time of weaning at Lago Preto, but uacaris have a specialist diet unlike that of other Neotropical primates and many of the fruits eaten are hard-shelled and difficult to open (Ayres 1986; Kinzey 1992; Boubli 1999; Bowler 2007; Norconk 2011; Barnett *et al.*, Chapter 16). It is therefore possible that weaning occurs at a later stage than in other species, to allow uacari young to develop sufficiently strong jaws to open these fruits. The birthing season may therefore be timed to coincide weaning with peak food availability, as with capuchins.

There are two possible explanations for the high proportion of females carrying young, seen in three alternating years at Lago Preto (2003, 2005, and 2007), and the lack of births in 2004. First, some females may give birth every year, and 2004 could have been a relatively unusual bad year. Alternatively, females could give birth on average every 2 years, and may have become synchronized or partially synchronized in their breeding cycles. This could in turn be connected to fruiting cycles of key resources, with particularly bad years leading to low birth rates or high infant mortality and subsequent high birth rates the following year, as a result of more females coming into season. A degree of synchronization could continue into following years.

The only previous data on interbirth intervals in uacaris came from a 30-month study of a captive population (Fontaine 1981). Suckling in this population continued until 22 months of age. Of the similarly sized New World monkeys, *Cebus capucinus* has a median interbirth interval in the wild of 26.4 months (Fedigan & Rose 1995), *Cebus apella* 22 months (Robinson & Janson 1987), *Cebus olivaceus* 26 months (Robinson & Janson 1987), *Cebus albifrons* 18 months (Kappeler & Pereira 2003) and *Pithecia pithecia* 21.9 months (Norconk in press, cited in

Norconk 2011). For *Chiropotes*, the most closely related genus to *Cacajao*, Peetz (2001) estimated a birth interval of at least 2 years. Given the evidence for interbirth periods of captive uacaris and wild related primates, it appears most likely that uacaris have an interbirth period of around 2 years.

Reasons for the lack of births in 2004 are uncertain, but it is possible that environmental conditions during the preceding year affected birth rates. Reduced food intake can result in reproductive suppression (Whitten 1983; van Schaik & van Noordwijk 1985). Between May and August, the 4 months preceding the birth season for uacaris at LPCC, *Mauritia flexuosa* palm fruits make up over 50% of the diet (Bowler 2007). Annual variation in *Mauritia flexuosa* productivity is a possible explanation for the lack of infants observed in 2004, but no firm link could be established. If red uacari reproduction is affected by annual variation in the production of *Mauritia flexuosa* or other key resources, this will need to be taken into account when modeling the demography of these populations.

The density of *Cacajao calvus ucayalii* recorded at Lago Preto increased dramatically between 2001 and 2007, and particularly between 2005 and 2006. It is unlikely that a habituation process is responsible for the observed density increases, because uacari groups are conspicuous and were not wary from the start of the study period. However, the differences seen could be partially due to the large group sizes and large home ranges observed. Uacaris at Lago Preto ranged over at least 2200 ha, and groups would often use one part of the range for several weeks before moving to a new area (Bowler 2007). Because the range of uacaris at the LPCC study site includes some areas outside the area surveyed, it is possible that one or more large groups of uacaris were using these areas for the duration of the census period in some years, while in other years groups were using areas that did contain census trails. Alternatively, some practical restrictions of the methods used could account for exaggerated differences in densities recorded between years. The practice of recording the perpendicular distance of the first individual sighted rather than the centre of the group at the time of the first sighting, although far more achievable in the field, is likely to lead to overestimation of densities (Marshall *et al.* 2008). If group sizes have increased at LPCC following increases in uacari density, the difference between the distances from the trail of the first individual sighted and the centre of the group is likely to be greater, further exaggerating density estimations. Nevertheless, density estimates are calculated in the same way for each year, and increased between every pair of consecutive censuses, demonstrating a real increase in the size of the uacari population.

The increase in population density of *Cacajao calvus ucayalii* at Lago Preto may be a result of changes in number of animals hunted. Hunting may have been reduced in the concession as a result of community-based conservation in the village of Carolina close to the reserve, and by the almost constant presence of researchers and WCS-Peru staff since April 2003. The most dramatic increase occurred between 2005 and 2006. This is consistent with the protection starting in 2003, but followed by a bad year for uacari births in 2004.

The population growth of *Cacajao calvus ucayalii* at LPCC appears very fast. It remains to be seen if this population will stabilize as it reaches the carrying capacity of the area, as it would in a density-dependent population, or if the population will fluctuate or crash. White-lipped peccaries cycle between high and low densities when hunting pressure is low, and this may be typical of species that forage in large groups (Fang *et al.* 2008). Although *Cacajao calvus ucayalii* forages in groups of variable sizes, it spends much of the time in large groups, especially when feeding in large patches of key resources such as *Mauritia flexuosa* (Bowler 2007). The possibility that the study population could crash once it reaches the environmental carrying capacity must therefore be considered. A tendency for populations of *Cacajao calvus ucayalii* to crash could explain the patchy distribution recorded for *Cacajao calvus ucayalii* by Salovaara *et al.* (2003) on the Yavarí and Yavarí–Mirín Rivers, although sensitivity to hunting may be a more parsimonious explanation. To make confident predictions about these possibilities, a more complete understanding of the life history characteristics of *Cacajao calvus* is required, with good data on population changes at Lago Preto and in the rest of the uacaris range. Further studies are required on population dynamics in response to food availability, as well as more accurate information on birth rates, mortality rates and group composition. These data would enable modeling of population viability, using software such as Vortex, inside and outside the reserve. The area surrounding LPCC consists largely of logging concessions (Figure 15.1), which make up a considerable part of the geographical range of *Cacajao calvus ucayalii*. Uacaris in these concessions are vulnerable to hunting by concession workers and will need to be included in any study on the viability of populations of *Cacajao calvus ucayalii*.

Acknowledgments

We are grateful for the assistance provided by research staff and students from WCS, DICE, UNAP and WWF, especially Miguel Antunez, Claudia Rios, Alfonso Mendez, Alfredo Dos Santos, Jorge Flores, Pedro Perez, Annie Escobedo, Zina Valverde, Maribel Recharte, Marcos Rios, Ricardo Zarate, Mary Inga, Lucho Moya, Kelly Moya, Lourdes Ruck, and the numerous research, volunteer students and Earthwatch volunteers who participated with the projects. We are also grateful to the boat crew, field assistants, and nurses, who provided dedicated service. Lesly Sanchez and Roxana Pezo are thanked for the administrative assistance. The people of Carolina and Nuevo Esperansa are thanked for their hospitality and their dedication for a sustainable future. The project was conducted in collaboration with the Wildlife Conservation Society, the Durrell Institute of Conservation and Ecology, the National University of the Peruvian Amazon (UNAP), the National Institute of Natural Resources – Peru (INRENA) and the Earthwatch Institute. The behavioral data collection was funded by Rufford Small Grants, the LA Zoo, Conservation International and Primate Conservation Inc.

References

Aquino, R. (1988). Preliminary survey on the population densities of *Cacajao calvus ucayalii*. *Primate Conservation*, **9**, 24–26.

Aquino, R. (1999). Observaciones preliminares sobre la dieta de *Cacajao calvus ucayalii* en el Nor-Oriente Peruano. *Neotropical Primates*, 7(1), 1-5.

Aquino, R., Alvarez, J. & Mulanovich, A. (2005). Diversidad y estado de conservacion de primates en las Sierras de Contamanna, Amazonia Peruana. *Revista Peruana de Biologia*, **12**(3), 427–434.

Ayres, J.M. (1986). The white uakaris and the Amazonian flooded forests. Unpublished PhD thesis, Cambridge University.

Ayres, J.M. & Johns, A.D. (1987). Conservation of white uacaries in Amazonian *várzea*. *Oryx*, **21**(2), 74–80.

Bodmer, R.E., Eisenberg, J.F. & Redford K.H. (1997). Hunting and the likelihood of extinction of Amazonian mammals. *Conservation Biology*, **11**, 460–466.

Bodmer, R.E., Puertas, P. & Antúnez, M. (2003). Use and sustainability of wildlife hunting in and around the proposed Yavarí reserved zone. In: *Peru: Yavari. Rapid Biological Inventories Report 11*, ed. N.C. Pitman, C. Vriesendorp & D. Moskovits). Chicago, IL: The Field Museum.

Boubli, J.P. (1999). Feeding ecology of black-headed uakaris (*Cacajao melanocephalus melanocephalus*) in the Pico de Neblina National Park, Brazil. *International Journal of Primatology*, **20**, 719–749.

Bowler, M. (2007). The ecology and conservation of the red uakari monkey on the Yavarí River, Peru. Unpublished PhD thesis, University of Kent, Canterbury.

Buckland, S.T., Anderson, D.R., Burnham, K.P., *et al.* (1993). *Distance Sampling: Estimating Abundance of Biological Populations*. New York, NY: Chapman and Hall.

Di Bitetti, M.S. & Janson, C.H. (2000). When will the stork arrive? Patterns of birth seasonality in Neotropical primates. *American Journal of Primatology*, **50**(2), 109–130.

Fang, T.G., Bodmer, R.E., Puertas, P.E., *et al.* (2008). *Certificación de pieles de pecaríes en la Amazonía Peruana. Una estrategia para la conservación y manejo de fauna silvestre en la Amazonía Peruana*. WCS, DICE, Darwin Initiative, INRENA, Fundamazonía. Lima, Peru.

Fedigan, L.M. & Rose, L.M. (1995). Interbirth interval variation in three sympatric species of Neotropical monkey. *American Journal of Primatology*, **37**, 9–24.

Fontaine, R. (1981). The uakaris, genus *Cacajao*. In *Ecology and Behavior of Neotropical Primates*, ed. A.F. Coimbra-Filho & R.A. Mittermeier. Rio de Janeiro: Academia Brasileira de Ciencias.

Kappeler, P.M. & Pereira, M.E. (2003). *Primate Life Histories and Socioecology*. Chicago, IL: University of Chicago.

Kinzey, W.G. (1992). Dietary and dental adaptations in the Pitheciinae. *American Journal of Physical Anthropology*, **88**, 499–514.

Marshall, A.R., Lovett, J.C. & White, P.C.L. (2008). Selection of line-transect methods for estimating the density of group-living animals: lessons from the primates. *American Journal of Primatology*, **70**(5), 452–462.

Norconk, M.A. (2011). Sakis, uakaris and titi monkeys. Behavioural diversity in a radiation of seed predators. In *Primates in Perspective* (2nd edn), ed. C.J. Campbell, A. Fuentes, K.C. MacKinnon, S.K. Beader & R.M. Stumpf. New York, NY: Oxford University Press, pp. 122–139.

Peetz, A. (2001). *Ecology and Social Organization of the Bearded Saki* Chiropotes satanas chiropotes *(Primates: Pitheciinae) in Venezuela (Ecotropical Monographs, No. 1)*. Bonn, Germany: Society of Tropical Ecology.

Peres, C.A. (2000). Evaluating the impact and sustainability of subsistence hunting at multiple Amazonian forest sites. In: *Hunting for Sustainability in Tropical Forests*, ed. J.G. Robinson & E.L. Bennett, pp. 31–57. New York, NY: Columbia University Press.

Puertas, P. & Bodmer, R.E. (1993). Conservation of a high diversity primate assemblage. *Biodiversity and Conservation*, **2**, 586–593.

Robinson J.G. & Janson, C.H. (1987). Capuchins, squirrel monkeys and atelines: Sociological convergence with Old World primates. In: *Primate Societies*, ed. B.B. Smuts, D.L. Cheney, R.M. Seyfarth, R.W. Wrangham & T.T. Struhsaker. Chicago, IL: University of Chicago Press, pp. 69–82.

Salovaara, K., Bodmer, R.E., Recharte, M., *et al.* (2003). Diversity and abundance of mammals. In: *Peru: Yavari. Rapid Biological Inventories Report 11*, ed. N.C. Pitman, C. Vriesendorp & D. Moskovits. Chicago, IL: The Field Museum.

Soini, P. (1986). A synecological study of a primate community in the Pacaya Samiria National Reserve, Peru. *Primate Conservation*, **7**, 63–71.

van Roosmalen, M.G.M., Mittermeier, R.A. & Milton, K. (1981). The bearded sakis, genus *Chiropotes*. In: *Ecology and Behavior of Neotropical Primates, Vol 1*, ed. A.F. Coimbra-Filho & R.A. Mittermeier. Rio de Janeiro: Academia Brasileira de Ciencias.

van Schaik, C.P. & van Noordwijk, M.A. (1985). Interannual variability in fruit abundance and the reproductive seasonality in Sumatran long-tailed macaques (*Macaca fascicularis*). *Journal of Zoology*, **4**, 533–549.

Whitten, P.L. (1983). Diet and dominance among female vervet monkeys (*Cercopithecus aethiops*). *American Journal of Primatology*, **5**(2), 139–159.

Chapter 16

Cacajao ouakary in Brazil and Colombia: patterns, puzzles and predictions

Adrian A. Barnett, Thomas R. Defler, Marcela Oliveira, Helder Queiroz & Bruna M. Bezerra

Introduction

The majority of taxa in the genus *Cacajao* have restricted ranges (Silva Junior *et al.*, Chapter 4), with *C. calvus calvus* having one of the smallest ranges of any of the larger Neotropical primates (an area some 100 × 230 km: the area of the US state of New Hampshire, and 80% the size of Belgium). While the geographic distribution of the golden-backed uacari, *Cacajao ouakary*, is not large in comparison to some other Amazonian primate taxa (namely *Alouatta seniculus*, *Cebus albifrons*, *C. macrocephalus*: Goldani *et al.* 2006), it is the largest of any member of the genus (only *C. c. ucayalii* coming close, see Silva Junior *et al.*, Chapter 4). Covering an estimated area of some 480 × 1225 km (580,000 km^2: an area about 25% bigger than California, and slightly bigger than France), the range of *C. ouakary* is sufficiently large to embrace some of the gradients in soil fertility, duration of dry season, plant community composition and life history that are reported to run either west to east, or northeast to southwest across the Amazon basin (Clinebell *et al.* 1995; ter Steege *et al.* 2006). Such variations have been shown to underlie differences in seasonality (phenology), quality (alkaloid load) and quantity (crop volume) of fruits (Mattson 1980; Gentry & Emmons 1987; Hammond & Brown 1995; Forget *et al.* 1999; Nascimento *et al.* 2005). As in the Paleotropics (e.g. Brugiere *et al.* 2002), Neotropical primates are acutely sensitive to differences in spatial and temporal patterns of resource availability (e.g. Peres 1994, 1997b; Emmons 1999; Defler 2001b; Haugaasen & Peres 2005; Bowler & Bodmer 2010), and consequent differences in diet, ranging and social behavior between conspecific populations are well documented (e.g. Hanya *et al.* 2002, 2004, 2008; Fashing *et al.* 2007). In this chapter we investigate the extent and nature of such variation in *C. ouakary* between three lowland Amazonian sites (Jaú and Amanã in Brazil, and the lower Río Apaporis in Colombia). Our goal is to relate our findings to physical, meteorological and biological variables that exist among the sites.

Black-backed and golden-backed uacaris were formerly described as subspecies of the black-headed uacari (*C. melanocephalus*: *C. m. melanocephalus* and *C. m. ouakary*, respectively) by Hershkovitz (1987), but were recently elevated to full species status following the genetic studies of Figueiredo (2006: see also Figueiredo *et al.*, Chapter 3 and Silva Junior *et al.*, Chapter 4): that arrangement is followed here. The biology of the two species has been summarized by Barnett (2005, 2010). Boubli *et al.* (2008) furnish what is known about the ecology of *C. ayresi*. Boubli and Tokuda (2008) have provided additional information on the diet and social biology of *C. melanocephalus*. Boubli and Lima (2009) provide computer models of the fundamental niche of *Cacajao*.

Golden-backed uacaris have been studied in Brazil in Jaú National Park on the lower Rio Negro (Rylands 1992; Barnett *et al.* 2002, 2005, 2011, 2012; Bezerra *et al.* 2010a, 2010b, 2011) and in Colombia on the lower Río Apaporis (Defler 1999, 2001a, 2004: Figure 16.1). In addition, there have been brief surveys on affluents of the upper Rio Negro, the Curicuriari and Uaupes rivers (Barnett & da Cunha 1991; da Cunha & Barnett 1990), and in the Amanã Sustainable Development Reserve (de Souza *et al.* 2004) within the Negro–Solimões interfluve (see Figure 16.1). The species appears to use mostly *terra firme* and *igapó* (black-water seasonally flooded forest), although visits to palm swamp occur when fruit are in season (Barnett 2005).

The golden-backed uacari is essentially limited to the Rio Negro–Solimes/Japurá interfluve (da Silva Junior *et al.*, Chapter 4 and Figure 16.1). Historical records of the nineteenth century suggest that the eastern limit has always been the eastern end of this interfluve (see Hershkovitz 1987). While uacaris are still there, the modern distribution has fragmented, probably from human impacts such as hunting and timber extraction (W. Spironello, A. Barnett, unpubl. data). Historically, the western limits may have been the Llanos de Yarí just south of the Serranía de la Macarena in Colombia (Guaviare). Klein and Klein (1976) received reports that individuals of this species occurred south of the Rio Guaviare as recently as the 1960s (Figure 16.1). On the northern bank of the Rio Negro, the species is replaced by both the black-backed uacari, *Cacajao melanocephalus* (see Silva Junior *et al.*, Chapter 4), and by *C. ayresi*, newly described by Boubli *et al.* (2008; see also Silva Junior *et al.*, Chapter 4).

Evolutionary Biology and Conservation of Titis, Sakis and Uacaris, eds. Liza M. Veiga, Adrian A. Barnett, Stephen F. Ferrari and Marilyn A. Norconk. Published by Cambridge University Press.

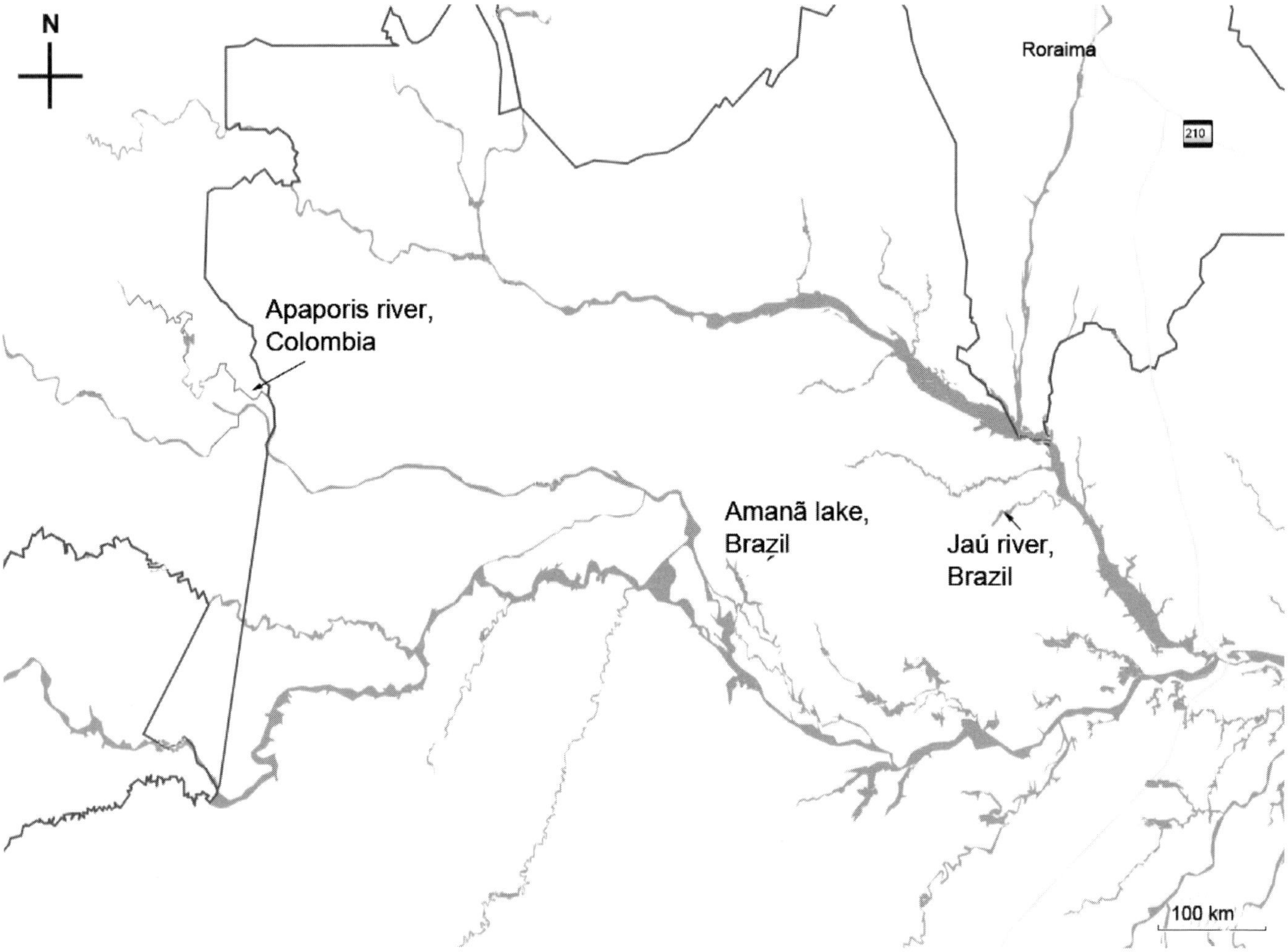

Figure 16.1 Map showing the three study sites: Jaú and Amanã (Brazil) and Caparú (Colombia).

Unfortunately, *C. ouakary* studies in Brazil and Colombia have not been conducted as part of a coordinated research program. Hence there are some methodological differences between sites and some aspects (e.g. the quantitative study of vocalizations: see Bezerra *et al.*, Chapter 30) have been studied only at one site. However, it has proved possible to make some comparisons in five areas of knowledge (diet – fruit; diet – non-fruit items; group size and fission–fusion activity; seasonal movements; reproductive seasonality and breeding behavior). These are considered below.

Study sites

Physical factors such as soil fertility and annual precipitation profiles have been shown to be important in determining the finer-grained aspects of a primate species' behavior between localities (see Panger *et al.* 2002; Peres 1993, 1997a, 1997b). Forest structure, species composition and phenology of the regional plant community (Gentry & Emmons 1987) influence the composition of macro-frugivore guilds of which primates are a prominent part (Dew & Boubli 2005). We begin by presenting a summary of the three major study sites.

Jaú (01°53′ 568″ S, 61°41′ 842″ W)

Jaú National Park is a 2,700,000 ha area between the Unini and Caribinani rivers on the south side of the Rio Negro, some 220 km west of Manaus. The geological formations underlying the park are mostly from the relatively recent (5.6–1.6 mya) and fertile Solimões Formation (Pinheiro & Borges 2004). However, the park's northwestern corner (where uacari studies have been carried out), is underlain by laterized sandstones of the Prosperança Formation, formed some 570–510 mya (Pinheiro & Borges 2004). Soils derived from such rocks are notably poor in nutrients required for plant growth (do Nascimento *et al.* 2004). Overlying these along river banks are locally derived re-worked alluvial deposits from the Holocene (Pinheiro & Borges 2004). Nutrient-poor, these soils underlie the *igapó*, the region's seasonally flooded river margin forest.

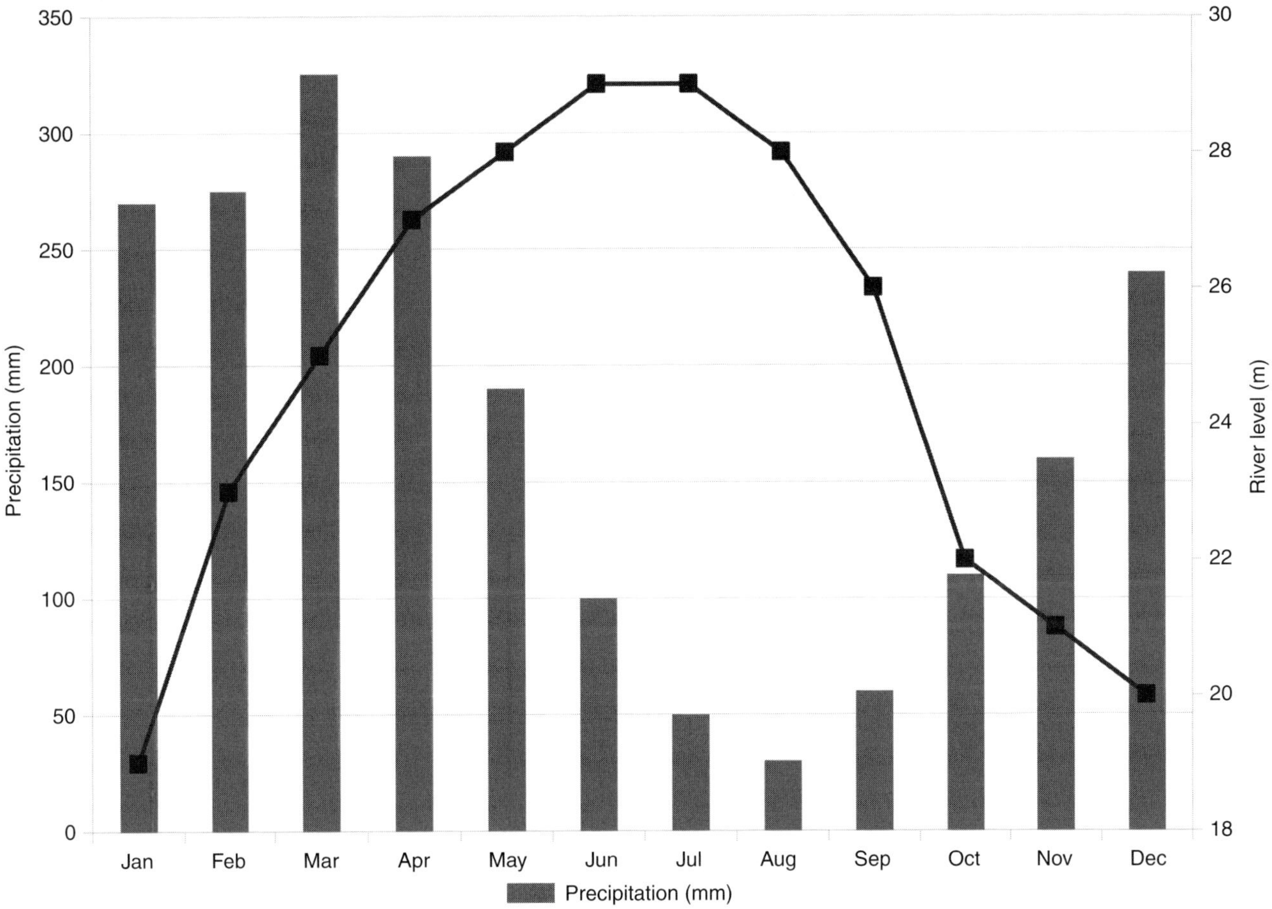

Figure 16.2 River levels and precipitation at Manaus (average of 10 years of data, redrawn from Ferreira 1997). (Bars = precipitation, line = river level.)

Annual rainfall peaks between December and April followed by a 5-month dry season (May to September, see Figure 16.2). Data have not been published for the park, but Ferreira (1997) collated data from Manaus (220 km to the west) that showed peaks of precipitation (275–325 mm) between January and April, compared with 40–60 mm between July and September (see Figure 16.2). Total annual rainfall in the Jaú–Manaus region is some 2200 mm (Goulding *et al.* 2003). Fruit is maximally available between January and June, with a minimum in August and October. About 80% of Jaú is rainforest, followed by *igapó* (12%), white-sand vegetation types (2%), palm and aroid swamps (2%), and disturbed land (< 1%: FVA-IBAMA 1998). At 0.4/km^2 (Pinheiro & Macedo 2004), the average human density in Jaú is substantially lower than the average for rural Amazonia (1.6/km^2: see Chapman & Peres 2001). Although it was substantial up until the 1950s (Leonardi 1996, 1999), hunting pressure and anthropogenic habitat destruction was minimal in the 1990s (FVA-IBAMA, 1998), and this appears to still be the case (A. Barnett, pers. obs.)

In addition to *Cacajao ouakary*, the local primate community consists of eight confirmed species (Barnett *et al.* 2002: Table 16.1).

Amanã (2°26′33″S, 64°43′30″W)

Positioned within the Solimões–Negro interfluve, the 2,350,000 ha Amanã Sustainable Development Reserve lies to the south of Jaú. Amanã's own southern border abuts the Mamirauá Sustainable Reserve (Figure 16.1). Soils are underlain by Holocene lake deposits of the Solimões Formation (RadamBrasil 1978; Frailey *et al.* 1988). These are more fertile than those in the parts of Jaú in which uacari studies have been undertaken (Moran *et al.* 2000).

Rainfall peaks occur between March and June and the peak of the dry season occurs between August and October. The average annual rainfall is 2910 mm. Fruit is maximally available between January and June, with an August–October minimum. Habitats include rainforest (circa 70%), *igapó* (10%), *várzea* (20%), and less than 0.5% disturbed land (Queiroz & Fernandes 2001). The average human density at Amanã is about 0.001 per hectare, due to a very small human population of some 2100 (Banco de Dados Demográficos do IDSM, unpubl. data). Hunting pressure occurs as part of the area's sustainable management program. Monitoring from five communities across a 7-year period recorded 45 uacaris killed by hunters, and it has been estimated that

Table 16.1 Primate species present at the three sites in Amazonia (x = present at the site; variation in taxonomic identity is specified when it varies among sites).

Primate species	Amanã (9 species)	Jaú (9 species)	Caparú (8 species)
Alouatta seniculus	x	x	x
Aotus	x Aotus sp.	x *A. trivirgatus* (*sensu* Hershkovitz 1983)	x *Aotus vociferans* (*sensu* Hershkovitz 1983)
Cacajao ouakary	x	x	x
Callicebus torquatus	x	x	x
Cebus albifrons	x	x	x
C. macrocephalus	x	x	x
Lagothrix lagothricha			x
Pithecia	x *Pithecia* sp.	x *P. pithecia*	
Saimiri sciureus	x	x	x
Saguinus inustus	x	x	

Note: the identity of the *Aotus* at Caparú is based on biogeographic data and requires specimen confirmation. The *Aotus* at Amanã are most probably *A. trivirgatus*, but this is supposition. We list *Cebus macrocephalus* according to the authority of Rylands and Mittermeier (2009), who, based on the unpublished thesis of Silva Junior (2001) eliminated all subspecies for the tufted capuchins. According to Silva Junior (2001, 2002), Colombian capuchins are *Cebus fatuellus*, which Rylands and Mittermeier (2009) list as *C. apella fatuellus*. Rylands and Mittermeier (2009) list the *Pithecia* species for the Jaú-Amanã region as *P. pithecia chrysocephala*. While *S. cassiquiarensis* has been used for squirrel monkeys from the Negro-Solimões interfluve, this taxon is not recognized by Rylands and Mittermeier (2009) and so we use *Saimiri sciureus* in preference. At all three sites the taxon would be *Saimiri sciureus cassiquiarensis*.

around 120 uacaris are actually taken each year (IDSM, unpubl. data). This means that *C. ouakary* is the third most hunted primate species (after *Alouatta seniculus* and *Cebus macrocephalus*). However, hunters at Amanã show a strong preference for ungulates and caviomorphs, and these are hunted much more frequently (IDSM, unpubl. data).

The primate community consists of nine known species (Amaral *et al.* 2005: Table 16.1). The presence of *Ateles* and *Lagothrix* in this part of the Negro–Solimões interfluve, tentatively suggested by de Souza *et al.* (2004), has not been confirmed. Barnett *et al.* (2002) considered that, if these large atelines ever did occur in this region, they were probably hunted out by the 1950s, and have not recolonized.

Caparú (3°29′20.8″S, 69°49′12.6″W)

Studies were concentrated in a 10 km^2 zone along Lago Taraira, a 24-km long oxbow lake situated on the lower Apaporis river, southern Vaupés Department, eastern Colombian Amazon, close to the Brazilian border (Figure 16.1). Forest topography is a combination of *igapó*, Pleistocene terrace and inland undulating Plio-Pleistocene hilly terrain, all with very infertile soils.

Local soils of Pleistocene terraces lie well above present-day highest flooding, consisting of both fluvial *aquepts* and *aquents* (poorly drained silty clay soils and infertile soils characteristic of seasonal vegetation, respectively). Deposited at the end of the last ice age as a result of the melting of Andean glaciers, they are extremely infertile soils, of blue–gray clay in texture and typical of inundated terrains subject to heavy precipitation. The other major soil type, that of the undulating hilly terrain of Plio-Pleistocene origin, has fine-grained sediments with petroferrous concretions and some sand and gravel. The latter has eroded from nearby ancient igneo-metamorphic rocky formations belonging to the Yupati Association, a southern prolongation of the Guyana Shield (Ibarra *et al.* 1979).

The 10 km^2 study area centered at Lago Taiaira, and has a very extensive floodable *igapó* on the south side. *Igapó* habitats make up 50% of Caparú, with lake surface 5%; transition forest up to Pleistocene terrace 10%; Pleistocene terrace 15%; Plio-Pleistocene hills 20%.

The *igapó* has low plant diversity, while the inland hills of red and darker clays have slightly more fertile soils and the region's most diverse plant communities. The extensive Pleistocene terraces are extremely infertile, being dominated by well-developed and very tall (30–35 m) *Micrandra spruceana* (Euphorbiaceae), a favorite food of *Cacajao* (soil data, Defler, unpubl.; botanical data, Defler & Defler 1996).

Rainfall averages 3836 mm ($n = 11$ years) with the average wettest and driest months being 319 mm (April) and 258 mm (September), respectively. There is, therefore, no dry season as generally defined (below 100 mm). The fruit peak, in terms of both number of species and volume of available fruit, occurs in April–May and availability then decreases to a low in October–November, before slowly beginning to rise again.

Hunting pressure was extremely low during a 17-year residence at this site. The local indigenous people consider the lake sacred, and only one local resident hunted there (*Agouti paca*). People do not usually hunt *Cacajao* in this part of Colombia, preferring *Lagothrix* which is still quite common (Defler *et al.* 2006). There is no nearby non-indigenous population, the closest regional inhabitants lived on the banks of the nearby Rio Caquetá, several hours' travel away. The site continues to have very low impact from human beings as reported by researchers coordinated by Conservación Internacional Colombia.

The local primate community consists of eight species (Defler 2004). As shown in Table 16.1, the species composition of the primate communities in the three study areas is very similar: Amanã and Jaú have the same species and share a further two genera that Caparú lacks (*Pithecia* and *Saguinus*), while Caparú possesses *Lagothrix lagotrichia*, a species not recorded for the more eastern sites. Where full species

Table 16.2 Primate densities at Caparú, based on three sequential years of surveys.

Species	Density (no. individuals per km²)
Alouatta seniculus	too low to register
Aotus (vociferans)	not measured
Cacajao ouakary	12
Callicebus torquatus	6.08
Cebus albifrons	too low to register
C. macrocephalus	8
Lagothrix lagothricha	13.1
Saimiri sciureus	8.8

determinations have been realized, the primate taxa at all three sites are the same at the species level. Densities for primates at Caparú are given in Table 16.2. The lack of quantitative data on relative primate species abundance at the other two sites does not allow a more functional comparison of the community dynamics.

Methodologies

At Jaú, Brazil, studies of uacaris have taken place since August 1999, but data presented here were collected from October 2006 to September 2007 (A. Barnett and B. Bezerra). In *igapó* animals were followed by canoe or on foot depending on the season and level of inundation. In *terra firme* animals were surveyed with two 5-km transects, one on either side of the Rio Jaú. Quantitative plant community data came from eight 0.25-ha quadrats in which all trees above 20 cm diameter breast height (dbh) have been marked, numbered and identified. All palms and lianas were also included. Phenology data were recorded monthly beginning in November 2006 (mature leaves, new leaves, flowers, fruit). Fruit hardness was recorded with a commercial fruit penetrometer, and the thickness of the pericarp (the layers of the fruit external to the seed) measured (Barnett, 2010).

At Amanã, Brazil, some information presented here comes from a study of hunting patterns of local communities, which has been operating since 2001 as part of a long-term study. The behavioral observations of *Cacajao* were collected during a year-long study during 2008 by M. Oliveira. Coming from the initial phase of a long-term study, conducted on a partially habituated group, these data may be regarded as preliminary. As at Jaú, animals were followed in a canoe when the *igapó* was flooded and on foot in the dry season.

At Caparú, Colombia, primate studies by T. Defler began in December 1982. Studies specifically on *Cacajao ouakary* took place from January 1994 up to June 1998, when he was forced to abandon the station and flee from FARC (Fuerzas Armadas Revolucionarias de Colombia) guerrilla demands that the researchers leave (Semple 2000). During the research period, both canoe and foot travel were used on an extensive marked trail system that eventually totaled 125 km. Behavioral data were gathered using both ad libitum and scan samples, although most of the quantitative data were burned by the FARC guerillas. Phenological data were gathered for 4.5 years, and published by Defler and Defler (1996). Most *Cacajao* food specimens collected for identification had to be abandoned in 1998.

Results

Diet – fruit

Between November 2006 and September 2008, some 189 plant diet items were registered for the golden-backed uacaris of Jaú, involving 136 identified taxa in 87 genera in 44 families. A further 12 diet items remained unidentified even to family level. Of the 189, 115 (60.8%) were fruits, of which 96 (50.8% of all items; 83.5% of fruits) were unripe when eaten. Of the 115 fruits, 85 (73.9%) had hard pericarps. All unripe fruits were eaten for their seeds or arils. Of the 19 species eaten when ripe, 10 were either ingested whole or eaten for their pulp or aril. Within the 44 plant families in the diet, Fabaceae, Sapotaceae and Lecythidaceae had the most species eaten (19, 19 and 10 species, respectively). Fruits eaten ranged in weight from 460 g (*Aldinia latifolia*: Fabaceae) to less than 3 g (several species including *Mabea nitida*, Euphorbiaceae and *Ternstroemia candolleana*, Theaceae). The largest fruit handled to extract a seed was 23 cm long × 11 cm wide (*Aldinia latifolia*: Fabaceae). The smallest fruits eaten by the uacaris at Jaú came from Myrtaceae genera *Blepharocarpus*, *Calyptranthes*, *Eugenia* and *Myrcia* with spherical fruits < 1 cm in diameter. The three most frequently eaten fruits were *Micropholis venulosa* (Sapotaceae: 19.7% of records), *Eschweilera tenuifolia* (Lecythidaceae: 11.6%) and *Buchenavia ochrograma* (Combretaceae: 7.1%).

At Amanã, golden-backed uacaris have been observed feeding on 24 plant species, of which 13 were identified. These are given in Table 16.3. Seven species (*Macrolobium acacifolium*, *Eschweilera* sp1, sp2 and sp3, *Crudia amazonica*, *Piranhea trifoliata*, *Tabebuia barbata* and *Copaifera* sp.) possessed hard pericarps.

In Colombia, only 20 *Cacajao* food items were identified (Table 16.3). *Eschweilera tenuifolia* was the most frequently eaten single food item, comprising some 10% of registered food choices, followed by immature seeds of *Micrandra spruceana*. *Cacajao* competed directly with *Lagothrix lagothricha* for young *Micrandra* seeds, which on maturing develop concentrated tannin deposits, and so are no longer attractive to monkeys. Defler (2004) suggests that high levels of tannins may make mature *Micrandra* seeds inedible. *Minquartia guianensis* and *M. punctata* (Olacaceae) were also shared by uacaris and woolly monkeys – although in this case, woollies generally ate flesh of the fruit and the *Cacajao* ingested young seeds. Species of *Brosimum* (Moraceae), *Mauritia* (Arecaceae) and *Pouteria* (Sapotaceae) were also eaten by both primates and in the case of *Pouteria* and *Brosimum* the *Lagothrix* always were interested in the flesh and *Cacajao* were interested in the seeds. Both primates ate the pulp of *Mauritia* palm fruits.

Table 16.3 Plant species eaten by *Cacajao ouakary* from three sites in Amazonia. (for Jaú data represents a subsample, data for the other two sites represents all known diet records. Hardness and thickness data based on collections at Jaú. All records from *igapó* unless otherwise stated).

Diet species (Family), pericarp hardness and thickness	Amanã	Jaú	Caparú
Abuta sp. (Menispermaceae) Not hard, 2 mm.			seed, *terra firme*
Amanoa oblongifolia (Euphorbiaceae) Hard, 1 mm		immature seeds, young leaves	
Amphizoma coriaceae (Celastraceae) Not hard, 2 mm			seed, *terra firme* and *igapó*
Anacardium sp. (Anacardiaceae) Not hard			flesh, *terra firme*
Aspidospermum oblongum (Apocynaceae) Hard, 3 mm		immature seed, *terra firme*	seed, *terra firme*
Astrocaryum juari (Arecaceae)		entire young fruits	
Attalea maripa (Arecaceae)		pulp (*terra firme*)	
Bromeliaceae		leaf bases *Aechma mertensii*	seeds Bromeliad sp.
Brosimum utile (Moraceae)			fruits, *terra firme*
Buchenavia sp. (Combretaceae) Not hard, 2 mm		young leaves *B. ochrogramma*	seeds
Burdachia prismatocarpa (Malpighiaceae) Hard, 2 mm		immature seeds	
Chaunochiron lauranthoides (Olacaceae) Not hard		immature seeds (+ insect larvae therein)	
Clusia sp. (Clusiaceae) Hard, 4–5 mm		insect larvae in petioles	aril, seed
Codonanthe crassifolia (Gerneriaceae)		flowers, leaves	
Copaifera sp. (Fab-Caes.) Hard, 1.5 mm	seeds		
Crudia amazonica (Fab-Caes.) Hard, 3 mm	seeds		
Eschweilera spp. (Lecythidaceae) Hard, > 4 mm	Sp. 1, 2, 3: seeds and flowers,	flowers, seeds, germinating seeds, young leaves *E. tenuifolia*	flowers, fruits (young entire), seeds, germinating seeds *E. tenuifolia*
Eleoloma glabrescens (Sapotacae) Not hard, 3 mm		imature seeds	
Gomphiluma gomphiifolia (Sapotaceae) Not hard, 3 mm		immature seeds	immature seed
Gymnoluma glabrescens (Sapotaceae) Not hard, 3 mm	seeds		
Hevea spp. (Euphorbiaceae) Hard, > 8 mm		immature seeds, pith from young shoots *H. spruceana*	immature seeds, *terra firme H. brasiliensis*
Hydrochorea marginata (Fab-Pap)		young leaves, flowers	
Inga sp(p). (Fab-Mim.) Not hard, 1 mm		aril, seed, (*terra firme*)	aril, seed (*terra firme* and *igapó*)
Licania heterophylla (Chrysobalenaceae) Not hard, 3 mm		pulp	
Mabea sp(p.) (Euphorbiaceae) Hard, 1 mm		seeds, *Mabea nitida*	seeds, *Mabea* sp.
Macrolobium acaciifolium (Fab-Caes.) Hard, > 1 mm	seeds	seeds	seeds

Table 16.3 *(cont.)*

Diet species (Family), pericarp hardness and thickness	Amanã	Jaú	Caparú
Mauritia flexuosa (Arecaceae) Not hard, 4 mm	pulp	pulp	pulp
Micrandra spruceana (Euphorbiaceae) Hard, 1 mm			immature seed, *terra firme*
Micropholis venusta (Sapotaceae) Not hard (fruits)		immature seeds, young leaves	
Minquartia punctata (Olacaceae) Not hard, 2 mm			pulp, seeds, *terra firme*
Monopteryx uaucu (Fab-Caes.) Not hard, 1mm			seed, *terra firme*
Ormosia paranaoides (Fab-Pap.) na		young leaves	
Paramachaerium ormosioides (Fab-Pap.) Not hard, 1 mm	seeds		
Parkia bicolor (Fab-Mim.) Not hard, 2 mm		unripe seeds	
Piranhea trifoliata (Euphorbiaceae) Hard, 2 mm	seeds		
Pouteria (Sapotaceae) Not hard, 3 mm		immature seed *P. excelsa*	immature seed, *terra firme* *P. laevigata*
Rheedia (Garcinia) acuminata (Guttiferae) Not hard, 2 mm	seeds		
Sloanea (Eleocarpaceae)	Seeds *S. porphyrocarpa*	seeds (and insect larva within) *Sloanea* sp.	
Swartzia laevicarpa (Fab-Pap) Not hard, 4 mm		aril	
Swartzia acuminata (Fab-Pap) Not hard, 4 mm		young fruits entire; older fruits, aril only	
Tabebuia barbata (Bignoniaceae) Not hard, 1 mm	seeds	unripe seeds	
Ternstroemia candolleana (Theaceae) Not hard, 1 mm		unripe seeds	

During the dry season at Caparú, flowers of *Eschweilera* were a very popular item, and the leaf bases of bromeliads were also used. During the dry season small groups of uacaris were twice observed traveling over the ground gleaning the sprouting seedlings of an unidentified large-seeded fruit from the *igapó* forest floor and eating them *in situ*.

Diet – non-fruit items

At Jaú, uacaris feed on a variety of non-fruit items, including: pith from the stems of *Hevea spruciana* (Euphorbiaceae) and *Endlicheria chalisa* (Lauraceae), flowers of *Eschweilera* (Lecythidaceae), *Codonanthe crassifolia* (Gesneriaceae) and *Tabebuia barbata* (Bignoniaceae), flower buds of the mistletoe *Phoradendron poeppigii* (Viscaceae), inflorescences of *Philodendron* (*megallophyllum*?) (Araceae), young leaves of *Buchenavia oxycarpa* and *B. ochrogramma* (Combretaceae), *Eschweilera* and *Amanoa oblongifolia* (Euphorbiaceae), and leaf-bases of *Aechmea mertensii* (Bromeliaceae). Together they comprised 21.3% of diet items recorded during the study period. While each was of minor importance overall, they were significant in some months (see Table 16.4). See Photo 16.1.

Invertebrates eaten included: the whip-scorpion *Heterophrynus* (probably *H. batesii*, Phrynidae: an individual taken after being disturbed by a foraging uacari), larvae of *Polybia* wasps (Polistidae, Hymenoptera: larvae extracted from raided nests by adult uacaris), Micro-lepidopteran (possibly Tortricidae) and Coleopteran (possibly Curculionidae) shoot-borers extracted from both stems and leaf pedicels of species of Clusiaceae, Fabaceae, Lauraceae, and Sapotaceae, microlepidopteran larvae extracted from the rolled leaflets of *Swartzia acuminata* (Fab.: Pap.), and Isoptera (probably *Nasutitermes*, Nasutitermitinae) picked individually from their broken on-branch mud-trails by a solitary female uacari. In addition, initial analysis of fecal material shows considerable volumes of insect material (mostly Isoptera, Coleoptera and Hymenoptera). Uacaris at Jaú were reported by local guides to eat Glossoscolecidae earthworms that seasonally migrate to the frass of palm frond bases to escape inundation. Although this has been observed by A. Barnett for *Cebus albifrons*, it has yet to be confirmed for *C. ouakary*.

In addition to direct micro-carnivory, a number of opened fruits showed high levels of insect infestation. These include *Eschweilera* (infestation by Micro-lepidoptera, Micro-hymenoptera), *Sloania* (Eleocarpaceae: Coleoptera larvae and adults), *Macrolobium* (Fabaceae: Coleoptera larvae), *Chaunochiron* (Olacaceae: Diptera larvae) and various Myrtaceae (Coleoptera larvae). In some species the infestation rates exceed 70% of the collected fruits, and insects living within constitute up to 17% of the wet weight of the fruit, and up to 30% of that of the seed(s). See Photo 16.2 below.

Table 16.4 Mean annual and maximum monthly percentages of non-fruit items in the diet of *Cacajao ouakary* at Jaú.

Diet item	Annual average (%)	Maximum monthly value (%)
Flowers (nectar and pollen)	7.2	45.9
Insect larvae (nests, leaves, shoot borers)	2.4	4.3
Leaf bases	2.3	5.1
Leaves (young)	10.8	64.3
Pith (from fresh shoots)	0.8	20.2

Observations from Amanã include the consumption of small insects (Orthoptera), and insect parts in fecal samples. The only observations of invertebrates being consumed at Caparú were of very small unidentified invertebrates occasionally picked from branch surfaces.

Group size

At Jaú, observed group sizes ranged from 2 to 100+, with singletons also being observed. In the fruit-rich parts of the year, groups of some 30–50 animals were common, although these frequently consisted of two to three subgroups of 15 or so individuals ($N = 63$). In the dry season groups rarely had more than 15 animals and subgroups of 3–5 animals were common ($N = 86$).

At Amanã, group sizes also varied seasonally. Here, uacaris occupied *terra firme* from September to December or January, the local low-water season. During this period, group size did not exceed 25 animals. During the period of rising floodwaters (December–May), group size gradually increased, to a counted maximum of 157 individuals. By May, once full flooding was achieved in *igapó*, the study group divided into two large subgroups, each with some 70 individuals. This is the peak period for fruit availability in *igapó* at Amanã (Luciane Lopes, pers. commun. to H. Queiroz). As the year continued, the progressive fission continued with the 70-member-strong groups subdividing further into subunits of 20+ individuals. Subgroups appeared to be of stable composition and occupied defined areas. However, although vocal contact was constant, agonistic encounters were not observed. Within the subunits, it was common to observe foraging parties of three individuals, which would forage within 3–50 m of each other. The

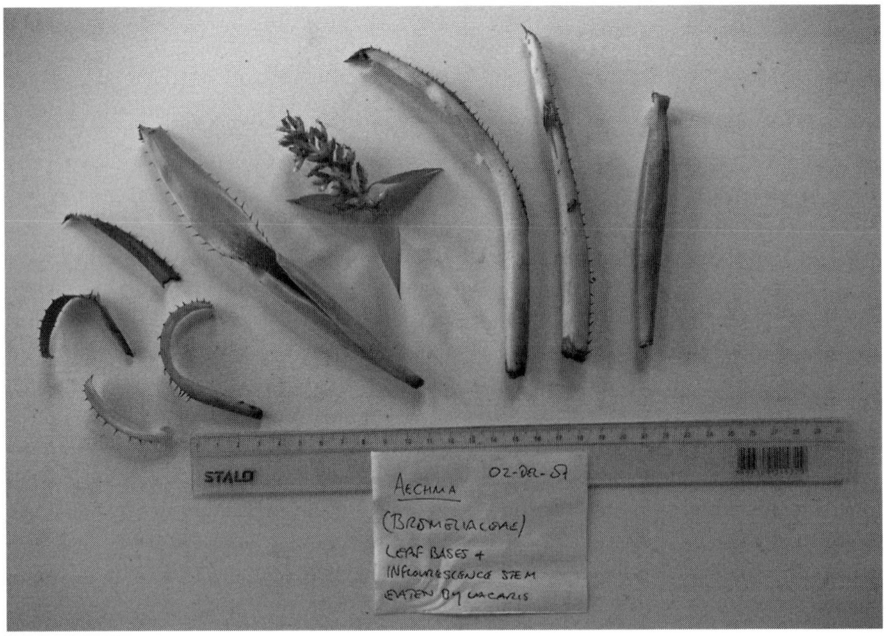

Photo 16.1 *Aechmea mertensii* (Bromeliaceae): golden-backed uacaris eat the succulent leaf bases as humans would eat the tegules of an artichoke (although without the butter). Photo: Adrian Barnett.

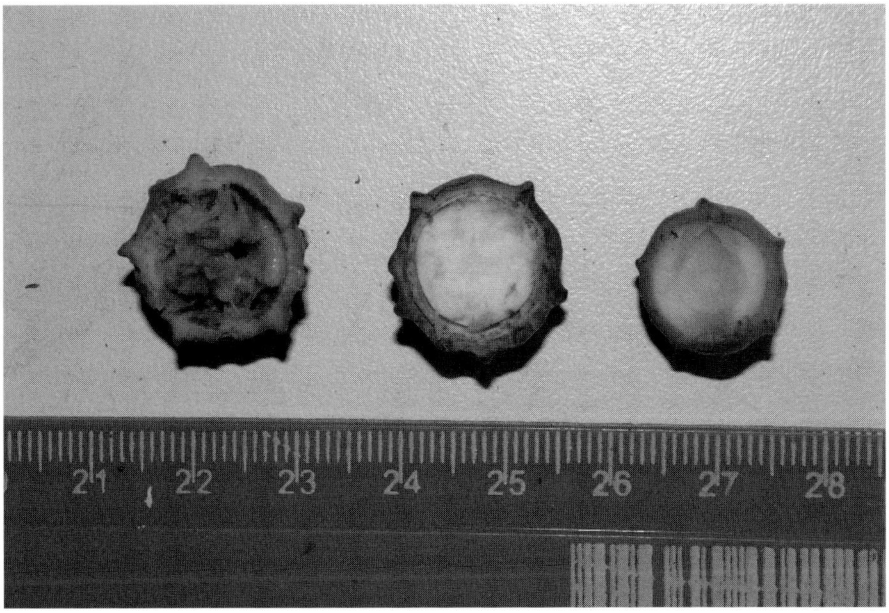

Photo 16.2 Fruits of *Chaunochiron loranthoides* (Olacaceae): fruit on far left is infested by larvae of *Anastrepha* sp. nov. (Diptera: Tephritidae). Golden-backed uacaris preferentially choose infested fruits, ingesting both seed and enclosed larvae. Photo: Adrian Barnett.

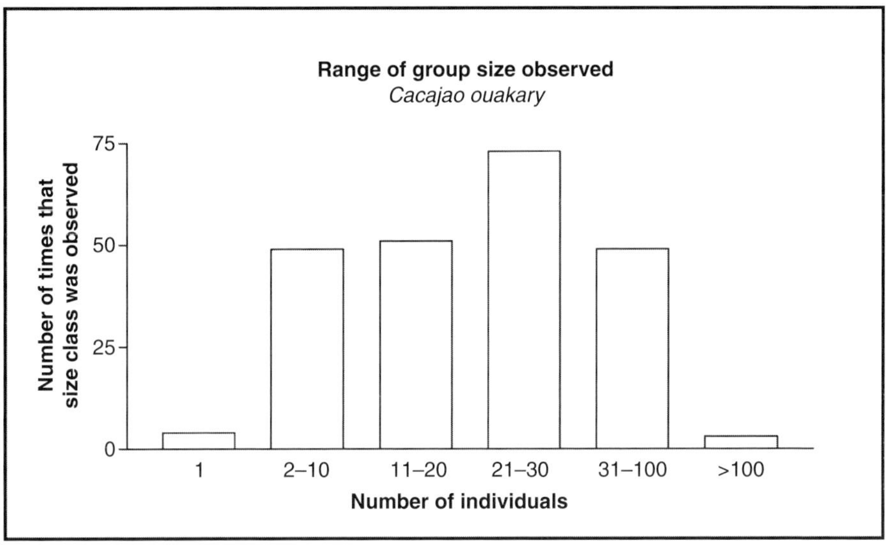

Figure 16.3 The range of group sizes observed for *Cacajao ouakary* at the Estación Biológica Caparú, Rio Apaporis, Colombia (adapted from Defler 1999).

compositional constancy of these small groups is not yet known. However, they were most frequently formed by two adult females and their offspring, or one adult female, an infant and a juvenile (male or female). Small all-male groups were also observed.

In Colombia, group sizes ranged from 2 to 200. One count in Pleistocene terrace forest was of about 190 individuals, a further 4–5 (probably incomplete) counts recorded groups of around 100 animals in *igapó* vegetation at the lake margin. Groups approaching 100 tended to be rare (see Figure 16.3). Group sizes tend to be larger during those parts of the year when heavy crop volumes, especially, were available. This was especially true of *Eschweilera*, the number one diet choice item for uacaris at Caparú. Annual distribution of group sizes at Caparú is given in Figure 16.3.

Densities

Defler (2001a) reported an effective ecological density of 12 animals/km^2 at Caparú, making it the second most abundant primate species at the site (see Table 16.2). Quantitative density estimations have yet to be undertaken at Jaú, but, in *igapó*, uacaris were encountered with much greater frequency overall than either *Cebus albifrons* or *Saimiri sciureus*, the other two diurnal primate species commonly encountered in this habitat. In the flooded season at Jaú, up to four groups of uacaris appeared to be using an area some 10 km^2, with group size varying between 10 and 40 animals, giving a rough maximum density of some 16 animals/km^2. A group of 5–9 uacaris inhabited year-round a 74-ha island of *igapó* in the middle of the Rio Jaú. These figures appear to confirm *C. ouakary*'s capacity for seasonally high densities. However, this must be offset against the large-scale seasonal movements (see above) which considerably reduce this density on an annual basis.

Species associations

In the over 80 encounters with uacaris at Jaú, multispecies groups containing uacaris and other primates have been observed just three times. In each case this has been with *Saimiri*, and in each instance the association did not appear to last more than one day. It is not clear if this is the result of actual avoidance or aggression on the part of the uacaris, but on at least one occasion, a troop of *Cebus albifrons* passing close to a troop of uacaris was subject to hissing vocalizations and aggressive posturing that included piloerection. Although occasional associations have been reported with anis (*Crotophaga* spp.: Curculidae) by Barnett *et al.* (2005), no regular associations were observed at Jaú between uacaris and birds, with only transitory associations occurring as uacaris moved through the territories of various insectivorous bird species (see Barnett *et al.*, Chapter 14).

At Amanã no associations were observed between uacaris and other primates between February to July (the period of maximum *igapó* flooding); however, agonistic vocal interactions were recorded between *Saimiri* and *Cacajao* at a large fruiting *Rheedia acuminata* tree (Clusiaceae). At full flooding, when, because some parts of the forest are fully submerged and the extent of the *igapó* is at a minimum, interactions with *Alouatta*, *Saimiri* and *Cebus macrocephalus* were observed, including feeding assemblages in the canopies of fruiting *Ficus* trees that lasted for up to 1 h. In addition, *Cacajao* and *Saimiri* were seen moving and foraging together for up to four consecutive hours. No bird associations were recorded.

At Caparú, multispecies groupings appear to be much more common than at either Jaú or Amanã. *Cacajao* was sighted over 20 times in association with *Cebus albifrons*, *Lagothrix lagothricha* and *Saimiri sciureus*. During these events the animals were within 20 m, overlapping and intermixed, and feeding in the same tree. In one spectacular sighting in flooded forest the following species were observed feeding in association for several hours: *Cacajao ouakary*, *Saimiri sciureus*, *Cebus albifrons*, Greater Anis (*Crotophagus major*: Cuculidae), pink river dolphins (*Inia geoffrensis*: Iniidae) and various species of fishes. The disturbance of the larger primates feeding on fruits (1) disturbed insects and small vertebrates that attracted the ani (and *Saimiri*) and (2) falling fruit fragments attracted feeding fish that in turn attracted the dolphin. Associations made up of some of these species were commonly observed at Caparú. Anis were commonly attracted to any primate feeding near the edge of the lake (this being the ecotone preferred by this bird). Pink dolphin and fish were commonly attracted to any fruit feeding that occcurred within flooded forest, including that by uacaris. The small hawks *Leucopternis albicollis* and *Harpagus bidentatus* commonly accompanied the *C. ouakary* (as well as all other large primates), to catch fleeing katydids.

During the period of *igapó* inundation, foraging associations of woolly monkeys and *Cacajao* were commonly observed. They were only observed feeding together in the *terra firme* forest on sweet-tasting ripe fruit such as *Brosimum utile* (Moraceae: a family near-absent from *igapó* at Jaú) and *Anacardium* sp. (Anacardiaceae: where both ate pulp). However, during the season of fruit scarcity uacaris dispersed throughout the inland forest and were rarely seen. During this period, uacaris were twice seen moving with woolly monkeys – the first observation was of an individual, the second of a pair. One of these associations lasted for at least 2 days.

Many fruit species eaten by *Cacajao* are more bitter (to human taste) than those ingested by *Lagothrix* and *Saimiri* (Defler, pers. obs.). Although *Cacajao* open hard-husked fruits using dental specializations (Kinzey 1992), *Lagothrix* are large and powerful animals and their diet also includes a number of hard-husked fruits (e.g. Lecythidaceae; Defler & Defler 1996). Woolly monkeys opened hard fruits with a combination of teeth, manual attrition and knocking against branches (similar to the suite of actions reported by Peres 1991 for *Cebus*, and for *Alouatta* by Oliveira-Filho & Galetti 1996).

Seasonal movements

At Jaú there is a seasonal dearth of fruit in the late dry season between late August and late November. At this time $< 1\%$ of the individual trees in the study quadrats bore fruit. Of a total of 317 plant species recorded for the area (palms, trees, lianas, epiphytes and shrubs combined), only 15 had fruit in this period compared with 28 in December–January, 36 in February–March, 72 in April–May, 52 in June–July. Of these 15, 11 were in *terra firme* forests.

In Jaú, study groups in areas with adjacent *terra firme* available migrated to this forest type during the period of fruit dearth in *igapó*. *Terra firme* also has a fruit dearth between late September and December (the figures given above are for both habitats combined). The migration from *igapó* to *terra firme* occurred only once fruits were available in *terra firme* (late October), a period when there was still very little fruit available in *igapó*. In December, uacaris inhabiting areas with contiguous *igapó* and *terra firme* had moved into *terra firme* and were eating seeds from unripe fruits of Apocynaceae and Sapotaceae. Groups were observed to move back into *igapó* by February, when floodwaters covered the forest floor and trees (especially *Eschweilera*) were beginning to produce flowers and fruits.

Seasonal shifts were also observed in group sizes, which also altered as resource availability changed. Observed group sizes in *terra firme* ranged from 6 to 22 animals. Although seasonal use of the contiguous habitats of *igapó* and *terra firme* appears to follow the peaks of fruit production in these habitats, at Jaú *igapó* does seem to be a preferred habitat, as uacaris in *terra firme* at Jaú were never observed further than 0.5 km from *igapó* (Barnett *et al.* 2005). In addition, Peres and Nascimento (2005) walked 342.3 km along transects in five locations in the park. The survey was made during the period of *igapó* inundation, and of 97 contacts with primate groups, not one was with *C. ouakary*, although Carlos Peres (pers. commun., 2000) reported that uacaris were abundant in adjacent *igapó* at the time. *Cebus albifrons* and *Saimiri sciureus* were regularly encountered in *igapó* at Jaú. *Alouatta* was encountered only very occasionally there, and *C. macrocephalus* was recorded only at the *terra firme–igapó* margin.

At Amanã, October is the peak of the dry season and there is little fruit in any of the surveyed forest types, including the still-flooded *igapó*. During this month, hunting-related censuses recorded uacaris in *terra firme* in both 2007 and 2008. Animals were some 4 km from the nearest body of open water, in a swampy area dominated by the palms *Mauritia flexuosa* and *M. huebneri* (Arecaceae). *Alouatta seniculus*, *Cebus albifrons*, *Cebus macrocephalus* and *Saimiri sciureus* have all been recorded in Amanã *igapó* during the dry season, the period of the year when the *igapó* is flooded.

At Caparú, all primates (except for *Cebus macrocephalus*) exploit the *igapó* in April, because the rising water level causes a huge pulse of fruit production (much higher than inland). During May–July *Cacajao* are commonly in the *igapó*, but in August–September food becomes scarce and they disappear from the *igapó*, spreading out inland (Defler 1999). Animals may even travel singly or in pairs during this time, making it very difficult to find them. The largest groups were sighted during the high water season April–August, when fruit (especially *Eschweilera*) are most abundant (Defler & Defler 1996; Defler 1999).

Reproductive seasonality and breeding behavior

Rylands (1992), Barnett (2010), Barnett *et al.* (2005) and Bezerra *et al.* (2011) all report young *C. ouakary* in March and April, the beginning of peak fruit availability. However, some out-of-season births evidently occur, as young infants have been observed in December (Barnett 2010).

At Amanã, infants have been recorded at the end of January and the beginning of February (coinciding with the onset of fruit production by *Rheedia acuminata*, a major diet item). Copulations were observed in November and July.

There seems to be a narrow birth season at Caparú (March–April). As at Jaú, birth seasons coincide with the local maxima of fruit availability in *igapó*. However, out-of-season births have been recorded: at Caparú; for example, a female died giving birth in November – the nadir of fruit availability.

A birth season, with the majority of births occurring at the onset of the rainy season fruiting period, appears to be common in black-water uacaris. Boubli (1999), working in Pico de Neblina National Park, Brazil, observed adult female *C. melanocephalus* with very small infants in March, the beginning of the rainy season and the period of peak fruit availability there. At a site nearby in Venezuela, lactating female *C. melanocephalus* were seen in March and April by Charles Handley (pers. commun.).

Social organization

At Jaú, the animals seem to have a fission–fusion social organization (i.e. subgroups may be separated by hundreds of meters, e.g. Silva & Ferrari 2009). Group size and composition varied from single individuals to approximately 100 individuals, and groups were composed of multiple males and females. A. Barnett observed no fighting or active aggression between October 2006 and April 2008; however, other social behaviors were also infrequent. Allo-grooming was very rarely observed ($N = 3$), though up to three individuals were involved in the activity. Interindividual distances during traveling frequently exceed five body lengths. Unless it is a mother-and-infant (i.e. when infants were carried at an early age), it is uncommon for foraging or resting uacaris to be closer than three body lengths ($n = 101$). In 47 of 74 feeding observations (63.5%), individuals foraged in a tree canopy alone (or with infant). In a further 9 (12.2%) there were two juveniles of the same social group in the same tree, and 12 consisted of adult and one or two juveniles (16.2%). Two or more adults were observed foraging together only on 6 occasions (8.1%: 2 adults in 4 instances, 3 adults in 2 instances). There is, however, sometimes the uttering of twittering, "tchó" and sharp whistle calls (Bezerra *et al.* 2010a, 2011; Bezerra *et al.*, Chapter 30), which seem to serve as contact vocalizations between dispersed individuals. At Caparú, upright tail wagging seems to signal well-being to other group members. Lowering the tail, or placing it to one side, occurs following situations which could have elicited feelings of being frightened or disturbed in the operant individual.

Data from Caparú indicate that uacaris may mate promiscuously, as there are always many adult males and females present in groups or subgroups. However, detailed research is required on subgroup composition, particularly the relationship of the male that often is seen to accompany what appears to be the core social unit of female–infant–juvenile–adolescent. Is this animal the sire of the young?

Given the apparent lack of aggression, it is interesting to note that young male golden-backed uacari appear to show delayed and cryptic social maturity. Two young males raised by T. Defler had no visible testicles until 4 years of age. This is also the case with *Chiropotes*. According to Dixson (1998), postnatal surges of testosterone have been correlated with testicular descent in some species of monkeys and humans, but we currently have no information on testicular function in young uacaris and bearded sakis. As secondary sexual characteristics develop quickly after testes descend, we presume that this is a case of delayed maturation and not an alternative reproductive tactic (*sensu* Gross 1996). Because young uacari and bearded saki males appear feminized, they may escape conflict from adult males until they have reached a body mass that is approaching the adult male size. That is to say, this feature is the cause, rather than the consequence, and may contribute to the low aggression levels witnessed in uacari society.

Discussion

Between-site differences in diet, social organization and group size have been recorded for a number of species (e.g. Defler 1982, *Cebus* spp.; de Moraes *et al.* 1998, *Brachyteles arachnoides*; Julliot 1992, *Alouatta* spp.; Boinski *et al.* 2005, *Saimiri* spp.).

Factors affecting these aspects include hunting (de Thoisy *et al.* 2005; Peres 1997a), habitat quality (Boinski *et al.* 2005; Peres 1997b), forest structure (Lemos de Sá & Strier 1992; Pinto *et al.* 1993), predation intensity (Di Fiore 2002), seasonality of fruit availability and the patch size at which diet items are available (Branch 1983; de Morais *et al.* 1998).

Cultural transmission can also be important (see Changeux 2005; Sapolsky 2006 for reviews of cultural transmission in non-ape primates, and Ottoni & Izar 2005; Schaffer & Aurelli 2005; Dindo *et al.* 2008 for specific Neotropical examples), and this may explain the curious fact that at none of the study sites have uacaris been seen to rob the nests of river turtles even though this has been filmed at a location on the Rio Negro (Barnett 2005). The use of pith at Jaú may also fall into this category, while removal of insect larvae from inside petioles simply may not yet have been observed at other sites.

Beyond these caveats, there are, nevertheless, some interesting differences between the three sites, one of the most obvious being the extent to which members of the different populations appear to penetrate *terra firme* during the dry season *igapó* fruit dearth. Defler (1999) reported that animals migrated up to at least 2.5 km into the *terra firme*. Animals at Amanã appear to penetrate further still. These are not unique events, since, elsewhere in Colombia, Hernández-Camacho and Defler (1985) reported observing the species up to 1 km from *igapó* during the dry season, while in Brazil, Palacios and Peres (2005) also found it in *terra firme* in the late dry season on the Caquetá. In contrast, individuals at Jaú seemed rarely to move more than 0.5 km from the *igapó–terra firme* ecotone. In addition, the extent to which uacaris associate with other species (both primates and birds) varies greatly between sites. Adequate interpretation of this must await quantitative data on patch size and crop volume. Furthermore, while group size and composition appear very similar between the three sites, only at Amanã have *C. ouakary* been observed to form regular bachelor groups.

A comparison of the three populations of *C. ouakary* appears to offer more similarities than differences. For example, uacaris at all sites travel in very large groups that fragment when resources are scarce, break into subgroups that have similar compositions (female, and young of several years, plus an associated male in proximity), move between habitats as resources dictate and breed at the onset of the fruiting season. They also have a diet in which hard-husked unripe fruits are important and (for two populations at least) species of the genus *Eschweilera* are the pre-eminent diet species. At none of the localities are the animals totally dependent on fruits and seeds. Rather, they incorporate other diet items (plant and invertebrates) as availability permits (or, possibly, as nutritional needs demand).

None of the populations surveyed in this chapter seems to have a society with high levels of aggression. All may have social groups that are composed of many males and females with (despite the group composition mentioned above) no obvious pairings or consortships during the mating season.

Similarities in our findings may be due to the relatively restrictive nature of the *igapó* habitat, the limited number of ways in which it may be exploited, and the limited suite of options available to a medium-sized primate seed predator.

The danger of sampling error, particularly with regard to species included in the diet, must also be considered. Some of the differences may be more apparent than real: for example, *Crudia* has been recorded as a diet item at Amanã, but never at Jaú. However, it was only recorded twice at Amanã, and so it is either of very little significance or it is a species that future fieldwork will register for Jaú. The same is true for *Rheedia* and *Copaifera*, both present at Jaú, but not yet recorded in the diet as they have been at Amanã (where *Rheedia* is a major diet item).

It is, perhaps, telling that while some behaviors are comparable among these three sites, they are also the ones that are the easiest to observe (diet, group size, presence/absence of fission–fusion, presence/absence of seasonal between-habitat migration), and hence about which we are currently best informed. Given the apparent flexibility of uacari behaviors, the failure to observe a particular comportment or sequence at one locality may be, given our current state of knowledge, due to sampling error more than any other single factor. Other behaviors (e.g. social ones related to mate choice, dispersal patterns and kin relations) must await more complete data.

The existence of variation in diet, habitat use and ranging behavior between populations is not unique to *C. ouakary* – variation in these factors also occurs in other *Cacajao* taxa. For example, different populations of *C. c. ucayalii* show great variations in these aspects, as a comparison of studies by Aquino and coworkers (e.g. Aquino 1998; Aquino & Encarnación, 1994, 1999), Bowler (2007) and Bowler & Bodmer (2009, 2010) shows. Comparing Lehman and Robertson (1994a, 1994b) with Boubli (1997) also shows this for *C. melanocephalus*. There also seems to be variation between species: with much higher levels of aggression being observed in *C. c. ucayalii* (Bowler 2007), than in *C. ouakary* (Bezerra *et al.* 2011). There are parallel behaviors in all taxa – the tracking of diet resources and the use of the most common species, for example (Ayres 1986; Boubli 1997; Bowler 2007), as well as small-scale similarities, such as terrestrial foraging to exploit seeds germinating in the period when seasonally inundated forests are dried out [observed both for *C. ouakary* and by Ayres (1986) for *C. calvus calvus*].

There are, however, some intriguing differences between the *C. ouakary* populations and those of its closest relative, *C. melanocephalus*. For example, Boubli (1999) reported that group sizes remained constant throughout the year at about 70 individuals and observed neither fission–fusion nor seasonal migration between habitats. Although this species does occur in *igapó* (Lehman & Robertson 1994a, 1994b), those studied by Boubli occupied Amazonian *caatinga* (*sensu* Whitmore 1998; Prance 2001), a forest habitat which is rarely affected by flooding. The extreme nutrient poverty of the soil (which is poor even by Amazonian standards – see Boubli 2002), produces a tree species community dominated by *Eperua*. The formations at Boubli's study site are extensive, making seasonal between-habitat migration challenging for a 4 kg primate. As noted by Boubli and Tokuda (2008), many of the most important tree species exploited by *C. melanocephalus* in the Amazonian *caatinga* of the Neblina region have an extended phenology and a non-clumped distribution. This makes the distribution of resources predictable in both space and time in a manner that does not encourage group fragmentation. In this context the social ecology of the *C. melanocephalus* population studied by Boubli (although not necessarily the whole species) should be viewed in context the larger pattern of edaphically constrained mammalian densities that occurs across the Amazon basin (see Emmons 1984, 1999; Coomes 1992; Peres & Janson 1999; Haugaasen & Peres 2005; Lambert *et al.* 2006).

In some cases, similarities might be more apparent than real: fission–fusion is easy to observe remotely, but difficult to understand. Is the membership the same or different? If the same, who are those individuals – kin, reproductive partners, allies? These salient questions cannot currently be answered, and more work is needed to discover if the apparent possession of the same social organization in fact conceals some locality-specific nuances in social structure that reflect variations in key environmental parameters (e.g. Andelman 1986).

What is certainly becoming clear is that *C. ouakary* is not a simple crusher of hard-husked seeds, but, like most animals, has a suite of foraging and social strategies. *Cacajao ouakary* is capable of switching between a variety of diet items and nutritive modes of folivory, granivory and insectivory. This, and a highly flexible social system, mean that the ecology of any species of *Cacajao* probably will not be something that can be readily pieced together from the ecological snapshots of medium-term studies widely separated in time and space. Integrated and long-term studies of several populations using coordinated methodologies are required if we are to not only record the extent and nature of behavioral variation in this fascinating primate, but to understand the proximate and ultimate causes that underlie them. This knowledge is key, because within these factors lie the species' vulnerabilities and it is only through a thorough knowledge of these that we can hope for the effective conservation of these remarkable animals.

Acknowledgments

A. Barnett and B. Bezerra's studies were funded by: Akodon Ecological Consulting, American Society of Primatologists, Columbus Zoo & Aquarium Conservation Fund, Servicios FORT (Manaus), LSB Leakey Foundation, Leakey Trust (UK), Marapanim (Manaus), Ms. Rosie MIL Fund, Primate Action Fund (via the Margot Marsh Biodiversity Foundation), Percy Sladen Memorial Fund, Pittsburgh Zoo Conservation Fund, Primate Conservation Inc., Roehampton University, Sophie Danforth Conservation Biology Fund, Wildlife Conservation Society (grants to Barnett). Bezerra received support from Programme Alban, ORS Award, Bristol University, Amazon Ecopark Lodge, Living Rain Forest Foundation (Manaus) and IdeaWild. Barnett and Bezerra would also like to thank the staff of Nucleo do Unidades de Conservação (NUC) of IBAMA, SISBIO, CNPq and Fundação Vítoria Amazônica (especially Sérgio Borges and Ricardo Almeida), and our field guides, Eduadro Elizio de Souza, Roberto da Silva and Maria Souza de Bom Jesus. Welma de Souza Silva (Univ. Federal de Amazonas & INPA) identified many of the plants. AB also thanks Caroline Ross, Ann MacLarnon and Charles Handley; BB thanks Gareth Jones and Antonio de Souto. Lucy Todd redrew one of the graphs, all other maps and figures were done by Bruna Bezerra.

T. Defler's studies have been financed by Conservation International, National Geographic Society, and the Margot Marsh Biodiversity Foundation. He thanks José Vicente Rodríguez Mahecha of Conservación Internacional Colombia, the Universidad Nacional de Colombia, its staff and administrators, the people of La Pedrera, Amazonas, Colombia, field assistant Angel Yukuna and Sara Bennett.

H. Quieroz and M. Oliveira thank Brazilian National Council for the Development of Science and Technology (CNPq), Mamirauá Institute for Sustainable Development (IDSM), and its terrestrial vertebrate ecology research group (particularly João Valsecchi, Tatiana Vieira and Nayara Cardoso), field assistants Francisco Santos, Edu Carmo, Francisco Carmo and Eriney Carmo, and the inhabitants of São Jesus do Baré, Amanã Reserve.

References

Amaral, J.V., Muniz, I.C.M., Avelar, A.A., *et al.* (2005). Primatas da Reserva de Desenvolvimento Sustentavial Amaná, Amazonas. In *Programa e Livro de Resumos XI Congresso Brasileiro de Primatologia*, ed. J.C. Bicca-Marques. Porto Alegre: Sociedade Brasileira de Primatologia, p. 70.

Andelman, S. (1986). Ecological and social determinants of cercopithecine mating patterns. In: *Ecological Aspects of Social Evolution*, ed. D.I. Rubenstein & R.W. Wrangham. Princeton, NJ:

Princeton University Press, pp. 201–216.

Aquino, R. (1998). Some observations on the ecology of *Cacajao calvus ucayalii* in the Peruvian Amazon. *Primate Conservation*, **18**, 21–24.

Aquino, R. & Encarnacion, F. (1994). Primates of Peru. *Primate Report*, **40**, 1–127.

Aquino, R. & Encarnacion, F. (1999). [Preliminary observations on the diet of *Cacajao calvus ucayalii* in northeast Peru.] *Neotropical Primates*, **7**, 1–5.

Ayres, J.M. (1986). Uakaris and Amazonian flooded forest. PhD dissertation, University of Cambridge.

Barnett, A.A. (2005). *Cacajao melanocephalus. Mammalian Species*, **776**, 1–6.

Barnett, A.A. (2010). Diet, habitat use and conservation ecology of the golden-backed uacari, *Cacajao ouakary*, in Jaú National Park, Amazonian Brazil. PhD thesis, Roehampton University, London. http://roehampton.openrepository.com/roehampton/

Barnett, A. & da Cunha, A.C. (1991). The golden-backed uacari of the upper Rio Negro Brazil. *Oryx*, **25**, 80–88.

Barnett, A.A., Borges, S., de Castilho, C.V., *et al.* (2002). Primates of Jaú National Park, Amazonas, Brazil. *Neotropical Primates*, **10**, 65–70.

Barnett, AA., de Castilho, C.V., Shapley, R.L., *et al.* (2005). Diet, habitat selection and natural history of *Cacajao melanocephalus ouakary* in Jau National Park, Brazil. *International Journal of Primatology*, **26**, 949–969.

Barnett, A.A., Schiel, V., Deveny, A., *et al.* (2011). Predation on *Cacajao ouakary* and *Cebus albifrons* (Primates: Platyrrhini) by harpy eagles. *Mammalia*, **75**, 169–172.

Barnett, A.A., Shaw, P., Spironello, W.R., *et al.* (2012). Sleeping site selection by golden-backed uacaris, *Cacajao melanocephalus ouakary* (Pitheciidae), in Amazonian flooded forests. *Primates*, **53**, 273–285.

Bezerra, B.M., Barnett, A.A., Souto A., *et al.* (2011). Ethogram, time budget and ecology of the Golden-backed uacari. *International Journal of Primatology*, **32**, 46–68.

Bezerra, B.M., Souto, A.S. & Jones, G. (2010a). Vocal repertoire of golden-backed uacaris (*Cacajao melanocephalus*): call structure and context. *International Journal of Primatology*, **31**, 79–778.

Bezerra, B.M., Souto, A.S. & Jones, G. (2010b). Responses of golden-backed uacaris, *Cacajao melanocephalus*, to call playback: implications for surveys in the flooded igapó forest. *Primates*, **51**, 327–336.

Boinski, S., Kauffman, L., Ehmke, E., *et al.* (2005). Dispersal patterns among three species of squirrel monkeys (*Saimiri oerstedii, S. boliviensis*, and *S. sciureus*): I. Divergent costs and benefits. *Behaviour*, **142**, 525–632.

Boubli, J.P. (1997). Ecology of the black uakari monkey, *Cacajao melanocephalus melanocephalus*, in Pico da Neblina National Park, Brazil. PhD dissertation, University of California, Berkeley.

Boubli, J.P. (1999). Feeding ecology of black-headed uacaris (*Cacajao melanocephalus melanocephalus*) in Pico da Neblina National Park. *International Journal of Primatology*, **20**, 719–749.

Boubli, J.P. (2002). Lowland floristic assessment of Pico da Neblina National Park, Brazil. *Vegetatio*, **160**, 149–167.

Boubli, J-P. & de Lima, M.G. (2009). Modeling the geographical distribution and fundamental niches of *Cacajao* spp. and *Chiropotes israelita* in northwestern Amazonia via a maximum entropy algorithm. *International Journal of Primatology*, **30**, 217–228.

Boubli, J.P. & Tokuda, M. (2008). Socioecology of black uakari monkeys, *Cacajao hosomi*, in Pico da Neblina National Park, Brazil: the role of the peculiar spatial–temporal distribution of resources in the Neblina forests. *Primate Report*, **75**, 3–10.

Boubli, J.P., Silva, M.N.F., Amado, M.V., *et al.* (2008). A taxonomic reassessment of *Cacajao melanocephalus* Humboldt (1811), with the description of two new species. *International Journal of Primatology*, **29**, 723–741.

Bowler, M. (2007). The ecology and conservation of the red uakari monkey on the Yavari River, Peru. PhD dissertation, University of Kent, Canterbury.

Bowler, M. & Bodmer, R. (2009). Social behavior in fission–fusion groups of red uakari monkeys (*Cacajao calvus ucayalii*). *American Journal of Primatology*, **71**, 1–12.

Bowler, M. & Bodmer, R. (2010). Diet and food choice in Peruvian red uakaris (*Cacajao calvus ucayalii*): selective or opportunistic seed predation? *International Journal of Primatology*, **32**, 1109–1122.

Branch, L.C. (1983). Seasonal and habitat differences in the abundance of primates in the Amazon (Tapajos) National Park, Brazil. *Primates*, **24**, 424–431.

Brugiere, D., Gautier, J-P., Moungazi, A., *et al.* (2002). Primate diet and biomass in relation to vegetation composition and fruiting phenology in a rain forest in Gabon. *International Journal of Primatology*, **23**, 999–1024.

Changeux, J.P. (2005). Genes, brains, and culture: from monkey to human. In *From Monkey Brain to Human Brain. A Fyssen Foundation Symposium*, ed. S. Dehaene, J.R. Duhamel, M.D. Hauser & G. Rizzolatti. Cambridge, MA: MIT Press, pp. 73–94.

Chapman, C.A. & Peres, C.A. (2001). Primate conservation in the new millennium: the role of scientists. *Evolutionary Anthropology*, **10**, 16–33.

Clinebell, R.R., Phillips, O.L., Gentry, A.H., *et al.* (1995). Prediction of neotropical tree and liana species richness from soil and climatic data. *Biodiversity and Conservation*, **4**, 56–90.

Coomes, O. (1992). Making a living in the Amazon rainforest: peasants, land, and economy in the Tahuayo River basin of northeastern Peru. Doctoral dissertation, University of Wisconsin-Madison.

da Cunha, A.C. & Barnett, A.A. (1990). Sightings of the golden-backed uacari, *Cacajao melanocephalus ouakary*, on the Upper Rio Negro, Amazonas, Brazil. *Primate Conservation*, **11**, 8–11.

Defler, T.R. (1982). A comparison of intergroup behavior in *Cebus albifrons* and *C. apella. Primates*, **23**, 385–392.

Defler, T.R. (1999). Fission–fusion behavior in *Cacajao melanocephalus ouakary. Neotropical Primates*, **7**, 5–8.

Defler, T.R. (2001a). *Cacajao melanocephalus ouakary* densities on the lower Apaporis River, Colombian Amazon. *Primate Report*, **61**, 31–36.

Defler, T.R. (2001b). Densidad de especies y organización espacial de una comunidad de primates: Estación Biológica Caparú, Departamento del Vaupés, Colombia. In: *Primatología del Nuevo Mundo*, ed. V. Pereira-Bengoa, F. Nassar-Montoya & A. Savage. Bogotá: Centro de Primatología Araguatas Ltda, pp. 23–39.

Defler, T.R. (2004). *Primates of Colombia.* Bogotá: Conservación Internacional Colombia.

Defler, T.R. & Defler, S.B. (1996). The diet of a group of woolly monkeys (*Lagothrix lagothricha lagothricha*) in southeastern Colombia. *International Journal of Primatology*, **17**, 161–190.

Defler, T.R., Palacios, E., Rodríguez, A., *et al.* (2006). *Lagothrix lagothricha*. In *Libro Rojo De Los Mamiferos de Colombia*, ed. J.V. Rodriguez-Mahecha, M. Alberico, F. Trujillo & J. Jorgensen. Bogota: Conservacion Internacional Colombia, pp. 220–225.

de Moraes, P.L.R., de Carvalho, O. & Strier K.B. (1998). Population variation in patch and party size in muriquis (*Brachyteles arachnoides*). *International Journal of Primatology*, **19**, 325–337.

de Souza, L.L., Queiroz, H.L. & Ayres, J.M. (2004). The mottled-face tamarin, *Saguinus inustus*, in the Amana Sustainable Development Reserve, Amazonas, Brazil. *Neotropical Primates*, **12**, 121–122.

de Thoisy, B., Renoux, F. & Julliot, C. (2005). Hunting in northern French Guiana and its impact on primate communities. *Oryx*, **39**, 149–157.

Dew, L.J. & Boubli, J-P. (2005). *Tropical Fruits and Frugivores: the search for strong interactors*. Basel: Karger.

Di Fiore, A. (2002). Predator sensitive foraging in ateline primates. In: *Eat or Be Eaten: Predator Sensitive Foraging Among Primates*, ed. L.E. Miller. Cambridge: Cambridge University Press, pp. 242–267.

Dindo, M., Thierry, B. & Whiten, A. (2008). Social diffusion of novel foraging methods in brown capuchin monkeys (*Cebus apella*). *Proceedings of the Royal Society B, Biological Sciences*, **275**, 187–193.

Dixson, A.F. (1998). *Primate Sexuality: Comparative Studies of the Prosimians, Monkeys, Apes, and Human Beings*. Oxford: Oxford University Press.

Do Nascimento, N.R., Bueno, G.T., Fritsch, E., *et al.* (2004). Podzolization as a deferralitization process: a study of an Acrisol-Podzol sequence derived from Palaeozoic sandstones in the northern upper Amazon Basin. *European Journal Soil Science*, **55**, 523–538.

Emmons, L.H. (1984). Geographic variation in densities and diversities of non-flying mammals in Amazonia. *Biotropica*, **16**, 210–222.

Emmons, L.H. (1999). Of mice and monkeys: primates as predictors of mammal community richness. In *Primate Communities*, ed. J.G. Fleagle, C.H. Janson & K.E. Reed. Cambridge: Cambridge University Press, pp. 171–188.

Fashing, P.J., Dierenfeld, E.S. & Mowry, C.B. (2007). Influence of plant and soil chemistry on food selection, ranging patterns, and biomass of *Colobus guereza* in Kakamega Forest, Kenya. *International Journal of Primatology*, **28**, 673–703.

Ferreira, L.V. (1997). Effects of the duration of flooding on species richness and floristic composition in three hectares in the Jaú National Park, in floodplain forest in Central Amazonia. *Biodiversity and Conservation*, **6**, 1353–1363.

Figueiredo, W. (2006). Estimativas de tempos de divergência em platirrinos e filogenia molecular e filogeografia dos uacaris, parauacus e cuxius. PhD thesis, Programa de Pós-Graduação em Genética e Biologia Molecular, Universidade Federal do Pará.

Forget, P-M., Kitajima, K. & Foster, R.B. (1999). Pre- and post-dispersal seed predation in *Tachigali versicolor* (Caesalpiniaceae): effects of timing of fruiting and variation among trees. *Journal of Tropical Ecology*, **15**, 61–81.

Frailey, C.D., Lavina, E.L., Rancy, A., *et al.* (1988). A proposed Pleistocene/Holocene lake in the Amazon basin and its significance to Amazonian geneology and biogeography. *Acta Amazonic*, **18**, 119–143.

FVA-IBAMA. (1998). *Parque Nacional do Jaú. Plano de Manejo – Vercão 8*. Manaus, Brazil: FundaçãoVitória Amazônica and Instituto Brasiliero do Meio Ambiente e dos Recursos Naturais Renováveis.

Gentry, A.H. & Emmons, L.H. (1987). Geographical variation in fertility, phenology and composition of understory of Neotropical forests. *Biotropic*, **19**, 216–227.

Goldani, A., Carvalho, G.S. & Bicca-Marques, J.C. (2006). Distribution patterns of Neotropical primates (Platyrrhini) based on parsimony analysis of endemicity. *Brazilian Journal of Biology*, **66**, 61–74.

Goulding, M., Barthem, R. & Ferreira, E. (2003). *The Smithsonian Atlas of the Amazon*. Washington, DC: Smithsonian Institution Press.

Gross, M.R. (1996). Alternative reproductive stategies and tactics: diversity within sexes. *Trends in Ecology and Evolution*, **11**, 92–98.

Hammond, D.S. & Brown, V.K. (1995). Seed weights of woody plants in relation to disturbance, dispersal, soil type in wet Neotropical forests. *Ecology*, **76**, 2544–2561.

Hanya, G., Matsubara, M., Hayaishi, S., *et al.* (2008). Food conditions, competitive regime, and female social relationships in Japanese macaques: within-population variation on Yakushima. *Primates*, **49**, 116–125.

Hanya, G., Noma, N. & Agetsuma, N. (2002). Altitudinal and seasonal variations in the diet of Japanese macaques in Yakushima. *Primates*, **44**, 51–59.

Hanya, G., Yoshihiro, S., Zamma, K., *et al.* (2004). Environmental determinants of the altitudinal variations in relative group densities of Japanese macaques on Yakushima. *Ecological Research*, **19**, 485–493.

Haugaasen, T. & Peres, C.A. (2005). Mammal assemblage structure in Amazonian flooded and unflooded forests. *Journal of Tropical Ecology*, **21**, 133–145.

Hernández-Camacho, J. & Defler, T.R. (1985). Some aspects of the conservation of non-human primates in Colombia. *Primate Conservation*, **6**, 42–50.

Hershkovitz, P. (1983). Two new species of night monkeys, genus *Aotus* (Cebidae: Platyrrhini): a preliminary report on *Aotus* taxonomy. *American Journal of Primatology*, **4**, 209–244.

Hershkovitz, P. (1987). Uacaries: the new world monkeys of the genus *Cacajao* (Cebidae, Platyrrhini): a preliminary taxonomic review, with a description of a new sub-species. *American Journal of Primatology*, **12**, 1–53.

Ibarra, C.J., Morelo, J., Briceño, A., *et al.* (1979). Los suelos. Grupo técnico del Proyecto Radargramétrico del Amazonas, La Amazonia Colombiana y sus Recursos. Bogotá: República de Colombia, pp. 97–216.

Julliot, C. (1992). Diet diversity and habitat of howler monkeys. In *Current Primatology, Volume I: Ecology and Evolution*, ed. B. Strasbourg Thierry, J.R. Anderson, J.J. Roeder & N. Herrenschmidt. Paris: Université Louis Pasteur, pp. 67–78.

Kinzey, W. (1992). Dietary and dental adaptations in the Pitheciinae. *American Journal of Physical Anthropology*, **88**, 499–514.

Klein, L.L. & Klein, D.J. (1976). Neotropical primates: aspects of habitat usage, population density, and regional distribution in La Macarena, Colombia. In *Neotropical Primates: Field Studies and Conservation*, ed. R.W. Thorington, Jr. & P.G. Heltne. Washington, DC: National Academy of Sciences, pp. 70–78.

Lambert, T.D., Malcolm, J.R. & Zimmerman, B.L. (2006). Amazonian small mammal abundance in relation to habitat structure and resource abundance. *Journal of Mammalogy*, **87**, 766–776.

Lehman, S.M. & Robertson, K.L. (1994a). Preliminary survey of *Cacajao melanocephalus melanocephalus* in Southern Venezuela. *International Journal of Primatology*, **15**, 927–934.

Lehman, S.M. & Robertson, K.L. (1994b). Survey of Humboldt's black head uakari (*Cacajao melanocephalus melanocephalus*) in southern Amazonas, Venezuela. *American Journal of Primatology*, **33**, 223.

Lemos de Sá, R.M. & Strier, K.B. (1992). A preliminary comparison of forest structure and use by two isolated groups of woolly spider monkeys, *Brachyteles arachnoides*. *Biotropica*, **24**, 455–459.

Leonardi, V. (1996). *Entre árvores e esquecimentos: história social nos sertões da Brasil.* Editoria Universidade de Brasília.

Leonardi, V. (1999). *Os Historiadores e os Rios: natureza e ruína na Amazônia brasiliera*. Editoria Universidade de Brasília.

Mattson, W.J. Jr. (1980). Herbivory in relation to plant nitrogen content. *Annual Review of Ecology and Systematics*, **11**, 119–161.

Moran, E.F., Brondizio, E.S., Tucker, J.M., *et al.* (2000). Effects of soil fertility and land-use on forest succession in Amazônia. *Forest Ecology and Management*, **139**, 93–108.

Nascimento, H.E.M., Laurance, W.F., Condit, R., *et al.* (2005). Demographic and life-history correlates for Amazonian trees. *Journal of Vegetation Science*, **16**, 625–634.

Oliveira-Filho, A.T. & Galetti, M. (1996). Seed predation of *Cariniana estrellensis* (Lecythidaceae) by black howler monkeys, *Alouatta caraya*. *Primates*, **37**, 87–90.

Ottoni, E.B. & Izar, P. (2005). Tool use by semifree and wild capuchin monkeys (*Cebus apella*): from developmental to inter-populational studies. [Abstract]. In: *Culture Conference*, ed. S. Perry & C. Boesch. Leipzig: Max Planck Institute for Evolutionary Anthropology, p. 29.

Palacios, E. & Peres, C.A. (2005). Primate population densities in three nutrient-poor Amazonian terra firme forests of south-eastern Colombia. *Folia Primatologica*, **76**, 135–145

Panger, M.A., Perry S., Rose R., *et al.* (2002). Cross-site differences in foraging behavior of white-faced capuchins (*Cebus capucinus*). *American Journal of Physical Anthropology*, **119**, 52–66.

Peres, C.A. (1991). Seed predation of *Cariniana micrantha* (Lecythidaceae) by brown capuchin monkeys in Central Amazonia. *Biotropica*, **23**, 262–270

Peres, C.A. (1993). Structure and spatial organization of an Amazonian terra firme forest primate community. *Journal of Tropical Ecology*, **9**, 259–276.

Peres, C.A. (1994). Primate responses to phenological changes in an Amazonian terra firme forest. *Biotropica*, **26**, 98–112.

Peres, C.A. (1997a). Effects of habitat quality and hunting pressure on arboreal folivore densities in neotropical forests: a case study of howler monkeys (*Alouatta* spp.). *Folia Primatologica*, **68**, 199–222.

Peres, C.A. (1997b). Primate community structure at twenty western Amazonian flooded and unflooded forests. *Journal of Tropical Ecology*, **13**, 381–405.

Peres, C.A. & Janson, C.H. (1999). Species coexistence, distribution, and environmental determinants of neotropical primate richness: a community-level zoogeographic analysis. In *Primate Communities*, ed. J.G. Fleagle, C.H. Janson & K.E. Reed. Cambridge: Cambridge University Press, pp. 55–74.

Peres, C.A. & Nascimento, M. (2005). Unpublished report to Fundaçăi Vitória Amazônica on survey of large mammals in upper Rio Jaú. 14 pages. FVA, Manaus.

Pinheiro, M.R. & Borges, S.H. (2004). Caracterização das Janela para a Biodiversidade do Parque Nacional do Jaú. In *Janelas para a Biodiversidade no Parque Nacional do Jaú: uma estratégia para o estudo da biodiversidade na Amazônia*, ed. S.H. Borges, S. Iwanaga, C.C. Durigan & C.C. Pinheiro. Manaus, AM: WWF-FVA-IBAMA-Prensa Ipiranga, pp. 19–40.

Pinheiro, M.R. & Macedo, A.B. (2004). Dinâmica da população humana nos rios do Parque Nacional do Jaú. In *Janelas para a Biodiversidade no Parque Nacional do Jaú: uma estratégia para o estudo da biodiversidade na Amazônia*, ed. S.H. Borges, S. Iwanaga, C.C. Durigan & C.C. Pinheiro. Manaus, AM: WWF-FVA-IBAMA-Prensa Ipiranga, pp. 43–61.

Pinto, L.P.S., Costa, C.M.R., Strier, K.B., *et al.* (1993). Habitat density and group size of primates in a Brazilian tropical forest. *Folia Primatologica*, **61**, 135–43.

Prance, G. (2001). Amazon ecosystems. In *Encylopedia of Biodiversity, Vol. 1*, ed. S.A. Levin. San Diego, CA: Academic Press, pp. 145–157.

Queiroz, H.L. & Fernandes, M.E.B. (2001). *A New World Heritage Site: The Central Amazon Protected Areas: A Proposal for the Establishment of a Natural World Heritage and Cultural Landscape World Heritage Site*, Volumes I and II. Brasília: The Brazilian Ministry of Environment and UNESCO.

RadamBrasil. (1978). *Sheet SA.20/Manaus. Levantamento de Recursos Naturais, vol. 18: Departamento Nacional da Produção Mineral.* Rio de Janeiro: Ministério das Minas e Energia.

Rylands, A.B. (1992). Relatorio: Viagem ao Parque Nacional de Jau. Unpublished report to Fundacao Vitoria Amazonica, Manaus, p. 7.

Rylands, A.B. & Mittermeier, R.A. (2009). The diversity of the New World primates (Platyrrhini): an annotated taxonomy. South American Primates: Comparative Perspectives. In *The Study Of Behavior, Ecology, And Conservation*, ed. P.A. Garber, A. Estrada, J.C. Bicca-Marques, E.W. Heymann & K.B. Strier. New York, NY: Springer, pp. 23–54.

Sapolsky, R.M. (2006). Culture in animals: the case of a non-human primate culture of low aggression and high affiliation. *Social Forces*, **85**, 217–233.

Schaffer, C.M. & Aurelli, F. (2005). Preliminary evidence for cultural variation in wild spider monkeys. [Abstract.] In *Culture Conference*, ed. S. Perry & C. Boesch. Leipzig: Max Planck Institute for Evolutionary Anthropology, p. 35.

Semple, K. (2000). A habitat held hostage. *Audubon Magazine*, Oct–Dec, 82–103.

Silva, S.S.B. & Ferrari, S.F. (2009). Behavior patterns of southern bearded sakis (*Chiropotes satanas*) in the fragmented landscape of eastern Brazilian Amazonia. *American Journal of Primatology*, **71**, 1–7.

Silva Jr., J. de S. (2001). Especiação nos Macacos-prego e Caiararas, Gênero *Cebus* Erxleben, 1777 (Primates, Cebidae). Doctoral thesis, Universidade Federal do Rio de Janeiro, Rio de Janeiro.

Silva, Jr., J. de S. (2002). Sistemática dos macacos-prego e caiararas, gênero *Cebus* Erxleben, 1777 (Primates, Cebidae). In *Livro de Resumos, X Congresso Brasileiro de Primatologia : Amazônia – A última Fronteira*, November 10–15. Belém: Sociedade Brasileira de Primatolgia, p. 35.

ter Steege, H., Pitman, N.C.A., Phillips, O.L., *et al.* (2006). Continental-scale patterns of canopy tree composition and function across Amazonia. *Nature*, **443**, 444–447.

Whitmore, T.C. (1998). *An Introduction to Tropical Rain Forests*. Oxford: Oxford University Press.

Chapter 17

Ecology and behavior of titi monkeys (genus *Callicebus*)

Júlio César Bicca-Marques & Eckhard W. Heymann

Introduction

Titi monkeys represent one of the most diverse radiations of New World primates, comprising at least 29 species (van Roosmalen *et al.* 2002; Wallace *et al.* 2006; Rylands & Mittermeier 2009), distributed into five clades: *Callicebus donacophilus* group, *Callicebus moloch* group, *Callicebus cupreus* group, *Callicebus torquatus* group and *Callicebus personatus* group (van Roosmalen *et al.* 2002). *Callicebus* spp. are found in the forests of the Amazon and Orinoco basins, the Chaco and dry forests of Paraguay and Bolivia, and the Brazilian Atlantic forest (van Roosmalen *et al.* 2002).

Despite this high diversity of forms and ample geographic distribution in South America and the extraordinary increase in Platyrrhine field and captive research over the last three decades, our knowledge on these small- to medium-sized (adult body mass 0.8–1.4 kg: Smith & Jungers 1997), shy primates is still very scanty. Knowledge of *Callicebus* has not grown at the same rate as that of other cebid genera, such as *Saguinus*, *Alouatta* and *Cebus*, since the publication of the most comprehensive review on their ecology and behavior by Warren Kinzey in 1981. More recent reviews of *Callicebus* include Robinson *et al.* (1987), Kinzey (1997) and Norconk (2011). In this chapter we review the current literature on titis and discuss how new findings have improved our understanding of their ecology and behavior, helped us to address unresolved issues, challenge previous interpretations, and provide guidelines for future research.

Habitat

Titis are found in a variety of forest types: primary and secondary, deciduous, semi-deciduous and evergreen, *terra firme*, gallery and flooded, lowland and upland, dry and moist, among others (e.g. Bennett *et al.* 2001; Buchanan-Smith *et al.* 2000; Defler 2004; Marinho-Filho & Veríssimo 1997; Martins *et al.* 1988; van Roosmalen *et al.* 2002). Annual rainfall in *Callicebus* spp. habitats ranges from as low as 500 mm in Brazilian *Caatinga* (*C. barbarabrownae*, Marinho-Filho & Veríssimo 1997), to as much as 3800 mm in Colombian Amazon (*C. lugens*, Palacios *et al.* 1997). Altitudinally, titis range from about sea level to at least 1000 m a.s.l. (Kinzey 1981; van Roosmalen *et al.* 2002).

In areas of sympatry in Colombia, *C. ornatus* uses mainly the poorly stratified and low bushy forest, whereas *C. lugens* is found primarily in high and well-stratified forest (Defler 1994). van Roosmalen *et al.* (2002) contend that members of the *C. torquatus* group are *terra firme* rainforest specialists, but Defler (1994, 2004) observed *C. lugens* entering *várzea* (white-water flooded) forest. If their association with undisturbed forest, as described by van Roosmalen *et al.* (2002), is confirmed as more studies are conducted and the alteration and fragmentation of their habitat increase, this would qualify as an actual ecological distinction between the members of the *C. torquatus* group and all other titis. The latter have been observed in disturbed and/or fragmented forests (e.g. Bicca-Marques & Garber 2004; Heiduck 1997; Mark 2003; Mason 1968; Müller 1995; Wright 1985). Michalski and Peres (2005), for instance, recorded *C. moloch* inhabiting 108 out of 129 forest patches ranging in size from 0.5 to 13,551 ha in southern Brazilian Amazon.

Population ecology

Estimates of abundance or population density vary considerably (Table 17.1). Data on population sizes are still too scarce to determine whether this wide range reflects species-, population- or site-specific characteristics. There is evidence, however, to suggest that density estimates from behavioral studies are probably more reliable than those from surveys (and even then, they may underestimate density if range overlap is not taken into account). Their secretive lifestyle (Kinzey 1981; Terborgh 1983) makes titis less detectable during censuses than more active and noisy species like capuchin or squirrel monkeys. Although this apparent difference between methods may be flawed if high-density sites are chosen for behavioral studies, data from a 100-ha forest fragment in the Brazilian Amazon where Bicca-Marques and colleagues (unpublished

Evolutionary Biology and Conservation of Titis, Sakis and Uacaris, eds. Liza M. Veiga, Adrian A. Barnett, Stephen F. Ferrari and Marilyn A. Norconk. Published by Cambridge University Press.

Table 17.1 Group size and population density of *Callicebus* spp.

Species	MGS[1]	N[2]	Abundance/density	Method[3]	Source
C. cupreus group					
C. cupreus	7	6	0.6–5.7 grp/10 km; 7.2–> 30 ind/km^2	S/B	Peres (1988); Cameron *et al.* (1989); Knogge & Heymann (1995); Bicca-Marques *et al.* (2002); Heymann *et al.* (unpubl. data)
C. caligatus	4	5		S	Ferrari *et al.* (2000)
C. ornatus	4	14	400–450 ind/km^2	B	Mason (1968); Robinson (1979); Porras (2000)
	2	43	60–369 ind/km^2	S	Wagner *et al.* (2009)
C. moloch group					
C. moloch			0.9 grp/10 km	S	Martins *et al.* (1988)
C. bernhardi	4	15		S	Ferrari *et al.* (2000)
C. brunneus	7	>169	2.8–26 ind/km^2	S/B	Wright (1985); Encarnación & Castro (1990); Buchanan-Smith *et al.* (2000); Ferrari *et al.* (2000); Bossuyt (2002); Lawrence (2003)
C. aureipalatii[4]		30	6.2 ind/km^2	S	Wallace *et al.* (2006)
C. torquatus group					
C. lugens	5	14	32 ind/km^2; ~5 grp/km^2	S/B	Defler (1983); Palacios *et al.* (1997); Polanco-Ochoa (1993); Palacios & Rodriguez (Chapter 20)
C. lucifer	5	3	20 ind/km^2	B	Kinzey *et al.* (1977); Kinzey & Wright (1982); Easley (1982)
	3	39	2.5–2.8 ind/km^2; 0.8–1.2 grp/km^2	S	Aquino *et al.* (2008)
C. personatus group					
C. personatus	6	3	5.5–7.6 ind/km^2	S/B	Kinzey & Becker (1983); Chiarello (2000); Price & Piedade (2001b)
C. melanochir	6	>1	17 ind/km^2	B	Heiduck (2002); Müller (1996a)
C. nigrifrons	6	>16	7–10 ind/km^2; 1.6–4.3 ind/10 km	S	Torres de Assumpção (1983); Kimura (pers. commun. cited in Kinzey & Becker 1983); Stallings & Robinson (1991); Neri (1997)

[1] Maximum group size; [2] sample size (number of sightings or study groups); [3] S = population survey, B = behavioral study; [4] the authors that described this species could not determine whether it belongs to the *C. cupreus* or the *C. moloch* group.

data) carried out both line transect censuses and behavioral studies on *C. cupreus* strongly support the contention that these methods actually differ in their sensitivity to detect titi groups. Whereas walking along a 2.7-km transect produced an estimate of only 4 groups (although the frequency and spatial distribution of groups calling almost simultaneously in a given morning suggested a higher density), a detailed behavioral study (Bicca-Marques 2000; Bicca-Marques *et al.* 2002) estimated more than 30 individuals in at least 8 groups in the same area.

Titis live in family groups often containing an adult pair and 1–3 offspring (Table 17.2). Groups containing a third adult are rare and have been observed so far in only four species (*C. cupreus*, *C. lugens*, *C. modestus* and *C. personatus*). Bicca-Marques *et al.* (2002; see also Silveira *et al.* 1998) described a group temporarily containing three adult males (including at least two brothers) after the death of the resident adult male. The follow-up of this group for a period of about 3 years allows the suggestion that the presence of more than one adult of each sex in a group may represent a particular phase of group dynamics. Defler (1983) cites one group of *C. lugens* (out of 10 studied) that comprised 3 adults (sex unknown), 1 subadult and 1 newborn. A *C. personatus* study group followed for over 100 h by Price and Piedade (2001b) had a similar composition (1 breeding female, 2 other adults of unknown sex, 1 juvenile, and 1 infant). Groups with more than two adults might result from delayed migration by offspring, but could also represent deviations from pair living. An intriguingly large number of adults apparently belonging to a single group of *C. modestus* was cited by Felton *et al.* (2006). One of the two groups recorded by these authors during a quick survey in Bolivia contained three adults out of five individuals, whereas the other had an amazing composition of six adults and one infant. This striking discrepancy group composition urges for longer-term research on *C. modestus* to determine whether this species deviates from the standard titi social organization. The available data do not allow us to rule out of the possibility that this group composition was a consequence of limited dispersal opportunities in a fragmented landscape (although locals report that titis may cross distances of 300–400 m between forest patches on the ground). Alternatively, it may have represented a transient stage in the group's history, an encounter between two groups or a misclassification of individuals into age classes (among other possibilities).

However, it seems that under certain conditions, such as a low availability of reproductive positions and/or of unoccupied habitat at high-density sites, individuals may delay dispersal from their natal groups upon reaching adulthood (Defler 1983; Bossuyt 2002). According to Bossuyt (2002), the death of a parent may also play a key role by increasing the advantages of

Table 17.2 Activity budget and use of space by *Callicebus* spp.

Species	Activity budget (% time)				Use of space[1]		Duration (months)	Source
	Rest	Feed/forage	Move/travel	Social	Home range (ha)	Day range (m)		
C. donacophilus group								
C. oenanthe	31	35	22	10	2	605–722[2]	8	DeLuycker (2007)
C. cupreus group								
C. ornatus	48	20	20	11	14	852 ($N = 4$)	6	Polanco-Ochoa (1993)
						[578–1152]		
					< 1	568 ($N = 12$)	1	Mason (1968)
						[315–870]		
C. moloch group								
C. brunneus					7 ($N = 6$)	637 ($N = 89$)[5]	15	Wright (1985, 1986)
					[6–8]	[150–1450]	?	
					12			Bossuyt (2002)
					[6–18]			
					< 2 ($N = 1$)		1	Lawrence (2003)
	59	13	18	10	3 ($N = 5$)		10	Lawrence (2007)
C. torquatus group								
C. lugens	54	18	23	5	22		12	Palacios & Rodriguez
					14 ($N = 5$)		16?	(1995, cited in Defler
					[9–22]			2004), Palacios *et al.*
								(1997)
								Defler (1983)
					15		?	Campos (1990)
C. lucifer	55[3]	30[3]	15[3]	–[3]	20	684 ($N = 6$)	3	Kinzey (1977), Kinzey
						[466–919]		*et al.* (1977)
	63	16	16	5	18	819 (± 37;	14	Easley (1982)
						$N = 22$)		
C. personatus group								
C. personatus		17			12 ($N = 2$)	1004 ($N = 8$)	6	Price & Piedade (2001a)
					[11–12]	[816–1332]		
	62[4]	21[4]	15[4]	2[4]	5	695 ($N = 6$)	2	Kinzey & Becker (1983)
						[523–801]		
C. melanochir	40	27	32	1	24	1008	11	Müller (1995, 1996a),
					22	[532–1539]	12	Müller & Pissinatti
								(1995)
								Heiduck (1997, 2002)
C. nigrifrons	35	26	33	6			12	Souza *et al.* (1996)
	29	40	25	6	48	1270 ($N = 42$)	5	Neri (1997)
					28 ($N = 4$)	[740–2697]	13?	Torres de Assumpção
					[21–38]			(1983)

Obs.: Percentages were rounded to the nearest integer.
[1] Data provided are mean home range and day range, sample size in parentheses, and range in brackets; [2] average of the dry and wet seasons, respectively; [3] recalculated based on the original data to reach 100% after the exclusion of records "unknown". Although Kinzey *et al.* (1977) do not cite social behavior, Kinzey and Wright (1982) report a significant involvement in grooming in a later study; [4] recalculated as described in [3] considering also authors' statement that individuals were involved in grooming at least 1.3% of the time, but were scored as resting; [5] average for two groups.

subadult (especially male) permanence at the natal group because of its potential aid in infant care and protection against predators.

The release of subadults from competition with same-sex adults is a weak candidate to explain such longer natal residence because of the apparent lack of social pressure from parents (at least in the form of agonistic interactions, for triggering offspring dispersal; Robinson *et al.* 1987; Müller 1996a; Bossuyt 2002; but see Assunção 2005 for a report on an adult male chasing and evicting a subadult in a *C. nigrifrons*

Plate 1 *Pithecia irrorata.* Photo by Francisco Fonseca.

Plate 2 Young male white-faced saki (*Pithecia pithecia*) from Brownsberg Nature Park Suriname. Photo by Nick Robl.

Plate 3 *Chiropotes Sagulatus* male in feeding tree. Brownsberg Nature Park, Suriname. Photo: Tremaine Gregory.

Plate 4 *Chiropotes Sagulatus* male feeding. Brownsberg Nature Park, Suriname. Photo: Tremaine Gregory.

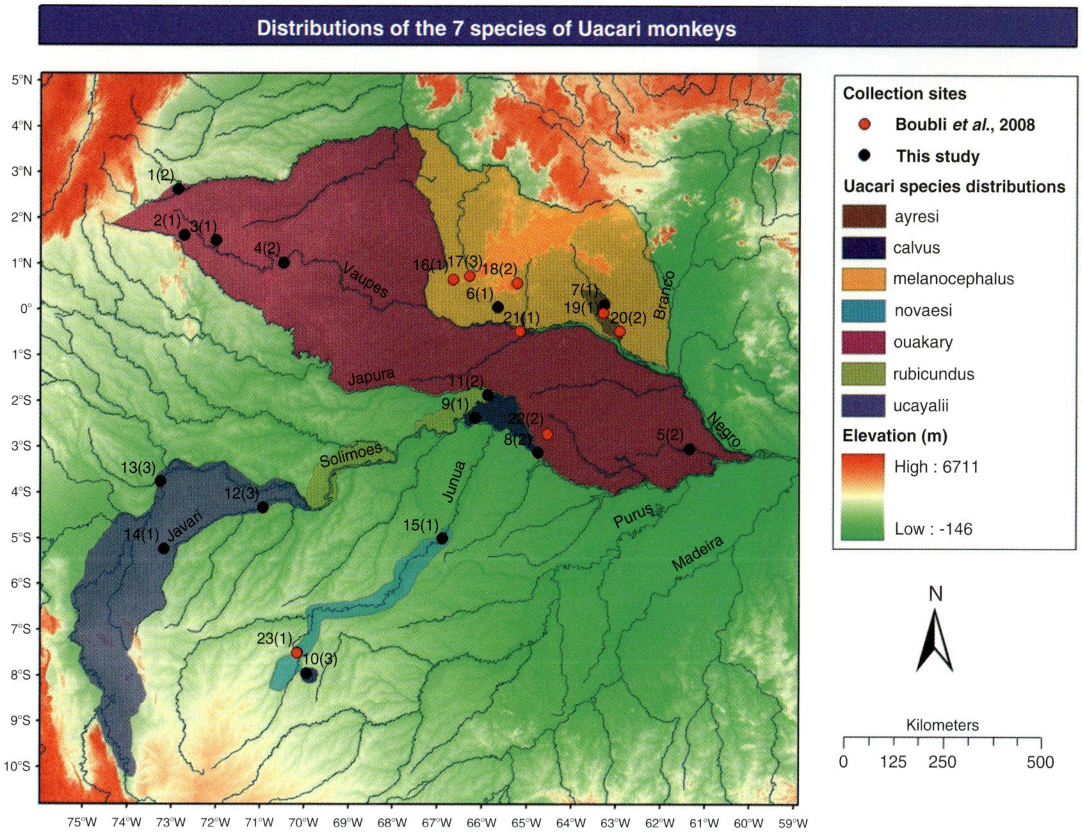

Figure 3.1 Distribution of the collecting localities of the *Cacajao* specimens included in the present analysis, and approximate limits of the geographic ranges of the different species and subspecies.

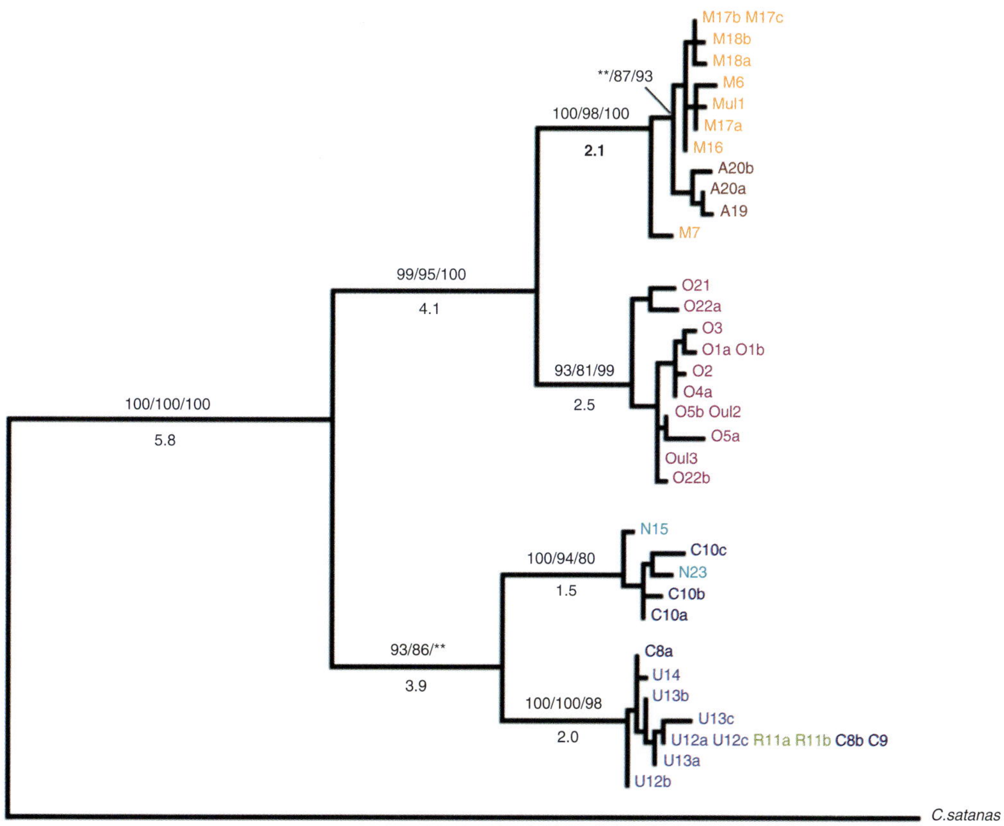

Figure 3.2 Phylogenetic tree constructed in the MrBayes program from cytochrome *b* in *Cacajao*. Numbers above branches are MP and ML bootstrap values, and Bayesian posterior probabilities, respectively, and those below the branches are estimated divergence times. Specimen codes follow Appendix 3.1, and are color-coded according to their geographic distribution (see Figure 3.1).

Photo 6.1 Young male white-faced saki using procumbent lower incisors to extract mesocarp from fruit. Brownsberg Nature Park, Suriname. Photo: Nick Robl.

Photo 6.2 While most seeds eaten by golden-backed uacaris are unripe, the fruits containing them differ greatly in size and shape. Uacaris also eat flowers, insects and pith. Photo: Adrian Barnett.

Photo 12.3 *Lecythis zabucajo* is one of the largest of the Lecythidaceae fruit in the New World. Fruit dimensions: approximately 12.0 cm wide and 11.5 cm long and fruit weight is more than 570 g (wet weight). Brownsberg Nature Park, Suriname. Photo: Marilyn Norconk. See Mori *et al.* (2010, http://sweetgum.nybg.org/lp/index.php).

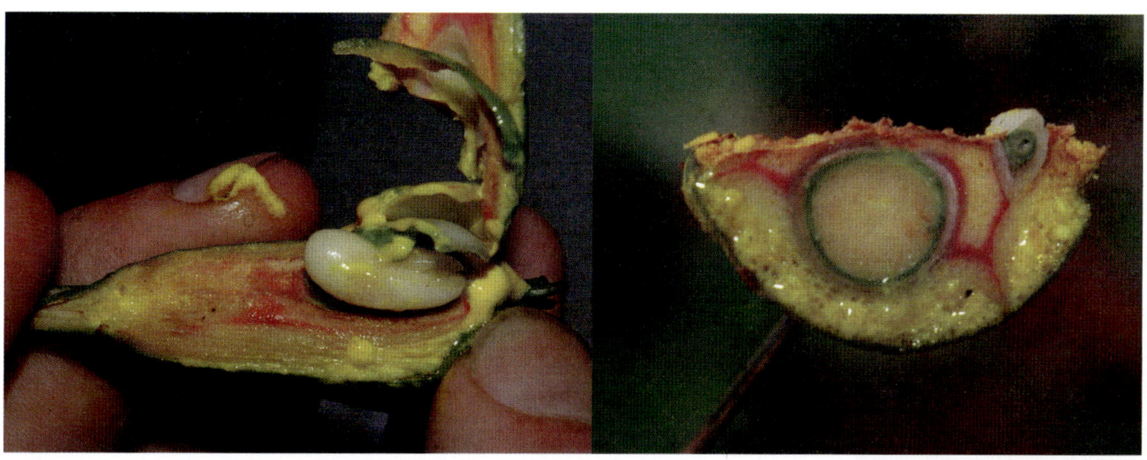

Photo 14.5 The pericarp of *Lorostemon negrense* (Clusiaceae) fruits are rich in canals. From these, a very sticky latex rushes abundantly when the fruit is broken. Taking hours to harden, it appears insoluble in spittle, alcohol or water. Despite this, golden-backed uacaris fed intensively on *L. negrense* fruit during their brief period of availability. Photo: Adrian Barnett.

Plate 5

Photo 14.1 Juvenile *Cacajao calvus novaesi* kept as a pet by members of the Morro Alto community. Uacari Sustainable Development Reserve on the Rio Jurua, Brazil. Photo: Waldener Endo.

Photo 14.2 *Cacajao calvus calvus* feeding. Photo: Rachel Acosta.

Photo 14.3 (1) Leaves are seasonally important for golden-backed uacaris, especially in the dry season when there is little fruit available. (2) The variety of size, shape and hardness of uacari diet fruits is impressively varied. (3) Samples of fruits eaten, rejected or ignored by golden-backed uacaris were studied to determine foraging rules underlying their preferences. Flowers were most prevalent in the diet during those periods of the year when they were numerically dominant over fruits and young leaves. The flowers of *Eschweilera tenuifolia* (Lecythidaceae) dominated the diet on some days. Flowers were destructively consumed both for anthers (pollen) and nectar. Photo by Adrian A. Barnett

Photo 15.1 Red uacari male (*Cacajao calvus ucayalii*), Yavari River, Peru. Photo: Mark Bowler.

Photo 15.2 Adult male uacari (*Cacajao calvus ucayalii*) in aguajal palm (*Mauritia flexuosa*), Yavari River, Peru. Photo: Mark Bowler.

Photo 17.1 *Callicebus caquetensis* feeding on guava from S. Caquetá, Colombia. Photo: Javier Garcia.

Photo 19.1 *Callicebus cupreus* grooming. Photo: Mirjam Nadjafzadeh.

Photo 25.1 Female white-faced saki (*P. pithecia*) feeding. Brownsberg Nature Park, Suriname. Photo: Marilyn Norconk.

Photo 25.2 *Pithecia albicans*. Photo: Francisco Fonseca.

Photo 27.1 *Pithecia irrorata* male (left) and female (right). Note the low level of sexual dimorphism in this taxon from southwestern Brazil, northern Bolivia, and southeastern Peru. Photo: Francisco Fonseca.

Photo 27.2 Adult male white-faced saki grooming an adult female, *Pithecia pithecia*, Brownsberg Nature Park, Suriname. Compare the sex-specific white face and black body hair of the adult male in this photo with the adult male from the western Amazon Basin in Photo 27.1. Photo: Marilyn Norconk.

Photo 27.3 Vigilant adult male white-faced saki, Brownsberg Nature Park, Suriname. Photo: Marilyn Norconk.

Photo 27.4 Adult male white-faced saki (*Pithecia pithecia*) exhibiting sternal scent gland. Brownsberg Nature Park, Suriname. Photo: Nick Robl.

Photo 28.1 Adult white-faced saki female feeding in Brownsberg Nature Park, Suriname. Photo: Marilyn Norconk.

Photo 28.2 Adult white-faced saki male feeding in Brownsberg Nature Park, Suriname. Photo: Marilyn Norconk.

Photo 28.4 Adult female white-faced saki (*Pithecia pithecia*) grooming younger female. Infant male in foreground. Brownsberg Nature Park, Suriname. Photo: Marilyn Norconk.

Photo 32.2 *Callicebus caquetensis* in S. Caquetá, Colombia. Photo: Javier Garcia.

Photo 32.3 *Callicebus cupreus ornatus*, Colombia. Photo: Thomas Defler.

Photo 35.1 *Callicebus caquetensis* moving between forest fragments. Photo: Javier Garcia.

group). Müller (1996b) contends that "group size in *Callicebus* is evidently regulated by the offspring" (p. 20), whereas Bossuyt (2002) states that females appear to disperse at a younger age than do males (see also Hoffman 1998 for similar results in captive groups). Bicca-Marques *et al.* (2002) suggest that titi monogamy may have evolved, at least in part, through female–female intolerance. No data are available regarding mating in these groups containing more than one adult couple to challenge the monogamous system, and only one case of a female mating with more than one male is reported in the literature (Mason 1968). Modern techniques of non-invasive genetic analyses have much to contribute.

Titis seen alone during surveys may be either dispersing individuals (mainly subadults of both sexes) or members of groups for which other group mates were not detected. Most authors state that dispersal takes place around 2–4 years of age, often after or soon before the birth of its second or third sibling (Defler 1983; Wright 1985; Neri 1997; Palacios *et al.* 1997; Bossuyt 2002). Wright (1985) reports 10 subadult *C. brunneus* dispersing from their natal groups mostly at the second half of the year. Most births in wild groups occur from July to January (Defler 1983, 2004; Kinzey & Becker 1983; Knogge & Heymann 1995; Wright 1985; Heiduck 1997; Palacios *et al.* 1997; Price & Piedade 2001b; Bicca-Marques *et al.* 2002; Tirado-Herrera & Heymann 2004), with a peak around September in *C. brunneus* (Wright 1985) or January in *C. lugens* (Defler 1983). On the other hand, births in a captive population in the Northern Hemisphere occurred throughout the year (Welker *et al.* 1998b).

Activity budget

Few studies have focused on the time allocated by wild titis to different activities and no long-term research has been published showing how these are distributed over the day. A considerable variation in the time invested in all major activities by different species has been reported (Table 17.2), but the extent to which methodological differences, particularly in the way authors have defined behavioral categories, and species-, habitat-, population-, site- and/or group-specific characteristics contribute to this variability is unknown.

Resting is often the most frequently observed behavior, followed by similar investments in foraging/feeding and moving/traveling, whereas social behavior represented at most about 10% of the daily active period (see Table 17.2). Interstudy differences may be a consequence of population- and/or group-specific traits (such as population density and home-range size) that potentially influence the kinds of social interactions each study group was involved in, but methodological differences may also be partially responsible. For example, activity budgets of *C. lucifer* (Kinzey *et al.* 1977) and *C. personatus* (Kinzey & Becker 1983) do not include social activities (they are probably incorporated in other categories, mainly "resting"). Evidence that social interactions represent an important element of the daily activities of these species is provided by Kinzey and Wright (1982). They observed that grooming accounted for about 10% of *C. lucifer*'s time during the active period and as much as 19% if grooming sessions in the sleeping tree are also considered. Late afternoon grooming sessions may last more than 3 h in this species (but only a few minutes in *C. personatus*: Kinzey & Becker 1983, and *C. nigrifrons*: Neri 1997), although individual grooming bouts last on average about 3 min.

The main function of grooming among titis appears to be the reinforcement of social bonds, with associated hygienic services probably playing a secondary role. In addition, no relationship seems to exist between grooming and sexual behavior (Kinzey & Wright 1982). All group members take an active part during these sessions, but there are clear age–sex-related differences. Adults groom more than immature individuals, and the adult male was the principal groomer of the infant in the wild (Kinzey & Wright 1982) but not in captivity (Welker *et al.* 1998a). In captivity, at least, infants may groom the father (Jantschke *et al.* 1995; Welker *et al.* 1998a). The father is also the main infant carrier both in the wild and in captivity (Kinzey *et al.* 1977; Kinzey & Becker 1983; Wright 1985; Mendoza & Mason 1986; Welker *et al.* 1998a).

Other social behaviors, such as play or aggression, are rare (Mason 1968; Kinzey 1981; Robinson *et al.* 1987; Defler 2004; Assunção 2005). Infants and older siblings are the most frequent playmates (Kinzey & Wright 1982; Jantschke *et al.* 1995; Knogge & Heymann 1995; Welker *et al.*, 1998a), but the father may also play with the youngest individual, especially in the absence of other immature group members (Kinzey *et al.* 1977; Wright 1986; Müller 1996b; Assunção 2005). In *C. nigrifrons*, play occurs most frequently just prior to night sleep (Neri 1997). A case of maternal infanticide (Cäsar *et al.* 2008) and a case of adoption (Cäsar & Young 2008) were reported in this species.

The father may share food with the infant (Kinzey 1977; Wright 1986) and immatures were observed taking away food from others in captive groups (Welker *et al.* 1998a). Neri (1997) observed an adult male finding a dove nest, driving the incubating bird away, and then letting the female eat the egg. This report is suggestive of adult male deference to the adult females at feeding sites. Bicca-Marques *et al.* (2003) report that dominance plays a minor role in determining access to resources even when the amount of food is small and not sufficient to satiate all family members.

Foraging ecology

Titi diet consists mainly of fruits with animal prey (especially insects) and leaves being the most important supplements (Table 17.3). The contention by van Roosmalen *et al.* (2002) that the diet of the members of *C. torquatus* group "includes insects as well, whereas the other groups are partial folivores" (p. 5) indicates a degree of dietary diversity which can only be appropriately tested with additional long-term studies on a more diverse array of species. Current data suggest titis from *C. personatus* group include only small amounts of animal prey or none at all in their diets. The only species from the *C. moloch* group for which there is published data available (*C. brunneus*) also eats more leaves than prey. Members of the *C. torquatus* and *C. cupreus* groups appear to include small amounts of both leaves and prey in their diets.

Photo 17.1 *Callicebus caquetensis* feeding on guava from S. Caquetá, Colombia. Photo: Javier Garcia. (See color plate section.)

In fact, because of the great importance of seeds in the diet, Palacios *et al.* (1997) suggested *C. lugens* should be considered a frugivorous–granivorous species, while Kinzey (1977) observed that seed predation accounted for 37% of *C. lucifer*'s diet (seeds = 4% + nuts = 33%). On the other hand, *C. brunneus* does not appear to qualify as a seed predator, as, according to Wright (1985), brown titis ingest and defecate small seeds and spit out medium and large ones. Given the diet compositions considered above, it is possible that at least Amazonia's *C. torquatus* group and the Atlantic forest's *C. melanochir* (Table 17.3) could be labeled secondarily granivorous instead of folivorous or insectivorous. In addition, for both *C. lugens* and *C. melanochir* seeds represent an important resource in periods of low food availability (Heiduck 1997; Palacios *et al.* 1997). Unfortunately, most researchers did not distinguish the fruit part ingested, hampering any evaluation of the appropriateness of Palacios *et al.*'s (1997) classification to other titis. Recent long-term studies, however, show that Kinzey's (1992) statement that titis consume very few seeds (see also Norconk 2011) do not apply for at least some *Callicebus* species. See Photos 17.1 and 17.2.

Fruit eating, the ingestion of entire fruits, or pulp, arils, or seeds only (Heiduck 1997; Palacios *et al.* 1997; Palacios & Rodriguez, Chapter 20), occurs throughout the year (Wright 1985; Müller 1996a; Souza *et al.* 1996; Heiduck 1997; Palacios & Rodriguez, Chapter 20). Given a choice, *C. brunneus* prefers ripe over unripe fruit. Chosen items are often small ($\leq$ 2 cm), and taste sweet or bland and mealy. Color is variable (Wright 1985; see also Palacios & Rodriguez, Chapter 20). Daily consumption of leaves is

Photo 17.2 *Callicebus torquatus lucifer* feeding at Ome Ecological Station, Amazonas Colombia between the Putumayo and Amazonas Rivers. Photo: Thomas Defler.

Table 17.3 Contribution of food types and species richness of the diet of *Callicebus* spp.

Species	Contribution to diet (%)					Diet richness			Duration (months)	Source
	Fruit[1]	Leaves[2]	Flowers	Prey	Other/unknown	PS[3]	FS[4]	LS[5]		
C. donacophilus group										
C. oenanthe	36 [1]	8	3	49	3	> 57	39	10	8	DeLuycker (2007)
C. cupreus group										
C. ornatus	81 [?]	8	2	9	–	50	46		6	Polanco (1992, cited in Defler 2004)
C. moloch group										
C. brunneus	54 [?]	28	(2)[7]	17	1	96[8]	81[8]	14[8]	15	Wright (1985)[6]
	50 [?]	39	–	11	–				5	Crandlemire-Sacco (1988)
	70 [?][9]	14[9]	< 1[9]	8[9]	7	35[10]	28[10]		10	Lawrence (2007)
C. torquatus group										
C. lugens	86 [27]	6	4	3	–	62	45		6	Palacios *et al.* (1997)
	82 [48]	10	3	4	1	92			12	Palacios & Rodriguez (Chapter 20)
C. lucifer	67 [37]	13	–	14	6	35	26	9	3	Kinzey (1977)
	74 [?]	9	< 1	16	–		> 57		14	Easley (1982)
C. personatus group										
C. personatus	55 [?]	18	22	–	5		28[11]	18[11]	6	Price & Piedade (2001a)
	57 [?]	26	2	–	15				6	Price & Piedade (2001a)
	81 [?]	18	1	–	–		27		2	Kinzey & Becker (1983)
C. melanochir	77 [34?]	17	–	–	6	> 91	69	30	11	Müller (1996a)
	85 [26]	14	< 1	< 1	–	85			12	Heiduck (1997)
C. nigrifrons	46 [?]	33	11	10	–				12	Souza *et al.* (1996)
	64 [?]	5	24	3	4	42	28	19	5	Neri (1997)

Obs.: Percentages were rounded to the nearest integer.
[1] Includes ingestion of whole fruit or fruit parts (when the % accounted for by the ingestion of seeds only is informed, it is cited in brackets); [2] includes ingestion of leaves or leaf parts in general; [3] number of plant species used as food sources; [4] number of fruit sources; [5] number of leaf sources; [6] summary data on the total contribution of each item to the diet were not provided in Wright's thesis. Therefore, they were obtained from Robinson *et al.* (1987); [7] flower consumption by *C. brunneus* was not mentioned in the table of Robinson *et al.* (1987), but represented 2% of the diet according to the original thesis; [8] number of species eaten over the course of one year; [9] average from five groups calculated from table 2.12 of Lawrence (2007); [10] maximum numbers from five study groups; [11] data from the two groups pooled together by the authors.

observed in the more folivorous titi species (Wright 1985). A preference for young leaves has been reported for many species, although mature ones and petioles may also be ingested (Kinzey 1977; Wright 1985; Crandlemire-Sacco 1988; Müller 1996a; Heiduck 1997; Price & Piedade 2001a; Palacios & Rodriguez, Chapter 20). The prey spectrum exploited by *Callicebus* spp. is discussed in Heymann and Nadjafzadeh (Chapter 19).

Other food items include flowers (petals, nectar and pollen), stems, tendrils and bark (Wright 1985; Müller 1996a; Souza *et al.* 1996; Neri 1997; Palacios *et al.* 1997; Price & Piedade 2001a; Palacios & Rodriguez, Chapter 20). The intake of soil, especially from ant mounds (*Atta* sp.), was observed 14 times in *C. melanochir* (Müller 1996a; Müller *et al.* 1997). Müller *et al.* (1997) did not find a relationship between the frequency of geophagy and the monthly ingestion of leaves. They also failed to identify significant clues for the function of this behavior when comparing ingested and not ingested soil samples, but concluded that the chemical and physical properties of the soils eaten are compatible with a toxin adsorption function. *Callicebus cupreus* sporadically consumes soil from arboreal termitaria (Tirado Herrera & Heymann, unpubl. data). No reports of drinking are found in the literature (Kinzey 1977, 1981).

Vegetal foods are exploited from trees, vines, parasites, and epiphytes (Wright 1985) from a large number of species (Table 17.3). The most important plant families in terms of percentage contribution to the diet were Myrtaceae, Sapotaceae and Moraceae for *C. melanochir* (Müller 1996a) and Myristicaceae, Euphorbiaceae and Moraceae for *C. lugens* (Palacios *et al.* 1997). Moraceae was often the most diverse food source for titis, with 13 species exploited by *C. brunneus* (Wright 1985), eight by *C. lugens* (Palacios *et al.* 1997), and seven by *C. personatus* (Price & Piedade 2001a). Other important plant families are Fabaceae (10 species, *C. brunneus*; Wright 1985), Myristicaceae (7 species, *C. lugens*; Palacios *et al.* 1997), Myrtaceae (5 species, *C. personatus*; Price & Piedade 2001a), Euphorbiaceae and Rubiaceae (4 species each, *C. melanochir*; Heiduck 1997), and Fabaceae, Rubiaceae and Sapindaceae (4 species each, *C. nigrifrons*; Neri 1997). According to Müller (1996a), 13 species of Myrtaceae and 10 species of Sapotaceae were exploited as fruit sources by *C. melanochir*, whereas Rubiaceae and Euphorbiaceae (4 species each) were the most diverse sources of leaves. Bignoniaceae, Leguminosae and Convolvulaceae are important sources of flowers (Wright 1985; Neri 1997; Price & Piedade 2001a).

Feeding trees exploited by titis are often relatively small (height < 20 m; crown diameter < 10 m) and produce fruits in a piecemeal fashion over a long fruiting season (2–6 months: Wright 1985, 1986; Müller 1996a; Palacios *et al.* 1997). Given the opportunity, however, titis also feed in larger trees (Heiduck 1997; Müller 1996a; Wright 1985). The apparent preference for small feeding trees reported by Müller (1996a) seems to be better explained by a lower availability of large trees at some study sites and/or by interference, contest competition with other primates for feeding sites that produce larger amounts of fruit over a shorter time interval (Wright 1985).

Researchers have reported that titis are chased away from feeding sites by larger monkeys (*Ateles* spp., *Lagothrix* spp. and *Cebus* spp.) and monkeys living in larger social groups (*Saimiri* spp.), and therefore avoid them at more productive fruiting sites (Wright 1985, 1986; Kinzey 1992; Defler 2004; but see Neri 1997, for the apparent lack of competition between *C. nigrifrons* and *Cebus libidinosus*). For instance, Kinzey (1992), based on Wright's data, showed that *C. brunneus* fed on *Ficus erythrosticta* mainly before and after peak production probably as a means of avoiding meeting competitors at this attractive resource. Wright (1985) considered that awaking earlier in the morning and arriving prior to competitors at large fruit sources is a strategy likely to counterbalance *C. brunneus*' competitive disadvantage during the wet season in Manu National Park.

The small feeding trees visited by *Callicebus* spp. do not produce sufficient fruit for all group members of many of the potentially competing primates and are often ignored by them (Wright 1986; Palacios *et al.* 1997). Callitrichines, however, also feed on such small trees (Terborgh 1983; Garber 1993), indicating the possibility for competition between them and titis. Although Kinzey and Becker (1983) observed *C. personatus* and *Callithrix geoffroyi* resting, traveling and even feeding together for 15 min in the same tree without signs of antagonism, and Crandlemire-Sacco (1988; see also C.H. Janson, pers. commun, cited in Kinzey 1981) and Bicca-Marques (unpublished data) observed *C. brunneus* and *C. cupreus*, respectively, frequently associating with *Saguinus fuscicollis* or *Saguinus imperator* at feeding sites with apparently minimal mutual interference, Neri (1997) reported that *C. nigrifrons* is less comfortable in the presence of *Callithrix penicillata* than of *Cebus libidinosus*.

Photo 17.3 *Callicebus barbarbrownae*, adult female carrying offspring, Lamarão, Bahia state, Brazil. Photo by Antonio Estrela.

Use of space

Family groups may range over an area of less than 1 ha in highly disturbed and fragmented forests (Mason 1968) up to about 50 ha (Neri 1997), although home range rarely exceeds 25 ha (Table 17.2). Despite being considered territorial animals that maintain intergroup spacing or defend range boundaries or resources through genus-specific long calls emitted especially in early morning (Mason 1968; Robinson 1979, 1981; Kinzey & Robinson 1983; Wright 1985; Melo & Mendes 2000; Porras 2000), varying degrees of home-range overlap were reported by several authors (< 10% in *C. lucifer*, Kinzey *et al.* 1977; 13% in *C. melanochir*, Müller 1995; 18–19% in *C. personatus*, Price & Piedade 2001b; 5–40% in *C. lugens*, Defler 1983). In addition, there are reports of adult offspring forming new families and setting up overlapping or adjacent home ranges with their natal groups' range (*C. cupreus*, Bicca-Marques *et al.* 2002) or probably occupying part of it (Easley & Kinzey 1986; see also Defler 2004).

Day range varied from 150 m in *C. brunneus* to 2697 m in *C. nigrifrons*. Mean day ranges were quite similar among species from the *C. cupreus*, *C. moloch* and *C. torquatus* groups, often about 600–700 m, but appear to be relatively greater (> 1000 m) among members of the *C. personatus* group (Table 17.2). As most studies were short term, these differences may be a result of the duration and time of the year in which the research was conducted. Wright (1985), for example, reported monthly mean path lengths ranging from 349 to 1108 m for the same *C. brunneus* study group.

The average distance traveled between consecutive feeding sites is short (57 m, *C. brunneus*, Wright 1986; 109 m, *C. melanochir*, Müller 1996a), and may involve the use of habitual routes (Wright 1985; Neri 1997; a trend of using the same pathways also was reported in captivity, Fragaszy 1980). According to Wright (1985, p. 86), "*Callicebus* often traplines many small-crowned fruit trees of a single species." Considering that the utility of visual cues within the forest canopy is limited to about 20–30 m (Dominy *et al.* 2001), and that olfactory cues may play only a minor role in both between- and within-patch foraging decisions by diurnal monkeys, including *C. cupreus* (Garber & Hannon 1993; Dominy *et al.* 2001; Bicca-Marques & Garber 2004), it is very likely that titis rely on spatial information to navigate between feeding patches.

Supporting evidence that they are capable of perceiving, storing and retrieving information on the spatial distribution of food rewards in their foraging decision-making process comes from experimental field and laboratory studies (Andrews 1988; Bicca-Marques 2000, 2005; Bicca-Marques & Garber 2004). Red titis successfully adopted a win–return strategy in their decision-making process when the spatial information was predictable over time, but promptly stopped applying it when the location of food rewards changed randomly across experimental trials (Bicca-Marques 2005). In addition, Andrews (1988) observed his study subjects discriminated among feeding sites varying widely in food availability.

A preference for the lowest forest strata, often below 10 m, has been reported for many species (*C. brunneus*, *C. caligatus*, *C. cinerascens*, *C. cupreus*, *C. donacophilus*, *C. moloch* and *C. nigrifrons*; Bicca-Marques unpubl. data; Terborgh 1983; Wright 1985; Cameron *et al.* 1989; Neri 1997; Buchanan-Smith *et al.* 2000; Ferrari *et al.* 2000), whereas *C. lucifer* appears to use higher forest levels (Lawler *et al.* 2006). Wright (1985) relates the greater time spent resting and at lower levels as a predator avoidance strategy. Titis sporadically descend to the ground (Kinzey *et al.* 1977; Kinzey 1981; Müller *et al.* 1997; Neri 1997; Felton *et al.* 2006).

While daytime activities are concentrated at the lowest forest levels, night sleep takes place at higher strata. Kinzey *et al.* (1977) reported the use of emergent trees that were located both in the centre and along borders of *C. lucifer*'s home range. Their study subjects often selected one of the primary horizontal branches of the tree crown at an average height of 25 m (range: 17–33 m, $N = 12$), located slightly higher than the surrounding closed canopy, but protected from raptor detection by the branches above. A similar selection of sleeping trees was observed in *C. personatus* (mean height = 32 m, range: 27–38 m, $N = 9$; Kinzey & Becker 1983). Protection against predator detection in general also may be achieved by selecting sleeping sites in dense foliage or vine tangles. There is no report, however, of titis using of holes for sleeping (Wright 1985). Both *C. nigrifrons* and *C. brunneus* slept in vine tangles and relatively exposed trunks (Wright 1985; Neri 1997). Mean sleeping tree height used by *C. brunneus* was 21 m (range: 12–30 m, $N = 38$; Wright 1985). All sleeping trees used by two *C. brunneus* study groups were located at least 100 m from the border of their home ranges, allowing Wright (1985) to suggest that the location of neighboring groups appears to have a greater influence on sleeping site selection than protection from predators or food distribution. Thirty-three different trees often exceeding 10 m in height were used by *C. nigrifrons* during 42 nights (Neri 1997). *Callicebus melanochir*, on the other hand, was more selective and used only six emergent trees (five in undisturbed forest and one in selectively logged forest) characterized by dense foliage for sleeping over 47 nights (Heiduck 2002).

Predators

As mentioned before, several aspects of titi monkey ecology and behavior have been related to a predator avoidance strategy. Potential predators of titis include diurnal birds of prey, felids, snakes and capuchin monkeys (*Cebus* spp.) (Kinzey 1981; Wright 1985; Defler 2004; Cisneros-Heredia *et al.* 2005). Cisneros-Heredia *et al.* (2005) observed a boa (*Boa constrictor*) preying upon a *C. discolor* at the Tiputini Biodiversity Station in the Ecuadorian Amazon. Wright (1985) cited hawk eagles (*Spizaetus tyrannus* and *S. ornatus*) preying upon titis at Cocha Cashu. Unsuccessful attacks by three raptor species (*Accipiter bicolor*, *Morphnus guianensis* and *Spizaetus ornatus*) and one felid (*Felis pardalis*) were also reported by her. A successful raptor attack was also observed by C. Freese (pers. commun. to Kinzey *et al.* 1977). Lawrence (2003) observed a female being killed by a *Cebus* sp. in Peru, and an infant *C. moloch* was killed and eaten by an adult male *Cebus apella* on an island produced by the flooding of Tucuruí dam in the State of Pará in Brazil (Sampaio & Ferrari 2005). The latter authors considered that the event was facilitated by relatively high primate population densities produced by the isolation of the study site.

Conservation

As with primates in general, including other pitheciines, the conservation of titis is hampered by habitat destruction, hunting and the pet trade, whose importance may vary geographically and culturally. According to the 2010 IUCN Red List, the most threatened species live in the Atlantic forest: *Callicebus barbarabrownae* is Critically Endangered (Veiga *et al.* 2008e; see Photo 17.3), *C. coimbrai* is Endangered (Veiga *et al.* 2008f) and *C. personatus* and *C. melanochir* are Vulnerable (Veiga *et al.* 2008b and 2008d, respectively), whereas *C. nigrifrons* is Near Threatened (Veiga *et al.* 2008c) and *C. stephennashi* is Data Deficient (Veiga 2008). *Callicebus modestus* (Veiga *et al.* 2008g), *C. oenanthe* (Veiga *et al.* 2008a) and *C. olallae* (Veiga *et al.* 2008h) are also Endangered, whereas *C. medemi* and *C. ornatus* are Vulnerable (Veiga & Palacios 2008a and 2008b, respectively). The remaining species are classified as Least Concern (IUCN 2010).

Because of their widespread ability to thrive in forest fragments and anthropogenically disturbed environments (see "Habitat" above), titis may reach their highest densities in these habitats (when not hunted by local people (see van Roosmalen *et al.* 2002; and also "Population ecology" above). According to van Roosmalen *et al.* (2002), these areas may offer optimal habitat and permanent food supply for titis. Although this hypothesis may hold true, the success of titis in forests near human settlements possibly results from a combination of this adaptability and a release from (or reduction of) the competition with larger (hunted) primates (see "Foraging ecology" above) through a process of density compensation (see Peres & Dolman 2000, for weak evidence for the influence of this community-level phenomenon on primate assemblages).

Acknowledgments

We thank the editors for inviting us to contribute to this volume. JCBM also thanks his wife and children for their support and for understanding his absence during some family time while writing this chapter and the Brazilian National Research Council for financial support (CNPq # 306090/2006–6 and 303154/2009–8).

References

Andrews, M.W. (1988). Selection of food sites by *Callicebus moloch* and *Saimiri sciureus* under spatially and temporally varying food distribution. *Learning and Motivation*, **19**, 254–268.

Aquino, R., Terrones, W., Cornejo, F., *et al.* (2008). Geographic distribution and possible taxonomic distinction of *Callicebus torquatus* populations (Pitheciidae: Primates) in Peruvian Amazônia. *American Journal of Primatology*, **70**, 1181–1186.

Assunção, M.L. (2005). Estudo preliminar da freqüência e duração do comportamento vocal e social em *Callicebus nigrifrons* – Spix, 1823 na RPPN Serra do Caraça-MG. Unpublished Bachelor's dissertation, Pontifícia Universidade Católica de Minas Gerais.

Bennett, C.L., Leonard, S. & Carter, S. (2001). Abundance, diversity, and patterns of distribution of primates on the Tapiche river in Amazonian Peru. *American Journal of Primatology*, **54**, 119–126.

Bicca-Marques, J.C. (2000). Cognitive aspects of within-patch foraging decisions in wild diurnal and nocturnal New World monkeys. Unpublished PhD thesis, University of Illinois at Urbana-Champaign.

Bicca-Marques, J.C. (2005). The win–stay rule in within-patch foraging decisions in free-ranging titi monkeys (*Callicebus cupreus cupreus*) and tamarins (*Saguinus imperator imperator* and *S. fuscicollis weddelli*). *Journal of Comparative Psychology*, **119**, 343–351.

Bicca-Marques, J.C. & Garber, P.A. (2004). Use of spatial, visual, and olfactory information during foraging in wild nocturnal and diurnal anthropoids: a field experiment comparing *Aotus*, *Callicebus*, and *Saguinus*. *American Journal of Primatology*, **62**, 171–187.

Bicca-Marques, J.C., Garber, P.A. & Azevedo-Lopes, M.A.O. (2002). Evidence of three resident adult male group members in a species of monogamous primate, the red titi monkey (*Callicebus cupreus*). *Mammalia*, **66**, 138–142.

Bicca-Marques, J.C., Garber, P.A. & Azevedo-Lopes, M.A.O. (2003). An experimental field study of social foraging in wild titi monkeys (*Callicebus cupreus*) under conditions of changing patch quality. *Revista de Etologia*, **Supplement**, 35.

Bossuyt, F. (2002). Natal dispersal of titi monkeys (*Callicebus moloch*) at Cocha Cashu, Manu National Park, Peru. *American Journal of Physical Anthropology*, **Supplement 34**, **47**.

Buchanan-Smith, H.M., Hardie, S.M., Caceres, C., *et al.* (2000). Distribution and forest utilization of *Saguinus* and other primates of the Pando Department, Northern Bolivia. *International Journal of Primatology*, **21**, 353–379.

Cameron, R., Wiltshire, C., Foley, C., *et al.* (1989). Goeldi's monkey and other primates in northern Bolivia. *Primate Conservation*, **10**, 62–70.

Campos, Y.F. (1990). Preferencias de habitat, aspectos reproductivos y comportamiento de canto, como factores determinantes en la territorialidad de *Callicebus torquatus*, en la Amazonia Ecuatoriana. Tesis de Licenciatura, Pontifica Universidad Católica del Ecuador, Quito.

Cäsar, C. & Young, R.J. (2008). A case of adoption in a wild group of black-fronted titi monkeys (*Callicebus nigrifrons*). *Primates*, **49**, 146–148.

Cäsar, C., Franco, E.S., Soares, G.C.N., *et al.* (2008). Observed case of maternal infanticide in a wild group of black-fronted titi monkeys (*Callicebus nigrifrons*). *Primates*, **49**, 143–145.

Chiarello, A.G. (2000). Density and population size of mammals in remnants of Brazilian Atlantic forest. *Conservation Biology*, **14**, 1649–1657.

Cisneros-Heredia, D.F., León-Reyes, A. & Seger, S. (2005). *Boa constrictor* predation on a titi monkey, *Callicebus discolor*. *Neotropical Primates*, **13**, 11–12.

Crandlemire-Sacco, J. (1988). An ecological comparison of two sympatric primates – *Saguinus fuscicollis* and *Callicebus moloch* of Amazonian Peru. *Primates*, **29**, 465–475.

Defler, T.R. (1983). Some population characteristics of *Callicebus torquatus lugens* (Humboldt, 1812) (Primates: Cebidae) in eastern Colombia. *Lozania*, **38**, 1–9.

Defler, T.R. (1994). *Callicebus torquatus* is not a white-sand specialist. *American Journal of Primatology*, **33**, 149–154.

Defler, T.R. (2004). *Primates of Colombia*. Bogotá: Conservation International.

DeLuycker, A.M. (2007). The ecology and behavior of the Rio Mayo titi monkey (*Callicebus oenanthe*) in the Alto Mayo, northern Peru. Unpublished PhD thesis, Washington University.

Dominy, N.J., Lucas, P.W., Osório, D., *et al.* (2001). The sensory ecology of primate food perception. *Evolutionary Anthropology*, **10**, 171–186.

Easley, S.P. (1982). Ecology and behavior of *Callicebus torquatus*, Cebidae, Primates. Unpublished PhD thesis, Washington University.

Easley, S.P. & Kinzey, W.G. (1986). Territorial shift in the yellow-handed titi monkey (*Callicebus torquatus*). *American Journal of Primatology*, **11**, 307–318.

Encarnación, F. & Castro, N. (1990). Informe preliminar sobre censo de primates no humanos em el sur oriente peruano: Iberia y Iñapari (Departamento de Madre de Dios), mayo 15–junio 14, 1978. In *La Primatologia en el Peru*, ed. N.E. Castro-Rodríguez. Lima: Proyecto Peruano de Primatologia, pp. 57–67.

Felton, A., Felton, A.M., Wallace, R.B., *et al.* (2006). Identification, behavioral observations, and notes on the distribution of the titi monkeys *Callicebus modestus* Lönnberg, 1939 and *Callicebus olallae*, Lonnberg, 1939. *Primate Conservation*, **20**, 41–46.

Ferrari, S.F., Iwanaga, S., Messias, M.R., *et al.* (2000). Titi monkeys (*Callicebus* spp., Atelidae: Platyrrhini) in the Brazilian State of Rondônia. *Primates*, **41**, 229–234.

Fragaszy, D.M. (1980). Comparative studies of squirrel monkeys (*Saimiri*) and titi monkeys (*Callicebus*) in travel tasks. *Zeitschrift für Tierpsychologie*, **54**, 1–36.

Garber, P.A. (1993). Seasonal patterns of diet and ranging in two species of tamarin monkeys: stability vs. variability. *International Journal of Primatology*, **14**, 145–166.

Garber, P.A. & Hannon, B. (1993). Modeling monkeys: a comparison of computer-generated and naturally occurring foraging patterns in two species of Neotropical primates. *International Journal of Primatology*, **14**, 827–852.

Heiduck, S. (1997). Food choice in masked titi monkeys (*Callicebus personatus melanochir*): selectivity or opportunism? *International Journal of Primatology*, **18**, 487–502.

Heiduck, S. (2002). The use of disturbed and undisturbed forest by masked titi monkeys *Callicebus personatus melanochir* is proportional to food availability. *Oryx*, **36**, 133–139.

Hilton-Taylor, C., Rylands, A.B. & Aguiar, J.M. (2004). IUCN red list – Neotropical primates. *Neotropical Primates*, **12**, 33–35.

Hoffman, K.A. (1998). Transition from juvenile to adult stages of development in titi monkeys (*Callicebus moloch*). Unpublished PhD thesis, University of California at Davis.

IUCN. (2010). *2010 IUCN Red List of Threatened Species*. <www.iucnredlist.org>. Downloaded on 26 May 2010.

Jantschke, B., Welker, C. & Klaiber-Schuh, A. (1995). Notes on breeding of the titi monkey *Callicebus cupreus*. *Folia Primatologica*, **65**, 210–213.

Kinzey, W.G. (1977). Diet and feeding behaviour of *Callicebus torquatus*. In *Primate Ecology: Studies of Feeding and Ranging Behaviour in Lemurs, Monkeys and Apes*, ed. T.H. Clutton-Brock. London: Academic Press, pp. 127–151.

Kinzey, W.G. (1981). The titi monkeys, genus *Callicebus*. In *Ecology and Behavior of Neotropical Primates*, vol. **1**, ed. A.F. Coimbra-Filho & R.A. Mittermeier. Rio de Janeiro: Academia Brasileira de Ciências, pp. 241–276.

Kinzey, W.G. (1992). Dietary and dental adaptations in the Pitheciinae. *American Journal of Physical Anthropology*, **88**, 499–514.

Kinzey, W.G. (1997). *Callicebus*. In *New World Primates: Ecology, Evolution, and Behavior*, ed. W.G. Kinzey. New York, NY: Aldine de Gruyter, pp. 213–221.

Kinzey, W.G. & Becker, M. (1983). Activity pattern of the masked titi monkey, *Callicebus personatus*. *Primates*, **24**, 337–343.

Kinzey, W.G. & Robinson, J.G. (1983). Intergroup loud calls, range size, and spacing in *Callicebus torquatus*. *American Journal of Physical Anthropology*, **60**, 539–544.

Kinzey, W.G. & Wright, P.C. (1982). Grooming behavior in the titi monkey (*Callicebus torquatus*). *American Journal of Primatology*, **3**, 267–275.

Kinzey, W.G., Rosenberger, A.L., Heisler, P.S., *et al.* (1977). A preliminary field investigation of the yellow-handed titi monkey, *Callicebus torquatus torquatus*, in northern Peru. *Primates*, **18**, 159–181.

Knogge, C. & Heymann, E.W. (1995). Field observation of twinning in the dusky titi monkey, *Callicebus cupreus*. *Folia Primatologica*, **65**, 118–120.

Lawler, R.R., Ford, S.M., Wright, P.C., *et al.* (2006). The locomotor behavior of *Callicebus brunneus* and *Callicebus torquatus*. *Folia Primatologica*, **77**, 228–239.

Lawrence, J.M. (2003). Preliminary report on the natural history of brown titi monkeys (*Callicebus brunneus*) at the Los Amigos Research Station, Madre de Dios, Peru. *American Journal of Physical Anthropology*, **36** (Suppl.), 136.

Lawrence, J.M. (2007). Understanding the pair bond in brown titi monkeys (*Callicebus brunneus*): male and female reproductive interests. Unpublished PhD thesis, Columbia University.

Marinho-Filho, J. & Veríssimo, E.W. (1997). The rediscovery of *Callicebus personatus barbarabrownae* in northeastern Brazil with a new western limit for its distribution. *Primates*, **38**, 429–433.

Mark, M.M. (2003). Some observations on *Callicebus oenanthe* in the upper Río Mayo valley, Peru. *Neotropical Primates*, **11**, 183–187.

Martins, E.S., Ayres, J.M. & Valle, M.B.R. (1988). On the status of *Ateles belzebuth marginatus* with notes on other primates of the Iriri river basin. *Primate Conservation*, **9**, 87–91.

Mason, W.A. (1968). Use of space by *Callicebus* groups. In *Primates: Studies in Adaptation and Variability*, ed. P.C. Lee. New York, NY: Holt, Rinehart and Winston, pp. 200–216.

Melo, F.R. & Mendes, S.L. (2000). Emissão de gritos longos por grupos de *Callicebus nigrifrons* e suas reações a playbacks. In *A Primatologia no Brasil, vol. 7*, ed. C. Alonso & A. Langguth. João Pessoa: SBPr & Editora Universitária, pp. 215–222.

Mendoza, S.D. & Mason, W.A. (1986). Parental division of labour and differentiation of attachments in a monogamous primate (*Callicebus moloch*). *Animal Behaviour*, **34**, 1336–1347.

Michalski, F. & Peres, C.A. (2005). Anthropogenic determinants of primate and carnivore local extinctions in a fragmented forest landscape of southern Amazonia. *Biological Conservation*, **124**, 383–396.

Müller, K.-H. (1995). Ranging in masked titi monkeys (*Callicebus personatus*) in Brazil. *Folia Primatologica*, **65**, 224–228.

Müller, K.-H. (1996a). Diet and feeding ecology of masked titis (*Callicebus personatus*). In *Adaptive Radiations of Neotropical Primates*, ed. M.A. Norconk, A.L. Rosenberger & P.A. Garber. New York, NY: Plenum Press, pp. 383–401.

Müller, K.-H. (1996b). Emigration of a masked titi monkey (*Callicebus personatus*) from an established group and the foundation of a new group. *Neotropical Primates*, **4**, 19–21.

Müller, K.-H. & Pissinatti, A. (1995). Ecology and feeding behavior of masked titi monkeys. *Neotropical Primates*, **3**(2), 51–52.

Müller, K.-H., Ahl, C. & Hartmann, G. (1997). Geophagy in masked titi monkeys (*Callicebus personatus melanochir*) in Brazil. *Primates*, **38**, 69–77.

Neri, F.M. (1997). Manejo de Callicebus personatus, Geoffroy 1812, resgatados: Uma tentativa de reintrodução e estudos ecológicos de um grupo silvestre na Reserva do Patrimônio Natural Galheiro – Minas Gerais. Unpublished Master's dissertation, Universidade Federal de Minas Gerais.

Norconk, M.A. (2011). Sakis, uakaris, and titi monkeys: behavioral diversity in a radiation of primate seed predators. In *Primates in Perspective*, 2nd edn, ed. C.J. Campbell, A. Fuentes, K.C. MacKinnon, S.K. Bearder & R.M. Stumpf. New York, NY: Oxford University Press, pp. 122–139.

Palacios, E., Rodriguez, A. & Defler, T.R. (1997). Diet of group of *Callicebus torquatus lugens* (Humboldt, 1812) during the annual resource bottleneck in Amazonian Colombia. *International Journal of Primatology*, **18**, 503–522.

Peres, C.A. (1988). Primate community structure in western Brazilian Amazônia. *Primate Conservation*, **9**, 83–87.

Peres, C.A. & Dolman, P.M. (2000). Density compensation in neotropical primate communities: evidence from 56 hunted and nonhunted Amazonian forests of varying productivity. *Oecologia*, **122**, 175–189.

Polanco-Ochoa, R. (1993). Use of space by *Callicebus cupreus ornatus* (Primates: Cebidae) in Macarena, Colombia. *Field Studies of New World Monkeys, La Macarena, Colombia*, **8**, 19–31.

Porras, M. (2000). Comunicación vocal y su relación com las actividades, estructura social y contexto comportamental em *Callicebus cupreus ornatus*. In *A Primatologia no Brasil, vol. 7*, ed. C. Alonso & A. Langguth. João Pessoa: SBPr e Editora Universitária, pp. 265–274.

Price, E.C. & Piedade, H.M. (2001a). Diet of northern masked titi monkeys (*Callicebus personatus*). *Folia Primatologica*, **72**, 335–338.

Price, E.C. & Piedade, H.M. (2001b). Ranging behavior and intraspecific relationships of masked titi monkeys (*Callicebus personatus personatus*). *American Journal of Primatology*, **53**, 87–92.

Robinson, J.G. (1979). Vocal regulation of use of space by groups of titi monkeys *Callicebus moloch*. *Behavioral Ecology and Sociobiology*, **5**, 1–15.

Robinson, J.G. (1981). Vocal regulation of inter- and intragroup spacing during boundary encounters in the titi monkey, *Callicebus moloch*. *Primates*, **22**, 161–172.

Robinson, J.G., Wright, P.C. & Kinzey, W.G. (1987). Monogamous cebids and their relatives: Intergroup calls and spacing. In *Primate Societies*, ed. B.B. Smuts, D.L. Cheney, R.M. Seyfarth, R.W. Wrangham & T.T. Struhsaker. Chicago, IL: The University of Chicago Press, pp. 44–53.

Rylands, A.B. & Mittermeier, R.A. (2009). The diversity of the New World primates (Platyrrhini): an annotated taxonomy. In *South American Primates: Comparative Perspectives in the Study of Behavior, Ecology, and Conservation*, ed. P.A. Garber, A. Estrada, J.C. Bicca-Marques, E.W. Heymann & K.B. Strier. New York, NY: Springer, pp. 23–54.

Sampaio, D.T. & Ferrari, S.F. (2005). Predation of an infant titi monkey (*Callicebus moloch*) by a tufted capuchin (*Cebus apella*). *Folia Primatologica*, **76**, 113–115.

Silveira, G., Bicca-Marques, J.C. & Nunes, C.A. (1998). On the capture of titi monkeys (*Callicebus cupreus*) using the Peruvian method. *Neotropical Primates*, **6**, 114–115.

Smith, R.J. & Jungers, W.L. (1997). Body mass in comparative primatology. *Journal of Human Evolution*, **32**, 523–559.

Souza, S.B., Martins, M.M. & Setz, E.Z.F. (1996). Activity pattern and feeding ecology of sympatric masked titi monkeys and buffy tufted-ear marmosets. In *Abstracts of the XVIth Congress of the International Primatological Society/XIXth Conference of the American Society of Primatologists*, abstract #155.

Stallings, J.R. & Robinson, J.G. (1991). Disturbance, forest heterogeneity and primate communities in a Brazilian Atlantic forest park. In *A Primatologia no Brasil, vol. 3*, ed. A.B. Rylands & A.T. Bernardes. Belo Horizonte: Fundação Biodiversitas, pp. 357–368.

Terborgh, J. (1983). *Five New World Primates*. Princeton, NJ: Princeton University Press.

Tirado-Herrera, E.R. & Heymann, E.W. (2004). Does mom need more protein? Preliminary observations on differences in diet composition in a pair of red titi monkeys (*Callicebus cupreus*). *Folia Primatologica*, **75**, 150–153.

Torres de Assumpção, C. (1983). An ecological study of the primates of Southeastern Brazil, with a reappraisal of *Cebus apella races*. Unpublished PhD thesis, University of Edinburgh.

van Roosmalen, M.G., van Roosmalen, T. & Mittermeier, R.A. (2002). A taxonomic review of the titi monkeys, genus *Callicebus* Thomas, 1903, with the description of two new species, *Callicebus bernhardi* and *Callicebus stephennashi*, from Brazilian Amazonia. *Neotropical Primates*, **10** (Suppl.), 1–52.

Veiga, L.M. (2008). *Callicebus stephennashi*. In *2010 IUCN Red List of Threatened Species*. <www.iucnredlist.org>. Downloaded May 26, 2010.

Veiga, L.M. & Palacios, E. (2008a). *Callicebus medemi*. In *2010 IUCN Red List of Threatened Species*. <www.iucnredlist.org>. Downloaded May 26, 2010.

Veiga, L.M. & Palacios, E. (2008b). *Callicebus ornatus*. In *2010 IUCN Red List of Threatened Species*. <www.iucnredlist.org>. Downloaded May 26, 2010.

Veiga, L.M., DeLuycker, A.M., Bóveda-Penalba, A.J., *et al.* (2008a). *Callicebus oenanthe*. In *2010 IUCN Red List of Threatened Species*. <www.iucnredlist.org>. Downloaded May 26, 2010.

Veiga, L.M., Ferrari, S.F., Kierulff, C.M., *et al.* (2008b). *Callicebus personatus*. In *2010 IUCN Red List of Threatened Species*. <www.iucnredlist.org>. Downloaded May 26, 2010.

Veiga, L.M., Kierulff, C.M., Oliveira, M.M., *et al.* (2008c). *Callicebus nigrifrons*. In *2010 IUCN Red List of Threatened Species*. <www.iucnredlist.org>. Downloaded May 26, 2010.

Veiga, L.M., Printes, R.C., Ferrari, S.F., *et al.* (2008d). *Callicebus melanochir*. In *2010 IUCN Red List of Threatened Species*. <www.iucnredlist.org>. Downloaded May 26, 2010.

Veiga, L.M., Printes, R.C., Rylands, A.B., *et al.* (2008e). *Callicebus barbarabrownae*. In *2010 IUCN Red List of Threatened Species*. <www.iucnredlist.org>. Downloaded May 26, 2010.

Veiga, L.M., Sousa, M.C., Jerusalinsky, L., *et al.* (2008f). *Callicebus coimbrai*. In *2010 IUCN Red List of Threatened Species*. <www.iucnredlist.org>. Downloaded May 26, 2010.

Veiga, L.M., Wallace, R.B. & Martinez, J. (2008g). *Callicebus modestus*. In *2010 IUCN Red List of Threatened Species*. <www.iucnredlist.org>. Downloaded May 26, 2010.

Veiga, L.M., Wallace, R.B. & Martinez, J. (2008h). *Callicebus olallae*. In *2010 IUCN Red List of Threatened Species*. <www.iucnredlist.org>. Downloaded May 26, 2010.

Wagner, M., Castro, F. & Stevenson, P.R. (2009). Habitat characterization and population status of the dusky titi (*Callicebus ornatus*) in fragmented forests, Meta, Colombia. *Neotropical Primates*, **16**, 18–24.

Wallace, R.B., Gómez, H., Felton, A., *et al.* (2006). On a new species of titi monkey, genus *Callicebus* Thomas (Primates, Pitheciidae), from Western Bolivia with preliminary notes on distribution and abundance. *Primate Conservation*, **20**, 29–39.

Welker, C., Jantschke, B. & Klaiber-Schuh, A. (1998a). Behavioural data on the titi monkey *Callicebus cupreus* and the owl monkey *Aotus azarae boliviensis*. A contribution to the discussion on the correct systematic classification of these species. Part III: Living in family groups. *Primate Report*, **51**, 29–42.

Welker, C., Jantschke, B. & Klaiber-Schuh, A. (1998b). Behavioural data on the titi monkey *Callicebus cupreus* and the owl monkey *Aotus azarae boliviensis*. A contribution to the discussion on the correct systematic classification of these species. Part IV: Breeding biology. *Primate Report*, **51**, 43–53.

Wright, P.C. (1985). The costs and benefits of nocturnality for *Aotus trivirgatus* (the night monkey). Unpublished PhD thesis, City University of New York.

Wright, P.C. (1986). Ecological correlates of monogamy in *Aotus* and *Callicebus*. In *Primate Ecology and Conservation*, ed. J.G. Else & P.C. Lee. Cambridge: Cambridge University Press, pp. 159–167.

Chapter 18

Costs of foraging in the Southern Bahian masked titi monkey (*Callicebus melanochir*)

Stefanie Heiduck

Introduction

Food abundance varies in time for most primate species (Hladik 1988; van Schaik *et al.* 1993; van Schaik & Pfannes 2005) and there are different strategies to cope with fluctuating abundances of food resources (Hemingway & Bynum 2005). Strategies to overcome food shortage are of special interest for behavioral ecologists because times of food scarcity may have powerful influences on foraging adaptations and diet, and thus the ecology of a species (Rosenberger 1992; Lambert 2007).

Classical Optimal Foraging Theory (OFT) assumes that natural selection favors the efficiency of foraging behavior so as to maximize net energy gain, i.e. the benefit/costs ratio. Within this framework "benefit" is defined as the energy gained by food intake. In contrast, "costs" are defined as the energy invested to obtain food and include search and handling costs (Pyke *et al.* 1977; Stephens & Krebs 1986; Krebs & Davies 1993). Food resources for frugivorous primates have been described as patchily distributed (e.g. Oates 1987). Therefore, search cost is the energy invested in travel between patches, whereas handling cost is the energy spent to obtain food within the patches (MacArthur & Pianka 1966; NcNair 1983). In a seasonal environment there are two major strategies for primates to overcome times of food shortage: (1) a dietary switch to more profitable food items, and (2) a variation of energy investment in foraging. While the first would improve the income or the benefit side of the foraging balance, the second strategy would effect the cost side of the balance (Hemingway & Bynum 2005; Lambert 2007).

In this chapter I will analyze the costs of foraging behavior for the Southern Bahian masked titi monkey (*Callicebus melanochir*). Titi monkeys, genus *Callicebus*, are small arboreal primates who live in groups of 2–5 individuals, consisting of an adult pair and its non-reproductive offspring. Reproduction is seasonal with one offspring per year. Titi monkeys are territorial and highly frugivorous (Kinzey 1981). *C. melanochir* is one of the five masked titi monkeys (*personatus* group) living in the Atlantic rain forest of eastern Brazil (van Roosmalen *et al.* 2002). There are only few studies on the ecology and behavior of the *personatus* group (Kinzey & Becker 1983; Price & Piedade 2001a, 2001b) with only two focusing particularly on *C. melanochir* (Müller 1995, 1996; Heiduck 1997, 2002). The diet of *C. melanochir* consists of fleshy fruit parts (51.6%), seeds (26.4%) and young leaves (14.1%) (Heiduck 1997). Phytochemical analysis showed no major differences in chemical composition of the food types. Food selection was not affected by food chemicals or energy content of food species. The main factor of food choice was the abundance of the plant species. Thus, *C. melanochir* was classified as a non-selective or opportunistic forager (Heiduck 1997).

The aim of this study is to test hypotheses from two Optimal Foraging (OF) models using variable regimes, high and low food availability, to assess the importance of handling and search costs for foraging patterns of *C. melanochir*.

The Marginal Value Theorem of Charnov (1976) makes predictions about the time an animal should spend within a food patch, called patch residence time. This model assumes that food intake rate decreases with the time an animal spends within a food patch because of patch depletion. An animal should leave a depleting food resource when its actual energy intake rate drops to the potential average energy intake rate for the environment. In a seasonal habitat with fluctuating food resources, therefore, an increase in patch residence times for patches of the same quality when overall food availability is low would be expected.

The second model makes predictions concerning search costs between patches (Norberg 1977). Under the assumption that patch use patterns remain the same, the model predicts that in times of lower food availability, less energy should be invested in searching. It would therefore be expected that *C. melanochir* would invest less energy in travel between food patches when food is scarce if patch use strategies do not vary.

Methods

Study site

The study site is located in the Atlantic rain forest near the town of Una, in the southern part of the state of Bahia, Brazil (15°20′S, 39°05′W). It is an isolated forest fragment of 80 ha

Evolutionary Biology and Conservation of Titis, Sakis and Uacaris, eds. Liza M. Veiga, Adrian A. Barnett, Stephen F. Ferrari and Marilyn A. Norconk. Published by Cambridge University Press.

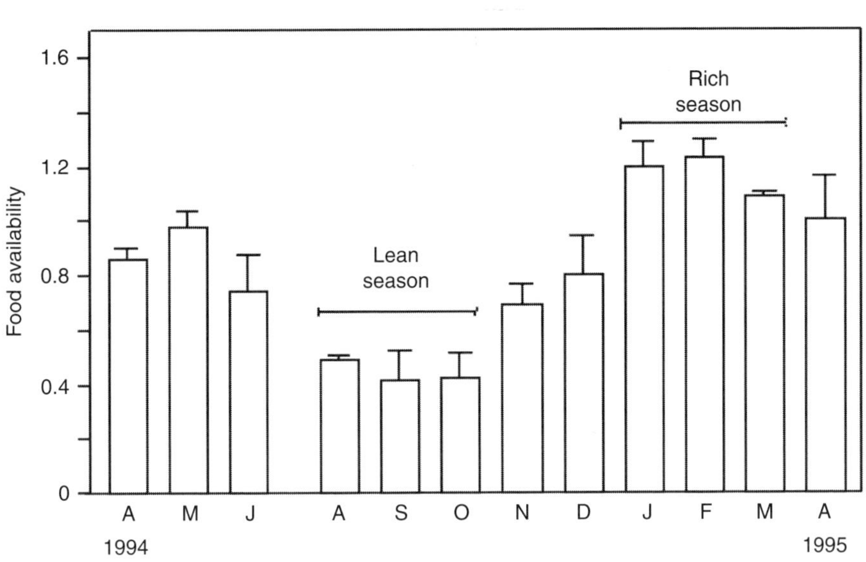

Figure 18.1 Monthly variation in food availability. Means and standard deviations from 2–3 phenological observation days per month ($n = 26$) on 75 trees. Availability scores: 0 = none to 4 = many food items per tree. Lean and rich season distinguished with posthoc Scheffé test after one-way ANOVA.

with a high degree of anthropogenic disturbance (for details, see Heiduck 2002) located at the Lemos Maia Experimental Station of the local cocoa growing authority Commissão Executiva do Plano da Lavoura Cacaueira (CEPLAC). Two other primate species, the golden-headed lion tamarin (*Leontopithecus chrysomelas*) and Wied's tufted-eared marmoset (*Callithrix kuhlii*), were seen at the site. During the study, four groups of *C. melanochir* lived in the fragment.

Temporal variation in food availability was determined by phenological observations of plant species known to be used by *C. melanochir* (for details, see Heiduck 1997). Food availability varied between months and a lean season of low food availability could be distinguished from a rich season of higher food availability (Figure 18.1). There is no seasonal variation in precipitation rates in the study area, but a clear seasonal shift in temperature which is positively correlated with food availability (Heiduck 1997).

The study focused on one group of *C. melanochir* consisting of 3–4 individuals (one adult male, one adult female, one juvenile and one infant born in September 1994). During the study period the group used a home range of about 22 ha, which was mapped with 1500 labeled trees.

Behavioral observations

I observed the study group during one year from April 1994 through April 1995, except for July 1994. Behavioral data were collected one day per week from dawn to dusk for a total of 47 observation days (4 days per month, but 3 in April 1995). Using instantaneous scan sampling (Altmann 1974), I recorded the activity of each group member every 5 min as well as the location of the group within the home range.

Following the standard procedures in behavioral ecology, time and travel distances are used as a substitute for actual energy expenses (Krebs & Davies 1993). As a substitute for handling costs, I measured feeding time, which is the time an animal spends manipulating and ingesting food items. I distinguished between the feeding time within one food patch, called patch residence time, and daily feeding time, irrespective of how many patches the animals visited. Feeding times were extrapolated from the instantaneous sampling scores by multiplying them by five minutes and represent group means. Daily means were calculated for patch residence times. Other behavioral activities were calculated in the same way as daily feeding time. Search costs were measured by the daily distances travelled by the study group. The daily distance traveled is the sum of distances between all 5-min instantaneous sampling locations per day. They were calculated with Pythagoras' theorem using the coordinates of the mapped trees.

I defined one food tree as a patch. Food trees (i.e. any tree that was entered and eaten in) were marked and measured for patch size, which is the diameter in breast height (dbh) as this correlates with crown volume ($r = 0.78$, $p < 0.001$, $n = 474$). Whenever an animal entered a food tree the amount of food within the patch relative to the patch size was estimated by a five score scheme (0 = none; 4 = many food items). I calculated means for patch size and amount of food within a patch used by the study group on one observation day as daily means.

I analyzed the behavior of the study group and the characteristics of the food patches in relation to the two seasons. Each season of 3 months was represented by 12 observation days ($n = 24$). Data were tested for normal distribution with Kolmogorov–Smirnov and as there was no deviation from normality parametric statistics were performed with Statistica 4.5 (Stat. Soft Inc., 1993).

Results

Patch use

I compared the number of patches used per day, patch size, amount of food within patches and patch residence times between the lean and the rich season to assess information on different patch use strategies in different seasons. During the entire study period the animals used on average 21.9 (± 5.6) food patches per day ($n = 47$ days). In the lean season the study group used fewer patches (19.3 ± 4.7) per day than in the rich season (27.3 ± 4.7; $t = -4.20$, $p < 0.001$, $n = 24$) (Figure 18.2a). The mean patch size, i.e. mean dbh of the trees used by the group during the study period, was 21.6 cm (± 6.2 cm) and patch size did not vary between seasons ($t = -0.83$, n.s., $n = 24$). The amount of food within patches was significantly lower in the lean season than in the rich season ($t = -2.90$, $p < 0.01$, $n = 24$) (Figure 18.2b). Mean patch residence time was 5.5 min (± 2.0 min) per patch during the study period. During the lean season mean patch residence time was significantly higher (6.5 ± 2.4 min) than during the rich season (4.2 ± 1.3 min; $t = 2.86$, $p < 0.01$, $n = 24$) (Figure 18.2c).

Activity budgets and travel distances

To assess information on general energy expenditure, I compared the energy-consuming activities feeding, playing and traveling in different seasons. In total, the group spent 4 h 20 min (± 40 min) per day for those activities and no difference was found between seasons ($t = 0.91$, n.s., $n = 24$).

However, a detailed analysis of the allocation of the activities revealed some differences between seasons. During the study period the animals spent 2 h 30 min (± 40 min) feeding, 1 h 33 min (± 28 min) traveling and 3.7 min (± 4.0 min) playing per day on average ($n = 47$). In the lean season, daily feeding time (3 h 09 min ± 26 min) increased significantly compared to the rich season (2 h 10 min ± 26 min; $t = 5.13$, $p < 0.001$, $n = 24$) (Figure 18.3a) while playing time decreased (lean: 0.1 ± 0.1 min; rich: 6.5 ± 3.3 min; $t = -6.68$, $p < 0.001$, $n = 24$). However, travel time did not vary between seasons ($t = -1.40$, n.s., $n = 24$) (Figure 18.3b). Mean daily distance traveled was 1015 m (± 168 m) during the study period. In the lean season daily distances traveled (897 ± 186 m) were significantly shorter than during the rich season (1144 ± 149 m; $t = -3.60$, $p < 0.001$, $n = 24$) (Figure 18.3c). The average distance animals traveled between food patches, calculated as the ratio between daily travel distance and number of patches visited, remained similar in both seasons (49 ± 13 m; $t = 1.25$, n.s., $n = 24$).

Discussion

I investigated costs of foraging for a small frugivorous pitheciine, the Southern Bahian masked titi monkey, under two conditions with different food availabilities: lean and rich season. In the lean season the study group traveled shorter distances and spent more time feeding than in the rich season. They visited fewer patches and spent more time in each patch. These patches were of the same size as those used in the rich season but contained a smaller amount of food.

To assess the importance of handling and search costs I tested predictions derived from two OF models. According to the Marginal Value Theorem of Charnov (1976), I expected an increase in patch residence times for patches of the same quality when food availability is low. Patch quality is conventionally defined as a function of the potential net energy gain and depends on patch size, amount of food within the patch, energy content and handling time among others (Stephens & Krebs 1986; Krebs & Davies 1993). Energy calculations could not be made on a patch basis for this study, but there was no difference in energy content or other phytochemical properties of fleshy fruit parts and seeds consumed during lean and rich seasons (Heiduck 1997). Patches used in the lean season contained a smaller amount of food and might therefore be regarded as patches with slightly lower quality than those used in the rich season. Hence, I conclude that the Marginal Value Theorem was supported by this study.

Norberg's model (1977) predicts a reduction of travel distances when food is scarce under the assumption that patch use patterns remain the same. The study group reduced their travel distances during the lean season. On first sight this seems to support Norberg's model. However, the model assumes that energy investment in travel depends on the potential net energy gain within food patches. If diet and/or patch utilization vary, a change in travel must be regarded as a secondary effect. Norberg's model is therefore not appropriate to explain the monkeys' foraging behavior in the present study.

The Southern Bahian masked titi monkey copes with times of food scarcity by modifying its patch-use strategy. When food availability is low, patch residence time increases. Given the fact that patches in the lean season were of slightly lower quality and energy intake rate decreases with time because of patch depletion (Chapman 1988; Snaith & Chapman 2005), *C. melanochir*'s net energy gain is presumably lower in the lean season. To meet their energetic needs the animals in this study spent more time feeding and reduced their travel distances, i.e. they increased their handling costs while reducing their search costs. This pattern was also found in other primate species (Milton 1980; Zhang 1995; Doran 1997; Agetsuma & Nakagawa 1998) and is described as typical for energy minimizers (Schoener 1971).

Seasonality in food availability is one of the driving forces in the evolution of specific morphological and behavioral patterns (Rosenberger 1992; Hemingway & Bynum 2005; Lambert 2007). In particular, the use of so-called fallback foods during times of scarcity seems to have a major impact on feeding adaptations (Lambert *et al.* 2004; Marshall & Wrangham 2007). The genus *Callicebus* is classified as an ecological generalist (Youlatos 2004; Lawler *et al.* 2006) in contrast to other pitheciines, which include high proportions of seeds in their diet (Norconk 2011).

C. melanochir's increase in feeding time during the lean season was mainly caused by an increase in feeding on seeds, while time spent feeding on fleshy fruit parts and leaves remained the same as in the rich season (Heiduck 1997). The relative proportion of time spent feeding on seeds comprised 18.6% in the rich season and 30.5% in the lean season. Although seeds consumed by *C. melanochir* did not differ substantially in their nutritional composition from fleshy fruit

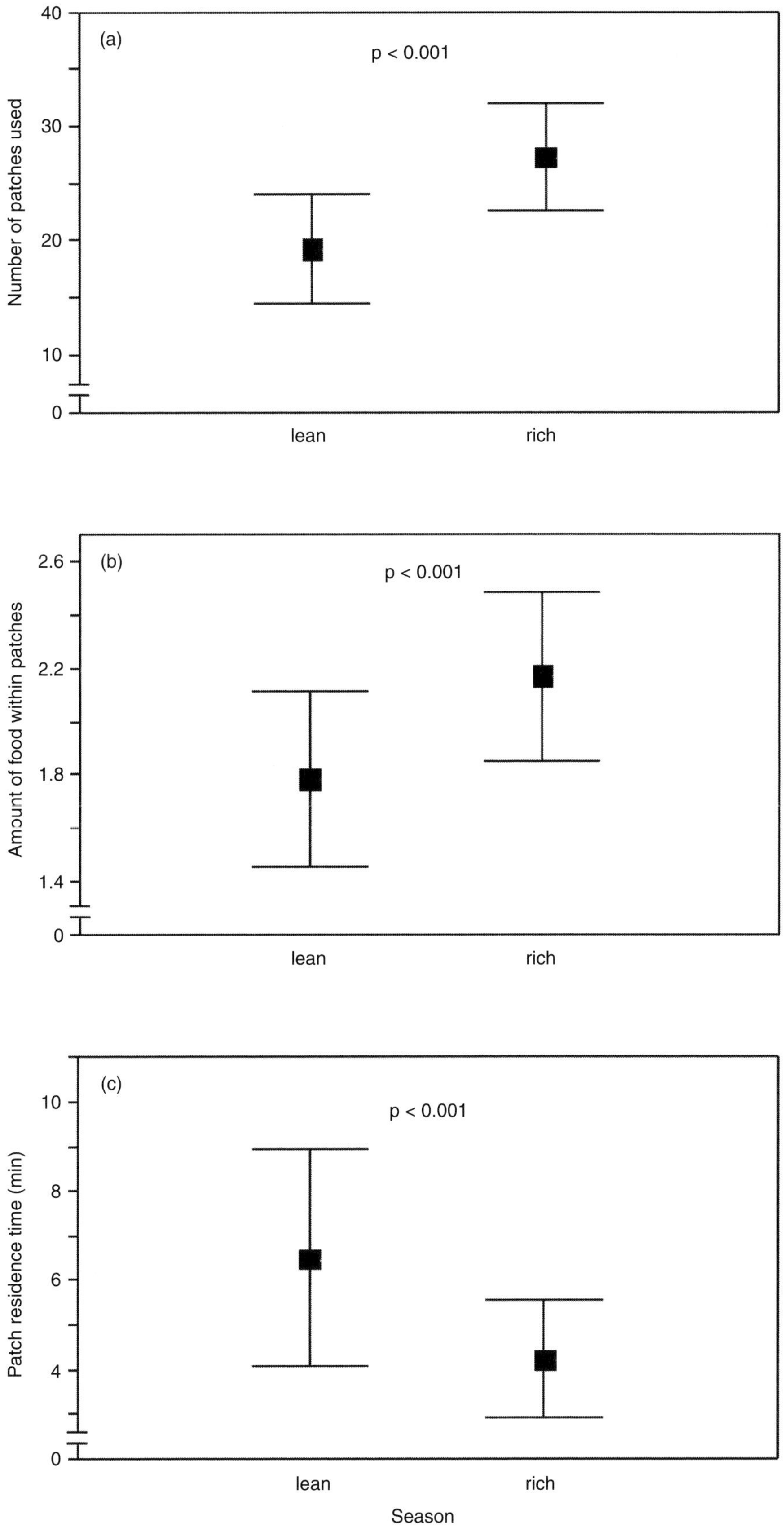

Figure 18.2 (a) Number of patches used, (b) mean amount of food within patches and (c) mean patch residence time per day in lean and rich season. (Means and standard deviations from 12 observation days per season; *t*-test, $n = 24$.)

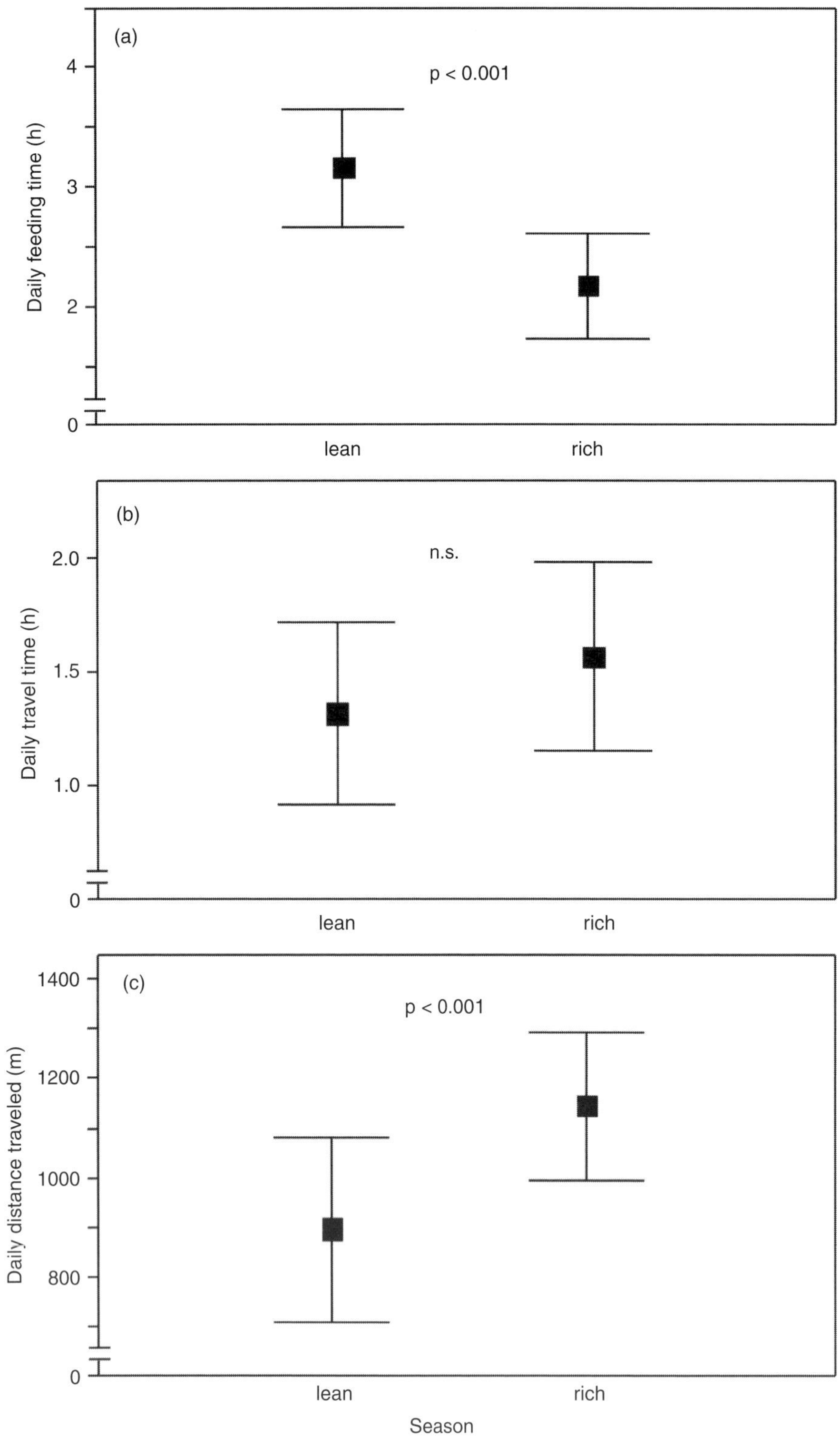

Figure 18.3 (a) Daily feeding time, (b) daily travel time and (c) daily distance traveled in the lean and the rich season. (Means and standard deviations from 12 observation days per season; *t*-test, $n = 24$.)

parts eaten (Heiduck 1997), preyed seeds are mostly protected by hard pericarps and therefore need extended handling time (Kinzey & Norconk 1993).

The foraging strategy of *C. melanochir* consists of a reduction of search costs and an increase in handling effort when food is scarce. This might be due to a tendency of using seeds as fallback food. Unfortunately, there is insufficient data available on seed predation in the genus *Callicebus* to test whether increased use of seeds during food scarcity is a general pattern within the genus or if it is due to the extreme rarity of animal prey in the food spectrum of the *personatus* group (see Chapter 24). More studies on different *Callicebus* species with detailed analysis of quantity and quality of food resources available and used are necessary to evaluate the foraging strategies of titi monkeys and the selective forces that shaped them.

Acknowledgments

This study was part of a cooperation agreement between the Deutsches Primatenzentrum Goettingen and the Centro de Primatologia do Rio de Janeiro. Logistic support at the field site was provided by the Commissão Executiva do Plano da Lavoura Cacaueira (CEPLAC). The research was financially supported by Deutscher Akademischer Austauschdienst (DAAD HSP II 516 503 501 4) and Deutsche Forschungsgemeinschaft (DFG GA 342/5-1). Field work would not have been possible without the help of the field assistants, with special thanks to Geomário Santos Souza. I am grateful to Jörg Ganzhorn for invaluable help and discussions through all stages of the project. For helpful comments on the manuscript I thank Eckhard W. Heymann, Dietmar Zinner and two anonymous reviewers.

References

Agetsuma, N. & Nakagawa, N. (1998). Effects of habitat differences on feeding behaviors of Japanese monkeys: comparison between Yakushima and Kinkazan. *Primates*, **39**, 275–289.

Altmann, J. (1974). Observational study of behavior: sampling methods. *Behaviour*, **49**, 227–267.

Chapman, C. (1988). Patch use and patch depletion by spider and howling monkeys of Santa Rosa National Park, Costa Rica. *Behaviour*, **105**, 99–116.

Charnov, E.L. (1976). Optimal foraging, the marginal value theorem. *Theoretical Population Biology*, **9**, 129–136.

Doran, D. (1997). Influence of seasonality on activity patterns, feeding behavior, ranging, and grouping patterns in Tai chimpanzees. *International Journal of Primatology*, **18**, 183–206.

Heiduck, S. (1997). Food choice in masked titi monkeys (*Callicebus personatus melanochir*): selectivity of opportunism? *International Journal of Primatology*, **18**, 487–502.

Heiduck, S. (2002). The use of disturbed and undisturbed forest by masked titi monkeys *Callicebus personatus melanochir* is proportional to food availability. *Oryx*, **36**, 133–139.

Hemingway, C.A. & Bynum, N. (2005). The influence of seasonality on primate diet and ranging. In *Seasonality in Primates: Studies of Living and Extinct Human and Non-Human Primates*, ed. D.K. Brockman & C.P. van Schaik. Cambridge: Cambridge University Press, pp. 57–104.

Hladik, C.M. (1988). Seasonal variations in food supply for wild primates. In *Coping with Uncertainty in Food Supply*, ed. J. Garine & G.A. Harrison. Oxford: Clarendon Press, pp. 1–25.

Kinzey, W.G. (1981). The titi monkey, genus *Callicebus*. In *Ecology and Behavior of Neotropical Primates*, ed. A.F. Coimbra-Filho & R.A. Mittermeier. Rio de Janeiro: Academia Brasileira de Ciencias, pp. 241–276.

Kinzey, W.G. & Becker, M. (1983). Activity patterns of the masked titi monkey, *Callicebus personatus*. *Primates*, **24**, 337–343.

Kinzey, W.G. & Norconk, M.A. (1993). Physical and chemical Properties of fruit and seeds eaten by *Pithecia* and *Chiropotes* in Surinam and Venezuela. *International Journal of Primatology*, **14**, 207–227.

Krebs, J.R. & Davies, N.B. (1993). *An Introduction to Behavioural Ecology*, 3rd edn. Oxford: Blackwell Science.

Lambert, J.E. (2007). Primate nutritional ecology: feeding biology and diet at ecological and evolutionary scales. In *Primates in Perspective*, ed. C.J. Campbell, A. Fuentes, K.C. MacKinnon, M. Panger & S. Bearder. Oxford: Oxford University Press, pp. 482–495.

Lambert, J.E., Chapman, C.A., Wrangham, R.W., *et al.* (2004). Hardness of Cercopithecine foods: implications for the critical function of enamel thickness in exploiting fallback foods. *American Journal of Physical Anthropology*, **125**, 363–368.

Lawler, R.R., Ford, S.M., Wright, P.C., *et al.* (2006). The locomotor behavior of *Callicebus brunneus* and *Callicebus torquatus*. *Folia Primatologica*, 77, 228–239.

MacArthur, R.H. & Pianka, E.R. (1966). On optimal use of a patchy environment. *American Naturalist*, **100**, 603–609.

Marshall, A.J. & Wrangham, R.W. (2007). Evolutionary consequences of fallback foods. *International Journal of Primatology*, **28**, 1219–1235.

McNair, J.N. (1983). A class of patch-use strategies. *American Zoologist*, **23**, 303–313.

Milton, K. (1980). *The Foraging Strategy of Howler Monkeys: A Study in Primate Economics*. New York, NY: Columbia University Press.

Müller, K.-H. (1995). Ranging in masked titi monkeys (*Callicebus personatus*) in Brazil. *Folia Primatologica*, **65**, 224–228.

Müller, K.-H. (1996). Diet and feeding ecology of masked titis (*Callicebus personatus*). In *Adaptive Radiations of Neotropical Primates*, ed. M.A. Norconk, A.L. Rosenberger & P.A. Garber. New York, NY: Plenum Press, pp. 383–401.

Norberg, R.A. (1977). An ecological theory on foraging time and energetics and choice of optimal food-searching method. *Journal of Animal Ecology*, **46**, 511–529.

Norconk, M.A. (2011). Sakis, uakaris, and titi monkeys: behavioural diversity in a radiation of primate seed predators. In *Primates in Perspective* (2nd edn), ed. C.J. Campbell, A. Fuentes, K.C. MacKinnon, S.K. Bearder & R.M. Stumpf. Oxford: Oxford University Press, pp. 122–139.

Oates, J.F. (1987). Food distribution and foraging behavior. In *Primate Societies*, B. Smuts, D. Cheney, R. Seyfarth, R. Wrangham & T. Struhsaker. Chicago, IL: University of Chicago Press, pp. 197–209.

Price, E.C. & Piedade, H.M. (2001a). Ranging behavior and intraspecific relationships of masked titi monkeys (*Callicebus personatus personatus*). *American Journal of Primatology*, **53**, 87–92.

Price, E.C. & Piedade, H.M. (2001b). Diet of northern masked titi monkeys (*Callicebus personatus*). *Folia Primatologica*, **72**, 335–338.

Pyke, G.H., Pulliam, H.R. & Charnov E.L. (1977). Optimal foraging: a selective review of theory and tests. *Quaterly Review of Biology*, **52**, 137–154.

Rosenberger, A.L. (1992). Evolution of feeding niches in New World monkeys. *American Journal of Physical Anthropology*, **88**, 525–562.

Schoener, T.W. (1971). Theory of feeding strategies. *Annual Review of Ecology and Systematics*, **2**, 369–404.

Snaith, T.V. & Chapman, C.A. (2005). Towards an ecological solution to the folivore paradox: patch depletion as an indicator of within-group scramble competition in red colobus monkeys (*Piliocolobus tephrosceles*). *Behavioral Ecology and Sociobiology*, **59**, 185–190.

Stephens, D.W. & Krebs, J.R. (1986). *Foraging Theory*. Princeton, NJ: Princeton University Press.

van Roosmalen, M.G.M., van Roosmalen, T. & Mittermeier, R.A. (2002). A taxonomic review of the titi monkeys, genus *Callicebus* Thomas, 1903, with the description of two new species, *Callicebus bernhardi* and *Callicebus stephennashi*, from Brazilian Amazonia. *Neotropical Primates*, **10** (Suppl.), 1–52.

van Schaik, C.P. & Pfannes, K.R. (2005). Tropical climates and phenology: a primate perspective. In *Seasonality in Primates: Studies of Living and Extinct Human and Non-Human Primates*, ed. D.K. Brockman & C.P. van Schaik. Cambridge: Cambridge University Press, pp. 23–54.

van Schaik, C.P., Terborgh, J.W. & Wright, S.J. (1993). The phenology of tropical forests: adaptive significance and consequences for primary consumers. *Annual Review of Ecology and Systematics*, **24**, 353–377.

Youlatos, D. (2004). Multivariate analysis of organismal and habitat parameters in two neotropical primate communities. *American Journal of Physical Anthropology*, **123**, 181–194.

Zhang, S.Y. (1995) Activity and ranging patterns in relation to fruit utilization by brown capuchins (*Cebus apella*) in French Guiana. *International Journal of Primatology*, **16**, 489–507.

Chapter 19

Insectivory and prey foraging techniques in *Callicebus* – a case study of *Callicebus cupreus* and a comparison to other pitheciids

Eckhard W. Heymann & Mirjam N. Nadjafzadeh

Introduction

For primates, animal prey and leaves represent alternative sources of protein, and no primate species combines large amounts of prey and leaves in the diet (Chivers & Hladik 1980; Martin 1990). These alternative dietary strategies can be explained by body-size constraints on the efficiency of prey foraging and by physiological constraints imposed on the digestion of leaves (Terborgh 1992). The absolute amount of obtainable prey does not increase with body size, thus making prey foraging unprofitable for large primate species, unless prey occurs in large clusters (like ant or termite mounds) (Terborgh 1983). The digestion of leaves through bacterial fermentation requires large amounts of space, which is not available in animals of small body size (Martin 1990). Consequently, primate diets show a predictable relationship with body mass (Martin 1990): while most extant primate species actually include fruit pulp in their diet, smaller species (below ca. 1–2 kg) usually supplement their diet with animal prey, while larger species (above ca. 1–2 kg) supplement with leaves (Harding 1981; Terborgh 1992). Additional factors such as the quality of leaves in the habitat or prey capture efficiency may, however, determine whether prey or leaves are used as source of protein. Kinzey (1978, 1997) compared the diets of two species of titi monkeys, and attributed the comparatively high amount of prey in the diet of *Callicebus lucifer* to leaves being very sclerophyllous and not readily available as a source of protein in its white-sand habitat, while very low capture efficiency was assumed to limit the amount of prey in *Callicebus brunneus*. He also suggested that "a 'critical function' … such as reducing the dietary protein source" (Kinzey 1978, p. 383) influences the molar structure of these otherwise basically frugivorous titi monkey species (Rosenberger & Kinzey 1996). Similarly, Rosenberger (1992) and Lambert (2007) emphasized that behavioral, morphological and physiological adaptations to the acquisition and consumption of minor food categories which may be vital for supplying certain essential nutrients, either throughout the year or as a fallback resource during critical periods of food scarcity, may be under strong selective pressure.

Members of the Pitheciidae feed either principally on fruit pulp (titi monkeys) or seeds (sakis and uacaris), and are classified as "sclerocarpic frugivores" (Kinzey 1981; Norconk 2011; Rosenberger 1992). Neither fruit pulp nor seeds usually provide significant amounts of protein, so pitheciids have to exploit either prey or leaves to obtain proteins. In light of Kinzey's (1978, 1997) hypothesis on the source of interspecific variation in prey consumption in titi monkeys, it is thus timely to review the available evidence for insectivory in the Pitheciidae. In this chapter we therefore examine (a) the contribution of prey to the diet and (b) the prey spectrum and prey foraging techniques, and analyze (c) whether a relationship exists between body mass and the contribution of prey to the diet, as predicted from theory. As a starting point for these comparisons and analyses, we provide detailed information on prey foraging in red titi monkeys, *Callicebus cupreus*.

Methods

The data on prey foraging in *C. cupreus* stem from a study by the second author on two groups of this species at the Estación Biológica Quebrada Blanco (EBQB) in northeastern Peru (Nadjafzadeh 2005). Detailed information on study methods are provided by Nadjafzadeh and Heymann (2008). While Nadjafzadeh (2005) and Nadjafzadeh and Heymann (2008) reported data for the two study groups separately, for the purpose of this chapter we combined these data, as the two groups did not differ in any respect.

For the comparison with other Pitheciidae, we searched the literature for data on the amount of prey in the diet (in terms of feeding time or of stomach contents), prey spectrum, prey search and capture techniques, and body mass. The relationship between body mass and feeding time on prey was examined through regression analyses with Statistica® 6.0.

Evolutionary Biology and Conservation of Titis, Sakis and Uacaris, eds. Liza M. Veiga, Adrian A. Barnett, Stephen F. Ferrari and Marilyn A. Norconk. Published by Cambridge University Press.

Animal prey in the diet of *Callicebus* and other pitheciids

Callicebus cupreus

The two study groups of *Callicebus cupreus* at EBQB dedicated between 11.7% and 14.8%, respectively, of feeding time on prey. These values fall into the range of these values from two earlier studies at EBQB which reported 11.1% and 18.0% (Table 19.1).

Comparison with other pitheciids

Based on evidence from direct feeding observations or on the examination of stomach contents, all other titi monkey species except for *Callicebus personatus* also include prey in their diet (Table 19.1). The proportion of time spent consuming prey varies between 0.3% in *Callicebus melanochir* and 22% in *Callicebus oenanthe*. Notably, proportions are similarly high in the closely related *C. cupreus* and *discolor* and in *C. brunneus*, very divergent for the closely related *C. lucifer* and *lugens*, and similarly low in the closely related *C. personatus*, *C. melanochir* and *nigrifrons*. Proportions of time spent consuming prey by eastern Brazilian species of *Callicebus* (*C. personatus*, *C. melanochir C. nigrifrons*) are more similar to other pitheciids, where the proportion varies between 0.4% in *Pithecia albicans* and 5.2% in *Cacajao calvus calvus* (Table 19.1). *Pithecia irrorata* in southeastern Peru "fed extensively on invertebrates" (Palminteri *et al.* 2005, p. 157), but quantitative data are not available.

Callicebus cupreus exploited prey from at least five orders of insects and from the Araneida; almost half of all prey feeding corresponded to Hymenoptera, mainly socially living ants (Figure 19.1). Unidentified prey mainly involved small items that were grabbed and directly put into the mouth, without providing the observers an opportunity for identification. This unidentified portion probably includes tiny hymenopterans, small enough to be put straight into the mouth without further handling. The detection of residuals of small hymenopterans in feces supports this assumption (Nadjafzadeh 2005).

In other species of *Callicebus* and in the pitheciines, animal prey also comprises diverse orders of insects, and spiders (Table 19.1). Ants are the prey category that is exploited by the majority of species and populations, followed by caterpillars, beetles and orthopterans (katydids and grasshoppers). Except for a single observation of a *C. nigrifrons* feeding on a pigeon egg (Neri 1997) and very sporadic bird egg consumption by *P. pithecia* (0.4% of prey feeding time; Homburg 1998), there is a notable absence of predation on vertebrates. Prey consumed by *Callicebus* and other pitheciids includes both taxa that live socially (many ant species) or can be found in large aggregations (many caterpillars), and taxa that usually occur solitarily (e.g. orthopterans).

Quantitative data on the amount of different prey taxa are lacking for other *Callicebus* species, but are available for some pitheciines. The prey spectrum of *Chiropotes satanas* is dominated by caterpillars, which represented 70% of all prey in a group of *C. s. chiropotes* studied by Peetz (2001), and 60.5% and 25.2%, respectively, in two groups of *C. s. satanas* studied by Veiga (2006) and Veiga and Ferrari (2006). Caterpillars also dominate in the prey spectrum of *C. c. calvus* and *Cacajao melanocephalus* as determined from stomach contents (Ayres 1986). Termites represented more than 50% of prey feeding records in *Pithecia pithecia*, with ants ranking second (Homburg 1998). However, termites were only taken on two observation days during their mass swarming.

Photo 19.1 *Callicebus cupreus* grooming. Photo: Mirjam Nadjafzadeh. (See color plate section.)

Table 19.1 Contribution of prey to the diet (as % of feeding time or % of stomach content) and prey items of *Callicebus* and other pitheciids.

Species	Prey in diet (%)	Prey items	Data type	Reference
Callicebus brunneus	< 1	n.a.[a]	o[b]	Kinzey 1978
Callicebus brunneus	(~15%)[c]	Moths, caterpillars, chrysalises, ants, beetles, grasshoppers, katydids, larvae, spiders	o	Wright 1985, 1986
Callicebus brunneus	8.0	n.a.	o	Lawrence 2007
Callicebus brunneus	11.0	Spiders, caterpillars, moths, butterflies, ants	o	Crandlemire-Sacco 1988
Callicebus cupreus	11.1	Ants, katydids	o	Tirado Herrera & Heymann 2004
Callicebus cupreus	18.0	Ants, katydids, spiders, mantis, butterflies, termites	o	Pérez Yamacita 2010
Callicebus cupreus, Group W Group N	11.7 14.8[d]	Ants, beetles, katydids, grasshoppers, caterpillars, wasps, mantis, spiders	o	Nadjafzadeh 2005
Callicebus discolor	13.0	n.a.	o	Youlatos & Pozo Rivera 1999
Callicebus lucifer	n.a.	Orthopterans, hymenopterans, beetles	sc[e]	Izawa 1975
Callicebus lucifer	20.0[f]	Ants, insect galls, beetles, hymenopteran larvae	o	Kinzey 1978
Callicebus lucifer	15.8	Grasshoppers, praying mantis, spiders, larvae	o	Easley 1982
Callicebus lugens	n.a.	Caterpillars, wasps, ants, beetles, flies	sc	Milton & Nessimian 1984
Callicebus lugens	3.4[g]	Ants and other hymenopterans, beetles, orthopterans, spiders, galls	o	Palacios *et al.* 1997
Callicebus lugens	19.5[h]	Katydids, ants, spiders	o	Alvarez & Heymann 2009
Callicebus melanochir	0.0	–	o	Müller 1996
Callicebus melanochir	0.3	n.a.	o	Heiduck 1997
Callicebus oenanthe	22.0	Ants, spiders, cicada, grasshoppers, butterflies, beetles, bees/wasps, caterpillars, walking sticks, galls, unidentified small flying insects	o	DeLuycker 2007, 2009
Callicebus ornatus	n.a.	Beetles, orthopterans	sc	Hernández Camacho & Cooper 1976
Callicebus ornatus	9.4	n.a.	o	Polanco 1992 quoted in Defler 2004
Callicebus nigrifrons	3.4	Caterpillars, galls, chrysalises, bird eggs	o	Neri 1997
Callicebus nigrifrons	10.0	n.a.	o	Souza *et al.* 1996
Callicebus personatus	0.0	–	o	Kinzey & Becker 1983
Callicebus personatus	0.0	–	o	Price & Piedade 2001
Cacajao calvus calvus	5.2	Caterpillars, orthopterans, beetles, spiders, butterflies, flies, cicadas, homopteran nymphs	o sc	Ayres 1986, 1989

Table 19.1 (*cont.*)

Species	Prey in diet (%)	Prey items	Data type	Reference
Cacajao calvus ucayalii	2.8	Ants, caterpillars, termites, mantids, katydids		Bowler 2007
Cacajao melanocephalus melanocephalus	2.0	Ants, spiders, grasshoppers, katydids, roaches	o	Boubli 1999
Cacajao melanocephalus ouakari	n.a.	Caterpillars, hymenopterans, beetles, cicadas, orthopterans, spiders	sc	Ayres 1986, 1989
Cacajao melanocephalus ouakari	n.a.	Wasps	o	Barnett *et al.* 2005
Chiropotes albinasus	n.a.	Unidentified larvae(?)	o	Ayres 1981
Chiropotes satanas chiropotes	n.a.	Caterpillars, ants, beetles	sc	Ayres & Nessimian 1982
Chiropotes satanas chiropotes	n.a.	Ants, flies, beetles	sc	Mittermeier *et al.* 1983
Chiropotes satanas chiropotes	n.a.	Caterpillars, ants, beetles	sc	Frazão 1991
Chiropotes satanas chiropotes	0.5	Caterpillars	o	Kinzey & Norconk 1993; Norconk 1996
Chiropotes satanas chiropotes	3.9	Caterpillars, cicadas, ants, orthopterans, hymenopterans (nests, galls), spiders	o	Peetz 2001
Chiropotes satanas satanas, mainland group	4.6	Caterpillars, hymenopterans, termites, cicadas, beetles, spiders	o	Veiga 2006; Veiga & Ferrari 2006
island group	3.7			
Pithecia albicans	n.a.	Caterpillars, spiders, hymenopterans	sc	Johns 1986
Pithecia albicans	0.4	n.a.	o	Peres 1993
Pithecia irrorata	n.a.[i]	n.a.	o	Palminteri *et al.* 2005
Pithecia monachus	n.a.	Ants, butterflies	sc	Izawa 1975
Pithecia monachus	~3.5[j]	Ants	o	Soini 1987
Pithecia pithecia	2.3	Ants, spiders, caterpillars	o	Kinzey & Norconk 1993; Norconk 1996
Pithecia pithecia	3.7	Ants, caterpillars, termites, hymenopterans, orthopterans	o	Homburg 1998
Pithecia pithecia	1.8	n.a.	o	Norconk & Conklin-Brittain 2004
Pithecia pithecia	<1	n.a.	o	Norconk & Setz (Chapter 25)

[a] n.a.: no information available. [b] o: direct observation. [c] Estimated from figure 3b in Wright (1986), includes search time. [d] Rainy season only. [e] sc: stomach content. [f] Different figures are provided in Kinzey (1977: 14%) and Kinzey *et al.* (1977: 16%). [g] During period of low fruit and leaf abundance. [h] During period of high fruit and leaf abundance. [i] "extensively" (Palminteri *et al.* 2005, p. 157). [j] Calculated from data in Soini (1987).

These findings suggest that for *Callicebus* and the other pitheciids it may be profitable to exploit prey that occurs in larger patches (social ants, caterpillars, swarming termites). However, on a cautionary note, it must be taken into consideration that feeding on prey in larger patches – which can involve all or most of the group members – may be more conspicuous and thus be more easily recorded than feeding on solitary prey items that is done individually.

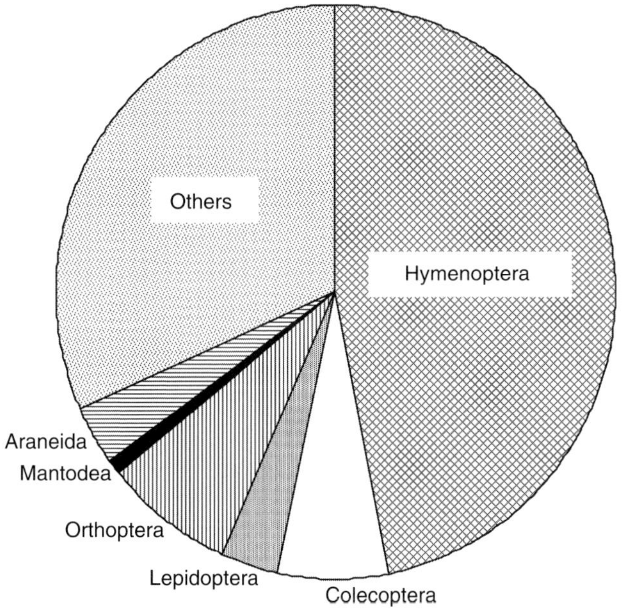

Figure 19.1 The proportion of different prey taxa in the diet of *Callicebus cupreus* (from data in Nadjafzadeh 2005 Nadjafzadeh & Heymann 2008).

Prey capture techniques

Callicebus cupreus

Red titi monkeys mainly forage for prey in the lower strata of the forest (Figure 19.3), and occasionally obtain prey even from the ground. While searching and capturing prey they are usually sitting or standing on branches (48.6% of records) or lianas (28.6%), but also clinging to trunks (19.8%); other substrates like palm leaves, epiphytes and ground are only sporadically used. Red titi monkeys are positioned on horizontal (0–15°) and diagonal (15–75°) supports during 40.6% and 36.4% of prey searches and captures, respectively, while 23% take place while the titi monkeys are positioned on a vertical substrate (75–90°) (Nadjafzadeh 2005).

Red titi monkeys search and find prey most often on living leaves. Dead leaves, trunks, branches, and gleaning insects from the air are used with more or less similar frequencies (Figure 19.4). Open microhabitats (i.e. microhabitats where prey items are exposed to view) account for 81.8% of all prey searches and captures. From these open microhabitats prey is grabbed directly with one or both hands, or is first fixed with the hand(s) and subsequently transferred to the mouth. Closed microhabitats that require slight manipulation comprise enrolled leaves, mainly dead, but also living. Only on four occasions was extractive foraging in microhabitats observed that require intensive manipulation (bark, dead branch).

Comparison with other pitheciids

Information provided by Alvarez and Heymann (2009), DeLuycker (2007), Easley (1982), Kinzey (1977), Neri (1997) and Wright (1985) indicate that *C. oenanthe*, *C. brunneus*,

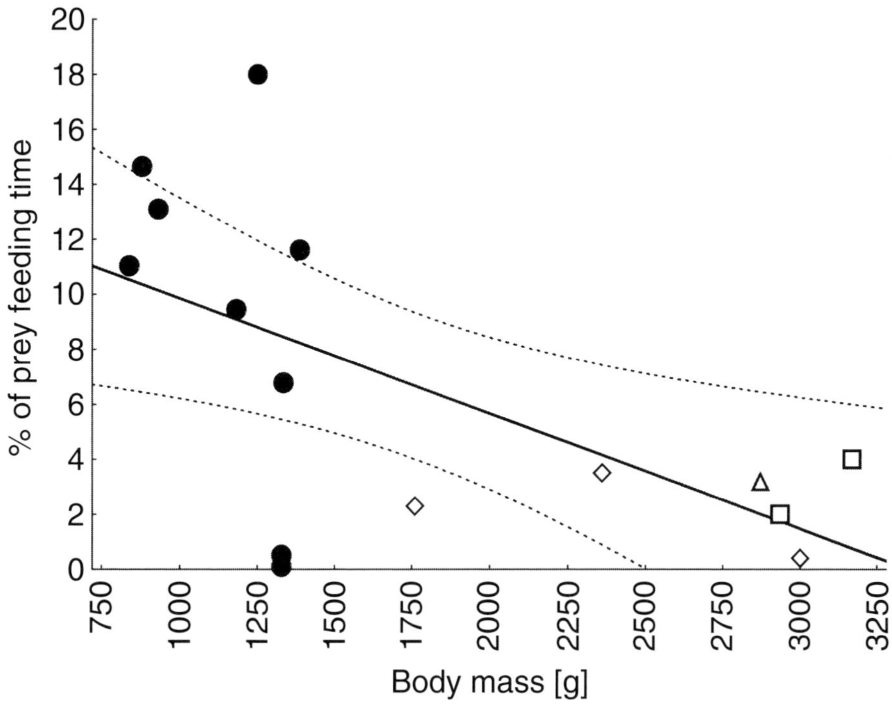

Figure 19.2 The proportion of feeding time on prey in *Callicebus* (●), *Pithecia* (◇), *Chiropotes* (△) and *Cacajao* (□) in relation to body mass. ——: regression line (% prey feeding time = 14.045 – 0.004 * body mass); ……: 95% confidence intervals. For body mass, midpoints of values for males and females were calculated from data in Smith and Jungers (1997), Hernández Camacho and Defler (1985, for *C. lugens*), Hershkovitz (1990, for *C. discolor* and *C. ornatus*), and Bicca-Marques *et al.* (2002, for *C. cupreus*). Body mass data from *C. personatus* were used for *C. melanochir* and *C. nigrifrons*, since head–body length data in Hershkovitz (1990) indicate similar size. If more than one data point was available for the proportion of feeding time on prey for one species (see Table 19.1), the midpoint of values was calculated, except for *C. brunneus* (values provided by Kinzey [1978] and Wright [1985] for this species were not used because the former was based on a very small observation time, and the latter included search time).

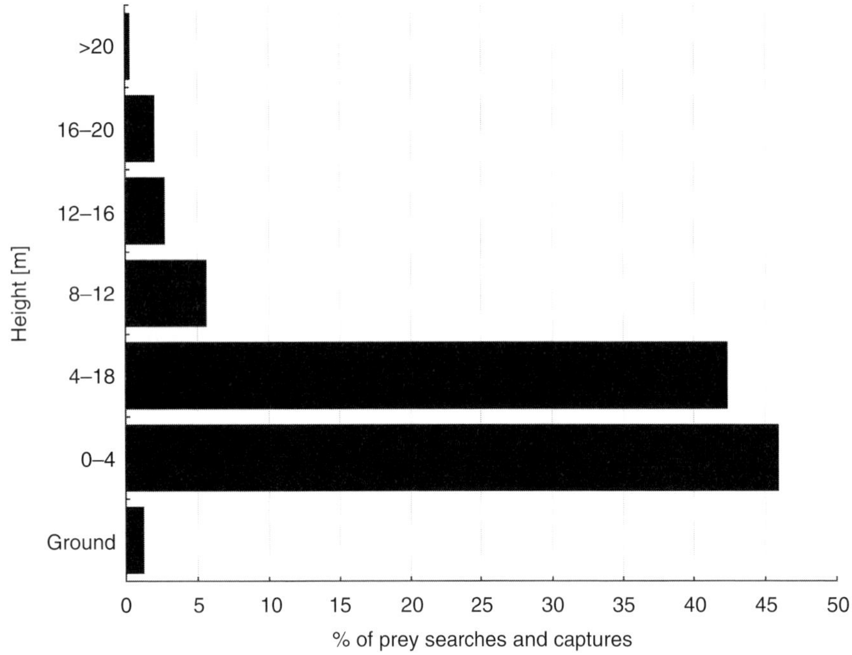

Figure 19.3 Height use during prey search and capture by *Callicebus cupreus* (mean values for two study groups, total $N = 938$ records) (from data in Nadjafzadeh 2005 Nadjafzadeh & Heymann 2008).

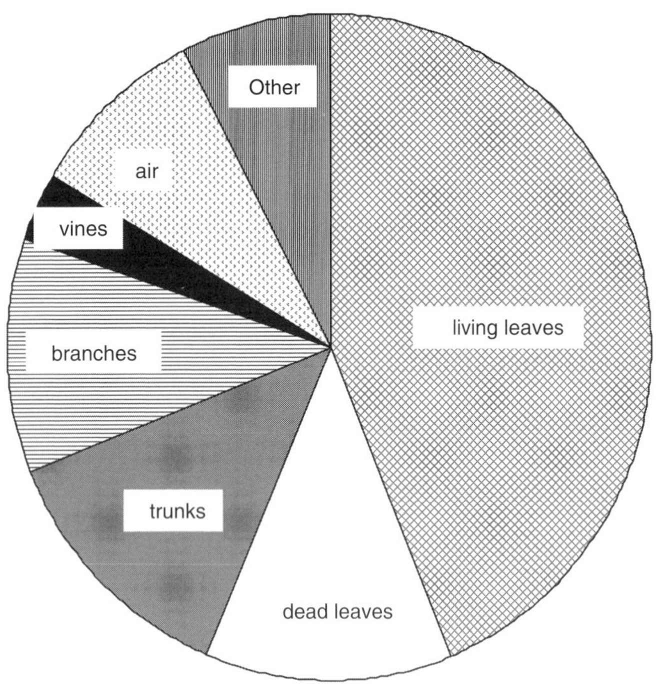

Figure 19.4 Microhabitats on which prey is searched and captured by *Callicebus cupreus* (from data in Nadjafzadeh 2005 and Nadjafzadeh & Heymann 2008).

C. lucifer, *C. lugens* and *C. nigrifrons* search for and capture prey on trunks and branches, unroll dead and green leaves, grab flying insects, but do not to tear open bark or branches, or grab into holes. This coincides with observations on *C. cupreus*, although it remains to be determined for most pitheciids whether different microhabitats are also used to the same or similar degree to obtain prey (but see Alvarez & Heymann 2009 for comparative quantitative information on *C. lugens*).

Chiropotes locates prey on living leaves and dry fruits. They also extract prey from dead branches, and have been observed to strip off bark and search beneath (Peetz 2001; Veiga 2006; Veiga & Ferrari 2006). *Cacajao calvus ucayalii* search on dead and living leaves, they split open the end of branches that contained chambers with ants, and were once seen to capture termites on the wing (Bowler 2007). *Cacajao melanocephalus* unrolls dry leaves, strips off bark from dead branches, and scrapes ant nests off trees (Boubli 1999). Similar to titi monkeys, they capture flying insects directly from the air (Boubli 1999). *Pithecia pithecia* also unrolls dry leaves, and from the descriptions in Homburg (1998) it can be deduced that prey is also commonly taken from open microhabitats such as leaf surfaces or branches.

There is also some evidence for "indirect prey foraging", i.e. uptake of insects infesting the fruit pulp or seeds, in *C. c. calvus* (Ayres 1989) and possibly *C. lugens* (Alvarez Vargas 2007).

Animal prey and body size in pitheciids

Nutritional ecology theory predicts that the amount of prey in the diet should decrease with increasing body mass (Chivers & Hladik 1980; Martin 1990; see "Introduction" above). Examination of the proportion of feeding time on prey in *Callicebus* and other pitheciids in relation to body mass reveals a significant negative correlation ($r = -0.61$, $R^2_{adj} = 0.33$, $F_{1,13} = 7.83$, $p < 0.05$; Figure 19.2). This correlation is likely to be biased because no phylogenetic correction was applied. Nevertheless, it indicates that the difference in the amount of time dedicated to prey consumption between *Callicebus* as a genus on the one hand and *Pithecia, Chiropotes and Cacajao* on the other hand might be accounted for by differences in body size. However, there is considerable intrageneric variation in *Callicebus* that cannot be explained by body-size differences. The highest (17.9%) and lowest (0%) proportions of feeding time on prey are found in the larger *Callicebus* species (Figure 19.1). Potentially, this variation may represent biases arising from different lengths of study periods, from the coverage of different seasons, or from group idiosyncrasies. Nadjafzadeh (2005) noted that there was no difference in the amount of prey between wet and dry season in *C. cupreus*. However, for the same species, there was some variation between studies made at different times and of different duration (Perez Yamacita 2010; Tirado Herrera & Heymann 2004; see Table 19.1). In contrast, two independent studies of *C. melanochir* in subsequent years consistently reported very little or no prey consumption (Heiduck 1997; Müller 1996). Highly contrasting proportions of feeding time on prey were found in two groups of *C. lugens*: one studied during the period of low fruit and leaf abundance showed low proportions of prey in the diet (Palacios *et al.* 1997), the other during the period of high fruit and leaf abundance high proportions (Alvarez Vargas 2007; Alvarez & Heymann 2009). Lawrence (2007) reported little overall variation (6–11%) between groups of *C. brunneus*, but marked seasonal variation, with higher proportions of insect consumption during the season with low fruit availability. The available data do not allow drawing a conclusion on what is causing variation within the genus *Callicebus*.

Conclusions and perspectives

It is clear from our review that animal prey is an integral component of the diet of almost all *Callicebus* and most other pitheciids, although the proportion of prey varies considerably. For some of the *Callicebus* species, time spent feeding on prey matches or even exceeds values from other Platyrrhines that are often classified as frugivores–insectivores or gummivores-insectivores, like the callitrichids (e.g. Digby *et al.* 2007). All pitheciid species studied so far exploit a variety of behaviors to search for and capture prey. From the available descriptions, these behaviors seem to be relatively uniform within *Callicebus* and mainly involve the search for and capture of prey exposed to view, being superficially similar to many tamarins (see Garber 1993), while *Chiropotes* and *Cacajao* also employ "destructive" foraging in the sense of Terborgh (1983). The difference between foraging techniques of *Callicebus* and the other pitheciids, along with the relationship between body mass and time spent feeding on insects, support Terborgh's (1983) considerations on the role of body size for prey foraging and consumption.

Red titi monkeys differ in their prey foraging techniques and consequently in prey spectrum from sympatric tamarin monkeys: while tamarins (*Saguinus mystax, Saguinus fuscicollis*) mainly capture prey that is occurring solitarily, *C. cupreus* mainly takes prey living socially (Nadjafzadeh 2005; Nadjafzadeh & Heymann 2008). Given the almost complete lack of quantitative data on foraging techniques and prey spectrum in other *Callicebus* and in pitheciine species, it is unclear at the moment how far they differ from or overlap with other sympatric insect-feeding primates such as *Cebus, Saimiri, Saguinus* or *Callimico*. Quantitative analysis of patterns of prey foraging will help to define the feeding niches of pitheciids in more detail and to understand whether and how far prey foraging contributes to niche differentiation and the avoidance of interspecific competition, as suggested by Terborgh (1983). The inclusion of prey foraging techniques in interspecific comparisons will also allow obtaining a deeper understanding of the evolution of feeding niches in platyrrhines in general (Rosenberger 1992).

It is unclear for the moment whether the interspecific variation of the amount of prey within *Callicebus* relates to species-specific traits (e.g. dental, postcranial or gut morphology) or population/habitat-specific factors such as the temporal and spatial availability and distribution of prey. Interestingly, Kinzey's (1978) hypothesis on dental differences between *C. brunneus* and *C. lucifer* does not hold up in light of the available data. Kinzey had argued that a longer shearing blade (cristid oblique) in the former species and a larger crushing and grinding surface (talonid basin) in the latter is related to their different emphasis on leaves and insects, respectively, as sources of protein. Nevertheless, given the nutritional importance of protein, foraging for and consumption of animal prey must be taken into consideration in the interpretation of the feeding adaptations - be they morphological, behavioral, or physiological - of *Callicebus* and the other pitheciids.

Note added in proof

After the manuscript had been accepted for publication, additional information on prey feeding in *Callicebus* emerged.

In a 1-year study, a group of *Callicebus coimbrai* included 2.6% insects in its diet (Souza-Alves *et al.* 2011). Prey consumption peaked in the dry season months November–December, when the titi monkeys fed on agglomerations of the caterpillar *Pseudosphinx tetrio* (Lepidoptera: Sphingidae). Other prey items were Phasmatodea and Homoptera.

In a 9-month study, a group of *Callicebus nigrifrons* included 11% insects in its diet (Barbosa Caselli & Setz 2011).

In a 4-month study (March–June, the end of the rainy season) of two groups of *Callicebus cupreus* at EBQB, Kulp (2011) reported 8.0% and 8.1% of prey in the diet, respectively.

The general conclusions of the chapter are not altered by this additional information.

Acknowledgments

We are thankful to the editors for inviting us to contribute this chapter. Field work on *Callicebus cupreus* was performed under authorization no. 88-2004-INRENA-IFFS-DCB from the Instituto Nacional de Recursos Naturales (INRENA) in Lima, and would have been impossible without the help of our field assistants Camilo Flores Amasifuén, Ney Shahuano Tello, Manuel Shahuano Tello and Jeisen Shahuano Tello. During field work MNN was supported by a grant from the Christian-Vogel-Fonds of the Gesellschaft für Primatologie. We are grateful to Stefanie Heiduck and two anonymous reviewers for very helpful comments on previous versions of this chapter.

References

Alvarez, S.J. & Heymann, E.W. (2009). Forrajeo de insectos por la viudita (*Callicebus lugens*). In *Estación Biológica Mosiro Itajura-Caparú – Biodiversidad en el Territorio del Yaigojé-Apaporis*, ed. G. Alarcón-Nieto & E. Palacios. Bogotá: Conservación Internacional, pp. 187–192.

Alvarez Vargas, S.J. (2007). Fruit choice by white-handed titi monkeys (*Callicebus lugens*) at the Apaporis river, Colombia. Unpublished Master's thesis, University of Göttingen, Göttingen.

Ayres, J.M. (1981). Observações sobre a ecologia e o comportamento dox cuxiús (*Chiropotes albinasus* and *Chiropotes satanas*, Cebidae: Primates). Master's thesis, Universidade do Amazonas, Manaus.

Ayres, J.M. (1986). Uakaris and Amazonian flooded forest. PhD thesis, University of Cambridge, Cambridge.

Ayres, J.M. (1989). Comparative feeding ecology of the uakari and bearded saki, *Cacajao* and *Chiropotes*. *Journal of Human Evolution*, **18**, 697–716.

Ayres, J.M. & Nessimian, J.L. (1982). Evidence for insectivory in *Chiropotes satanas*. *Primates*, **23**, 458–459.

Barbosa Caselli, C. & Setz, E.Z.F. (2011). Feeding ecology and activity pattern of black-fronted titi monkeys (*Callicebus nigrifrons*) in a semideciduous tropical forest of southern Brazil. *Primates*, **52**, 351–359.

Barnett, A.A., Volkmar de Castilho, C., Shapley, R.L., *et al.* (2005). Diet, habitat selection and natural history of *Cacajao melanocephalus ouakary* in Jaú National Park, Brazil. *International Journal of Primatology*, **26**, 949–969.

Bicca-Marques, J.C., Garber, P.A. & Azevedo-Lopes, M.A.O. (2002). Evidence of three resident adult male group members in a species of monogamous primates, the red titi monkey (*Callicebus cupreus*). *Mammalia*, **66**, 138–142.

Boubli, J.P. (1999). Feeding ecology of black-headed uacaris (*Cacajao melanocephalus melanocephalus*) in Pico da Neblina National Park, Brazil. *International Journal of Primatology*, **20**, 719–749.

Bowler, M. (2007). The ecology and conservation of the red uakari monkey on the Yavarí river, Peru. PhD thesis, University of Kent, Canterbury.

Chivers, D.J. & Hladik, C.M. (1980). Morphology of the gastrointestinal tract in primates. *Journal of Morphology*, **166**, 337–386.

Crandlemire-Sacco, J.L. (1988). An ecological comparison of two sympatric primates: *Saguinus fuscicollis* and *Callicebus moloch* of Amazonian Peru. *Primates*, **29**, 465–475.

Defler, T.R. (2004). *Primates de Colombia*. Bogotá: Conservación Internacional.

DeLuycker, A.D. (2007). The ecology and behavior of the Río Mayo titi monkey (*Callicebus oenanthe*) in the Alto Mayo, northern Peru. PhD thesis, Washington University, St. Louis.

DeLuycker, A.D. (2009). Insect foraging strategies of the Rio Mayo titi monkey *Callicebusoenanthe*) in a fragmented forest, northern Peru. *American Journal of Primatology*, **71** (Suppl.), 100.

Digby, L.J., Ferrari, S.F. & Saltzman, W. (2007). Callitrichines: the role of competition in cooperatively breeding species. In *Primates in Perspective*, ed. C.J. Campbell, A. Fuentes, K.C. Mackinnon, M. Panger & S.K. Bearder. New York, NY: Oxford University Press, pp. 85–105.

Easley, S.P. (1982). Ecology and behavior of *Callicebus torquatus*, Cebidae, Primates. PhD thesis, Washington University, St. Louis.

Frazão, E. (1991). Insectivory in free-ranging bearded saki (*Chiropotes satanas chiropotes*). *Primates*, **32**, 243–245.

Garber, P.A. (1993). Feeding ecology and behaviour of the genus *Saguinus*. In *Marmosets and Tamarins: Systematics, Behaviour, and Ecology*, ed. A.B. Rylands. Oxford: Oxford University Press, pp. 273–295.

Harding, R.S.O. (1981). An order of omnivores: nonhuman primate diets in the wild. In *Omnivorous Primates: Gathering and Hunting in Human Evolution*, ed. R.S.O. Harding & G. Teleki. New York, NY: Columbia University Press, pp. 191–214.

Heiduck, S. (1997). Food choice in masked titi monkeys (*Callicebus personatus melanochir*): selectivity or opportunism? *International Journal of Primatology*, **18**, 487–502.

Hernández Camacho, J. & Cooper, R.W. (1976). The nonhuman primates of Colombia. In *Neotropical Primates. Field Studies and Conservation*, ed. R.W. Thorington, Jr. & P.G. Heltne. Washington, DC: National Academy of Sciences, pp. 35–69.

Hernández Camacho, J. & Defler, T. (1985). Some aspects of the conservation of non-human primates in Colombia. *Primate Conservation*, **6**, 42–50.

Hershkovitz, P. (1990). Titis, New World monkeys of the genus *Callicebus* (Cebidae, Platyrrhini): a preliminary taxonomic review. *Fieldiana Zoology*, **55**, 1–109.

Homburg, I. (1998). *Ökologie und Sozialverhalten von Weißgesicht-Sakis. Eine Freilandstudie in Venezuela.* Göttingen: Cuvillier Verlag.

Izawa, K. (1975). Foods and feeding behaviour of monkeys in the upper Amazon basin. *Primates*, **16**, 295–316.

Johns, A. (1986). Notes on the ecology and current status of the buffy saki, *Pithecia albicans*. *Primate Conservation*, **7**, 26–29.

Kinzey, W.G. (1977). Diet and feeding behavior in *Callicebus torquatus*. In *Primate Ecology. Studies of Feeding and Ranging Behaviour in Lemurs, Monkeys and Apes*, ed. T.H. Clutton-Brock. London: Academic Press, pp. 127–151.

Kinzey, W.G. (1978). Feeding behaviour and molar features in two species of titi monkey. In *Recent Advances in Primatology 1. Behaviour*, ed. D.J. Chivers & J. Herbert. London: Academic Press, pp. 373–385.

Kinzey, W.G. (1981). The titi monkeys, genus *Callicebus*. In *Ecology and Behavior of Neotropical Primates*, ed. A.F. Coimbra-Filho & R.A. Mittermeier. Rio de Janeiro: Academia Brasileira de Ciências, pp. 241–276.

Kinzey, W.G. (1992). Dietary and dental adaptations in the Pitheciinae. *American Journal of Physical Anthropology*, **88**, 499–514.

Kinzey, W.G. (1997). *New World Primates. Ecology, Evolution, and Behavior*. New York, NY: Aldine de Gruyter.

Kinzey, W.G. & Becker, M. (1983). Activity patterns of the masked titi monkey, *Callicebus personatus*. *Primates*, **24**, 337–343.

Kinzey, W.G. & Norconk, M.A. (1993). Physical and chemical properties of fruit and seeds eaten by *Pithecia* and *Chiropotes* in Surinam and Venezuela. *International Journal of Primatology*, **14**, 207–227.

Kinzey, W.G., Rosenberger, A.L., Heisler, P.S., *et al.* (1977). A preliminary field investigation of the yellow-handed titi monkey, *Callicebus torquatus torquatus*, in northern Peru. *Primates*, **18**, 159–181.

Kulp, J. (2011). Activity budgets and diet composition of red titi monkeys (*Callicebus cupreus*) in primary forest and forest edge. Masters' thesis, University of Göttingen, Göttingen.

Lambert, J.E. (2007). Primate nutritional ecology. Feeding biology and diet at ecological and evolutionary scales. In *Primates in Perspective*, ed. C.J. Campbell, A. Fuentes, K.C. Mackinnon, M. Panger & S.K. Bearder. New York, NY: Oxford University Press, pp. 482–495.

Lawrence, J.M. (2007). Understanding the pair bond in brown titi monkeys (*Callicebus brunneus*): male and female reproductive interests. PhD thesis, Columbia University, New York.

Martin, R.D. (1990). *Primate Origins and Evolution*. London: Chapman and Hall.

Milton, K. & Nessimian, J.L. (1984). Evidence for insectivory in two primate species (*Callicebus torquatus lugens* and *Lagothrix lagothricha lagothricha*) from northwestern Amazonia. *American Journal of Primatology*, **6**, 367–371.

Mittermeier, R.A., Konstant, W.R., Ginsberg, H., *et al.* (1983). Further evidence of insect consumption in the bearded saki monkey, *Chiropotes satanas chiropotes*. *Primates*, **24**, 602–605.

Müller, K.-H. (1996). Diet and feeding ecology of masked titis (*Callicebus personatus*). In *Adaptive Radiations of Neotropical Primates*, ed. M.A. Norconk, A.L. Rosenberger & P.A. Garber. New York, NY: Plenum Press, pp. 383–401.

Nadjafzadeh, M.N. (2005). Strategien und Techniken des Beuteerwerbs von Kupferroten Springaffen, *Callicebus cupreus*, im Vergleich zu den sympatrischen Tamarinarten *Saguinus mystax* und *Saguinus fuscicollis* im Nordosten Perus. Diploma thesis, University of Bochum, Bochum.

Nadjafzadeh, M.N. & Heymann, E.W. (2008). Prey foraging of red titi monkeys, *Callicebus cupreus*, in comparison to sympatric tamarins, *Saguinus mystax* and *Saguinus fuscicollis*. *American Journal of Physical Anthropology*, **135**, 56–63.

Neri, F.M. (1997). Manejo de *Callicebus personatus*, Geofroy 1812, resgatados: uma tentativa de reintrodução e estudos ecológicos de um grupo silvestre na Reserva Particular do Partrimônio Natural Galheiro – Minas Gerais. Master's thesis, Universidade Federal de Minas Gerais, Belo Horizonte.

Norconk, M.A. (1996). Seasonal variation in the diets of white-faced and bearded sakis (*Pithecia pithecia* and *Chiropotes satanas*) in Guri Lake, Venezuela. In *Adaptive Radiations of Neotropical Primates*, ed. M.A. Norconk, A.L. Rosenberger & P.A. Garber. New York, NY: Plenum Press, pp. 403–423.

Norconk, M.A. (2011). Sakis, uakaris, and titi monkeys: behavioral diversity in a radiation of primate seed predators. In *Primates in perspective* (2nd edn), ed. C.J. Campbell, A. Fuentes, K.C. Mackinnon, S.K. Bearder & R.M. Stumpf. New York, NY: Oxford University Press, pp. 122–139.

Norconk, M.A. & Conklin-Brittain, N. (2004). Variation on frugivory: the diet of Venezuelan white-faced sakis. *International Journal of Primatology*, **25**, 1–25.

Palacios, E., Rodríguez, A. & Defler, T. (1997). Diet of a group of *Callicebus torquatus lugens* (Humboldt, 1812) during the annual resource bottleneck in Amazonian Colombia. *International Journal of Primatology*, **18**, 503–538.

Palminteri, S., Powell, G. & Collado, E. (2005). Ecology and habitat use of the monk saki (*Pithecia monachus*) in southeastern Peru: a dietary generalist with a perplexing distribution. In *Abstracts of the 19th Annual Meeting of the Society for Conservation Biology, Brasília,* July 15–19, 2005, pp. 156–157.

Peetz, A. (2001). Ecology and social organization of the bearded saki *Chiropotes satanas chiropotes* (Primates: Pitheciinae) in Venezuela. *Ecotropical Monographs*, **1**, 1–170.

Peres, C.A. (1993). Notes on the ecology of buffy saki monkeys (*Pithecia albicans*, Gray 1860): a canopy seed-predator. *American Journal of Primatology*, **31**, 129–140.

Pérez Yamacita, J.G. (2010). Ecología alimenticia y uso de hábitat del "tocón colorado" (*Callicebus cupreus*, Pitheciidae) en la Estación Biológica Quebrada Blanco, Río Tahuayo, Loreto – Perú. Licenciatura thesis, Universidad Nacional de la Amazonía Peruana, Iquitos.

Price, E.C. & Piedade, H.M. (2001). Diet of northern masked titi monkeys (*Callicebus personatus*). *Folia Primatologica*, **72**, 335–338.

Rosenberger, A.L. (1992). Evolution of feeding niches in New World monkeys. *American Journal of Physical Anthropology*, **88**, 525–562.

Rosenberger, A.L. & Kinzey, W.G. (1996). Functional patterns of molar occlusion in Platyrrhine primates. *American Journal of Physical Anthropology*, **45**, 281–298.

Smith, R.J. & Jungers, W.L. (1997). Body mass in comparative primatology. *Journal of Human Evolution*, **32**, 523–559.

Soini, P. (1987). La dieta del mono huapo (*Pithecia monachus*). *Informe de Pacaya*, **25**, 1–12.

Souza, S.B., Martins, M.M. & Setz, E.Z.F. (1996). Activity pattern and feeding ecology of sympatric masked titi monkeys and buffy tufted-ear marmosets. In *Abstracts of the XVIth Congress of the International Primatological Society/ XIXth Conference of the American Society of Primatologists*, abstract #155.

Souza-Alves, J.P., Fontes, I.P., Chagas, R.R. D., *et al.* (2011). Seasonal versatility in the feeding ecology of a group of titis (*Callicebus coimbrai*) in the northern Brazilian Atlantic Forest. *American Journal of Primatology*, **73**, 1199–1209.

Terborgh, J. (1983). *Five New World Primates*. Princeton, NJ: Princeton University Press.

Terborgh, J. (1992). *Diversity and the Tropical Rain Forest*. New York, NY: Scientific American Library.

Tirado Herrera, E.R. & Heymann, E.W. (2004). Does mom need more protein? Preliminary observations on differences in diet composition in a pair of red titi monkeys, *Callicebus cupreus*. *Folia Primatologica*, **75**, 150–153.

Veiga, L.M. (2006). Ecologia e comportamento do cuxiú-preto (*Chiropotes satanas*) na paisagem fragmentada da Amazônia orienta. Dcotoral dissertation, Universidade Federal do Pará, Belém.

Veiga, L.M. & Ferrari, S.F. (2006). Predation of arthropods by southern bearded sakis (*Chiropotes satanas*) in Eastern Brazilian Amazonia. *American Journal of Primatology*, **68**, 209–215.

Wright, P.C. (1985). The costs and benefits of nocturnality for *Aotus trivirgatus* (the night monkey). PhD thesis, City University of New York.

Wright, P.C. (1986). The nocturnal primate niche in the New World. *Journal of Human Evolution*, **18**, 635–658.

Youlatos, D. & Pozo Rivera, W. (1999). Preliminary observations on the songo songo (dusky titi monkey, *Callicebus moloch*) of northeastern Ecuador. *Neotropical Primates*, **7**, 45–46.

Chapter 20

Seed eating by *Callicebus lugens* at Caparú Biological Station, on the lower Apaporis River, Colombian Amazonia

Erwin Palacios & Adriana Rodríguez

Introduction

Titi monkeys (genus *Callicebus*) comprise upwards of 30 species found mainly in the tropical forests of the Amazon and Orinoco basins (van Roosmalen *et al.* 2002; Wallace *et al.* 2006). Despite the relative ease with which they may be observed, *Callicebus* species have been the subject of limited studies, and consequently our knowledge of many facets of their ecology is still poor. Feeding ecology is one of the aspects for which data are still scarce, despite the seminal work of the late Warren Kinzey in Peru and of the studies his work prompted (e.g. Easley 1982; Wright 1986; Müller 1996; Heiduck 1997; Palacios & Rodríguez 1995; Palacios *et al.* 1997; Price & Piedade 2001; Hernández & Castillo 2002; Alvarez, in progress). Recent studies of titis have revealed that ecological traits such as group size, home range size, feeding ecology and positional behavior differ greatly between species (Müller 1995, 1996, *C. personatus*; Price & Piedade 2001, *C. personatus*; Bicca-Marques *et al.* 2002, *C. cupreus*; Bossuyt 2002, *C. moloch*; Lawler *et al.* 2005, *C. brunneus* and *C. torquatus*). In the same fashion, earlier studies of *C. torquatus* (*C. regulus sensu* van Roosmalen *et al.* 2002) and *C. moloch* (*C. cupreus sensu* van Roosmalen *et al.* 2002) in Peru revealed some striking differences between the diets of these two species, classifying the latter as a frugivore–folivore, and *C. regulus* as frugivore–insectivore (Kinzey 1977, 1978). Such distinctions have been broadly used when describing the diet of the *Callicebus torquatus* group (see, for, example van Roosmalen *et al.* 2002; Norconk 2011). Nevertheless, in the light of some recent studies (e.g. Heiduck 1997; Palacios *et al.* 1997; Hernández & Castillo 2002), it is becoming clear that *Callicebus* species show a range of dietary preferences, for example including in their diets variable amounts of unripe seeds. Seeds make up a substantial portion of the diet of pitheciids (Norconk *et al.* 1998), this being a prime characteristic of this primate family. As such, seed consumption is better understood for *Chiropotes* and *Pithecia*, but not for *Callicebus* species, partially because of the lack of studies mentioned above (*Cacajao* is also poorly known). As a contribution to a better understanding of the feeding ecology of *Callicebus* species in particular, and their position as seed predators within the pitheciid ecological continuum in general, we present data on seed consumption by *C. lugens* during a one-year period in southeastern Colombia.

Methods

Study site

This study was conducted at Estación Biológica Caparú (EBC) (1°09.25′S, 69°50.02′W) (www.teamnetwork.org/en/field_stations/caparu-biological-station) in the lower Apaporis River, Colombian Amazon (Figure 20.1). The forest at EBC corresponds to lowland (200 m a.s.l.) undisturbed forest, where the three major habitats include (1) low hills ("*colinas*") or *terra firme* forest with clayey soils supporting the greatest floristic diversity (~160 species/ha: Defler & Defler 1996); (2) Pleistocene river terraces (Carvajal *et al.* 1979) with an intermediate floristic diversity (~90 species/ha), growing on a predominantly sandy substrate mixed with clays, and (3) *igapó* or seasonally (April–December) flooded blackwater forest, where plant species diversity is the lowest (~ 60 species/ha: Defler & Defler 1996; Palacios & Rodríguez 2001). Mean annual temperature and rainfall at EBC are 25.1°C and 3836 ± 486.3 mm, respectively (n = 11 years) (Defler & Defler 1996, p. 164). Rainfall at EBC follows a seasonally bimodal distribution, and there is no dry season (i.e. monthly precipitation ≤ 100 mm). The driest month (September) averages 258 mm (Defler 1996; Defler & Defler 1996, p. 164) and the wettest period of the year occurs between April and June.

Subjects and data collection

Callicebus lugens is one of the five *Callicebus* species found in Colombia. Widely distributed within Colombia, Brazil and Venezuela (see Hershkovitz 1990; Porter *et al.*, Chapter 32; Ferrari *et al.*, Appendix B), titis are small-sized (ca. 1300 g), monogamous primates. The study group was the same as that studied in 1993–1994 by Palacios and Rodríguez (1995), and

Evolutionary Biology and Conservation of Titis, Sakis and Uacaris, eds. Liza M. Veiga, Adrian A. Barnett, Stephen F. Ferrari and Marilyn A. Norconk. Published by Cambridge University Press.

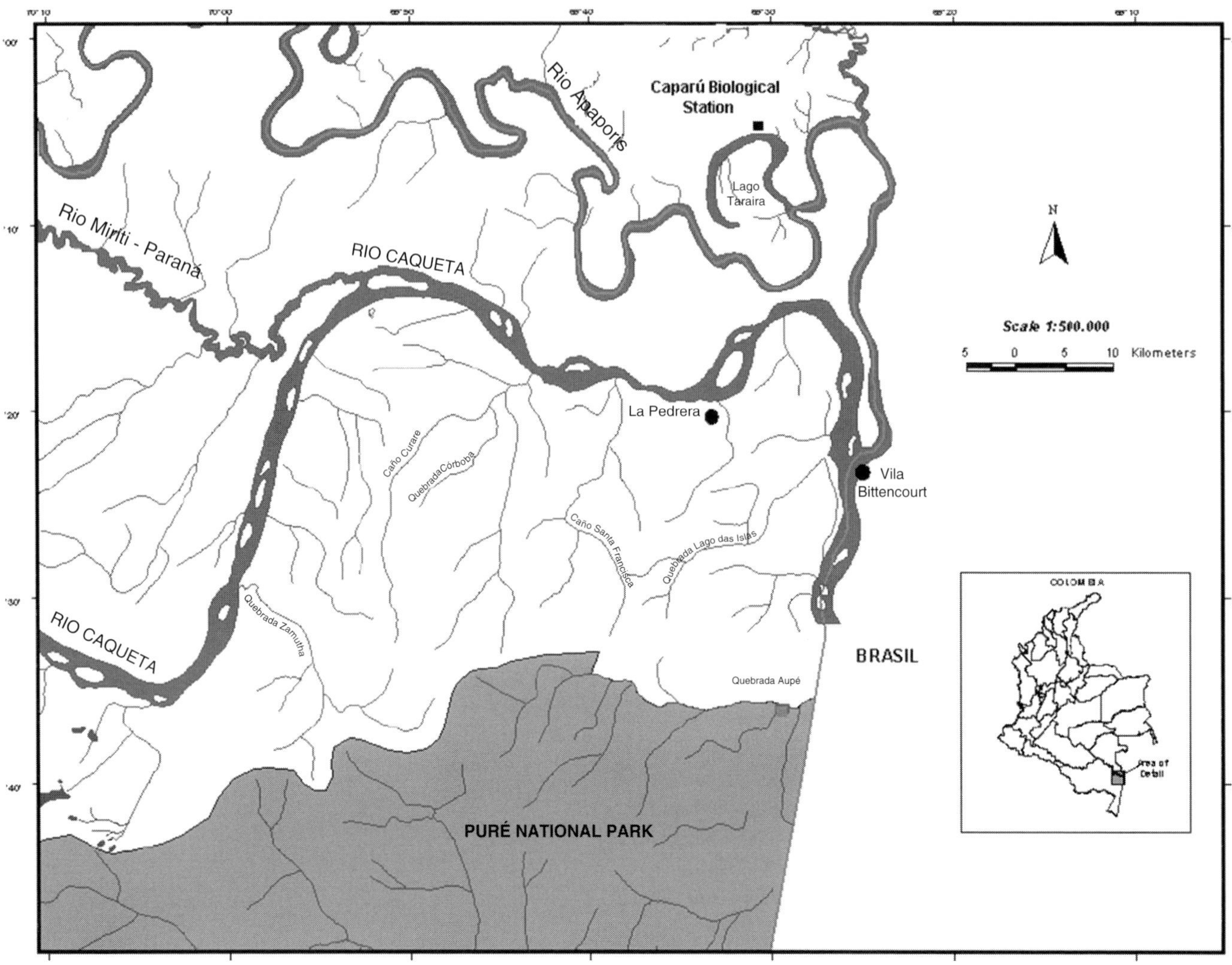

Figure 20.1 Map of the lower Caquetá and Apaporis rivers region showing the location of Caparú Biological Station.

has been habituated to observers for more than 15 years. During 12 consecutive months, we observed the study group, an adult pair and a juvenile, for 4 complete days per month (but 5 days in August and November, and 3 in June), a total of 49 observation days.

Observations began at 05.30 h and usually ended 11.30 h later. Data were gathered using a slow-scan method (Altmann 1974; Robinson 1981), assessing the behavioral state of all visible animals every 15 min. Because the titis carried out all their activities as a cohesive social unit, we were able to observe all three animals during almost every sample. When feeding occurred, we timed individual feeding bouts to obtain time scores. A bout started when any individual began to feed and ended when the animal left the feeding tree, and/or clearly stopped feeding (handling, holding, chewing or ingesting food items). We noted the plant species and major food categories (e.g. ripe fruits, young seeds, young leaves, flowers, insects) of feeding trees. Individual time scores were summed to represent the group's feeding time from any given resource.

Seed ingestion by *C. lugens* was considered under two categories: (i) when the animals consumed seeds on their own (Palacios & Rodríguez 1995; Palacios *et al.* 1997) and (ii) when they were eaten together with fruit flesh (Kinzey & Norconk 1993). We collected plant voucher specimens on a monthly basis. They were later identified and deposited at the Instituto de Investigaciones Amazónicas (SINCHI), Bogotá D.C., and Royal Botanical Gardens, Kew.

By conducting a phenological survey we estimated overall year-round resource availability. Monthly, we observed a total of 360 trees (180 $\geq$ 10 cm DBH and 180 $\leq$ 10 cm DBH) in Pleistocene terrace habitat (Palacios & Rodríguez 1997), and for each tree we registered its phenophase (new leaves, flowers, and mature and immature fruits) and assigned an abundance value on a scale from 0 to 4, where 0 = 0%, 1 = 1–25%, 2 = 26–50%, 3 = 51–75% and 4 = 76–100%. Following Peres (1994), an index of the monthly availability was calculated as (100 × [sum of all abundance scores]/# of trees in a monthly sample).

Results

Overall diet

The study group spent most of its feeding time (94.8%) consuming resources of plant origin. Plant resources in the titi diet were mostly fruits and seeds, which accounted for 33.9% and 48.3% of the feeding time, respectively (Figure 20.2). Consumption of fruit items included mature arils, whole ripe fruits, as well as mesocarps and pericarps of ripe and unripe fruits. The latter accounted for 8.2% of the time spent feeding on fruits. Non-reproductive plant parts represented 9.5% of the feeding time, while flowers were consumed during 3.1% of this time. New leaves were by far the most frequently consumed (90.2%) non-fruit item, with leaf buds comprising 4.8%, and liana tips and young stems just 0.5%. The titis were never seen eating mature leaves during the current study, although they exploited this food source during 1993–1994 (Palacios & Rodríguez 1995; Palacios *et al.* 1997). The titis spent just 3.8% of the feeding time consuming invertebrate prey (insects and small spiders), the former mainly represented by ants and their larval stages (Figure 20.2). Searching for and capturing prey hidden in rolled leaves (both old and dry, and new and green) was a common behavior adopted to gain access to small spider colonies and caterpillars.

Seed consumption and floristic composition

A total of 92 plant species representing 43 genera in 31 families comprised the dietary choices of the titi group. Seeds were obtained from 23 plant species, which included mostly trees and some lianas (Table 20.1). Fruit flesh and seeds were eaten together during 8.2% of the seed eating time, and were obtained from 4 species: *Helycostilis tomentosa* (Moraceae), *Coussarea* sp. (Rubiaceae), *Eugenia griseiflora* (Myrtaceae) and an unidentified Clusiaceae. The remaining 19 species were the sources of seeds which the titis consumed during the remaining 40.1% of time, after separating them from fruit pericarps, and pulp or arils (Table 20.1).

Seeds were predominantly harvested from unripe fruits, except those of an unidentified Annonaceae (Annonaceae sp. 1) extracted from mature fruits. For two *Brosimum* species and for *Helyanthostilis sprucei* (Moraceae) and *Heterostemon conjugatus* (Caesalpinioideae), seeds were extracted from both ripe and near-unripe fruits.

Only two species were used for their seeds, new leaves and flowers, *Anaxagorea brachycarpa* (Annonaceae) and *H. conjugatus*, while one *Brosimum* (Moraceae) species was used for their new leaves and seeds, three species of *Iryanthera* (Myristicaceae) for their unripe seeds and mature arils, and one unidentified Annonaceae (sp. 1) for its mature seeds and pulp (Table 20.1). Seeds of two species dispersed from the tree by anaemophily, *Odontadenia macrantha* (Apocynaceae), and *Aspidosperma marcgravianum* (Apocynaceae) were picked up from the forest floor and were then eaten after the animal climbed back into a nearby tree.

The top ten species used by the study group represented 77.7% of the time for seed consumption, and 45.5% of the total feeding time. Five of these species were included among the 10 most important species consumed by the group during a previous 6-month study of the same group in 1993–1994 (Palacios & Rodríguez 1995; Palacios *et al.* 1997). Seeds were found and eaten between 1 and 12 m above the forest floor during 68.7% of the seed eating time, while 30.9% of this feeding time took place between heights of 13 and 30 m above the ground.

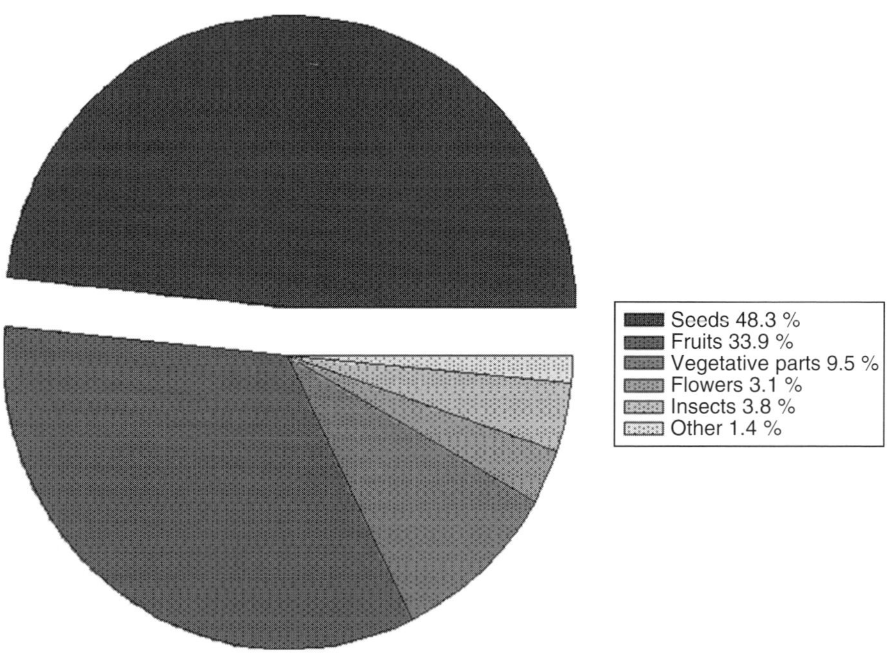

Figure 20.2 Proportion of different food items in the diet of *Callicebus lugens*.

Table 20.1 Plant species used for their seeds by *Callicebus lugens*. Months during which any food item (flowers, arils, pulps, and new leaves) of the species was eaten are shown. Months in bold-faced type represent those when seeds were eaten (in some cases, during these months other food sources of the species could also take place). Life form: t = tree; l = liana.

Family and species	Life form	Period taken
Annonaceae		
Anaxagorea brachycarpa	t	**Jan–Mar**, **May–Jun**, **Aug,** Sep, Oct, **Nov–Dec**
Annonaceae sp.	t	**Mar**
Clusiaceae		
Clusia sp.	l	**Aug**
Convolvulaceae		
Maripa sp.	l	**Jan**, **Mar**, **May**, **Jun**
Euphorbiaceae		
Conceveiba guianensis	t	**Jan**
Sandwithia heterocalyx	t	**Jan–Dec**
Leguminosae Caesalpinioideae		
Heterostemon conjugatus	t	**Jan–Feb**, Apr–**Dec**
Monimiaceae		
Siparuna descipiens	t	**Mar**, **Jul**, Nov–**Dec**
Moraceae		
Brosimum rubescens	t	**Aug**, **Nov–Dec**
Brosimum lactescens	t	**Nov–Dec**
Brosimum potabile	t	**Jan–Mar**
Brosimum utile	t	**May, Nov–Dec**
Brosimum sp.	t	**Nov–Dec**
Helianthostylis sprucei	t	**Feb–Jun**, **Aug–Oct**
Helicostylis tomentosa	t	**Oct**
Myristicaceae		
Iryanthera crassifolia	t	Jan, **Feb**, **Mar**, **Jul–Aug**, **Nov–Dec**
Iryanthera juruensis	t	Jan–Feb, **Nov**
Iryanthera ulei	t	**Jan**, **Mar**
Iryanthera elliptica	t	Feb, **Apr–May**
Myrtaceae		
Eugenia griseiflora	t	**Apr**, **Dec**
Rubiaceae		
Coussarea sp.	t	**Jul–Sep**
Apocynaceae		
Odontadenia macrantha	l	**Dec**
Aspidosperma marcgravianum	t	**Mar**

Seasonal variation in seed consumption

Although seeds were consumed year round, monthly consumption varied greatly, ranging from 19.7% (October) to 79.8% (July) of the monthly feeding time. Seed consumption was correlated with neither the availability of immature fruits ($r_s = -0.28$; $P > 0.1$), nor mature fruit ($r_s = -0.29$; $P > 0.1$). Seeds exceeded the proportion of ripe fruit during 8 months of the year (Figure 20.3), and feeding on seeds was strongly negatively correlated with other fruit parts ($r_s = -0.85$; $P < 0.01$). Some species were used over extensive periods of time: the seeds of *S. heterocalyx* and *H. sprucei* were used by the titis for 12 and 10 months, respectively, while the seeds and flowers of *A. brachycarpa* and *H. conjugatus* were used for more than 10 months each, and seeds and ripe arils of *Iryanthera crassifolia* were used by the titis during 7 months (Table 20.1). Remaining sources of seeds were used for between 1 and 3 months.

Discussion

Seeds are the most important food item in the diet of *C. lugens* in southeastern Colombia on a year-round basis. Seeds also represented a considerable proportion (26.9%) of the diet of the same group during a previous 6-month study (Palacios *et al.* 1997). The previous study mainly covered the lean season (August–November) when the availability of many resources (especially fleshy fruits) sharply

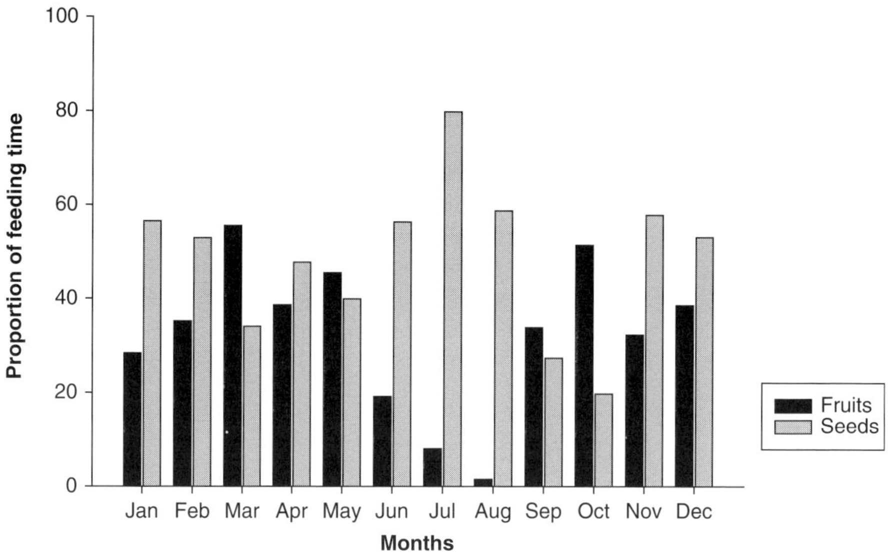

Figure 20.3 Monthly variation in the feeding time devoted to seeds and fruits by *Callicebus lugens*.

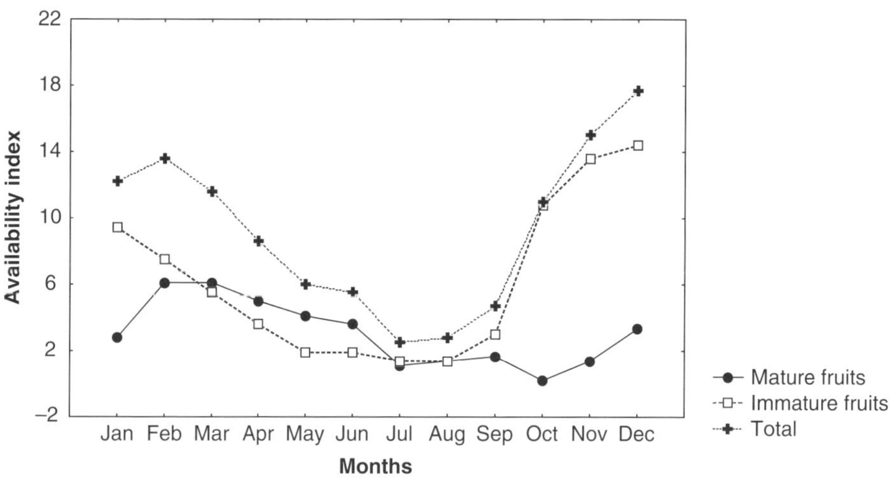

Figure 20.4 Monthly variation in mature, immature and total fruits availability in the Pleistocene terraces habitat during the study period.

drops (Palacios & Rodríguez, 1997). Based on such results, we concluded (Palacios *et al.*, 1997) that seed eating in *C. lugens* could be a strategy to face periods of scarcity of food sources, especially because of the protein-rich resource which seeds represent (Janzen 1981; Waterman 1984; Heiduck 1997). The results of the current study emphasize that seeds are indeed a very important food source for *C. lugens* during low food availability periods, but that they are also a key resource during months when fruit availability peaks (Figure 20.4). For example, the time the titis devoted to seed consumption during February, November and December, when total fruit production reached the highest levels, was higher than for fruit eating. High levels of seed consumption have been also documented for *C. melanochir* in Brazil's Atlantic forest by Heiduck (1997: 26.4% of the total feeding time). Kinzey's studies on the feeding behavior of *C. torquatus* in Perú (Kinzey 1977, 1978, 1981) did not mention the consumption of seeds, but it is likely they were included in their diet, as his food plant list included species such as *Anaxagorea* sp. (Annonaceae), *Hevea* sp. (Moraceae), and *Alchornea* sp. (Euphorbiaceae), which lack either pulp or arils (Palacios *et al.* 1997).

The continued consumption of seeds during the year by the titis is favored by the availability of small, but temporally extended, crops of species such as *S. heterocalyx* and *A. brachycarpa*. Tree species that fruit asynchronously, such as *H. sprucei* and *I. crassifolia*, provide more abundant sources of seeds for extended periods. The seeds of the remaining set of species are harvested by the titis especially during the first and last months of the year. Like bearded sakis and uacaris (Norconk *et al.* 1998), *C. lugens* consumes young seeds and mesocarps from the same plant species. Four species of the genus *Iryanthera* provided these important food sources in the lower Apaporis River. The strategy of *C. lugens* to exploit seeds in the lower strata of the canopy allows them to avoid competition with other sympatric primate species, especially when using seeds from the small trees of *A. brachycarpa*.

Martin *et al.* (2003) stated that the lack of crack-resistant properties in the enamel of *Callicebus* correlated with the lower proportion of seeds in their diet. At least for *C. lugens*, it correlates with the softer seeds they consume, compared with those eaten by other pitheciines (Kinzey & Norconk 1990, 1993), but not necessarily with the amount of seeds in the diet. As Kinzey (1992) and Rosenberger (1992) have discussed, the dentition of *Callicebus* is best interpreted as part of the hard-fruit and seed-adapted continuum. *Callicebus lugens* makes use of its canines to open small, multilocular fruits such as *S. heterocalyx* and *Conceveiba guianensis* to extract their small seeds, as does *P. pithecia* to extract tiny seeds of many species of Euphorbiaceae (Norconk 2011). The titis were observed to use their canines to puncture the pericarps of immature *Iryanthera* spp. fruits, and broke down the relatively hard immature seeds with their molars, after having scraped the endocarp away with their incisors. The canines of *C. lugens* also play an important role in processing the seeds of *H. sprucei*, seeds which titis are unable to directly masticate because of their large size (mean = 1.7 cm, $n = 15$). These seeds are punctured and broken up with the canines, so that the resultant smaller pieces are easily processed with the molars. In the same fashion, opening the hard pods of *Heterostemon conjugatus* is facilitated by the use of canines. At the time *C. lugens* exploits immature seeds, these pods have a very leathery texture that is difficult to rip apart to get access to two or three seeds. Using their canines they were able to perforate the pods to extract the seeds – usually with the fruits still attached to the tree.

The high proportions of seeds in the diet of *C. lugens*, and their morphological and behavioral adaptation for eating seeds, suggests that this species could be considered specialized seed predators. This may well also be true for some other species in the genus *Callicebus*. More studies focusing on the diet of the various species of the genus (especially the *torquatus* group) are needed in order to better document and understand the particularities of the feeding ecology of the titis, and their position within the pitheciine dietary continuum.

Acknowledgments

Thanks to COLCIENCIAS (Colombia) and Fundación para la Promoción de la Investigación y la Tecnología – Banco de la República for funding this study. Thanks also to Dairon Cárdenas, César Marín, René López, Diego Giraldo and Alberto Posada of Instituto de Investigaciones Amazónicas SINCHI, and to Sir Ghillean T. Prance and T. D. Pennington of the Royal Botanical Gardens, Kew for helping us with plant specimen identification. Two anonymous reviewers provided critical comments on early drafts of the manuscript. Angel and Lucas Yucuna, and Raúl Cubeo provided invaluable help with the collection of botanical specimens.

References

Altmann, J. (1974). Observational study of behaviour: sampling methods. *Behavior*, **49**, 277–265.

Bicca-Marques, J.C., Garber, P.A. & Azvedo-Lopez, M.A.O. (2002). Evidence of three resident adult male group members in a species of monogamous primate, the red titi monkey (*Callicebus cupreus*). *Mammalia*, **66**, 138–142.

Bossuyt, F. (2002). Natal dispersal of titi monkeys (*Callicebus moloch*) at Cocha Cashu, Manu National Park, Peru. *American Journal of Physical Anthropology*, **93**, 505–524.

Carvajal, L.F.J., Posada, A.F.N., Molina, M.L.C., *et al.* (1979). Bosques. In *La Amazonía Colombiana y sus Recursos (Proyecto Radargramétrico del Amazonas, IGAC.*, Bogotá: República de Colombia, pp. 217–322.

Defler, T.R. (1996). Aspects of the ranging pattern in a group of wild woolly monkeys (*Lagothrix lagothricha*) in the NW Amazon of Colombia. *American Journal of Primatology*, **38**, 289–302.

Defler, T.R. & Defler, S. (1996). Diet of a group of *Lagothrix lagothricha* in southeastern Colombia. *International Journal of Primatology*, **17**, 161–189.

Easley, S.P. (1982). The ecology and behaviour of *Callicebus torquatus* (Cebidae: Primates). Unpublished PhD dissertation, Washington University, St. Louis.

Heiduck, S. (1997). Food choice in masked titi monkeys (*Callicebus personatus melanochir*): selectivity or opportunism? *International Journal of Primatology*, **18**, 487–502.

Hernández, B. & Castillo, C. (2002). Uso del espacio por *Callicebus torquatus lugens* en la Estación Biológica Caparú. Bachelor's thesis, Universidad Nacional de Colombia, Santafé de Bogotá.

Hershkovitz, P. (1990). Titis, New World monkeys of the genus *Callicebus* (Cebidae: Platyrrhini): a preliminary taxonomic review. *Fieldiana Zoology* new series no. **55**, 1–109.

Janzen, D.H. (1981). Digestive seed predation by a Costa Rican baird's tapir. *Biotropica*. **13** (Supplement), 59–63.

Kinzey, W.G. (1977). Diet and feeding behaviour of *Callicebus torquatus*. In *Primate Ecology: Studies of feeding and ranging behaviour in Lemurs, Monkeys and Apes*, ed. T.H. Clutton Brock. London: Academic Press, pp. 127–151.

Kinzey, W.G. (1978). Feeding behavior and molar features in two species of titi monkey. In *Recent Advances in Primatology*, Vol. **1**, ed. D.J. Chivers & J. Herbert. New York, NY: Academic Press.

Kinzey, W.G. (1981). The titi monkeys, genus *Callicebus*. In *Ecology and behaviour of Neotropical primates*, ed. A.F. Coimbra-Filho & R. Mittermeier. Rio de Janeiro: Academia Brasileira de Ciencias, pp. 241–276.

Kinzey, W.G. (1992). Dietary and dental adaptations in the Pitheciinae. *American Journal of Physical Anthropology*, **88**, 499–514.

Kinzey, W.G. & Norconk, M.A. (1990). Hardness as a basis for fruit choice in two sympatric primates. *American Journal of Physical Anthropology*, **81**, 5–15.

Kinzey, W.G. & Norconk, M.A. (1993). Physical and chemical properties of fruit and seeds eaten by *Pithecia* and *Chiropotes* in Surinam and Venezuela. *International Journal of Primatology*, **14**, 207–227.

Lawler, R.R., Ford, S.M., Wright, P.C., *et al.* (2005). The locomotor behavior of

Callicebus brunneus and *Callicebus torquatus*. *Folia Primatologica*, 77, 228–239.

Martin, L.B., Olejniczak, A.J. & Maas, M.C. (2003). Enamel thickness and microstructure in pitheciine primates, with comments on dietary adaptations of the middle Miocene hominoid *Kenyapithecus*. *Journal of Human Evolution*, **45**, 351–367.

Müller, K.H. (1995). Ranking in masked titi monkeys (*Callicebus personatus*) in Brazil. *Folia Primatologica*, **65**, 224–228.

Müller, K.H. (1996). Feeding ecology of masked titis, *Callicebus personatus*. In *Adaptive Radiations of Neotropical Primates*, ed. M.A. Norconk, A.L. Rosenberger & P.A. Garber. New York, NY: Plenum, pp. 383–401.

Norconk, M.A. (2011). Sakis, uakaris and titi monkeys: behavioral diversity in a radiation of primate seed predators. In *Primates in Perspective* (2nd end), ed. C.J. Campbell, A. Fuentes, K.C. MacKinnon, S.K. Bearder & R.M. Stumpf. New York, NY: Oxford University Press, pp. 122–139.

Norconk, M.A., Grafton, B.W. & Conklin-Brittain, N.L. (1998). Seed dispersal by Neotropical seed predators. *American Journal of Primatology*, **45**, 103–126.

Palacios, E. & Rodríguez, A. (1995). Caracterización de la dieta y Comportamiento Alimentario de *Callicebus torquatus lugens*. Bachelor's thesis, Universidad Nacional de Colombia, Santafé de Bogotá.

Palacios, E. & Rodríguez, A. (1997). *Abundancia de frutos y patrones de producción en tres hábitats de bosque primario en la Amazonía Colombiana*. Final report to Fundación para la Promoción de la Investigación y la Tecnología – Banco de La República, Santafé de Bogotá.

Palacios, E. & Rodríguez, A. (2001). Ranging pattern and use of space in a group of red howler monkeys (*Alouatta seniculus*) in a southeastern Colombian rainforest. *American Journal of Primatology*, **55**, 233–251.

Palacios, E., Rodríguez, A. & Defler, T. (1997). Diet of a group of *Callicebus torquatus lugens* (Humboldt, 1812) during the annual resource bottleneck in Amazonian Colombia. *International Journal of Primatology*, **18**, 503–522.

Peres, C.A. (1994). Primate responses to phenological changes in an Amazonian terra firme forest. *Biotropica*, **26**, 285–294.

Price, E. & Piedade, H.M. (2001). Diet of northern masked titi monkeys (*Callicebus personatus*). *Folia Primatologica*, **72**, 335–338.

Robinson, J.G. (1981). Vocal regulation of inter- and intragroup spacing during boundary encounters in the Titi monkey, *Callicebus moloch*. *Primates*, **22**, 161–172.

Rosenberger, A.L. (1992). Evolution of feeding niches in New World monkeys. *American Journal of Physical Anthropology*, **88**, 525–562.

van Roosmalen, M.G.M., van Roosmalen, T. & Mittermeier, R.A. (2002). A taxonomic review of the titi monkeys, genus *Callicebus* Thomas, 1903, with the description of two new species, *Callicebus bernhardi* and *Callicebus stephennashi*, from Brazilian Amazonia. *Neotropical Primates*, **10**(Suppl.), 1–52.

Wallace, R.B., Gomez, H., Felton, A., *et al.* (2006). On a new species of titi monkey, genus *Callicebus* Thomas (Primates, Pitheciidae), from western Bolivia with preliminary notes on distribution and abundance. *Primate Conservation*, **20**, 29–39.

Waterman, P.G. (1984). Food acquisition and processing as a function of plant chemistry. In *Food Acquisition and Processing in Primates*, ed. D.J. Chivers, B.A. Wood & A. Bilsborough. New York, NY: Plenum, pp. 177–211.

Wright, P.C. 1986. Ecological correlates of monogamy in *Aotus* and *Callicebus*. In *Primate Ecology and Conservation*, ed. J.G. Else & P.C. Lee. Cambridge: Cambridge University Press, pp. 159–167.

Chapter 21

Callicebus in Manu National Park: territory, resources, scent marking and vocalizations

Patricia C. Wright

Introduction

Callicebus brunneus live in discrete nuclear family units, with each territorial group bordering several other groups (Wright 1989). Six to eight groups form a "neighborhood" with intergroup communication accomplished via loud duets or family choruses. As in gibbons, siamangs, and indri, the *Callicebus* adult pairs that hold territories duet loudly in the morning (Mason 1966, 1968; Marshall & Marshall 1973; Mitani 1984, 1985; Pollock 1986). The *Callicebus* calls project over 1 km, incorporate a series of chirrups, pants, and whoop-gobbles, and have been described as "operatic turkey gobble" (Kinzey 1981, 1997; Kinzey & Robinson 1983). Male and female duet components differ in tone and timing (Robinson 1979, 1981; Robinson *et al.* 1987). Several hypotheses have been suggested regarding the function of loud duets in primates, including defense of territory (Robinson *et al.* 1987; Cowlishaw 1992), resource (Kinzey 1997) and mate (Tilson & Tenaza 1976). Duets may advertise the health and strength of the caller, and hence the threat to competitors of the same sex (Gittens 1984; Mitani 1984).

In this study I predicted that if calling is used as territorial defense, then dawn chorusing would occur throughout the home range periodically to declare this property occupied. This chorusing system would act as auditory "no trespassing" signs posted throughout the year as the food sources within the territory are crucial over an annual cycle (Emlen & Oring 1977). If more discrete resource defense is driving the system, morning duets should occur near large fruit trees with ripe fruit. If mate defense is a primary selective force, duetting should occur more frequently before and during the mating season than at other times of the year. Furthermore, considering that subadults join in the morning loud calls (Robinson *et al.* 1987), they may be advertising for mates. Thus, I predicted that calling by subadults would increase during the months before natal dispersal.

Methods

Study site

The study was conducted in the trail system of Estación Biológica Cocha Cashu (Cocha Cashu Biological Research Station) in Manu National Park, southeastern Peru (11°51′S, 71°19′W) from August 1980 to August 1981 and August–December 1982. The park, established in 1973, is one of the largest protected areas of rain forest in the world, encompassing 15,000 km^2 on the western edge of the Amazon basin. Even before the establishment of the park, animal and plant populations were unaffected by hunting or logging because of the low human population density in the region and difficult accessibility. Annual rainfall is 2000 mm, with most of the rain falling from October to April (Terborgh 1983). More than 28 km of trail covers a 5–7 km^2 area bordering the Manu River. Manu National Park has 12 sympatric species of primates; of which, 8, including *C. brunneus* have been the subject of a dissertation or book (Terborgh 1983; Janson 1985; Symington 1988; Wright 1989; Goldizen & Terborgh 1989; Mitchell *et al.* 1991; Rodman & Bossuyt 2007). This is one of the highest primate community biomasses documented in Amazonia (Janson & Emmons 1990). *Callicebus brunneus* is monomorphic, weighs approximately 1000 g, lives in discrete monogamous pairs, and engages in extensive paternal care (Kinzey 1981, 1997; Wright 1984, 1986, 1989, 1990).

Censuses

Censuses were conducted at the beginning of this study within the Cocha Cashu trail system. I walked all the main trails at Cocha Cashu at a speed of approximately 1 km/h. Diurnal censuses began at 06:00 h and ended at 12:00 h. The sighting distance from the trail was determined to be 15 m. *Callicebus* individuals were detected visually, by hearing vocalizations, or by hearing movement of branches. For each individual encountered, I noted species, time, location on the trail,

Evolutionary Biology and Conservation of Titis, Sakis and Uacaris, eds. Liza M. Veiga, Adrian A. Barnett, Stephen F. Ferrari and Marilyn A. Norconk. Published by Cambridge University Press.

perpendicular distance from trail to first individual seen, height from ground, calls given and interindividual distance of group members. I also estimated age according to body size, and determined sex by observing the angle of urination. Censuses were conducted twice monthly from January to August 1981 and September to December 1982.

Focal groups

Two focal *Callicebus* groups were followed continuously from exiting until entering the sleep tree for 5 days per month. Activity such as feeding, resting and traveling was sampled every 5 min on all individuals in the group. Ad libitum data were taken during these 5-day samples on vocalizations, scent marking, fighting, playing and copulating. Aggression included chasing, loud vocalizations and branch shaking which generated attention-getting noise.

During focal group follows, morning censuses of neighboring group locations were accomplished by taking the angle and time of all dawn calls emitted that morning, using the focal group's location as a base point. Triangulating the calls with a second observer was possible for about 6 months of the study.

No animals were captured in this study. Individual *Callicebus* were distinguished by size (adults were larger than one or two year olds), and amounts of white on tail tips or above the lip.

Home-range sizes of the two focal groups were determined after plotting the 16 months of 5 day or night samples and measuring the area within the minimum convex polygon. Home-range sizes for the remaining groups were estimated by plotting locations obtained both visually and by hearing during censuses over the 16-month period.

Fruit abundance

One hundred fruit traps were placed 25 m apart and collected bimonthly. All fruits of species eaten by the monkeys were counted in the traps. Percentage of traps containing monkey fruit was calculated for each month (fruit score) (Terborgh 1983).

Results

Home-range size and species densities

Censuses and focal group follows suggested that *Callicebus brunneus* at Cocha Cashu Research Station live in monogamous pairs throughout the year, and occupy exclusive home ranges from which invaders of the same species, whether solitary or paired, are chased. Population density was 25–30 individuals/km^2. Home-range size was 6–8 ha ($N = 6$), mean of 6.8 ha (Figure 21.1). (Home ranges bordered the river or lake and extended about 100–200 m into the interior.) *C. brunneus* were rarely seen in fig swamps or higher ground. Group size ranged from 2 to 5 (mean = 4.16; $N = 6$), and births took place in May, July, September and November ($N = 8$ births).

Intragroup communication

Callicebus brunneus have a short warble call to alert to aerial predators and a longer warble call to mob felid predators. Contact calls directed by the male towards the whole group include a high-pitched whistle which functions to reunite him with his group, and a louder low gobble for focused traveling to the next tree.

Intergroup communication

Loud call chorusing

Callicebus brunneus pairs gave synchronized dawn duets at 5–15 min intervals, on average once every other morning in different areas of their home ranges (Figure 21.2). Pairs called from a fixed point each day and did not move until the calling bout was over, after approximately 5–15 min. In a 5-day sample ($n = 26$), calls came from about three different points, usually one at the north, south and middle of the home range. Calls were contagious and elicited response from 4 to 6 groups, many located across the lake. Male and female calls were different in pitch and tone, with males slightly lower as described for other *Callicebus* species (Robinson 1981). The subadult in focal group II did not call in October; however, he joined his parents in the loud call in November before dispersing in December. The juvenile did not join in any duets until April, and then called every time with parents thereafter, refraining in October when his group did not give any loud calls.

Loud calls and energy minimization

The frequency of loud call chorusing varied seasonally with frequent loud calls during the season of high fruit abundance (November–May; 3–4 times per morning every day) and rare chorusing during June–October, the season of scarce resources (Figure 21.3).

During June–August *Callicebus* minimised energy expended on travel and activity patterns. Mean daily path length in June was only 349 ± 143.6 m, compared with March's mean daily path length which was 1108 ± 214.9, nearly four times farther. Activity was reduced from 12 h/day from September–March to 4–5 h/day in June–August.

Copulations and communication

In focal Group II mating was observed on only two days: 7 April (15:21 h) and 8 April (11:50 h). Time between copulations, assuming that no nocturnal copulations occurred, was 20 h 29 min. I monitored the female continually from 4 to 10 April. Loud call choruses occurred between the hours of 5:50 and 7:04; once on 4 April (duet), twice on 7 April (trio) the day of copulation, none on 8 April the second day of copulation and twice on 9 April (trio), when a lone male was chased away by the subadult. By observing newborn infants, we could estimate that matings occurred from February to June for five of the six groups, and therefore females were gestating during the season of food scarcity.

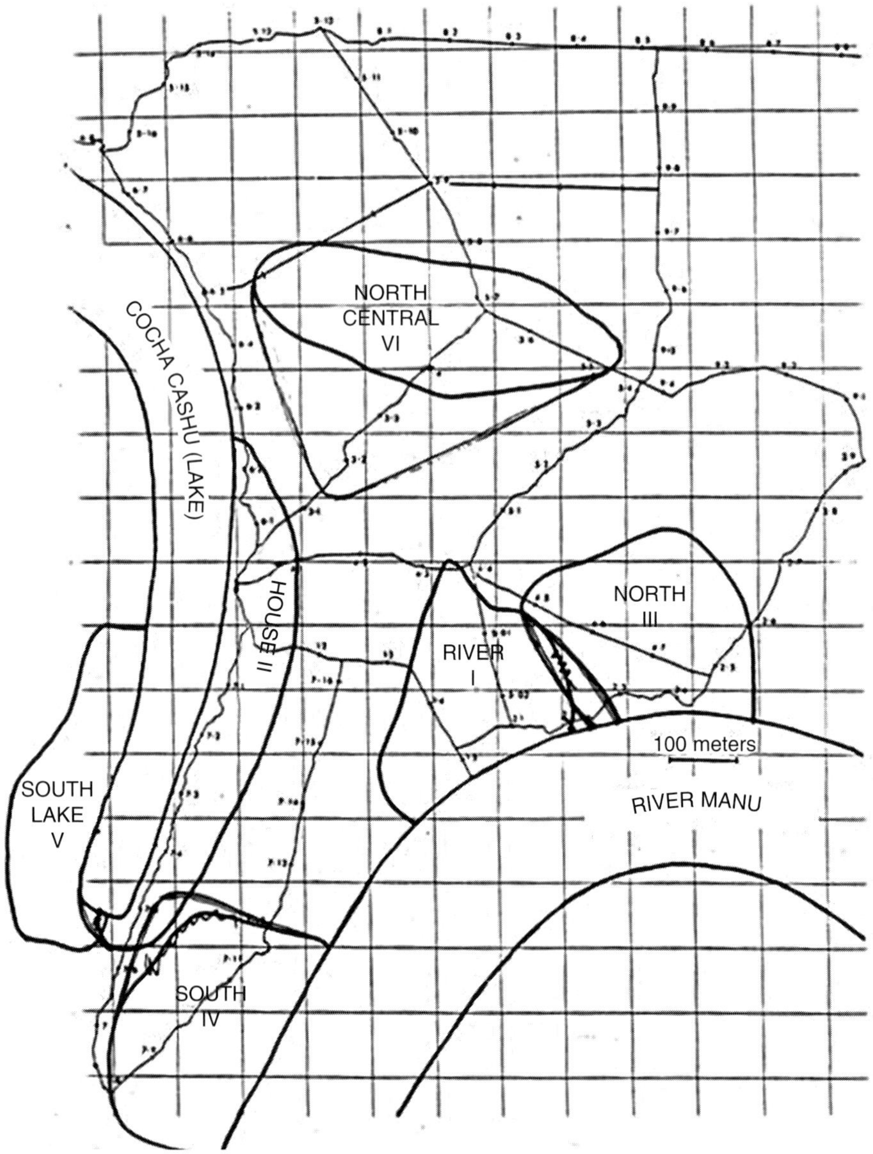

Figure 21.1 Territories of *Callicebus brunneus* groups from censuses 1980–1982 at Cocha Cashu Research Station, Manu National Park, Peru.

Scent-marking

Callicebus brunneus did not rely on scent-marking to demarcate territories. There were 43 observations of chest marking by male monkeys within the 16 months of observing two groups: a subadult male chest marked a horizontal branch for 1.6 min 3 days before his mother copulated with the adult male in the group; the adult male chest marked branches for less than a minute after copulations with his mate; and this same adult male rubbed his chest onto the female's back before copulation. Neither copulations nor scent-marking were observed in Group I.

Intergroup agonistic contests (IACs)

From September 1980 to August 1981, only four IACs were observed in focal follows of Group II and two in Group I. All IACs occurred at territorial borders when a large fruit tree at the border was at peak fruiting (Figure 21.2). Duration of encounters ranged from 5 to 30 min. In the IACs observed, the mated pair, both male and female, responded with a series of loud calls and the subadult male (if present) joined the calling. Calls were interspersed with charges toward the border and chasing the opposing group. In all cases, females remained 25–30 m behind her mate, quietly sitting with infants and juveniles while the adult male and subadult male called and chased. After repeating the call, chasing the opponent, stopping to call again, and then continuing the chase, the retreating group was chased by either the subadult (if a subadult was present), or by the adult male. The "winner" gave the final loud call, and remained at the territorial boundary for 10–15 min, often feeding in the fruit tree, after the other group had retreated. I never observed any direct combat, only chases. Several aggressive interactions occurred at dawn as a solitary individual approached the sleep tree. The subadult male from the resident family attacked and chased the intruder for over 100 m. This behavior suggests that both territorial defense and mate defense may be important functions of IACs, with the adult male guarding his mate, while the subadult male defended the territory. Females did not actively participate in territory or mate defense during the observed IACs.

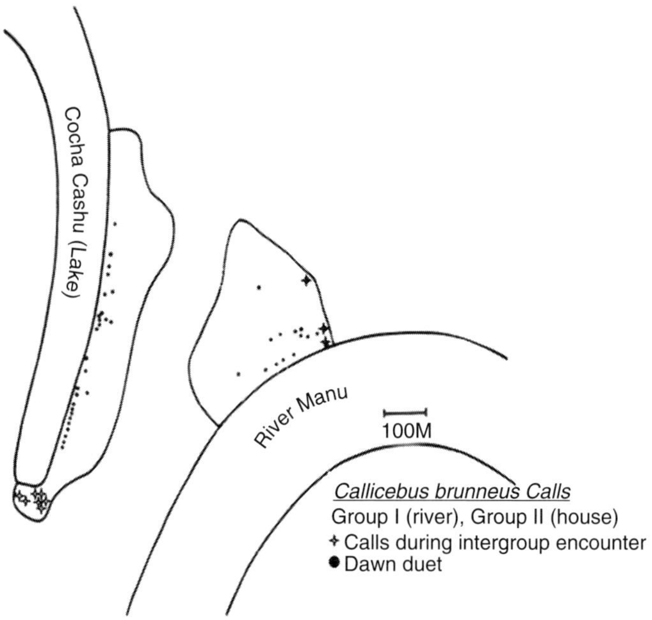

Figure 21.2 Map of Intergroup Agonistic Contests (IAC) and dawn choruses of *Callicebus* Group I and Group II. This is a composite of 16 months (n = 16 IAC).

Dispersal patterns and behavior of solitary individuals

I observed the dispersal of a 3-year-old subadult *C. brunneus* male from Group II when the infant of the group was 5 months old and starting to become independent. The subadult became more distant from other group members during the month before dispersal. Although he lagged as much as 25 m behind the group, he did not exit the group, or cross boundaries, before dispersal. No aggression was seen between the father and subadult male.

Encounters between solitary *C. brunneus* and a resident group were observed five times. The group responded immediately to sightings of solitary individuals by giving a series of loud duets and chasing the intruder. In two cases, the solitary individuals could be identified as male (by observing angularity of urination), and were chased by the resident adult male or subadult male. One solitary individual, which approached Group II in April, was an older male, easily distinguished because half his tail was missing. He had been seen 4 years earlier, holding a territory with a female and offspring over 2 km from focal Group II. An additional solitary male approached Group I in November. This solitary male had a complete tail, and was sighted and chased by the male and female in Group I intermittently for 4 consecutive days. Solitary males lingered near territorial boundaries between

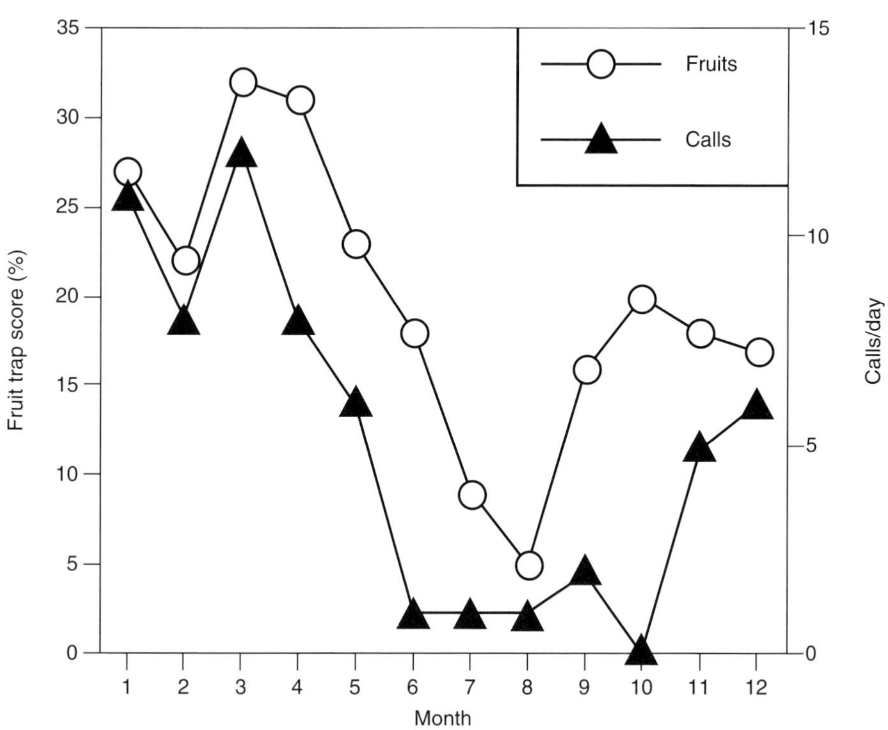

Figure 21.3 Availability of fruits, measured by fruit traps over an annual cycle at Manu National Park, Peru compared to loud calls per month for *Callicebus*.

groups, slept alone in trees at the boundaries, moved low and quietly in the forest, and did not call. In each case, after a few days the solitary individual disappeared and was observed alone on the other side of the lake a few weeks later.

Dispersal by subadults was observed in four groups during this study. These dispersals occurred during the months of July ($N = 1$), September ($N = 1$), and December ($N = 2$). Censuses suggested that all subadults had dispersed from all eight groups before January.

Outcomes of predictions

Calling and IACs are associated with territorial defense. I predicted that if calling was used as territorial defense, that calling and IACs by both breeding territory-holders would be distributed equally throughout all seasons, as territorial boundaries and the food within it are crucial over an annual cycle (Emlen & Oring, 1977), and that calling and IACs would be restricted to territorial boundaries. *Callicebus brunneus* calling frequency was significantly positively correlated with high fruit abundance as indicated by fruit traps (Spearman Rank correlation $P = 0.012$, based on data in Figure 21.3). IACs were not evenly distributed throughout the seasons, and all IACs occurred during the season of high ripe fruit abundance. Scent-marking was not observed at territorial borders, but ranging patterns indicated that all borders were checked within a 5- to 6-day period. All IACs occurred at territorial boundaries, as predicted for the territorial defense hypothesis.

Loud calls and IACs are associated with resource defense. I predicted that if resource defense was driving the system, loud calls and IACs would occur near large fruit trees with ripe fruit. *Callicebus brunneus* loud choruses were given during months with high fruit abundance (Figure 21.3), but not in the proximity of fruit trees. IACs occurred at large trees with ripe fruit at the border of the territory. As predicted, IACs were associated with resource defense in *C. brunneus*, but loud calls were not.

Loud calls and IACs are associated with mate defense. I predicted that if mate defense was a primary selective force, loud call duetting would be more concentrated before and during the mating season than at other times of the year. I predicted that IACs would decrease during these months, to avoid cuckoldry, as extra-pair copulations have been observed during IACs in other monogamous species (Mason 1966; Hubrecht 1985; Garber *et al.* 1993; Palombit 1994). I predicted that IACs would decrease during the months when infants were < 3 months old to avoid possibilities of infanticide by neighboring groups (Cowlishaw 1992). Enthusiastic loud choruses were given during the days preceding and during breeding in *Callicebus*, but this behavior was no different than that seen when the female was not breeding. IACs were observed at peak mating months, allowing the possibility of extra-pair copulations, but in observations, the subadult rather than the mated pair chased the opposing pair.

Loud calls and IACs facilitate courtship and mate choice for subadults before dispersal. I predicted that if calling and IACs facilitated courtship and mate choice for subadults, that subadults would participate in both dawn loud call duets and IACs. In *Callicebus*, the subadults behaved as predicted by actively joining choruses the month before natal dispersal and chasing the neighbor deep inside its home range as part of the an IAC. No copulations were observed during IACs in this study.

Discussion

Defense of territory, discrete resources and mates

In general, aggressive primate intergroup interactions occur over access to food, territories or mates (Dunbar 1988; van Schaik *et al.* 1992). In obligate monogamy, the stakes are high for both the mated male and the mated female to guard a high-quality territory and a productive mate against intruders (Kleiman 1977). However, IACs at borders are energetically expensive and can lead to severe wounds. Loud dawn duets or scent-marking may function as a less-expensive form of mate, resource and territory defense (Lewis 2005, Pochron *et al.* 2005; Powzyk & Mowry 2006; Overdorff & Tecot 2006), but threats and advertisements are not effective without enforcement (Oates 1995). Intergroup encounters may provide a challenge for the mated pair, since extra-pair copulations have been observed to occur during interactions of *Callicebus lugens* or *C. medemi* (Mason 1966, van Roosmalen *et al.* 2002) and other territorial primates (Garber *et al.* 1993; Palombit 1994). The best strategy for the territory-holding male might be to avoid cuckoldry by avoiding IACs when his partner is receptive. Also, the mated pair with offspring might avoid IACs because of the possibility of infanticide (van Schaik & Dunbar 1990; Palombit 1999). The subjects of this study solved this "conflict" of resource and territory defense versus mate defense by allowing the subadult to chase intruders, while the paired male remained with his mate (mate defense).

Courting, mate choice and dispersal

Dispersal at puberty may be a requirement of both male and female offspring born into monogamous groups (Kleiman 1977; Clutton-Brock 1989; Wittenberger & Tilson 1980). High-quality mate selection is crucial to high lifetime reproductive success, because mates can remain together until the death of a partner. Territory quality is an additional factor in fitness (Wittenberger & Tilson 1980). Successful decision-making for a subadult would be assisted by knowledge of location and quality of available mates. For maturing monogamous primates, territory and mate acquisition are essential prerequisites for successful reproduction (Overdorff 1996). Early reproduction is the best strategy for high lifetime reproductive output (Altmann *et al.* 1988), and tactics for subadults to acquire a high-quality mate and territory earlier in life will be favored. In addition to the territorial and mate defense components, important to the adult pair, this study presents some evidence that the morning loud choruses in *Callicebus* offer an opportunity for a subadult to advertise his/her availability as a mate and assess potential mates.

In addition to vocal communication, visual assessment of other subadults is restricted to intergroup encounters, most of which are aggressive. Observations of tamarins suggest that subadults use IACs to court, copulate with and facilitate transfer of potential mates (Hubrecht 1985; Goldizen 1987; Goldizen *et al.* 1996; Buchanan-Smith & Jordan 1992; Garber *et al.* 1993; Savage *et al.* 1996). In this study, no matings were observed during IACs, but visual inspection and chasing by subadults of other groups was observed. Mason (1966) documented an extra-pair copulation during an IAC in Colombia, and W. Kinzey (pers. commun.) observed a subadult *C. torquatus* disperse during an IAC.

In long-term relationships, defending a mate may not be restricted to estrous seasons (Fernandez-Duque *et al.* 1997). The increase of IACs in the breeding months indicates that testing pair bonds may be a good strategy, but these data are not complete enough to test that possibility. IACs did occur when infants were 3 months and younger and thus did not support an infanticide avoidance strategy. However, in three instances, mothers did not participate in the encounters and nursed the infants 25–50 m from the IACs. This may be a good strategy to protect against the risk of infanticide.

Conclusions

My findings in Peru suggest that *C. brunneus* uses morning choruses to advertise its territory. This small territory contains enough food to feed a family (3–5 individuals) over the annual cycle, and therefore it is worth defending. Over an annual cycle, significantly more dawn choruses were given when fruit was abundant, suggesting a resource defense function. *Callicebus brunneus* relies on loud calls for territorial advertisement, and unlike *Aotus*, *Saguinus* and many prosimian primates (Wright 1999), does not use scent-marking to define territorial boundaries. The fact that *C. brunneus* does not call during the period of scarce resources suggests that this species minimizes energy at this time by not broadcasting labor-intensive calls when there are no fruits over which to fight or guard. When fruits were available at territorial borders, *Callicebus* often showed aggressive behavior against neighboring groups. During several contests, subadults charged towards the hostile groups, chasing the intruders, while the adult male guarded his mate and the adult female guarded her infant from potentially infanticidal stranger males. Subadults may use encounters to assess mating opportunities in these neighboring groups. Dispersal is seasonal and subadults have been seen to transfer into neighboring groups (Wright 1989; Rodman & Bossuyt 2007). Loud advertisement calling by either *Callicebus* males or females was not associated with mating. Pair-bonded primates are in close proximity and may not need to search for one another when the female is in estrous. Recent fieldwork on wild *Callicebus* has brought a fresh look at the associations between loud calling, resource defense, and mate defense and these new data will soon be available (Lawrence 2007).

Acknowledgments

The Ministry of Agriculture of Peru is acknowledged for permission to do research in the Manu National Park. This research was funded in part by Wenner-Gren Foundation Grant 4282 and a NSF Doctoral Dissertation Improvement Grant. Many thanks to John Terborgh, Robin Foster, Louise Emmons, John Allman, Peg Stern and Patrick Daniels who all contributed to this project. Dave Sivertson assisted with recording and triangulating the calls. Thanks to Stacey Tecot and three reviewers for their wise comments, and to Marilyn Norconk, Adrian Barnett and Liza Veiga for inviting me to write this chapter.

I also would like to acknowledge the work of Francis Bossuyt, graduate student at the University of California, Davis who followed-up on this work on *Callicebus* at Cocha Cashu from 1996 to 2000. His advisor Peter Rodman followed through with analyzing these data after the untimely death of Francis in the field, and thank you Peter for that.

References

Altmann, J., Hausfater, G. & Altmann, S.A. (1988). Determinants of reproductive success in savannah baboons, *Papio cynocephalus*. In *Reproductive Success*, ed. T.H. Clutton-Brock. Chicago, IL: Chicago University Press, pp. 403–418.

Buchanan-Smith, H.M. & Jordan, T.R. (1992). An experimental investigation of the pair bond in the callitrichid monkey, *Saguinus labiatus*. *International Journal of Primatology*, **13**, 95–99.

Clutton-Brock, T.H. (1989). Female transfer and inbreeding avoidance in social mammals. *Nature*, **337**, 70–72.

Cowlishaw, G. (1992). Song function in gibbons. *Behaviour*, **121**, 131–153.

Dunbar, R.I.M. (1988). *Primate Social Systems*. New York, NY: Cornell University Press.

Emlen, S.T. & Oring, L.W. (1977). Ecology, sexual selection, and the evolution of mating systems. *Science*, **197**, 215–223.

Fernandez-Duque, E., Mason, W.A. & Mendoza, S.P. (1997). Effects of duration of separation on responses to mates and strangers in the monogamous titi monkey (*Callicebus moloch*). *American Journal of Primatology*, **43**, 225–237.

Garber, P.A., Pruetz, J.D. & Isaacson, J. (1993). Patterns of range use, range defense and intergroup spacing in moustached tamarin monkeys (*Saguinus mystax*). *Primates*, **34**, 11–25.

Gittens, S.P. (1984). Territorial advertisement and defense in gibbons. In *The Lesser Apes*, ed. H. Preuschoft. Edinburgh: Edinburgh University Press.

Goldizen, A.W. (1987). Facultative polyandry and the role of infant-carrying in the wild saddle-back tamarins (*Saguinus fuscicollis*). *Behavioral Ecology and Sociobiology*, **20**, 99–109.

Goldizen, A.W. & Terborgh, J.T. (1989). Demography and dispersal patterns of a tamarin population: possible causes of delayed breeding. *American Naturalist*, **134**, 208–224.

Goldizen, A.W., Mendelson, J., San Vlaardingen, M., *et al.* (1996). Saddle-back tamarin (*Saguinus fuscicolis*) reproductive strategies: evidence from a thirteen-year study of a marked population. *American Journal of Primatology*, **38**, 57–83.

Hubrecht, R.C. (1985). Home-range and use and territorial behavior in the common marmoset, *Callithrix jacchus jacchus*, at the Tapacura Field Station, Recife, Brazil. *International Journal of Primatology*, **6**, 553–550.

Janson, C.H. (1985). Aggressive competition and individual food competition in wild brown capuchin monkeys (*Cebus apella*). *Behavioral Ecology and Sociobiology*, **18**, 125–138.

Janson, C.H. & Emmons, L.H. (1990). The ecological structuring of the nonflying mammals communities at Cocha Cashu Biological Station, Manu National Park, Peru. In *Four Neotropical Rainforests*, ed. A.H. Gentry. New Haven, CT: Yale University Press, pp. 314–338.

Kinzey, W.G. (1981). The titi monkeys, genus *Callicebus*. In *Ecology and Behavior of Neotropical Primates*, ed. A.F. Coimbra-Filho & R.A. Mittermeier. Rio de Janeiro: Brasilian Academy of Sciences, pp. 241–276.

Kinzey, W.G. (1997). *The New World Primates Ecology, Evolution, and Behavior*. New York, NY: Aldine de Gruyter.

Kinzey, W.G. & Robinson, J.G. (1983). Intergroup loud calls, range size and spacing in *Callicebus torquatus*. *American Journal of Physical Anthropology*, **60**, 539–544.

Kleiman, D.G. (1977). Monogamy in mammals. *Quarterly Review of Biology*, **52**, 39–69.

Lawrence, J.M. (2007). Understanding the pair bond in brown titi monkeys (*Callicebus brunneus*): male and female reproductive interests. PhD dissertation, Columbia University, New York, NY

Lewis, R.J. (2005). Sex differences in scent-marking in Sifaka: mating conflict or mate guarding. *American Journal of Physical Anthropology*, **128**, 388–398.

Marshall, J.T. & Marshall, E.R. (1973). Gibbons and their territorial songs. *Science*, **193**, 235–237.

Mason, W. (1966). Social organization of the South American monkey, *Callicebus moloch*: a preliminary report. *Tulane Studies in Zoology*, **13**, 23–28.

Mason, W.A. (1968). Use of space by *Callicebus* groups. In *Primates: Studies in Adaptation and Variability*, ed. P.C. Jay. New York, NY: Holt, Rinehart & Winston, pp. 200–216.

Mitani, J.C. (1984). The behavioral regulation of monogamy in gibbons (*Hylobates muelleri*). *Behavioral Ecology and Sociobiology*, **15**, 225–229.

Mitani, J.C. (1985). Location specific responses of gibbons (*Hylobates muelleri*) to male songs. *Journal of Comparative Ethology*, **70**, 219–224.

Mitchell, C.L., Boinski, S. & van Schaik, C.P. (1991). Competitive regimes and female bonding in two species of squirrel monkeys (*Saimiri oerstedii* and *S. sciureus*). *Behavioral Ecology and Sociobiology*, **28**, 55–60.

Oates, J.F. (1995). The dangers of conservation by rural development: a case study from the forests of Nigeria. *Oryx*, **29**, 252–263.

Overdorff, D.J. (1996). Ecological correlates to social structure in two lemur species in Madagascar. *American Journal of Physical Anthropology*, **100**, 487–506.

Overdorff, D.J. & Tecot, S.R. (2006). Social pair-bonding and resource defense in wild red-bellied lemurs (*Eulemur rubriventer*). In *Lemurs: Ecology and Adaptation*, ed. L. Gould & M. Sauther. New York, NY: Springer Press, pp. 1–36.

Palombit, R.A. (1994). Dynamic pair bonds in hylobatids: implications regarding monogamous social systems. *Behaviour*, **128**, 65–101.

Palombit, R.A. (1999). Infanticide and the evolution of pair bonds in nonhuman primates. *Evolutionary Anthropology*, **7**, 117–129.

Pochron, S.T., Morelli, T.L., Terranova, P., *et al.* (2005). Patterns of male scentmarking in *Propithecus edwardsi* of Ranomafana National Park, Madagascar. *American Journal of Primatology*, **65**, 103–115.

Pollock, J. (1986). The song of the indris (*Indri indri*; Primates: Lemuroidea): natural history, form and function. *International Journal of Primatology*, **7**, 225–229.

Powzyk, J.A. & Mowry, C.B. (2006). The ecology of *Indri indri*: Madagascar's ghost of the forest. In *Madagascar's Lemurs: Ecology and Adaptation on an Island of Diversity*, ed. L. Gould & M. Sauther. New York, NY: Springer/Kluwer Press, pp. 353–368.

Robinson, J.G. (1979). Vocal regulation of use of space by groups of titi monkeys *Callicebus moloch*. *Behavioral Ecology and Sociobiology*, **5**, 1–15.

Robinson, J.G. (1981). Vocal regulation of inter- and intragroup spacing during boundary encounters in the titi monkey, *Callicebus moloch*. *Primates*, **22**, 161–172.

Robinson, J.G., Wright, P.C. & Kinzey, W.G. (1987). Monogamous cebids and their relatives: intergroup calls and spacing. In *Primate Societies*, ed. B.B. Smuts, D.L. Cheney, R.M. Seyfarth, R.W. Wrangham & T.T. Struhsaker. Chicago, IL: Chicago University Press, pp. 44–53.

Rodman, P. & Bossuyt, F. (2007). Fathers and stepfathers: familial relations of old and new males within groups of *Callicebus brunneus* in southeastern Peru. *American Journal of Physical Anthropology*, **44**, 201.

Savage, A., Giraldo, L.H., Soto, L.H., *et al.* (1996). Demography, group composition and dispersal in wild cotton-top tamarins (*Saguinus oedipus*) groups. *American Journal of Primatology*, **38**, 85–100.

Symington, M.M. (1988). Demography, ranging patterns, and activity budgets of black spider monkeys (*Ateles paniscus chameck*) in the Manu National Park, Peru. *American Journal of Primatology*, **15**, 45–67.

Terborgh, J.W. (1983). *Five New World Primates: A Study in Comparative Ecology*. Princeton, NJ: Princeton University Press.

Tilson, R. & Tenaza, R. (1976). Monogamy and duetting in an Old World monkey. *Nature*, **263**, 320–321.

van Roosmalen, M., van Roosmalen, T. & Mittermeier, R.A. (2002). A taxonomic view of the titi monkey Genus *Callicebus*, Thomas 1903 with the description of two new species *Callicebus bernhardi* and *Callicebus stephennashi* from Brasilian Amazonia. *Neotropical Primates*, **10**, 1–35.

van Schaik, C.P. & Dunbar, R.I.M. (1990). The evolution of monogamy in large primates: a new hypothesis and some crucial tests. *Behaviour*, **115**, 30–42.

van Schaik, C.P., Assink, P.R. & Salafsky, N. (1992). Territorial behavior in Southeast

Asian langurs: resource defense or mate defense? *American Journal of Primatology*, **26**, 233–242.

Wittenberger, J.F. & Tilson, R.L. (1980). The evolution of monogamy: Hypotheses and evidence. *Annual Review in Ecology and Systematics*, **11**, 197–232.

Wright, P.C. (1984). Biparental care in *Aotus trivirgatus* and *Callicebus moloch*. In *Female Primates: Studies by Women Primatologists*, ed. M.E. Small. New York, NY: Alan R. Liss, pp. 59–75.

Wright, P.C. (1986). Ecological correlates to monogamy. In *Primate Ecology and Conservation*, ed. J.C. Else & P.C. Lee. Cambridge: Cambridge University Press, pp. 159–168.

Wright, P.C. (1989). The nocturnal primate niche in the New World. *Journal of Human Evolution*, **18**, 635–646.

Wright, P.C. (1990). Patterns of paternal care in primates. *International Journal of Primatology*, **11**, 89–102.

Wright, P.C. (1999). Lemur traits and Madagascar ecology: coping with an island environment. *Yearbook of the Journal of Physical Anthropology*, **42**, 31–72.

Wright, P.C., Toyoma, L.M. & Simons, E.L. (1987). Courtship and copulation in *Tarsius bancanus*. *Folia Primatologica*, **46**, 142–148.

Chapter 22

Ecology and behavior of bearded sakis (genus *Chiropotes*)

Liza M. Veiga & Stephen F. Ferrari

Introduction

The genus *Chiropotes* includes five species (*C. albinasus*, *C. chiropotes*, *C. sagulatus*, *C. satanas* and *C. utahickae*) distributed throughout the Amazon and Orinoco basins east of the rivers Madeira, Jiparaná and Negro in the Guianas, Suriname, Venezuela and Brazil (see Silva Júnior *et al.*, Chapter 4). The northern, southern, and western limits of the geographic range of these primates appear to be determined by ecological factors, principally the availability of appropriate habitats (Ayres 1989; Norconk 2011; Ayres & Prance, Chapter 12).

However, with the exception of a possible zone of sympatry in the northern Amazon basin (Boubli 2002), the geographic ranges of *Chiropotes* and *Cacajao* are almost entirely allopatric. This may be related to the broad morphological and ecological similarities of these two genera, which may thus be too competitive to coexist in the same habitat (Ayres & Prance, Chapter 12).

Bearded sakis are intriguing monkeys in both their appearance (Figure 22.1) and their behavior. They are medium-sized platyrrhines, with adults measuring approximately 40–50 cm from the top of the head to the base of the tail and weighing between 2 and 4 kg (Ayres 1981; Hershkovitz 1985). Both sexes have long, bushy, fox-like tails, and sport a beard and two bulbous growths of hair on the top of their heads, which cover the powerful masticatory muscles used to open tough immature fruits (Kay *et al.*, Chapter 1). These primates are extremely agile and fast-moving, are superb climbers and leapers, and spend most of their time in the highest forest strata. Like other pitheciids, they rarely descend to the ground (Barnett *et al.* 2012). They are also extremely shy animals, and are difficult to observe or habituate (Pinto *et al.*, Chapter 13). This explains in part why, despite their ample geographic distribution, there have been so few detailed behavioral field studies of bearded sakis.

Bearded sakis are highly specialized morphologically for the consumption of immature seeds. In addition to powerful masseter and temporalis muscles, their dental anatomy (Figures 22.2 and 22.3) enables them to open extremely tough fruits, to access the highly protected seeds (Kinzey 1992; Rosenberger 1992; Lucas & Teaford 1994). They have huge, tusk-like, divergent canines to penetrate and open up hard husked fruits, procumbent incisors which are used to extract and scrape out the soft, pliable seeds, and relatively small, rounded molars with low relief, which are used to masticate the extracted material (van Roosmalen *et al.* 1981; Kinzey 1992).

Pioneering studies of *Chiropotes* were undertaken by Russell Mittermeier (Mittermeier & Coimbra-Filho 1981), Marcio Ayres (1981), and Marc van Roosmalen (van Roosmalen *et al.* 1981). While a small handful of studies were conducted in subsequent years, the detailed, long-term monitoring of habituated groups only began in earnest in the late 1990s. While these recent studies represent a quantum leap in our understanding of the ecology of bearded sakis, the available body of data is still greatly limited in comparison with the better-known platyrrhine genera, such as howlers (*Alouatta*), tamarins (*Saguinus*) and capuchins (*Cebus*). Van Roosmalen *et al.* (1981) provided the first detailed review of the biology of the genus, while Norconk (2011) summarized the principal ecological and behavioral data available by the beginning of the twenty-first century. This chapter reviews the current scientific knowledge of the behavior and ecology of the bearded sakis, discusses the contribution of the most recent fieldwork, and contemplates directions for future research.

Habitat

Bearded sakis prefer tall, lowland *terra firme* rainforest (van Roosmalen *et al.* 1981; Ayres 1981; Frazão 1992), but have also been found in montane forests, savannahs (Mittermeier & van Roosmalen 1981; Norconk *et al.* 2003) and flooded habitats, including *igapó* swamp (Mittermeier & Coimbra-Filho 1981; Ayres 1981; Wallace & Painter 1996) and mangrove (Silva Jr. *et al.* 1992). As bearded sakis are specialized frugivores, they require access to a year-round supply of fruits, which suggests that they would be unable to survive in mangrove or savannah habitats without having access to more productive and diverse forests in neighboring areas.

The earliest observations of bearded sakis in the wild found no evidence of their occurrence in secondary habitats, which resulted in the widespread belief that they were dependent on primary forest and were completely intolerant of habitat

Evolutionary Biology and Conservation of Titis, Sakis and Uacaris, eds. Liza M. Veiga, Adrian A. Barnett, Stephen F. Ferrari and Marilyn A. Norconk. Published by Cambridge University Press.

Figure 22.1 Male black-bearded saki at Carajás Zoo, Brazil. Photo: Eduardo. L. Paschoalini.

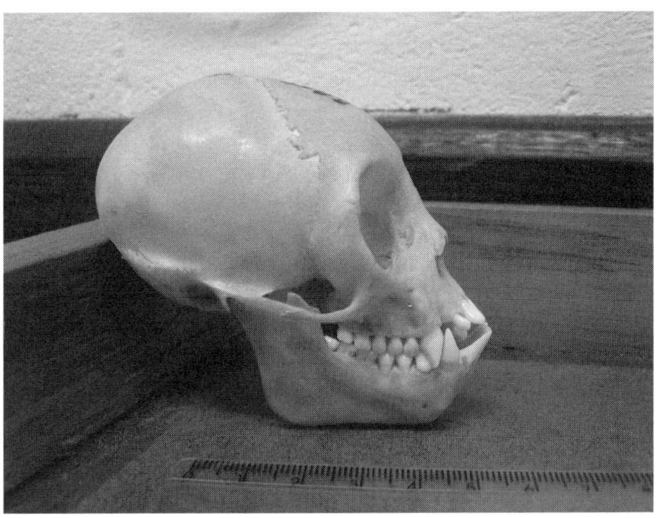

Figure 22.3 Skull (side view) of an adult male bearded saki, *Chiropotes satanas*, taken at the Emilio Goeldi Museum Mammal Collection, Brazil. Photo: Liza M. Veiga.

disturbance (Johns & Ayres 1987). Later fieldwork in the heavily fragmented landscape of southeastern Amazonia nevertheless recorded *C. satanas* in logged and disturbed forests (Silva Jr. 1991; Ferrari & Lopes 1996), and subsequent studies at a number of sites in the region and the central

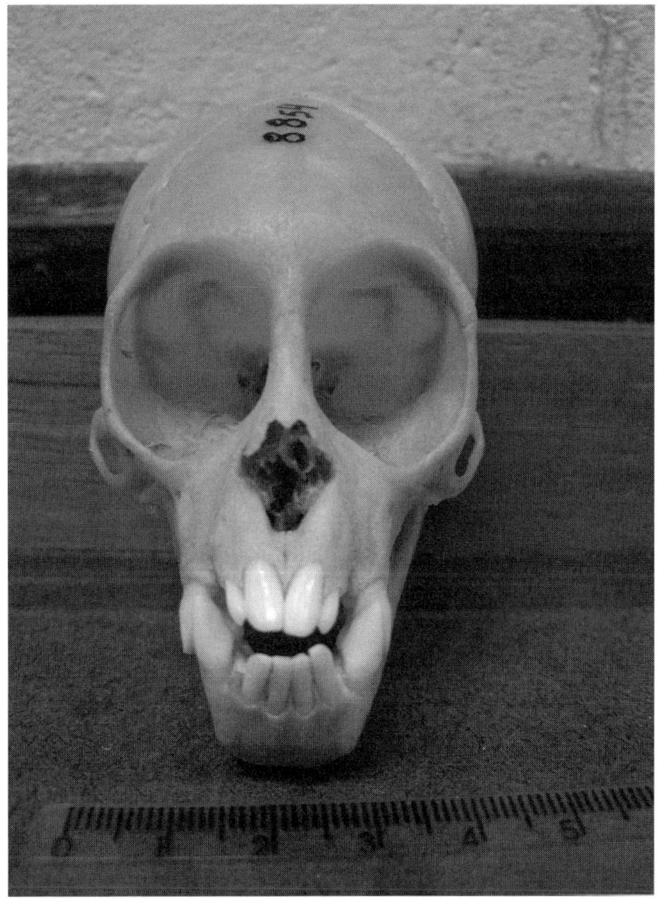

Figure 22.2 Skull (front view) of an adult male bearded saki, *Chiropotes satanas*, taken at the Emilio Goeldi Museum Mammal Collection, Brazil. Photo: Liza M. Veiga.

Amazon basin indicated that bearded saki populations are able to tolerate relatively intensive anthropogenic impact over the long term (Ferrari *et al.* 1999b; Bobadilla & Ferrari 2000; Lopes & Ferrari 2000; Pereira 2002; Port-Carvalho & Ferrari 2004; Boyle *et al.* 2009, Chapter 24).

Two especially important sites for the understanding of the tolerance of the bearded sakis towards habitat disturbance are the hydroelectric reservoirs at Guri in Venezuela (Peetz 2001) and Tucuruí, in the southeastern Brazilian Amazon basin (Veiga & Ferrari 2006; Silva & Ferrari 2008; Guimarães 2011; Santos *et al.*, Chapter 23). Among other findings, these studies have shown that populations of bearded sakis are able to survive over the long term - that is, at least three generations - on small islands, of the order of 20–180 ha.

Social structure and behavior

Bearded sakis form relatively large multimale–multifemale groups, with as many as 65 members (van Roosmalen *et al.* 1981; Ayres 1989; Frazão 1992; Peetz 2001; Santos 2002; Silva 2003; Norconk *et al.* 2003; Vieira 2005; Pinto 2008; Gregory 2011; Shaffer 2012; Gregory & Norconk Chapter 28), which may subdivide regularly into smaller social units. Early studies reported that the large groups traveled and slept together, and only fragmented temporarily into local foraging parties (Fleagle 1988; van Roosmalen *et al.* 1988; Ayres 1989; Kinzey & Norconk 1990; Frazão 1992; Peetz 2001). More recent studies have nevertheless recorded more fluid social systems, reminiscent of the fission–fusion societies of other specialized frugivores, such as spider monkeys (Symington 1988; Chapman *et al.* 1994) and chimpanzees (Goodall 1986).

Fission–fusion systems have evolved as a response to temporal variation in the distribution of feeding resources (primarily, fruiting trees), with the size and dispersal of foraging parties reflecting the size and spatial configuration of feeding patches. By adjusting party size to resource abundance, the primates are able not only to minimize intraspecific competition and potential conflicts among group members, but also maximize foraging efficiency (Leighton & Leighton 1983; White & Wrangham 1988; Chapman 1990; Overdorff 1996; Shaffer 2012).

In particular, the recent studies of *C. satanas*, *C. albinasus*, and *C. sagulatus* in Brazil (Veiga 2006; Veiga *et al.* 2006; Pinto 2008), Suriname (Gregory 2011) and Guiana (Shaffer 2012) have found conclusive evidence of complex fission–fusion dynamics in all three species, which is consistent with the pattern observed in other frugivorous primates, and indicates that this is standard behavior in the genus. In all three species, troop members rarely came together as a single social unit, but rather spent most of their time in subgroups of varying size and composition. However, no clear evidence was found of a systematic relationship between grouping patterns and the size or distribution of feeding patches. In *C. satanas*, Veiga (2006) noted that troop fusion appeared to be influenced in part by the availability of the immature seeds of *Simarouba amara*, a large tree species with clumped distribution and synchronized fruiting, which was the most important food item in the troop's diet.

The mating system of the bearded sakis is poorly understood. Based on preliminary observations of *C. albinasus*, Ayres (1981) believed that bearded sakis may form breeding pairs or nuclei within larger social units, a viewpoint which endured in the subsequent literature, for want of any more conclusive evidence. More recent observations nevertheless indicate a polygynandrous system, although the reliable interpretation of social behavior observed in the wild is hampered by the difficulties of identifying individual subjects in general, and within specific social contexts in particular. Ultimately, observations of reproductive behavior are extremely rare, and any interpretation of mating systems must be treated with caution.

Nevertheless, a polygynandrous system is supported indirectly by evidence of male social behavior. In a study of *C. satanas*, Veiga & Silva (2005) recorded a high degree of affiliative behavior among male group members, and low levels of agonism (0.1% of the activity budget). Male bearded sakis are also relatively gregarious in comparison with females, tend to groom each other more, rest together more, and engage in more affiliative behaviors, such as lining-up and hugging (Peetz 2001; Veiga *et al.* 2006; Silva & Ferrari 2009; Gregory 2011). The males are also extremely tolerant of juveniles, and will often allow them to sit on their backs during rest periods. This affiliative behavior among males suggests possible kinship ties, and by extension, male philopatry and female dispersal, but there is little concrete evidence to confirm such conclusions.

Social behavior is a relatively minor component of the activity of most groups (Table 22.1), and while physical contact (lining-up, hugging) may be an important social bonding mechanism, agonistic behaviors are rare in general. Older males appear to lead group movements in some cases, although there is little conclusive evidence of systematic social hierarchies in any species.

Activity budget

Activity budgets – the proportion of time dedicated to different behavioral categories – provide important insights into the ecological strategies and adaptability of primate species, in particular in the context of comparative analyses among periods, populations and species. Seasonal shifts in activity patterns can be especially informative with regard to the behavioral flexibility and ecological adaptations of a species.

To date, 16 long-term field studies (lasting 6 months or more) of habituated or semi-habituated groups of bearded sakis have taken place (Table 22.1). While the ground-breaking studies of Ayres (1981) and van Roosmalen *et al.* (1981) were the longest conducted to date, they provide the least reliable data, and emphasize the difficulties of monitoring these primates, especially in continuous forest habitats (see Pinto *et al.*, Chapter 13).

A key project, which marked the advent of detailed ecological studies of *Chiropotes*, was that of Peetz (2001), who monitored a group of 27 habituated *C. chiropotes* for 15 months on a 180-ha island in Guri Lake in Venezuela. This study provided a yardstick for all subsequent fieldwork, in terms of the detail of the data collected. Studies at a second reservoir, Tucuruí, in the southeastern Brazilian Amazon basin, have provided the bulk of the data on two other bearded saki species, *C. utahickae* (Santos 2002; Vieira 2005) and *C. satanas* (Santos 2002; Silva 2003; Veiga 2006; Guimarães 2011). Despite the challenges, the principal behavioral data on *C. albinasus* (Pinto 2008) and *C. sagulatus* (Gregory 2011; Shaffer 2012) are derived from studies in continuous forest.

In general, bearded sakis spend the bulk of their time foraging and moving, although considerable differences are found across species and study sites (Table 22.1). This variation is difficult to interpret, given the considerable differences among studies in variables such as habitat type (continuous forest, large or small fragments) and the study period. All the data available for *C. albinasus*, for example, are from continuous forest sites, whereas those for *C. chiropotes* were collected

Table 22.1 Activity budget and use of space by *Chiropotes* spp.

Species	Duration of study (months)	Habitat (ha)	Group size	Activity budget (% time)				Use of space[1]			Source
				Rest	Feed/ forage	Move/ travel	Social	Home range (ha)	Day range, Mean ± SD or min-max (m)	Mean ± SD height (m)	
C. albinasus	17	Continuous forest	22.5± 3.5 (*n* = 4)	–	–	–	–	250–350 (est.)	2500–4500	10–29	Ayres (1981)
	11	Continuous forest	56	27	24	35	9	1000 +	3667 ± 1687 1840–7809	–	Pinto (2008)
C. chiropotes	5	Island (180)	14					180 (est.)	1050 424–1780	–	Kinzey & Norconk (1993)
	15	Island (180)	22	21	47	19	11	122	1600 ± 550 500–2700	–	Peetz (2001)
C. sagulatus	3	Fragment (10)	2	–	–	–	–	10	1300	–	Ayres (1981)
	28	Continuous forest	15+	–	–	–	–	200–250 (est.)	2500 (est.)	–	van Roosmalen *et al.* (1981)
	6	Continuous forest	16	–	–	–	–	–	3200 ± 1100	–	Norconk & Kinzey (1994)
	12	Fragment (1100)	30+	–	–	–	–	–	1097 ± 590	–	Frazão (1992)
	13	Continuous forest	45 (18 ± 13)	48.5	20.3	31.1	–	742	2362 ± 821 809–3386	–	Gregory (2011)
	15	Continuous forest	65+	–	–	–	–	800	4000	–	Shaffer (2012)
		Continuous forest and fragments	8–35	–	–	–	–	1, 10, 100, 559	0.04–3000	–	Boyle *et al.* (2009)
C. satanas	7	Mainland (1300)	27	–	–	–	–	57	–	–	Santos (2002)
	6	Island (16.3)	7	16	25	56	3	16	–	15 ± 4	Silva (2003)
	6	Mainland (1300)	34	27	24	46	3	70	–	12 ± 4	Silva (2003)
	12	Mainland (1300)	39	26	29	35	9	99	4025 ± 994 1560–6270	17 ± 4	Veiga (2006)
	3	Fragment (63)	17	14	20	58	8	–	–	15 ± 5	Port-Carvalho & Ferrari (2004)
	12	Island (19.4)	8	23	34	26	15	17	2807 ± 289 1900–3680	19 ± 5	Veiga (2006)
C. utahickae	8	Continuous forest	–	–	–	–	–	–	–	18 ± 7	Bobadilla & Ferrari (1998)
	8	Island (129)	24	9	59	31	1	100 (est.)	–	–	Santos (2002)
	6	Island (129)	23	11	37	51	1	58	2530 ± 1 1940–4080	17 ± 11	Vieira (2005)

Obs.: Percentages were rounded to the nearest integer.

on a reservoir island, and while shorter-term studies may return similar activity budgets to those recorded over a full annual cycle (Silva 2003; Veiga 2006), this may be more coincidental than meaningful.

As bearded sakis' diets are composed mainly of seeds, which are relatively rich in energy (Norconk *et al.* 2009), it makes sense for them to adopt an energy-maximizing foraging strategy, and devote relatively more time to searching for and traveling between dispersed patches of foods and relatively less time resting and socializing, which is consistent with the general pattern observed in the data (Table 22.1). Activity budgets may also vary significantly over time in response to seasonal changes in the availability and distribution of feeding resources (Strier 2000). *Chiropotes chiropotes* in Venezuela and *C. sagulatus* in Suriname dedicated more time to feeding in the dry season months when seeds were the most important food source (Peetz 2001; Gregory 2011), although the results of other long-term studies are less conclusive.

Foraging ecology

Almost all field studies of bearded sakis have provided some estimate of the composition of the diet of the study group (Table 22.2). Once again, differences in habitat, methodological procedures and study period may be at least as important as any potential variation among species or populations, although the data classify *Chiropotes* emphatically as a specialized frugivore and seed predator. In the vast majority of studies, fruit makes up at least 80% of the diet, with seeds (primarily immature) typically accounting for a large part. While flowers may constitute a relatively important item at some sites, leaves and other non-reproductive plant parts (shoots and pith) are clearly an occasional and rare component of the diet of bearded sakis, even in highly impacted environments.

Seasonally, seeds appear to be more important during the dry season. However, bearded sakis are able to include seeds in their diets throughout the year due to the diversity of phenological cycles presented by different plant species in tropical forests, as well as the ability of the monkeys to exploit seeds at distinct stages of maturity, varying from very young to almost completely ripe. This allows them to avoid the resource bottlenecks faced by other highly frugivorous primates, particularly those which depend on ripe fruit pulp, such as spider monkeys.

A major reduction in the fruit component of the diet was observed only in *C. satanas* at highly impacted sites in southeastern Amazonia. In these cases, the diet was complemented with flowers, which might be seen as an extension of the seed predation strategy, with the seeds being consumed at an even earlier stage of development. While such a strategy may guarantee subsistence over the short term in areas with a reduced resource base (in particular on small islands), there may be negative long-term implications, given that these flowers will not develop into fruits, which may be a vital resource later in the year. On one island, in particular, Silva (2003) noted that the bearded sakis may have been malnourished. These island studies also provide the only known record of geophagy – consumption of earth from termite nests – although it is unclear what function this behavior may have (Veiga & Ferrari 2006).

Chiropotes diets are also characterized by a considerable diversity of plant species, which obviously reflects the marked biological diversity of the Amazonian forests they inhabit. While the bulk of these species may contribute only a small component of the diet, even over the short term, others may be considered dietary staples, and bearded sakis may depend heavily on certain specific resources at a given site. The most prominent plant families in bearded saki diets include the Lecythidaceae and Sapotaceae, the distribution of which in the Amazon basin may be fundamental to the zoogeography of *Chiropotes* (Ayres & Prance, Chapter 12). The diversity of Amazonian forests is far greater, however, and simultaneous studies of *C. satanas* at neighboring sites (separated by around 6 km) in Tucuruí found relatively little overlap in the most important plant species used by the different study groups (Silva 2003; Veiga 2006).

Use of space

Most detailed monitoring of habituated groups has taken place on islands or in forest fragments, and estimates of home-range size derived from these studies are influenced fundamentally by the configuration of the habitat available to the animals (Table 22.1). The first estimates of home-range size in bearded sakis in continuous forest (Ayres 1981; van Roosmalen *et al.* 1981) indicated natural ranges much larger than even the largest of these fragments, although even these studies may represent underestimates, possibly related to the monitoring of subgroups. All the more recent field studies in continuous forest have recorded estimates of well over 500 ha, although these studies also involved relatively large groups of bearded sakis, but this would appear to be consistent with the natural characteristics of the genus, that is, groups of 30–60 individuals inhabiting home ranges of the order of hundreds or even thousands of hectares.

Curiously, however, the long-term monitoring of a relatively large *C. satanas* group in an ample fragment of forest (1300 ha) recorded a home range of less than 100 ha (Santos 2002; Silva 2003; Veiga 2006). While the presence of neighboring groups within the same area may have contributed to the reduced size of this home range, this finding does raise a number of questions with regard to the factors that may determine range size in these primates. While one possibility is species-level differences in ecological adaptations, it seems more likely that the underlying factors may be related to habitat composition or productivity. Unfortunately, there are still too few comparative data to support a more conclusive evaluation of these contrasts.

As might be expected from their large home ranges, bearded sakis also travel relatively long distances each day (Table 22.1) in comparison with the smaller pitheciids. It is interesting to note that groups inhabiting relatively small islands or forest fragments travel similar or even greater distances each day than

Table 22.2 Contribution of food types and the species richness of the diet of *Chiropotes* species.

Species	Contribution to diet (%)					Diet richness			Duration of the study (months)	Source
	Fruit[1]	Leaves[2]	Flowers[3]	Prey	Other/ unknown	PS[6]	FS[7]	LS[8]		
Chiropotes chiropotes	90 [36]		3		7[5]	51			17	Ayres (1981)
	93 [54]	1	5	1		125			11	Pinto (2008)
Chiropotes chiropotes	96 [75]		1	1		39			5	Kinzey & Norconk (1993),
	83 [58]	4	1	12[4]		73			17	Norconk (1996)
	92 [51]	3	1	4	1	134	23	8	15	Peetz (2001), Peetz & Homburg (1998)
Chiropotes sagulatus	73 [63]	11	11		5	18			3	Ayres (1981)
	96 [66]	1	3			85			28	van Roosmalen *et al.* (1981)
	96 [86]	1	1			50+			6	Kinzey & Norconk (1990)
	93 [72]	1	1	4	1	112+			12	Frazão (1992)
	95 [85]	0.4	4	0.5	95	91	1		7	Gregory (2011)
Chiropotes satanas	99 [63]		1			37			3	Port-Carvalho & Ferrari (2004)
	50 [41]		50			45			6	Santos (2002), mainland
	80 [38]		20			40+			6	Silva (2003), mainland
	80 [54]	3	12	5	1	149+			12	Veiga (2006), mainland
	42 [30]		58	1		21			6	Silva (2003), island
	74 [60]	4	17	4	1	119			12	Veiga (2006), island
						66			7	Guimarães (2011)
Chiropotes utahickae	81 [76]	1	19			87			6	Santos (2002)
	80 [36]	4	17	1	1	119			6	Vieira (2005)

Obs.: Percentages were rounded to the nearest integer.
[1] Includes ingestion of whole fruit or fruit parts [% of seeds]; [2] includes ingestion of non-reproductive plant parts such as shoots, young leaves and pith; [3] includes whole flowers, buds, petals or flower reproductive parts; [4] includes searching for arthropods; [5] includes vegetative material; [6] number of plant species used as food sources; [7] number of fruit sources; [8] number of leaf sources.

those in large home ranges in continuous forest. This similarity reflects distinct patterns of ranging behavior, however, with much less linear daily paths being observed in groups with smaller ranges, with a greater tendency for overlap. In a systematic comparison of ranging behavior in fragments of different sizes, in fact, Boyle *et al.* (2009) confirmed that the bearded sakis in smaller fragments revisited feeding trees far more frequently than those in larger fragments.

Predators

As for other platyrrhines (Ferrari 2009), the principal predators of bearded sakis appear to be raptors, although there are too few recorded events to allow anything more than cautious speculation on this fundamentally important aspect of the ecology of the genus. The preference of these monkeys for the highest forest strata almost certainly precludes frequent attacks by terrestrial predators (snakes and felids), and their typically rapid movements through the forest canopy probably make them difficult targets for raptors. This is supported indirectly by studies of prey remains under Harpy Eagle (*Harpia harpyja*) nests in Guyana (Rettig 1978) and Suriname (Ford & Boinski 2007). Whereas sloths (*Bradypus* and *Choloepus*) and capuchins (*Cebus*) – which are similar in size and habitat preferences to *Chiropotes* – were common prey items at both sites, bearded sakis were rare prey in Guyana, and were not recorded at all in Suriname.

Harpy eagles have been observed attacking both *C. utahickae* (Martins *et al.* 2005) and *C. albinasus* (Rafaela Soares, pers. comm.), but no other raptors are known to prey on bearded sakis. The only other observed event involved a boa (*Boa constrictor*) attacking an adult *C. utahickae* (Ferrari *et al.*

2004). While boas are typically terrestrial predators, this attack took place in the forest canopy.

Other aspects of the behavior of these primates – such as the use of sleeping sites – also appear to reflect antipredator strategies. Bearded sakis typically sleep in tall trees, often hidden among lianas. The males, which are the most prominent social players in bearded saki groups (see above), also appear to play an important role in antipredator defense. During perceived threats from raptors, *C. satanas* males were seen piloerected, bounding along branches, and even leaping in the direction of the potential predator, while females and young sought cover (Veiga 2006).

Conservation

The geographic ranges of all five species of bearded saki cover vast tracts of northern South America, of the order of hundreds of thousands of square kilometers, although these ranges vary considerably in their human population density and degree of anthropogenic impact. Logging, deforestation and habitat fragmentation are the principal threats to all *Chiropotes* species, although these monkeys are also hunted in many areas for both their meat and their bushy tails, which are used as adornments by some indigenous peoples or as house dusters by colonists.

The species can be divided into two distinct groups, those distributed north of the Amazon River (*C. chiropotes* and *C. sagulatus*), which occupy relatively isolated and sparsely populated areas, and those south of the Amazon (*C. albinasus*, *C. satanas*, and *C. utahickae*), which face varying degrees of anthropogenic impact from the ongoing advance of the agricultural frontiers of the southern Amazon basin, known as the "arc of deforestation" (Fearnside *et al.* 2009). The three southern species are all endemic to Brazil, where the intensity of deforestation increases from west to east, which also coincides with decreasing geographic ranges. In other words, *C. albinasus* is not only distributed over a much larger area than *C. satanas*, this area is more isolated and less deforested in general than the southeastern Amazonian range of *C. satanas*.

Whereas the two northern species are under no immediate risk of extinction, then, the three southern species are all considered to be Endangered by the IUCN. In fact, *C. satanas* is classified as Critically Endangered (CR) (Veiga *et al.* 2008a), due primarily to the widespread deforestation within its natural range, where more than half the original forest cover has been lost. Johns & Ayres (1987) had in fact predicted that this species would have been extinct by the turn of the twentieth century, although this conclusion was derived from a paucity of data on surviving populations, and a now superseded understanding of the species' tolerance of habitat disturbance. While the species is now known to be more widespread and abundant than it was thought to be 25 years ago, it is endemic to the most densely populated corner of the Amazon basin, which has a long tradition of deforestation, ongoing development and few protected areas. One positive aspect, however, is that the Brazilian federal environment institute (ICMBio) is currently developing an action plan for the species, which may result in more effective conservation measures.

Further west, *C. utahickae* faces similar but generally less-intense problems, which are mediated by the much larger geographic range of the species and a more consolidated network of protected areas. The species was nevertheless considered to be Endangered with extinction in the most recent review of its IUCN category (Veiga *et al.* 2008b). While it has an even larger geographic range, and was not considered to be in any danger of extinction in previous classifications, *C. albinasus* has also now been listed as Endangered (Veiga *et al.* 2008c), due primarily to the ongoing advance of agricultural frontiers in the southern Amazon basin, characterized by deforestation for cattle ranching and the installation of plantations of soybeans and other cash crops. Nevertheless, this species can be considered to be at less risk than *C. utahickae* –at least for the time being – due to its much larger and more isolated geographic range.

As shown in this chapter, the recent ecological studies of *Chiropotes* indicate that bearded sakis are far more tolerant of habitat disturbance and alterations of the resource base than was originally thought. While this means that populations of these primates may be able to survive over the long term in many impacted areas, this apparently favorable ecological characteristic may have negative long-term implications for the ecosystem as a whole and, ultimately, the bearded saki populations. The increasingly intense predation of seeds (and in some cases, flowers) in small fragments will likely have an adverse effect on the recruitment of many tree species, altering the composition and structure of the forest over the long term and, in turn, the resource base available to the bearded sakis. Understanding such processes will obviously depend on the collection of more systematic, long-term ecological data covering a number of generations.

Avenues for future research

Moving rapidly in the upper reaches of the rainforest canopy, bearded sakis are notoriously difficult to monitor in the wild, even when the study subjects are habituated to the presence of human observers. These difficulties are exacerbated by the fission–fusion system of these primates, which, together with their relatively cryptic coloration, hampers the reliable identification of individuals under field conditions. The fact that most of the more detailed studies of the ecology of *Chiropotes* have been conducted in relatively small patches of forest is thus understandable, but introduces a range of potential bias for the analysis of ecological patterns within or between populations or species, especially as some species have only been studied under these circumstances and others only in continuous forest (Table 22.1). These problems are exacerbated by the fact that, for even the best-studied species, data are available from only a small number of sites.

What is patently clear from the overview of the ecological data presented in this chapter is that, despite the considerable advances in the scientific understanding of the behavior and ecology of the bearded sakis derived from fieldwork conducted over the past 20 years, there are still many more questions to be asked. The available data (Tables 22.1 and 22.2) reveal ecological patterns that appear to be typical of the genus *Chiropotes*, such as a highly frugivorous/granivorous diet, relatively large social groups and ample home ranges, and associated behavior, such as fission–fusion social organization. However, the data are also characterized by considerable variation in many aspects, only some of which can be linked systematically to specific ecological factors, such as habitat disturbance. In particular, it is still unclear if any significant degree of ecological differentiation exists among the species (or between regions, populations, and so on). Obviously, a much larger body of data from a wider range of study sites should be the major priority for future research into this genus.

Considering the ecological role of bearded sakis in Amazonian forests, in particular as specialized seed predators, and their vulnerability to ongoing deforestation, another, more specific priority for future research should be the understanding of the influence of the foraging behavior of these primates on the recruitment of key tree species. As mentioned in the previous section, increasing predation pressure on resource plants in fragmented habitats may have highly deleterious long-term effects on both *Chiropotes* populations and the ecosystems they inhabit. This may mean that the tolerance of habitat disturbance may be a short-term phenomenon limited by eventual modifications in the composition and structure of the forest. Studies of seedling recruitment patterns at sites with different densities of bearded sakis (see, for example, Lopes & Ferrari 1994) might provide valuable insights into the effects of seed predation, with important practical applications for the development of effective conservation and management programs.

Given the difficulties of identifying subjects reliably under field conditions, the social behavior of the bearded sakis, and in particular their mating system, remains largely unknown. While appropriate tracking technology exists, most fieldworkers have been reluctant to consider using such an approach due to the potential risks of capturing such elusive animals. It is unclear whether technical advances will provide any less-risky options in the foreseeable future, but one viable approach would be the collection of fecal samples for DNA analyses, which could provide insights into the genetic relationships among group members, and in particular, the relative reproductive success of different male group members.

One other potentially lucrative approach to the analysis of the available data on the ecology of the bearded sakis may be ecological modeling, which may play an increasingly important role in conservation planning. Approaches such as Population Viability Analysis (Lacy *et al.* 2009) may be essential to the development of effective conservation and management strategies, especially for the most endangered species, although reliable analyses will still depend on the collection of trustworthy long-term data on demographic parameters.

Acknowledgments

The authors are grateful to Eletronorte S.A., CAPES, CNPq, and Primate Conservation Inc. (LMV) for their support.

References

Ayres, J.M. (1981). Observações sobre a ecologia e o comportamento dos cuxiús (*Chiropotes albinasus* e *Chiropotes satanas*, Cebidae, Primates). Masters' dissertation, Instituto Nacional de Pesquisas da Amazônia e Fundação Universidade do Amazonas, Manaus.

Ayres, J.M. (1989). Comparative feeding ecology of the uakari and bearded saki, *Cacajao* and *Chiropotes*. *Journal of Human Evolution*, **8**, 697–716.

Ayres, J.M. & Milton, K. (1981). Primate and habitat survey in Tapajos River. *Zoologia*, **111**, 1–11.

Barnett, A.A., Boyle, S.A., Norconk, M.A., *et al.* (2012). Terrestrial activity in Pitheciins (*Cacajao*, *Chiropotes*, and *Pithecia*). *American Journal of Primatology*, **74**, xxx.

Bobadilla, U.L. & Ferrari, S.F. (1998). First detailed field data on *Chiropotes satanas utahicki* Hershkovitz, 1985. *Neotropical Primates*, **6**, 17–18.

Bobadilla, U.L. & Ferrari, S.F. (2000). Habitat use by *Chiropotes satanas utahicki* and syntopic platyrrhines in eastern Amazonia. *American Journal of Primatology*, **50**, 215–224.

Boubli, J.P. (2002). Western extension of the range of bearded sakis: a possible new taxon of *Chiropotes* sympatric with *Cacajao* in the Pico da Neblina National Park, Brazil. *Neotropical Primates*, **10**, 1–4.

Boyle, S.A., Lourenco, W.C., da Vilva L.R., *et al.* (2009). Travel and spatial patterns change when *Chiropotes satanas chiropotes* inhabit forest fragments. *International Journal of Primatology*, **30**, 515–531.

Branch, L.C. (1983). Seasonal and habitat differences in the abundance of primates in the Amazon (Tapajos) National Park, Brazil. *Primates*, **24**, 424–431.

Chapman, C.A. (1990). Ecological constraints on group size in three species of Neotropical primates. *Folia Primatologica*, **55**, 1–9.

Chapman, C.A., Wrangham, R.W. & Chapman, L.J. (1994). Indices of habitat-wide fruit abundance in tropical forests. *Biotropica*, **26**, 160–171.

Fearnside, P.M., Righi, C.A., Graça, P.M.L.A., *et al.* (2009). Biomass and greenhouse gas emissions from land-use changes in Brazil's Amazonian "arc of deforestation": the states of Mato Grosso and Rondônia. *Forest Ecology and Management*, **9**, 1968–1978.

Ferrari, S.F. (2009). Predation risk and anti-predator strategies. In *South American Primates: Comparative Perspectives in the Study of Behavior, Ecology and Conservation*, ed. P.A. Garber, A. Estrada, J.C. Bicca-Marques, E.W. Heymann & K.B. Strier. New York, NY: Springer, pp. 251–277.

Ferrari, S.F. & Lopes, M.A. (1996). Primate population in eastern Amazonia. In *Adaptive Radiations of Neotropical Primates*, ed. M.A. Norconk, A.L. Rosenberger & P.A. Garber. New York, NY: Plenum Press, pp. 53–68.

Ferrari, S.F., Emidio-Silva, C., Lopes, M.A., *et al.* (1999b). Bearded sakis in south-eastern Amazonia – back from the brink? *Oryx*, **33**, 346–351.

Ferrari, S.F., Iwanga, S., Coutinho, P.E.G., *et al.* (1999a). Zoogeography of *Chiropotes albinasus* (Platyrrhini,

Atelidae) in southwestern Amazonia. *International Journal of Primatology*, **20**, 995–1004.

Ferrari, S.F., Iwanaga, S., Ravetta, A.L., *et al.* (2003). Dynamics of primate communities along the Santarem–Cuiaba Highway in south-central Brazilian Amazonia. In *Primates in Fragments: Ecology and Conservation*, ed. L.K. Marsh. New York, NY: Kluwer Academic/ Plenum, pp. 123–144.

Ferrari, S.F., Pereira, W.L.A., Santos R.R., *et al.* (2004). Fatal attack of a boa constrictor on a bearded saki (*Chiropotes satanas utahicki*). *Folia Primatologica*, **75**, 111–113.

Fleagle, J.G. (1988). *Primate Adaptation and Evolution*. London: Academic Press.

Ford, S.M. & Boinski, S. (2007). Primate predation by harpy eagles in the Central Surinam Nature Reserve. *American Journal of Physical Anthropology*, Suppl. **44**, 109.

Frazão, E. (1992). Dieta e Estratégia de Forragear de *Chiropotes satanas chiropotes* (Cebidae: Primates) na Amazônia Central Brasileira. Masters' dissertation, INPA/FUA, Manaus.

Goodall, J. (1986). *The Chimpanzees of Gombe: Patterns of Behavior*. Cambridge, MA: Harvard University Press.

Gregory, L.T. (2011). Socioecology of the Guianan bearded saki, *Chiropotes sagulatus*. Unpublished dissertation, Kent State University, Kent OH, USA.

Guimarães, A.C.P. (2011). Ecologia e dieta de *Chiropotes satanas* (Hoffmannsegg, 1807) em fragmento florestal na área de influência da UHE de Tucuruí-Pará. Masters' dissertation, UFPA/Goeldi Museum, Belém.

Hershkovitz, P. (1985). A preliminary taxonomic review of the South American bearded saki monkeys genus *Chiropotes* (Cebidae, Platyrrhini), with the description of a new subspecies. *Fieldiana Zoology*, **27**, 1–46.

Johns, A.D. & Ayres, J.M. (1987). Bearded sakis beyond the brink. *Oryx*, **21**, 164–167.

Kinzey, W.G. (1992). Dietary and dental adaptations in the Pitheciinae. *American Journal of Physical Anthropology*, **88**, 499–514.

Kinzey, W.G. & Norconk, M.A. (1990). Hardness as a basis of fruit choice in two sympatric primates. *American Journal of Physical Anthropology*, **81**, 5–15.

Kinzey, W.G. & Norconk, M.A. (1993). Physical and chemical properties of fruit and seeds eaten by *Pithecia* and *Chiropotes* in Surinam and Venezuela. *International Journal of Primatology*, **14**, 207–227.

Lacy, R.C., Borbat, M. & Pollak, J.P. (2009). *Vortex: A stochastic simulation of the extinction process*. Version 9.99. Chicago Zoological Society, Brookfield, Illinois, USA (http://www.vortex.org/vortex.html).

Leighton, M. & Leighton, D. (1983). Vertebrate responses to fruiting seasonality within a Bornean rain forest. In *Tropical Rain Forest: Ecology and Management*, ed. S.L. Sutton, T.C. Whitemore & A.C. Chadwick. Oxford: Blackwell Scientific Publications, pp. 181–196.

Lopes. M.A. & Ferrari, M.A. (1994). Differential recruitment of *Eschweilera albiflora* (Lecythidaceae) seedlings at two sites in western Brazilian Amazonia. *Tropical Ecology*, **35**, 25–34.

Lopes. M.A. & Ferrari, M.A. (2000). Effects of human colonization on the abundance and diversity of mammals in eastern Brazilian Amazonia. *Conservation Biology*, **14**, 1658–1665.

Lucas, P.W. & Teaford, M.F. (1994). Functional morphology of colobine teeth. In *Colobine Monkeys: Their Ecology, Behavior and Evolution*, ed. A.G. Davies & J.F. Oates. Cambridge: Cambridge University Press, pp. 173–203.

Martins, E.S., Ayres, J.M. & Ribeiro do Valle, M.B. (1988). On the status of *Ateles belzebuth marginatus* with notes on other primates of the Iriri River basin. *Primate Conservation*, **9**, 87–91.

Martins, S.S., Lima, E.M. & Silva Jr, J.S. (2005). Predation of bearded saki (*Chiropotes utahicki*) by a harpy eagle (*Harpia harpyja*). *Neotropical Primates*, **13**, 7–10.

Mittermeier, R.A. & Coimbra Filho, A.F. (1981). *Ecology and Behaviour of Neotropical Primates*, Vol. 1. Rio de Janeiro: Academia Brasileira de Ciências.

Mittermeier, R.A. & van Roosmalen, M.G.M. (1981). Preliminary observations on habitat utilization and diet in eight Surinam monkeys. *Folia Primatologica*, **36**, 1–39.

Norconk, M.A. (1996). Seasonal variation in the diets of white-faced and bearded sakis (*Pithecia pithecia* and *Chiropotes satanas*) in Guri Lake, Venezuela. In *Adaptive Radiations of Neotropical Primates*, ed. M.A. Norconk, A.L. Rosenberger & P.A. Garber. New York, NY: Plenum Press, pp. 403–423, 547–548.

Norconk, M.A. (2011). Sakis, uakaris, and titi monkeys: behavioral diversity in a radiation of primate seed predators. In *Primates in Perspective*, ed. C.J. Campbell, A. Fuentes, K C. MacKinnon, S.K. Bearder & R.M. Stumpf. Oxford: Oxford University Press, pp. 122–139.

Norconk, M.A. & Kinzey, W.G. (1994). Challenge of neotropical frugivory: travel patterns of spider monkeys and bearded sakis. *American Journal of Primatology*, **34**, 171–183.

Norconk, M.A., Raghanti, M.A., Martin, S.K., *et al.* (2003). Primates of Brownsberg Natuurpark, Suriname, with particular attention to the Pitheciins. *Neotropical Primates*, **11**, 94–100.

Norconk, M.A., Wright, B.W., Conklin-Brittain, N.L., *et al.* (2009). Mechanical and nutritional properties of food as factors in platyrrhine dietary adaptations. In *South American Primates: Testing New Theories in the Study of Primate Behavior, Ecology, and Conservation*, ed. P.A. Garber, A. Estrada, C. Bicca-Marques, E. Heymann & K. Strier. New York, NY: Springer Science, pp. 279–319.

Overdorff, D.J. (1996). Ecological correlates to social structure in two lemur species in Madagascar. *American Journal of Physical Anthropology*, **100**, 487–506.

Peetz, A. (2001). Ecology and social organisation of the bearded saki *Chiropotes satanas chiropotes* (Primates: Pitheciinae) in Venezuela. *Ecotropical Monographs*, **1**, 1–170.

Peetz, A. & Homburg, I. (1998). Seasonal variation in the feeding ecology of bearded sakis (*Chiropotes satanas chiropotes*) in relation to their activity budget. *Folia Primatologica*, **69**, 217.

Pereira, A.P.C.P. (2002). Ecologia alimentar do cuxiú-preto (*Chiropotes satanas satanas*) na Fazenda Amanda, Pará. Masters' dissertation, Universidade Federal do Pará, Belém. Programa de Pós-graduação em Teoria e Pesquisa do Comportamento.

Pinto, L.P. (2008). Ecologia alimentar do cuxiú-de-nariz vermelho *Chiropotes albinasus* (Primates: Pitheciidae) na floresta Nacional do Tapajós, Pará. Unpublished doctoral dissertation, Universidade Estadual de Campinas, Brazil.

Port-Carvalho, M. & Ferrari, S.F. (2004). Occurrence and diet of the black bearded saki (*Chiropotes satanas satanas*) in the fragmented landscape of western Maranhão, Brazil. *Neotropical Primates*, **12**, 17–21.

Rettig, N.L. (1978). Breeding behavior of the harpy eagle (*Harpia harpyja*). *Auk*, **95**, 629–643.

Rosenberger, A.L. (1992). Evolution of feeding niches in New World monkeys. *American Journal of Physical Anthropology*, **88**, 525–562.

Santos, R.R. (2002). Ecologia de cuxiús (*Chiropotes satanas*) na Amazônia Oriental: Perspectivas para a conservação de populações fragmentadas. Masters' dissertation, Museu Paraense Emílio Goéldi e Universidade Federal do Pará, Belém.

Shaffer, C.A. (2012). GIS analysis of the ranging behavior, group cohesiveness, and patch use of bearded sakis (*Chiropotes sagulatus*) in the upper Essequibo Conservation Concession, Guyana. *American Journal of Physical Anthropology*, **147**(S54), 267.

Silva Jr., J.S. (1991). Distribuição geográfica do cuxiú-preto (*Chiropotes satanas satanas* Hoffmansegg. 1807) na Amazônia maranhense (Cebidae: Primates). In *A Primatologia no Brasil – 3*, ed. A.B. Rylands & A.T. Bernardes. Belo Horizonte: Sociedade Brasileira de Primatologia, pp. 275–284.

Silva Jr., J.S., Queiroz, H.L. & Fernandes, M.E.B. (1992). Primatas do Maranhão: dados preliminares (Primates: Platyrrhini). In: XIX Congresso Brasileiro de Zoologia, Belém, PA. Anais do XIX Congresso Brasileiro de Zoologia, p. 173.

Silva, S.S.B. (2003). Comportamento Alimentar do cuxiú-preto (*Chiropotes satanas*) na área de influência do reservatório da usina hidrelétrica de Tucuruí–Pará. Masters' dissertation, Museu Paraense Emílio Goéldi e Universidade Federal do Pará, Belém.

Silva, S.S.B. & Ferrari, S.F. (2009). Behavior patterns of southern bearded sakis (*Chiropotes satanas*) in the fragmented landscape of eastern Brazilian Amazonia. *American Journal of Primatology*, **71**, 1–7.

Strier, K.B. (2000). *Primate Behavioral Ecology*. Boston, MA: Allyn and Bacon.

Symington, M.M. (1988). Food competition and foraging party size in the black spider monkey (*Ateles paniscus chamek*). *Behaviour*, **105**, 117–134.

Symington, M.M. (1990). Fission–fusion social organisation in Ateles and Pan. *International Journal of Primatology*, **11**, 47–61.

van Roosmalen, M.G.M., Mittermeier, R.A. & Milton, K. (1981). The bearded saki, genus *Chiropotes*. In *Ecology and Behavior of Neotropical Primates Vol. 1*, ed. A.F. Coimbra-Filho & R.A. Mittermeier. Rio de Janeiro: Academia Brasileira de Ciências, pp. 419–441.

van Roosmalen, M.G.M., Mittermeier, R.A. & Fleagle, J.G. (1988). Diet of the northern bearded saki (*Chiropotes satanas chiropotes*): a neotropical seed predator. *American Journal of Primatology*, **14**, 11–35.

Veiga, L.M. (2006). Ecologia e comportamento do cuxiú-preto (*Chiropotes satanas*) na paisagem fragmentada da Amazônia oriental. Unpublished doctoral dissertation, Universidade Federal do Pará, Brazil.

Veiga, L.M. & Ferrari, S.F. (2006). Predation of arthropods by southern bearded sakis (*Chiropotes satanas*) in eastern Brazilian Amazonia. *American Journal of Primatology*, **68**, 209–215.

Veiga, L.M. & Ferrari, S.F. (2008). Coping with habitat fragmentation: flowers as an alternative resource for bearded sakis (*Chiropotes* spp.). *Primate Eye*, **96**, 104–105.

Veiga, L.M. & Silva, S.S.B. (2005). Relatives or just good friends? Affiliative relationships among male southern bearded sakis (*Chiropotes satanas*). *Livro de Resumos, XI Congresso Brasileiro de Primatologia*, Porto Alegre, 13–18 February 2005, p. 174.

Veiga, L.M., Pinto, L.P. & Ferrari, S.F. (2006). Fission–fusion sociality in bearded sakis (*Chiropotes albinasus* and *Chiropotes satanas*) in Brazilian Amazonia. *International Journal of Primatology*, **27**(Suppl 1), #224.

Veiga, L.M., Pinto, L.P., Ferrari, S.F., *et al.* (2008c). *Chiropotes albinasus. IUCN Red List of Threatened Species*. www.iucnredlist.org.

Veiga, L.M., Silva Jr., J.S., Ferrari, S.F., *et al.* (2008a). *Chiropotes satanas. IUCN Red List of Threatened Species*. www.iucnredlist.org.

Veiga, L.M., Silva Jr., J.S., Ferrari, S.F., *et al.* (2008b). *Chiropotes utahickae. IUCN Red List of Threatened Species*. www.iucnredlist.org.

Vieira, T. (2005). Aspectos da ecologia do cuxiú de Uta Hick, *Chiropotes utahickae* (Hershkovitz, 1985), com ênfase na exploração alimentar de espécies arbóreas da ilha de Germoplasma, Tucuruí-PA, Masters' dissertation, Museu Paraense Emílio Goéldi e Universidade Federal do Pará, Belém.

Wallace, R.B. & Painter, R.L.E. (1996). Notes on a distribution river boundary and southern range extension for two species of Amazonian primates. *Neotropical Primates*, **4**, 149–151.

White, F.J. & Wrangham, R.W. (1988). Feeding competition and patch size in chimpanzee species *Pan paniscus* and *Pan troglodytes*. *Behaviour*, **105**, 148–164.

Chapter 23

Feeding ecology of Uta Hick's bearded saki (*Chiropotes utahickae*) on a man-made island in southeastern Brazilian Amazonia: seasonal and longitudinal variation

Ricardo R. Santos, Tatiana M. Vieira & Stephen F. Ferrari

Introduction

Chiropotes utahickae – which is endemic to Brazil – is one of the least known of the five species of bearded sakis recognized by Silva Jr. *et al.* (Chapter 4). Bobadilla and Ferrari (1998, 2000) and Ferrari *et al.* (1999) provide some general information, and confirm considerably greater densities in disturbed habitat in comparison with continuous forest. More recently, Ferrari *et al.* (2002) surveyed the Tucuruí Reservoir on the Tocantins River in the south of the state of Pará, where remnant populations of *C. utahickae* are found on a number of relatively small islands, including the site of the present study.

While the pioneering studies of Ayres (1981, 1989), van Roosmalen *et al.* (1981), Johns and Ayres (1987) and Frazão (1992) indicated that bearded sakis depend on relatively large areas of undisturbed forest, more recent research (Peetz 2001; Port-Carvalho & Ferrari 2004; Veiga 2006; Silva & Ferrari 2009) has provided a very different perspective on the ecology of *Chiropotes*. In particular, whereas early estimates of home-range size varied from 200–500 ha, and Pinto (2008) recorded a range of approximately 1000 ha for *Chiropotes albinasus* in continuous forest, some groups are now known to survive in much smaller fragments, some less than 20 ha, although this tolerance of habitat disturbance may be mediated by significant shifts in resource exploitation.

The bearded saki group that was monitored in the present study is a case in point. In addition to providing the first detailed data on many aspects of the ecology of *C. utahickae*, this study provides a number of insights into the species' ability to tolerate habitat fragmentation. Despite the small size of the island, study group members were able to maintain a diet relatively rich in fruit and seeds, although alternative resources, in particular flowers, were also relatively important, especially during some months.

Methods

The study took place on the 129-ha Germoplasma Island (3°52′S, 49°38′W), located close to the left bank of the Tucuruí Reservoir on the Tocantins River in the Brazilian state of Pará. The island is covered predominantly by primary *terra firme* forest, with a central plantation of native tree species (18 ha). Climate is humid equatorial, with a marked wet season between December and June, and a dry season from July to November. Six other platyrrhine species are found on the island (*Alouatta belzebul*, *Aotus infulatus*, *Callicebus moloch*, *Cebus apella*, *Saguinus niger* and *Saimiri sciureus*).

The resident group of *C. utahickae* had 24 members in April 2001, increasing to at least 28 individuals by the end of 2003, although from March 2004 onwards no more than 23 animals were seen together at a given time. The latter may have been a subgroup, however, considering that it used no more than half of the island during monitoring in 2004 (Vieira 2005). Preliminary observations and habituation were carried out between April and June 2001, and continuous monitoring was conducted in two periods – July to November 2001, and March to August 2004 – which correspond approximately to dry and wet season samples, respectively.

Quantitative behavioral data were collected in instantaneous scan samples (Martin & Bateson 1993), with scans being carried out at 5-min intervals throughout the daily activity period (following Ayres 1981 and Peetz 2001). In 2001, data were collected on 5 days each month, except August (4 days), whereas in 2004, the monthly sample was 8 days. During a scan, the activity state of each visible subject was recorded. Four main behavioral categories were used (rest, locomotion, feeding and "miscellaneous": primarily sociosexual behavior), and activity budgets were obtained from the relative frequency of records of each category. In the case of feeding records, details of the item or plant part being consumed were recorded, and whenever possible, sources were marked for later identification. The composition of the diet was estimated from the relative frequency of different items: $n_i/N \times 100$, where n_i = number of scan records of item i, and N = the total number of feeding records (identified items) collected during the period.

Evolutionary Biology and Conservation of Titis, Sakis and Uacaris, eds. Liza M. Veiga, Adrian A. Barnett, Stephen F. Ferrari and Marilyn A. Norconk. Published by Cambridge University Press.

Results

General behavioral characteristics

The behavior of study group members was typical of that of bearded sakis at other sites, being characterized by rapid movements in the uppermost forest strata and the frequent formation of subgroups, some with as few as three individuals. The group's general activity budget (n = 16,767 scan sample records) was composed of 47.96% locomotion, 41.06% feeding, 10.51% resting and 0.47% miscellaneous activities. While this feeding rate is relatively high in comparison with most other platyrrhines, it appears to be typical of the pitheciines in general (Ayres 1986; Boubli 1997; Homburg 1998; Peetz 2001; Pinto 2008; Silva & Ferrari 2009).

It nevertheless seems unlikely that these animals do in fact devote such a large proportion of their activity time to feeding. What seems more likely here is that other categories, in particular rest, are relatively underestimated due to differences in their visibility. The relative cohesion of groups in feeding trees may reinforce this trend, which may be exacerbated by intrinsic observer bias towards feeding behavior, the observation of which tends to be prioritized in studies of this kind.

The proportion of time spent feeding varied considerably among months, from a low of 20.78% in March to a high of 62.67% in November. Increasing time spent feeding during the course of the year reflected a general trend of an increasing proportion of seeds in the diet (see below).

Feeding behavior and diet

Study group members were frequently observed feeding in relatively small subgroups, although Santos (2002) did not find any systematic relationship between the size of foraging parties and that of feeding trees. While the diet encompassed a wide variety of items, including non-reproductive plant parts such as leaves and cambium, the major components – seeds and fruit – were typical of *Chiropotes* (Figure 23.1). The relative importance of flowers is less characteristic, and may reflect a reduced availability of preferred items at the study site.

The group's diet varied considerably among months (Figure 23.2), with a general tendency for the consumption of seeds to increase during the dry season, and that of fruit and flowers to decrease. However, the most frugivorous diet was recorded in August, at the beginning of the dry season, although the highest proportion of seeds was recorded at the end of the dry season (November). Flowers were most important in September, although 87.1% of feeding records referred to a single species, *Alexa grandiflora* (Fabaceae).

Most (89.42%) seeds were immature when consumed, and this pattern predominated in all months (Figure 23.3a). The consumption of fruit was more balanced, however, with only 54.40% of records involving unripe items, although there was a clear seasonal pattern here (Figure 23.3b). Ripe fruit accounted for as many as half of all feeding records during the wet season months between March and May, but fell below 10% in the remaining months. The peak of fruit feeding in August was thus due to the consumption of the immature mesocarp of *Inga alba* (Mimosaceae) in 2004, when it accounted for all but one of the records of the consumption of immature fruit, and 60.74% of the group's diet.

Overall, at least 161 different plant species from 44 families contributed to the study group's diet (see Santos 2002; Vieira 2005). However, only six species (*Alexa grandiflora*, *Annona tenuipes* (Annonaceae), *Bertholletia excelsa* (Lecythidaceae), *Eschweilera* sp. (Lecythidaceae), *Inga alba* and *Newtonia suavolens* [Mimosaceae]) provided more than half (50.3%) of

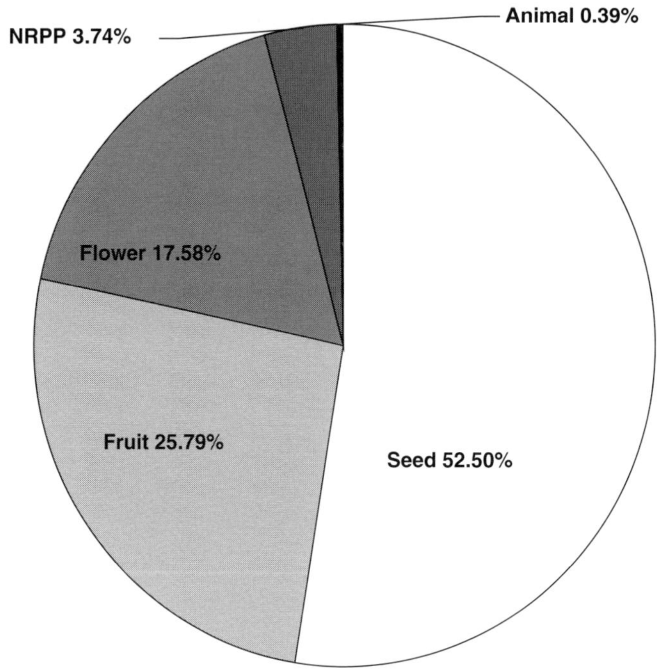

Figure 23.1 General diet of the *C. utahickae* study group, based on the percentages of scan sample records (n = 6655, excluding unidentified items, water and suckling). NRPP, non-reproductive plant parts (leaves, leaf shoots, cambium, palm heart).

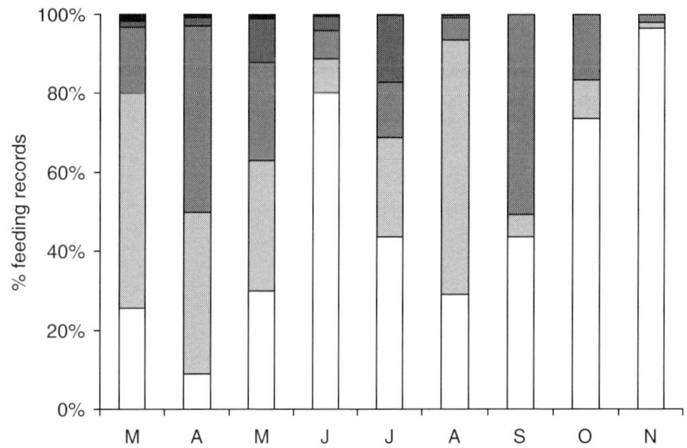

Figure 23.2 Seasonal variation in the diet of the *C. utahickae* study group, based on the percentages of scan sample records (for legend, see Figure 23.1).

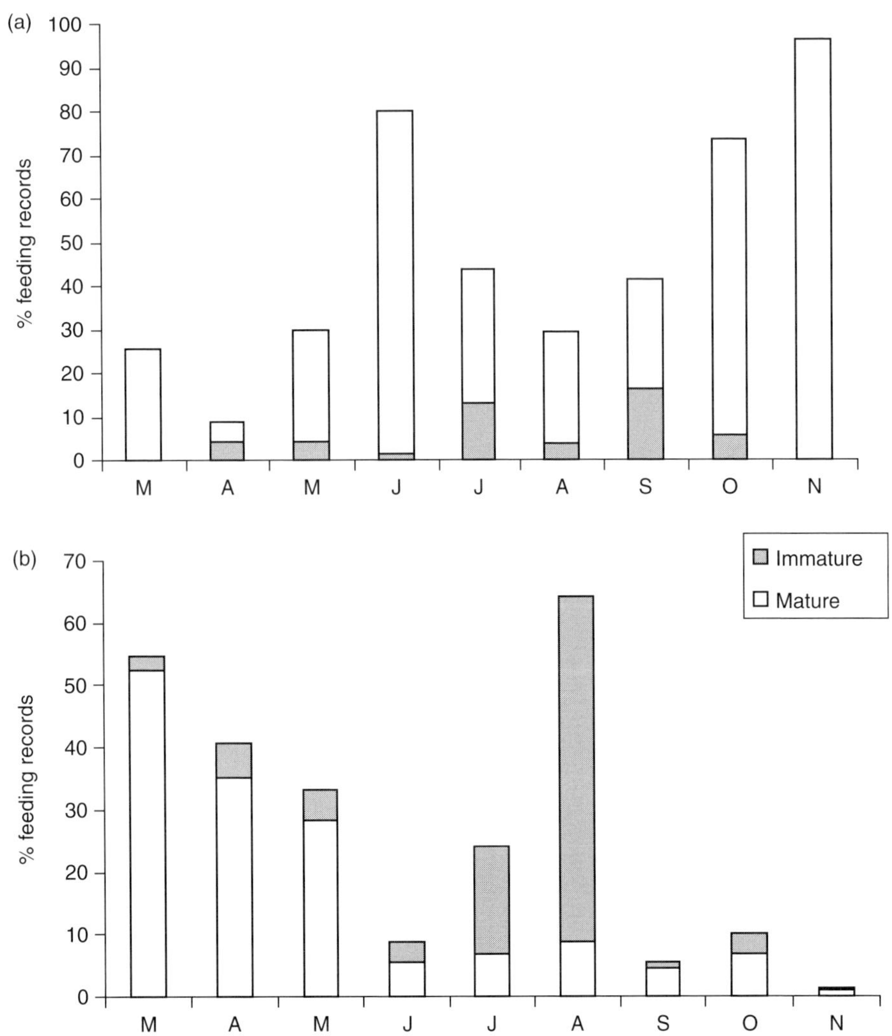

Figure 23.3 Seasonal variation in the relative contribution of (a) immature and mature seeds and (b) immature and mature fruit to the diet of the *C. utahickae* study group.

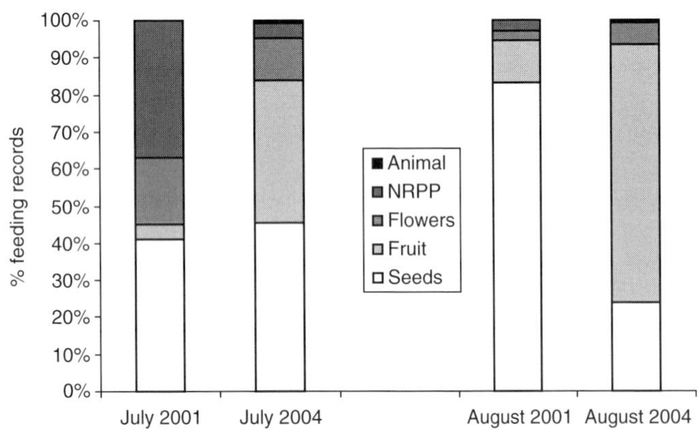

Figure 23.4 Longitudinal variation in the composition of the study group's diet, comparing July and August of 2001 and 2004.

feeding records (identified items), whereas 34 species were recorded only once. Similarly, almost half (46.1%) of identified species belonged to one of the top six families (Annonaceae, Bignoniaceae, Caesalpinaceae, Mimosaceae, Moraceae, and Sapotaceae), while 18 families (40.9%) were represented by only one species.

The Lecythidaceae was represented by only six species, despite the apparent importance of this family to the pitheciines (see Ayres & Prance, Chapter 12), although two species – *Bertholletia excelsa*, *Eschweilera* sp. – were among the top six. *Bertholletia* – the Brazil nut tree – provided almost a third of the flowers consumed which together with flowers from *A. grandiflora* provided 13.5% of the diet. By contrast, *Eschweilera* sp. and *N. suavolens* provided mostly seeds, while *Annona tenuipes* and *I. alba* were exploited for their fruit (mesocarp).

Longitudinal variation

Two months were sampled in both years of monitoring, permitting a tentative evaluation of longitudinal variation. Perhaps surprisingly, there was almost as much variation between months of the same year as between years, starting with time spent feeding, which declined from 75.24% of activity time in July 2001 to 29.27% in August of the same year, whereas in 2004, it increased from 39.09% in July to 50.72% in August.

While the value for July 2001 is almost certainly a substantial overestimate of the actual time spent ingesting food, this higher feeding rate does correlate with the unusual diet recorded in this month (Figure 23.4). Non-reproductive plant parts were a negligible component of the study group's diet in all other months, whereas in July 2001 they contributed 36.9% of feeding records. This was due primarily to the consumption of the leaf buds of *Newtonia suaveolans*. In July 2004, only a small amount of plant parts were consumed, with a corresponding increase in the consumption of fruit. This was due in part to the exploitation of the immature mesocarp of three species of *Inga* (*Inga alba*, *Inga falcistipula* and *Inga* sp.), which together contributed 21.0% of feeding records in this month.

In August, the contribution of the immature mesocarp of *I. alba* to the group's diet increased to 60.7%. This contrasts with the marked predominance of seeds (47.1% of which were provided by *Enterolobium schomburgkii*, Mimosaceae) in the same month of 2001.

Overall, the comparison points to a major shift in the availability of resources between the two years, probably as a result of the varying phenological patterns, primarily, the availability of *Inga* fruit in 2004, when it was consumed from March onwards. This genus was not recorded at all in 2001 (July to November), however. *Inga* may thus have either fruited earlier in 2001 or, possibly, not at all. The data from July 2001 suggest that fruit of any kind may have been relatively scarce during this month.

This contrast is further reinforced by the much smaller number of plant species exploited during 2001 in comparison with 2004. To standardize sampling effort for this comparison, 5 days were selected randomly from July 2004 and 4 from August of this year. Whereas only 12 species were recorded during July 2001 and 17 in August, 26 and 28 species were recorded in the respective months in 2004.

There was relatively little overlap in the species exploited in the same month in the two years (Jaccard's index of similarity: $J = 0.188$ for July, and $J = 0.125$ for August), although the similarity between months of the same year was also lower than might be expected ($J = 0.304$ for 2001 and $J = 0.385$ for 2004). These values reflect the overall variability in the composition of the diet. In addition, while diversity was relatively low in July 2001 (Shannon–Wiener's index of diversity: $H' = 1.287$) in comparison with the same month of 2004 ($H' = 2.675$), the opposite trend was observed for August ($H' = 2.124$ in 2001; $H' = 1.731$ in 2004). The lower values were the result of the marked predominance of a single species in the diet (*Newtonia suaveolans* in July 2001, and *Inga alba* in August 2004). While less pronounced, the other months were also characterized by the typical pattern of intensive exploitation of a few species seen throughout the study.

Discussion

Despite inhabiting a relatively small, isolated area of forest, the members of the *C. utahickae* study group presented patterns of behavior similar to those of bearded sakis at other sites. Perhaps the most valid comparisons can be made with the study of Peetz (2001), in which a group of 22 *Chiropotes chiropotes* was monitored on a reservoir island (Danto Machado) of roughly similar size to Germoplasma.

As in many other pitheciines (Ayres 1986; Boubli 1997; Homburg 1998; Peetz 2001; Pinto 2008; Silva & Ferrari 2009), group members spent an unexpectedly large proportion of their time feeding. Peetz (2001) suggested that high feeding rates may be related to the exploitation of seeds, which requires more time for manipulation and the extraction of nutrients than unspecialized frugivory. In this case, a certain proportion of the feeding records would probably best be classified as foraging, i.e. the capture, rather than the ingestion of nutrients. Interestingly, in the present study, there was a strong positive correlation between time spent feeding each month and the proportion of seeds in the diet (Spearman's $r_s = 0.700$, $p < 0.05$, $n = 9$).

Whereas the exploitation of immature seeds is typical of *Chiropotes*, the secondary importance of atypical items (flowers and leaf buds) at Germoplasma, rather than fruit, may point to resource deficiencies at the site. Flowers were also a relatively important component of the diet of *Chiropotes satanas* on islands at the right bank of the reservoir (Silva 2003; Veiga 2006) and Setz (1993) recorded a similar pattern in a fragment-dwelling group of *Pithecia pithecia chrysocephala*. At Germoplasma, flowers were an important rcsourcc in some months, although they were amply spaced (April, July 2001, and September), and did not follow a well-defined seasonal pattern. Rather, the longitudinal analysis indicates an unpredictable pattern of resource availability, with the sakis presumably exploiting the most abundant resources available at any given time.

While a surprisingly large number of plant species were exploited, considering the size of the island, a small number provided the bulk of the feeding records, as on Danto Machado (Peetz 2001), where the five top-ranking species provided three-quarters of feeding records. The two groups consumed seeds in similar proportions, but flowers, in particular, and non-reproductive plant parts were much less important on Danto Machado. By contrast, whereas Germoplasma sakis were rarely observed preying on insects, this resource was relatively important in some months on Danto Machado, as it was on the right bank of the reservoir (Veiga & Ferrari 2006). These contrasts almost certainly reflect local differences in the configuration of resources, rather than interspecific differences in foraging adaptations.

By the end of the present study period, the Germoplasma sakis had been isolated on the island for almost 20 years. Despite the island's small size, the results of this study indicate that the sakis have been able to survive over at least two generations without any major shifts in their behavior or ecology. However, the topical importance of atypical resources indicates that this situation may be dependent on alternative, possibly suboptimal foraging strategies, although post-mortem examination of an adult study group member (Ferrari *et al.* 2004) revealed no evidence of nutritional stress or deficiencies. While *C. utahickae* may be relatively tolerant of the effects of anthropogenic habitat fragmentation, remnant populations in isolated fragments such as Germoplasma island will almost certainly require active management over the long term (Turner & Corlett 1996).

Acknowledgments

Fieldwork was supported by Eletronorte S.A. and the Kapok Foundation. We are grateful to CAPES and CNPq for funding this research, and to Rubens Ghilardi Jr, Tacachi Hatanaka and Edilene Nunes for logistical support in the field. We especially thank Jucelino Rodrigues Leal Chaves and Antônio Vieira de Macedo for their field assistance.

References

Ayres, J.M.C. (1981). Observações sobre a Ecologia e o Comportamento dos Cuxiús (Chiropotes albinasus e Chiropotes satanas, Cebidae: Primates). Unpublished Master's dissertation, Fundação Universidade do Amazonas.

Ayres, J.M.C. (1986). Uakaris and Amazonian flooded forests. Unpublished PhD thesis, Cambridge University.

Ayres, J.M. (1989). Comparative feeding ecology of the uakari and bearded saki, *Cacajao* and *Chiropotes*. *Journal of Human Evolution*, **18**, 697–716.

Bobadilla, U.L. & Ferrari, S.F. (1998). First detailed field data on *Chiropotes satanas utahicki* Hershkovitz, 1985. *Neotropical Primates*, **6**, 17–18.

Bobadilla, U.L. & Ferrari, S.F. (2000). Habitat use by *Chiropotes satanas utahicki* and syntopic platyrrhines in eastern Amazonia. *American Journal of Primatology*, **50**, 215–224.

Boubli, J.P. (1997). A study of the black uakari, *Cacajao melanocephalus melanocephalus*, in the Pico da Neblina National Park, Brazil. *Neotropical Primates*, **5**, 113–115.

Ferrari, S.F., Ghilardi Jr., R., Lima, E.M., *et al.* (2002). Mudanças a longo prazo nas populações de mamíferos da área de influência da Usina Hidrelétrica de Tucuruí, Pará. *Resumos do XXIV° Congresso Brasileiro de Zoologia*, 540–541.

Ferrari, S.F., Pereira, W.P.A., Santos, R.R., *et al.* (2004). Fatal attack of a *Boa constrictor* on a bearded saki (*Chiropotes satanas utahicki*). *Folia Primatologica*, **75**, 111–113.

Ferrari, S.F., Silva, C.E., Lopes, M.A., *et al.* (1999). Bearded sakis in southeastern Amazonia – back from the brink? *Oryx*, **33**, 346–351.

Frazão, E.R. (1992). Dieta e estratégia de forragear de *Chiropotes satanas chiropotes* (Cebidae: Primates) na Amazônia central brasileira. Unpublished Masters' dissertation, Instituto Nacional de Pesquisa da Amazônia.

Homburg, I. (1998). Ökologie und Sozialverhalten einer Gruppe von Weiβgesicht-sakis (*Pithecia pithecia pithecia* Linnaeus 1766) im Estado Bolívar, Venezuela. Unpublished PhD thesis, Universität Bielefeld.

Johns, A.D. & Ayres, J.M. (1987). Southern bearded sakis beyond the brink. *Oryx*, **21**, 164–167.

Martin, P. & Bateson, P. (1993). *Measuring Behaviour: An Introductory Guide*. Cambridge: Cambridge University Press.

Peetz, A. (2001). Ecology and social organization of the bearded saki *Chiropotes satanas chiropotes* (Primates: Pitheciinae) in Venezuela. *Ecotropical Monographs*, **1**, 1–170.

Pinto, L.P. (2008). Ecologia alimentar de um grupo de cuxiús-de-nariz-vermelho *Chiropotes albinasus* (Primates: Pitheciidae) na Floresta Nacional do Tapajós, Pará. Unpublished PhD thesis, Universidade Estadual de Campinas.

Port-Carvalho, M. & Ferrari, S.F. (2004). Occurrence and diet of the black bearded saki (*Chiropotes satanas satanas*) in the fragmented landscape of western Maranhão, Brazil. *Neotropical Primates*, **12**, 17–21.

Santos, R.R. (2002). Ecologia de cuxiús (*Chiropotes satanas*) na Amazônia oriental: perspectivas para a conservação de populações fragmentadas. Unpublished Masters' dissertation, Museu Paraense Emílio Goeldi.

Setz, E.Z.F. (1993). Ecologia alimentar de um grupo de parauacus (*Pithecia pithecia chrysocephala*) em um fragmento florestal na Amazônia central. Unpublished PhD thesis, Universidade Estadual de Campinas.

Silva, S.S.B. (2003). Comportamento alimentar de cuxiú-preto (*Chiropotes satanas*) na área de influência do reservatório da usina hidrelétrica de Tucuruí-PA. Unpublished Master's dissertation, Museu Paraense Emílio Goeldi.

Silva, S.S.B. & Ferrari, S.F. (2009). Behavior patterns of southern bearded sakis (*Chiropotes satanas*) in the fragmented landscape of eastern Brazilian Amazonia. *American Journal of Primatology*, **71**, 1–7.

Turner, I.M. & Corlett, R.T. (1996). The conservation value of small, isolated fragments of lowland tropical rain forest. *Trends in Ecology & Evolution*, **11**, 330–333.

van Roosmalen, M.G.M., Mittermeier, R.A. & Milton, K. (1981). The bearded sakis, genus *Chiropotes*. In *Ecology and Behavior of Neotropical Primates*, Volume 1, ed. A.F. Coimbra-Filho & R.A. Mittermeier. Rio de Janeiro: Academia Brasileira de Ciências, pp. 419–441.

Veiga, L.M.V. (2006). Ecologia e organização social de cuxiús, Chiropotes satanas satanas, em Tucuruí, Pará. Unpublished PhD thesis, Universidade Federal do Pará.

Veiga, L.M.V. & Ferrari, S.F. (2006). Predation of arthropods by southern bearded sakis (*Chiropotes satanas*) in eastern Brazilian Amazonia. *American Journal of Primatology*, **68**, 209–215.

Vieira, T.M. (2005). Aspectos da ecologia do Cuxiú de Uta Hick, *Chiropotes utahickae* (Hershkovitz, 1985), com ênfase na exploração alimentar de espécies arbóreas da Ilha de Germoplasma, Tucuruí – PA. Unpublished Master's dissertation, Museu Paraense Emílio Goeldi.

Chapter 24

The behavioral ecology of northern bearded sakis (*Chiropotes satanas chiropotes*) living in forest fragments of Central Brazilian Amazonia

Sarah A. Boyle, Andrew T. Smith, Wilson R. Spironello & Charles E. Zartman

Introduction

Habitat fragmentation is an increasing global trend that impacts the sustainability of ecosystems worldwide (Lovejoy 2006). In the case of forest fragmentation, areas of contiguous forest are cleared, and a mosaic of patches surrounded by non-forested matrix remains. Forest fragmentation can affect local climate (Bierregaard Jr *et al.* 1992; Achard *et al.* 2002), habitat suitability (Gascon *et al.* 2000; Laurance *et al.* 2000), species richness (Bierregaard Jr *et al.* 1992; Malcolm 1997), seed dispersal (Chapman & Onderdonk 1998; Estrada *et al.* 1999), and predator–prey interactions (Asquith *et al.* 1997).

Primates vary in their diet composition and specializations for obtaining resources, and large-bodied primates that eat fruit usually require large home ranges (Clutton-Brock & Harvey 1977; Johns & Skorupa 1987; Onderdonk & Chapman 2000). Yet, studies relating degree of frugivory and species presence within forest fragments have yielded different results. Currently there is no consensus regarding the use of primate characteristics to predict the vulnerability of a species in fragmented habitat (Johns & Skorupa 1987; Estrada & Coates-Estrada 1996; Onderdonk & Chapman 2000; Marsh *et al.* 2003; Boyle & Smith 2010a). Therefore, further data are necessary to determine whether a general pattern exists between diet specialization and survival in fragmented habitats.

We focused our research on the northern bearded saki (*Chiropotes satanas chiropotes*). Bearded sakis (*Chiropotes* spp.) are specialized seed predators that derive approximately 90% of their diet from fruit (Ayres 1981; Peetz 2001). Although some studies have examined the genus *Chiropotes* in human-altered habitats (Ferrari *et al.* 1999; Peetz 2001; Port-Carvalho & Ferrari 2004; Veiga & Ferrari 2006), much of this research has focused on forested islands that were formed by the construction of and subsequent flooding by hydroelectric dams (Peetz 2001; Santos 2002; Veiga & Ferrari 2006).

Bearded sakis (*Chiropotes* spp.) typically live in large social groups that separate and rejoin throughout the day, have large home ranges, and travel long daily distances (Norconk 2011). Previous studies of bearded sakis have found that the species' home range varies from 200 to 250 ha (Ayres 1981; van Roosmalen *et al.* 1981) in continuous forest. Research in Pará, Brazil (Silva 2003; Veiga 2006) and Venezuela (Peetz 2001) revealed that bearded sakis on human-created islands (see above) occupy areas of 16–250 ha. Thus, bearded saki home range size appears to be flexible.

The characteristic home-range size of bearded sakis in continuous forest is larger than any of the forest fragments in our study area. Therefore, our goals were to determine the minimal fragment size needed to sustain northern bearded saki populations, and also to determine how groups inhabiting fragmented forests differed behaviorally and ecologically from groups in undisturbed habitats. Here we present a review of our findings.

Methods

Study site

We conducted this research in the Biological Dynamics of Forest Fragments Project (BDFFP) reserve, located approximately 80 km north of Manaus, in the Brazilian state of Amazonas. BDFFP is facilitated by the Instituto Nacional de Pesquisas da Amazônia (INPA) and the Smithsonian Tropical Research Institute (STRI). The project, initially called the Minimal Critical Size of Ecosystems project, began in 1979. Studies prior to and after fragmentation have provided a catalogue of changes in species distribution and loss (Lovejoy *et al.* 1984; Rylands & Keuroghlian 1988; Ferraz *et al.* 2003), forest composition and structure (Ferreira & Laurance 1997; Laurance *et al.* 1997; Nascimento *et al.* 2006) and microclimate (Kapos 1989; Murcia 1995).

The study site is tropical moist *terra firme* forest that receives 1900–3500 mm of rain annually with a dry season from June to October (Gascon & Bierregaard 2001; Laurance 2001). There are three fragment size classes (1 ha, 10 ha and 100 ha), as well as continuous primary forest that serves as a control for comparison (Gascon & Lovejoy 1998; Figure 24.1). The degree of isolation (i.e. distance to nearest forest patch, condition of the surrounding matrix) varies among the BDFFP fragments (Bierregaard & Stouffer 1997; Boyle & Smith

Evolutionary Biology and Conservation of Titis, Sakis and Uacaris, eds. Liza M. Veiga, Adrian A. Barnett, Stephen F. Ferrari and Marilyn A. Norconk. Published by Cambridge University Press.

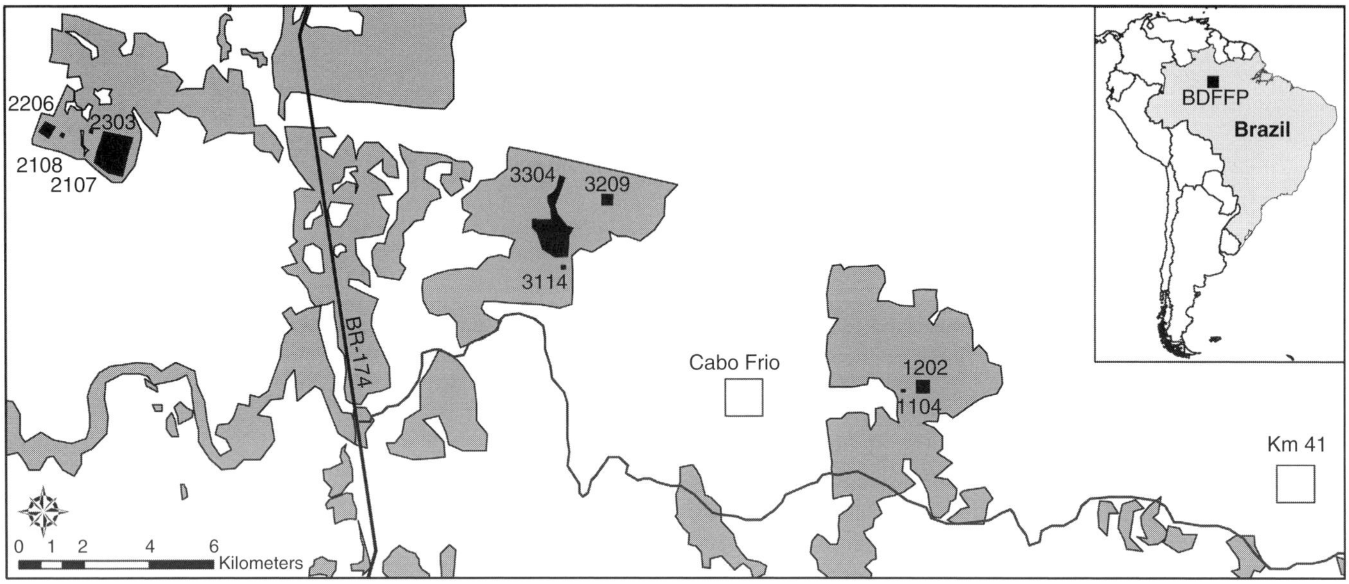

Figure 24.1 BDFFP study site. Nine forest fragments (black polygons) and two areas of continuous forest (white polygons outlined in black) were surveyed. Gray polygons represent the matrix. Figure modified from Boyle (2008a).

2010a), and in 2006 there was more secondary growth forest in the matrix than in previous years (Boyle & Smith 2010a). In addition, there have been changes to the forest surrounding the study area, due to increased human pressures (Laurance & Luizão 2007; Boyle 2008a).

Although research at BDFFP has been ongoing for more than three decades, research involving bearded sakis had been minimal, consisting of sporadic annual censuses in the forest fragments (Gilbert 2003), and behavioral research in the continuous forest (Frazão 1992). Our research is the first behavioral ecology study of bearded sakis that takes into account each of the nine fragments in addition to areas of continuous forest.

The presence of bearded sakis in the BDFFP fragments has been variable since the reserves were first isolated. Upon isolation, bearded sakis left the isolated reserve areas (Rylands & Keuroghlian 1988). Additional censuses found bearded sakis to be absent from the fragments until the mid 1990s when an adult male spent five months in one of the 10-ha fragments (#2206). In 1995, an adult male, adult female and infant were spotted in another 10-ha fragment (#1202) during a two-week period (Gilbert 1994; Gilbert & Setz 2001). An adult male bearded saki was present in one of the 10-ha BDFFP fragments (#1202) in 1997, and was joined by an adult female bearded saki by 2001. The third 10-ha fragment (#3206) has had no history of bearded saki inhabitants. The presence of bearded sakis in the two 100-ha BDFFP forest fragments (#2303 and #3303) was not noted until a census in 2000 (Gilbert 2003). Prior to 2003, bearded sakis had not been spotted in any of the four 1-ha BDFFP fragments.

Five other primate species (*Alouatta macconnelli*, *Ateles paniscus*, *Cebus apella*, *Pithecia pithecia* and *Saguinus midas*) reside in the BDFFP study area. Some species (e.g. *Alouatta*) frequently occupy forest fragments in all three size classes (1 ha, 10 ha and 100 ha), while other species (e.g. *Ateles*) are rarely present (Gilbert 2003; Boyle & Smith 2010a).

Sampling methods

We collected data during a preliminary study in July–August 2003, and then from January 2005 to June 2006. We surveyed nine forest fragments at the Dimona, Esteio and Porto Alegre ranches – four 1-ha, three 10-ha, and two 100-ha fragments – and two areas of continuous forest.

On the first day in each study area we conducted a primate census by walking line transects along already established trails, following the past methods used at BDFFP (Rylands & Keuroghlian 1988; Gilbert 2003). If bearded sakis were encountered during the primate census (Boyle & Smith 2010a), they were designated the focal group for that fragment's data cycle, and subsequently relocated and followed for 3 consecutive days. One cycle through the nine fragments and two areas of continuous forest lasted approximately 56 days for a total of four cycles annually.

Upon locating a group of bearded sakis, we tracked the same group from the time they awoke in the morning until the time they settled down for the night (approximately 0530–1730), for a total of 604 behavioral contact hours. Due to the presence of researchers in the study area for nearly 30 years, the bearded sakis were habituated to human observers. Using group scan sampling techniques (Altmann 1974), we recorded the group's GPS coordinates using a handheld GPS receiver every 5 min, as well as group size and composition, and the activity of each individual in sight. Activities included eating, foraging, resting, moving, traveling, social interactions and "other." If individuals were eating fruit, flowers or leaves, we marked the tree or liana with plastic flagging and assigned it a unique number. We returned later to each feeding location for floral identification. We noted the tree's GPS coordinates, diameter at breast height (DBH), and the type of food (fruit, flower, leaf).

Later, we relocated the flagged trees, epiphytes and lianas using maps produced from the GPS data, and we collected plant material samples in order to identify food sources. Identifications were conducted both in the field and the INPA herbarium in Manaus, Brazil.

Spatial analyses

Based on the GPS data, we mapped the daily routes of the bearded saki groups and their feeding sites using ArcView 3.3. We determined daily distance traveled and home range (Boyle *et al.* 2009a) using the Home Range Extension (Rodgers & Carr 1998) for ArcView. We measured home-range size using both Minimum Convex Polygon (Odum & Kuenzler 1955) and Adaptive Kernel (Worton 1987) methods due to differences in home-range estimates when sample size was small (Boyle *et al.* 2009b).

Summary of findings

Presence in fragments

Bearded saki groups were present in five of the nine forest fragments (one 1-ha, two 10-ha and two 100-ha fragments); however, only two of these groups were permanent residents during the study period (Table 24.1; Boyle & Smith 2010b). There were no bearded sakis in any of the four 1-ha fragments, with the exception of the presence of one subadult male in the 1-ha fragment (#2107) in July–August 2003. We do not know if this individual left the forest fragment or died.

Two of the 10-ha fragments (#1202 and #2206) had one bearded saki group each. These two groups were consistently present in the fragments in July–August 2003 and January 2005–June 2006. We never witnessed either group leaving the forest fragment. We never spotted bearded sakis in the 10-ha fragment (#3209) at Porto Alegre.

Each 100-ha fragment hosted one bearded saki group; however, these groups did not remain in their respective forest fragments (Boyle & Smith 2010b). These fragments were not fully isolated as forest "islands" (Boyle & Smith 2010a), and the bearded sakis used forested corridors to travel in and out of these fragments.

Group size

Bearded saki group size ranged from 1 to 35 individuals, and as forest patch size increased, so did group size (Boyle & Smith 2010b). Average group size (with standard error given in parentheses) was 1 individual in the 1-ha fragment, 3.79 (± 0.21) individuals in the 10-ha fragments, 12.15 (± 1.44) individuals in the 100-ha fragments and 22.89 (± 5.09) individuals in the continuous forest (Figure 24.2). Group size was constant at four individuals during the 2003 and 2005–2006 study periods for the group inhabiting a 10-ha fragment (#2206). Group size was four individuals for the group inhabiting another 10-ha fragment (#1202) until the disappearance of an adult female in October 2005. We do not know whether

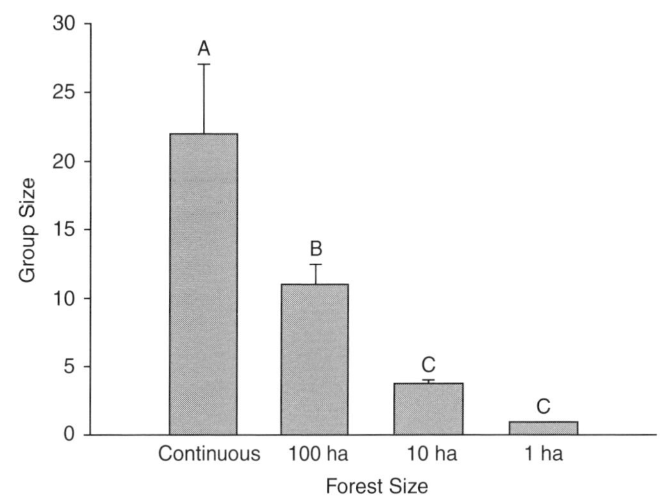

Figure 24.2 Average group size. Standard error bars are present and A, B and C represent differences among the bearded saki groups. Data are from Boyle and Smith (2010b).

Table 24.1 Northern bearded saki presence in nine forest fragments (2003–2006).

Size	Present		Status	Absent	
1 ha	Dimona	(#2107)	2003 only	Dimona	(#2108)
				Esteio	(#1104)
				Porto Alegre	(#3114)
10 ha	Dimona	(#2206)	Permanent	Porto Alegre	(#3209)
	Esteio	(#1202)	Permanent		
100 ha	Dimona	(#2303)	Nomadic		
	Porto Alegre	(#3304)	Nomadic		

Table 24.2 Combined northern bearded saki diet across all BDFFP study sites.

Family	Total individuals	# Genera	# Species	Family	Total individuals	# Genera	# Species
Anacardiaceae	2 (0.22%)	1 (0.87%)	2 (0.82%)	Humiriaceae	10 (1.09%)	3 (2.61%)	3 (1.23%)
Anisophylleaceae	1 (0.11%)	1 (0.87%)	1 (0.41%)	Icacinaceae	1 (0.11%)	1 (0.87%)	1 (0.41%)
Annonaceae	41 (4.49%)	6 (5.22%)	11 (4.51%)	Lauraceae	3 (0.33%)	1 (0.87%)	1 (0.41%)
Apocynaceae	20 (2.19%)	4 (3.48%)	5 (2.05%)	Lecythidaceae	132 (14.44)	4 (3.48%)	16 (6.56%)
Araceae	2 (0.22%)	2 (1.74%)	2 (0.82%)	Leguminosae	55 (6.02%)	14 (12.17%)	24 (9.84%)
Arecaceae	1 (0.11%)	1 (0.87%)	1 (0.41%)	Loganiaceae	6 (0.66%)	1 (0.87%)	3 (1.23%)
Bignoniaceae	13 (1.42%)	5 (4.35%)	8 (3.28%)	Malpighiaceae	7 (0.77%)	1 (0.87%)	2 (0.82%)
Bombacaceae	5 (0.55%)	2 (1.74%)	3 (1.23%)	Marcgraviaceae	2 (0.22%)	1 (0.87%)	1 (0.41%)
Burseraceae	28 (3.06%)	1 (0.87%)	6 (2.46%)	Melastomataceae	18 (1.97%)	2 (1.74%)	2 (0.82%)
Caryocaraceae	5 (0.55%)	1 (0.87%)	3 (1.23%)	Memecylaceae	2 (0.22%)	1 (0.87%)	1 (0.41%)
Cecropiaceae	38 (4.16%)	1 (0.87%)	8 (3.28%)	Menispermaceae	13 (1.42%)	3 (2.61%)	6 (2.46%)
Celastraceae	2 (0.22%)	1 (0.87%)	1 (0.41%)	Moraceae	39 (4.27%)	6 (5.22%)	12 (4.92%)
Chrysobalanaceae	71 (7.77%)	3 (2.61%)	18 (7.38%)	Myristicaceae	12 (1.31%)	2 (1.74%)	3 (1.23%)
Clusiaceae	15 (1.64%)	6 (5.22%)	8 (3.28%)	Myrtaceae	1 (0.11%)	1 (0.87%)	1 (0.41%)
Combretaceae	1 (0.11%)	1 (0.87%)	1 (0.41%)	Olacaceae	5 (0.55%)	3 (2.61%)	3 (1.23%)
Convolvulaceae	3 (0.33%)	1 (0.87%)	2 (0.82%)	Passifloraceae	1 (0.11%)	1 (0.87%)	1 (0.41%)
Cucurbitaceae	1 (0.11%)	1 (0.87%)	1 (0.41%)	Polygalaceae	19 (2.08%)	2 (1.74%)	3 (1.23%)
Dilleniaceae	6 (0.66%)	3 (2.61%)	4 (1.64%)	Quiinaceae	1 (0.11%)	1 (0.87%)	1 (0.41%)
Duckeodendraceae	1 (0.11%)	1 (0.87%)	1 (0.41%)	Rubiaceae	1 (0.11%)	1 (0.87%)	1 (0.41%)
Ebenaceae	2 (0.22%)	1 (0.87%)	2 (0.82%)	Sapotaceae	207 (22.65%)	7 (6.09%)	47 (19.26%)
Elaeocarpaceae	3 (0.33%)	1 (0.87%)	3 (1.23%)	Simaroubaceae	3 (0.33%)	2 (1.74%)	2 (0.82%)
Euphorbiaceae	68 (7.44%)	4 (3.48%)	5 (2.05%)	Violaceae	1 (0.11%)	1 (0.87%)	1 (0.41%)
Flacourtiaceae	4 (0.44%)	1 (0.87%)	1 (0.41%)	Vochysiaceae	4 (0.44%)	3 (2.61%)	4 (1.64%)
Hippocrateaceae	39 (4.27%)	4 (3.48%)	8 (3.28%)				

the female left the fragment or died. During our study, no new individuals joined either of the bearded saki groups in the 10-ha fragments (Boyle & Smith 2010b).

Single offspring births occurred from mid October to early November 2005 in groups from the continuous forest, as well as groups that used the 100-ha fragments. Neither of the two 10-ha fragment bearded saki groups experienced births during the study period.

Diet

We tagged 993 trees, lianas and epiphytes that served as food for bearded sakis during our behavioral observations. These included 244 species in 115 genera in 14 different families (Table 24.2; Boyle 2008b; Boyle *et al.* 2012). We were unable to identify 19 of the 933 specimens.

The three families that were represented most often in the bearded saki diet were Sapotaceae (23%), Lecythidaceae (14%), and Chrysobalanaceae (8%). Overall, the most frequently used genera were *Eschweilera* (Lecythidaceae), *Pouteria* (Sapotaceae), and *Licania* (Chrysobalanaceae). The most prominent plant species were *Micrandropsis scleroxylon* (Euphorbiaceae), *Eschweilera truncata* (Lecythidaceae) and *Ecclinusa guianensis* (Sapotaceae). Lianas and epiphytes represented 17% of the overall bearded saki diet, and there was no difference in the proportion of lianas and epiphytes in the diet across forest sizes (Boyle 2008b; Boyle *et al.* 2012).

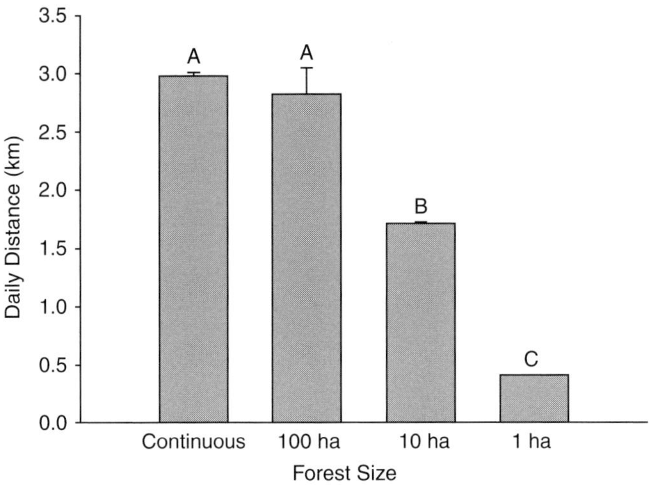

Figure 24.3 Average daily distance traveled. Standard error bars are present and A, B and C represent differences among the bearded saki groups. Data are from Boyle *et al.* (2009a).

Average DBH of feeding trees was 37.9 cm (± 5.90). Of the 933 trees, epiphytes and lianas that were consumed by the bearded sakis, 94% served as a fruit resource for the monkeys, 5% as a flower resource and less than 1% as a leaf resource (Boyle 2008b).

There was little dietary species overlap across all six inhabited sites (four forest fragments and two areas of continuous forest) in 2005–2006. Bearded sakis consumed only four species *Ecclinusa guianensis*, *Eschweilera truncata*, *Hevea guianensis* and *Micropholis guyanensis* (1.5 percent of the pooled flora) in all six study areas.

Travel patterns

Home-range size varied from 12 ha in a 10-ha fragment (actual fragment size was 13 ha) to 559 ha in the continuous forest (Boyle *et al.* 2009a). Bearded saki monkeys travelled greater daily distances in the continuous forest and 100-ha fragments than in the smaller fragments (Figure 24.3; Boyle *et al.* 2009a). Average daily distance was 2.99 km (± 0.02) in the continuous forest, 2.83 km (± 0.22) in the 100-ha fragments, 1.72 km (± 0.09) in the 10-ha fragments and 0.41 km in the 1-ha fragment. Bearded sakis living in the smaller fragments travelled in more circular patterns, and they revisited feeding trees more often throughout a day than did monkeys living in the larger forested areas (Boyle *et al.* 2009a). Furthermore, the monkeys did not concentrate their efforts in low-lying, riparian areas, yet they did travel through these areas to reach other areas of the forest.

Discussion

Northern bearded sakis show extreme flexibility in their behavioral ecology, as individuals in fragmented areas were found to have smaller group sizes, shorter daily distances traveled and smaller home ranges than individuals in continuous forest. Additionally, northern bearded sakis were found to exhibit variability in their diets across fragments. However, it is unclear whether this behavioral flexibility is sufficient for the long-term viability of this species in fragmented habitats.

We found northern bearded saki groups permanently inhabiting forest fragments as small as 2–3% the size of their home range in continuous forest. It may seem counterintuitive that "permanent" groups existed in two of the 10-ha fragments, while the groups in the 100-ha fragments left and re-entered the fragments frequently. We would have expected that the monkeys would have moved in and out of the smaller fragments more often than the larger fragments; however, one possible explanation for these results is that neither 100-ha fragment was completely isolated, but both 10-ha fragments were isolated. Northern bearded sakis in the 100-ha fragments nevertheless comprised smaller groups and traveled shorter daily distances than individuals in the continuous forest.

Flexibility was also evident in the species eaten by the northern bearded sakis, as less than half of consumed plant species were eaten at more than one study area. These results indicate that northern bearded sakis are capable of consuming a variety of plant species. When the plant species were separated by type (fruit, flower, leaf), we found that the northern bearded sakis consistently ate fruit (primarily seeds).

Although northern bearded sakis were present in forest fragments that were fractions of their characteristic home range in continuous forest, we found substantial differences in the behavioral ecology of northern bearded sakis in forest fragments and those in continuous forest. Two 10-ha fragments in this study contained northern bearded saki groups, but it is unknown how long these groups of three and four individuals can remain in an area as small as 10 ha, especially as there was no noted travel in and out of these fragments, and the monkeys in the 10-ha fragments failed to breed through the duration of the study. Copulations were witnessed in June and August 2005 in a 100-ha fragment and continuous forest site, and infants were seen in both 100-ha groups and both continuous forest groups in November and December 2005 (the early wet season). Bearded saki monkeys (*Chiropotes* spp.) reach sexual maturity at approximately 36 months, and interbirth interval is greater than 24 months (Peetz 2001). There have been no young juveniles in the 10-ha fragments since this study began in 2003; therefore, it is possible to infer that there had not been a successful birth in either group residing in the 10-ha fragments for at least 3.5 years (November 2002–June 2006). Therefore, we suggest continued monitoring of the forest fragments to document patterns of use and movement between forest areas, as well as changes in group composition.

Acknowledgments

We thank BDFFP, CNPq and IBAMA for support and permission to conduct this research. Waldete C. Lourenço, Lívia R. da Silva, Alaercio M. dos Reis, Osmaildo F. da Silva, and

Alexandro E. dos Santos provided invaluable field assistance, and Paulo A. C. L. Assunção and Ana Andrade assisted with plant identifications. John Alcock, David Kabelik, and Leanne Nash provided helpful feedback on the manuscript. Funding was provided by Arizona State University, BDFFP, Organization for Tropical Studies and Instituto Nacional de Pesquisas da Amazônia, Fulbright, Margot Marsh Biodiversity Foundation, American Society of Primatologists, Primate Conservation Inc., and Idea Wild. This is publication number 474 in the BDFFP Technical Series.

References

Achard, F., Eva, H.D., Stihig, H.-J., *et al.* (2002). Determination of deforestation rates of the world's humid tropical forests. *Science*, **297**, 999–1002.

Altmann, J. (1974). Observational study of behavior: sampling methods. *Behaviour*, **48**, 227–265.

Asquith, N.M., Wright, S.J. & Clauss, M.J. (1997). Does mammal community composition control recruitment in neotropical forests? Evidence from Panama. *Ecology*, **78**, 941–946.

Ayres, J.M. (1981). Observações sobre a ecologia e o comportamento dos cuxiús (*Chiropotes albinasus* e *Chiropotes satanas*, Cebidae: Primates). Unpublished Master's thesis, Instituto Nacional de Pesquisa da Amazônia and Fundação Universidade do Amazonas (FUA), Manaus, Amazonas.

Bierregaard Jr., R.O. & Stouffer, P.C. (1997). Understory birds and dynamic habitat mosaics in Amazonian rainforests. In *Tropical Forest Remnants: Ecology, Management, and Conservation of Fragmented Communities*, ed. W.F. Laurance & R.O. Bierregaard Jr. Chicago, IL: University of Chicago Press, pp. 138–155.

Bierregaard Jr., R.O., Lovejoy, T.E., Kapos, V., *et al.* (1992). The biological dynamics of tropical rain-forest fragments. *Bioscience*, **42**, 859–866.

Boyle, S.A. (2008a). Human impacts on primate conservation in central Amazonia. *Tropical Conservation Science*, **1**, 1–12.

Boyle, S.A. (2008b). The effects of forest fragmentation on primates in the Brazilian Amazon. Unpublished PhD dissertation, Arizona State University, Tempe, Arizona.

Boyle, S.A. & Smith, A.T. (2010a). Can landscape and species characteristics predict primate presence in forest fragments in the Brazilian Amazon? *Biological Conservation*, **143**, 1134–1143.

Boyle, S.A. & Smith, A.T. (2010b). Behavioral modifications in northern bearded saki monkeys (*Chiropotes satanas chiropotes*) in forest fragments of central Amazonia. *Primates*, **51**, 43–51.

Boyle, S.A., Lourenço, W.C., da Silva, L.R., *et al.* (2009a). Travel and spatial patterns change when northern bearded saki monkeys (*Chiropotes satanas chiropotes*) live in forest fragments. *International Journal of Primatology*, **30**, 515–531.

Boyle, S.A., Lourenço, W.C., da Silva, L.R., *et al.* (2009b). Home range estimates vary with sample size and methods. *Folia Primatologica*, **80**, 33–42.

Boyle, S.A., Zartman, C.E, *et al.* (2012). Implications of habitat fragmentation on the diet of bearded saki monkeys in central Amazonian forest. *Journal of Mammalogy*, **93**, 959–976.

Chapman, C.A. & Onderdonk, D.A. (1998). Forests without primates: primate/plant codependency. *American Journal of Primatology*, **45**, 127–141.

Clutton-Brock, T.H. & Harvey, P.H. (1977). Primate ecology and social organization *Journal of Zoology*, **183**, 1–39.

Estrada, A. & Coates-Estrada, R. (1996). Tropical rain forest fragmentation and wild populations of primates at Los Tuxtlas, Mexico. *International Journal of Primatology*, **17**, 759–783.

Estrada, A., Anzures, A. & Coates-Estrada, R. (1999). Tropical rain forest fragmentation, howler monkeys (*Alouatta palliata*), and dung beetles at Los Tuxtlas, Mexico. *American Journal of Primatology*, **48**, 253–262.

Ferrari, S.F., Emidio-Silva, C., Lopes, M.A., *et al.* (1999). Bearded sakis in south-eastern Amazonia – back from the brink? *Oryx*, **33**, 346–351.

Ferraz, G.C., Russell, G.J., Stouffer, P.C., *et al.* (2003). Rates of species loss from Amazonian forest fragments. *Proceedings of the National Academies of Sciences of the United States of America*, **100**, 14069–14073.

Ferreira, L.V. & Laurance, W.F. (1997). Effects of forest fragmentation on mortality and damage of selected trees in Central Amazonia. *Conservation Biology*, **11**, 797–801.

Frazão, E.R. (1992). Comportamento alimentar de *Chiropotes satanas chiropotes* (CEBIDAE: PRIMATAS) na Amazônia Central Brasileira. Unpublished Master's thesis, Instituto Nacional de Pesquisas da Amazônia, Manaus, Amazonas.

Gascon, C. & Bierregaard, Jr., R.O. (2001). The Biological Dynamics of Forest Fragments Project: the study site, experimental design, and research activity. In *Lessons from Amazonia: The Ecology and Conservation of a Fragmented Forest*, ed. R.O. Bierregaard Jr., C. Gascon, T.E. Lovejoy & R. Mesquita. New Haven, CT: Yale University Press, pp. 31–42.

Gascon, C. & Lovejoy, T.E. (1998). Ecological impacts of forest fragmentation in central Amazonia. *Zoology – Analysis of Complex Systems*, **101**, 273–280.

Gascon, C., Williamson, G.B. & da Fonseca, G.A.B. (2000). Ecology – receding forest edges and vanishing reserves. *Science*, **288**, 1356–1358.

Gilbert, K.A. (1994). Endoparasitic infection in red howling monkeys (*Alouatta seniculus*) in the central Amazonian basin: a cost of sociality? Unpublished PhD thesis, Rutgers University, New Brunswick, New Jersey.

Gilbert, K.A. (2003). Primates and fragmentation of the Amazon forest. In *Primates in Fragments: Ecology and Conservation*, ed. L.K. Marsh. New York, NY: Kluwer Academic, pp. 145–157.

Gilbert, K.A. & Setz, E.Z. (2001). Primates in a fragmented landscape. Six species in Central Amazonia. In *Lessons From Amazonia: The Ecology and Conservation of a Fragmented Forest*, ed. R.O. Bierregaard Jr, C. Gascon, T.E. Lovejoy & R. Mesquita. New Haven, CT: Yale University Press, pp. 262–270.

Johns, A.D. & Skorupa, J.P. (1987). Responses of rain forest primates to habitat disturbance: a review. *International Journal of Primatology*, **8**, 157–191.

Kapos, V. (1989). Effects of isolation on the water status of forest patches in the Brazilian Amazon. *Journal of Tropical Ecology*, **5**, 173–185.

Laurance, W.F. (2001). The hyper-diverse flora of the central Amazon. In *Lessons From Amazonia: The Ecology and Conservation of a Fragmented Forest*, ed. R.O. Bierregaard Jr., C. Gascon, T.E. Lovejoy & R. Mesquita. New Haven, CT: Yale University Press, pp. 47–53.

Laurance, W.F. & Luizão, R. (2007). Driving a wedge into the Amazon. *Nature*, **448**, 409–410.

Laurance, W.F., Laurance, S., Ferreira, L.V., *et al.* (1997). Biomass collapse in Amazonian forest fragments. *Science*, **278**, 1117–1118.

Laurance, W.F., Vasconcelos, H.L. & Lovejoy, T.E. (2000). Forest loss and fragmentation in the Amazon: implications for wildlife conservation. *Oryx*, **34**, 39–45.

Lovejoy, T.E. (2006). Protected areas: a prism for a changing world. *Trends in Ecology and Evolution*, **21**, 329–333.

Lovejoy, T.E., Rankin, J.M., Bierregaard, Jr., R.O., *et al.* (1984). Ecosystem decay of Amazon forest remnants. In *Extinctions*, ed. M.H. Nitecki. Chicago, IL: University of Chicago Press, pp. 295–325.

Malcolm, J.R. (1997). Biomass and diversity of small mammals in Amazonian forest fragments. In *Tropical Forest Remnants: Ecology, Management, and Conservation of Fragmented Communities*, ed. W.F. Laurance & R.O. Bierregaard, Jr. Chicago, IL: University of Chicago Press, pp. 207–221.

Marsh, L.K., Chapman, C.A., Norconk, M.A., *et al.* (2003). Fragmentation: specter of the future or the spirit of conservation? In *Primates in Fragments: Ecology and Conservation*, ed. L.K. Marsh. New York, NY: Kluwer Academic, pp. 381–398.

Murcia, C. (1995). Edge effects in fragmented forests: implications for conservation. *Trends in Ecology and Evolution*, **10**, 58–62.

Nascimento, H.E.M., Andrade, A.C.S., Camargo, J.L.C., *et al.* (2006). Effects of the surrounding matrix on tree recruitment in Amazonian forest fragments. *Conservation Biology*, **20**, 853–860.

Norconk, M.A. (2011). Sakis, uakaris, and titi monkeys: behavioural diversity in a radiation of primate seed predators. In *Primates in Perspective*, 2nd edn, ed. C.J. Campbell, A. Fuentes, K.C. MacKinnon, M. Panger & S.K. Bearder. New York, NY: Oxford University Press, pp. 123–138.

Odum, E.P. & Kuenzler, E.J. (1955). Measurement of territory and home range size in birds. *Auk*, **72**, 128–137.

Onderdonk, D.A. & Chapman, C.A. (2000). Coping with forest fragmentation: the primates of Kibale National Park, Uganda. *International Journal of Primatology*, **21**, 587–611.

Peetz, A. (2001). Ecology and social organization of the bearded saki *Chiropotes satanas chiropotes* (Primates: Pithecinae) in Venezuela. *Ecotropical Monographs*, **1**, 1–170.

Port-Carvalho, M. & Ferrari, S.F. (2004). Occurrence and diet of the black bearded saki (*Chiropotes satanas satanas*) in the fragmented landscape of western Maranhão, Brazil. *Neotropical Primates*, **12**, 17–21.

Rodgers, A.R. & Carr, A.P. (1998). *HRE: The Home Range Extension for ArcView.* Thunder Bay, Ontario, Canada: Ontario Ministry of Natural Resources, Centre for Northern Forest Ecosystem Research.

Rylands, A.B. & Keuroghlian, A. (1988). Primate populations in continuous forest and forest fragments in central Amazonia. *Acta Amazônica*, **18**, 291–307.

Santos, R.R. (2002). Ecologia de cuxiús (*Chiropotes satanas*) na Amazônia Oriental: Perspectivas para a conservação de populaces fragmentadas. Unpublished Master's thesis, Universidade Federal do Pará, Belém, Pará.

Silva, S.S.B. (2003). Comporamento alimentar do cuxiú-preto (*Chiropotes satanas*) na area de influência do reservatório da usina hidreléctrica de Tucuruí-Pará. Unpublished Master's thesis, Universidade Federal do Pará, Belém, Pará.

van Roosmalen, M.G.M., Mittermeier, R.A. & Milton, K. (1981). The bearded sakis, Genus *Chiropotes*. In *Ecology and Behavior of Neotropical Primates*, ed. A.F. Coimbra-Filho & R.A. Mittermeier. Rio de Janeiro: Academia Brasileira de Ciências, pp. 419–442.

Veiga, L.M. (2006). A ecologia e o comportamento do cuxiú-preto (*Chiropotes satanas*) na paisagem fragmentada da Amazônia Oriental. Unpublished PhD dissertation, Universidade Federal do Pará, Belém, Pará.

Veiga, L.M. & Ferrari, S.F. (2006). Predation of arthropods by southern bearded sakis (*Chiropotes satanas*) in eastern Brazilian Amazonia. *American Journal of Primatology*, **68**, 209–215.

Worton, B.J. (1987). Kernel methods for estimating the utilization distribution in home-range studies. *Ecology*, **70**, 164–168.

Chapter 25

Ecology and behavior of saki monkeys (genus *Pithecia*)

Marilyn A. Norconk & Eleonore Z. Setz

Introduction

Sakis are widely distributed across northern South America and the Amazon Basin (Figure 25.1). They are currently divided into five species, but the taxonomy of the western Amazon species is in the process of revision (Marsh, pers. commun.). Hershkovitz (1987) divided the sakis into two groups: the *Pithecia pithecia* group from the eastern Amazon Basin and the Guiana Shield consisting of two subspecies largely distinguished on the basis of coloration of the adult male facial mask, and the *P. monachus* group, south of the Amazon River in central and western Amazonia. At the time of Hershkovitz's (1987) revision, the *P. monachus* group consisted of four species: *P. albicans*, *P. aequatorialis*, with two subspecies each of *P. irrorata* and *P. monachus*. The validity of dividing the genus into two species groups was recently demonstrated by Marroig and Cheverud (2004) using discriminant function techniques to analyze the crania of all taxa except *P. monachus milleri*. Using a larger data set as a guide to assess meaningful morphological distances among the 16 genera of platyrrhines, they recognized two valid species of *P. irrorata* (*P. irrorata* and *P. vanzolinii*) (Marroig & Cheverud 2004). Further, they recommended retaining subspecies designation for the two *P. pithecia* subspecies, but lacked sufficient samples to evaluate *P. monachus* taxa in detail.

Nearest-neighbor distances using a variety of data sets support the view that *Pithecia* clusters with and is primitive to bearded sakis (*Chiropotes*) and uacaris (*Cacajao*) (morphological data: Rosenberger 1984; Ford 1986; Kay 1990; Marroig & Cheverud 2001; Horovitz & Meyer 1997; nDNA: Schneider *et al.* 1993; Harada *et al.* 1995; Schneider & Rosenberger 1996;

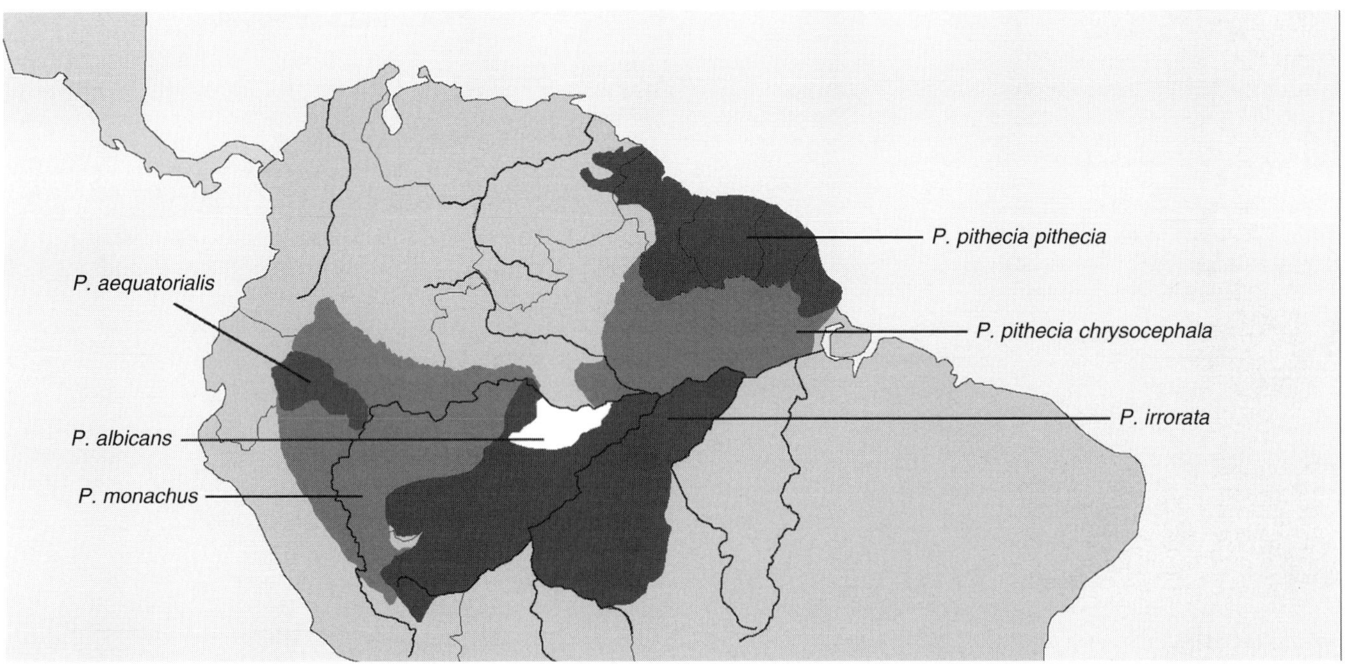

Figure 25.1 Map showing the relative distributions of five species of *Pithecia* and two subspecies of *P. pithecia*. The boundaries of species' distributions remain in flux with recent revisions proposed by Aquino *et al.* (2009) and Oliveira *et al.* (2009). Map based on distribution data from Hershkovitz (1987) and configured by E. Bailey.

Evolutionary Biology and Conservation of Titis, Sakis and Uacaris, eds. Liza M. Veiga, Adrian A. Barnett, Stephen F. Ferrari and Marilyn A. Norconk. Published by Cambridge University Press.

mitochrondrial genes: Chatterjee *et al.* 2009). Yet as Ford (1986, p. 114) pointed out, "*Pithecia* is itself specialized and derived from the common ancestor of this group, if not dentally, certainly postcranially and karyotypically." (See Kay 1990; Meldrum & Kay 1997; Rosenberger 2002; and Kay *et al.*, Chapter 1, and Rosenberger & Tejedor, Chapter 2 for views on the fossil history of the pitheciine clade.) Morphological data sets provide a larger variety of interpretations of relationships between titi monkeys and the sakis and uacaris (see Ford 1986; Kay 1990).

Habitat

Sakis occupy a wide range of habitats from lowland tropical rainforests in Ecuador (Di Fiore *et al.* 2007), Colombia (Defler 2004), Peru (Palminteri *et al.*, Chapter 11), and Brazil (Peres 1993a, 1993b) to montane forests (c. 500 m) on bauxite mountains (Mittermeier 1977; Norconk *et al.* 2003; Oliveira *et al.* 2009). They also inhabit regions of relatively low rainfall as well as impoverished forests, e.g. dry tropical forests, gallery forests and small islands in Venezuela (Kinzey *et al.* 1988; Norconk *et al.* 2003; Defler 2004); gallery and *terra firme* forests in Guiana (Lehman 2006; Lehman *et al.* 2006); and sandy plains and upland forests in Suriname (Mittermeier 1977; Hammond 2005).

Figure 25.2 Juvenile male *Pithecia pithecia* in Brownsberg Nature Park, July 2006. Photo: Tremaine Gregory.

Like all platyrrhines, sakis are primarily arboreal, but species appear to vary in the height of forest used during travel. Walker (1996, 2005) and Vié *et al.* (2001) found that *P. pithecia* used the lower to middle canopy at higher frequencies than the upper canopy levels and this finding correlates well with their predominant vertical clinging (Figure 25.2) and leaping locomotion (75% of their locomotor pattern; Oliveira et al. 1985). Happel (1982) found that *P. monachus* travelled in lower levels of the forest than when they rested and fed, and leaping represented 33% of locomotor samples in her study. *Pithecia albicans* and *P. irrorata* use of upper levels of the forest during travel (Peres 1993a; Jacobs *et al.* 2008, respectively). See Photos 25.1 and 25.2.

The intermembral index of *P. pithecia* (two averages reported: 74.2 and 75) is slightly lower than *P. monachus* (76 and 77.9) (Hershkovitz 1987, table VII) and Walker (2005) argues that limb proportions and femoral morphology of *P. pithecia* is distinctive relative both to other *Pithecia* species and to *Chiropotes* spp. Indeed, common names in some countries capture the distinctiveness of their locomotion – "volador" for *P. monachus* in Colombia and "flying jack" for *P. pithecia* in Guiana. In contrast, the higher intermembral index of *Chiropotes* spp. (four individuals range from 82 to 85: Hershkovitz 1985, table 1) reflects their above-branch quadrupedalism.

Photo 25.1 Female white-faced saki (*P. pithecia*) feeding. Brownsberg Nature Park, Suriname. Photo: Marilyn Norconk. (See color plate section.)

Photo 25.2 *Pithecia albicans*. Photo: Francisco Fonseca. (See color plate section.)

Population ecology

Estimates of saki population densities calculated from censuses range from < 1 individual/km^2 to 36 individuals/km^2 and only one or a few groups have been reported in km^2 or 10 km of linear distance (Table 25.1). Most data are from censuses and are likely to be minimum estimates of population sizes. Sakis are very cryptic monkeys, difficult to habituate to human observers and scatter when encountered. Thus obtaining an accurate group count is difficult unless a group is seen multiple times and censuses are repeated. Indeed, de Thoisy *et al.* (2008) found that the number of sakis observed in multiple census transects began to stabilize only after 90 km of sampling. The good news is that there are a number of long-term studies under way in *terra firme* forests that will provide a better estimate of population sizes. Palminteri *et al.* (Chapter 11) are studying several habituated groups of *P. irrorata* and report that this species occurs in high densities locally in southeastern Peru and Norconk *et al.* (2003; Norconk & Grafton 2003) report high densities of *P. pithecia* in Brownsberg Nature Park, Suriname (Table 25.1). Di Fiore and colleagues have also habituated groups of wild sakis in Ecuador.

Lehman *et al.* (2001, 2006) in Guiana and Soini (1986) and Aquino *et al.* (2009) in Peru conducted comprehensive surveys of sakis and found that group size ranged from two to eight individuals. Until recently, the mating system of sakis was thought to be monogamous (Robinson *et al.* 1987) with observations typically of an adult pair and offspring. More recent data from censuses support that view (Aquino *et al.* 2009), but long-term studies have begun to report that groups contain multiple breeding females and/or multiple males (Setz 1993; Setz & Gaspar 1997; Norconk 1996; Thompson *et al.*, 2011). We suspect that as in the case of population densities, censuses underestimate group size in sakis. Using average group size data from Table 25.1, we found that censuses reported significantly smaller group sizes than long-term studies (Mann–Whitney $U = 21.0$, one-way $P < 0.05$). Group sizes averaged 3.75 from censuses (range 1.9–5.5; $n = 19$) and 4.98 (range 2.6–6.9; $n = 5$) from long-term studies.

Extra adult males (presumed sons of the breeding male) are tolerated in free-ranging saki groups in Suriname (Norconk, pers. obs.) and may engage in minimally contested breeding within the natal group (Thompson 2010). With the exception of births and deaths, groups are often stable over several years of observations, but in one case a group collapsed for unknown reasons and a 4-year-old male inherited his father's territory and subsequently attracted a non-group female (Thompson *et al.*, 2011). Thus, we suspect that both males and females have a range of breeding options that include dispersal, moving into breeding positions upon the death or disappearance of an older adult and cooperative breeding in the presence of an older adult. Di Fiore *et al.* (2007) provided data on mate replacement in *P. aequatorialis* following the death of the breeding male. They also documented the presence of males in the population that could detect and respond quickly to an available breeding position – similar to a situation reported by Saulo R. Silva (pers. commun.) for titis. In the *P. aequatorialis* case, there were no offspring of breeding age to move into a breeding position.

Table 25.1 Group size and population density of *Pithecia* spp.

Species	Group size					Group density[4]	Method[5]	Source
	Mean ± SD	Min–max	*N*[1]	Extra?[2]	Method[3]			
P. monachus group								
Pithecia aequatorialis		1–7	4		S			Heymann *et al.* 2002
		2–8	123	no	S		T	Aquino *et al.* 2009
	4		1	yes	B			Di Fiore *et al.* 2007
		2–5	6–7		S			Marsh 2004
P. albicans	4.6 ± 1.5	3–7	5		S	0.9/km^2	T	Peres 1993a
	5.1	3–8	6		S	1.92/km^2	T	Peres 1993b
	5.2[6]		39		S	0.9/10 km	T	Johns 1986
	3.9[7]		18		S	1.2/10 km	T	Johns 1986
	2.5[8]		6		S	0.1/10 km	T	Johns 1986
	3.7[9]		3		S	0.3/10 km	T	Johns 1986
P. irrorata	2.78 ± 1.37				S	0.56[10]/10 km	T	Ferrari *et al.* 1999
	3.5	2–5	6		S			Buchanan-Smith *et al.* 2000
P. monachus	3.75	2–5	16		S	1.4[11]/10 km	T	Bennett *et al.* 2001
	3.8	2–8	23		S	3.7[12]/km^2	T	Soini 1986
	3.1 ± 1.6	1–6	10	yes	S		T	Izawa 1976
	1.9	1–4	51		S		T	Izawa & Yoneda 1981
P. pithecia group								
P. pithecia chrysocephala	6	6–7	1	yes[14]	B	10/km^2		Setz & Gaspar 1997; Gilbert & Setz 2001
	2.6 ± 0.5	2–3	3	yes	B	15/km^2		Oliveira *et al.* 1985
						15.5 ind/ km^2 ± 10.32	T	Andrade 2007
P. pithecia pithecia	4.8 ± 2.4	2–12	21	yes	S			Lehman *et al.* 2001
	3.3 ± 1.7	1–5	10		S			Muckenhirn *et al.* 1975
	2.7 ± 0.8	2–4	9		S			Mittermeier 1977
	3.7 ± 1.2	2–6	10	yes[14]	S	6.1/10 km	T	Norconk *et al.* 2003
	2.3 ± 1.2	1–5	35	yes	S	0.14[13]/10 km	T	Vié *et al.* 2001
	6.9 ± 1.4	5–9	1	yes[14]	B			Norconk 2006
	5.4 ± 0.7	2–7	3	yes[14]	B			Thompson 2010
	5	4–6	2		S			Schwarzkopf & Rylands 1989
	5.5 ± 2.5	3–8	2		S			Kinzey *et al.* 1988

[1] Sample size (number of sightings or study groups); [2] are there more than one adult male and female?; [3] S, population survey; B, behavioral study; [4] densities shown either as group sightings per 10 km walked, groups per km^2; [5] T, transect census; B, behavioral study; [6] primary forest (estimated individual density of 9/km^2); [7] logged forest (estimated individual density of 18/km^2); [8]cultivated mosaic (estimated individual density of.9/km^2); [9] forest island (estimated individual density of 2.4/km^2); [10] group density calculated from individual density estimate of 1.56/10 km surveyed; [11] data reported for east side of Tapiche River; west side group size was 2.0; estimated individual density = 17.21 individuals/km^2 on east side and 9/km^2 on west side; group size range reported combines both east and west side of river; [12]calculated from individuals/km^2 density (14.3/km^2) (was *P. hirsuta* at the time of publication); [13] group density = 0.28/km^2; 0.64 individuals/km^2; [14]two breeding females.

Generally, genetic data are critical to teasing out the interaction between behavior and kinship.

Activity budgets

As in *Callicebus* (Bicca-Marques and Heymann, Chapter 17), there are few reports of activity budgets in *Pithecia* and they are only from *P. pithecia* populations (Table 25.2). Gregory's study (2006) was on a habituated group in *terra firme* forest, Setz's (1993) study was in a forest fragment, and Homburg's (1998) study was on a land-bridge island. As yet, sample sizes are too low for accurate assessments.

Males (*P. pithecia*) are the main participants in intertroop encounters (ITEs) at Brownsberg Nature Park in Suriname (Thompson 2006) compared with reports of female participation in ITEs in Lago Guri (Venezuela) sakis (Norconk 2006). See Thompson and Norconk (Chapter 27) for changes in male–male proximity before and after ITEs in Suriname.

Table 25.2 Activity budget and use of space by *Pithecia* spp.

Species	Activity budget (% time)				Use of space[1]		Duration (months)	Source
	Rest	Feed/ forage	Move/ travel	Social	Home range (ha)	Day range (m)		
P. monachus group								
P. albicans					172.4 (5) [147–204]		7[2]	Peres 1993b
P.monachus					24.9 (3) [9.7–42]		13	Soini 1986
P. pithecia group								
P. pithecia chrysocephala	30	25	43	2		1003 ±278 (30) [420–1416]	2 (38 days)	Setz 1993
P. pithecia pithecia	38	43	16	3	12.8 (1)	1104 ± 255 (59)	12	Homburg 1998[3]
					12.8 (1)	1772 ± 274 (47)	3	Cunningham 2003[3]; Cunningham & Janson 2007[3]
	44	26	30			1000[4] (1)	3	Gregory 2006[5]
					107, 287[6]	1880 ± 520[7] (3) [300–2800]	10	Vié *et al.* 2001
					10.3 (1)			Norconk *et al.* 2003[5]
					c. 28 (3) [18.9, 24.7, 40.6]	c. 700– > 1000 (3)	7	Thompson, pers. commun.[5, 8]

Obs.: Percentages were rounded to the nearest integer.
[1] Data provided are mean home range and day range, sample size in parentheses, and range in brackets; [2] represents 146 independent sightings; groups were not habituated; [3] Lago Guri, Venezuela; [4] estimated day range based on travel speed; [5] Brownsberg Nature Park, Suriname; [6] minimum convex polygon technique; translocated individuals; [7] range: translocated and resident sakis; figure reported is the average path length of a pair of resident (non-migrating) sakis; [8] study in progress.

Foraging ecology

Pithecia spp. ingest seeds, fleshy fruit, whole fruit (including pulp and seeds), young leaves, flowers from trees, shrubs, lianas and hemiparasites (Setz 1993; Homburg 1998). Using traditional categories, *Pithecia* are strong frugivores, with seeds representing between 26 and 64% of the annual diet (Table 25.3). Although sakis ingest seeds year round, seeds were ingested in higher proportions in the dry season (60% to > 80% of monthly diet) than the wet season (17–40%) at Lago Guri, Venezuela (Homburg 1998, table 4.10; also see Palminteri *et al.*, Chapter 11). The seasonal difference in seed ingestion at the Lago Guri site was related to higher availability of younger, maturing seeds in the dry season, but sakis ingested mature seeds in both periods and only mature seeds were available in the wet season (Norconk 1996).

Seed eating occurred at a lower frequency in the Colosso fragment in Brazil (Setz 1993), with a higher proportion of seeds ingested in the wet season (33%) than the dry season (20%). On the basis of her data, Setz characterized sakis as primarily ripe fruit eaters. The difference between Colosso and Lago Guri may be related to rainfall and forest type, but results at Brownsberg Nature Park are in line with those at Lago Guri (Table 25.3). Thus, all studies report that sakis are strongly frugivorous and some studies report seeds as the major component of the diet.

Sakis also ingest a high diversity of foods that make up smaller, and perhaps nutritionally important, parts of their diet (Norconk & Conklin-Brittain 2004; Norconk 2011). For example, arils of plant species from the Connaraceae family represented an important part of saki diets at both Lago Guri and Brownsberg. Norconk and Conklin-Brittain (2004, table II) reported that average lipid values of arils (64.2% by dry matter, DM) were more than twice the value of lipids in immature whole fruit (29.9% DM), and both young (13.2% DM) and mature seeds (24.8% DM). They also noted that the average values of crude protein in young seeds, young whole fruit, flowers and young leaves ranged by plant part from 14.9% DM to 18.4% DM, but that the calculated intake of protein from the plant portion of the diet was relatively low in all seasons, particularly during the late wet season (4% DM) (Norconk & Conklin-Brittain 2004, table III). Insects may take up the slack in protein intake during low periods of protein availability in the plant portion of the diet. Homburg found that animal prey from at least 5 insect taxa (Hymenoptera, Orthopteroidea, Isoptera, Lepidoptera and Formicoidea) made up at least 3% of their annual diet (1998, p. 74), but thus far we

Table 25.3 Contribution of food types and species richness of the diet of *Pithecia* spp.

Species	% contribution to diet					Diet richness			Duration (months)	Source
	Fruit[1]	Leaves[2]	Flowers	Prey	Other/ Unknown	PS[3]	FS[4]	LS[5]		
P. monachus group										
P. albicans	80 [47]	10	7	<1	3	93	81	7	7	Peres 1993b
P. irrorata	98 [83]	1–2	0.5			168+			18	Palminteri *et al.*, Chapter 11
P. monachus	93 [38]	4	3			64			13	Soini 1986
P. pithecia group										
P. pithecia chrysocephala	81 [26]	11	8	0	0.3[6]	188	169	15	3[7]/72[8]	Setz 1993, 1999
P. pithecia pithecia	88 [61]	7	2	2	1				12	Norconk 1996; Norconk & Conklin-Brittain 2004
	84 [53]	10	2	4	-	80	69	22	17	Homburg 1998
						21	21		3	Gregory 2006
	95 [64]	4	1	<1	<1	63	63	?	4	Norconk, unpubl.

Obs.: Percentages were rounded to the nearest integer.
[1] Includes ingestion of whole fruit or fruit parts (the % accounted for by the ingestion of seeds only is cited in brackets); [2] includes ingestion of leaves or leaf parts in general; [3] number of plant species used as food sources; [4] number of fruit species; [5] number of leaf species; [6] termitaria earth; [7] for dietary composition; [8] for species in the diet.

Table 25.4 Top five plant families in the diets of *P. pithecia pithecia* (Lago Guri, Venezuela & Brownsberg Nature Park, Suriname), *P. pithecia chrysocephala* (Amazonia, Brazil), and *P. irrorata* (Los Amigos, Peru).

Lago Guri, Venezuela[1]	% diet	Brownsberg Nature Park, Suriname[2]	% diet	Amazonia, Brazil[3]	% diet	Los Amigos, Peru[4]	% diet
Connaraceae	20.7	Euphorbiaceae	18.5	Fabaceae (mim)	12.5	Moraceae	22.7
Erythroxylaceae	15.0	Fabaceae (mim[5])	17.3	Moraceae	11.6	Sapotaceae	16.2
Rubiaceae	9.5	Lecythidaceae	12.7	Sapotaceae	7.4	Fabaceae (mim)	8.5
Chrysobalanaceae	8.5	Araceae	9.6	Melastomataceae	6.8	Arecaceae	5.8
Loganiaceae	6.6	Melastomataceae	3.8	Malpighiaceae	5.7	Myristicaceae	5.5

[1] Homburg 1998 (annual cycle: 1991–1992); [2] Norconk, unpubl. data (4 months: short dry and early long wet seasons, 2007); [3] Setz 1993 (3 months: dry, wet and transition seasons; 2152 feeding records); [4] Palminteri *et al.*, Chapter 11 (18 months: 2004–2005); [5] = Mimosaceae.

do not have a good estimate of nutrients gained from ingesting insects for sakis.

Intakes of insect prey and leaves are difficult to estimate and undoubtedly are often underestimated, but Setz (1993) suggests that leaves (and flowers) increased in importance in the dry season (17%) compared to the wet season (4%) at the Colosso site. Norconk (1996) found that young leaves were not a high proportion of the monthly diet, but the sakis ingested some leaves almost daily at the Lago Guri site.

Data from four studies demonstrate the diversity of resources used by *Pithecia*. Given the disparate locations of these field sites, it is not surprising that there was almost no overlap in the top five plant families. However, a relatively strong degree of similarity emerged by comparing entire food lists. Jaccard Indexes of Diversity at the plant family level range from 39% (dry tropical forest of Lago Guri and wet Los Amigos forest in southeast Peru) to 87.3% (Los Amigos, Peru and Brownsberg Nature Park, Suriname, a wet montane forest in Suriname) (Table 25.4). Fourteen plant families were the same for all sites. At the species level, intersite comparabilities are much lower (1% between Lago Guri (Homburg 1998) and Petit Saul (French Guiana: Vié *et al.* 2001) to a maximum of 7.8% between two Brazilian sites, Colosso (Setz 1993) and Urucu (Brazil: Peres 1993a).

Some of the plant families commonly used by sakis produce fruit that are widely used by other frugivorous primates (e.g. Rubiaceae, Moraceae, Passifloraceae and Sapotaceae, and *Inga* at the genus level), but others (Euphorbiaceae and Bignoniaceae seeds and arils from Connaraceae species) appear to be more specific to sakis. Connaraceae arils and Euphorbiaceae seeds were among the highest sources of lipids for the Lago Guri sakis and young Bignoniaceae seeds were high in crude protein and lipids (Norconk & Conklin-Brittain 2004).

Use of space

Estimates of home-range size for *Pithecia* spp. vary by an order of magnitude from 10.3 ha to > 200 ha (Table 25.2). In an early study, Izawa (1976) had the impression that *P. monachus* groups were nomadic. Even now, it is difficult to assess whether some species of *Pithecia* have absolutely smaller home ranges than others or whether populations vary based on differences in habitat quality. Once again, there are too few data to distinguish between these possibilities.

Populations of at least two saki species (*P. aequatorialis* and *P. pithecia*) are territorial (Norconk 2006, 2011; Di Fiore *et al.* 2007) and preliminary data suggest that both species give specific long calls related to territorial behavior (D. Hurst, pers. commun., 2008). Daily path lengths average at least 1 km, longer than most estimates for *Callicebus* (Bicca-Marques & Heymann, Chapter 17; C. Caselli, pers. commun.), but shorter than bearded sakis and uacaris (Norconk 2011; Barnett *et al.*, Chapter 14; Veiga and Ferrari, Chapter 22).

Predators

There are no accurate assessments of predation rates for sakis, but in the course of monitoring remains of prey dropping beneath a harpy eagle nest between 2002 and 2005 in the Central Suriname Nature Reserve (CSNR), Boinski and colleagues estimated that more than 20 sakis were prey to a pair of eagles (S. Boinski, pers. commun., 2007; also see Rettig 1978; Izor 1985; Ford & Boinski 2007). This statistic is striking given the difficulty of censusing sakis and often reported low population densities (Table 25.1). It is also striking that sakis ranked second, after howlers, among harpy eagle primate prey in the CSNR study. Even smaller birds of prey like the crested eagle, *Morphnus guianensis*, are known to attack young sakis (Gilbert 2000) and hunting is known to occur among the Nambiquara people in Brazil (Setz 1991). Bezerra *et al.* (2009) also reported saki predation by harpy eagles.

Several morphological and behavioral traits may have evolved to minimize predation risk in sakis. First, in several species of sakis, both males and females are cryptically colored. Second, sakis respond in a stereotypical manner to potential birds of prey, dropping to within a meter of the ground and remaining silent and motionless for > 1 h (Gleason & Norconk 2002). Third, sakis appear to choose their sleeping trees with care and exhibit several stereotypical behaviors when entering a sleeping tree. Observations at both Lago Guri, Venezuela and Colosso, Brazil documented that sakis approach the vicinity of sleeping trees up to 3 h before nightfall, i.e. between 1500 and 1600 h, and remained inactive in the lower to middle portions of the forest for up to an hour before the group dispersed into 1–4 sleeping trees (Setz 1999). Setz (1999) found that the saki group she observed rarely slept in the same trees on consecutive nights (only 4 in 128 pairs of nights). Her study group used 33 different sleeping sites in 156 nights. Four trees were used about half the nights in one year, but only 17 trees were used between years (Setz 1999). Working in Lago Guri, Venezuela, Brush (2000) suggested that choice of sleeping area was related to finding a sufficient number of large trees to sleep within easy vocalization (and perhaps visual) distance. Sleeping trees were characterized on the basis of tree size (DBH), crown height and breadth, and number of supports over a 10-week period and compared to non-sleeping trees. Variance in sleeping tree architecture loaded heavily on aspects of tree size (49.5% of variance) – number of available perches ($t = 4.36$, df $= 76$, $p < 0.0001$) and availability of suitable trees within 16 m radius ($t = 7.23$, df $= 76$, $p < 0.0001$) distinguished sleeping trees from non-use trees (Brush 2000, p. 44).

Conservation

As we still do not have good estimates of primate densities in any *Pithecia* species, it is difficult to ascertain the conservation risk for these species. All species are labelled as Least Concern by the IUCN redlist of threatened species (http://www.iucnredlist.org/), perhaps more on the basis of their broad geographic distribution than on known population statistics. The North American captive population of 130 white-faced sakis is housed in 36 institutions (Frampton 2004). The taxonomy of the genus *Pithecia* is currently under revision by L. Marsh and colleagues (pers. commun., 2008).

Avenues for future research

There is still much to be discovered about saki behavior and ecology and we do not fully understand the similarities and differences between them and the larger-bodied bearded sakis and uacaris. All three pitheciine genera share highly derived dentition and seed predation feeding strategies, but differ in many other parameters (group size, home-range size, degree of territoriality, sexual dimorphism; Norconk 2011). A single taxon, *Pithecia pithecia*, has more or less carried the standard of genus identity, but Peres' (1993a) study on habitat use of *P. albicans*, observations on territoriality and mate replacement by Di Fiore *et al.* (2007), and comparative vocalizations by D. Hurst (pers. commun., 2008) suggest that *P. pithecia* may represent a departure from Amazonian sakis in ecology and social behavior. As more populations are studied, it may turn out that *P. pithecia* is the odd one out, with the highest degree of sexual dichromatism and smallest body mass, specialized use of low forest levels, vertical clinging and leaping locomotion and low intermembral index, tolerance of dry habitats, and perhaps a higher tendency toward polygyny than other saki species. Thus, as in all the pitheciine taxa, there are surprises yet to be discovered in *Pithecia* spp.

Acknowledgments

MN is grateful to EDELCA (Electrificación del Caroní) for permission to conduct research at Lago Guri, Venezuela, and to the Foundation for Nature Conservation in Suriname (Stinasu) for permission to conduct research at Brownsberg Nature Park. She is particularly grateful to Marie Djosetro (Director of Stinasu), Reggie Slijngard, Bart Di Dijn, Iwan Molgo, Gisla Sabitrie Doerga, and Muriel Djaspan. MN would like to thank field assistants and graduate students at both the Lago Guri and Brownsberg field sites, particularly Avis Anzelc, Jason Brush, Terry Gleason, Brian Grafton, Tremie Gregory, Delanie Hurst, Eliza Little, Nick Robl, Kate Talbot, Cynthia Thompson, Ari Vreedzaam, Suzanne Walker and Claire Wigmore. MN's funding for research reviewed or reported in this chapter was provided by the University Research Council and Graduate Student Senate, Kent State University, National Science Foundation and Wenner Gren Foundation for Anthropological Research.

References

Andrade, P.S. (2007). *Estudos Populacionais dos Primatas em duas florestas Nacionais do Oeste do Pará, Brasil. Tese em Recursos Florestais, com opção em Conservação de Ecossistemas Florestais.* ESALQ/USP. Piracicaba: SP.

Aquino, R., Cornejo, F.M., Lozxano, E.P., *et al.* (2009). Geographic distribution and demography of *Pithecia aequatorialis* (Pitheciidae) in Peruvian Amazonia. *American Journal of Primatology*, **71**, 964–968.

Bennett, C.L., Leonard, S. & Carter, S. (2001). Abundance, diversity, and patterns of distribution of primates on the Tapiche River in Amazonian Peru. *American Journal of Primatology*, **54**, 119–126.

Bezerra, B.M., Barnett, A.A., Souto, A., *et al.* (2009). Predation by the tayra on the common marmoset and the pale-throated three-toed sloth. *Journal of Ethology* , **27**, 91–96.

Brush, J.A. (2000). *Sleeping ecology of white-faced saki monkeys in Lago Guri, Venezuela.* Unpublished Master's thesis, Kent State University.

Buchanan-Smith, H.M., Hardie, S.M., Caceres, C., *et al.* (2000). Distribution and forest utilization of *Saguinus* and other primates of the Pando Department, northern Bolivia. *International Journal of Primatology*, **21**, 353–373.

Chatterjee, H.J., Ho, S.Y.W., Barnes, I., ***et al.*** (2009). Estimating the phylogeny and divergence times of primates using a supermatrix approach. *BMC Evolutionary Biology*, **9**, 259.

Cunningham, E.P. (2003). *The use of memory in Pithecia pithecia 's foraging stategy.* Unpublished PhD thesis, City University of New York.

Cunningham, E.P. & Janson, C. (2007). Integrating information about location and value of resources by white-faced saki monkeys. *Animal Cognition*, **10**, 293–304.

Defler, T.R. (2004). *Primates of Colombia. Bogatá*, Colombia: Conservation International.

Di Fiore, A., Fernandez-Duque, E. & Hurst, D. (2007). Adult male replacement in socially monogamous equatorial saki monkeys (*Pithecia aequatorialis*). *Folia Primatologica*, **78**, 88–98.

Ferrari, S.F., Iwanaga, S., Coutinho, P.E.G., *et al* . (1999). Zoogeography of *Chiropotes albinasus* (Platyrrhini, Atelidae) in southwestern Amazonia. *International Journal of Primatology*, **20**, 995–1004.

Ford, S.M. (1986). Systematics of the New World monkeys. In *Comparative Primate Biology*, ed. D.R. Swindler & J. Erwin. New York, NY: Liss, pp. 73–135.

Ford, S. & Boinski, S. (2007). Primate predation by harpy eagles in the Central Suriname Nature Reserve. *American Journal of Physical Anthropology*, **S44**, 109.

Frampton, T. (2004). *White-faced saki (Pithecia pithecia) studbook update.* Melbourne, FL: Brevard Zoo.

Gilbert, K.A. (2000). Attempted predation on a white-faced saki in the Central Amazon. *Neotropical Primates*, **8**, 103–104.

Gilbert, K.A. & Setz, E.Z.F. (2001). Primates in a fragmented landscape: six species in a fragmented landscape. In *Lessons from Amazonia: The Ecology and Conservation of a Fragmented Forest*, ed. R.O. Bierregaard Jr., C. Gascon, T.E. Lovejoy & R.C.G. Mesquita. New Haven, CT: Yale University Press, pp. 262–270.

Gleason, T.M. & Norconk, M.A. (2002). Predation risk and antipredator adaptations in white-faced sakis, *Pithecia pithecia*. In *Eat or Be Eaten: Predator Sensitive Foraging Among Primates*, ed. L. E. Miller. New York, NY: Cambridge University Press, pp. 169–184.

Gregory, L.T. (2006). *Comparative socioecology of sympatric, free-ranging white-faced and bearded sakis in Brownsberg Nature Park, Suriname.* Unpublished Master's thesis, Kent State University.

Hammond, D.S. (2005). Biophysical features of the Guiana Shield. In *Tropical Forests of the Guiana Shield*, ed. D.S. Hammond. Cambridge, MA: CABI, pp. 15–194.

Happel, R.E. (1982). Ecology of *Pithecia hirsuta* in Peru. *Journal of Human Evolution*, **11**, 581–590.

Harada, M.L., Schneider, H., Schneider, M.P. C., *et al.* (1995). DNA evidence on the phylogenetic systematics of New World monkeys: support for the sister-grouping of *Cebus* and *Saimiri* from two unlinked nuclear genes. *Molecular Phylogenetics and Evolution*, **4**, 331–349.

Hershkovitz, P. (1985). A preliminary taxonomic review of the South American Bearded Saki Monkeys genus *Chiropotes* (Cebidae, Platyrrhini), with the description of a new subspecies. *Fieldiana*, **27**(1363), 1–46.

Hershkovitz, P. (1987). The taxonomy of South American sakis, genus *Pithecia* (Cebidae, Platyrrhini): a preliminary report and critical review with the description of a new species and a new subspecies. *American Journal of Primatology*, **12**, 387–468.

Heymann, E.W., Encarnación C.F. & Canaquin Y.J.E. (2002). Primates of the Río Curaray, northern Peruvian Amazon. *International Journal of Primatology*, **23**, 191–201.

Homburg, I. (1998). *Ökologie and sozialverhalten einer gruppe von weissgesicht-sakis (Pithecia pithecia pithecia Linnaeus 1766) im Estado Bolívar, Venezuela*. Unpublished PhD thesis, Universität Bielefeld, Germany.

Horovitz, I. & Meyer, A. (1997). Evolutionary trends in the ecology of New World monkeys inferred from a combined phylogenetic analysis of nuclear, mitochrondrial, and morphological data. In *Molecular Evolution and Adaptive Radiation*, ed. T.J. Givnish & K.J. Sytsma. Cambridge: Cambridge University Press, pp. 189–221.

Izawa, K. (1976). Group sizes and compositions of monkeys in the upper Amazon Basin. *Primates*, **17**, 367–399.

Izawa, K. & Yoneda, M. (1981). Habitat utilization of nonhuman primates in a forest of western Pando, Bolivia. *Kyoto University Overseas Research Reports of New World Monkeys*, **2**, 13–22.

Izor, R.J. (1985). Sloths and other mammalian prey of the Harpy eagle. In *Ecology of Armadillos, Sloths, and Vermilinguas*, ed. G.G. Montgomery. Washington, DC: Smithsonian Institution Press, pp. 343–346.

Jacobs, J., Nole, I., Palminteri, S., *et al.* (2008). First come, first serve: "sit and wait" behavior in dung beetles at the source of primate dung. *Neotropical Entomology* , **37**, 641–645.

Johns, A. (1986). Notes on the ecology and current status of the buffy saki, *Pithecia albicans*. *Primate Conservation*, **7**, 26–29.

Kay, R.F. (1990). The phyletic relationships of extant and fossil Pitheciinae (Platyrrhini, Anthropoidea). *Journal of Human Evolution*, **19**, 175–208.

Kinzey, W.G., Norconk, M.A. & Alvarez-Cordero, E. (1988). Primate survey of eastern Bolívar, Venezuela. *Primate Conservation*, **9**, 66–70.

Lehman, S.M. (2006). Nested distribution patterns and the historical biogeography of the primates of Guyana. In *Primate Biogeography: Progress and Prospects*, ed. S.M. Lehman & J.G. Fleagle. New York, NY: Springer, pp. 63–80.

Lehman, S.M., Prince, W. & Mayor, M. (2001). Variations in group size in white-faced sakis (*Pithecia pithecia*): evidence for monogamy or seasonal congregations? *Neotropical Primates*, **9**, 96–101.

Lehman, S.M., Sussman, R.W., Phillips, C.J., *et al.* (2006). Ecological biogeography of primates in Guyana. In *Primate Biogeography: Progress and Prospects*, ed. S.M. Lehman & J.G. Fleagle. New York, NY: Springer, pp. 105–133.

Marroig, G. & Cheverud, J.M. (2001). A comparison of phenotypic variation and covariation patterns and the role of phylogeny, ecology, and ontogeny during cranial evolution of New World monkeys. *Evolution*, **55**, 2576–2600.

Marroig, G. & Cheverud, J.M. (2004). Cranial evolution in sakis (*Pithecia*, Platyrrhini) I: Interspecific differentiation and allometric patterns. *American Journal of Physical Anthropology*, **125**, 266–278.

Marsh, L.M. (2004). Primate species at the Tiputini Biodiversity Station, Ecuador. *Neotropical Primates*, **12**, 75–78.

Meldrum, D.J. & Kay, R.F. (1997). *Nuciruptor rubricae*, a new pitheciin seed predator from the Miocene of Colombia. *American Journal of Physical Anthropology*, **102**, 407–427.

Mittermeier, R.A. (1977). *Distribution, synecology and conservation of Surinam monkeys*. Unpublished PhD thesis, Harvard University.

Muckenhirn, N.A., Mortensen, B.K., Vessey, S., *et al.* (1975). *Report on a Primate Survey in Guyana*. Washington, DC: Pan American Health Organization, pp. 1–49.

Norconk, M.A. (1996). Seasonal variation in the diets of white-faced and bearded sakis (*Pithecia pithecia* and *Chiropotes satanas*) in Guri Lake, Venezuela. In *Adaptive Radiations of Neotropical Primates*, ed. M.A. Norconk, A.L. Rosenberger & P.A. Garber. New York, NY: Plenum, pp. 403–423.

Norconk, M.A. (2006). A long-term study of group dynamics and female reproduction in Venezuelan *Pithecia pithecia*. *International Journal of Primatology*, **27**, 653–674.

Norconk, M.A. (2011). Sakis, uakaris, and titi monkeys: behavioral diversity in a radiation of primate seed predators. In *Primates in Perspective*, ed. C.J. Campbell, A. Fuentes, K.C. MacKinnon, S.K. Bearder & R.M. Stumpf. Cambridge: Cambridge University Press, pp. 122–139.

Norconk, M.A. & Conklin-Brittain, N.L. (2004). Variation on frugivory: the diet of Venezuelan white-faced sakis. *International Journal of Primatology*, **25**, 1–26.

Norconk, M.A. & Grafton, B.W. (2003). Changes in forest composition and potential feeding tree availability on a small land-bridge island in Lago Guri, Venezuela. In *Primates in Fragments*, ed. L. Marsh. New York, NY: Kluwer Academic, pp. 211–227.

Norconk, M.A., Raghanti, M.A., Martin, S. K., *et al.* (2003). Primates of Brownsberg Natuurpark, Suriname, with particular attention to the pitheciins. *Neotropical Primates*, **11**, 94–100.

Oliveira, J.M.S., Lima, M.C., Bonvincino, C., *et al.* (1985). Preliminary notes on the ecology and behavior of the Guianan saki (*Pithecia pithecia*, Linnaeus 1766; Cebidae, Primate). *Acta Amazonica*, **15**, 249–263.

Oliveira, L.C., Loretto, D., Viana, L.R., *et al* . (2009). Primate community of the tropical rain forests of Saraca-Taquera National Forest, Para, Brazil. *Brazilian Journal of Biology* , **69**, 1091–1099.

Peres, C.A. (1993a). Notes on the ecology of buffy saki monkeys (*Pithecia albicans*, Gray 1860): a canopy seed-predator. *American Journal of Primatology*, **31**, 129–140.

Peres, C.A. (1993b). Structure and spatial organization of an Amazonian terre firme forest primate community. *Journal of Tropical Ecology*, **9**, 259–276.

Rettig, N.L. (1978). Breeding of the harpy eagle (*Harpia harpyja*). *Auk* , **95**, 629–643.

Robinson, J.G., Wright, P.C. & Kinzey, W.G. (1987). Monogamous cebids and their relatives: intergroup calls and spacing. In *Primate Societies*, ed. B.B. Smuts, D.L. Cheney, R.M. Seyfarth, R.W., Wrangham & T.T. Struhsaker. Chicago, IL: University of Chicago Press, pp. 44–53.

Rosenberger, A.L. (1984). Fossil New World monkeys dispute the molecular clock. *Journal of Human Evolution*, **13**, 737–742.

Rosenberger, A.L. (2002). Platyrrhine paleontology and systematics: the paradigm shifts. In *The Primate Fossil Record*, ed. W.C. Hartwig. Cambridge: Cambridge University Press, pp. 151–159.

Schneider, H. & Rosenberger, A.L. (1996). Molecules, morphology, and platyrrhine systematics. In *Adaptive Radiations of Neotropical Primates*, ed. M.A. Norconk, A.L. Rosenberger & P.A. Garber. New York, NY: Plenum, pp. 3–19.

Schneider, H., Schneider, M.P.C., Sampaio, I., *et al.* (1993). Molecular phylogeny of the New World monkeys (Platyrrhini, Primates). *Molecular Phylogenetics and Evolution*, **2**, 225–242.

Schwarzkopf, L. & Rylands, A.B. (1989). Primate species richness in relation to habitat structure in Amazonian rainforest fragments. *Biological Conservation*, **48**, 1–12.

Setz, E.Z.F. (1991). Animals in the Nambiquara diet: methods of collection and processing. *Journal of Ethnobiology*, **11**, 1–22.

Setz, E.Z.F. (1993). *Ecologia alimentar de um grupo de parauacus (Pithecia pithecia chrysocephala) em um fragmento florestal na Amazónia Central.* Unpublished PhD thesis, Inst. Biol. UNICAMP, Campinas, SP.

Setz, E.Z.F. (1999). Sleeping habits of golden-faced sakis in a forest fragment in the central Amazon. *American Journal of Primatology*, **49** ,100.

Setz, E.Z.F. & Gaspar, D.d.A. (1997). Scent-marking behaviour in free-ranging golden-faced saki monkeys, *Pithecia pithecia chrysocephala*: sex differences and context. *Journal of Zoology, London*, **241**, 603–611.

Soini, P. (1986). A synecological study of a primate community in Pacaya-Samiria National Reserve, Peru. *Primate Conservation*, **7**, 63–71.

de Thoisy, B., Brosse, S. & Dubois, M.A. (2008). Assessment of large-vertebrate species richness and relative abundance in Neotropical forest using line-transect census: what is the minimal effort required? *Biodiversity and Conservation* , **17**, 2627–2644.

Thompson, C.L. (2006). *Sex biases and inter- and intragroup social behaviour in white-faced saki monkeys (Pithecia pithecia).* Unpublished Master's thesis, Kent State University.

Thompson, C.L. (2010). Monogamous and non-monogamous mating behavior in white-faced saki monkeys (*Pithecia pithecia*). *American Journal of Physical Anthropology* , **50(Suppl.)**, 229.

Thompson, C.L., Norconk, M.A., Anzelc, A., *et al.* (2011). Group dynamics of free-ranging white-faced sakis (*Pithecia pithecia*) at Brownsberg Nature Park. *American Journal of Primatology*, **73**, 1051–1061.

Vié, J.C., Richard-Hansen, C. & Fournier-Chambrillon, C. (2001). Abundance, use of space, and activity patterns of white-faced sakis (*Pithecia pithecia*) in French Guiana. *American Journal of Primatology*, **55**, 203–221.

Walker, S.E. (1996). The evolution of positional behavior in the saki-uakaris (*Pithecia, Chiropotes*, and *Cacajao*). In *Adaptive Radiations of Neotropical Primates*, ed. M.A. Norconk, A.L. Rosenberger & P.A. Garber. New York, NY: Plenum, pp. 335–367.

Walker, S.E. (2005). Leaping behavior of *Pithecia pithecia* and *Chiropotes satanas* in Eastern Venezuela. *American Journal of Primatology*, **66**, 369–387.

Chapter 26

Finding the balance: optimizing predator avoidance and food encounters through individual positioning in *Pithecia pithecia* during travel

E.P. Cunningham, A.L. Harrison-Levine & R.G. Norman

Introduction

The relative positions of members of a group during travel can affect their fitness. Individuals in peripheral positions (DeVore & Washburn 1963; Hamilton 1971; Rhine 1975), particularly in the front of a moving group (Rhine 1975; Bumann *et al.* 1997; Di Blanco & Hirsch 2006) have the greatest risk of predation. Leading individuals may, however, have greater foraging success (Robinson 1981; Hall & Fedigan 1997; Janson 1990a, 1990b; Hirsch 2007). Being in front is most advantageous when the group is foraging for quickly depleted resources: by the time the rest of the group arrives, many of the resources may have been consumed (Robinson 1981; Janson 1990a, 1990b; Hall & Fedigan 1997; Hirsch 2007). A similar trade-off between increased predation risk and reduced foraging efficiency exists in relation to nearest-neighbor distances. More isolated individuals are more likely to be taken by a predator as they benefit less from the vigilance and the dilution effect of neighbors. Increasing nearest-neighbor distance, however, sometimes lessens feeding interference (Janson 1990b).

In this chapter, we examine the configuration of a group of six white-faced sakis (*Pithecia pithecia*) during travel. Our goal is to determine whether they have a defined structure, and to study the influences of foraging competition and predation risk on group geometry during travel during contrasting periods of abundance and scarcity. The small size of *P. pithecia* groups, combined with their small body size [approximately 2 kg (Robinson *et al.* 1987)], makes them highly vulnerable to predators. At the study site, sakis frequently alarm-call in response to large birds and have remained motionless in the understory for longer than an hour when they have felt threatened. Other potential saki predators in the area include jaguars (*Panthera onca*), and red-tailed boas (*Boa constrictor*) (Gleason & Norconk 2002). Sakis have very low rates of aggression during feeding. In a concurrent study, aggressive interactions occurred at a rate of 0.02/min of observation in feeding trees (Harrison-Levine, unpubl. MA thesis, 1998). During the present study, the number of large fruiting trees declined. As a result, the sakis relied more on leaves, insect material and small trees and shrubs with fewer fruit (Cunningham 2003; Cunningham & Janson 2006).

If intragroup feeding competition influences group structure during travel, then adult females, the age–sex class with the highest nutritional need, should be in the front of the group or in other vulnerable positions, where they will suffer less from the depletion of resources. Adult male positional behavior is harder to interpret: males could position themselves in the front and other peripheral positions to increase their personal foraging success or act as a predation buffer. The presence of juveniles in vulnerable positions indicates the influence of feeding competition because it is predicted that they would occupy a protected position given their small body size and relative inexperience.

Photo 26.1 Breeding female (Ma) in Isla Redonda group of white-faced sakis (*P. pithecia*), Lago Guri, Venezuela. Photo: Elena Cunningham.

Evolutionary Biology and Conservation of Titis, Sakis and Uacaris, eds. Liza M. Veiga, Adrian A. Barnett, Stephen F. Ferrari and Marilyn A. Norconk. Published by Cambridge University Press.

Photo 26.2 Breeding male (Big Daddy) in Isla Redonda group of white-faced sakis (*P. pithecia*), Lago Guri, Venezuela. Photo: Elena Cunningham.

Methods

Study site

The study site is a 12.8-ha island in Guri Lake, Venezuela (7.45°N, 62.52°W). We observed six sakis: two adult males, two adult females, one subadult female and one juvenile male. Data were collected in three observation periods between March and May 1996 (20 days in March, 21 in April, and 21 in May). We collected continuous focal animal data and conducted group scans every 30 min. Data collected during group scans included location (recorded as distances and bearings from previously mapped trails and feeding trees), individual identity, activity (feed, move, rest, other) and time of sighting of the individuals. See Cunningham (2003) and Cunningham and Janson (2006) for additional details on the study site and data collection.

Analysis

Three age–sex classes were used in analyses: adult male, adult female and juvenile. The trail system of the island, the daily paths of the focal animals, and the locations of sakis that were recorded during group scans were plotted on Tracker, public-domain software created by Janson and Lee (1995, unpubl.). Of the 776 group scans conducted, only 40 (March: $n = 14$, April: $n = 10$, May: $n = 16$) met the stringent criteria of a travel scan described below. First, when more than one scan from a single day was analyzed, the scans were separated by at least 1 h. An hour is probably long enough to ensure independence between samples (following Robinson 1981). Second, scans were eliminated if all the individuals were not found (a rare event), if the recorded time for the location of the leading individual was later than the time for the individuals in the rear (a later time for the leading individual suggests that the individuals might not have been in front, but passed other individuals' positions in the intervening time). Third, scans were eliminated if the group was not traveling. As individuals often had brief feeding and resting bouts as other members of the group continued to travel, there was no clear demarcation between feeding, resting and traveling.

A travel bout was defined when the focal animal traveled at least 20 m in the 10-min period starting 5 min before the beginning of the scan and if the focal animal traveled at least 75 m in 30 min preceding the beginning of the scan. The 40 scans included in this analysis averaged 3.5 min in duration due to the time it took to locate and identify six traveling monkeys.

For each scan, the relative location of every group member was plotted on a five-by-six grid (Table 26.1 is an example of a plot of one scan). We compared the observed positions of the monkeys to positions expected by chance. In this study, chance is defined as the random placement of six monkeys. Randomly occupied cells were shifted so that a row could not be occupied

Table 26.1 Sample plot illustrating the relative positions of each group member during scans, where M1 & M2 = adult males, F1 & F2 = adult females, J1 = juvenile male and J2 = juvenile female. Recorded on 16 March 1996, 13:00 h.

Leading edge (front) of traveling group				
M1				M2
	F1	J1	F2	
		J2		

unless every row in front of it had at least one occupied cell, peripheral columns could not be occupied unless columns closer to the center had at least one occupied cell. Monte Carlo simulation of 100,000,000 replications was performed to generate expected cell proportions which were then scaled to six monkeys observed 40 times.

We also assigned a vulnerability score to each individual in each scan. We assumed (cf. Hamilton 1971; Bumann *et al.* 1997) that predators attack the first prey they encounter. We considered each saki vulnerable from eight directions (front, front right diagonal, right, etc.). An individual who traveled without any neighbours within 15 m would be given a score of eight. For an individual in a given cell, each direction in which at least one cell was occupied by a saki less than 15 m away functioned to reduce the vulnerability score by one. For example, for the scan that was plotted on Table 26.1, the nearest-neighbor for M1 was 31 m away and the nearest neighbor for M2 was 48 m away. Therefore, M1 and M2 both had vulnerability scores of 8. The other members of the group were within 15 m of each other. Their vulnerability scores were: F1 = 6, F2 = 6, J1 = 5, J2 = 5. If a member of the group was surrounded by the five other members of the group in five directions, he or she would have a vulnerability score of three, the lowest possible score in this study.

Although, the seeds of *Licania discolor* (Chrysobalanaceae) made up the sakis' most important food source (accounting for 80% or more of their feeding minutes each month of the study), *L. discolor* seeds ingested by the sakis dropped 25% between March and April, and an additional 51% between April and May. As highly productive *Licania* trees, which could support feeding bouts of over an hour in duration, became scarce, short feeding bouts (< 5 min) accounted for a higher percentage of feeding time (March: 17%, April: 28%, May: 33%) (Cunningham & Janson 2006). Based on these data, we considered March the period of abundance and May the period of scarcity. April, a transitional period in terms of fruit abundance, with the smallest sample size (10 scans), was excluded from this analysis.

We used SPSS versions 15 and 16 to conduct statistical analysis. Probability levels are 2-tailed and significant at < 0.05.

Table 26.2 The relative positions of individuals in each age-sex class, recorded during 40 analyzed scans. Counts for adult male positions are in **bold face**, adult female counts are *italicized and underlined*, and juvenile counts are in regular font. (*N* = 240 recorded positions.)

Leading edge (front) of traveling group				
2	**4**	**13**	**7**	**5**
3	*7*	*2*	*5*	*2*
1	4	5	5	0
1	**4**	**4**	**6**	**3**
2	*9*	*11*	*8*	*1*
2	6	9	6	1
1	**1**	**7**	**3**	**3**
2	*2*	*6*	*2*	*0*
1	10	9	7	1
0	**1**	**4**	**1**	**2**
2	*4*	*4*	*2*	*1*
0	1	6	1	1
0	**1**	**3**	**1**	**1**
1	*2*	*2*	*0*	*0*
1	0	0	1	0
0	**1**	**0**	**0**	**1**
0	*0*	*0*	*0*	*0*
0	0	1	1	0

Results

Average group spread parallel to the line of travel (26.2 ± 24.2 m) was greater than average perpendicular spread (18.3 ± 14.1 m). The differences were not statistically significant (p = .095, Wilcoxon signed ranks test), suggesting that group geometry in sakis is highly variable. Table 26.2 is a composite of all the relative positions of each member of the group in the 40 samples. There was no significant difference between observed and expected positions (exact p value = 0.09).

We could not statistically analyze the placement of age-sex classes on the grid because many of the cells would have expected counts less than five. However, we conducted two analyses that indicate that the placement of age–sex classes in the group is not random. Adult males were most often in the front row (48.5%), followed by females (26.8%) and then juveniles (22.7%) ($\chi^2 = 9.91$, $p = 0.01$, $N = 240$).

We also considered the vulnerability scores of individual positions relative to rank and age–sex class. High vulnerability scores indicate peripheral positioning. Adult male vulnerability scores were highest (5.86 ± 1.27), adult female scores were intermediate (5.45 ± 1.33) and juvenile scores were lowest (5.25 ± 0.92) ($F_{2,221} = 7.226$, $p = 0.001$). Posthoc tests revealed that adult males differed significantly from juveniles (Tukey HSD adult males – adult females, p = 0.077; adult males – juveniles, p =.011; adult females – juveniles p = 0.748).

The differences in the means of nearest-neighbor distances for the three age–sex classes was significant (adult males: 8.8 ± 9.6 m; adult females: 6.5 ± 7.9 m; juveniles: 4.6 ± 3.0 m, $F_{2,237} = 8.066$, $p < 0.001$). Posthoc tests (Tukey HSD) showed that the significant difference was between adult males and juveniles. These results suggest that minimizing the risk of predation for vulnerable group members has the strongest influence on positioning behavior during travel.

Effect of resource abundance on group structure

Resource abundance did not significantly affect vulnerability scores, nearest-neighbor distances, or the number of occupied cells in the first row. There was, however, a significant change in the number of individuals in the forward 10 m of the group. As resources became scarcer, significantly more individuals moved in towards the front of the group (Figure 26.1). While adult males had little change in percentage of their samples that were scored in the forward 10 m (68–71%), the increase was greater for adult females (57–71%) and for juveniles (43–73%).

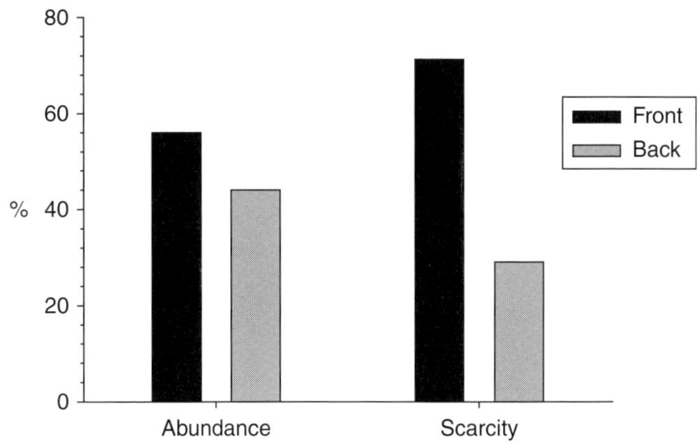

Figure 26.1 Percentage of individuals in the front and the back of the group during periods of abundance and scarcity. Front: within 10 m of the leading individual and parallel to the line of travel. Back: > than 10 m from the leading individual and parallel to the line of travel. [χ^2 (1) = 4.30, p =.038, N = 180.]

Discussion

Positions of the sakis were variable during travel; the overall configuration of the group did not vary from expected. Age-sex classes, however, were not randomly distributed. Adult males occupied the most vulnerable positions, including the leading edge of the group. Juveniles were in the most protected positions. During March, when large *L. discolor* trees were most abundantly in fruit, group geometry while traveling likely functioned to protect the youngest members of the group from predation. During May, when highly productive *L. discolor* trees were scarcer and smaller, scattered resources played a more important role in the sakis' diet and females and juveniles spent more time in the forward 10 m of the group. By moving to the front of the group, females and juveniles increased their chances of seeing resources first. As sakis display very low levels of interindividual aggression while feeding (Harrison-Levine, unpubl. MA thesis, 1998), the individual that sees a resource first will most likely be able to exploit it. Because moving towards the front of the group increases the risk of predation, we might expect such a move to be accompanied by a decrease in nearest-neighbor distances (Bumann *et al.* 1997). Saki females and juveniles, however, did not reduce nearest-neighbor distances. It appears that when resources became scarcer, females and juveniles were willing to increase predation risk to increase their foraging success.

The study group's configuration during travel was distinct from the pattern observed during feeding bouts by Harrison-Levine (unpubl. MA thesis, 1998). When feeding in large trees, an adult male had the largest number of neighbors feeding within 2 m, suggesting that he occupied a central position. In contrast, the juvenile female, was most often in a tree by herself and had the greatest nearest-neighbor distances, indicating that she occupied peripheral positions. The configuration of saki groups seems to be flexible, changing with activity and responding to predation risk as well as the abundance of resources.

Many aspects of pitheciine positioning behavior remain to be explored. A larger sample size would make it possible to determine whether rank affects saki position during travel. Studies of pitheciines, such as *Chiropotes* and *Cacajao* who live in large, flexible groups (Norconk 2011), could show how group size influences positional behavior. Analysing group structure during travel in three dimensions would provide valuable information, particularly for platyrrhines that all share arboreality as a common feature.

References

Bumann, D., Krause, J. & Rubenstein, D. (1997). Mortality risk of spatial positions in animal groups: the danger of being in the front. *Behaviour*, **134**, 1063–1076.

Cunningham, E.P. (2003). The use of memory in *Pithecia pithecia's* foraging strategy. Unpublished PhD thesis, City University of New York.

Cunningham, E.P. & Janson, C.H. (2006). *Pithecia pithecia*'s behavioral response to decreasing fruit abundance. *American Journal of Primatology*, **68**, 491–497.

DeVore, I. & Washburn, S. L. (1963). Baboon ecology and human evolution. *Viking Fund Publications in Anthropology*, **36**, 335–367.

Di Blanco, Y. & Hirsch, B.T. (2006). Determinants of vigilance behavior in the ring-tailed coati (*Nasua nasua*): the importance of within-group spatial position. *Behavioral Ecology and Sociobiology*, **61**, 173–182.

Gleason, T.M. & Norconk, M.A. (2002). Predation risk and antipredator adaptations in white-faced sakis, *Pithecia pithecia*. In *Eat or Be Eaten: Predation Sensitive Foraging Among Primates*, ed. L.E. Miller. New York, NY: Cambridge University Press, pp. 169–184.

Hall, C.L. & Fedigan, L.M. (1997). Spatial benefits afforded by high rank in white-faced capuchins. *Animal Behaviour*, **53**, 1069–1082.

Hamilton, W.D. (1971). Geometry for the selfish herd. *Journal of Theoretical Biology*, **31**, 295–311.

Hirsch, B.T. (2007). Costs and benefits of within-group spatial position: a feeding competition model. *Quarterly Review of Biology*, **82**, 9–27.

Janson, C.H. (1990a). Social correlates of individual spatial choice in foraging groups of brown capuchin monkeys, *Cebus apella*. *Animal Behaviour*, **40**, 910–921.

Janson, C.H. (1990b). Ecological consequences of individual spatial choice in foraging groups of brown capuchin monkeys, *Cebus apella*. *Animal Behaviour*, **40**, 922–934.

Norconk, M.A. (2011). Sakis, uakaris, and titi monkeys: behavioral diversity in a radiation of primate seed predators. In *Primates in Perspective* (2nd edn), ed. C.J. Campbell, A. Fuentes, K.C. MacKinnon, S.K. Bearder & R.M. Stumpf. New York, NY: Oxford University Press, pp. 122–139.

Rhine, R.J. (1975). The order of movement of yellow baboons (*Papio cynocephalus*). *Folia Primatologica*, **23**, 72–104.

Robinson, J.G. (1981). Spatial structure in foraging groups of wedge-capped capuchin monkeys *Cebus nigrivittatus*. *Animal Behaviour*, **29**, 1036–1056.

Robinson, J.G., Wright, P.C. & Kinzey, W.G. (1987). Monogomous cebids and their relatives: intergroup calls and spacing. In *Primate Societies*, ed. B.B. Smuts, D.L. Cheney, R.M. Seyfarth, R.W. Wrangham & T.T. Struhsaker. Chicago, IL: University of Chicago Press, pp. 44–53.

Chapter 27

Testing models of social behavior with regard to inter- and intratroop interactions in free-ranging white-faced sakis

Cynthia L. Thompson & Marilyn A. Norconk

Introduction

Since males and females use different strategies to maximize their reproductive success (e.g. Clutton-Brock 1988), they are also expected to use disparate means when navigating their social environment. The striking sexual dichromatism of white-faced sakis (*Pithecia pithecia pithecia*), a potential indicator of male–male competition or female choice (both of which may lead to very different intra- and intertroop social dynamics), make sakis a particularly interesting case to examine sex-biased differences in social behavior.

All sakis (*Pithecia* spp.) form small social groups consisting of 2–12 individuals (Buchanan *et al.* 1981; Lehman *et al.* 2001; Vié *et al.* 2001; Norconk *et al.* 2003; Norconk 2006, 2011), but some groups are too large to fit a traditional definition of monogamy (e.g. Fuentes 2002). Although some observers of larger groups have considered the possibility that multiple smaller units coalesce into larger ones on a temporary basis (Buchanan *et al.* 1981; Lehman *et al.* 2001), our studies of wild, habituated white-faced sakis in Suriname and Venezuela suggest that this is unlikely given the intense aggression during intergroup interactions. However, the social and/or ecological conditions that underlie group size expansion in sakis are not completely clear. While older information and census data lent support to a one-male, one-female social unit (Buchanan *et al.* 1981; Oliveira *et al.* 1985; Lehman *et al.* 2001; Norconk 2011, Norconk & Setz, Chapter 25), longer studies on habituated individuals suggest a more complex troop configuration/social organization (Rosenberger *et al.* 1996; Kinzey 1997; Norconk 2006; Di Fiore *et al.* 2007; Thompson & Norconk 2011). In this chapter, we describe data collected on social relationships in two free-ranging troops of white-faced sakis at Brownsberg Nature Park, Suriname, and place the study in the context of existing models of social behavior.

Current models explaining inter- and intrasexual social relationships in primates fall into three main types: (1) socioecological models, (2) infanticide avoidance models, and (3) monogamy models. We will focus on infanticide and monogamy models in this study because socioecological models (e.g. Wrangham 1980; van Schaik 1989; Sterck *et al.* 1997) all share the common perspective that resource distribution dictates the competitive regime among females. Competition, in turn, is predicted to shape social relationships. As troops of *P. pithecia* usually contain no more than two adult females, and are unlikely to be based on a "core network" of females, socioecological models are difficult to apply.

The infanticide avoidance model predicts that the threat of male infanticide (and female avoidance of it) is the main evolutionary force shaping social relationships among adults (van Schaik & Dunbar 1990; van Schaik 1996, 2000). Relationships among females are expected to be weak as females are non-bonded (*sensu* Wrangham 1980). The primary social bond is expected to be between males and females, with females being more heavily invested in maintaining the relationship (Table 27.1), with the exception of monogamous mating systems, where males and females face similar risks from infanticide (van Schaik & Dunbar 1990). In short, it is expected that females respond to potentially heavy reproductive losses via infanticide by developing close associations with troop males.

Monogamy models are also potentially relevant to white-faced sakis. These models explain why a male would maintain an exclusive relationship with a single female, due to dispersion of resources that females need for reproduction or dispersion of the females themselves (Table 27.1). Unlike the infanticide model, male–female relationships are expected to be relatively weak or maintained by the male.

The dearth of studies on wild, habituated, free-ranging sakis has made it impossible to go beyond numerical data on troop size and composition provided by census data. We provide data on social interactions among adult white-faced sakis that will provide the first step to test evolutionary models characterizing social relationships in primates and other vertebrates. In particular, we (1) quantify patterns of proximity and proximity maintenance behaviors in two

Evolutionary Biology and Conservation of Titis, Sakis and Uacaris, eds. Liza M. Veiga, Adrian A. Barnett, Stephen F. Ferrari and Marilyn A. Norconk. Published by Cambridge University Press.

Table 27.1 Interpreting social relationships in primates: expectations of the infanticide and three monogamy models (adapted from Fuentes 1999, 2002).

Model	Intergroup aggression	M–F bond	Other relevant expectations	References
Infanticide avoidance	Male→male ---- Females non-participants in ITEs	Permanent M;F associations, maintained by the female (but see text)	Females should be non-bonded	van Schaik & Dunbar 1990; van Schaik 1996, 2000
Monogamy models				
1. Females as a limited or widely dispersed resource	Female→Both sexes	May be relatively weak	Exclusively two-adult groups	Wrangham 1980; van Schaik & van Hooff 1983
2. Mate guarding	Male→Male	Maintained by male	Females are dispersed	Palombit 1999
3. Male defense from predators and/or conspecifics	Both sexes →Both Sexes	May be weak	Intergroup competition for resources	Wittenberger & Tilson 1980; van Schaik & Dunbar 1990

wild white-faced saki troops, (2) examine the frequency of grooming interactions as an estimate of the intensity of social bonds among adults, and (3) characterize competitive interactions between troops via an analysis of intertroop encounters.

Methods

Study site and subjects

Research was conducted at Brownsberg Nature Park, Suriname (see Gregory & Norconk, Chapter 28, for a description of the site), from May 30 to August 5, 2005 during the long wet season and the beginning of the long dry season. Because an initial habituation period was necessary for the two troops, the data analyzed below were collected from June 21 to August 5, 2005. Two social troops with adjoining home ranges were habituated to human observers: Peach troop contained five individuals; Junco troop had four (Table 27.2). Each troop had one breeding male and one or two breeding females.

Table 27.2 Age-sex composition of *P. pithecia* study troops at Brownsberg Nature Park and estimated ages for subadults.

Troop	Individual code	Description[1]
Peach	PEA	Adult male
	FTS[2]	Subadult male (born late 2002/early 2003)
	SSO[3]	Subadult male (born late 2003/early 2004)
	MIR	Adult female
	TIN	Adult female
Junco	JUC	Adult male
	MAM	Adult female (mother of BEL)
	SCR	Subadult female
	BEL[4]	Juvenile female (about 5–6 months old)

[1] Subadult males were smaller than adults, but exhibited adult coloration (see Norconk 2006). Estimated ages are based on the observations of previous years. Peach group has been observed since 2003; Junco group since 2005. Despite the absence of observers from August 2004 to May 2005, it is assumed that troop membership was stable based on number and age composition of the individuals.
[2] FTS was first observed as a self-locomoting infant in June 2003 and as a juvenile one year before this study (summer 2004).
[3] SSO was observed as a young juvenile based on pelage and facial coloration in summer 2004.
[4] BEL was observed nursing, and was not included in data analysis.

Data collection

Data were collected 6 days a week, following troops from sleeping tree to sleeping tree, for a total of 204.6 h: 145.2 h from Peach troop and 43.8 h from Junco troop. Instantaneous scan samples were conducted every 15 min on a focal animal, followed by a 5-min continuous sample. Focal animals were often selected based on visibility, with an effort made to achieve equal sampling time for each individual. Sampling focal animals for a maximum of four consecutive samples avoided biases toward particularly visible individuals. Focal animals were not reselected for at least 1 h.

Two types of data were collected during scan samples: (1) distance (proximity) of the focal animal to all other visible troop members, which was estimated by one observer after undergoing an initial training period, and (2) activity of the focal animal, classified into three mutually exclusive categories: resting, feeding or traveling. This classification of activity patterns was used to subdivide the proximity analysis into categories that may display different social patterns (for example, proximity to individuals may be less constrained while resting than feeding). During continuous samples that followed instantaneous scan samples (on the same individual), we recorded approaches and leaves (within 1 m) to/from the focal and by the focal that lasted at least 5 s, and grooming episodes. Continuous sampling yielded 176 social interactions for Peach troop, and 21 for Junco troop, from over 6.32 h of sampling.

All-occurrence sampling (Altmann 1974) was used to describe behaviors during intertroop encounters (ITEs). Data collected during these events included (1) troops involved in the encounter (including both study troops that occupied contiguous home ranges as well as less habituated troops), (2) participation (active or non-active) of each troop member in the encounter, (3) distance and membership of agonistic chases during ITEs, (4) context-specific and generalized alarm vocalizations (Buchanan 1978) given by individuals either immediately before, during, or after an ITE, and (5) scent-marking activities. Active participation was defined as being at the leading edge of the encounter (nearest to the neighboring troop), engaging in displays (branch-shaking) and piloerection. Non-active troop members remained in the background and fed or rested while active members engaged in ITEs.

Data analysis

Proximity distances among all sex categories (male–male, male–female, female–female) were regressed on time of day, to determine if time needed to be controlled for in subsequent analyses. There was no significant relationship between time of day and proximity for any of the sex categories (MF: $F = 0.618$, $p = 0.432$, $R^2 = 0.002$; MM: $F = 0.243$, $p = 0.623$, $R^2 = 0.002$; FF: $F = 1.599$, $p = 0.210$, $R^2 = 0.019$), and hence data collected over the course of the day were analyzed together. When the proximity data were divided into subsets by activity, the assumption of equal variances was violated, and we accordingly conducted non-parametric tests for these data. Inter-individual distances between sex categories were compared with Kruskal–Wallis and Mann–Whitney tests for Peach and Junco troops, respectively, separately for the three activity categories.

Approaches and grooming interactions were analyzed using chi-square tests, with the categories being both males and females approaching/grooming the same and opposite sex. Expected values were based on troop composition. Approaches and withdrawals were also used to calculate Hinde's Index, a measure of responsibility for association, between males and females:

$$H = (A_f/A_f + A_m) - (W_f/W_f + W_m)$$

where A_f = approaches of females to males, A_m = approaches of males to females, W_f = withdrawal of females from males and W_m = withdrawal of males from females. The Hinde Index ranges from −1 to 1, where −1 reflects males being completely responsible for proximity maintenance, 1 reflects complete responsibility by the female, and 0 reflects equal levels of proximity maintenance by both sexes (Hinde & Atkinson 1970; Martin & Bateson 1986).

Individual participation and frequency of behaviors for 10 ITEs are reported. On days in which ITEs occurred, the proximity distance data taken before the ITE were compared to those after the ITE, via *t*-tests for each sex category.

Results

Sex differences in proximity

White-faced saki troops in this study were overall relatively cohesive (Table 27.3); proximity ranged from within 1 to 30 m. In Peach troop, the closest proximity of sex categories was between males and females during resting ($p = 0.03$) and while traveling (ns). While male–female distances in Junco troop were comparable to Peach troop ($t = -0.213$, ns), the overall proximity pattern between sex categories differed between the troops due to variation in female–female relationships. In Junco troop, MAM and SCR were in extremely close association; it is possible that SCR was a recent offspring of MAM, whereas the genetic relationship between the adult females in Peach troop was unknown. Alternately, this could represent differences in adult female relationships based on age. Female–female social distances in Peach troop were generally further than male–female distance, except when feeding, in which

Table 27.3 Sex-biased differences in proximity, by group and activity.

Activity	Male–female distance (m) ($\bar{x} \pm$ SD)	Male–male distance (m) ($\bar{x} \pm$ SD)	Female–female distance (m) ($\bar{x} \pm$ SD)	*p*
Peach Group				
feeding (*n* = 124)	6.32 ± 4.69	6.51 ± 3.80	6.29 ± 4.48	0.651
resting (*n* = 238)	4.67 ± 4.93	6.21 ± 5.05	5.66 ± 6.34	0.032*
traveling (*n* = 46)	8.08 ± 7.49	8.33 ± 4.45	15.83 ± 12.35	0.281
Junco Group				
feeding (*n* = 28)	4.47 ± 2.20	–	2.56 ± 3.25	0.050*
resting (*n* = 35)	6.33 ± 4.64	–	2.50 ± 2.71	0.014*
traveling (*n* =12)	6.57 ± 1.62	–	4.20 ± 1.48	0.039*

Notes:
Kruskal–Wallis test was conducted for Peach group and Mann–Whitney for Junco group. All distances are in meters. *$P \leq 0.05$.

differences between sex categories were not significant. Lastly, male–male proximity was the furthest while resting and feeding, but closer than female–female distance when traveling. On average, males were in closer proximity to females than to other males.

Time spent in close proximity differed between sexes as well. For Peach troop, 39.5% and 31.5% of all male–female and female–female distances, respectively, were $\leq$ 2 m; this was true for only 19% of male–male distances. This confirms the findings of the Kruskal–Wallis test in Peach troop: males spent more time in close proximity to females than to other males, and females spent more time closer to males than to other females. However, the reverse was true for Junco troop, in which 29.8% of male–female distances and 67.9% of female–female distances were $\leq$ 2 m. Time spent in close proximity collaborated the patterns of overall proximity distances. In sum, for Peach troop, proximity was closer between the sexes rather than within (MF < FF/MM), whereas the reverse was true for Junco troop (FF < MF).

Photo 27.1 *Pithecia irrorata* male (left) and female (right). Note the low level of sexual dimorphism in this taxon from southwestern Brazil, northern Bolivia, and southeastern Peru. Photo: Francisco Fonseca. (See color plate section.)

Behavioral interactions

Junco troop did not have enough observed interactions to analyze separately, so the Peach troop was analyzed alone. Frequencies of observed approaches differed significantly from expected (χ^2 = 12.89, p = 0.002; Figure 27.1a). Approaches uniting male–female dyads were more often initiated by females, whereas male approaches to females did not deviate from chance. Despite this, Hinde's Index between males and females of Peach troop was $H = 0.045$, indicating equal responsibility for proximity maintenance between the sexes. Withdrawals were not analyzed alone, as they are contingent upon approaches (that is, the number of withdrawals of a male from a female will be limited by approaches uniting males and females). Grooming interactions also significantly differed from expectations ($\chi^2 = 114.53$, $p < 0.001$; figure 27.1b). Males rarely groomed other troop members regardless of sex, while females groomed males almost three times more than expected. These results suggest that while responsibility for proximity maintenance was equal between the sexes, grooming was heavily biased toward female investment in the inter-sexual bond (also see Thompson & Norconk 2011). See Photos 27.1 and 27.2.

Photo 27.2 Adult male white-faced saki grooming an adult female, *Pithecia pithecia*, Brownsberg Nature Park, Suriname. Compare the sex-specific white face and black body hair of the adult male in this photo with the adult male from the western Amazon Basin in Photo 27.1. Photo: Marilyn Norconk. (See color plate section.)

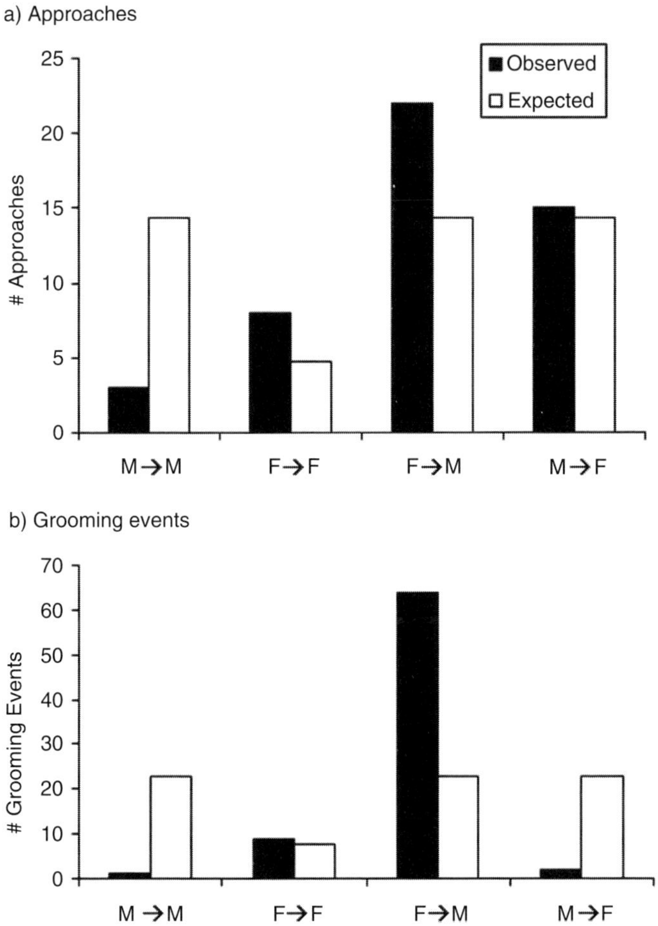

Figure 27.1 Peach group social interaction frequencies (filled bars), compared to expected values (open bars) for (a) approaches and (b) grooming events. M, male; F, female; arrows indicate direction of approach/groom. Expected values are based on group composition.

Intertroop encounters

Ten ITEs were observed, seven of which were between Peach and Junco troops, while the other three involved unhabituated neighboring troops (two with Peach troop, one with Junco). Typical ITEs were initiated by a loud vocalization upon detection of another troop (Norconk 2006). Characteristic behaviors (Norconk 2006) then ensued, with males (adult and subadults) of each troop rushing toward the neighboring troop and engaging in a "stand-off" or period of time during which neither male advanced nor retreated. They displayed, vocalized, became piloerected, scent-marked, and at times engaged in chases with the neighboring troop's males. In Peach troop, the two eldest males, PEA and FTS, often jointly engaged in these activities, generally while in close proximity (Table 27.4). Females remained behind their troop's male(s) during all encounters, never approached nor chased individuals in the neighboring troop, but did vocalize and scent marked. At the cessation of the ITE, troops would retreat into their respective territories – displacement of the "losing" troop was not observed, although ITEs generally

Table 27.4 Individual participation in intertroop encounters in Peach and Junco troops at Brownsberg Nature Park, Suriname. Values are the percentage of encounters in which behaviors were displayed. "Active" is defined in Methods. Due to wide dispersion of groups during ITEs, only subsets of individuals could be observed at a single time (usually the active individuals). ? = instances where an individual's identity was strongly inferred (by sex or process of elimination), but could not be confirmed, thus percentages are considered minimum values. Detailed descriptions of each encounter are given in Thompson (2006) and Thompson *et al.* (2012).

	Active (%)	Chasing (%)	Being chased (%)	Vocalize (%)	Scent-mark (%)
Peach					
PEA	100	66.7	0	100	37.5
FTS	100	77.8	11.1	100	0 (12.5?)
SSO	12.5 (37.5?)	0	0	100	0
MIR	0	0	0	100	0
TIM	0	0	0	100	0
Junco					
JUC	100	0	66.7	100	44.4
MAM	0	0	0	77.8 (100?)	22.2
SCR	0	0	0	0 (55.6?)	0

occurred along what was assumed to be territory boundaries or overlap zones.

ITEs also generated sex-specific changes in proximity: male–male proximity became significantly closer after ITEs (on average 3.4 m closer after ITEs, $p = 0.008$); no other sex category displayed a significant difference (M–F: 1.5 m closer after, $p = 0.157$, F–F: 3.8 m further after, $p = 0.08$).

Discussion

These results indicate that white-faced sakis have stronger intersexual than intrasexual bonds as well as stronger affiliation among females than males, with the caveat that troop males jointly engage in aggressive behaviors during territorial defense. See Photos 27.3 and 27.4.

Using the criteria of the social models in Table 27.1, what do these preliminary findings suggest regarding the selection pressures influencing white-faced saki social behavior? As several of the major tenets of the monogamy models are violated (Table 27.5), infanticide avoidance is the only explanation that cannot be rejected at this time. Females clearly associate with males, who play the "protector" role by actively engaging in ITEs (but see Thompson *et al.* 2012). Aggressive behaviors during ITEs were directed at extra-troop males, but not females. Indeed, females did not approach the neighboring troop in any ITE. Thus, females were never directly involved in intertroop encounters, nor were they in proximity to non-troop males. This was true for MAM, who had a nursing juvenile, as well as MIR and TIN, who did not have dependent offspring. Although responsibility for proximity maintenance was equal between the sexes, the asymmetry in grooming relationships demonstrates that investment in the intersexual bond is female-biased. The postulation that females are non-bonded

Photo 27.3 Vigilant adult male white-faced saki, Brownsberg Nature Park, Suriname. Photo: Marilyn Norconk. (See color plate section.)

Photo 27.4 Adult male white-faced saki (*Pithecia pithecia*) exhibiting sternal scent gland. Brownsberg Nature Park, Suriname. Photo: Nick Robl. (See color plate section.)

requires further study, but is suggested in the data collected during this study: adult females in Peach troop did not appear to prefer other females as resting or grooming partners. The reverse pattern displayed between MAM and SCR in Junco troop is potentially due to a mother–daughter relationship. It is also possible that variability in female–female relationships (due to kinship, troop history, social preferences, etc.) is representative of the species as a whole, as is true for other species that fit the infanticide models, like gorillas (Harcourt 1979; Stewart & Harcourt 1987; Watts 2001; Stokes 2004; Robbins 2007).

White-faced saki male–male relationships observed in this study also mimic those found in other infanticidal species. Peach troop, with more than one adult-sized male, was almost always the chaser of the one-male troop (Junco). This is consistent with observations from Lago Guri, Venezuela, where of two troops, the larger one (in number of males and total troop size) always chased the smaller one (Norconk 2006).

Multimale troops may have an advantage in intertroop encounters, as has been demonstrated in red howlers, Thomas's langurs and gorillas, where the presence of multiple males prevents takeovers by extra-troop males and ultimately reduces the risk of infanticide (red howlers: Pope 1990; Thomas's langurs: Steenbeck 1999; Steenbeck *et al.* 2000; gorillas: Watts 1989, 2000; Robbins, 1995; Bradley *et al.* 2005).

While our preliminary data fit the model, infanticide has never been observed in wild sakis. It is appropriate to make two points here. First, it is not surprising that such a rare behavior has never been documented in white-faced sakis given the paucity of long-term, free-ranging studies on this species. Most long-term studies have taken place either on

Table 27.5 Tentative differential support for infanticide and monogamy models in white-faced sakis (✓ = supported; x = not supported; ? = see text for discussion).

Model	Intergroup aggression		M–F bond		Other relevant expectations	
Infanticide avoidance	Male→male ---- Females hang-back during ITEs	✓ ✓	Permanent M–F associations, maintained by the female (but see text)	✓	Females should be non-bonded	?[a]
Monogamy models:						
1. Females as a limited or widely dispersed resource	Female→Both sexes	x	May be relatively weak	x	Species occurs in two-adult groups exclusively	x[b]
2. Mate guarding	Male→Male	✓	Maintained by male	x	Females are dispersed	x[c]
3. Male defense from predators and/or conspecifics	Both sexes →Both Sexes	x	May be weak	x	Intergroup competition for resources	?[d]

Notes.
[a] See discussion in text.
[b] Buchanan *et al.* 1981; Gleason & Norconk 1995; Lehman *et al.* 2001; Vié *et al.* 2001; Norconk *et al.* 2003.
[c] As the study groups (and other documented groups) contained more than one adult female, we do not consider females to be dispersed.
[d] The data presented here suggest intergroup aggression (and thus competition) is high; some evidence exists that this aggression reflects competition for food resources (Thompson *et al.* 2012).

islands or in forest fragments, where interactions with other troops and the possibility of immigrating males were minimal (Setz & Gaspar 1997; Norconk 2011). Continued research on habituated, known individuals may elucidate such behavior. Indeed, Di Fiore *et al.* (2007) observed a transition period following the death of the resident male in a *Pithecia aequatorialis* troop, in which the incoming male directed aggression toward the juvenile and infant. Second, van Schaik and Dunbar (1990) argue that the ghost of infanticide past, present in the suite of behaviors examined above, is sufficient to invoke it as a former selection pressure shaping current behavior.

Hence, while infanticide has not yet been documented in white-faced sakis, the notion that infanticide prevention has shaped their social behavior cannot be rejected at this time. The model's expectations are indeed met, and appear to better explain evolutionary pressures affecting this species than either the socioecological or monogamy models. While caution is certainly warranted when invoking models to explain a data set of only two troops, we feel these data are indeed noteworthy as (1) there are currently limited data on social behavior in *Pithecia pithecia* in a free-ranging population (but see Thompson & Norconk 2011), and (2) the composition of the study troops reflects the norm for white-faced sakis as a species. Thus, rather than viewing the incongruities between troops as a limitation, we believe that our data provide important information on natural variation in troop organization. In sum, although more long-term research is needed to elucidate the ultimate causes of white-faced saki social behavior, we advocate that the infanticide model should remain under serious consideration.

References

Altmann, J. (1974). Observational study of behavior: sampling methods. *Behaviour*, **49**, 227–265.

Bradley, B.J., Robbins, M.M., Williamson, E.A., *et al.* (2005). Mountain gorilla tug-of-war: silverbacks have limited control over reproduction in multimale groups. *Proceedings of the National Academy of Sciences*, **102**(26), 9418–9423.

Buchanan, D.B. (1978). Communication and ecology of pithecine monkeys with special reference to *Pithecia pithecia*. *Dissertation Abstracts International*, **B39**(3), 1159–1160.

Buchanan, D.B., Mittermeier, R.A. & van Roosmalen, M.G.M. (1981). The saki monkeys, genus *Pithecia*. In *Ecology and Behavior of Neotropical Primates*, ed. A.F. Coimbra-Filho & R.A. Mittermeier. Rio de Janeiro: Academia Brasileira de Ciências, pp. 391–417.

Clutton-Brock, T.H. (1988). *Reproductive Success: Studies of Individual Variation in Contrasting Breeding Systems*. Chicago, IL University of Chicago Press.

Di Fiore, A., Fernandez-Duque, E. & Hurst, D. (2007). Adult male replacement in socially monogamous equatorial saki monkeys (*Pithecia aequatorialis*). *Folia Primatologica*, **78**, 88–89.

Fuentes, A. (1999). Re-evaluating primate monogamy. *American Anthropologist*, **100**(4), 890–907.

Fuentes, A. (2002). Patterns and trends in primate pair bonds. *International Journal of Primatology*, **23**(5), 953–978.

Gleason, T.M. & Norconk, M.A. (1995). Intragroup spacing and agonistic interactions in white-faced sakis. *American Journal of Primatology*, **36**, 125.

Harcourt, A.H. (1979). Social relationships among adult female mountain gorillas. *Animal Behavior*, **27**, 251–264.

Hinde, R.A. & Atkinson, S. (1970). Assessing the roles of social partners in maintaining mutual proximity, as exemplified by mother–infant relations in rhesus monkeys. *Animal Behaviour*, **18**, 169–176.

Kinzey, W.G. (1997). *New World Primates: Ecology, Evolution, and Behavior*. New York, NY: Aldine de Gruyter.

Lehman, S.M., Prince, W. & Mayor, M. (2001). Variations in group size in white-faced sakis (*Pithecia pithecia*):

evidence for monogamy or seasonal congregations? *Neotropical Primates*, **9**(3), 96–101.

Martin, P. & Bateson, P. (1986). *Measuring Behavior: An Introductory Guide*, 2nd edn. Cambridge: Cambridge University Press.

Norconk, M.A. (2006). Long-term study of group dynamics and female reproduction in Venezuelan *Pithecia pithecia*. *International Journal of Primatology*, **27**(3), 653–674.

Norconk, M.A. (2011). Sakis, uakaris, and titi monkeys. In *Primates in Perspective*, eds. C.J. Campbell, A. Fuentes, K.C. MacKinnon, S.K. Bearder & R.M. Stumpf. New York, NY: Oxford University Press, pp. 122–139.

Norconk, M.A., Raghanti, M.A., Martin, S.K., *et al.* (2003). Primates of Brownsberg Natuurpark, Suriname, with particular attention to the pitheciins. *Neotropical Primates*, **11**, 94–100.

Oliveira, J.M.S., Lima, M.G., Bonvincin, C., *et al.* (1985). Preliminary notes on the ecology and behavior of the Guianan saki (*Pithecia pithecia*, Linnaeus 1766; Cebidae, Primate). *Acta Amazonica*, **15**(1–2), 249–263.

Palombit, R.A. (1999). Infanticide and the evolution of pair bonds in nonhuman primates. *Evolutionary Anthropology: Issues, News, and Reviews*, **7**(4), 117.

Pope, T.R. (1990). The reproductive consequences of male cooperation in the red howler monkey: paternity exclusion in multi-male and single-male troops using genetic markers. *Behavioral Ecology and Sociobiology*, **27**(6), 439–446.

Robbins, M.M. (1995). A demographic analysis of male life history and social structure of mountain gorillas. *Behaviour*, **132**(1–2), 21–47.

Robbins, M.M. (2007). Gorillas: diversity in ecology and behavior. In *Primates in Perspective*, ed. C.J. Campbell, A. Fuentes, K.C. MacKinnon, M. Panger & S.K. Bearder. New York, NY: Oxford University Press, pp. 305–320.

Rosenberger, A.L., Norconk, M.A. & Garber, P.A. (1996) New perspectives on the pitheciines. In *Adaptive Radiations of Neotropical Primates*, ed. M.A. Norconk, A.L. Rosenberger & P.A. Garber. New York, NY: Plenum Press, pp. 427–432.

Setz, E.Z.F. & Gaspar, D.A. (1997). Scent-marking behaviour in free-ranging golden-faced saki monkeys, *Pithecia pithecia chyrysocephala*: sex differences and context. *Journal of Zoology, London*, **241**, 603–611.

Steenbeek, R. (1999). Tenure related changes in wild Thomas's langurs: I: between-group interactions. *Behaviour*, **136**(5), 595–625.

Steenbeek, R., Sterck, E.H.M., De Vries, H., *et al.* (2000). Costs and benefits of the one-male, age-graded, and all male phases in wild Thomas's langur groups. In *Primate Males: Causes and Consequences of Variation in Group Composition*, ed. P.M. Kappeler. Cambridge: Cambridge University Press, pp. 130–145.

Sterck, E.H.M., Watts, D.P. & van Schaik, C.P. (1997). The evolution of female social relationships in nonhuman primates. *Behavioral Ecology and Sociobiology*, **41**, 291–309.

Stewart, K.G. & Harcourt, A.H. (1987). Gorillas: variation in female relationships. In *Primate Societies*, ed. B.B. Smuts, D.L. Cheney, R.M. Seyfarth, R.W. Wrangham & T.T. Struhsaker. Chicago, IL: University of Chicago Press, pp. 155–164.

Stokes, E.J. (2004). Within-group social relationships among females and adult males in wild western lowland gorillas (*Gorilla gorilla gorilla*). *American Journal of Primatology*, **64**(2), 233–246.

Thompson, C.L. (2006). Sex biases in inter- and intra-group social behavior of wild white-faced saki monkeys (*Pithecia pithecia*). Unpublished Master's thesis, Kent State University.

Thompson, C.L. & Norconk, M.A. (2011). Within-group social bonds in white-faced saki monkeys (*Pithecia pithecia*) display male–female pair preference. *American Journal of Primatology*, **73**, 1051–1061.

Thompson, C.L., Norconk, M.A. & Whitten, P.L. (2012). Why fight? Selective forces favoring between-group aggression in a variably pair-living primate, the white-faced saki (*Pithecia pithecia*). *Behaviour*, **149**, 795–820.

van Schaik, C.P. (1989). The ecology of social relationships amongst female primates. In *Comparative Socioecology: The Behavioural Ecology of Humans and Other Mammals*, ed. V. Standen & R.A. Foley. Oxford: Blackwell Scientific Publications, pp. 195–218.

van Schaik, C.P. (1996). Social evolution in primates: the role of ecological factors and male behaviour. *Proceedings of the British Academy*, **88**, 9–31.

van Schaik, C.P. (2000). Social counterstrategies against infanticide by males in primates and other mammals. In *Primate Males: Causes and Consequences of Variation in Group Composition*, ed. P.M. Kappeler. Cambridge: Cambridge University Press, pp. 34–52.

van Schaik, C.P. & Dunbar, R.I.M. (1990). The evolution of monogamy in large primates: a new hypothesis and some crucial tests. *Behaviour*, **115**(1–2), 30–62.

van Schaik, C.P. & van Hooff, J.A.R.A.M. (1983). On the ultimate causes of primate social systems. *Behaviour*, **85**, 91–117.

Vié, J., Richard-Hansen, C. & Fournier-Chambrillon, C. (2001). Abundance, use of space, and activity patterns of white-faced sakis (*Pithecia pithecia*) in French Guiana. *American Journal of Primatology*, **55**(4), 203–221.

Watts, D.P. (1989). Infanticide in mountain gorillas: new cases and a reconsideration of the evidence. *Ethology*, **81**, 1–18.

Watts, D.P. (2000). Causes and consequences of variation in male mountain gorilla life histories and group membership. In *Primate Males: Causes and Consequences of Variation in Group Composition*, ed. P.M. Kappeler. Cambridge: Cambridge University Press, pp. 169–180.

Watts, D.P. (2001). Social relationships of female mountain gorillas. In *Mountain Gorillas: Three Decades of Research at Karisoke*, ed. M.M. Robbins, P. Sicotte & K.J. Stewart. Cambridge: Cambridge University Press, pp. 215–240.

Wittenberger, J.F. & Tilson, R.L. (1980). The evolution of monogamy: hypotheses and evidence. *Annual Review of Ecology and Systematics*, **11**, 197–232.

Wrangham, R.W. (1980). An ecological model of female-bonded primate groups. *Behaviour*, **75**, 262–299.

Chapter 28

Comparative socioecology of sympatric, free-ranging white-faced and bearded saki monkeys in Suriname: preliminary data

L. Tremaine Gregory & Marilyn A. Norconk

Introduction

Bearded and white-faced sakis (*Chiropotes* spp. and *Pithecia pithecia*) are distributed across northern Amazonia and the Guianas, and are sympatric throughout much of this region (Fleagle & Mittermeier 1980; Norconk *et al.* 1996; Gilbert & Setz 2001; Lehman 2004). Although some seed dispersal occurs (Norconk *et al.* 1998), these species are known to be seed predators (Kinzey 1992; Kinzey & Norconk 1990; Norconk 1996, 2011; Norconk *et al.*, Chapter 6) and show significant similarities in dentition (Kinzey 1992; Teaford & Runestad 1992; Martin *et al.* 2003), but there have been few studies to determine specific mechanisms that enable what appear to be near dental equivalents to exist in sympatry.

The most striking dietary similarity among all pitheciines – their preference for young and mature seeds – can be paired with an equally unusual dental complex that enables them to gain access to seeds even if well protected by tough or brittle pericarps (Kinzey 1992; Rosenberger *et al.* 1996; Norconk *et al.* 2009). Broader ecological similarities between bearded sakis and uacaris, and smaller-bodied *Pithecia pithecia* largely end with feeding adaptations. For example, Mittermeier (1977) found that bearded sakis and white-faced sakis in Suriname were stratified vertically, with bearded sakis moving through upper levels of the forest and white-faced sakis occupying middle to lower levels. Similarly, Oliveira *et al.* (1985) observed differences in the way that bearded and white-faced sakis use vertical space, and they suggested

Photo 28.1 Adult white-faced saki female feeding in Brownsberg Nature Park, Suriname. Photo: Marilyn Norconk. (See color plate section.)

Evolutionary Biology and Conservation of Titis, Sakis and Uacaris, eds. Liza M. Veiga, Adrian A. Barnett, Stephen F. Ferrari and Marilyn A. Norconk. Published by Cambridge University Press.

Photo 28.2 Adult white-faced saki male feeding in Brownsberg Nature Park, Suriname. Photo: Marilyn Norconk. (See color plate section.)

that spatial use was not the only difference between the species – they believed the species also differed in body size, social organization, habitat preferences and locomotion.

In addition to body mass differences and preferences for lower forest levels during travel, white-faced sakis are sexually dichromatic, live in smaller, more cohesive groups, have smaller home ranges, are territorial, and use vertical clinging and leaping locomotion within the lower levels of the forest (Table 28.1). While multimale–multifemale bearded saki groups can exceed 40 individuals (Norconk *et al.* 2003), white-faced saki groups have been described as "monogamous" (Robinson *et al.* 1987) or as small family units (Fuentes 1999), consisting of one breeding male (Lehman *et al.* 2001; Norconk 2011; also see Thompson & Norconk, Chapter 27).

In spite of these studies, knowledge of pitheciine behavior and ecology has lagged behind that of other platyrrhines. Long-term data are lacking particularly with regard to intra- and extra-group social dynamics, and transfer or dispersal of individuals. We present the results of a preliminary study in Suriname that focuses on the ecology and social behavior of sympatric *Chiropotes sagulatus* and *Pithecia pithecia.* We then introduce a descriptive model to compare the two species and provide some insight into where we think our knowledge of these species is lacking. This explanatory model was designed to integrate behavior and ecology with possible anatomical correlates and characterize "adaptive suites" (*sensu* Bartholomew 1972; Pianka 1994). Thus, we extend our descriptive field study to illustrate how different lines of research may inform us about the evolutionary strategies of pitheciines.

Photo 28.3 *Chiropotes sagulatus* male feeding, Brownsberg Nature Park, Suriname. Photo: Tremaine Gregory.

Table 28.1 Comparison of morphological and socioecological characteristics of white-faced and bearded sakis.

Trait	White-faced sakis	Bearded sakis	Source
Dental morphology	Procumbent incisors; massive, laterally projecting canines; little molar cusp relief; crenulated enamel; thin enamel with decussating prisms	Same	Ayres 1989; Hershkovitz 1985, 1987; Kinzey 1992; Martin *et al.* 2003
Body mass (size dimorphism: ♂ : ♀)	♂: 1900 g ♀: 1515 g (Minimal: 1.25)	♂: 3100 g ♀: 2600 g (Minimal: 1.19)	Ford 1994; Glander & Norconk, unpubl.
Coloration and sexual dichromatism	♂: Black pelage with white facial "mask"; ♀: Gray/brown pelage with white or ochre facial (malar) stripes. See Photos 28.1 and 28.2	♂ and ♀: Black with brown to rufous-colored back; ♂♂ with larger facial features and pendulous testes	Hershkovitz 1979, 1985, 1987
Group size[a]	Throughout their range: 2–12; BNP[b]: 5.7 (3–8, $n = 10$)	Brazil: 22.89 ± 5.09; BNP: 37.4 (22–44, $n = 3$)	Boyle 2008; Lehman *et al.* 2001; Norconk *et al.* 2003; Oliveira *et al.* 1985
Subgrouping[a]	None	Extensive	Ayres 1981, 1986; Kinzey & Cunningham 1994; Norconk & Kinzey 1994; Veiga *et al.* 2006
Day range; home range	1.9 km; 68–148 ha	3.2 km; 200–559 ha	Boyle 2008; Norconk & Kinzey 1994; van Roosmalen *et al.* 1981; Vié *et al.* 2001
Locomotion	Mostly VCL[c], some quadrupedal walking	Mostly quadrupedal walking, some generalized leaping	Fleagle & Mittermeier 1980; Walker 2005
Territorial defense	Frequent intergroup encounters	None observed	Peetz 2001; Thompson 2006; Veiga *et al.* 2005
Percent fruit in diet; preferred plant families[a]	84.3–93.3%; Connaraceae, Erthroxylaceae, Rubiaceae, Chrysobalanaceae, Lecythidaceae, Loganiaceae, Leguminosae	72.6–97.1%; Sapotaceae, Lecythidaceae, Euphorbiacae, Burseraceae, Chrysobalanaceae, Moraceae, Leguminosae, Loranthaceae, Meliaceae, Simaroubaceae	Ayres 1981; Boyle 2008; Homburg 1997; Kinzey & Norconk 1990; Mittermeier 1977; Norconk 1996; Norconk & Conklin-Brittain 2004; Peetz 2001; van Roosmalen *et al.* 1981; Veiga 2006
Tree level preferred[a]	Low–mid canopy	Upper canopy	Fleagle & Mittermeier 1980; Mittermeier 1977

Notes: [a] Also investigated in the present study. [b] BNP = Brownsberg Nature Park, Suriname. [c] VCL = vertical clinging and leaping.

Methods

Situated in the Brownsberg Mountain Range of the Guiana Shield of northeastern South America, Brownsberg Nature Park (5°01′N, 55°34′W) encompasses 8418 ha (Fitzgerald *et al.* 2002). Suriname's only nature park, the Brownsberg consists of a 530-m high lateritic plateau and forested slopes (Fitzgerald *et al.* 2002). The range in elevation allows for substantial floral and faunal diversity in the park, including eight primates (*Alouatta seniculus*, *Cebus apella*, *C. olivaceus*, *Ateles paniscus*, *Saguinus midas*, *Saimiri sciureus* and the two saki species; de Dijn *et al.* 2007; Fitzgerald *et al.* 2002; Norconk *et al.* 2003).

Data were collected (by LTG) from May to August 2005 and from May to July 2006, the long wet season and early portion of the long dry season, on two well-habituated groups of white-faced sakis (WFS) and multiple groups of bearded sakis (BDS). Because bearded sakis are known to have very large home ranges (Ayres 1981; Norconk 2011; Pinto 2008; Veiga 2006), home ranges overlap with other groups (Ayres 1989) and bearded saki groups are known to fragment (Veiga *et al.* 2006), data were collected whenever the monkeys could be located.

White-faced saki group size was stable within each study period and changed as a result of three births and the loss of one adult in the second year. Group size ranged from 5 to 6 in Group 1 and 4 to 5 in Group 2 over the duration of the study. Bearded saki group size and composition was estimated every 10 min during sampling periods, but their height in the forest precluded accurate and complete group counts.

One-minute group scans were taken on all individuals that were visible every 10 min to document activity (traveling, feeding, or resting), forest height/level used, group size and composition (for BDS), and location of the group relative to the nearest trail marker. Because we did not have access to reliable GPS signals at the time of this study, linear distance from the previous scan was estimated (in meters) using a Freestyle® digital pedometer. Data are reported in monkey minutes (i.e. frequency data were weighted by the total number of individuals visible in each scan) to reflect the possibility that individuals in a group were engaged in different activities.

We used a five-category scale instead of absolute tree heights to accommodate variation in the crown dimensions of different tree species (i.e. upper-level canopy users may travel at 30 m or 15 m depending on the height of the tree). This system allowed us to document forest use, not simply height.

To determine the distance traveled/hour, we summed six, 10-min pedometer readings recorded for each hour of the day. White-faced sakis were followed for a total of 10 days (286 scan samples) and bearded sakis were followed for 9 days (221 scan samples) in 2005.

All-occurrence feeding data (Altmann 1974) were taken to record the duration of feeding bouts (first individual in to last individual out of tree). Data were converted to total feeding min/fruit species for each sample period. The number of individuals feeding and the diameter at breast height (DBH) of feeding trees were also recorded. This was particularly important in the case of bearded sakis, as feeding bouts were brief, group members often fed in multiple trees, and travel took them to remote locations in the park. In 2005, we recorded 187 feeding bouts for white-faced sakis and 50 feeding bouts for bearded sakis; in 2006, 229 bouts for white-faced sakis and 23 for bearded sakis. Differences in sample sizes reflect the relative difficulty of finding and collecting data on bearded sakis compared with white-faced sakis.

Mann–Whitney U tests and associated natural estimators (Hollander & Wolfe 1973, p. 76) were used to test for interspecific differences in travel rates and to compare the DBH of feeding trees between the two sakis. We used a nonparametric analysis of variance (Kruskal–Wallis test) to analyze differences in forest levels used between species. Interspecific differences in activity patterns were tested using chi-square tests. As the null hypothesis for activity patterns was rejected, the strength of association was calculated using Cramer's V.

Niche overlap in the plant species used for feeding was calculated as a percentage using the following statistic from Krebs (1989):

$$P_{jk} = [\Sigma(\text{minimum } p_{ij}, p_{ik})]100$$

where p_{ij} is the proportion of the diet represented by the plant species for one monkey species (or one field season), and p_{ik} is the proportion of the diet represented by the plant species for the other monkey species.

Results

Interspecific differences in activity patterns, travel speed and habitat use

The smaller social groups of white-faced sakis were stable within the two sample periods, but the larger groups of bearded sakis were flexible, splitting into subgroups that ranged from 3 to 10 individuals both between and within sample periods. We also found significant differences in activity patterns, with bearded sakis resting less and traveling more than white-faced sakis ($\chi^2 = 118.6$, $p < 0.001$, Cramer's V = 0.24; WFS: $n = 553$, BDS: $n = 1495$ monkey minutes, Figure 28.1).

Bearded sakis travelled significantly faster than white-faced sakis: BDS = 75 ± 82 m/10 min, $n = 221$; WFS = 20 ± 32 m/10 min, $n = 286$, Mann–Whitney U, $z = 7.720$, $p < 0.001$), had higher variance in travel rate, and had a higher maximum distance travelled/10 min (Figure 28.2). Ten-minute samples used to estimate hourly travel distance yielded a similar trend (Figure 28.3): bearded sakis have a faster, more variable travel pattern (Mann–Whitney U, $z = 3.761$, $p > 0.001$, WFS: $n = 9$ intervals, BDS: $n = 11$ intervals). Furthermore, while white-faced

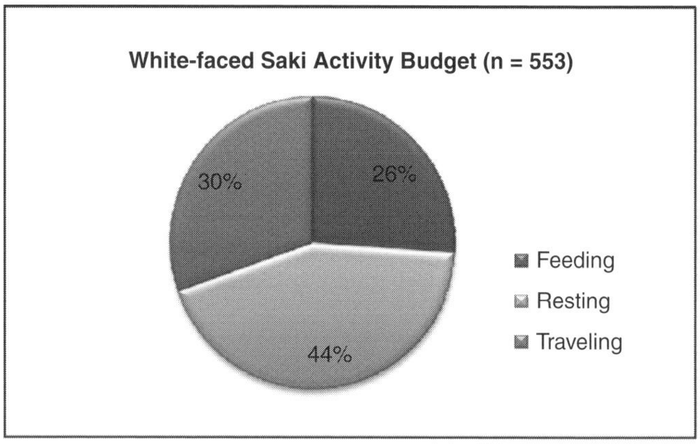

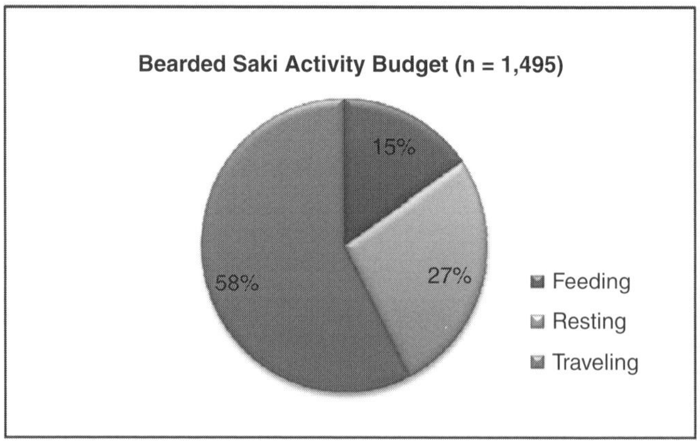

Figure 28.1 Activity budgets of white-faced sakis and bearded sakis. Sample sizes are in monkey minutes (see Methods).

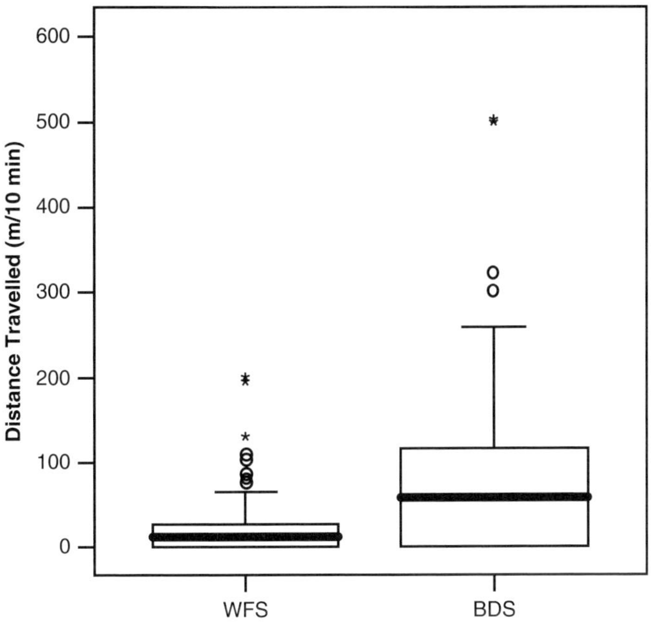

Figure 28.2 Box plot of travel rates (m/10 m) of white-faced sakis (WFS, $n = 286$) and bearded sakis (BDS, $n = 221$, $z = 7.720$, $p < 0.001$). Notes: box = 25th–75th percentile, * = extremes, O = outliers, tail = largest observed value (not an outlier), dark bar = mean.

sakis stopped traveling and were inactive after 16:00 h, bearded sakis continued to travel until at least 17:50 h.

White-faced sakis used a broader range of, and generally lower, forest levels (Kruskal–Wallace $H = 21.2$, $p < 0.001$, Figure 28.4) during any activity and used smaller trees than bearded sakis when feeding (Mann–Whitney U, $z = 6.292$, $p < 0.001$, Figure 28.5). White-faced sakis were also scored more frequently on the bole, while bearded sakis were recorded more frequently on terminal branches. This reflects the high frequency of vertical clinging and leaping locomotor behavior in white-faced sakis and upper crown use combined with leaping in bearded sakis (Walker 2005; Davis & Walker-Pacheco, Chapter 8).

Dietary overlap

A total of 51 species in 28 plant families were used as feeding sources for both species, with 30.8% overlap between the two years (Krebs $P_{2005,2006}$, 11 spp.). Interestingly, within-year plant species overlap between white-faced sakis and bearded sakis was quite low, < 20% (2005: $P_{WFS,BDS} = 19.0\%$, 4 spp. and 2006: $P_{WFS,BDS} = 15.8\%$, 3 spp.). During these wet-to-early-dry-season samples, the three most important plant families for

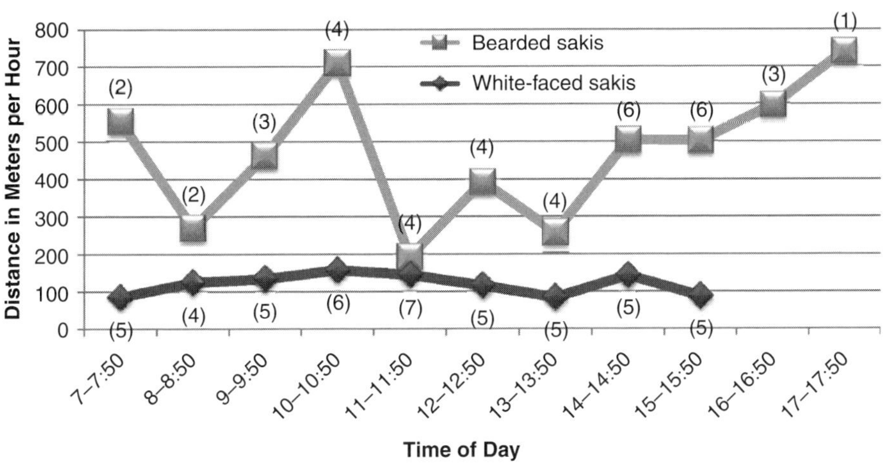

Figure 28.3 Travel distance per 60-min interval for white-faced and bearded sakis. Sample sizes for each hour are in parentheses.

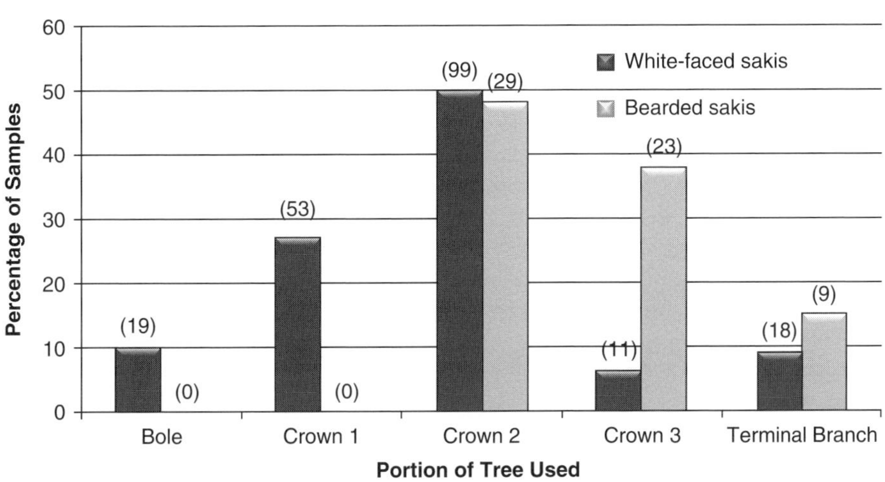

Figure 28.4 Comparison of tree use by the two saki species during travel, feeding, and resting activities. Instead of estimating height of feeding animals in a tree, trees were divided into functional areas: bole, three vertical portions of the crown (low, middle and high), and terminal branches, regardless of portion of crown used. Sample sizes for each area are indicated in parentheses.

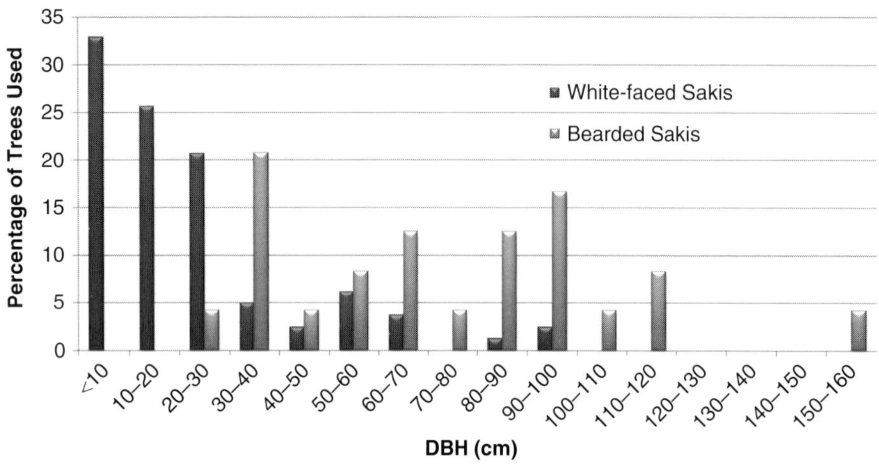

Figure 28.5 Proportion of trees used for feeding by the two saki species sorted by tree size (DBH) (white-faced sakis: $n = 82$ and bearded sakis: $n = 24$). Data are combined for two sample periods: 2005 and 2006. Bearded sakis appear to prefer large trees, 40% of which were > 90 cm DBH in this preliminary data set.

Table 28.2 Top-ranked feeding species used by white-faced sakis in 2005 and 2006 study periods. Monkey feeding minutes = the sum of the duration of each feeding bout multiplied by the number of monkeys feeding (h:min). Rank of a plant species is determined by the number of monkey feeding minutes. "Percent of total" indicates the percent of the total feeding minutes and bouts represented by the 10 most important plant species.

2005					2006				
Rank	**Plant species**	**Plant family**[a]	**Monkey feeding h:min**	**No. of bouts**	**Rank**	**Plant species**	**Plant family**	**Monkey feeding h:min**	**No. of bouts**
1	*Salacia cordata*	CELA	24:50	13	1	*Cheiloclinium cognatum*	HIPR	33:17	31
2	*Calycorectes cf grandifolius*	MYRT	12:55	11	2	*Talisia micrantha*	SAPI	28:39	14
3	*Vouacapoua americana*[b]	CAES	12:40	19	3	*Heteropsis flexuosa*	AR++	9:06	41
4	*Micranda brownsbergensis*	EUPH	7:17	4	4	*Salacia cordata*	CELA	8:50	14
5	*Pausandra martinii*	EUPH	6:55	32	5	*Pausandra martinii*	EUPH	6:33	60
6	*Heteropsis flexuosa*	AR++	6:22	41	6	*Micranda brownsbergensis*	EUPH	2:56	7
7	*Bellucia grosslarioides*	MELA	4:49	9	7	*Psiguria triphylla*	CUCU	2:47	9
8	*Inga* sp.	MIMO	4:20	11	8	*Rheedia benthamiana*	CLUS	2:40	8
9	*Eugenia* sp. 1	MYRT	1:54	2	9	*Tovomita cf grata*	CLUS	1:48	2
10	*Philodendron* sp. 1	AR++	1:44	7	10	*Inga* spp.	MIMO	1:30	9
Total feeding minutes and feeding bouts			84:37	149	Total feeding minutes and feeding bouts			98:19	199
Percent of total			89.4%	79.7%	Percent of total			92.8%	86.7%

Notes
Abbreviations of plant families: [a] APOC, Apocynaceae; AR++, Araceae; CAES, Caesalpinaceae; CELA, Celastraceae; CHRY, Chrysobalanaceae; CLUS, Clusiaceae; CUCU, Cucurbitaceae; EUPH, Euphorbiaceae; HIPR, Hippocrataceae; LECY, Lecythidaceae; MELA, Melastomataceae; MIMO, Mimosaceae; MYRT, Myrtaceae; OLAC, Olacaceae; SAPI, Sapindaceae; SAPO, Sapotaceae; VOCH, Vochysiaceae.
[b] Plant species consumed by both saki species.

white-faced sakis were Celastraceae, Hippocrateaceae, and Sapindaceae, representing 43% of feeding minutes (Table 28.2). A largely different set of resources from the Vochysiaceae, Olacaceae, and Caesalpiniaceae families represented 33% of bearded saki feeding minutes (Table 28.3). The top 10 plant species represented more than 85% of total feeding minutes for both primates. Euphorbiaceae species of *Pausandra martini* and *Micranda brownsbergensis*, *Inga* spp. (Mimosaceae), and *Salacia cordata* (Celastraceae) may provide reliable interannual feeding sources for white-faced sakis. Only *S. cordata* was used between years for bearded sakis.

Discussion

Consistent with predictions from the literature, we found clear differences between white-faced and bearded sakis in social and ecological characteristics (Table 28.1). White-faced saki

Table 28.3 Top-ranked feeding species used by bearded sakis in 2005 and 2006 study periods. (See Table 28.2 for explanation of data cited.)

2005					2006				
Rank	**Plant species**	**Plant family**[a]	**Monkey feeding h:min**	**No. of bouts**	**Rank**	**Plant species**	**Plant family**	**Monkey feeding h:min**	**No. of bouts**
1	*Minquartia guianensis*	OLAC	6:25	3	1	*Ruizterania albiflora*	VOCH	8:06	3
2	*Vouacapoua americana*[b]	CAES	5:42	11	2	*Pouteria guianensis*	SAPO	4:53	6
3	Unknown 4		4:48	1	3	*Parahancornia fasciculata*	APOC	3:14	4
4	*Salacia cordata*[b]	CELA	4:47	7	4	*Pouteria* spp.	SAPO	2:11	5
5	*Tontelea coriacea*	HIPR	4:05	4	5	*Licania* sp. 1	CHRY	0:45	1
6	Unknown 2		3:54	1	6	*Chrysophyllum eximium*	SAPO	0:32	1
7	*Qualea coerulea*	VOCH	3:50	3	7	*Qualea rosea*	VOCH	0:21	1
8	Unknown 1		0:54	1	8	*Hevea guianensis*	EUPH	0:15	1
9	*Couratari stellata*	LECY	0:53	3	9	*Inga cf capitata*	MIMO	0:12	1
10	Unknown 14	EUPH	0:44	1	10	*Salacia cordata*[b]	CELA	0:05	1
Total feeding minutes and feeding bouts			36:02	35	Total feeding minutes and feeding bouts			20:29	23
Percent of total			89.9%	70%	Percent of total			100%	100%

Notes:
Abbreviations of plant families: [a] APOC, Apocynaceae; AR++, Araceae; CAES, Caesalpinaceae; CELA, Celastraceae; CHRY, Chrysobalanaceae; CLUS, Clusiaceae; CUCU, Cucurbitaceae; EUPH, Euphorbiaceae; HIPR, Hippocrataceae; LECY, Lecythidaceae; MELA, Melastomataceae; MIMO, Mimosaceae; MYRT, Myrtaceae; OLAC, Olacaceae; SAPI, Sapindaceae; SAPO, Sapotaceae; VOCH, Vochysiaceae.
[b] Plant species consumed by both saki species.

Photo 28.4 Adult female white-faced saki (*Pithecia pithecia*) grooming younger female. Infant male in foreground. Brownsberg Nature Park, Suriname. Photo: Marilyn Norconk. (See color plate section.)

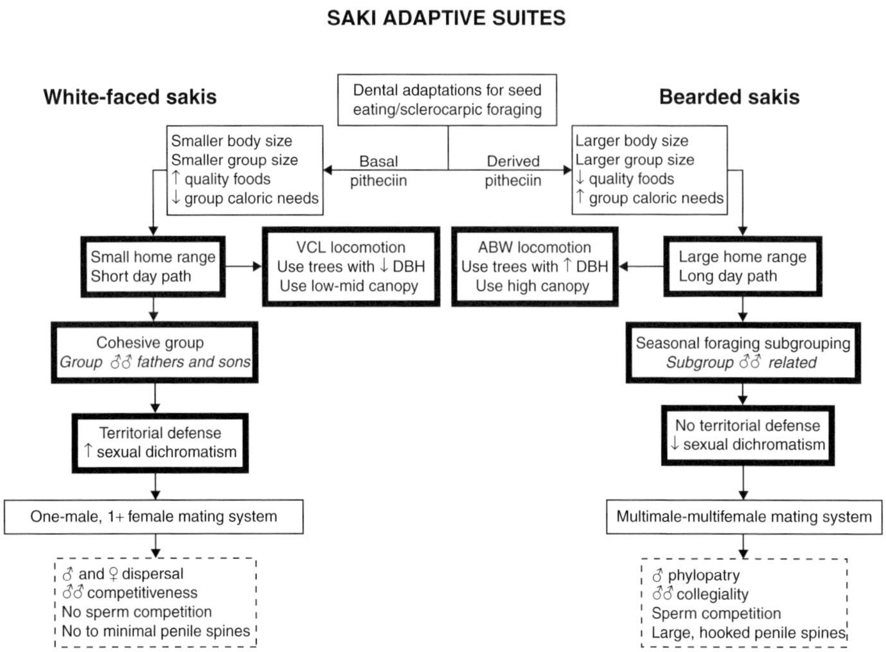

Figure 28.6 We drew on data from the literature, previous experience and from this preliminary study to construct paired adaptive suites of behaviors exhibited by white-faced and bearded sakis. Boxes with a bold outline indicate traits that were investigated in this study. Inferences are indicated by boxes with dashed lines or italicised words. Beginning with their common dental morphology, the two species subsequently diverge in many aspects of social, ecological, and perhaps reproductive behaviors. See Discussion.

groups are small and cohesive, whereas bearded sakis have large groups that fission into subgroups. Bearded sakis travel faster and farther than white-faced sakis, and this rapid pace influences their activity patterns – bearded sakis spend more time traveling than feeding or resting. In contrast, white-faced sakis spent an average of 44% of their active day resting.

Preference for feeding tree size (DBH) by bearded sakis was tested by comparing our data with a random sample of 2847 trees measured by Hans ter Steege and colleagues at Brownsberg (pers. commun.). We found that bearded sakis have a preference for rarer, larger trees (normal approximation of Mann–Whitney U, $z = 7.432$, $p < 0.001$, $\mathrm{pr}(x > y) = 94\%$). Trees smaller than 10 cm DBH were not measured or counted in the ter Steege sample, and therefore cannot be compared with trees used by white-faced sakis. However, it is clear that white-faced and bearded sakis often feed in trees of different sizes.

Ayres' (1986) assertion that *Chiropotes* spp. and *Pithecia* spp. have divergent diets seems to hold true in Brownsberg (Figure 28.5). Despite similarities in dental morphology, we found very low overlap in feeding species. In fact, we found higher overlap between sympatric brown capuchins (BC) and both saki species in the 2005 sample ($P_{WFS,BC} = 33.0\%$, 7 spp. and $P_{BDS,BC} = 21.1\%$, 4 spp.) than between the two saki species (Gregory 2006). While both saki species share the ability to open young, well-protected fruit and masticate seeds (Kinzey 1992), perhaps differences in other ecological factors as well as different body sizes have contributed to dietary niche diversification (Oliveira *et al.* 1985; Rosenberger 1992).

While highly derived dental traits characterize both pitheciines in this study (Kinzey 1992; Martin *et al.* 2003; Norconk *et al.* 2009; Rosenberger 1992; Rosenberger *et al.* 1996; Teaford & Runestad 1992; Figure 28.6), divergent pathways have apparently led to differences in ecology and behavior between bearded and white-faced sakis. Smaller body size, group size, home-range size, as well as shorter daily path lengths may have facilitated increased group cohesion for white-faced sakis compared with bearded sakis. We propose that territorial defense and male–male intergroup aggression expressed in white-faced sakis is enhanced by sexual dichromatism with the white faces and black body pelage of males providing a striking contrast to the cryptic coloration of females. Bearded sakis appear to lack territorial behavior in the form of aggression towards extra-group males (Peetz 2001; Veiga *et al.* 2005; Gregory 2011). Within groups, adult male bearded sakis are not only tolerant of each another, but actively seek other males as social partners (Peetz 2001; Veiga *et al.* 2005; Gregory 2011).

As in other multimale primate species (e.g. muriquis: Strier 1997), bearded sakis and uacaris have developed affiliative mechanisms among males that may reduce within-group male–male competition (Bowler & Bodmer 2009; Peetz 2001; Veiga *et al.* 2005; Gregory 2011). This scenario would facilitate male philopatry and male–male tolerance, but, as has been suggested for muriquis and chimpanzees, a more subtle form of competition in the form of sperm competition may have evolved in bearded sakis (muriquis: Milton 1985; chimpanzees: Harcourt 1995). Males could reap the benefits of aggregating with kin (Hamilton 1964), and yet compete covertly. Based on a study of pitheciine genital morphology, particularly the presence of penile spines, Hershkovitz (1993) predicted that sperm competition would be found in both *Cacajao* and *Chiropotes*, yet be less likely in *Pithecia*. Although socioecological differences between white-faced and bearded sakis are becoming clearer, future studies should focus on understanding bearded saki group composition, male–male relations, dispersal patterns and mating patterns, as well as an understanding of the effects of ecological conditions on the social behavior of both species.

Acknowledgments

We would like to thank STINASU (Foundation for Nature Conservation in Suriname) for permission to conduct the study. We are also grateful to Bart de Dijn for providing maps of the site and for the field assistance of Cyndie Thompson, Terry Gleason, and Ari Vreedzaam. Funding to LTG was provided by the Graduate Student Senate, Kent State University.

References

Altmann, J. (1974). Observational study of behavior: sampling methods. *Behaviour*, **49**, 227–267.

Ayres, J.M. (1981). Observações sobre a ecologia e o comportamento dos cuxiús (*Chiropotes albinasus* e *Chiropotes satanas*, Cebidae: Primates). Unpublished MA thesis, Fundação Universidade do Amazonas.

Ayres, J.M. (1986). Uakaris and Amazonian flooded forest. Unpublished PhD thesis, Cambridge, UK.

Ayres, J.M. (1989). Comparative feeding ecology of the uakari and bearded saki, *Cacajao* and *Chiropotes*. *Journal of Human Evolution*, **18**, 697–716.

Bartholomew, G.A. (1972). Body temperature and energy metabolism. In *Animal Physiology: Principles and Adaptations*, ed. M.S. Gordon. New York, NY: Macmillan, pp. 364–449.

Bowler, M. & Bodmer, R. (2009). Social behavior in fission–fusion groups of red uakari monkeys (*Cacajao calvus ucayalii*). *American Journal of Primatology*, **71**, 976–987.

Boyle, S.A. (2008). The effects of forest fragmentation on primates in the Brazilian Amazon. PhD Dissertation, Arizona State University, Tempe.

de Dijn, B.P.E., Molgo, I., *et al.*(2007). The biodiversity of the Brownsberg. In *A Rapid Biological Assessment of the Lely and Nassau Plateaus, Suriname (with Additional Information on the Brownsberg Plateau). RAP Bulletin of Biological Assessment 43*, ed. L.E. Alonso & J.H. Mol. Arlington, VA: Conservation International.

Fitzgerald, K.A., de Dijn, B.P.E., and Mitro, S. (2002). *Brownsberg Nature Park: Ecological Research and Monitoring Program 2001–2006*. Paramaribo, Suriname: STINASU, Foundation for Nature Conservation in Suriname.

Fleagle, J.G. & Mittermeier, R.A. (1980). Locomotor behavior, body size, and comparative ecology of seven Surinam monkeys. *American Journal of Physical Anthropology*, **52**, 301–314.

Ford, S.M. (1994). Evolution of sexual dimorphism in body weight in platyrrhines. *American Journal of Primatology*, **34**, 221–244.

Fuentes, A. (1999). Re-evaluating primate monogamy. *American Anthropologist*, **100**, 890–907.

Gilbert, K.A. & Setz, E.Z.F. (2001). Primates in a fragmented landscape: Six species in a fragmented landscape. In *Lessons from Amazonia: The Ecology and Conservation of a Fragmented Forest*, ed. R.O. Bierregaard, Jr., C. Gascon, T.E. Lovejoy & R.C.G. Mesquita. New Haven, CT: Yale University Press, pp. 262–270.

Gregory, T.L. (2006). Comparative socioecology of white-faced and bearded saki monkeys. Unpublished MA thesis, Kent State University, USA.

Gregory, T.L. (2011) Socioecology of the Guianan bearded saki, *Chiropotes sagulatus*. Unpublished Doctoral dissertation, Kent State University, USA.

Hamilton, W.D. (1964). The genetical evolution of social behavior. *Journal of Theoretical Biology*, **122**, 95–121.

Harcourt, A.H. (1995). Sexual selection and sperm competition in primates: what are male genitalia good for? *Evolutionary Anthropology*, **4**, 121–129.

Hershkovitz, P. (1979). The species of sakis, genus *Pithecia* (Cebidae, Primates), with notes on sexual dichromatism. *Folia Primatologica*, **31**, 1–22.

Hershkovitz, P. (1985). A preliminary taxonomic review of the South American bearded saki monkeys genus *Chiropotes* (Cebidae, Platyrrhini) with the description of a new subspecies. *Fieldiana: Zoology*, **27**(1363), 1–46.

Hershkovitz, P. (1987). The taxonomy of South American sakis, genus *Pithecia* (Cebidae, Platyrrhini): a preliminary report and critical review with the description of a new species and subspecies. *American Journal of Primatology*, **12**, 387–468.

Hershkovitz, P. (1993). Male external genitalia of non-prehensile tailed South American monkeys. Part 1: Subfamily Pitheciinae, Family Cebidae. *Fieldiana: Zoology New Series*, **73**, 1–17.

Hollander, M. & Wolfe, D. (1973). *Nonparametric Statistical Methods*. New York, NY: John Wiley & Sons, Inc.

Homburg, I. (1997). *Ökologie and sozialverhalten einer gruppe von weissgesicht-sakis (Pithecia pithecia pithecia Linnaeus 1766) im Estado Bolívar*, Venezuela. Universität Bielefeld, Germany.

Kinzey, W.G. (1992). Dietary and dental adaptations in the Pitheciinae. *American Journal of Physical Anthropology*, **88**, 499–514.

Kinzey, W.G. & Cunningham, E.P. (1994). Variability in platyrrhine social organization. *American Journal of Physical Anthropology*, **34**, 185–198.

Kinzey, W.G. & Norconk, M.A. (1990). Hardness as a basis of fruit choice in two sympatric primates. *American Journal of Physical Anthropology*, **81**, 5–15.

Krebs, C.J. (1989). *Ecological Methodology*. New York, NY: Harper and Row.

Lehman, S.M. (2004). Distribution and diversity of primates in Guyana: species–area relationships and riverine barriers. *International Journal of Primatology*, **25**, 73–95.

Lehman, S.M., Prince, W. & Mayor, M. (2001). Variations in group size in white-faced sakis (*Pithecia pithecia*): evidence for monogamy or seasonal aggregations? *Neotropical Primates*, **9**, 96–101.

Martin, L.B., Olejniczak, A.J. & Maas, M.C. (2003). Enamel thickness and microstructure in pitheciin primates, with comments on dietary adaptations of the middle Miocene hominoid *Kenyapithecus*. *Journal of Human Evolution*, **45**, 351–367.

Milton, K. (1985). Mating patterns of wooly spider monkeys *Brachyteles arachnoides*: implications for female choice. *Behavioral Ecology and Sociobiology*, **17**, 53–59.

Mittermeier, R.A. (1977). Distribution, synecology, and conservation of Suriname monkeys. Unpublished PhD thesis, Harvard University.

Norconk, M.A. (1996). Seasonal variation in the diets of white-faced and bearded sakis (*Pithecia pithecia* and *Chiropotes satanas*) in Guri Lake, Venezuela. In *Adaptive Radiations of Neotropical Primates*, ed. M.A. Norconk, A.L. Rosenberger & P.A. Garber. New York, NY: Plenum Press, pp. 403–423.

Norconk, M.A. (2011). Sakis, uakaris, and titi monkeys: behavioral diversity in a radiation of primate seed predators. In *Primates in Perspective*, 2nd edn, ed. C.J. Campbell, A. Fuentes, K.C. MacKinnon, S.K. Bearder, & R.M Stumpf. New York, NY: Oxford University Press, pp. 122–139.

Norconk, M.A. & Conklin-Brittain, N.L. (2004). Variation on frugivory: the diet of Venezuelan white-faced sakis. *International Journal of Primatology*, **25**, 1–26.

Norconk, M.A. & Kinzey, W.G. (1994). Challenge of neotropical frugivory: travel patterns of spider monkeys and bearded sakis. *American Journal of Primatology*, **34**, 171–183.

Norconk, M.A., Grafton, B.W. & Conklin-Brittain, N.L. (1998). Seed dispersal by neotropical seed predators. *American Journal of Primatology*, **45**, 103–126.

Norconk, M.A., Raghanti, M.A., Martin, S.K., *et al.* (2003). Primates of Brownsberg Natuurpark, Suriname, with particular attention to the Pitheciins. *Neotropical Primates*, **11**, 94–99.

Norconk, M.A., Sussman, R.W. & Phillips-Conroy, J.E. (1996) Primates of Guayana Shield Forests. In *Adaptive Radiations of Neotropical Primates*, ed. M.A. Norconk, A.L. Rosenberger & P.A. Garber. New York, NY: Plenum, pp. 69–83.

Norconk, M.A., Wright, B.W., Conklin-Brittain, N.L., *et al.* (2009). Mechanical and nutritional properties of foods as factors in platyrrhine dietary adaptations. In *South American Primates: Testing New Theories in The Study of Primate Behavior, Ecology, and Conservation*, ed. P.A. Garber, A. Estrada, C. Bicca-Marques, E. Heymann & K. Strier. New York,NY: Springer Science, pp. 279–319.

Oliveira, J.M.S., Lima, M.G., Bonvincino, C., *et al.* (1985). Preliminary notes on the ecology and behavior of the Guianan saki (*Pithecia pithecia*, Linnaeus 1766; Cebidae, Primate). *Acta Amazonica*, **15**, 249–263.

Peetz, A. (2001). Ecology and social organization of the bearded saki *Chiropotes satanas chiropotes* (Primates: Pitheciinae) in Venezuela. *Ecotropical Monographs*, **1**, 1–170.

Pianka, E.R. (1994). *Evolutionary Ecology*. New York, NY: HarperCollins College Publishers.

Pinto, L.P. (2008). Ecologia alimentar do cuxiú-de-nariz-vermelho *Chiropotes albinasus* (Primates: Pitheciidae) na Floresta Nacional do Tapajós, Pará. PhD dissertation, Instituto de Biologia, Universidade Estadual de Campinas, Campinas, SP, Brazil.

Robinson, J.G., Wright, P.C. & Kinzey, W.G. (1987). Monogamous cebids and their relatives: intergroup calls and spacing. In *Primate Societies*, ed. B.B. Smuts, D.L. Cheney, R.M. Seyfarth, R.W. Wrangham & T.T. Struhsaker. Chicago, IL: University of Chicago Press, pp. 44–53.

Rosenberger, A.L. (1992). Evolution of feeding niches in New World monkeys. *American Journal of Physical Anthropology*, **88**, 525–562.

Rosenberger, A.L., Norconk, M.A. & Garber, P.A. (1996). New perspectives on the pitheciines. In *Adaptive Radiations of Neotropical Primates*, ed. M.A. Norconk, A.L. Rosenberger & P.A. Garber. New York, NY: Plenum Press, pp. 329–33.

Shaffer, C.A. (2012). Ranging behavior, group cohesiveness, and patch use in northern bearded sakis (*Chiropotes sagulatus*) in Guyana. Unpublished doctoral dissertation, Washington University.

Strier, K.B. (1997). Subtle cues of social relations in male muriqui monkeys. In *New World Primates: Ecology, Evolution, and Behavior*, ed. W.G. Kinzey. New York, NY: Aldine de Greyter, pp. 109–118.

Teaford, M.F. & Runestad, J.A. (1992). Dental microwear and diet in Venezuelan primates. *American Journal of Physical Anthropology*, **88**, 347–364.

Thompson, C. (2006). Sex biases in inter and intra-group social behavior of wild white-faced saki monkeys (*Pithecia pithecia*). Unpublished MA thesis, Kent State University.

van Roosmalen, M.G.M., Mittermeier, R.A. & Milton, K. (1981). The bearded sakis, genus *Chiropotes*. In *Ecology and Behavior of Neotropical Primates*, ed. A.F. Coimbra-Filho & R.A. Mittermeier. Rio de Janeiro: Academia Brasileira de Ciências, pp. 419–441.

Veiga, L.M. (2006). *Ecologia e comportamento do cuxiú-preto (Chiropotes satanas) na paisgem fragmentada de Amazônia oriental, Brasil.* Belem, Pará, Universidade Federal do Pará.

Veiga, L.M., Pinto, L.P. & Ferrari, S.F. (2006). *Fission–fusion sociality in bearded sakis* (Chiropotes albinasus and Chiropotes satanas) *in Brazilian Amazonia*. Proceedings of the XXIst Congress of the International Primatological Society, Uganda, p. 261.

Veiga, L.M., Silva, S.S.B. & Ferrari, S.F. (2005). Relatives or just good friends? Affiliative relationships among male southern bearded sakis (*Chiropotes satanas*). Livro de Resumos, XI Congresso Brasileiro de Primatologia, Porto Alegre, p. 174.

Vié, J.C., Richard-Hansen, C. & Fournier-Chambrillon, C. (2001). Abundance, use of space, and activity patterns of white-faced sakis (*Pithecia pithecia*) in French Guiana. *American Journal of Primatology*, **55**, 203–221.

Walker, S.E. (2005). Leaping behavior of *Pithecia pithecia* and *Chiropotes satanas* in eastern Venezuela. *American Journal of Primatology*, **66**, 369–387.

Chapter 29

Pair-mate relationships and parenting in equatorial saki monkeys (*Pithecia aequatorialis*) and red titi monkeys (*Callicebus discolor*) of Ecuador

Eduardo Fernandez-Duque, Anthony Di Fiore & Ana Gabriela de Luna

Introduction

Socially monogamous primates are often described as displaying a suite of behavioral characteristics that includes a high degree of affiliation and social tolerance among pair-mates with both individuals being responsible for maintaining their close spatial association (Kleiman 1977; Anzenberger 1988; Palombit 1996; Fernandez-Duque *et al.* 2000; Fuentes 2002; Reichard 2003).

Among the pitheciins, titis (*Callicebus* spp.) conform to this "classic" pattern of social monogamy. They are always encountered in small groups of 2–5 individuals, the nucleus of which is an adult male and an adult female (Kinzey 1981; Wright 1985; Robinson *et al.* 1987; Defler 2004; Carrillo *et al.* 2005; Norconk 2011). In wild groups, pair-mates typically stay within a few meters of each other during feeding, traveling and resting periods, and show coordinated activities (Mason 1966; Robinson 1979, 1981; Kinzey & Wright 1982; Wright 1985; Mendoza & Mason 1986a; Price & Piedade 2001). Based on the high degree of intimacy, coordination, interdependence between pair-mates and distress following separation, the existence of a strong and specific mutual attachment or "bond" is regularly inferred (Mason 1975; Mendoza & Mason 1986b; Anzenberger 1988; Fernandez-Duque *et al.* 1997).

Our understanding of pair-mate relationships and social behavior in sakis (*Pithecia* spp.) is far less clear than in titis because there have only been a handful of studies focused on identified and habituated individuals (Setz & Gaspar 1997; Norconk 2006; Di Fiore *et al.* 2007). Sakis have also been reported to live in small social groups that include a single breeding pair and several young. Although there have also been studies reporting larger groups (Norconk 2011), most of those groups were found in island habitats that limit the dispersal possibilities of individuals (Setz & Gaspar 1997; Vié *et al.* 2001; Norconk 2006), or during censuses of non-habituated individuals that limit the possibility of precise group identification (Lehman *et al.* 2001).

Coordinated displays and joint participation in territory defense are also thought to reflect the existence of a pair bond between the male and the female of a socially monogamous group (Robinson 1979, 1981; Mitani 1984; Raemaekers & Raemaekers 1985). Titis use ranges with relatively little overlap, and routinely perform behaviors at the borders of those ranges that include duetting and joint visual displays (Mason 1968; Robinson 1981; Wright 1985; Price & Piedade 2001; Dacier *et al.* in press). For sakis, on the other hand, the specific contributions of male and female pair-mates to intergroup interactions or territory defense remain unexamined. The limited data available on use of space by sakis in non-island habitats indicate that their ranges may be somewhat exclusive and defended (Vié *et al.* 2001; Norconk *et al.* 2003).

Finally, extensive male involvement in infant care is also commonly associated with social monogamy (Kleiman 1981; Kleiman & Malcolm 1981; Palombit 1999; Maestripieri 2002; Fernandez-Duque *et al.* 2009). Among the pitheciins, the male titi monkey carries the infant most of the time and also plays, grooms and shares food with the infant, suggesting that the infant may be primarily attached to the putative father rather than the mother (Fragaszy *et al.* 1982; Kinzey & Becker 1983; Wright 1984; Mendoza & Mason 1986b; Hoffman *et al.* 1995; Tirado-Herrera & Heymann 2004). In all of the other pitheciin genera, by contrast, direct paternal care is relatively absent. Male sakis do not routinely transport infants, although they may play and interact socially with older young (Norconk 2011).

The extent to which strong pair bonds characterize sakis like they do in titis has not been investigated. Nor have there been evaluations of paternal care in sakis that consider the services that males may provide to females and their offspring, such as anti-predator vigilance or territory defense. In the following paragraphs, we give a descriptive overview of the patterns of pair-mate relationships and paternal care seen in equatorial saki monkeys (*P. aequatorialis*) and red titi monkeys (*Callicebus discolor*), based on comparative data collected from two groups of titis and one group of sakis in western Amazonia.

Evolutionary Biology and Conservation of Titis, Sakis and Uacaris, eds. Liza M. Veiga, Adrian A. Barnett, Stephen F. Ferrari and Marilyn A. Norconk. Published by Cambridge University Press.

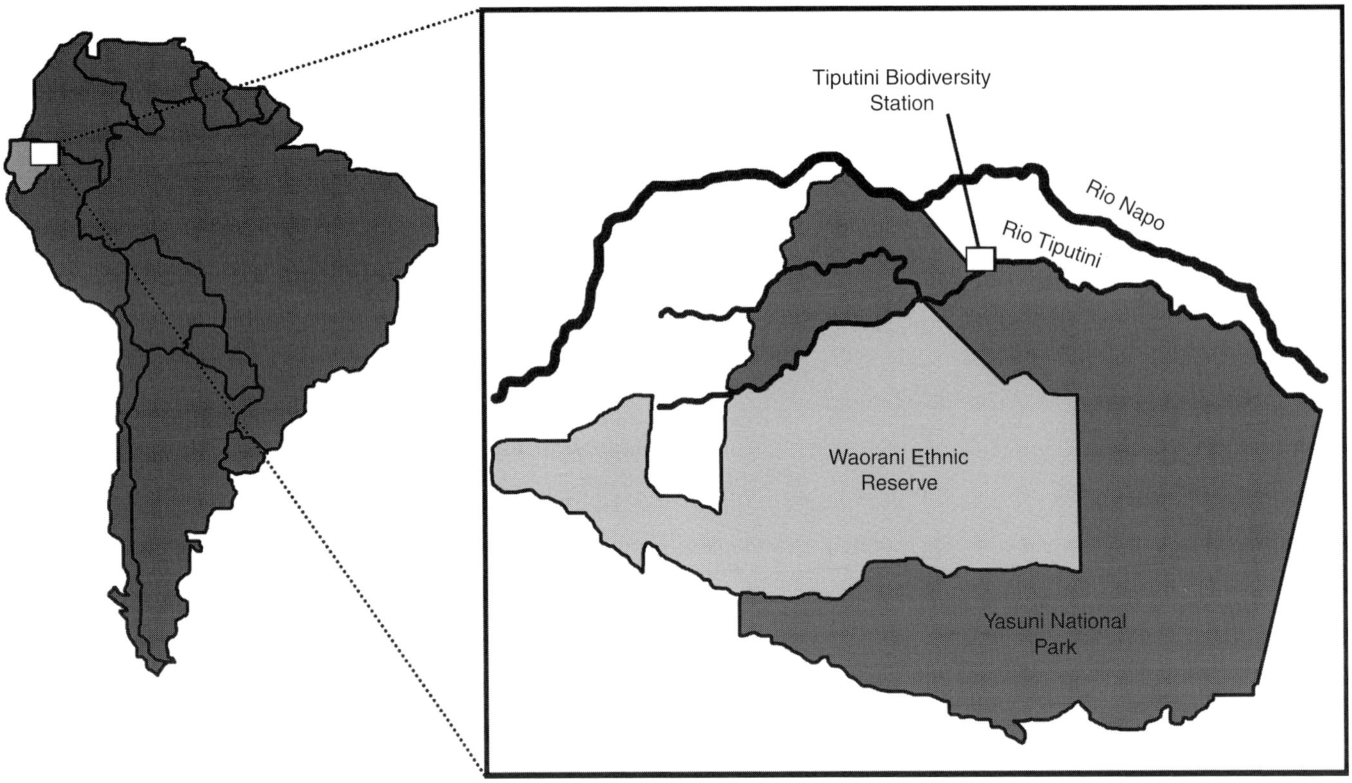

Figure 29.1 Location of the study site in the Yasuní National Park and Biosphere Reserve in the Ecuadorian Amazon.

Methods

Area of study

Since 2003 we have been studying three species of monogamous primates (owl monkeys: *Aotus vociferans*; titi monkeys: *Callicebus discolor*; saki monkeys: *Pithecia aequatorialis*) at the Tiputini Biodiversity Station (76°08′W, 0°38′S), located in the Yasuní National Park and Biosphere Reserve in Ecuador (Carrillo *et al.* 2005; Di Fiore *et al.* 2007; Fernandez-Duque *et al.* 2008, 2008b; Dacier *et al.* 2011).

Groups of study

The saki group has been monitored continuously since November 2003. During that month, we darted and captured the adult male and fitted him with a radio collar following procedures we have also used to capture owl monkeys (*Aotus azarai*) in Argentina (Fernandez-Duque & Rotundo 2003; Juárez *et al.* 2011), as well as owl monkeys (*Aotus vociferans*), titi monkeys (*Callicebus discolor*), capuchins (*Cebus albifrons*), squirrel monkeys (*Saimiri sciureus*), spider monkeys (*Ateles belzebuth*) and woolly monkeys (*Lagothrix lagotricha*) in Ecuador (Di Fiore & Fernandez-Duque, unpubl. data, 2007).

At the time of darting, the group consisted of an adult male, an adult female and a male juvenile of approximately 6 months of age. The original adult male was replaced by a new adult male in October 2004 (Di Fiore *et al.* 2007); the dependent juvenile dispersed in October 2007, when he was approximately 4.5 years of age; and two infants were born to the adult female in March 2005 and November 2006. The first infant disappeared in February 2006 when it was approximately 11 months old and the second one was in the group until December 2007 when the study reported here was completed.

Regarding the titis, we collected data from one fully habituated group studied since November 2003 and a second group added to the study in October 2006. The first group consisted initially of two animals, an adult male and an adult female. During these years, the female gave birth to two offspring. The first one was born in January 2004 and disappeared in November 2006 when it was almost 3 years of age. The second, born in January 2005, disappeared in February 2006 when it was 13 months old. The original adult female disappeared in March 2007, and she was replaced by a new adult female shortly thereafter. A new infant was born to the new female in December 2007. The second group consisted of an adult male–female pair when it was added to the study. Two infants were born in this group, one in November 2006 and the other in November 2007.

Data collection

We collected behavioral data during 20-min focal samples of all group members. Each day we sampled animals opportunistically based on visibility, but following the rule that successive focal samples of the same individual had to be separated by at least 20 min. Across days, we maintained an approximate balance in the number of focal samples collected per individual. Focal samples were collected in an approximately balanced schedule across months. During each focal sample, we recorded one of six basic behavioral states (resting, foraging, moving, social, other, and out of view) for the focal animal, and the identity of its nearest neighbor (or neighbors, if multiple group members were equidistant from the focal individual) as instantaneous sampling points every 2 min. Additionally, between the instantaneous sampling points, we continuously recorded all occurrences of select behaviors relevant to parental care and male–female relationships (e.g. grooming, food sharing, play, vigilance, nursing, rejection, infant transfers, and approach/leave interactions). Finally, we recorded ad libitum any conspicuous but rare events (e.g. intergroup encounters, fights among group members) that occurred too infrequently for our sampling protocol to yield an adequate assessment of their rate.

Data analysis

Between July 2004 and March 2007 we collected approximately 530 h of focal data on the saki group (63 h while the original adult male was in the group and the remainder with his replacement). We also collected approximately 330 h on two groups of titis; 320 h with the primary study group and 11 h with the second one. The limited data collected from the second group were not included in the analyses presented below, although they were in good agreement with the results obtained from the first group.

To characterize the social behavior of individual titis and sakis, we first calculated, for each focal sample, the proportion of instantaneous sampling points that focal individuals spent with each other group member as its nearest neighbor, the proportion of sampling points they spent giving and receiving grooming, and their rates of participation in food sharing, play, vigilance, nursing, rejection and infant transfers. We calculated rates as the number of occurrences of the behavior of interest per focal sample. We then computed a mean across focal samples for each individual.

To examine which individuals were more responsible for maintaining the observed spatial patterns, we calculated Hinde's indices for each pair of individuals within a group. The index (I) is calculated as the proportion of "approaches" (AP) between two individuals (A and B) that one of them is responsible for, minus the proportion of departures ("leaves", LV) that the same individual is responsible for, multiplied by 100.

$$I_{AB} = \left[\frac{(\#AP\,A \Rightarrow B)}{(\#APA \Rightarrow B + \#AP\,B \Rightarrow A)} - \frac{(\#LV\,A \Rightarrow B)}{(\#LVA \Rightarrow B + \#LVB \Rightarrow A)}\right] \times 100$$

Thus, the index ranges from 100 (if individual A is responsible for all approaches to B and no departures from B) to −100 (if A never approaches B and consistently leaves B). A value of zero would indicate that those two individuals approach and leave one another at similar rates.

Results

Species difference in pair-mate relationships

Titis and sakis differed in the spatial relationships between mates. Both male and female titis had their partner as the nearest neighbor more frequently than the sakis did (Figure 29.2).

Titi pair-mates also spent more time in social contact (i.e. resting in contact and tail twining) than did saki pair-mates (Figure 29.3a). Despite the difference in time spent in contact, the amount of grooming seen among pair-mates, at least following the replacement of the original resident male saki, was similar in the two species (Figure 29.3b). Titi pair-mates also approached each other more frequently than did saki pair-mates (Figure 29.4).

Both species showed sex differences in the relative contributions to proximity maintenance. Males were more responsible than females for maintaining proximity, as reflected in Hinde's indices of 25 and 24 for male titis and sakis,

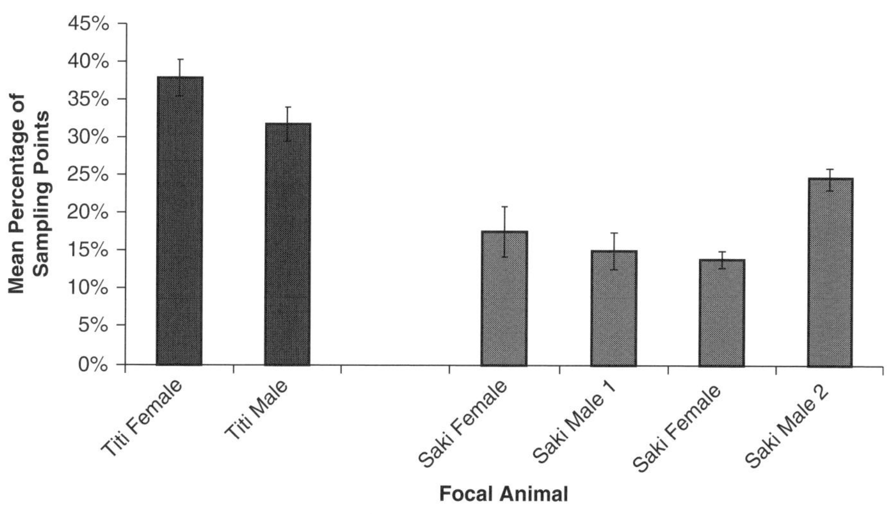

Figure 29.2 Differences between species in the nearest-neighbor relationships among pair-mates of titi and saki groups. Bars show the mean proportion of sampling points per focal sample (± SE) when the focal animal's pair-mate was the nearest neighbor.

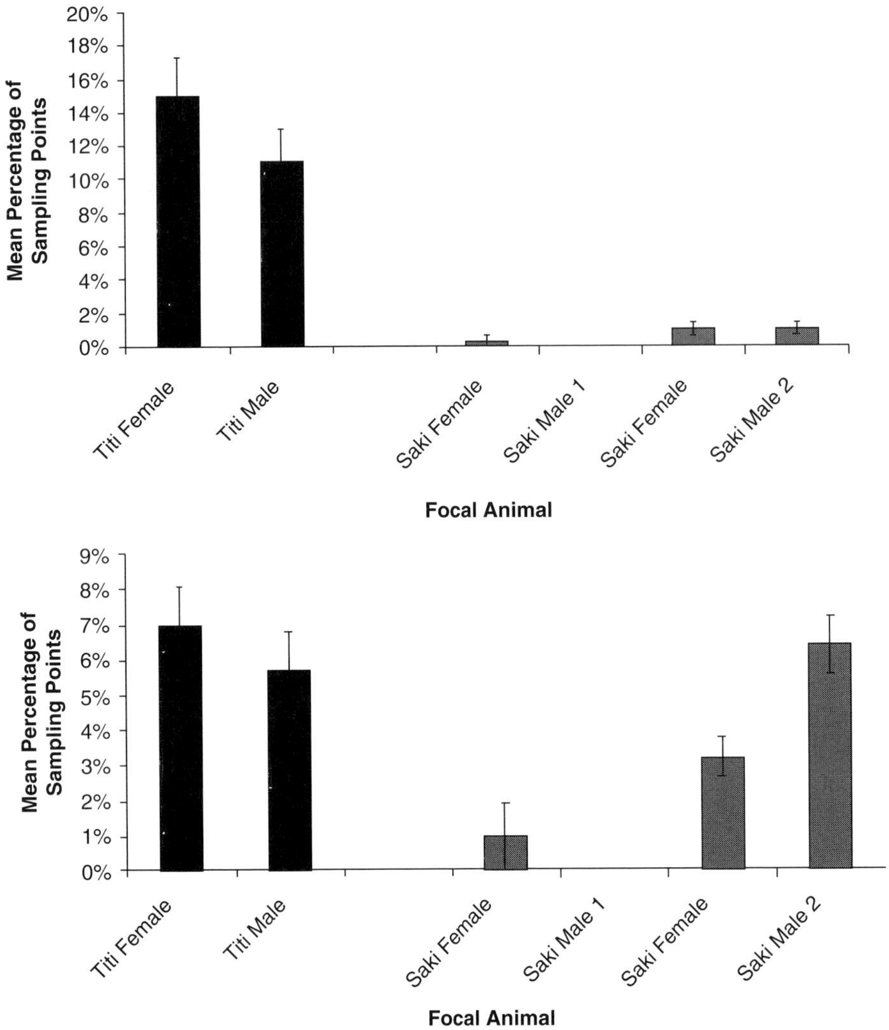

Figure 29.3 Differences between species in the time spent (a) in social contact and (b) grooming among pair-mates of titi and saki groups. Bars reflect the mean proportion of sampling points per focal sample (± SE) when the focal animal's pair-mate was in social contact or grooming/being groomed by the focal animal.

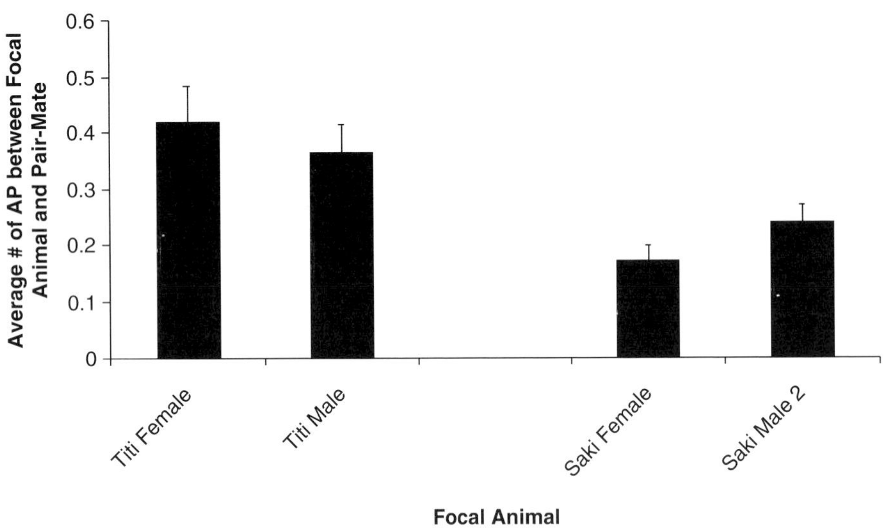

Figure 29.4 Differences between species and between the sexes in the rates of approach among pair-mates of titi and saki groups. Bars reflect mean number of approaches to partner plus approaches received from partner per focal sample (± SE). AP refers to approach.

respectively. In general, males of both species approached their partners more than females approached males, while females were more often responsible for breaking proximity from their partners (Table 29.1).

There were other aspects of saki and titi social behavior that may reflect additional differences in the nature of the bond between mates. The two titi pairs regularly performed coordinated duets, in good agreement with previous findings

(Mason 1966; Kinzey & Robinson 1983). On the other hand, no coordinated vocalizations were observed between saki pair-mates. Moreover, preliminary playback experiments to the study group showed that the male saki was the one responding to playbacks of territorial vocalizations, by vocalizing and approaching the speaker (Di Fiore & Fernandez-Duque, unpubl. data, 2006).

Species differences in parenting

Differences in social behavior between titis and sakis were most pronounced with regard to patterns of infant care. The male titis were heavily involved in infant care. They started carrying the infants the first week after birth, and during the first 4 months of life they carried them substantially more than the mothers did. Even when not carrying the infant, the adult male titi monkey was more often the nearest neighbor of infants than was the adult female. In fact, over the first 4 months of life, one titi infant whom we focused on collecting parental carrying data spent between 80 and 100% of its time either in contact with or closest to its putative father rather than its mother (Figure 29.5).

Male sakis, by contrast, were never observed carrying infants. When the saki infant was already independent, the offspring sometimes ($n = 7$) stayed next to the stationary male while the female moved off to forage. Additionally, when the infant was approximately 7 months of age, we observed the male sharing food with him, although this was not a common behavior.

There were also clear differences between titis and sakis in the spatial arrangements of adults and their weaned offspring. Saki juveniles had their mother as a nearest neighbor more frequently than they had the resident adult male, whereas the titi juveniles had the resident adult male as their nearest neighbor more frequently than they had their mother (Figure 29.6). In both species, the juveniles were more responsible than the adults for maintaining proximity, although the pattern was less skewed in the titis than the sakis (mean juvenile-to-adult Hinde Index, 20 vs. 44, respectively).

Table 29.1 Rates of approaches and leaves among titi and saki pair-mates (± SE).

	Male AP female	Female AP male	Male LV female	Female LV male
Titis	0.18 ± 0.002	0.11 ± 0.001	0.14 ± 0.001	0.28 ± 0.003
Sakis	0.08 ± 0.001	0.04 ± 0.001	0.08 ± 0.001	0.12 ± 0.001

AP, approach; LV, leave.
Rates calculated as number of occurrences of approaches or leaves per 20-min focal sample averaged across samples.

Discussion

Our data show that although both titis and sakis live in socially monogamous groups, there are important differences between the species in the nature of the relationship between males and females, as well as in the nature of the parent–offspring interactions. In this preliminary examination, those differences were reflected in: (1) the spatial relationships among group members, (2) the differences in sex-specific contributions to the maintenance of spatial proximity, and (3) the differences in the amount of direct care provided by males.

The spatial associations among group members were very different in the titi and saki groups. Titi pair-mates stayed in close proximity to each other most of the time, whereas the saki male and female tended to be further apart. Still, it is important to note that the incoming saki male spent substantially more time with the female as his nearest neighbor than did the initial resident male (Di Fiore *et al.* 2007), suggesting that a close spatial association may be important in saki monkeys during the process of pair formation.

Although sakis and titis differ in size and in the extent of physical dimorphism between the sexes, our data show that

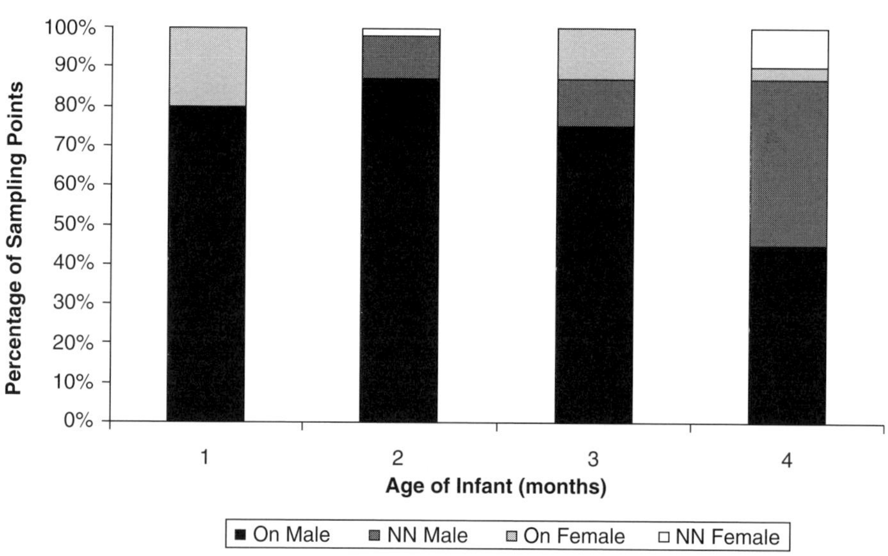

Figure 29.5 Patterns of carrying and nearest neighbor relationships between one titi monkey infant, its mother and the putative father during the first four months after birth. Bars reflect the mean proportion of sampling points per focal sample when the infant was being carried by the mother, the putative father or moving independently with the mother or putative father as its nearest neighbor.

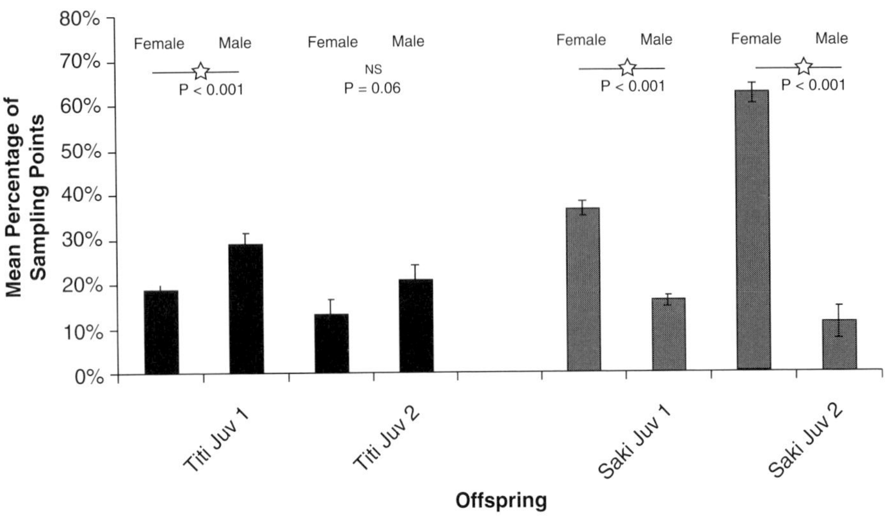

Figure 29.6 Species differences in the nearest-neighbor relationships between juveniles and their parents. Bars reflect the mean proportion of sampling points per focal sample (± SE) when the adult female and the adult male were the nearest neighbor of the juvenile. Stars indicate statistically significant differences (Wilcoxon tests for two related samples, $p < 0.001$).

males showed more interest in females than females did in males in both species. This was reflected in both the rate of approaches and in Hinde's Index, which summarizes both the "approach" and "leave" sides of proximity maintenance. Additionally, during preliminary playback experiments only saki males responded to the simulated presence of an intruder. And, although it has been reported that titi pair-mates regularly approach the territory border together and display in a coordinated fashion (Mason 1966; Robinson 1981), our own experience indicates that there may be important sex differences in the responses to intruders. During playback experiments and group territorial encounters males more frequently led the group towards the sound source, while displaying, standing on hind legs and chasing intruders (Di Fiore & Fernandez-Duque, unpubl. data). Finally, it was very clear that the two genera differed in the contribution of the two sexes to infant care.

In agreement with previous reports, the male titi was the main carrier of the infant from the first week of life (Mendoza & Mason 1986a; Norconk 2011). This is a most remarkable aspect of the behavioral biology of titis, only seen elsewhere in primates among owl monkeys (Wright 1984; Rotundo *et al.* 2005; Fernandez-Duque *et al.* 2009) and callitrichines (Digby *et al.* 2007). Given that paternal care is prevalent in those small-bodied taxa, but absent in the relatively larger sakis, it seems reasonable to suggest that social monogamy is maintained because the care provided by the male contributes to reducing the metabolic costs to the female of raising relatively large offspring.

Socially monogamous primates have historically been described as having a relatively prolonged and essentially stable relationship that included exclusive mating between them. This assumption, however, has appropriately been questioned (Palombit 1994a), and reports of extra-pair copulations among socially monogamous primates keep accumulating (Mason 1966; Palombit 1994b; Reichard 1995; Fietz *et al.* 2000; Oka & Takenaka 2001). Pair-mate turnover has been documented in all three socially monogamous genera we are investigating (Di Fiore *et al.* 2007; Rodman & Bossuyt 2007; Fernandez-Duque *et al.* 2008a, 2008b). In the future, ongoing genetic analyses of group structure will help us elucidate the temporal stability of these socially monogamous groups, as well as the relative importance of paternal care in favoring the evolution of monogamy.

In conclusion, social monogamy is not a unitary phenomenon in primates; it can exist without males and females being strongly pair-bonded, and it can exist in the absence of paternal care (Di Fiore & Fernandez-Duque 2007). Thus, in searching for explanations for social monogamy and paternal care in primates, we should expect that there may be different mechanisms at play for different taxa.

Acknowledgments

The authors are most grateful to the Wenner-Gren Foundation, the L.S.B. Leakey Foundation, the J. William Fulbright Scholar Program, Primate Conservation, Inc., Idea Wild, New York University, and the Zoological Society of San Diego for funding this research. Special thanks are also due to the Ecuadorian government – especially officials of the Ministerio de Ambiente – for their continued interest in our primate research, and to Dr. David Romo and Sr. Jaime Guerra of the Universidad San Francisco de Quito for scientific and logistical support at the Tiputini Biodiversity Station. The authors also wish to thank each of the many volunteers and students who made this research possible by spending long hours in the forest, including A. Dacier, Y. Di Blanco, D. Hurst, A. Larson, C. Schmitt, C. Sendall, M. Rotundo, and "los tigres" of the Tiputini Biodiversity Station. The research described here was done in full agreement with all Ecuadorian legislation and was approved by the IACUC committee of New York University.

References

Anzenberger, G. (1988). The pairbond in the titi monkey (*Callicebus moloch*): intrinsic versus extrinsic contributions of the pairmates. *Folia Primatologica*, **50**, 188–203.

Carrillo, G., Di Fiore, A. & Fernandez-Duque, E. (2005). Dieta, forrajeo y presupuesto de tiempo en cotoncillos (*Callicebus discolor*) del Parque Nacional Yasuní en la Amazonía Ecuatoriana. *Neotropical Primates*, **13**(2), 7–11.

Dacier, A., De Luna, A.G., Fernandez-Duque, E., *et al.* (2011). Estimating population density of Amazonian titi monkeys (*Callicebus discolor*) via playback point counts. *Biotropica*, **43**(2), 135–140.

Defler, T.R. (2004). *Titi monkeys. Primates of Colombia.* Bogotá: Conservation International, pp. 298–322.

Di Fiore, A. & Fernandez-Duque, E. (2007). A comparison of paternal care in three socially-monogamous neotropical primates. *American Journal of Physical Anthropology*, **132**(S44), 99.

Di Fiore, A., Fernandez-Duque, E. & Hurst, D. (2007). Adult male replacement in socially monogamous equatorial saki monkeys (*Pithecia aequatorialis*) *Folia Primatologica*, **78**, 88–98.

Digby, L.J., Ferrari, S.F. & Saltzman, W. (2007). Callitrichines. The role of competition in cooperatively breeding species. In *Primates in Perspective*, ed. C.J. Campbell, A. Fuentes, K.C. MacKinnon, M. Panger & S.K. Bearder. Oxford: Oxford University Press, pp. 85–105.

Fernandez-Duque, E. & Rotundo, M. (2003). Field methods for capturing and marking Azarai night monkeys. *International Journal of Primatology*, **24**(5), 1113–1120.

Fernandez-Duque, E., Di Fiore, A. & Carrillo-Bilbao, G. (2008a). Behavior, ecology and demography of *Aotus vociferans* in Yasuní National Park, Ecuador. *International Journal of Primatology*, **29**(2), 421–431.

Fernandez-Duque, E., Juárez, C. & Di Fiore, A. (2008b). Adult male replacement and subsequent infant care by male and siblings in socially monogamous owl monkeys (*Aotus azarai*). *Primates*, **49**, 81–84.

Fernandez-Duque, E., Mason, W.A. & Mendoza, S.P. (1997). Effects of duration of separation on responses to mates and strangers in the monogamous titi monkey (*Callicebus moloch*). *American Journal of Primatology*, **43**(3), 225–237.

Fernandez-Duque, E., Valeggia, C.R. & Mason, W.A. (2000). Effects of pair-bond and social context on male–female interactions in captive titi monkeys (*Callicebus moloch*, Primates: Cebidae). *Ethology*, **106**(12), 1067–1082.

Fernandez-Duque, E., Valeggia, C.R. & Mendoza, S.P. (2009). The biology of paternal care. *The Annual Review of Anthropology*, **38**, 115–130.

Fietz, J., Zischler, H., Schwiegk, C., *et al.* (2000). High rates of extra-pair young in the pair-living fat-tailed dwarf lemur, *Cheirogaleus medius. Behavioral Ecology and Sociobiology*, **49**(1), 8–17.

Fragaszy, D.M., Schwarz, S. & Shimosaka, D. (1982). Longitudinal observations of care and development of infant titi monkeys (*Callicebus moloch*). *American Journal of Primatology*, **2**, 191–200.

Fuentes, A. (2002). Patterns and trends in primate pair bonds. *International Journal of Primatology*, **23**, 953–973.

Hoffman, K.A., Mendoza, S.P., Hennessy, M.B., *et al.* (1995). Responses of infant titi monkeys, *Callicebus moloch*, to removal of one or both parents: evidence for paternal attachment. *Developmental Psychobiology*, **28**(7), 399–407.

Juárez, C.P., Rotundo, M., Berg, W.J., *et al.* (2011). Costs and benefits of radio-collaring on the behavior, demography and conservation of owl monkeys (*Aotus azarai*) in Formosa, Argentina. *International Journal of Primatology*, **32**, 69–820.

Kinzey, W.G. (1981). The titi monkeys, genus *Callicebus*. In *Ecology and Behavior of Neotropical Primates*, ed. A.F. Coimbra-Filho & R.A. Mittermeier. Rio de Janeiro: Academia Brasileira de Ciências, vol. **1**, pp. 241–276.

Kinzey, W.G. & Becker, M. (1983). Activity pattern of the masked titi monkey, *Callicebus personatus. Primates*, **24**(3), 337–343.

Kinzey, W.G. & Robinson, J.G. (1983). Intergroup loud calls, range size, and spacing in *Callicebus torquatus. American Journal of Physical Anthropology*, **60**, 539–544.

Kinzey, W.G. & Wright, P.C. (1982). Grooming behavior in the titi monkey (*Callicebus torquatus*). *American Journal of Primatology*, **3**, 267–275.

Kleiman, D.G. (1977). Monogamy in mammals. *The Quarterly Review of Biology*, **52**, 39–69.

Kleiman, D.G. (1981). Correlations among life history characteristics of mammalian species exhibiting two extreme forms of monogamy. In *Natural Selection and Social Behavior*, ed. R.D. Alexander & D.W. Tinkle. New York, NY: Chiron Press, pp. 332–344.

Kleiman, D.G. & Malcolm, J.R. (1981). The evolution of male parental investment in mammals. In *Parental Care in Mammals*, ed. D.G. Gubernick & P.H. Klopfer. New York, NY: Plenum Press, pp. 347–387.

Lehman, S.M., Prince, W. & Mayor, M. (2001). Variations in group size in white-faced sakis (*Pithecia pithecia*): evidence for monogamy or seasonal congregations? *Neotropical Primates*, **9**(3), 96–101.

Maestripieri, D. (2002). Parent–offspring conflict in primates. *International Journal of Primatology*, **23**(4), 923–951.

Mason, W.A. (1966). Social organization of the South American monkey, *Callicebus moloch*: a preliminary report. *Tulane Studies in Zoology*, **13**, 23–28.

Mason, W.A. (1968). Use of space by *Callicebus* groups. In *Primates: Studies in Adaptation and Variability*, ed. P.C. Jay. New York, NY: Holt, Rinehart & Winston, pp. 200–216.

Mason, W.A. (1975). Comparative studies of social behavior in *Callicebus* and *Saimiri*: strength and specificity of attraction between male–female cagemates. *Folia Primatologica*, **23**, 113–123.

Mendoza, S.P. & Mason, W.A. (1986a). Parental division of labour and differentiation of attachments in a monogamous primate (*Callicebus moloch*). *Animal Behaviour*, **34**, 1336–1347.

Mendoza, S.P. & Mason, W.A. (1986b). Constrasting responses to intruders and to involuntary separation by monogamous and polygynous New World monkeys. *Physiology and Behavior*, **38**, 795–801.

Mitani, J.C. (1984). The behavioral regulation of monogamy in gibbons (*Hylobates muelleri*). *Behavioral Ecology and Sociobiology*, **15**, 225–229.

Norconk, M.A. (2006). Long-term study of group dynamics and female reproduction in Venezuelan *Pithecia pithecia. International Journal of Primatology*, **27**(3), 653–674.

Norconk, M.A. (2011). Sakis, uakaris and titi monkeys: behavioral diversity in a radiation of seed predators. In *Primates in Perspective* (2nd end), ed. C.J. Campbell, A. Fuentes, K.C. MacKinnon, S.K. Bearder & R.M. Stumpf. Oxford: Oxford University Press, pp. 122–139.

Norconk, M.A., Raghanti, M.A., Martin, S.K., *et al.* (2003). Primates of Brownsberg Natuurpark, Suriname, with particular attention to the pitheciins. *Neotropical Primates*, **11**(2), 94–100.

Oka, T. & Takenaka, O. (2001). Wild gibbons' parentage tested by non-invasive DNA sampling and PCR-amplified polymorphic microsatellites. *Primates*, **42**(1), 67–73.

Palombit, R. (1994a). Dynamic pair bonds in hylobatids: implications regarding monogamous social systems. *Behaviour*, **128**, 65–101.

Palombit, R.A. (1994b). Extra-pair copulation in a monogamous ape. *Animal Behaviour*, **47**, 721–723.

Palombit, R.A. (1996). Pair bonds in monogamous apes: a comparison of the Siamang (*Hylobates syndactilus*) and the white-handed gibbon (*Hylobates lar*). *Behaviour*, **133**, 321–356.

Palombit, R.A. (1999). Infanticide and the evolution of pair bonds in nonhuman primates. *Evolutionary Anthropology*, **7**, 117–129.

Price, E.C. & Piedade, H.M. (2001). Ranging behavior and intraspecific relationships of masked titi monkeys (*Callicebus personatus personatus*). *American Journal of Primatology*, **53**, 87–92.

Raemaekers, J.J. & Raemaekers, P.M. (1985). Field playback of loud calls to gibbons (*Hylobates lar*): territorial, sex-specific and species-specific responses. *Animal Behaviour*, **33**, 481–493.

Reichard, U. (1995). Extra-pair copulations in a monogamous gibbon (*Hylobates lar*). *Ethology*, **100**, 99–112.

Reichard, U. (2003). Monogamy: past and present. In *Monogamy. Mating Strategies and Partnerships in Birds, Humans and Other Mammals*, ed. U.H. Reichard & C. Boesch. Cambridge: Cambridge University Press, pp. 3–25.

Robinson, J.G. (1979). Vocal regulation of use of space by groups of titi monkeys *Callicebus moloch. Behavioral Ecology and Sociobiology*, **5**, 1–15.

Robinson, J.G. (1981). Vocal regulation of inter- and intragroup spacing during boundary encounters in the titi monkey, *Callicebus moloch. Primates*, **22**(2), 161–172.

Robinson, J.G., Wright, P.C. & Kinzey, W.G. (1987). Monogamous cebids and their relatives: intergroup calls and spacing. In *Primate Societies*, ed. B.B. Smuts, D.L. Cheney, R.M. Seyfarth, R. Wrangham & T.T. Struhsaker. Chicago, IL: The University of Chicago Press, pp. 44–53.

Rodman, P.S. & Bossuyt, F.J. (2007). Fathers and stepfathers: familial relations of old and new males within groups of *Callicebus brunneus* in southeastern Perú. *American Journal of Physical Anthropology*, **132**(S44).

Rotundo, M., Fernandez-Duque, E. & Dixson, A.F. (2005). Infant development and parental care in free-ranging *Aotus azarai azarai* in Argentina. *International Journal of Primatology*, **26**(6), 1459–1473.

Setz, E.Z.F. & Gaspar, D.A. (1997). Scent marking behaviour in free-ranging golden-faced saki monkeys *Pithecia pithecia chrysocephala*: sex differences and context. *Journal of Zoology*, **241**, 603–611.

Tirado-Herrera, E.R. & Heymann, E.W. (2004). Does mom need more protein? Preliminary observations on differences in diet composition in a pair of red titi monkeys (*Callicebus cupreus*). *Folia Primatologica*, **75**(3), 150–153.

Vié, J.-C., Richard-Hansen, C. & Fournier-Chambrillon, C. (2001). Abundance, use of space, and activity patterns of white-faced sakis (*Pithecia pithecia*) in French Guiana. *American Journal of Primatology*, **55**, 203–221.

Wright, P.C. (1984). Biparental care in *Aotus trivirgatus* and *Callicebus moloch.* In *Female Primates: Studies by Women Primatologists*, ed. M. Small. New York, NY: Alan R. Liss, Inc., pp. 59–75.

Wright, P.C. (1985). *The costs and benefits of nocturnality for Aotus trivirgatus (the night monkey).* City University of New York.

Chapter 30

Vocal communication in *Cacajao*, *Chiropotes* and *Pithecia*: current knowledge and future directions

Bruna M. Bezerra, Adrian A. Barnett, Antonio S. Souto & Gareth Jones

Introduction

Animal communication can be defined as a process in which one individual (the signaller) influences the behavior of another (the receiver) through the transmission of signals (e.g. Krebs & Davies 1996; Shettleworth 1998; Bradbury & Vehrencamp 1998). It plays a vital role in many interactions among animals, including aggression, predation, feeding and mating (e.g. Marler 1961; Shettleworth 1998). Primates are known to communicate through chemical, visual and/or auditory signals (e.g. Napier & Napier 1967) and the way they communicate will depend on the restrictions imposed by their habits and habitats (Krebs & Davies 1996; Bradbury & Vehrencamp 1998; Brumm *et al.* 2003).

The acoustic signals considered here often constitute an important communication tool for arboreal primates, due to visual limitations imposed by dense foliage (Altmann 1967). Phylogeny, body weight and social context are known to influence vocalizations of nonhuman primates (e.g. Inoue 1988; Hauser 1993). The propagation of sound in habitats where primates live may be influenced by such factors as frequency of the vocal signal, the height above the ground at which it is emitted, meteorological conditions and time of day (Ingard 1953; Marten *et al.* 1977; Waser & Waser 1977; Morton 1975; Waser & Brown 1986; Brown & Waser 1988; de la Torre & Snowdon 2002; Sugiura *et al.* 2006). Nonhuman primates vocalize in varied contexts, and many calls are context-specific (e.g. Cheney & Seyfarth 1982; Roush & Snowdon 1999; Snowdon 2001) and consequently many primate species use a varied number of calls in social interactions. Thus, studying the vocal repertoire can provide key insights into the social ecology and behavior of the species concerned. In this chapter we (i) review the information on vocal communication currently available for *Cacajao*, *Chiropotes* and *Pithecia* species (e.g. Rosenberger 1981; Walker & Ayres 1996); (ii) consider how this informs our knowledge of pitheciine social systems and social ecology; and (iii) also take the opportunity to highlight those areas where our knowledge is weakest and suggest avenues for future research.

Vocal repertoires, sociality and the extent of current knowledge

Together with *Callicebus* (subfamily Callicebiidae) the subfamily Pitheciidae constitutes the family Pithecidae (Groves 2001). The vocal repertoire of *Callicebus* is the subject of another chapter of this book, and will not be considered further here. All members of the subfamily Pitheciidae are diurnal and arboreal. *Pithecia* species tend to have small social groups and home ranges in contrast with *Cacajao* and *Chiropotes* species (Norconk 2011). Groups can reach up to 200 individuals in some of the species, although, due to food resource-related fission–fusion sociality, encountered groups are often much (40+) to very much (2–15) smaller (e.g. Kinzey & Norconk 1990; Kinzey & Cunningham 1994; Rowe 1996; Defler, 1999, 2001, 2005; Norconk 2011). The social system varies greatly across species and even within the three genera, with monogamy, plurally breeding groups, multifemale–multimale groups and even solitary individuals having been observed in the subfamily Pitheciidae (Rowe 1996; Norconk 2011). Thus, for a group such as the pitheciids, it may be expected that vocal communication will be paramount because the monkeys dwell in the canopy, travel over large areas and may occur in large groups which are often widely dispersed and which may frequently fragment into subgroups when resource patchiness prompts it. Not only will vocalizations be important, but it is also likely that pitheciids will show patterns of vocal variation that will allow us to unravel the complex threads of cause-and-effect in a given species' social ecology. Furthermore, the vocalizations would predict types of vocal behavior that might occur in other species of the group yet to be studied.

The field of pitheciid vocal studies is in its infancy, with the real size of the vocal repertoires still remaining uncertain for all members of the subfamily (see Table 30.1). Even when the calls of a species have been investigated (e.g. Fontaine 1981; Buchanan *et al.* 1981; Fernandes 1991), it is rare for the studies to specify, describe physically via spectrograms

Evolutionary Biology and Conservation of Titis, Sakis and Uacaris, eds. Liza M. Veiga, Adrian A. Barnett, Stephen F. Ferrari and Marilyn A. Norconk. Published by Cambridge University Press.

Table 30.1 Vocal repertoire and sociality of primates in the subfamily Pitheciidae. Group sizes were taken mostly from Norconk (2011). Vocal repertoire sizes are not definitive for the species presented below as most of them are based on basic and not specific studies on vocal communication.

Species	Repertoire size Physically/ onomatopoetically	Group size	Social structure	Higher and lower call frequencies	Source
Cacajao calvus rubicundus	11/12	2–55 (up to 100)	Multimale–multifemale	1.2 to 10 kHz	Fontaine 1981; Rowe 1996
C. c. calvus		30–48			
C. c. ucayalii		8–70			
C. c. novaesi					
C. melanocephalus ouakary	9	1–30 (up to 200)	Multimale–multifemale group	0.77–4.14 kHz	Defler 2005; Bezerra *et al.* 2007; Bezerra 2010; Bezerra *et al.* 2010a
C. m. melanocephalus			1male–multifemale, multimale–multifemale		
C. ayresi					Boubli *et al.* 2008
Chiropotes satanas	/2	10–30	Multimale–multifemale, monogamy		van Roosmalen *et al.*, 1988; Rowe, 1996
C. chiropotes		8–44			
C. israelita					
C. utahickae	4				Fernandes 1991.
C. albinasus		2–26	Multimale–multifemale		van Roosmalen *et al.*, 1988
Pithecia monachus	1/5, 2	2.9 (2–9)	Monogamy	3 kH	Buchanan *et al.* 1981
P. irrorata	/1	1–3.5	Monogamy		Hill 1960; Rowe 1996
P. pithecia	12/5	1–12	Monogamy, solitary, plurally breeding groups		Buchanan *et al.* 1981; Kinzey 1997; Lehman *et al.* 2001; Norconk 2006; Heline 2007. Di Fiori *et al.* 2007
P. aequatorialis		1–7			Rowe 1996
P. albicans	/1	3–7	Monogamy		Norconk 2011
P. hirsuta		2–8			

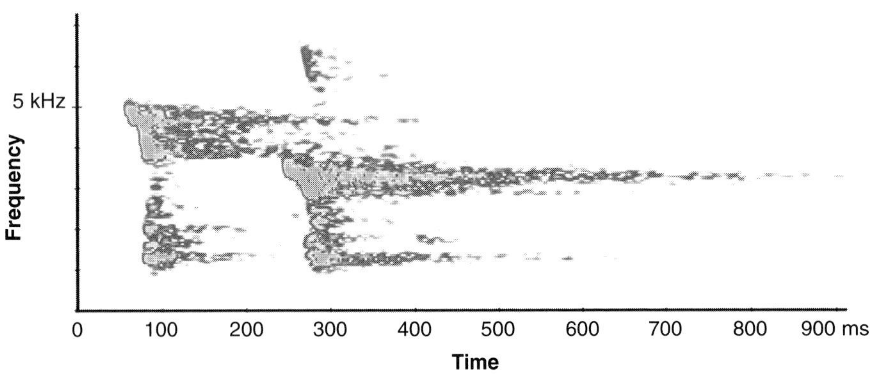

Figure 30.1 Spectrogram of '*tcho*' calls in *Cacajao m. ouakary* (time in ms, frequency in kHz). This call can be uttered singly or in series and in various behavioral contexts. The physical structure of this call varies according to behavioral context and individual signaller. For a complete description of the vocal repertoire of the golden-backed uacaris see Bezerra (2010) and Bezerra *et al.* (2010a).

(see Figure 30.1) and/or quantify properly the vocalizations (e.g. Bezerra 2010; Bezerra *et al.* 2010a). As a result, we often end up with only general and more superficial information such as: *Pithecia irrorata*, gray monkey saki, utter loud growls-grunt at human observers (Hill 1960); *Cacajao melanocephalus ouakary*, golden-backed uacari, presents a threat display that consists of piloerection associated with vocalization displays (Cunha & Barnett 1989); feeding *Chiropotes satanas*, bearded sakis, typically continue vocal communication after having split into subgroups (Kinzey & Norconk 1990); in *Cacajao* and *Chiropotes*, tail wagging frequently happens in parallel with vocalizations (Walker & Ayres 1996); *Pithecia albicans*, buffy sakis, have alarm calls as the most notable call (Rowe 1996); *C. m. ouakary* has a high-pitched call that has been compared to the call of an *Amazonas* parrot (Rowe 1996) and hunters and researchers both locate uacaris by their contact calls (Cunha & Barnett 1989, 1990; Rowe 1996); rates of contact calls and calls performed during intergroup interactions differed before and after adult male replacement in *Pithecia aequatorialis* (equatorial saki) groups (Di Fiore *et al.* 2007). These pieces of information are, without doubt, very valuable, but they make us realize how much still needs to be done regarding basic research on vocal communication system of *Cacajao*, *Pithecia* and *Chiropotes* species. For many species we are still at the early stages of data description and collection.

Among- and within-genera comparisons of the vocalizations of insects (e.g. Hartley *et al.* 1974; Nickle 1976; Heller & von Helversen 2004); amphibians (e.g. Roy 1994); reptiles (e.g. Höbel & Gerhardt 2003); birds (e.g. Hand 1981; Devoogd *et al.* 1993; Radford 2005; Girão & Souto 2005; Thomassen & Povel 2006), and of mammals including primates (e.g. Jones *et al.* 2000; Peters & Tonkin-Leyhausen 2004; McCombe & Semple 2005, Bezerra & Souto 2008) have shown patterns that permit reconstruction of the evolution of the vocal communication systems themselves and of the social systems with which they are associated. This, given sufficient data, should also be possible for pitheciid monkeys. There are other variables that should also be considered when taking into account the form and function of current vocalizations. One is with the influence of sound propagation and distortion in the habitats (Marten *et al.* 1977; Bradbury & Vehrencamp 1998) and how the effects are ameliorated given the ecological and social context in which the communications are produced. Another is the link between size of social group and size of the species' vocal repertoire. Underpinned by the potential influence of the vocal communication in the evolution of primate social behavior (McCombe & Semple 2005), an increase in the group size would be predicted to lead to an increase in the repertoire size (McCombe & Semple 2005) as vocalizations develop to facilitate social bonding once groups become too big to perform the tactile and visual communication that serves in species with small close-knit and highly proximate groups (Dunbar 2003). However, we must consider that for large groups, we must not necessarily have an increase on the vocal repertoire size. A more subtle communication system can arise, for instance, by having some calls that include characteristics that encode cues about individuals, groups or subgroups (e.g. Sproul *et al.* 2006). Unfortunately, such lines of inquiry are currently limited by a critical lack of information on operational group sizes and vocal repertoire sizes for most pitheciid species.

In addition, the lack of studies at more than one site means that we are currently unable to undertake comparisons between localities and so knowledge of the subtle differences in the communication systems and social ecology and behavior of the pitheciids is still lacking. This lack of knowledge of the nature and extent of within-species variation also hampers our ability to make informed interspecific comparisons of vocal communication from an evolutionary point of view. The current data available are insufficient to evaluate variation in both structure (e.g. number of calls) and function of the vocal repertoire. Future basic research on the social system and vocal communication is needed if we are to create a better understanding on the ecology, sociality and evolution of the monkeys of the pitheciid subfamily.

Importance of vocal communication for conservation of the subfamily Pitheciidae

As shown in Table 30.1, very little is known about the vocal communication of Pitheciidae monkeys. Only partial and generally onomatopoetic descriptions of the vocal repertoire have been published, and these only for a limited number of species. Many of the pitheciine primates are threatened with extinction (www.iucn.org 2010, IUCN Red List of Threatened Species), and thus, the precise description of the acoustic signals emitted by pitheciine primates is likely to become an important tool in future efforts for the conservation of these species. The vocal signals are likely to lead to a valuable method for surveying these rare monkeys from their vocalizations in habitats where visual surveys would be difficult (Sutherland 1996; Bradbury & Vehrencamp 1998; Bezerra *et al.* 2010b). This technique is used for bird species (e.g. Ratcliffe *et al.* 1998; Turcotte & Desrochers 2002; Ambagis 2004; Boscolo *et al.* 2006), and some researchers have used it for mammals, including primates (e.g. Estrada 1982; Ogutu & Dubin 1998; Mills *et al.* 2001; Downey *et al.* 2006). Furthermore, knowledge on the vocal repertoire of pitheciins may help us to solve taxonomic issues given the peculiarities of vocalizations among the different species (e.g. Struhsaker 1970; Quris 1980; Hodun *et al.* 1981; Haimoff *et al.* 1982; Snowdon *et al.* 1986; Zimmermann *et al.* 1988; Courtenay & Bearder 1989; Hohmann 1989; Snowdon 1993). We believe that knowledge of vocalizations can also help improve the welfare of captive populations of pitheciins. Environment enrichments can be created to allow captive animals to display vocal repertoires as similar as possible to those produced by wild animals. Also, knowing the context of the calls (e.g. calls used under stressful situations) will help keepers to ascertain the conditions of the animals in captivity.

Future studies

We still need a great amount of basic research to describe quantitatively and qualitatively the vocal communication system of the pitheciids. Studies focusing on the general social ecology are also lacking, and thus this should be a target to be pursued for this group of primates. Once social systems and vocal repertoires are well defined, a great range of comparative studies can be conducted. This may lead us to use the calls as tools for assessing welfare, solving taxonomic issues, conducting surveys in difficult environments, and finally, increasing our knowledge on the general biology of pitheciine primates.

Acknowledgments

BMB was supported by Programme Alban; ORS award – Faculty of Sciences, School of Biological Sciences, University of Bristol; a Rufford Small Conservation Grant; an IDEA WILD grant, Amazon Ecopark Lodge and Living Rain Forest foundation.

References

Altmann, S.A. (1967). The structure of primate social communication. In *Social Communication Among Primates*, ed. S.A. Altman. Chicago, IL: University of Chicago Press, pp. 325–336.

Ambagis, J. (2004). A comparison of census and monitoring techniques for Leach's storm petrel. *Waterbirds*, **27**, 211–215.

Bezerra, B.M. (2010). Behaviour and vocal communication in golden-backed uakaris, *Cacajao melanocephalus*. PhD thesis, University of Bristol, United Kingdom.

Bezerra, B.M. & Souto, A.S. (2008). The structure and usage of the vocal repertoire of common marmosets. *International Journal of Primatology*, **29**, 671–701.

Bezerra, B.M., Barnett, A.A., Souto, A.S., *et al.* (2007). *Preliminary recordings on the vocalisations in golden-backed uacari (Cacajao melanocephalus ouakary)*, in Jaú National Park, Amazon, Brazil. 12th Brazilian Primatological Conference, Belo Horizonte, p. 8.

Bezerra, B.M., Souto, A.S. & Jones, G. (2010a). Vocal repertoire of golden-backed uakaris (*Cacajao melanocephalus*): call structure and context. International *Journal of Primatology*, **5**, 759–778.

Bezerra, B.M., Souto, A.S. & Jones, G. (2010b). Responses of golden-backed uakaris, *Cacajao melanocephalus*, to call playbacks: implications for surveys in the flooded Igapó forest. *Primates*, **51**, 327–336.

Boscolo, D., Metzger, J.P. & Viellard, J.M.E. (2006). Efficiency of playback for assessing the occurrence of five bird species in Brazilian Atlantic Forest fragments. *Anais da Academia Brasileira de Ciências*, **78**, 629–644.

Boubli, J.P., Silva, M.N.F., Amado, M.V., *et al.* (2008). A taxonomic reassessment of black uacari monkey, *Cacajao melanocephalus*, Humboldt (1811), with the description of two new species. *International Journal of Primatology*, **29**, 723–741.

Bradbury, J.W. & Vehrencamp, S.L. (1998). *Principles of Animal Communication.* Sunderland, MA: Sinauer Associates, Inc.

Brown, C.H. & Waser, P.M. (1988). Environmental influences on the structure of primate vocalizations. In *Primate Vocal Communication*, ed. D. Todt, P. Goedeking & D. Symmes. London. Springer-Verlag, pp. 51–68.

Brumm, H., Voss, K., Köllmer, I., *et al.* (2003). Acoustic communication in noise: regulation of call characteristics in a New World monkey. *Journal of Experimental Biology*, **207**, 443–448.

Buchanan, D.B., Mittermeier, R.A. & van Roosmalen, M.G.M. (1981). The saki monkeys, genus *Pithecia*. In *Ecology and Behaviour of Neotropical Primates.* Vol. **1**, ed. R.A. Coimbra-Filho & R. Mittermeier. Rio de Janeiro, Academia Brasileira de Ciências, pp. 391–417.

Cheney, D.L. & Seyfarth, R.M. (1982). Recognition of individuals within and between groups of free-ranging vervet monkeys. *American Zoologist*, **22**, 519–529.

Courtenay, D.O. & Bearder, S.K. (1989). The taxonomic status and distribution of bushbabies in Malawi with emphasis on the significance of vocalizations. *International Journal of Primatology*, **10**, 17–34.

Cunha da, A.C. & Barnett, A. (1989). Project Uakari. First Report; The preliminary survey; Part One – Zoology. Unpublished report to WWF-Netherlands, Pronatura (Brazil) and Royal Geographical Society, London, UK.

Cunha da, A.C. & Barnett, A. (1990). Sightings of the golden-backed uacari (*Cacajao melanocephalus ouakary*) on the upper Rio Negro, Amazonas, Brazil. *Primate Conservation*, **11**, 8–11.

Defler, T.R. (1999). Fission–fusion in the black-headed uacari (*Cacajao melanocephalus*) in eastern Colombia. *Neotropical Primates*, **7**, 5–8.

Defler, T.R. (2001). *Cacajao melanocephalus ouakary* densities on the lower Apaporis River, Colombian Amazon. *Primate Report*, **61**, 31–36.

Defler, T.R. (2005). *Primates of Colombia.* Conservation International. Spring Press.

De la Torre, S. & Snowdon, C.T. (2002). Environmental correlates of vocal communication of wild pygmy marmosets, *Cebuella pygmaea. Animal Behaviour*, **63**, 847–856.

Devoogd, T.J., Krebs, J.R., Healy, S.D., *et al.* (1993). Relations between song repertoire size and the volume of brain nuclei related to song: comparative evolutionary analyses amongst oscine birds. *Proceedings of the Royal Society of London: Biological Sciences*, **254**, 75–82.

Di Fiore, A., Fernandez-Duque, E. & Hurst, D. (2007). Adult male replacement in socially monogamous equatorial saki monkeys (*Pithecia aequatorialis*). *Folia Primatologica*, **78**, 88–98.

Downey, B.A., Jones, P.F., Quinlan, R.W., *et al.* (2006). Use of playback alarm calls to detect and quantify habitat use by Richardson's ground squirrels. *Wildlife Society Bulletin*, **34**, 480–484.

Dunbar, R.J.M. (2003). The origin and subsequent evolution of language. In *Language Evolution*, ed. M.H. Christiansen & S. Kirby. Oxford: Oxford University Press, pp. 219–234.

Estrada, A. (1982). Survey and census of howler monkeys (*Alouatta palliata*) in the rain forest of "Los Tuxtlas," Veracruz, Mexico. *American Journal of Primatology*, **2**, 363–372.

Fernandes, M.E.B. (1991). Comunicação social dos cuxiús (*Chiropotes satanas uitahicki*, Cebidae, Primates). *APrimatologia do Brasil*, **3**, 297–305.

Fontaine, R. (1981). The uakaris, genus *Cacajao.* In *Ecology and Behavior of Neotropical Primates*, ed. A.F. Coimbra-Filho & R.A. Mittermeier. Rio de Janeiro: Academia Brasileira de Ciências, pp. 443–493.

Girão, W. & Souto, A. (2005). Breeding period of araripe manakin *Antilophia bokermanni* inferred from vocalisation activity. *Cotinga*, **24**, 35–37.

Groves, C.P. (2001). *Primate Taxonomy.* Smithsonian Series in Comparative Evolutionary Biology. Washington, DC: Smithonian Institution.

Haimoff, E.H., Chivers, D.J., Gittins, S.P., *et al.* (1982). A phylogeny of gibbons (*Hylobates* spp.) based on morphological and behavioural characters. *Folia Primatologica*, **39**, 213–237.

Hand, J.L. (1981). A comparison of vocalizations of western gulls (*Larus occidentalis occidentalis* and *L. o. livens*). *Condor*, **83**, 289–301.

Hartley, J.C., Robinson, D.J. & Warne, A.C. (1974). Female response song in the ephippigerines *Steropleurus stali* and *Platystolus obvius* (Orthoptera, Tettigoniidae). *Animal Behaviour*, **22**, 382–389.

Hauser, M. (1993). The evolution of non-human primate vocalization: effects of phylogeny, body weight and social context. *American Naturalist*, **142**, 528–542.

Heline, W.T. (2007). Vocal repertoire of white-faced sakis (*Pithecia pithecia*). MSc thesis. Eastern Kentucky University.

Heller, K.G. & von Helversen, D. (2004). Acoustic communication in phaneropterid bushcrickets: species-specific delay of female stridulatory response and matching male sensory time window. *Behavioural Ecology and Sociobiology*, **18**, 189–198.

Hill, W.C.O. (1960). *Primates: Comparative Anatomy and Taxonomy*, Vol. **IV**. Edinburgh: Edinburgh University Press.

Höbel, G. & Gerhardt, H.C. (2003). Reproductive character displacement in the acoustic communication system of green tree frogs (*Hyla cinerea*). *Evolution*, **57**, 894–904.

Hodun, A., Snowdon, C.T. & Soini, P. (1981). Subspecific variation in the long calls of the tamarin, *Saguinus fuscicollis*. *Zeitschrift für Tierpsychologie*, **57**, 97–110.

Hohmann, G. (1989). Comparative study of vocal communication in two Asian leaf monkeys, *Presbytis johnii* and *Presbytis entellus*. *Folia Primatologica*, **52**, 27–57.

Ingard, U. (1953). A review of the influence of meteorological conditions on sound propagation. *The Journal of the Acoustical Society of America*, **25**, 405–411.

Inoue, M. (1988). Age gradation in vocalization and body weight in Japanese monkeys (*Macaca fuscata*). *Folia Primatologica*, **51**, 76–86.

International Union for Conservation of Nature and Natural Resources. 2010. *IUCN Red List Of Threatened Species*. www.iucn.org.

Jones, G., Vaughan, N. & Parsons, S. (2000). Acoustic identification of bats from directly sampled and time expanded recordings of vocalizations. *Acta Chiropterologica*, **2**, 155–170.

Kinzey, W.G. (ed.) (1997). *New World Primates: Ecology, Evolution and Behavior*. New York, NY: Aldine de Gruyter.

Kinzey, W.G. & Cunningham, E.P. (1994). Variability in platyrrhine social organization. *American Journal of Primatology*, **34**, 185–198.

Kinzey, W.G. & Norconk, M.A. (1990). Hardness as a basis of fruit choice in two sympatric primates. *American Journal of Physical Anthropology*, **81**, 5–15.

Krebs, J.R. & Davies, N.B. (1996). *Introdução à ecologia comportamental*. São Paulo: Editora Atheneu.

Lehman, S.M., Prince, W. & Mayor, M. (2001). Variations in group size in white-faced sakis (*Pithecia pithecia*): evidence for monogamy or seasonal congregations? *Neotropical Primates*, **9**, 96–101.

Marler, P. (1961). The logical analyses of animal communication. *Journal of Theoretical Biology*, **1**, 295–317.

Marten, K., Quine, D. & Marler, P. (1977). Sound transmission and its significance for animal vocalization. *Behavioural Ecology and Sociobiology*, **2**, 291–302.

McCombe, K. & Semple, S. (2005). Coevolution of vocal communication and sociality in primates. *Biology Letters*, **1**, 381–385.

Mills, M.G.L., Juritz, J.M. & Zucchini, W. (2001). Estimating the size of spotted hyaena (*Crocuta crocuta*) populations through playback recordings allowing for non-response. *Animal Conservation*, **4**, 335–343.

Morton, E.S. (1975). Ecological sources of selection in avian sounds. *American Naturalist*, **109**, 17–34.

Napier, J.R. & Napier, P.H. (1967). *A Handbook of Living Primates*. London: Academic Press.

Nickle, A.D. (1976). Interspecific differences in frequency and other physical parameters of pair forming sounds of bush katydids (Orthoptera: Tettigoniidae: Phaneropterinae). *Annals of the Entomological Society of America*, **69**, 1136–1144.

Norconk, M.A. (2006). Long-term study of group dynamics and female reproduction in Venezuelan *Pithecia pithecia*. *International Journal of Primatology*, **27**, 1573–8604.

Norconk, M.A. (2011). Sakis, uakaris, and titi monkeys. In *Primates in Perspective*, ed. C.J. Campbell, A. Fuentes, K. MacKinnon, S.K. Bearder & R.M. Stumpf. New York, NY: Oxford University Press, pp. 122–13p.

Ogutu, J.O. & Dublin, H.T. (1998). The response of lions and spotted hyaenas to sound playbacks as a technique for estimating population size. *African Journal of Ecology*, **36**, 83–95.

Peters, G. & Tonkin-Leyhausen, B.A. (2004). Evolution of acoustic communication signals of mammals: friendly close-range vocalizations in Felidae (Carnivora). *Journal of Mammalian Evolution*, **6**, 129–159.

Quris, R. (1980). Emission vocale de forte intensité chez *Cercocebus galeritus agilis*: structure, caractéristiques spécifiques e individuelles, modes d'émission. *Mammalia*, **44**, 35–50.

Radford, A.N. (2005). Group-specific vocal signatures and neighbour–stranger discrimination in the cooperatively breeding green woodhoopoe. *Animal Behaviour*, **70**, 1227–1234.

Ratcliffe, N., Vaughan, D., Whyte, C., *et al.* (1998). Development of playback census methods for Storm Petrels *Hydrobates pelagicus*. *Bird Study*, **45**, 302–312.

Rosenberger, A.L. (1981). Systematics: the higher taxa. In *Ecology and Behavior of Neotropical Primates*, ed. A.F. Coimbra Filho & R.A. Mittermeier. Rio de Janeiro: Academia Brasileira de Ciências, pp. 419–442.

Roush, R.S. & Snowdon, C.T. (1999). The effect of social status on food-associated calling behaviour in captive cotton-top tamarins. *Animal Behaviour*, **58**, 1299–1305.

Rowe, N. (1996). *The Pictorial Guide To The Living Primates*. East Hampton, NY: Patagonias Press.

Roy, D. (1994). Development of hearing in vertebrates with special reference to anuran acoustic communication. *Journal of Biosciences*, **19**, 629–644.

Shettleworth, S.J. (1998). *Cognition, Evolution and Behaviour*. Oxford: Oxford University Press.

Snowdon, C.T. (1993). A vocal taxonomy of the callitrichids. In *Marmosets and Tamarins: Systematics, Behaviour and Ecology*, ed. A.B. Rylands. Oxford: Oxford University Press, pp. 78–94.

Snowdon, C.T. (2001). Social process in communication and cognition in callitrichid monkeys: a review. *Animal Cognition*, **4**, 247–257.

Snowdon, C.T., Hodun, A., Rosenberger, A.L., *et al.* (1986). Long-call structure and its relation to taxonomy in lion tamarins. *American Journal of Primatology*, **11**, 253–261.

Sproul, C., Palleroni, A. & Hauser, M.D. (2006). Cottontop tamarin, *Saguinus oedipus*, alarm calls contain sufficient information for recognition of individual identity. *Animal Behaviour*, **72**, 1379–1385.

Struhsaker, T.T. (1970). Phylogenetic implications of some vocalizations of *Cercopithecus* monkeys. In *Old World*

Monkeys: Evolution, Systematics, and Behavior, ed. J.R. Napier & P.H. Napier. New York, NY: Academic Press, pp. 365–444.

Sugiura, H., Tanaka, T. & Masataca, N. (2006). Sound transmission in the habitats of Japanese macaques and its possible effect on population differences in coo calls. *Behaviour*, **143**, 993–1012.

Sutherland, W.J. (1996). *Ecological Census Techniques. A Handbook*. Cambridge: Cambridge University Press.

Thomassen, H.A. & Povel, G.D.E. (2006). Comparative and phylogenetic analysis of the echo clicks and social vocalizations of swiftlets (Aves: Apodidae). *Biological Journal of the Linnean Society*, **88**, 631–643.

Turcotte, Y. & Desrochers, A. (2002). Playbacks of mobbing calls of black-capped chickadees help estimate the abundance of forest birds in winter. *Journal of Field Ornithology*, **73**, 303–307.

van Roosmalen, M.G.M., Mittermeier, R.A. & Fleagle, J.G. (1988). Diet of northern bearded saki (*Chiropotes satanus chiropotes*): a Neotropical seed predator. *American Journal of Primatology*, **14**, 11–35.

Walker, S.E. & Ayres, J.M. (1996). Positional behavior of the white uakari (*Cacajao calvus calvus*). *American Journal of Physical Anthropology*, **101**, 161–172.

Waser, P.M. & Brown, C.H. (1986). Habitat acoustics and primate communication. *American Journal of Primatology*, **10**, 135–157.

Waser, P.M. & Waser, M.S. (1977). Experimental studies of primate vocalization: specializations for long distance propagation. *Zeitschrift für Tierpsychologie*, **43**, 239–263.

Zimmermann, E., Bearder, S.K., Doyle, G.A., *et al.* (1988). Variations in vocal patterns in Senegal and South African lesser bushbabies and their implications for taxonomic relationships. *Folia Primatologica* **51**, 87–105.

Part

IV

Conservation of the Pitheciids

Liza M. Veiga & Anthony B. Rylands

Diverse challenges and opportunities are encountered by the pitheciid genera depending on their innate ecological and behavioral characteristics and the interrelationship of these with their geographical location. The first three chapters of this section examine the effects of location, specifically, country- and/or region-specific trends on pitheciid survival. The three remaining chapters discuss key conservation topics, namely, pitheciids in captivity, fragmented living and the experience of conservation initiatives.

Consistent themes pertinent to conservation emerge from these chapters: what constitute the main threats for these primates, what level of regional, national and international interest exists in the conservation of biodiversity, and how development activities impact them.

In the second half of the section, life in captivity is examined. The role of *ex situ* living is becoming increasingly important as natural habitat disappears and population viability in the wild decreases. Some species such as the black-bearded saki and the San Martan titi monkey are becoming locally extinct and restricted to habitat patches.

Sadly, many regions are increasingly fragmented, so it is important to understand the implications of "fragmented living" for the pitheciids.

Main threats

1. Loss of habitat and hunting pressure.
2. Impact of organized development activities.

Evolutionary Biology and Conservation of Titis, Sakis and Uacaris, eds. Liza M. Veiga, Adrian A. Barnett, Stephen F. Ferrari and Marilyn A. Norconk.
Published by Cambridge University Press.

Impact

Brazil is the only home for the vast majority of the 50 recognized taxa occur in Brazil and 20 species are endemic, so conservation efforts here will have a huge impact on the survival of the group as a whole.

Another recurring topic is the need for further research. Geographic distribution, population densities, preferred food resources, habitat requirements, home range or territory sizes are unknown for many populations of pitheciids. Until more is known about populations in the wild, it is difficult to provide adequately for species in captivity.

Clearly, many challenges still lie ahead, and perhaps one of the greatest will be the identification of priorities for future studies. Conservation is the most obvious of these, and in general, studies of endangered species should probably be prioritized over those of less-threatened ones, as far as possible. However, the understanding of broader ecological processes and evolutionary patterns may be at least as important for the development of effective management strategies. Ultimately, the long-term survival of pitheciid populations will depend on the protection of the ecosystems they inhabit, and all the case studies presented in this section will surely contribute to this process.

Chapter 31

The Guyana Shield: Venezuela and the Guyanas

Shawn M. Lehman, Jean-Christophe Vié, Marilyn A. Norconk, Carlos Portillo-Quintero & Bernardo Urbani

Introduction

The Guyana Shield is an ancient geological formation underlying Guyana, Suriname and French Guiana, as well as parts of Colombia, Venezuela and Brazil (Hammond 2005a). This region encompasses approximately 1.8 million km^2 of tropical rainforest, tepuis (tabletop mountains), palm swamps and *llanos* (Parry & Eden 1997). The Guyana Shield is also one of the few tropical regions in which the majority of the forests remain intact (ter Steege 1993). As a result of its unique historical biogeography, the Guyanas are characterized by a remarkable diversity of plants and animals (Lindeman & Mori 1989), many of which remain largely unstudied and unexploited. Researchers have described approximately 6000 species of vascular plants; 1004 birds and 282 mammals in the region (e.g. Boggan *et al.* 1997; Lim *et al.* 2005; Milensky *et al.* 2005). There are four genera of Pitheciinae in the Guyanas (*Callicebus*, *Pithecia*, *Chiropotes* and *Cacajao*). In this chapter, we describe the conservation status of the pitheciines that range into Venezuela, Guyana, Suriname and French Guiana.

Venezuela

Venezuela is a large (916,445 km^2) country characterized by considerable biodiversity ranging across xeric scrublands in the extreme northwest through broad inland plains to coastal mangrove forests in the northeast. The Orinoco River bisects the country into northwest and southeast regions. The northwest region includes the northern end of the Andes and parts of the *llanos*. The southern region also contains extensive *llanos* as well as the northern part of the Amazon Basin.

There are four species of pitheciines in Venezuela, all of which are distributed in the forested region south of the

Photo 31.1 The open area behind this male white-faced saki (originally part of the group's home range), was deforested to install a telecommunications tower. Photo: Marilyn Norconk.

Evolutionary Biology and Conservation of Titis, Sakis and Uacaris, eds. Liza M. Veiga, Adrian A. Barnett, Stephen F. Ferrari and Marilyn A. Norconk. Published by Cambridge University Press.

Orinoco River. This region includes the Amazonas, Bolívar and Delta Amacuro states. *Cacajao melanocephalus melanocephalus* is restricted to the southern part of the Amazonas state, south of the upper Orinoco River, and along the Casiquiare branch and Negro and Guainía river basins. *Callicebus lugens* (formerly recognized as *Callicebus torquatus lugens* in Venezuela) is known from the western Paragua River basin in central Bolívar state up to the Brazilian border in the southern Amazonas state. *Chiropotes satanas chiropotes* (aka *Chiropotes sagulatus*) is distributed west of the Caroní River basin to the upper Orinoco River basin. *Chiropotes israelita* has recently been reported from the northern Brazilian Negro River region (Bonvicino *et al.* 2003), and may also be present in the southern Amazonas state. *Pithecia pithecia* is restricted to the northeastern Bolívar state and southern part of the Delta Amacuro state. This species is located east of the lower Caroní River, south of the lower Orinoco River, and along the Cuyuní river basin.

Most of the current range of pitheciines remains unprotected. Preliminary Gap Analysis Program data of pitheciines in Venezuela indicate only 24% of the total ranges for *Callicebus*, 41% for *Cacajao*, 24% for *Chiropotes* and 0% for *Pithecia* fall inside existing protected areas. No systematic research has so far been conducted on pitheciine population densities, and only a few studies have evaluated mean group size. *Pithecia* groups range from 5 to 9 individuals in Lago Guri and the Cuyuní River, Bolívar state (Norconk 2006; Urbani 2006), *Chiropotes* range from 15 to 22 in Lago Guri (Norconk *et al.* 1997; Peetz 2001) and groups of around 10 individuals have been reported for *Cacajao* around San Fernando de Atabapo and San Carlos de Río Negro, Amazonas state (Bodini 1983; Lehman & Robertson 1994). Pitheciine population densities can only be calculated from the well-studied Guri sites; however, these are islands that were formed after the construction of a dam. The other sites were visited in rapid surveys. There are no available data for *Callicebus*.

Specific risks and trends

Although pitheciines are protected from hunting by the Law for the Protection of Wild Fauna (Rodríguez & Rojas-Suárez 1995), enforcement is rare throughout Venezuela. Furthermore, economic issues are forcing many Venezuelans into primate hunting as a means of acquiring much-needed protein. Legal and illegal mining and consequently hunting has expanded, reaching historically remote areas as far as the border with Brazil and Guyana. Cattle ranching and logging activities are moving into areas inhabited by *Pithecia*, *Chiropotes* and *Callicebus* (B. Urbani, pers. obs.). In the northern Amazonas and Bolívar states, where most protected areas are small and close to human populations, construction of new roads is resulting in new settlements. Thus, it is likely that pitheciines are experiencing population declines in Venezuela.

History of previous conservation efforts to manage wild populations

Specific conservation programs have not been implemented for the protection of pitheciines in Venezuela. However, during the last few decades, there has been an increased interest in conducting scientific explorations in Venezuela (Huber 1990). There is also a network of local scientists with research interests in nature conservation based in Venezuela. More recently, the long-term research program directed by M.A. Norconk (Kent University) and the late W.G. Kinzey (City University of New York) was successful in obtaining important socioecological data on *Pithecia* and *Chiropotes*. This program has also trained undergraduate and graduate students from Venezuela and other countries in ecological methods. Although the project has been completed, it offers a successful example of international cooperation between primate researchers and a national public company (Electrificación del Caroní, EDELCA).

Summary of current and planned conservation programs

Venezuela is at a critical point in that it must determine the future of forest development and conservation. As the country deals with serious national economic pressures and the desire to reduce its long-term dependence on oil, plans are currently being made to preserve forested areas south of the Orinoco River. In a report on sustainable development of this region, Miranda *et al.* (1998) reached the following conclusions: (1) the benefits from logging and mining are not being fully realized at the national or local levels, (2) logging and mining currently have serious negative impacts on the environment and society, and (3) expansion of logging and mining activities will result in high environmental and social costs. The recommendations of the authors are to: (1) maximize revenues from forest resources to ensure that benefits contribute to long-term forest conservation, (2) minimize the environmental and social impacts of mining and logging, and (3) employ public participation in creating sustainable use of forest products. In addition, we suggest that it is necessary to protect areas for strict conservation purposes; and probably one of the best-preserved examples is the middle-to-upper Caura River basin.

Assessment of the likelihood of success

One option to conserve pitheciines and other primates is to improve the quality of life of local people, with better education and health services as well as alternative economic activities (Carrillo & Perera 1995; Perera 1997). Concerns regarding corruption at all institutional levels need to be addressed to achieve successful protection of primate habitats (Wright *et al.* 2007). Priority should be given to raise the interest of governmental and private agencies in the application of conservation plans (Esteves & Dumith 1998), in addition to the necessity of basic research on the ecology and behavior of pitheciines. For example, research is still needed for unstudied species, such as *C. lugens*. As indicated by Bevilacqua *et al.* (2002), Venezuela has the largest protected area system in Latin America relative to its size. However, the legal status of some protected areas needs to be clarified and strengthened, with prioritization for improvement of the working conditions for park personnel. Finally, pitheciine conservation status will be improved by national awareness campaigns and educational strategies.

Guyana

The Co-operative Republic of Guyana is a small country of 214,969 km^2 situated on the northeastern coast of South America. Approximately 80% of the country is covered by forest habitats, which contain hundreds of species of birds, fish, reptiles and mammals. Nine primate species exist in Guyana (*Ateles paniscus*, *Alouatta seniculus*, *Cebus albifrons*, *Cebus apella*, *Cebus olivaceus*, *C. sagulatus*, *P. pithecia*, *Saguinus midas* and *Saimiri sciureus*), with anecdotal evidence for *Aotus* (Sussman & Phillips-Conroy 1995). Although *P. pithecia* is found throughout the forested regions of Guyana, *C. sagulatus* is limited to the eastern and southwestern portions of the country (Lehman 2004b). Sightings rates for Pitheciines and other primates have reduced markedly over the last 30 years (Muckenhirn *et al.* 1975; Sussman & Phillips-Conroy 1995; Lehman 2000), representing a serious conservation issue that has been linked to anthropogenic disturbances.

Specific risks and trends

The main anthropogenic threats to Pitheciines result from logging, mining and hunting (Lehman *et al.* 2006). Since 1980, the Guyana Forestry Commission has handed out 8.8 million ha of state forests as logging concessions to foreign companies (Parry & Eden 1997). For example, the Barama concession is 1.69 million ha in northwest Guyana. A recent estimate of the harvesting rate for tropical hardwoods is 350,000–400,000 m^3/year (FAO 2003). However, there is concern that logging companies are steadily increasing harvesting rates (Parry & Eden 1997). Legal and illegal mining for gold, diamonds and bauxite also have negative impacts on forests and primate populations (Comiskey *et al.* 1993; Bergquist 2004). The direct effects of mining are limited in scale to the excavation site (i.e. local deforestation), but the cumulative secondary effects of polluted runoff from tailings and catchment ponds into rivers are of considerable concern for long-term health of forest habitats (Richardson & Funk 1999; Mol & Ouboter 2004). For example, in 1995, three million cubic meters of cyanide-laden waste water was spilled into local rivers at the Omai Gold Mines Ltd. operations in central Guyana. Although there are no data on pitheciine responses to logging in large concessions, Lehman *et al.* (2006) reported that small-scale logging had no effect on sighting rates for either *P. pithecia* or *C. sagulatus*. Hunting directly impacts primate populations in Guyana, with *P. pithecia* and *C. sagulatus* being exploited as food by Wapishana/Arawak Amerindians in southern Guyana and by Bush Negroes in eastern parts of the country (Cormier 2006). Despite the complete loss of some primate populations to wildlife trappers (Lehman *et al.* 1995), few *P. pithecia* are legally exported. However, there is a thriving illegal wildlife trade that includes pitheciines (Lehman, pers. obs.).

History of previous conservation efforts to manage wild populations

Of the 21 protected areas in Guyana, only 2 are officially legislated: Kaieteur National Park (62,700 ha) and Iwokrama Rainforest Reserve (371,992 ha). The non-legislated protected areas exist only on paper or at the whim of private land owners. There is extensive habitat disturbance of local rivers and associated forest habitats by illegal miners in Kaieteur National Park (Barnett *et al.* 2000). Iwokrama contains large tracts of undisturbed forest habitats, making it seem an ideal site for primate studies. However, the Iwokrama management plan allows for some commercial and cultural exploitation of forest resources, including traditional subsistence hunting of primates by local Amerindians in the wildlife preserve (185,731 ha). There are few data on Pitheciinae beta diversity and densities for much of Guyana, including the two protected areas. At present, it can be stated that *P. pithecia* exists in Kaieteur and Iwokrama (Sussman & Phillips-Conroy 1995; Lehman 2000; Wright 2005) and that *C. sagulatus* does not range into either protected area (Lehman 2000).

Although researchers associated with the Biological Diversity of the Guiana Shield Program at the Smithsonian Institution and Tropenbos International have made enormous contributions to our understanding of plant and animal biodiversity in Guyana (ter Steege 1993; Funk *et al.* 2005), these programs have or will soon cease operations in the country. Efforts by researchers and non-governmental organizations (NGOs) to assess and manage wild primate populations is a new phenomenon in Guyana, with the majority of primate studies being conducted in Iwokrama. Local NGOs have so far focused their efforts on small private reserves and eco-tourist sites (e.g. Karanambo-Lodge, Shell Beach), which have only limited impacts on local primate populations. Among the international NGOs working in Guyana, Conservation International has made the most progress by developing plans for a series of community-owned conservation areas in southern Guyana.

Summary of current and planned conservation programs

Few conservation programs are specific to primates and none to Pitheciinae in Guyana. Most conservation efforts have been focused on biodiversity assessments and protection of areas of presumed biodiversity value (Funk *et al.* 2005). In 1994, the Government of Guyana received funding from the World

Bank to consider creating a national protected areas system. The Iwokrama International Centre for Rain Forest Conservation and Development Act was passed into law in 1996. Then in 1997, the Government created the Environmental Protection Agency to oversee and monitor environmental activities. In 1999, the Government of Guyana expanded the size of Kaieteur Falls National Park from 11 ha to 62,700 ha. In 2006, the Global Environment Facility provided funding to support development of management and legal frameworks for a national protected area system in Guyana. The five priority areas under this project are Shell Beach, the Kanuku Mountains, Orinduik, Mount Roraima and the southern Guyana region.

Assessment of the likelihood of success

Despite conservation benefits achieved by the creation of the Iwokrama Rainforest Reserve and expansion of Kaieteur National Park, little has been accomplished directly in terms of primate conservation in Guyana. Primate populations are decreasing throughout the country (Sussman & Phillips-Conroy 1995; Lehman *et al.* 2006), likely as a result of deforestation, hunting and trapping. The Government of Guyana has delayed legislation to create new protected areas in eastern rain forests, which contain the highest primate species richness (Lehman 2004a). Furthermore, the Government and its international donors have indicated that a major objective of any new protected areas system will be to increase access to and exploitation of timber, gold and diamonds (Richardson & Funk 1999). Because of the above issues and the fact that Guyana scores very poorly using corruption perception indices at the global level (Lambsdorff 2006), there are serious doubts regarding chances for successful conservation programs of long-term benefit for primates.

Suriname

Suriname (previously Dutch Guiana) is located in northern South America on the Caribbean coast between Guyana and French Guiana. The country has one major city, the capital city of Paramaribo, very few paved roads (mostly around the capital and connecting eastern and western parts of the country along the coast, and the lowest human population density (2.7 people/km^2) in South America (*Economist* 2003). Myers *et al.* (2000, p. 857) described the entire region of the Guianas as a "major wilderness area" or "good news" region and this label is well-supported by Hammond's (2005a, p. 485) estimate (using year 2000 data) that approximately 90% of Suriname's total land area consists of tropical forest cover. However, the future of Suriname's forests should be tempered with concerns about the extent of legal and illegal mining and unregulated hunting, particularly in mining areas, and poor governmental regulation and/or interest in taking an active role in conserving natural resources (Colchester 1995).

Like the other Guianas, as well as the southern states of Venezuela, Brazil north of the Amazon River, and southeastern Colombia, Suriname's forests sit atop the ancient, geologically complex and highly weathered, Guiana Shield (see Hammond (2005b) for a thorough discussion of the geology of the region). Formed in the early Proterozoic of the Precambrian (2.4–2.0 billion years ago), the Guiana Shield is the largest region of exposed rock in the world. It has undergone changes associated with tectonic, volcanic, sedimentary and erosional activities (Hammond 2005c) and the result, in Suriname, is a landscape that is largely low in elevation and relatively flat in the interior dotted with more than 100 isolated remnant structures or massifs ranging from 500 to 1300 m (e.g. Brownsberg, Tafelberg and Juliana Top: Huber *et al.* 2003; Hammond 2005c). Unlike the interior, the northern-most portion of Suriname is a construction of the alluvial fan action of the Amazon River during the Tertiary. Presently, this area consists of mixed vegetation coastal plain, beaches and mangrove forest succeeded by a mixture of dry land, (seasonally flooded) forest, swamp forest and marsh forest (Lindeman & Mori 1989). These habitats of the forested interior versus the relatively open northern region have affected the distribution of both people and nonhuman primates.

Suriname has always had a low human population density. The Guiana Shield was occupied much later than the coastal regions and central Amazonia where resources may have been more abundant for hunters and gatherers or small-scale horticulturalists (Hammond 2005c). The middle period of Suriname history (from the early seventeenth century to the early twentieth century) had a stronger influence on the configuration and size of the current population. The Dutch sphere of influence on trade and sugar cane plantation development was pre-eminent in the seventeenth and eighteenth centuries, in both Guiana and Suriname, but was dependent on slave labor from West Africa. Over a period of 200 years until their emancipation in 1863 (Price 1996), an estimated 66,000 slaves escaped into the interior to pursue a subsistence-level lifestyle on the rivers (Thoden van Velzen *et al.* 2004). Interestingly, with their intimate knowledge of the rivers and their navigational and boat-making skills, they entered into a profitable collaboration with the gold miners in the late nineteenth century, opening up the interior to early mining (Thoden van Velzen *et al.* 2004). Today, maroons are themselves becoming major players in small- and medium-scale gold mining.

Specific risks and trends

Suriname became independent of Holland in 1975 and has suffered a military coup and civil war since then, resulting in periods of near economic collapse in the twentieth century (Kroes 1982; Thoden van Velzen *et al.* 2004). Given severe economic constraints, Colchester (1995, p. 18) warned that "the sale of the country's natural resources to foreign companies seems to be its only option." Suriname has a long history of foreign interest in its minerals, the extraction of bauxite being the most productive. Bauxite has been extracted from coastal and near-interior sites since 1922 (Hammond 2005c). Historically, SURALCO (parent company: Alcoa, Pittsburgh, USA) has been the principal company involved in the extraction and exportation of processed bauxite or alumina, but BHP Billiton (AustraloBritish mining company) began an assessment of the quality of naturally occurring bauxite and the potential destruction to the environment in 2002, in preparation for extraction in the largely undisturbed Bakhuis Mountains of western Suriname. These mountains form the northwestern border of the Central Suriname Nature Reserve (CSNR).

Hammond (2005b, 2005c) calculated that Suriname has placed 10.5% of its total area under protection; however, there is very little active protection of any of these areas. The country has one very large nature reserve that was designated a Natural World Heritage Site in 2002 (Huber *et al.* 2003). Through the efforts of Conservation International and donors and supporters, including the Suriname government (CSNR), this remote reserve should protect 1.6 million ha of pristine lowland tropical rainforest including two pitheciins, *Pithecia pithecia* and *Chiropotes sagulatus.* The same two pitheciins are found at Brownsberg Nature Park (BNP), Suriname's only park that is easily accessible to national and international visitors and researchers. The mountain and the surrounding forest are, unfortunately, also accessible to gold miners. Small- and medium-scale mining has been increasing over the past 5 years at the base of the mountain within park boundaries and recently professional hunters have been employed to procure meat for the miners (Norconk, pers. obs.). Without protection, the future of this diverse (de Dijn *et al.* 2006) and beautiful forest and the animals that inhabit it is very bleak.

As of 1988, the government had granted gold mining concessions on all the major rivers (Coppename, Saramacca, Kabalebo, Suriname and Marowijne rivers; Suriname 1987), but there has been a recent surge of both legal and illegal mining as the price of gold on the world market has more than doubled from $300/oz to the current price of $653/oz from 2002 to 2007 (http://www.nymex.com/index.aspx). Legal concessions granted to international mining companies (e.g. IAMGOLD and Reunion Gold Corp, both of Québec, Canada) do not include the small- and medium-scale illegal mining operations that are having a devastating effect on the forests and streambeds of the interior. Providing support for Colchester's view that the sale of natural resources is the country's only current income option, the government has thus far been unwilling to step in to regulate the scope and location of habitat destruction due to illegal or legal gold mining. In addition, the government has not moved to regulate the use of mercury that is used in processing small amounts of gold in spite of increasingly frequent reports of streambed and fish contamination (Mol & Ouboter 2004) and documented effects on human health (de Kom *et al.* 1998; Vahter *et al.* 2002).

Mittermeier (1977, 1991) conducted the only country-wide primate survey in Suriname, including 11 sites in the three vegetation zones. Sakis were absent in coastal and savanna regions until the savanna forest gave way to higher riverine forest habitats upstream, but were found throughout the interior. A recent census of the northern plateau and upper slope forest of BNP documented both species of sakis to be present and abundant (Norconk *et al.* 2003). White-faced sakis were abundant in the sense that numerous groups were documented, inhabiting small home ranges (about 10 ha). Bearded sakis were abundant by virtue of their large group size (as many as 44 individuals), but given their large home ranges there may be as few as five or six groups inhabiting the Park. There has been no recent census of pitheciins in the Central Suriname Nature Reserve, but it is presumed that they are also abundant throughout this pristine region.

The Suriname game law of 1954 specifies hunting seasons, classifies animals that can be hunted or traded, and regulates the issuance of permits (Hofwijks & Madhar 2006), but it applies only to the northern part of the country and the area around the Brokopondo reservoir (see summary of the game law of 1954 of Suriname in Duplaix 2001). Mittermeier (1977) documented hunting in the Lely mountains and Kaiser mountains in eastern Suriname 30 years ago, and recent wildlife surveys in the Lely mountains suggests that hunting is severe (Alonso & Mol 2007). According to Hofwijks and Madhar (2006, p. 15) the "commercial bushmeat trade is a burgeoning industry in Suriname with professional hunters working year-round in the interior." Their study included a survey of markets in and around the capital, Paramaribo, and interviews of tourists departing the international airport for Holland. Although the study by Hofwijks and Madhar (2006) did not specifically list primates in their findings, Ouboter (2001) and Duplaix (2001) found that hunting pressure was high for spider monkeys, capuchins and howlers. Mittermeier (1991, p. 99), using data collected in interviews, found that 81% of rural Surinamers eat monkeys. *Pithecia* was ranked the most preferred species by 26.7% of those people who eat monkeys and *Chiropotes* by 10% (Mittermeier 1991, p. 100); nevertheless, the two pitheciins are near the bottom of the list of mammals that are routinely hunted/preferred. Conversations with people living in the village of Brownsweg near the BNP suggest that hunters will shoot anything they find, including howlers, spider monkeys, capuchins and white-faced sakis (R. Finesi, pers. commun.).

Bushmeat-eating is not restricted to the interior. Twenty percent of the people surveyed in the departure lounge of a KLM flight to Holland ($n = 46$) claimed to have eaten bushmeat during their stay in Suriname, although they did not always know, or indicate that they knew, what they were eating (Hofwijks & Madhar 2006). In concert with the bushmeat trade, Ouboter (2001) documented a lively, albeit mostly non-primate, international animal trade from Suriname to the Netherlands, Spain, Canada, and the USA.

History of previous conservation efforts to manage wild populations

The Ministry of Natural Resources oversees the management and designation of national parks and nature reserves in Suriname. The Nature Conservation Division of the Forest Suriname Forest Service (Dienst Lands Bosbeheer, or LBB) is charged with direct management of protected areas in Suriname. STINASU (Stichting Natuur Beheer, Suriname) is a semi-governmental foundation charged with assisting the forest service (research, education and nature tourism). While it would seem that forest protection is inherent in the guidelines of LBB, there are limited staff and operational funds to protect the borders of the designated areas and no permanent presence of guards in any of the parks. Increased numbers of roads related to gold mining will surely open areas up that have been protected in the past by virtue of their remoteness.

Summary of current and planned conservation programs

There are no conservation plans directed at monkeys, let alone sakis. Suriname has 11 nature reserves and 1 nature park (Huber *et al.* 2003, pp. 91–92). These include two sites where sakis (*Pithecia pithecia* and *Chiropotes sagulatus*) have or are being studied and where they are known to be abundant (CSNR and BNP). Highly visible non-governmental agencies, e.g. World Wildlife Fund and Conservation International, are focused on protecting habitat and globally endangered target species, like the giant otter or sea turtles. Even though saki populations are difficult to quantify in a systematic way, due to the cryptic behavior of white-faced sakis and the large home ranges of bearded sakis, they are not likely to be identified as species in need of special protection.

Assessment of the likelihood of success

Combining the "good news" of the extensive forests in the interior of Suriname with increasing mining activities and the unwillingness of the government to minimize the number or size of mining and logging concessions presents a very mixed, and potentially tragic, picture for Suriname. Legislation aimed at regulating hunting in the interior and illegal mining, particularly in nature reserves, would be a significant improvement, but the heart of the illegal mining problem is unemployment and relatively low levels of education (Heemskerk 2001). What is currently missing at many levels of Suriname society is an appreciation for nature and the willingness to envision habitats and wildlife as valuable beyond monetary gain. It will be interesting to see if the June 4, 2007 report from Conservation International on the discovery of 24 new species in the Nassau Mountains of eastern Suriname (http://www.conservation.org/xp/frontlines/2007/06040701.xml), the result of rapid assessment project (RAP) undertaken with the support of BHP Billiton and Suriname Aluminium Company (Suralco), will have any effect on redirecting or curtailing mining in that area.

French Guiana

French Guiana is an overseas territory of France. It covers approximately 84,000 km^2 of which 90% is covered by forest. Human population is low (around 200,000 individuals) but is increasing quickly. The geography of French Guiana is characterized by a coastal strip (where the majority of the people live), and tropical forest which gradually rises to modest peaks not exceeding 830 m. The dominant tree families are Lecythidaceae, Caesalpiniaceae, Chrysobalanaceae and Sapotaceae (de Granville 1988).

Of the eight primate species found in French Guiana, two are in the Pitheciinae (*Chiropotes sagulatus* and *Pithecia pithecia*; de Thoisy & Dewynter 2003). *Pithecia pithecia* has been the subject of a specific study (Vié 1998; Vié *et al.* 2001). *Pithecia pithecia* is widely distributed but due to its elusive behavior, very rare vocalizations, small group size (2.3 animals per group), low density (0.64 individuals /km^2), and large home range (> 140 ha), the status of this species is difficult to assess (Vié 1998; Vié *et al.* 2001). Conversely, *C. sagulatus* has a limited distribution in the extreme south of the country for reasons that are poorly understood. Therefore, information on this species is almost non-existent and its precise distribution remains unknown.

Specific risks and trends

Logging is the main threat to forest habitats in French Guiana, but it is currently limited to a 70-km wide strip along the coast. It is considered "selective" in that only four tree species account for 70% of timber volume. The direct impact of logging on the forest structure varies according to studies (de Thoisy & Vié 1998) and its effect varies from one species to the other. *Pithecia pithecia* appear more abundant in hilly areas with a high continuous canopy, but they are also one of the two dominant species (with tamarins) in more disturbed habitats (de Thoisy 2010). However, construction of new roads provides access to protected, remote areas. This incursion, associated with a lack of hunting regulation and law enforcement, largely contributes to the current unsustainable hunting practices. Hunting efforts are most intense for *A. paniscus*, *A seniculus* and *Cebus* spp., with pitheciines being hunted opportunistically. Traditionally, local Amerindians focused their hunting effort on largest non-primate mammals. Monkey biomass varies from 0.3 to 14% of the total game biomass collected by Amerindians (de Thoisy 2010). However, recent studies around some human settlements clearly show that harvesting levels are unsustainable. Although prohibited, the sale of monkey meat is still common. All human communities, either indigenous or not, hunt monkeys, either for consumption or trade (de Thoisy *et al.* 2005). Some large-bodied primate species, such as *A. paniscus*, have been extirpated by hunting in vast areas including those around main human settlements, along main roads and more generally in a large band along the coast. Hunting is also associated with gold mining. The recent rush and the very large increase in the number of illegal settlements along all main rivers represent a major environmental and security problem. Large areas of lowland rain forest have been cleared for agriculture or flooded following the building of a dam (Vié 1999). Forest roads increase habitat fragmentation. Although some primate species are able to cross these roads, they represent a real barrier for *P. pithecia*. The species is known to survive well in forest fragments and does not avoid edges but forest fragmentation is likely to represent a long-term threat by reducing opportunities for individuals to disperse (Vié 1998; Vié *et al.* 2001).

History of previous conservation efforts to manage wild populations

A few protected areas have been created during the last 15 years that should benefit pitheciine conservation. Four natural reserves have been created (Nouragues, Trinité, Kaw and Amana), but they still have limited means and lack clearly defined management plans. Illegal mining and hunting still occur in some of the reserves. The proposal to include the area around the Petit Saut dam into a protected area, where the only studies on pitheciines have been conducted in French Guiana, was not pursued by the government despite successful protection for a few years (Vié 1999).

Three nonhuman primate species (*A. paniscus* and the two pitheciines) have been entirely protected by a ministerial decree since 1986. This decree also prohibits trade in monkey meat and other primate species can be hunted only for subsistence. Despite this decree, the bushmeat trade is still common and the price of monkey meat is lower than the price of other wild species sold legally. Thus, commercial hunting still persists in French Guiana, although fortunately, primates are not the main target. The conservation community has repeatedly asked for a revision of the law to include total protection to all primate species but this has been rejected, in particular because of lobbying by hunting groups. There are no hunting regulations, no quotas for hunted species and controls are ineffective.

Summary of current and planned conservation programs

March 2007 represents a landmark for nature conservation in French Guiana with the long-awaited creation of the "Parc Amazonien de Guyane". This is the culmination of a 15-year process to create a national park covering 3.3 million hectares of lowland forest, including 2 million hectares as the core area of the Park where nature protection will be stronger. The core area will be likely to cover the entire range of *C. sagulatus* in French Guiana. It will complement a network of recently created natural reserves. There is now very solid and documented evidence that primate communities decline very quickly following unsustainable hunting (often associated with logging, illegal mining and habitat destruction). The conservation community has asked repeatedly to ban illegal mining activities, to protect additional vulnerable species, to create canopy corridors when building new roads, and to shorten logging periods to limit the impact on biodiversity; all with limited results. On the other hand, local NGOs have been very active on other fronts such as awareness raising, biodiversity surveys and monitoring the impacts of human activities.

Assessment of the likelihood of success

Nature conservation is quite a recent notion in French Guiana, but over the last two decades significant progress has been made. Despite these efforts, the threats to biodiversity have increased more quickly than measures to mitigate them have been put into place. The network of protected areas is not entirely satisfactory with very few staff in some reserves, illegal activities taking place in others, and very few enforcement measures overall. Nevertheless, with the recent addition of the National Park, the protected area network is becoming potentially stronger and offering great potential for conservation of pitheciines. The impact will depend on political will and funding from the central French government as well as an adequate recruitment of the staff in charge of its management. More importantly the network comprising the newly created *Parc Amazonien de Guyane*, the Tumucumaque National Park and the ecological station of Grão Para and the Maicuru reserve, forms the largest area of protected rainforest in the world, covering 12 million ha. There is now a good legal framework to operate and reverse trends in biodiversity loss, and enforcement of current legislation would be a great step forward. Increasing public awareness as to the importance of the country's biodiversity is also a major task, in order to change the public's perception that forest resources are unlimited and of limited value to the country. This task has been undertaken mostly by local NGOs, but greater support from the government administration will improve the effectiveness of their efforts.

Acknowledgments

We thank Adrian Barnett, Liza Veiga, Stephen Ferrari and Marilyn Norconk for their kind invitation to participate in this edited volume. We thank Liza Veiga and the reviewer for excellent comments made on earlier versions of our manuscript. S.M. Lehman gratefully acknowledges the support of V. Funk and C. Kelloff of the Biodiversity of the Guianas Program, Smithsonian Institution; and R.W. Sussman and J. Phillips-Conroy of Washington University. C. Portillo-Quintero is currently supported by a University of Alberta Assistantship; and B. Urbani by a UIUC Assistantship, and a former Fulbright-OAS Fellowship.

References

Alonso, L.E. & Mol, J.H. (2007). *A Rapid Biodiversity Assessment of the Lely and Nassau Mountains, Suriname.* Washington, DC: Conservation International.

Barnett, A., Shapely, B., Lehman, S.M., *et al.* (2000). Primate records from the Potaro Plateau, western Guyana. *Neotropical Primates*, **8**, 35–40.

Bergquist, T. (2004). Fish community structure and organization in neotropical clear and blackwater streams in Guyana. MS thesis, Florida Atlantic University.

Bevilacqua, M., Cárdenas, L., Flores, A.L., *et al.*(2002). *The State of Venezuela's Forest: A Case Study of the Guayana Region.* Caracas: Global Forest Watch/WRI/ACOANA/Universidad Nacional Experimental de Guayana/ Provita/Fundación Polar.

Bodini, R. (1983). Distribución geográfica y conservación de primates sub-humanos en Venezuela. In *La Primatología en Latinoamérica*, ed. C.J. Saavedra, R.A. Mittermeier & I. Bastos-Santos. Bairro Cincão (Brazil): WWF/Editora Littera Maciel Ltda, pp. 101–113.

Boggan, J., Funk, V., Kelloff, C., *et al.* (1997). *Checklist of the Plants of the Guianas (Guyana, Surinam, French Guiana).* Washington, DC: Biological Diversity of the Guianas Program.

Bonvicino, C.R., Boubli, J.P., Otazú, I.B., *et al.* (2003). Morphologic, karyotypic, and molecular evidence of a new form of *Chiropotes* (primates, pitheciinae). *American Journal of Primatology*, **61**(3), 123–133.

Carrillo, A. & Perera, M.A. (1995). *Amazonas: Modernidad en Tradición. Contribuciones al Desarrollo Sustentable en el Estado Amazonas, Venezuela.* Caracas: SADA-Amazonas/ORPIA/ CAIAH/GTZ-Venezuela.

Colchester, M. (1995). *Forest Politics in Suriname.* Utrecht, The Netherlands: International Books.

Comiskey, J., Dallmeier, F., Aymard, G., *et al.* (1993). *Biodiversity Survey of Kwakwani, Guyana.* Washington, DC: The Smithsonian Institution/Man and the Biosphere Biological Diversity Program, p. 200.

Cormier, L. (2006). A preliminary review of neotropical primates in the subsistence and symbolism of indigenous lowland South American peoples. *Ecological and Environmental Anthropology*, **2**(1), 14–32.

de Dijn, B.P.E., Molgo, I., Norconk, M.A., *et al.* (2006). *Biodiversity of the Brownsberg, Report to Conservation International.*

de Granville, J.-J. (1988). Phytogeographical characteristics of the Guianan forests. *Taxon*, **37**(3), 578–594.

de Kom, J.F.M., van der Voet, G.B. & de Wolff, F.A. (1998). Mercury exposure of maroon workers in the small scale gold mining in Suriname. *Environmental Research Section A*, **77**, 91–97.

de Thoisy, B. (2010). Monos en Guyana Francesa: diversidad y abundancia en relación con los hábitats y sus amenazas. In *Primatologia en Colombia: avances al principio del milenio*, ed. V. Pereira-Bengoa, P.R. Stevenson, M.L. Bueno Fernando Nassar-Montoya. Bogota: Fundación Universitaria San Martín.

de Thoisy, B. & Dewynter, M. (2003). *Les primates de Guyane.* Cayenne, French Guiana: Collection Nature Guyanaise, Association Kwata et Sepanguy.

de Thoisy, B. & Vié, J.-C. (1998). Faune sauvage et activités humaines: chasse et exploitation forestière en Guyane Française. *JATBA, Revue d'Ethnobiologie*, **40**(1–2), 103–120.

de Thoisy, B., Renoux, F. & Julliot, C. (2005). Hunting in northern French Guiana and its impact on primate communities. *Oryx*, **39**(2), 149–157.

Duplaix, N. (2001). *Evaluation of the Animal and Plant Trade in the Guayana Shield Eco-Region, Preliminary Findings. W. W. F. G. F. a. E. C. Project.* Georgetown, Guyana: WWF.

Economist. (2003). *Pocket World in Figures.* Vicenzia, Italy: LEGO S.p.a.

Esteves, J. & Dumith, D.A. (1998). *Diversidad Biológica en Amazonas. Bases para una Estrategia de Gestión.* Caracas: SADA-Amazonas/PNUD/Fundación Polar.

FAO. (2003). *The State of the World's Forests.* Rome: Food and Agricultural Organization, United Nations.

Funk, V., Richardson, K.S. & Ferrier, S. (2005). Survey-gap analysis in expeditionary research: where do we go from here? *Biological Journal of the Linnean Society*, **85**(4), 549–567.

Hammond, D.S. (2005a). Biophysical features of the Guiana Shield. In *Tropical Forests of the Guiana Shield: Ancient Forests in a Modern World*, ed. D.S. Hammond. Wallingford: CAB International, pp. 15–194.

Hammond, D.S. (2005b). Forest conservation and management in the Guiana Shield. In *Tropical Forests of the Guiana Shield: Ancient Forests in a Modern World*, ed. D.S. Hammond. Wallingford: CAB International, pp. 481–520.

Hammond, D.S. (2005c). Socio-economic aspects of Guiana Shield forest use. In *Tropical Forests of the Guiana Shield: Ancient Forests in a Modern World*, ed. D.S. Hammond. Wallingford: CAB International, pp. 381–480.

Heemskerk, M. (2001). Maroon gold miners and mining risks in the Suriname Amazon. *Cultural Survival Quarterly*, **25**(1), 25–29.

Hofwijks, S. & Madhar, R. (2006). *Assessment of the Bushmeat and Wildplant Trade in Suriname.* The Guianas Regional Programme: World Wildlife Fund Guianas.

Huber, O. (1990). Estado actual de los conocimientos sobre flora y vegetación de la región Guayana, Venezuela. In *El Río Orinoco como Ecosistema – The Orinoco River as an Ecosystem*, ed. F.H. Weibezahn, H. Álvarez & W.M. Lewis Jr. Caracas: EDELCA/Fondo Editorial Acta Científica Venezolana/CAVN/Universidad Simón Bolívar, pp. 337–386.

Huber, O., Foster, M.N. & Pires, T.C.A. (2003). *Conservation Priorities for the Guayana Shield: 2002 Consensus.* Conservation International, Center for Applied Biodiversity Science.

Kroes, R. (1982). The small-town coup: the NCO political intervention in Surinam. *Armed Forces & Society*, **9**(1), 115.

Lambsdorff, J.G. (2006). *Corruption Perceptions Index.* Berlin: Transparency International Secretariat, p. 13.

Lehman, S.M. (2000). Primate community structure in Guyana: a biogeographic analysis. *International Journal of Primatology*, **21**(3), 333–351.

Lehman, S.M. (2004a). Biogeography of the primates of Guyana: effects of habitat use and diet on geographic distribution. *International Journal of Primatology*, **25**(6), 1225–1242.

Lehman, S.M. (2004b). Distribution and diversity of primates in Guyana: species–area relationships and riverine barriers. *International Journal of Primatology*, **25**(1), 73–95.

Lehman, S.M. & Robertson, K.L. (1994). Preliminary survey of *Cacajao melanocephalus melanocephalus* in southern Venezuela. *International Journal of Primatology*, **15**(6), 927–934.

Lehman, S.M., Prince, W. & Taylor, L.L. (1995). Habitat disturbance, hunting pressures, and primate distribution in NE Guyana. *American Journal of Primatology*, **36**, 137.

Lehman, S.M., Sussman, R.W., Phillips-Conroy, J. & Prince, W. (2006). Ecological biogeography of primates in Guyana. In *Primate Biogeography*, ed. S.M. Lehman & J.G. Fleagle. New York, NY: Plenum/Kluwer Press, pp. 105–130.

Lim, B.K., Engstrom, M.D. & Ochoa, J.G. (2005). Mammals. In *Checklist of the Terrestrial Vertebrates of the Guiana Shield*, ed. T. Hollowell & R.P. Reynolds. Washington, DC: Bulletin of the Biological Society of Washington, no. **13**, pp. 77–92.

Lindeman, J.C. & Mori, S.A. (1989). The Guianas. In *Floristic Inventory of Tropical Countries: The Status of Plant Systematics, Collections, and Vegetation, plus Recommendations for the Future*, ed. D.G. Campbell & H.D. Hammond. New York, NY: New York Botanical Gardens, pp. 375–390.

Milensky, C., Hinds, W., Aleixo, A., *et al.* (2005). Birds. In *Checklist of the Terrestrial Vertebrates of the Guiana Shield*, ed. T. Hollowell & R.P. Reynolds. Washington, DC: Bulletin of the Biological Society of Washington, no. 13, pp. 43–76.

Miranda, M., Blanco-Uribe, A.Q., Hernandez, L., *et al.*(1998). *All that Glitters is Not Gold: Balancing Conservation and Development in Venezuela's Frontier Forests*. Washington, DC: World Resources Institute, Forest Frontiers Initiative.

Mittermeier, R.A. (1977). Distribution, synecology, and conservation of Surinam monkeys. Unpublished PhD dissertation, Harvard University.

Mittermeier, R.A. (1991). Hunting and its effects on wild primate populations in Suriname. In *Neotropical Wildlife Use and Conservation*, ed. K.H. Redford & J.G. Robinson. Chicago, IL: University of Chicago Press, pp. 93–107.

Mol, J.H. & Ouboter, P.E. (2004). Downstream effects of erosion from small-scale gold mining on the instream habitat and fish community of a small Neotropical rainforest stream. *Conservation Biology*, **18**(1), 201–214.

Muckenhirn, N.A., Mortenson, B.K., Vessey, S., *et al.* (1975). *Report of a Primate Survey in Guyana*. Pan American Health Organization.

Myers, N., Mittermeier, R.A., Mittermeier, C.G., *et al.* (2000). Biodiversity hotspots for conservation priorities. *Nature*, **403**, 853–858.

Norconk, M. (2006). Long-term study of group dynamics and female reproduction in Venezuelan *Pithecia pithecia*. *International Journal of Primatology*, **27**(3), 653–674.

Norconk, M.A., Raghanti, M.A., Martin, S.K., *et al.* (2003). Primates of Brownsberg Natuurpark, Suriname, with particular attention to the pitheciins. *Neotropical Primates*, **11**, 94–100.

Norconk, M.A., Sussman, R.W. & Phillips-Conroy, J. (1997). Primates of Guayana Shield Forests: Venezuela and the Guianas. In *Adaptive Radiations of Neotropical Primates*, ed. M.A. Norconk, A.L. Rosenberger & P.A. Garber. New York, NY: Plenum Press, pp. 69–86.

Ouboter, P. (2001). Assessment of traded wildlife species. Report #GFECP07 to WWF-Guianas, September 2001. PDF available on the WWF-Guianas website and CD: WWF-Guianas Wildlife Management.

Parry, J.T. & Eden, M.J. (1997). Monitoring and managing land degradation in Guyana: the future. In *Land Use, Land Degradation and Land Management in Guyana*, ed. P.E. Williams, J.T. Parry & M.J. Eden. Surrey: Commonwealth Geographical Bureau, University of London, pp. 93–103.

Peetz, A. (2001). Ecology and social organization of the bearded saki *Chiropotes satanas chiropotes* (Primates: Pitheciinae) in Venezuela. *Ecotropical Monographs*, **1**, 1–170.

Perera, M.A. (1997). *Salud y Ambiente. Contribuciones al Conocimiento de la Antropología Médica y Ecología Cultural en Venezuela*. Caracas: FACES-Universidad Central de Venezuela.

Price, R. (1996). *Maroon Societies: Rebel Slave Communities in the Americas, 3rd edn*. Baltimore, MD: The Johns Hopkins University Press.

Richardson, K.S. & Funk, V. (1999). An approach to designing a systematic protected area system in Guyana. *Parks*, **9**, 7–16.

Rodríguez, J.P. & Rojas-Suárez, F. (1995). *Libro Rojo de la Fauna Venezolana*. Caracas: Provita/Fundación Polar/ Wildlife Conservation Society/Profauna (MARNR)/UICN.

Suriname, P.A.o. (1987). *Land Use and Concessions: Map D1*. The National Planning Office of Suriname and the General Secretariat of the Organization of American States.

Sussman, R.W. & Phillips-Conroy, J. (1995). A survey of the distribution and diversity of the primates of Guyana. *International Journal of Primatology*, **16**(5), 761–792.

ter Steege, H. (1993). *Patterns in Tropical Rain Forest in Guyana*. Wageningen: Stichting Tropenbos.

Thoden van Velzen, H.U.E.,van Wetering, W. & Van Der Elst, D. (2004). *In the Shadow of the Oracle: Religion as Politics in a Suriname Maroon Society*. Long Grove, IL: Waveland Press, Inc.

Urbani, B. (2006). A survey of primate populations in northeastern Venezuelan Guayana. *Primate Conservation*, **20**, 47–52.

Vahter, M., Berglund, M., Akesson, A., *et al.* (2002). Metals and women's health. *Environmental Research Section A*, **88**(3), 145–155.

Vié, J.-C. (1998). Les effets d'une perturbation majeure de l'habitat sur deux espèces de primates en Guyane Française: translocation de singes hurleurs roux (*Alouatta seniculus*), et translocation et insularisation de sakis à face pâle (*Pithecia pithecia*). Unpublished PhD dissertation, Université Montpellier II, France.

Vié, J.-C. (1999). Wildlife rescues – the case of the Petit Saut hydroelectric dam in French Guiana. *Oryx*, **33**(3), 115–126.

Vié, J.-C., Richard-Hansen, C. & Fournier-Chambrillon, C. (2001). Abundance, use of space, and activity patterns of white-faced sakis (*Pithecia pithecia*) in French Guiana. *American Journal of Primatology*, **55**(4), 203–221.

Wright, B.W. (2005). Ecological distinctions in diet, food toughness, and masticatory anatomy in a community of six neotropical primates in Guyana, South America. Unpublished PhD dissertation, University of Illinois.

Wright, S.J., Sanchez-Azofeifa, G.A., Portillo-Quintero, C., *et al.* (2007). Poverty and corruption compromise tropical forest reserves. *Ecological Applications*, **17**, 1259–1266.

Chapter 32

Pitheciid conservation in Ecuador, Colombia, Peru, Bolivia and Paraguay

Leila Porter, Janice Chism, Thomas R. Defler, Laura Marsh, Jesús Martinez, Hope Matthews, Wynlyn McBride, Diego G. Tirira, Marianela Velilla & Rob Wallace

Introduction

The forests of Colombia, Ecuador, Peru, Bolivia and Paraguay are home to at least 24 taxa of pitheciids (Table 32.1). Unfortunately, the majority of these taxa have not been well studied and many details of their geographic range limits, population densities, behavior and ecology are unknown. In this chapter we outline the distribution and conservation status of the Pitheciidae in this region with the goal of identifying taxa of the greatest conservation priority and the research projects most needed to develop adequate conservation plans for their protection.

Major taxonomic revisions of *Callicebus* have recently been undertaken (e.g. van Roosmalen *et al.* 2002), and a revision of *Pithecia* is currently under way (e.g. L. Marsh, unpubl. data). As a result of these revisions, populations of *Callicebus* that had been thought to represent single species have been divided into multiple species in some regions, and the splitting of *Pithecia* into more species is likely to occur in the near future. As more data are gathered about the morphology, behavior, genetic profiles and distribution of different populations, it is likely that more new taxa will be identified. In this chapter, however, unless otherwise noted, we use species and subspecies names from published taxonomies, van Roosmalen *et al.* (2002) for *Callicebus* and Hershkovitz (1987a, 1987b) for *Pithecia* and *Cacajao*.

Overall, the main threats to the pitheciids in this region are ones familiar to primate conservation biologists working

Photo 32.1 A saki monkey (*Pithecia irrorata*) in northwestern Bolivia, Department of the Pando. Photo: Leila Porter.

Evolutionary Biology and Conservation of Titis, Sakis and Uacaris, eds. Liza M. Veiga, Adrian A. Barnett, Stephen F. Ferrari and Marilyn A. Norconk. Published by Cambridge University Press.

Table 32.1 Conservation status and occurrence of Pitheciids in Protected Areas in the Western Amazon.

Country	Species	Conservation status	National parks	Other governmental protected areas	Other protection	Sources
Colombia	*Callicebus caquetensis*	CR ^				Defler *et al.* 2010
Colombia	*Callicebus discolor*	LC*				Defler 2003, 2004
Colombia	*Callicebus ornatus*	VU*	Serra La Macarena, Tiniguas			Defler 2003, 2004
Colombia	*Callicebus torquatus lugens*	LC*	Pichachos (?)			Defler 2003, 2004
Colombia	*Callicebus torquatus lucifer*	LC*	Cahuinarí, Puré, Amacayacu			Defler 2003, 2004
Colombia	*Callicebus torquatus medemi*	LC*	La Paya			Defler 2003, 2004
Colombia	*Pithecia monachus monachus*	LC*	Cahuinaría, Amacayacu, Puré			Defler 2003, 2004
Colombia	*Pithecia monachus milleri*	DD*	La Paya			Defler 2003, 2004
Colombia	*Cacajao melanocephalus ouakary*	LC*		Guaviare FR, Guarinía FR		Defler 2003, 2004
Ecuador	*Callicebus discolor*	LC*	Sumaco-Napo Galeras, Yasuní	Limoncocha BR, Cayambe-Coca ER Cuyabeno RFP		Tirira 2007, 2011
Ecuador	*Callicebus lucifer*	LC*		Cuyabeno RFP		Tirira 2007, 2011
Ecuador	*Pithecia aequatorialis*	LC*	Yasuní			Tirira 2007, 2011
Ecuador	*Pithecia monachus*	LC^	Yasuní	Cuyabeno RFP		Tirira 2007, 2011
Peru	*Callicebus brunneus*	LC*	Manu	? Megantoni NS (titi sp.), Tambopata NR	Los Amigos Conservation Area (private concession)	Bunce, J. pers. commun.; Lawrence, J., pers. commun.; Phillips *et al.* 2004; Vriesendorp *et al.* 2004
Peru	*Callicebus caligatus*	LC*		Possibly in Tamshiyacu Tahuayo CR		Hershkovitz 1988
Peru	*Callicebus cupreus*	LC*	Cordillera Azul	Tamshiyacu Tahuayo CR		Aquino & Encarnación 1994
Peru	*Callicebus discolor*	LC*	Tingo Maria	Pacaya-Samiria NR		Aquino & Encarnación 1994
Peru	*Callicebus lucifer*	LC*			Proposed RZ and NP at Amiyacu, Apayacu, Yaguas, Medio Putumayo	Vriesendorp *et al.* 2004
Peru	*Callicebus oenanthe*	CR+		Alto Mayo PF		Rowe & Martinez 2003

Table 32.1 *(cont.)*

Country	Species	Conservation status	National parks	Other governmental protected areas	Other protection	Sources
Peru	*Pithecia aequitorialis*	LC*			Range possibly overlaps with the proposed RZ and NP at Amiyacu, Apayacu, Yaguas, Medio Putumayo	Vriesendorp *et al.* 2004
Peru	*Pithecia irrorata*	LC*	Manu			Aquino & Encarnación 1994
Peru	*Pithecia monachus*	LC*	Cordillera Azul, Tingo Maria	Pacaya-Samiria NR, Tamshiyacu Tahuayo CR, Megantoni NS		Aquino & Encarnación 1994
Peru	*Cacajao calvus ucayalii*	VU*		Tamshiyacu Tahuayo CR		Aquino & Encarnación 1994
Bolivia	*Pithecia irrorata*	LC*			Proposed NR Frederico Roman	Alverson *et al.* 2003
Bolivia	*Callicebus pallescens*	DD^	Kaa-Iya			Ayala, unpubl. data
Bolivia	*Callicebus donacophilus*	LC^	Amboro			Martinez & Wallace 2007, Anderson *et al.* 1993
Bolivia	*Callicebus modestus*	EN^			Santa Rosa Municipal Reserve	Martinez & Wallace 2007
Bolivia	*Callicebus olallae*	CR^			Santa Rosa Municipal Reserve	Martinez & Wallace 2007
Bolivia	*Callicebus aureiplatii*	VU^	Madidi			Wallace *et al.* 2006
Bolivia	*Callicebus* sp.	LC*		Manuripi NR		Alverson *et al.* 2000
Bolivia	*Chiropotes albinasus*		Noel Kempff Mercado			Lisandro Saucedo, pers. commun. to R. Wallace, May 2010; Diego Romero, pers. commun. to R. Wallace, March 2011
Paraguay	*Callicebus pallescens*	NT	Defensores del Chaco, Rio Negro, Cerro Chovoreca		PR The Paraguayan Pantanal, PR Fortin Patria Lodge	Stallings 1985; Mercolli *et al.*, unpubl. report, 1999; Brunson *et al.*, unpubl. data

Conservation Status:
* indicates it is from IUCN 2008 and ⁺2011; ^indicates assessment by the authors (see text for details).
LC, Least Concern; NT, Near Threatened; VU, Vulnerable; EN, Endangered; CR, Critically Endangered; DD, Data Deficient. See IUCN 2008 for definitions.
Key to Protected Areas:
BR, Biological Reserve; CR, Communal Reserve; ER, Ecological Reserve; FR, Faunal Reserve; NR, National Reserve; NS, National Sanctuary; PF, Protected Forest; RFP, Reserve for Faunal Production; RZ, Reserved Zone.

throughout the world: hunting and habitat destruction. Even primates that live in the most remote forests are facing increasing levels of habitat loss and disturbance. Development projects have created or threaten to create roads deep into the forests in this region, accelerating human encroachment into these areas. Deforestation due to agricultural projects, logging, ranching and oil surveying occurs throughout this region, and in some areas has reduced the primates' habitats to isolated fragments of forest.

Hunting is a major threat to some of the pitheciids, even though they are generally not preferred quarry (e.g. Puertas & Bodmer 1993). Heavy hunting pressure can result from a combination of factors including market hunting in areas accessible to cities, and the consumption of monkeys by hunters searching for larger prey to sell in markets (Puertas & Bodmer 1993; Matthews 2005). Furthermore, as larger species of monkeys become scarce, hunters may increasingly target even medium and small pitheciids (Aquino 1988). Given the pressures of habitat destruction and hunting, protected areas have been established in all the countries discussed in this chapter (Table 32.1). Unfortunately, not all pitheciid species are found within the confines of a protected area.

In the sections that follow, we provide an overview of the pitheciids by country. We outline (1) a brief history of past and current conservation efforts in each region, (2) taxonomic problems that need to be resolved, (3) threats facing each taxon, (4) research needed to make assessment of conservation status possible or more accurate, and (5) taxa that require greater conservation action either through the enforcement of rules in existing protected areas or the creation of new protected areas.

Colombia

Despite a plethora of strong environmental legislation in the past years, violations within the country are common. The large-scale export of primates that occurred 35 years ago has stopped, but there is still a small-scale illicit trade, as Colombian monkeys are sold in Miami for $1000 or more per animal (T.R. Defler, pers. obs.). National interest in effective conservation has grown mightily since the 1970s and many people work in this area as professionals, but there are many incidences of violations, and national trade in wildlife continues to be a huge problem. Generally, hunting primates for food is ignored by all authorities, as subsistence hunting is guaranteed in the new constitution (T.R. Defler, pers. obs.).

Colombia currently has nine recognized taxa of pitheciid primates (Table 32.1), although in the future this number is expected to increase as regions are further explored and taxonomic problems are resolved. Colombian pitheciids are distributed east of the Cordillera Oriental up to about 500 m. The taxa that live close to the Cordillera are classified as Vulnerable or Critically Endangered (IUCN 2008) due to extensive habitat destruction from colonization and industrial agriculture, coupled with their very small distributions within the country. Violence between guerillas and paramilitary forces generates extreme insecurity in some parts of the country and prevents scientific study of and conservation efforts for some pitheciids.

The newest addition to the pitheciids in Colombia is *Callicebus caquetensis* that we suggest is critically endangered (Defler *et al.* 2010). The species was first mentioned by Moynihan (1976) but was ignored by Hershkovitz (1990), surely for lack of information. This species has been found between the Orteguaza and Caquetá rivers in Caquetá department, and based on its coloration it is described as similar to the members of the *Callicebus cupreus* group. The results of genetic analyses along with its pelage, however, reveal that *Callicebus caquetensis* is distinct from *C. ornatus* to the north and *C. discolor* to the south. The species is found in a zone of intense agricultural activity involving both cattle and coca leaf production. It is probably exposed to glyphosate spraying, which is a tool of Plan Colombia to reduce coca plantings within the country, and may very well be the most endangered Colombian primate.

Photo 32.2 *Callicebus caquetensis* in S. Caquetá, Colombia. Photo: Javier Garcia. (See color plate section.)

Callicebus discolor and *C. ornatus* are both classified as Vulnerable (IUCN 2008), but *C. discolor* is of greater concern due to its small geographic area and the extensive degree of habitat disturbance occurring there. Unfortunately, it is found entirely in a zone of conflict in southwestern Putumayo, where field work is dangerous. Estimates of population size are not available for this species, and no conservation plans are in place in this region even though it probably has the highest biodiversity in Colombia.

Callicebus ornatus is endemic to Colombia and has a very small distribution. Although extensive industrial agriculture occurs within its range, many small patches of forest still remain. Fortunately, it can survive in these small forests, although individuals that disperse between patches likely suffer heavy predation. The ecology and behavior of this taxon is well studied (e.g. Mason 1966; Robinson 1981; Porras 2000), but no action plan has been formulated for it. Some populations are found within protected areas (Table 32.1), although guerilla activity makes it impossible to monitor these areas. Interest in protecting *Callicebus ornatus* has increased among educated people from Bogotá, but African oil palm production continues throughout this species' distribution range and threatens its future.

Callicebus torquatus is at low risk of extinction, but its taxonomic status is unclear. Defler (2010) recognizes two subspecies, *C. t. lucifer* and *C. t. medemi* (a Colombian endemic), whereas van Roosmalen *et al.* (2002) and Groves (2005) elevate these subspecies to species based on pelage differences. Barros *et al.* (2000) further suggest that animals east of the Rio Negro should be elevated to their own species, *C. lugens*, which karyologically is well-supported. Karyological data have been useful for determining whether reproductive barriers exist between *Callicebus* populations (Bueno *et al.* 2006), and should be used to determine how many species and subspecies are present in Colombia within this genus (Defler & Bueno 2007) as previous recognition of titi monkey species by phenotype seems suspect to error. The Hershkovitz (1990) taxon *Callicebus t. lugens* is not clearly distinguishable from *Callicebus t. lucifer* (T.R. Defler, unpubl. data). Hernández-Camacho and Cooper (1976) did not even recognize *Callicebus t. lucifer* for the country and there are no clear diagnostic characteristics to distinguish these two taxa. Therefore, although we have a karyotype for what is called *Callicebus t. lucifer* (*sensu* Hershkovitz 1990), which seems to agree with its assignment as a subspecies of *Callicebus torquatus*, we cannot be certain that *Callicebus t. lugens* (*sensu* Hershkovitz 1990) corresponds to *Callicebus lugens* without geo-referenced karyotypes to recognize it (T.R. Defler & M.L. Bueno, unpubl. data).

Cacajao melanocephalus ouakary (considered by some to be *Cacajao ouakary* from chromosomal evidence) is considered to be of Least Concern (IUCN 2008). It is minimally affected by human activities throughout most of its distribution as it inhabits remote areas. However, the western part of its range is undergoing extensive colonization and habitat disturbance. There is no action plan for this species, although it is protected in two immense faunal reserves (Table 32.1). These monkeys are, however, hunted in these reserves by Macuje indigenous peoples. No population evaluation has been undertaken, and no field work for this taxon exists in Colombia, aside from what has been reported in Defler (1999, 2001, 2004).

Pithecia monachus monachus is considered Data Deficient (IUCN 2008). As it inhabits remote Amazonian locations,

Photo 32.3 *Callicebus cupreus ornatus*, Colombia. Photo: Thomas Defler. (See color plate section.)

there are no population estimates for this subspecies aside from comments by Defler (2003, 2004, 2010), who describes it as being widespread and commonly observed. Comparative genetic studies need to be done of populations in adjacent watersheds to determine if there is more than one subspecies present. There is no species action plan for this saki monkey, but it is protected in some national parks (Table 32.1). It is likely the most secure of the pitheciid taxa in Colombia.

The endemic *Pithecia monachus milleri* is Vulnerable in Colombia (IUCN 2008) due to its congruence with much colonist activity, although some populations are legally protected in parks (Table 32.1). There have been neither evaluations of the status of its populations nor any field studies undertaken to determine its behavioral ecology. In addition, molecular and karyological studies are needed to compare populations on both sides of the large Caquetá River, as they may represent genetically distinct populations.

The present political situation in Colombia does not bode well for primate conservation, as authorities are often unwilling to enforce laws in the field, due to the danger of retaliation by armed groups. Enforcement at airports is more probable, as in Leticia, where the national police work well with the local semi-autonomous corporation in charge of environmental enforcement, especially as many infractions are by foreign tourists (T.R. Defler, pers. obs.). But local people are not often pressed to observe national faunal regulations, especially if they have political influence (T.R. Defler, pers. obs.). Recently the case of the Colombian medical researcher Manuel Elkin Patarroyo has received national and international press due to the thousands of illegally purchased *Aotus* (purchased without permission from Brazil, Peru and Colombia) that he has used in his research, but he has not been sanctioned by the government (Los micos de Patarroyo 2008; Maldonado *et al.* 2009).

Ecuador

In Ecuador there are two genera of pitheciids, *Callicebus* and *Pithecia*, and traditionally two species have been recognized in each genus (Tirira 2007). Titi and saki monkeys are not regularly hunted due to their small size, and the belief by the Quichua and Huaorani that saki monkeys are "toxic". However, the Shuar and Achuar kill sakis to make "shrunken heads" for tourists. Both genera are taken as pets, but *Pithecia* is regarded a "bad pet" by Quichua because of their large canines (L. Marsh, pers. obs.).

The conservation of primates in Ecuador has been focused entirely on the large-bodied species, particularly the spider monkeys (*Ateles* spp.). As a result, no conservation efforts exist for any pitheciid species. The studies of pitheciids in Ecuador have focused on the ecology and behavior of four taxa including: *Callicebus discolor* in the Yasuní National Park (Pozo 2004; Carrillo *et al.* 2005; Cisneros-Heredia *et al.* 2005; Pozo & Youlatos 2005; Bravo-Cabezas, 2010); *C. lucifer* in the Cuyabeno Faunal Production Reserve (Ulloa 1988; Campos 1991; De Vries *et al.* 1993); *Pithecia aequatorialis* in the Yasuní National Park (Di Fiore *et al.* 2007); and *P. monachus* in the Cuyabeno Faunal Production Reserve (Navarrete 2001) and the Yasuní National Park (Pozo 2004; Moreano 2005; Pozo & Youlatos 2005). In addition, L. Marsh (unpubl. data) has begun revising the taxonomy of *Pithecia* in Ecuador.

The biggest threat to all primates in Ecuador is the habitat loss associated with oil development. The Ecuadorian government generally favors the oil industry rather than conservation initiatives, as oil companies bring in immediate revenue. As a result, wide transects made by oil surveyors bite through forests, even within national parks. Indigenous uprisings have resulted in few if any reforms on behalf of their rights or the protection of the environment in the face of the multibillion-dollar oil industry (Kimerling 1993; Wunder 2000).

Callicebus discolor is found in the Amazon region of Ecuador in tropical rainforests. This species is well-documented between the Aguarico River (in the north) and Pastaza River (in the south), but its distribution outside of this area is uncertain. It is found within primary forest, secondary and disturbed forests, and can live close to human settlements (De la Torre 2000; Tirira 2007).

Callicebus discolor is found within various protected areas (Table 32.1). There are no conservation programs for this species, but it can adapt well to disturbed forests and inhabits areas close to small human settlements, and it is rarely hunted. It is therefore considered to be of Least Concern by the IUCN (2008), and is included in the Red Book of the Mammals of Ecuador as Near Threatened (Tirira 2011). Its conservation status is considered to be stable, but additional research should be done to determine its distribution limits. Furthermore, studies should be undertaken to determine whether populations to the north and south of the Napo River are isolated and, if so, whether they represent distinct species.

Callicebus lucifer inhabits tropical rainforest in the northern region of the Ecuadorian Amazonia, but the limits of its distribution have not been documented. It is unknown whether it is present in the upper part of the Aguarico River (west of the Cuyabeno River). It is found in *terra firme* forest up to the edges of flooded forest (van Roosmalen *et al.* 2002; Tirira 2007) and is found in one protected area (Table 32.1). There are no conservation projects for this species, and although its conservation status is not well known, it is thought to be stable (Least Concern: IUCN 2008) and is included in the Red Book of Mammals of Ecuador as Near Threatened (Tirira 2011). It is of little commercial interest, but it is captured occasionally by indigenous people for pets. As for *C. discolor*, the distribution of *C. lucifer* should be better established. In addition, comparisons should be made between populations of *C. lucifer* in Ecuador with those in Colombia, to determine if they represent different subspecies.

Pithecia monachus and *P. aequatorialis* are the names currently recognized for the sakis in Ecuador (Tirira 2007). Using this taxonomy, *Pithecia aequatorialis* is found in the lower Amazon of Ecuador, in tropical forests below 300 m in altitude, and on both sides of the Napo River (according to

Hershkovitz 1987a); however, newly available evidence shows that it can be found only on the southern side of the Napo River (Tirira 2011). The IUCN (2008) lists this species as of Least Concern; but the Ecuadorian Red Book (Tirira 2011) classifies it as Near Threatened. There is little known about this species, but the largest populations appear to exist inside the Yasuni National Park (Marsh 2004). *Pithecia monachus* (Lönnberg 1938) is also found in tropical forests of the lower Amazon. It prefers the interior of *terra firma* forests, although it can also be found in flooded forests and along the edges of rivers in dense vegetation (De la Torre 2000; Tirira 2007). It is found within two areas of protection (Table 32.1), and although the IUCN considers it to be Data Deficient (IUCN 2008), it is considered to be Near Threatened in the Red Book of the Mammals of Ecuador (Tirira 2011). Further studies are needed to determine its conservation outlook and distribution limits in more detail.

The activity of petroleum companies, deforestation, hunting of primates, the sale of primates, and human population increase, are all factors that suggest that the future of primates in eastern Ecuador is not good. For this reason, in 2007, 2009 and 2010 a new evaluation of the Red List of Ecuadorian mammals was put in place under which all pitheciid species present in the country are included in the category Near Threatened (Tirira 2011).

Peru

As is often the case in primate conservation, it is not possible to develop conservation plans for species until their ecology and distribution have been studied and documented. In Peru, this work has only been undertaken in recent years and in relatively few locations; thus, many of Peru's pitheciids have yet to be extensively studied. For example, the distribution and abundance of *Cacajao calvus ucayalii* until recently has been limited to only a few field studies, rapid biological inventories and descriptions of chance encounters from researchers studying other primates (e.g. Pitman *et al.* 2003; Ward & Chism 2003; Bartecki & Heymann 1987). One of the best-documented attempts to protect a highly diverse primate assemblage, which includes the threatened red uakari (*Cacajao calvus ucayalii*), was the establishment in northeastern Peru of the Reserva Comunal Tamshiyacu Tahuayo in 1991 (Bodmer & Puertas 2000; Pitman *et al.* 2003). In 2007 this reserve was granted protected status by the state of Loreto and is now the Area de Conservacíon Regional Comunal de Tamshiyacu-Tahuayo (ACRCTT).

In recent years, three Rapid Biological Inventories have been undertaken in Peru (Pitman *et al.* 2003, 2004; Vriesendorp *et al.* 2004) and many studies have been conducted in both the Manu National Park (Soini 1982; Terborgh *et al.* 1984) and the Pacaya Samiria National Reserve (Soini 1986; Bodmer *et al.* 1999). However, much work still needs to be done to accurately assess the conservation status of pitheciids in Peru, as well as to describe and identify which species are found there (e.g. L. Marsh, unpubl. data).

Peru is home to at least 10 species of pitheciids representing 3 genera: *Callicebus*, *Pithecia* and *Cacajao* (Table 32.1). Some pitheciids have ranges that overlap national borders and are threatened by habitat destruction in one country but presently appear to have relatively healthy populations in other countries. For example, *Callicebus discolor* is of Least Concern in Peru but is considered to be Vulnerable in Colombia, where its range is restricted (IUCN 2008).

Surveys of primate populations in Peru have identified hunting as a major threat to the pitheciids (Soini 1982; Freese *et al.* 1982). Habitat destruction resulting from human encroachment, logging and forest clearance for agriculture also affects pitheciids in Peru (e.g. Bennett *et al.* 2001; Pitman *et al.* 2003). Until recently, many species have been protected by the remoteness of their habitat; however, such protection is coming to an end in many places.

Cacajao, represented in Peru by one subspecies, *C. calvus ucayalii*, is classified as Vulnerable (IUCN 2008). At one time this subspecies may have ranged into Brazil, but its presence there is now questionable (IUCN 2008). While its distribution may have always been patchy and local (Bowler *et al.* 2006), by the late 1980s Aquino (1988) believed it to have been eliminated from the southernmost part of its range in Peru. Part of the impression of patchiness, however, might be the result of uacaris' very large home ranges (Leonard & Bennett 1996; Bowler, Chapter 15). As there have been no detailed studies of the uacaris' distribution and abundance, lack of evidence should not be taken as definitive evidence of its absence at a locality within the known area of occurrence (e.g. Ward & Chism 2003; Matthews 2005). Peru's uacari populations are under strong pressure from a combination of hunting and habitat destruction mainly from logging (Aquino 1988; Heymann & Aquino 1994; Aquino & Encarnación 1994). The distribution of uacaris in Peru includes one protected area (Table 32.1), and a recently established conservation concession in what is hypothesized to be the area with the densest uacari population in the Yavarí River area (Bowler, Chapter 15). A recently completed study of this species (Bowler & Bodmer 2009) has provided important information on its social structure and group dynamics which will aid in developing strategies for its conservation.

Three species of saki have been identified in Peru: *Pithecia monachus*, *P. irrorata* and *P. aequatorialis*. Several additional taxa of *Pithecia* in Peru have been proposed (L. Marsh, unpubl. data) and recent studies suggest a possible new species in the ACRCTT (J. Chism, unpubl. data). Sakis are threatened by hunting, capture of infants for the pet trade, habitat loss, and the use of body parts in handicrafts (Aquino & Encarnación 1994). For example, sakis' tails are made into dusters for the tourist trade.

Pithecia monachus, with a wide distribution in Peru and parts of Ecuador, Colombia and Brazil, is classified as Least Risk/Least Concern (IUCN 2008). Although monk sakis have the most official protection of any pitheciid in Peru, occurring in three protected areas (Table 32.1), the degree of protection may be limited. A study comparing mammal populations in a heavily hunted and unhunted areas within the Pacaya–Samiria Reserve found that *P. monachus* was being hunted at unsustainable levels (Bodmer *et al.* 1999). This is particularly disturbing given their low reproductive rates and low densities.

Pithecia irrorata irrorata has a restricted distribution in Peru, occurring only in the southeastern part of the country. Its range includes Manu National Park where its abundance is described as either moderate (Aquino & Encarnación 1994) or rare (Terborgh *et al.* 1984). Outside the park it is threatened by hunting for meat and habitat loss primarily due to agriculture. Its range extends into Bolivia and Brazil, where it may be more common as its conservation status overall is Least Risk/Least Concern (IUCN 2008).

Pithecia aequatorialis is reported to be sympatric with *P. monachus* in northern Peru (Aquino & Encarnación 1994), which may be responsible for confusion about its true conservation status. Aquino and Encarnación (1994) state that *P. aequatorialis* is subject to strong hunting pressure. While designated as Least Risk/Least Concern (IUCN 2008), it does not occur in any protected areas in Peru (Table 32.1).

The most recent IUCN Redlist identifies six species of *Callicebus* in Peru: *C. brunneus*, *C. caligatus*, *C. cupreus*, *C. discolor*, *C. lucifer* and *C. oenanthe* (IUCN 2008). A recent analysis identified the Peruvian species as *C. oenanthe*, *C. cupreus*, *C. discolor*, *C. lucifer* and *C. dubius*, while the status of *C. caligatus* as a separate species, in Peru at least, was questioned (van Roosmalen *et al.* 2002). In addition, it is unclear whether *C. brunneus* actually occurs in Peru; the animals so identified may actually be *C. caligatus*, *C. dubius* or *C. toppini* (see section on Bolivia for details). The IUCN (2008) classifies all *Callicebus* species in Peru as Least Risk/Least Concern except for *C. oenanthe* which is considered Endangered due to its very restricted geographic and altitudinal range (van Roosmalen *et al.* 2002). This species' range includes the Alto Mayo Protected Forest; however, a recent survey (Rowe & Martinez 2003) detected no titis within the protected area. Furthermore, the construction of an all-weather road between Lima and Tarapoto is increasing human migration to the area and exacerbating the threat to *C. oenanthe*, as much of its habitat, even on steep slopes, has been cleared for agriculture (Rowe & Martinez 2003).

Distributions of all pitheciids in Peru with the possible exception of *Pithecia aequatorialis* include at least one protected area (Table 32.1). Peru has a large number of protected areas ranging from the most-protected national parks and sanctuaries (IUCN categories II and III) to the least-protected communal reserves, national reserves, protected forests and hunting preserves (IUCN category VI) (UNEP-WCMC 2007).

Several studies have found that protected areas have at least some potential to be effective in protecting species and habitats. A comparison of primate populations in Peru and Bolivia found higher densities of monkeys in areas protected from hunting and other human disturbance (Freese *et al.* 1982). Paradoxically, however, a recent survey for *P. aequatorialis* in the Itaya, Tigre and Curarae river basins found the highest densities of this species in the most heavily hunted areas along the Itaya River (Aquino *et al.* 2009). An increasing body of work indicates that reserves may act as sources for species which can then disperse into adjacent areas (sinks) depleted by hunting and other human activity (e.g. Novaro *et al.* 2000). Protected areas, while potentially effective at protecting habitat (Naughton-Treves *et al.* 2005), only conserve wildlife if hunting and other human disturbance can be controlled as well. Future conservation programs in Peru should be directed toward both habitat protection and enlisting community support to curtail human disturbance in areas with high primate diversity (Bodmer & Puertas 2000).

Bolivia

Bolivia is home to one species of *Pithecia*, and six species of *Callicebus* (Martinez & Wallace 2010; Table 32.1). In 2010 one species of *Chiropotes* was added to this list although this record has yet to be published (L. Saucedo, pers. commun., 2010; D. Romero, pers. commun., 2011). Until very recently many of the habitats of these primates had very low human population densities; however, intensive human immigration into lowland regions, ranching, agriculture and logging are causing extensive forest destruction. Two species of titi monkeys endemic to Bolivia face the greatest threat of extinction due to hunting, habitat disturbance and planned development projects through their ranges.

Over the last 40 years Bolivia has established 22 National Protected Areas that cover almost 17% of the Bolivian territory, and many of these protected areas are large and of global importance. Past research and conservation efforts on Bolivian primates have tended to focus on protected areas (Wallace *et al.*, Appendix A); however, in the last 10 years efforts have began to establish additional protected areas at the department or municipal level in order to adequately protect taxa that are not found in any of the National Protected areas, for example in Pando Department and in several municipalities in southwestern Beni (Wallace *et al.*, Appendix A).

Pithecia irrorata (Figure 32.1) appears to be distributed across the entire northern region of the Department of the Pando. Although it has not been censused consistently across this area, it has been documented at localities in the east, west and center of this region, indicating it is likely to be present throughout this entire area (e.g. Porter 2006; Alverson *et al.* 2000, 2003; Cameron & Buchanan-Smith 1991–1992; Izawa & Bejarano 1981). Because of its broad distribution, the species is not likely to be under immediate threat. Hunting of *Pithecia* likely occurs at low levels by rural communities throughout the year; however, as the range of *Pithecia* overlaps with Brazil nut harvesting areas, hunting pressure probably increases during the seasonal harvesting period when many more people are working in the forest. There are no protected areas for this species, although work is under way to establish a reserve in the northeastern corner of the Pando (Alverson *et al.* 2003).

Figure 32.1 A saki monkey (*Pithecia* cf *monachus*) in the Area de Conservacíon Regional Comunal de Tamshiyacu–Tahuayo, Peru. Photo: Alfredo Dosantos Santillan.

Callicebus pallescens is exclusively found in the Chaco ecosystem of Bolivia, extending into Paraguay and the dry forests of the Brazilian Pantanal. This species has only recently been upgraded from subspecies status (van Roosmalen *et al.* 2002), and confirmed localities are few. Within the Chaco it may be more abundant in more humid forest types, particularly riverine forests (J. Ayala, pers. commun., 2006). A large portion of the current hypothesized distribution of this species is found within the Kaa-Iya National Park, although it is very occasionally hunted by Izoceno indigenous communities in this area (J. Ayala, pers. commun., 2006). Further data regarding the overall distribution of this species, including the limits between *C. pallescens* and *C. donacophilus*, is required before assessing its conservation status.

Callicebus donacophilus has the widest distribution in Bolivia stretching east from the Manique River in the Beni Biosphere Reserve into Brazil (Ferrari *et al.* 2000). Although found in tropical humid forests it seems to be restricted to the slightly drier forests of southern Amazonia. Despite its wide distribution it is only known from two national protected areas in the country (Table 32.1). Mechanized agriculture has resulted in massive habitat loss within the part of its range around the city of Santa Cruz. Nevertheless, it is one of three primate species that survive within the confines of the city, and it has also been observed on the outskirts of several rural communities (R. Wallace, pers. obs.). These observations suggest that it should not be considered an immediate conservation priority, as it has some resilience to human disturbance and a relatively large distributional range.

Callicebus modestus and *C. olallae* are Bolivian endemics exclusively found in the southwestern portion of the Beni Department in relatively dry forest patches within a forest–savanna mosaic (Lönnberg 1939; Anderson 1997; Felton *et al.* 2006; Martinez & Wallace 2007). Although distributional ranges overlap for these species, taxonomic studies have consistently considered them to be distinct species, and also to be distinct from the nearby populations of *C. donacophilus* (Lönnberg 1939; Hershkovitz 1988, 1990; Kobayashi 1995; Martinez & Wallace 2007, 2010). Genetic studies to confirm the taxonomic status of these endemics are currently under way (J. Barrera, pers. commun., 2007). The conservation status of these endemics is alarming given their extremely restricted range, estimated at just 1800 km^2 for *C. modestus* and 400 km^2 for *C. olallae* (Martinez & Wallace 2007) and the planned major improvement of a regional thoroughfare called the "Northern Corridor" project. This project will create a paved road from La Paz north to Pando Department and inevitably stimulate forest loss and increased hunting pressure in the immediate vicinity. The range of *C. modestus* just touches the Beni Biosphere Reserve (Table 32.1), while *C. olallae* is absent from national protected areas. Fortunately, the Santa Rosa Municipal Reserve has recently been established, and the planned establishment of a management plan and program should help protect *C. olallae* and *C. modestus*. In addition, conservation programs should work with local cattle owners in order to promote preservation of wildlife in portions of their land.

Callicebus aureipalatii was recently described from the western lowlands of Bolivia (Wallace *et al.* 2006) where the *pied mont* forest at the base of the Andes grades into lowland humid riverine and floodplain forests. The eastern and northern distributional limits of this species require further study, but the range of this species does stretch into southern Peru along the Heath and Tambopata rivers, south of the Madre de Dios. This species occurs in the Madidi protected area, and as long as this protected area remains intact, the conservation status of this species is not an immediate concern.

An as yet unidentified *Callicebus* species previously identified as *C. brunneus* occurs in the northern limits of Bolivia across the Pando Department (Anderson 1997), including the area of the large Manuripi Wildlife Reserve, and probably south into northern Beni Department (Rowe & Martinez 2003). According to Hershkowitz (1988, 1990), the distribution of *C. brunneus* extends west into southern Peru and east and north into Brazil. However, van Roosmalen and colleagues (2002) suggested that *C. dubius* occurs in northern Bolivia. Recent studies in Brazil (Ferrari *et al.* 2000; Rohe & Silva-Jr. 2009), observations and photographic evidence from Pando (R. Wallace, unpubl. data) and observations on the original description of *C. brunneus* (Vermeer 2009) strongly suggest that *C. brunneus* does not occur west of the Madeira River and may in fact be a Brazilian endemic. Although it seems likely that titi monkey populations in Pando are not *C. brunneus*, according to recent studies there are at least three possibilities as to the identity of *Callicebus* there: *C. caligatus* (Ferrari *et al.* 2000), *C. dubius* (van Roosmalen *et al.* 2002; Rohe & Silva-Jr. 2009) and *C. toppini* (R. Wallace, unpubl. data). As such, further surveys are urgently required to clearly establish the number and identity of *Callicebus* species in this region (Wallace *et al.* 2006; Martinez & Wallace 2010) in order to adequately assess their conservation status.

In summary, the National Protected Area system adequately protects two of the six species of *Callicebus* in Bolivia (*C. aureipalatii* and *C. pallescens*). The relative dearth of protected areas in northern Bolivia is worrying for the conservation of *Pithecia irrorata* and *Callicebus* sp., although there are at least two initiatives for protected areas currently under way for this region which would dramatically improve the likelihood of successful conservation of these species in Bolivia (Federico Roman, in northeastern Pando, and Aquicuana in northern Beni). *Callicebus donacophilus* is thought to occur in small portions of several protected areas; however, to ensure it is properly protected, further research is needed to determine its precise distribution within these protected areas. Along with focused local and national environmental education efforts, improving the management of the recently created Santa Rosa del Yacuma Municipal Reserve represents the most important conservation intervention for *Callicebus olallae* and *C. modestus*, two of the most threatened primate species in the Americas.

Finally, recent photographic evidence from Noel Kempff Mercado National Park in extreme eastern Santa Cruz Department of Bolivia has determined the presence of the red-nosed bearded saki monkey (*Chiropotes albinasus*) close to the headwaters of the Itenez River (L. Saucedo, pers. commun., 2010; D. Romero, pers commun., 2011). Previous studies had noted this species along with other primate species exclusively on the Brazilian side of Itenez River, although anecdotal reports suggested this species might spill over to Bolivia (Wallace *et al.* 1996). The number of groups and extent of its distribution has yet to be determined and given the massive deforestation on the Brazilian side of the river in the last two decades, it is also possible that this presence is a recent development as a response to habitat loss and fragmentation.

Paraguay

In Paraguay, no effective conservation programs for primates have been established. There are five primates in Paraguay and all lack sufficient population data and ecological information to develop conservation plans for their protection. Paraguay has only one pitheciid primate, *Callicebus pallescens*. It is found west of the Rio Paraguay in a region known as the Chaco. The Chaco covers more than 60% of Paraguay: in the north the vegetation includes continuous xeric forest, whereas the south has swampland, gallery forests, and palm savanna mixed with forest (Stallings 1985). *Callicebus pallescens* is distributed from the border with Bolivia south to approximately 23°S, and from the Rio Paraguay west to approximately 61°30′W (Stallings 1985; W. Brunson, M. Velilla & G. Zuercher, *Assessment of Infrastructure and Primate Status in Paraguay*, unpubl. report, 2007).

The only information on Paraguayan primates has been collected during assessments of their distribution and conservation status (Stallings 1985; W. Brunson, M. Velilla & G. Zuercher, unpubl. data). Based on this information, the last "Threatened species workshop" changed *Callicebus pallescens* from the category Least Concern to Near Threatened for Paraguay (J.L. Cartes, E. Gomez, M. Velilla, H. Del Castillo, R. Owen, G. D'Elia & C. Lopéz-González, *Taller de Mamíferos Amenazados*, unpubl. report, 2005; IUCN 2008).

The main threats to *C. pallescens* are the drastic changes that occur in the Chaco region, particularly in the central portion, due to deforestation and habitat loss for cattle ranching. The total measured deforestation in the Chaco region from May 2005 through May 2006 was 130,000 ha (Asociación Guyra Paraguay, *Base de Datos de Guyra Paraguay*, unpubl. report, 2007). Other minor threats to *Callicebus* are hunting and the pet trade. According to Stallings (1985), the Lengua-Mascoy Indians hunted the species in the southern Chaco. However, it appears that most indigenous groups in the Paraguayan Chaco do not frequently hunt primates. *Callicebus* are sometimes sold as pets because of their docility and ability to adjust to captivity (Neris *et al.* 2002).

In Paraguay there are 11 protected areas found in the Chaco, covering 8.2% of the region (Sistema Nacional de Áreas Silvestres Protegidas del Paraguay 2007) and *Callicebus* is found in several of these areas (Table 32.1). Seven of these protected areas were designated by the UNESCO in 2005 as core areas for The Chaco Biosphere Reserve, covering an area of 4,707,250 ha (UNESCO 2007). The Biosphere reserve provides forest corridors between protected areas and thus protects a large area in the north; the Dry Chaco and Pantanal ecoregions. Even though this seems to be very well protected, there are still conservation problems. The main issue is that some of the national parks do not have the appropriate infrastructure or personnel to patrol the areas.

There are three ecoregions in the Chaco: Humid Chaco, Dry Chaco and a small portion of Pantanal. The Humid Chaco does not have adequate protection: there is only one park and it is not well established. The Dry Chaco does have adequate protected areas, especially in the northern portion. The Pantanal ecoregion will have proper protection once the Rio Negro National Park is fully implemented. When considering the distribution of *Callicebus pallescens* in these regions, it is clear that the species is not well-protected in all of its range. In particular, in the southern part of its distribution range (in the Humid Chaco) there are no protected areas. We therefore recommend the establishment of protected areas in this portion of the Chaco.

Conclusions

Few of the pitheciids from Colombia south to Paraguay have been well-studied, and few have conservation action plans in place for their protection. Even the pitheciids found within protected areas often face increasing pressure, as laws that prevent hunting and habitat degradation within these areas are seldom enforced (Terborgh *et al.* 2002). These problems are further exacerbated in Colombia in politically unstable areas where it is dangerous for scientists and government employees to work due to guerilla activity.

Despite these problems, protected areas have the potential to be effective in protecting species and habitats. Therefore, it is essential to enforce the laws within existing protected areas in order to prevent hunting and deforestation, and to form new protected areas for species not currently within any type of reserve or park.

In addition, research is necessary to determine the taxonomy of pitheciids throughout this region. Documentation of the similarities and differences in the genes, anatomy, behavior and ecology of different populations of titi monkeys and saki monkeys should help clarify how many species and subspecies are present in each country. Accurate taxonomies are essential for determining if pitheciid taxa are properly protected in their ranges.

Although work on all the pitheciids is required, we conclude by listing some priorities for conservation initiatives in each country.

- In Colombia, additional surveys are needed to define the geographic range of *Callicebus caquetensis*, and given its Critically Endangered status, reserves to protect this species should be established immediately. *Callicebus discolor* and *C. ornatus* are vulnerable and in need of conservation action plans given the extensive fragmentation of their habitats. Additional genetic work is required to assess the taxonomic status of the different populations of *Callicebus torquatus* and to determine if *C. lugens* is the same as *C. torquatus lugens* (*sensu* Hershkovitz 1990): given the diagnostic characteristics of the phenotype as compared to *C. t. lucifer*, this is not straightforward and should not be assumed.
- In Ecuador, taxonomic diversity within *Pithecia* must be critically examined as studies in progress indicate that the number of species in the country have been underestimated. Further investigations of the range limits and population sizes of these taxa should be undertaken to determine their status and to develop strategies for their protection.
- In Peru, *Cacajao calvus ucayalii* requires further protection, as it continues to face hunting pressure and habitat loss. *Pithecia* needs to be evaluated in order to determine the taxonomic and conservation status of its populations. In addition, the endemic *Callicebus oenanthe* requires increased protection given its restricted range and disappearing habitat.
- In Bolivia, the newly created Santa Rosa Municipal Reserve needs to be monitored closely, and management capacity installed in order to ensure that the endemic species *C. modestus* and *C. ollalae* are adequately protected. In addition, titi monkeys in the north of Bolivia need to be evaluated in order to determine whether they represent one or multiple species.
- In Paraguay, *Callicebus pallescens* requires further protection particularly in the southern part of its distribution range in the Humid Chaco.

References

Alverson, W., Moskovits, D. & Halm, I. (eds.). (2003). Bolivia: Pando, Frederico Roman. *Rapid Biological Inventories* 6. Chicago, IL: Field Museum, Environmental and Conservation Programs.

Alverson, W., Moskovits, D. & Shopland, J. (eds.). (2000). Bolivia: Pando, Rio Tahuamanu. *Rapid Biological Inventories* 1. Chicago, IL: The Field Museum, Environmental and Conservation Programs.

Anderson, S. (1997). Mammals of Bolivia: taxonomy and distribution. *Bulletin of the American Museum of Natural History*, **231**, 1–652.

Anderson, S., Riddle, B.R., Yates, T.L., *et al.* (1993). Los mamíferos del Parque Nacional Amboro y la región de Sana Cruz de la Sierra, Bolivia. *Special Publication of the Museum of Southwestern Biology*, **2**, 58.

Aquino, R. (1988). Preliminary survey on the population densities of *Cacajao calvus ucayalii*. *Primate Conservation*, **9**, 24–26.

Aquino, R. & Encarnación, F. (1994). Primates of Peru. *Primate Report*, **40**, 1–127.

Aquino, R., Cornejo, F., Lozano, E., *et al.* (2009). Geographic distribution and demography of *Pithecia aequatorialis* (Pitheciidae) in Peruvian Amazonia. *American Journal of Primatology*, **71**, 1–5.

Barros, R.M.S., Pieczarka J.C., Brigido, M.C. O., *et al.* (2000). A new karyotype in *Callicebus torquatus* (Cebidae, Primates). *Hereditas*, **133**, 55–58.

Bartecki, U. & Heymann, E. (1987). Sightings of red uakaris, *Cacajao calvus rubicundus*, at the Rio Blanco, Peruvian Amazon. *Primate Conservation*, **8**, 34–36.

Bennett, C., Leonard, S. & Carter, S. (2001). Abundance, diversity and patterns of distribution of primates on the Tapiche River in Amazonian Peru. *American Journal of Primatology*, **54**, 119–126.

Bodmer, R. & Puertas, P. (2000). Community-based comanagement of wildlife in the Peruvian Amazon. In *Hunting for Sustainable Tropical Forests*, ed. J. Robinson & E. Bennett. New York, NY: Columbia University Press, pp. 395–409.

Bodmer, R., Allen, C., Penn, J., *et al.* (1999). *Evaluating the sustainable use of wildlife in the Pacaya–Samiria National Reserve, Peru. America Verde Working Papers* **4**. Arlington, VA: The Nature Conservancy.

Bowler, M. & Bodmer, R. (2009). Social behavior in fission–fusion groups of red uakari monkeys (*Cacajao calvus ucayalii*). *American Journal of Primatology*, **71**, 976–987.

Bowler, M., Queiroz, H., Bodmer, R., *et al.* (2006). Uacaris for real: conservation initiatives in Peru and Brazil. *International Journal of Primatology*, **Suppl 1**, abstract 521.

Bravo-Cabezas, J.J. (2010). Patrones de actividad y uso de estrato vertical de *Callicebus discolor* (Primates: Pitheciidae) en tierra firme en el Parque Nacional Yasuní. Unpublished Bachelor thesis, Pontificia Universidad Católica del Ecuador.

Bueno, M.L., Ramírez-Orjuela, M., Leiovici, M., *et al.* (2006). Información cariológica del género *Callicebus* en Colombia. *Revista de la Academia Colombiana de Ciencias Exactas, Físicas y Naturales*, **30**, 109–115.

Cameron, R. & Buchanan-Smith, H. (1991). Primates of the Pando, Bolivia. *Primate Conservation*, **12**–13, 11–14.

Campos, F. (1991). Preferencias de hábitat, aspectos reproductivos y comportamiento de canto como factores determinantes en el comportamiento reproductivo de *Callicebus torquatus* en la Amazonía ecuatoriana. Unpublished Bachelor thesis, Pontificia Universidad Católica del Ecuador.

Carrillo B.G.A., Di Fiore, A. & Fernández-Duque, E. (2005). Dieta, forrajeo y presupuesto de tiempo en cotoncillos (*Callicebus discolor*) del Parque Nacional Yasuní en la Amazonia Ecuatoriana. *Neotropical Primates*, **13**, 7–11.

Cisneros-Heredia, D.F., León-Reyes, A. & Seger, S. (2005). *Boa constrictor* predation on a Titi monkey, *Callicebus discolor*. *Neotropical Primates*, **13**, 11–12.

Defler, T.R. (1999). Fission–fusion behavior in *Cacajao melanocephalus ouakary*. *Neotropical Primates*, **7**, 5–8.

Defler, T.R. (2001). *Cacajao melanocephalus ouakary* densities on the lower Apaporis River,*Colombian Amazon. Primate Report*, **61**, 31–36.

Defler, T.R. (2003). *Primates de Colombia*. Bogotá: Conservación Internacional de Colombia.

Defler, T.R. (2004). *Primates of Colombia*. Bogotá: Conservación Internacional de Colombia.

Defler, T.R. (2010). *Historia Natural de los Primates Colombianos*. Bogotá: Universidad Nacional de Colombia.

Defler, T.R. & Bueno, M.L. (2007). *Aotus* diversity and the species question. *Primate Conservation*, **22**, 55–70.

Defler, T.R., Bueno, M.L. & Garcia, J. (2010). *Callicebus caquetensis*: a new and critically endangered titi monkey from southern Caquetá, Colombia. *Primate Conservation*, **25**, 1–9.

De la Torre, S. (2000). *Primates de la Amazonía del Ecuador/Primates of Amazonian Ecuador*. Quito: SIMBIOE.

De Vries T., Campos, F., De la Torre, S., *et al.* (1993). Investigación y conservación en la Reserva de Producción Faunística Cuyabeno. In *La Investigación para la Conservación de la Diversidad Biológica en el Ecuador*, ed. P.A. Mena & L. Suárez. Quito: EcoCiencia, pp. 167–221.

Di Fiore, A., Fernandez-Duque, E. & Hurst, D. (2007). Adult male replacement in socially monogamous equatorial saki monkeys (*Pithecia aequatorialis*). *Folia Primatologica*, **78**, 88–98.

Felton, A., Felton A.M., Wallace, R.B., *et al.* (2006). Identification, distribution and behavioural observations of the titi monkeys *Callicebus modestus* Lönnberg 1939, and *Callicebus olallae* Lönnberg 1939. *Primate Conservation*, **20**, 41–46.

Ferrari S.F., Iwanaga, S., Messias, M.R., *et al.* (2000) Titi monkeys (*Callicebus* spp., Atelidae: Platyrrhini) in Brazilian state of Rondonia. *Primates*, **41**, 229–234.

Freese, C., Heltne, P., Castro, N., *et al.* (1982). Patterns and determinants of monkey densities in Peru and Bolivia, with notes on distribution. *International Journal of Primatology*, **3**, 53–90.

Groves, C.P. (2005). *Mammal Species of the World: A Taxonomic and Geographic Reference*, 3rd edn, ed. D.E. Wilson & D.M. Reeder. Baltimore, MD: Johns Hopkins University Press.

Hernandez-Camacho, J. & Cooper, R.W. (1976). The nonhuman primates of Colombia. In *Neotropical Primates: Field Studies and Conservation*, ed. R.W. Thorington Jr. & P.G. Heltne. Washington, DC: National Academy of Sciences, pp. 35–69.

Hershkovitz, P. (1987a). The taxonomy of South American sakis, genus *Pithecia* (Cebidae, Platyrrhini): a preliminary report and critical review with the description of a new species and a new subspecies. *American Journal of Primatology*, **12**, 387–468.

Hershkovitz, P. (1987b) Uacaris, New World monkeys of the genus *Cacajao* (Cebidae, Platyrrhini): a preliminary taxonomic review with description of a new subspecies. *American Journal of Primatology*, **12**, 1–53.

Hershkovitz, P. (1988). Origin, speciation and distribution of South American titi monkeys, genus *Callicebus* (Family Cebidae, Platyrrhini). *Proceedings of the Academy of Natural Sciences of Philadelphia*. **140**, 240–272.

Hershkovitz, P. (1990). Titis, New World monkeys of the genus *Callicebus* (Cebidae, Platyrrhini): a preliminary taxonomic review. *Fieldiana Zoology, new series*, **55**, 1–109.

Heymann, E. & Aquino, R. (1994). Exploraciones primatologicas en las Quebradas Blanco, Blanquillo y Tangarana (Rio Tahuayo, Amazonia, Peruana). *Folia Amazonica*, **6**, 1–2.

International Union for the Conservation of Nature (2008). *2008 IUCN Red List of Threatened Species*. Version 2010.4, viewed 28 February, 2010, http://www.iucnredlist.org.

International Union for the Conservation of Nature (2011). *2011 IUCN Red List of Threatened Species*, version 2012.2, viewed 12 February, 2012, http://www.iucnredlist.org.

Izawa, K. & Bejarano, G. (1981). Distribution ranges and patterns of nonhuman primates in western Pando, Bolivia. *Kyoto University Overseas Research Reports of New World Monkeys*, **2**, 1–11.

Kimerling, J. (1993). *Crudo Amazonico*. FCUNAE: Coca, Ecuador and Green San Francisco: Ink, Inc.

Kobayashi, S. (1995). A phylogenetic study of titi monkeys, genus *Callicebus*, based on cranial measurements: I. Phyletic groups of *Callicebus*. *Primates*, **36**, 101–120.

Leonard, S. & Bennett, C. (1996). Associative behavior of *Cacajao calvus ucayalii* with other primate species in Amazonian Peru. *Primates*, **37**, 227–230.

Lönnberg, E. (1938). Remarks on some members of the genera *Pithecia* and *Cacajao* from Brasil. *Arkiv für Zoologi*, **30^{A}**, 1–25.

Lönnberg, E. (1939). Notes on some members of the genus *Callicebus*. *Arkiv für Zoologi*, **31^{A}**, 1–18.

Los micos de Patarroyo. (2008). Camibo.com, viewed 31 March, 2008, http://www.cambio.com.co/portadacambio/751/ARTICULO-WEB-NOTA_INTERIOR_CAMBIO-3825952.html.

Maldonado, A., Nijman, V. & Bearder, S.K. (2009). Trade in night monkeys *Aotus* spp. in the Brazil–Colombia–Peru tri-border area: international wildlife trade regulations are ineffectively enforced. *Endangered Species Research*, **9**, 143–149.

Marsh, L.K. (2004). Primate species at the Tiputini Biodiversity Station, Ecuador. *Neotropical Primates*, **12**, 75–78.

Martinez, J. & Wallace, R.B. (2007). Further notes on the distribution of the Bolivian endemic titi monkeys, *Callicebus modestus* and *Callicebus olallae*. *Neotropical Primates*, **14**, 47–54.

Martínez, J. & Wallace, R.B. (2010). Pitheciidae. In *Distribución, Ecología y Conservación de los Mamíferos Medianos y Grandes de Bolivia*, ed. R.B. Wallace, H. Gómez, Z.R. Porcel & D.I. Rumiz. Santa Cruz de la Sierra, Bolivia: Centro de Ecología Difusión Simón I. Patiño, pp. 305–330.

Mason, W.A. (1966). Social organization of the South American monkey, *Callicebus moloch*: a preliminary report. *Tulane Studies in Zoology*, **13**, 23–28.

Matthews, H. (2005). The effects of hunting on a primate community in the Peruvian Amazon. Unpublished Masters thesis, Winthrop University.

Moreano, M. (2005). Uso de hábitat y comportamiento de *Pithecia monachus* (Primates: Pitheciidae) en el Parque Nacional Yasuní. Unpublished Bachelor thesis, Pontificia Universidad Católica del Ecuador.

Moynihan, M. (1976). *The New World Primates*. Princeton, NJ: Princeton University Press.

Naughton-Treves, L., Holland, M. & Brandon, K. (2005). The role of protected areas in conserving biodiversity and sustaining local livelihoods. *Annual Review Environment and Resources*, **30**, 219–252.

Navarrete, L. (2001). Estudio ecológico de *Pithecia monachus monachus* (Primates: Cebidae), Cuyabeno, Amazonía ecuatoriana. Unpublished Bachelor thesis, Pontificia Universidad Católica del Ecuador.

Neris, N., Colman, F., Ovelar, E., *et al.* (2002). *Guía de Mamíferos Medianos y Grandes del Paraguay: Distribución, Tendencia Poblacional y Utilización.* Asunción: Secretaría del Ambiente (SEAM) y Agencia de cooperación Internacional del Japón (JICA).

Novaro, A., Redford, K. & Bodmer, R. (2000). Effect of hunting in source-sink systems in the Neotropics. *Conservation Biology*, **14**, 713–721.

Phillips, K.A., Haas, M.E., Grafton, B.W., *et al.* (2004). Survey of the gastrointestinal parasites of the primate community at Tambopata National Reserve, Peru. *Journal of Zoology*, **264**, 149–151.

Pitman, N., Smith, R., Vriesendorp, C., *et al.* (eds.). (2004). *Peru: Ampiyacu, Apayacu, Yaguas, Medio Putumayo. Rapid Biological Inventory*, **12**. Chicago, IL: The Field Museum, Environmental and Conservation Programs, viewed 15 January, 2007, http://fm2.fieldmuseum.org/rbi/results_per12.asp.

Pitman, N., Vriesendorp, D., & Moskovits, D. (eds.) (2003). Peru: Yavarí. *Rapid Biological Inventory*, **11**. Chicago, IL: The Field Museum, Environmental and Conservation Programs.

Porras, M. (2000). Comunicación vocal y su relación a actividades, estructura social y contexto comportamental en *Callicebus cupreus ornatus. A Primatologia no Brasil*, **7**, 265–274.

Porter, L.M. (2006). Distribution and density of *Callimico goeldii* in northwestern Bolivia. *American Journal of Primatology*, **68**, 235–243.

Pozo R., W.E. (2004). Preferencias de hábitat de seis especies de primates simpátricos del Yasuní, Ecuador. *Ecología Aplicada*, **3**, 128–133.

Pozo R., W.E. & Youlatos, D. (2005). Estudio sinecológico de nueve especies de primates del Parque Nacional Yasuní, Ecuador. *Politécnica*, **26**, *Biología* **6**, 83–107.

Puertas, P. & Bodmer, R. (1993). Conservation of a high diversity primate assemblage. *Biodiversity and Conservation*, **2**, 586–593.

Robinson, J.G. (1981). Vocal regulation of inter- and intragroup spacing during boundary encounters in the titi monkey, *Callicebus moloch. Primates*, **22**, 161–172.

Rohe, F. & Silva-Jr., J.S. (2009). Confirmation of *Callicebus dubius* (Pitheciidae) distribution and evidence of invasion into the geographic range of *Callicebus stephennashi. Neotropical Primates*, **16**, 69–71.

Rowe, N. & Martinez, W. (2003). *Callicebus* sightings in Bolivia, Peru and Ecuador. *Neotropical Primates*, **11**, 32–35.

Sistema Nacional de Áreas Protegidas del Paraguay (2007). *Áreas Silvestres Protegidas del Paraguay*, viewed 4 April, 2011, http://www.undp.org.py/images_not/triptico_asps_2007.pdf.

Soini, P. (1982). Primate conservation in Peruvian Amazonia. *International Zoo Yearbook*, **22**, 37–46.

Soini, P. (1986). A synecological study of a primate community in the Pacaya–Samiria National Reserve, Peru. *Primate Conservation*, 7, 63–71.

Stallings, J.R. (1985).Distribution and status of primates in Paraguay. *Primate Conservation*, **6**, 51–58.

Terborgh, J., Fitzpatrick, J. & Emmons, L. (1984). Annotated checklist of bird and mammal species of Cocha Cashu Biological Station. *Fieldiana, Zoology, new series*, **21**, 1–29.

Terborgh, J., van Schaik, C., Rao, M., *et al.* (eds.). (2002). *Making Parks Work: Strategies for Preserving Tropical Nature.* Washington, DC: Island Press.

Tirira, D.G. (2007). *Guía de Campo de los Mamíferos del Ecuador.* Publicación especial sobre los mamíferos del Ecuador **6**. Quito: Ediciones Murciélago Blanco.

Tirira, D.G. (ed.) (2011). *Libro Rojo de los Mamíferos del Ecuador. Fundación Mamíferos y Conservación, Pontificia Universidad Católica del Ecuador y Ministerio del Ambiente del Ecuador.* Publicación especial sobre los mamíferos del Ecuador **8**. Quito: Ediciones Murciélago Blanco.

Ulloa, R. (1988). Estudio sinecológico de primates en la Reserva de Producción Faunística Cuyabeno, Amazonía ecuatoriana. Unpublished Bachelor thesis. Pontificia Universidad Católica del Ecuador.

United Nations Educational, Scientific and Cultural Organization. (2007). *Paraguay: El Chaco. UNESCO – MAB Biosphere Reserves Directory. The MAB Programme*, viewed 5 April, 2011, http://www.unesco.org/mabdb/br/brdir/directory/biores.asp?mode=all&code=PAR+02.

United Nations Environmental Programme, World Conservation Monitoring Centre. (2007). *Protected areas national management categories*, viewed 4 January, 2008, http://www.unep-wcmc.org/wdpa.

van Roosmalen, M., van Roosmalen, T. & Mittermeier, R. (2002). A taxonomic review of the titi monkeys, genus *Callicebus* Thomas, 1903, with the description of two new species, *Callicebus bernhardi* and *Callicebus stephennashi*, from the Brazilian Amazon. *Neotropical Primates*, **10**, 1–52.

Vermeer, J. (2009). On the identification of *Callicebus cupreus* and *Callicebus brunneus*. *Neotropical Primates*, **16**, 69–71.

Vriesendorp, C., Rivera Chávez, L., Moskovits, D., *et al.* (eds.). (2004). Peru: Magantoni. *Rapid Biological Inventory* 15. Chicago, IL: The Field Museum, Environmental and Conservation Programs, viewed 15 January, 2007, http://fm2.fieldmuseum.org/rbi/results_per15.asp.

Wallace, R.B., Gomez, H., Felton, A., *et al.* (2006). On a new species of titi monkey, genus *Callicebus* Thomas, from western Bolivia (Primates, Cebidae) with preliminary notes on distribution and abundance. *Primate Conservation*, **20**, 29–39.

Wallace, R.B., Painter, R.L.E., Taber, A.B., *et al.* (1996). Notes on a distributional river boundary and southern range extension for two species of Amazonian primates. *Neotropical Primates*, **4**, 149–151.

Ward, N. & Chism, J. (2003). A report on a new geographic location of red uakaris (*Cacajao calvus ucayalii*) on the Quebrada Tahuaillo in Northeastern Peru. *Neotropical Primates*, **11**, 19–22.

Wunder, S. (2000). *The Economics of Deforestation: The Example of Ecuador*. London: Macmillan Press.

Chapter 33

Brazil

Stephen F. Ferrari, José S. Silva Júnior, Manuella A. de Souza, Ana Luisa K. Albernaz, Marcelo M. Oliveira & Leandro Jerusalinsky

Introduction

Brazil is a country of many biological superlatives, and its pitheciid fauna is no exception to the general rule. It is home to all four pitheciid genera, and more species than any other country. In fact, the geographic ranges of only a small handful of species are located wholly outside Brazilian territory, and a considerable proportion is endemic to this nation. Of the non-endemic species, in addition, only a few are more widespread outside Brazil than within the country.

This Brazil-centric distribution of pitheciids has a number of different, sometimes conflicting implications for the conservation of the group. Up to a point, it reflects the relatively ample geographic ranges of most species, which is normally a positive factor from a conservation viewpoint. However, for the endemic species in particular, it may also restrict management options, and even the potential number of protected areas, in comparison with species that occur in more than one country. Nevertheless, each region, local area and species has a unique set of characteristics, which demand a unique approach to the planning of conservation strategies.

In general terms, the Brazilian government's record on wildlife conservation is probably at least as good as that of the other countries in which pitheciids are found. In addition to the consolidation of an ample, and still expanding network of protected areas, specific initiatives include the Federal Environment Institute's Primate Protection Centre (Centro de Proteção de Primatas Brasileiros: ICMBIO/CPB), which supports and coordinates primate conservation programs throughout the country. Two other institutions are important nuclei of research and captive breeding. The National Primate Centre (Centro Nacional de Primatas) was established in eastern Amazonia by the federal health ministry in 1978, while the Rio de Janeiro Primatology Centre (Centro de Primatologia do Rio de Janeiro) was created in 1975 by the Rio de Janeiro state government.

Endangered Brazilian pitheciids are now included in the agendas of two national committees, established to coordinate primate conservation initiatives in the Brazilian Amazon (Brasil 2005) and the northern Atlantic Forest and Caatinga (Brasil 2006a). These committees, which involve both government bodies, such as the CPB, and non-governmental organizations, establish conservation initiatives and guidelines, in particular through regular workshops.

An important recent advance has been the establishment of the nationwide system (Sistema Nacional de Unidades de Conservação da Natureza or SNUC: Brasil 2000) which regulates the creation and management of protected areas at all levels of political organization, from federal, state and municipal government to private individuals. In addition to specific efforts, such as the implementation of protected areas, Brazilian pitheciids benefit, at least potentially, from a number of different federal laws. One is the Forestry Code (Brasil 1965), which regulates the exploitation of natural habitats on privately owned land. Deforestation is prohibited in habitats denominated permanent protection areas, which include riverbanks and hilltops, and all rural properties are obliged to maintain a legal reserve of natural habitat corresponding to a specific proportion of their total area, which varies depending on the biome.

This legislation was recently reinforced specifically for the Atlantic Forest (Brasil 2006b), where it is now illegal to deforest any primary or late successional habitat within the geographic distribution of endangered species. This has obvious potential benefits for the Atlantic Forest titis. In the Amazon region, the legal reserve must cover 80% of the property, although there is a strong parliamentary lobby for this to be reduced to 50% or less.

The Fauna Protection Act (Brasil 1967) prohibits the hunting or commercial exploitation of native fauna, while the Environmental Crimes legislation (Brasil 1998) establishes strict penalties for a range of offences, including animal cruelty, although it also decriminalized subsistence hunting. As hunting in Brazil has always been primarily for subsistence, this has probably had little practical effect on the intensity of the pressure experienced by pitheciid populations. In practice, *Callicebus* is protected by its relative small body size, whereas the white uacari, *Cacajao calvus*, is rejected by most local hunters because of its supposed resemblance to a human being (Ayres 1986). By contrast, in addition to their meat, both *Chiropotes* and *Pithecia* are valued for their bushy tails, which

Evolutionary Biology and Conservation of Titis, Sakis and Uacaris, eds. Liza M. Veiga, Adrian A. Barnett, Stephen F. Ferrari and Marilyn A. Norconk. Published by Cambridge University Press.

are made into dusters, and hunting pressure may contribute to reduced densities of sakis in some areas (Lopes & Ferrari 2000).

Brazil has one of the World's most active primatology societies – A Sociedade Brasileira de Primatologia (www.primatologia.org.br) – which has more than 150 members, including scientists from all four corners of the country, although the Amazon basin is under-represented considerably in relation to its geographic area. There are also relatively few researchers working on pitheciids, in comparison with some other groups, such as the lion tamarins, for example (Kleiman & Rylands 2002), or even the common marmoset. However, there have been considerable advances since the pioneering work of Ayres (1981, 1986) and Kinzey & Becker (1983), and at least some ecological data are now available for most Brazilian pitheciids.

An additional problem for the conservation of the pitheciids in general, but in Brazil in particular, is the definition of species-level diversity and zoogeography. By virtue of its dimensions and ecological diversity, Brazil has, once again, been at the forefront of the recent discoveries of platyrrhine species, with three new species of titi monkeys being described in the past few years (Kobayashi & Langguth 1999; van Roosmalen *et al.* 2002), although there are also important questions to be resolved with regard to the classification of the pitheciines (see Figueiredo *et al.*, Chapter 3 and Silva Júnior *et al.*, Chapter 4). These questions, together with the overall scarcity of data on ecological patterns and geographical distribution, underpin the difficulties of planning the conservation of Brazil's pitheciid species.

The conservation of Brazilian pitheciids is best discussed in the context of four major geographic divisions – the Atlantic Forest, the southern Amazon basin (east of the Rio Madeira), the western basin (between the Madeira and Negro rivers), and the northern Amazon (east of the Negro) – which reflect broad regional differences in biodiversity, levels of anthropogenic disturbance and conservation status. Whereas the Atlantic Forest is home to a single pitheciid genus, for example, all four genera occur in the northern Amazon basin. Similarly, while the former is characterized by critical levels of deforestation, the latter encompasses some of the largest remaining tracts of continuous tropical forest found anywhere in the World.

Atlantic Forest

The titi monkeys of the *Callicebus personatus* group are the only pitheciids that occur in the Brazilian Atlantic Forest, where they range from central São Paulo state in the south to the right margin of the São Francisco River in the states of Bahia and Sergipe, in the north. The distribution of the group also includes adjacent Cerrado (savanna) and Caatinga (scrubland) ecosystems in São Paulo–Minas Gerais, and Bahia–Sergipe, respectively. The group's range is completely isolated from those of all other pitheciids, including other *Callicebus*, and none of the species are found anywhere in the Amazon basin.

The species of the *C. personatus* group have relatively small geographic ranges in comparison with most other pitheciids, one factor which contributes to their comparatively unfavorable conservation status. More importantly, their distribution coincides with the most densely populated region of the country, which is also that with the longest history of European colonization (Dean 1995), which has resulted in widespread deforestation, with most estimates of the remaining forest cover varying between 5% and 10%. Unfortunately, the species with the smallest ranges also face the highest levels of habitat loss, and the least protection.

The two species most threatened with extinction – *C. barbarabrownae* and *C. coimbrai* – have been assigned to the IUCN categories "critically endangered" (Veiga *et al.* 2008a) and "endangered" (Veiga *et al.* 2008b), respectively, based on criteria that include current distribution, habitat fragmentation levels, and population size. *Callicebus coimbrai* has one of, if not the smallest geographic distribution of any titi species, restricted primarily to the coastal strip of Atlantic Forest between the São Francisco River in the north (Sergipe), and northern Bahia to the south. Only slightly larger, the range of *C. barbarabrownae* is located to the west of that of *C. coimbrai*, in the *Caatinga* of the right margin of the middle and lower São Francisco.

Recent studies of both species (see Printes *et al.*, Chapter 5) have expanded considerably our knowledge of their distribution, and the number and size of populations, providing some room for cautious optimism. This is reflected in the recent (Veiga *et al.*, 2008b) reclassification of *C. coimbrai* from "critically endangered" to "endangered", based on a strict interpretation of IUCN criteria B, C and D. Obviously, this should not be considered a motive for complacency with regard to the urgency of conservation measures, in particular the establishment of protected areas.

One very recent development has been the establishment of the Mata do Junco state wildlife refuge in eastern Sergipe, which protects a small population of *C. coimbrai.* The CPB has also submitted preliminary proposals for the implementation of a federal wildlife refuge in the south of the state, destined specifically for the conservation of this species. On the other hand, while Sergipe's Serra de Itaiana National Park appears to be within the geographic range of *C. coimbrai*, the species is also certainly absent from this site (Oliveira *et al.* 2006; Jerusalinsky *et al.* 2006). Another important recent advance has been the implementation of an ecological monitoring program at the Fazenda Trapsa in southern Sergipe (Souza-Alves 2010).

Further south, the species *C. melanochir* and *C. personatus* appear to be in less immediate threat of extinction, and their current conservation status is "vulnerable" (Veiga *et al.* 2008c, 2008d). Both species have relatively ample ranges in comparison with the previous two, and occur in a number of relatively large conservation units (Printes *et al.*, Chapter 5). *Callicebus melanochir* is found along the coast from the right margin of the Paraguaçu River in Bahia to the left bank of the Mucuri

River in northern Espírito Santo (van Roosmalen *et al.* 2002), whereas *C. personatus* ranges as far south as Rio de Janeiro, and westwards to the upper Jequitinhonha. Both species have been the subject of detailed ecological studies by Müller (1996) and Heiduck (2002), and Price and Piedade (2001), respectively.

The fifth species, *Callicebus nigrifrons* is distributed over a relatively wide area, including much of São Paulo, southern Minas Gerais and eastern Rio de Janeiro state (Printes *et al.*, Chapter 5), a region almost the size of France. Given its relatively ample distribution, and its presence in protected areas such as the Serra do Mar reserve complex, and even the Cantareira State Park in the centre of the city of São Paulo (Trevelin *et al.* 2007), the species is currently considered to be "near threatened" by the IUCN (Veiga *et al.* 2008e), even though it has been assigned to more critical categories in previous assessments. As for all the other species, the extensive fragmentation that is typical of remaining habitat throughout its range is the major potential problem for the long term, given the likely eventual need for metapopulation management.

In the Atlantic Forest, the first primate conservation programs date back to the 1970s, although the early projects concentrated on "flagship" species such as the golden lion tamarin (*Leontopithecus rosalia*) and the muriqui (*Brachyteles* spp.). Specific initiatives involving titi monkeys were slow to materialize, for a number of reasons, not least because, up until the end of the twentieth century, what is now considered to be the *C. personatus* group consisted of a single species, with four subspecies (Hershkovitz 1988, 1990). In their description of *Callicebus coimbrai*, Kobayashi and Langguth (1999) elevated the remaining taxa to the species level, thereby "creating" a more problematic group of five species, all of which are considered threatened to some degree. This has led to the current interest in species such as *C. coimbrai* and *C. barbarabrownae*, but research and conservation efforts are still relatively incipient, and there are as yet very few ecological data available to underpin management strategies.

Southern Amazon basin

The pitheciids of the southern Amazon basin present an interesting pattern of distribution, with increasing species richness from east to west, reflecting overall trends within the biome (Ferrari 2004). Whereas *Chiropotes* occurs throughout most of the region, for example, *Callicebus* is only found west of the Tocantins River, and *Pithecia* west of the Tapajós. Up to a point, this east–west continuum also reflects decreasing anthropogenic pressure, although the overall trend is reversed in Rondônia, in the western extreme of the region, where deforestation rates are similar to those in the extreme east.

Deforestation rates are also high in the southern rim of the basin – the "Arc of Deforestation" – although precious little information is available on the presence of pitheciids in Mato Grosso or southern Pará, which limits the analysis of distribution patterns and, ultimately, population size and conservation status. There are two complementary questions here, both related to the inadequacies of the available data on the distribution of platyrrhines in the Amazon basin. The role of rivers in the zoogeography of Amazonian primates has long been recognized (Wallace 1852), but most species are represented by little more than a handful of widely dispersed localities, which often leads to considerable, often very subjective extrapolations for the definition of range limits. Worse still, in some cases, e.g. Ferrari (1995), the mistaken assignment of a locality to the opposite margin of a river has led to a major but erroneous extension of a species' range.

In the case of *Chiropotes utahickae*, for example, the species is quite clearly limited to the Xingu-Tocantins interfluvium (Hershkovitz 1985; Ferrari & Lopes 1996; Silva Júnior *et al.*, Chapter 4), but there are no recorded localities any further south than Carajás, in Pará, whereas forest ecosystems appropriate for bearded sakis may extend hundreds of kilometres further south, as far as northern Mato Grosso. While Hershkovitz (1985, figure 1) presumed that the whole of this area is occupied by *C. utahickae*, this is a potentially dangerous assumption on which to base conservation planning.

The second question is the local distribution of each species within its range. Once again, it has been standard practice in the study of the zoogeography of Amazonian primates to assume a homogeneous distribution within range limits, although a number of recent studies have shown that this may often be erroneous (Peres 1997; Ferrari 2004). In the specific case of the pitheciids of the southern basin, Ferrari *et al.* (2007) have documented significant lacunae in the occurrence of *Callicebus moloch* in the same Xingu–Tocantins interfluvium, including the Caxiuanã National Forest, one of the potentially most important protected areas for the conservation of the species in this interfluvium. A good knowledge of the distribution of populations within a species' range is essential for the development of effective conservation strategies (see Printes *et al.*, Chapter 5), but the ample ranges and lack of data that are typical of most Amazonian species may be more problematic than they might seem in some cases.

The black-bearded saki, *Chiropotes satanas*, is the most threatened pitheciid of the Brazilian Amazon, and is classified as "critically endangered" by the IUCN (Veiga *et al.* 2008f), on the basis of ongoing habitat loss and hunting pressure. Endemic to the eastern extreme of the basin, east of the Tocantins River in the states of Maranhão and Pará, the distribution of this species coincides with the most densely populated region of Brazilian Amazonia, which has a long tradition of European colonization, and deforestation levels that begin to rival those of the Atlantic Forest (Ferreira *et al.* 2005). The distribution of this species is relatively well delineated by the transition of the Amazon Forest proper, in the western half of Maranhão, to the *caatinga*, in the east, with the Tocantins River to the south and west.

More than 20 years after Johns and Ayres (1987) predicted that *C. satanas* would be extinct by the turn of the twentieth century, our knowledge of the ecology and distribution of the species has grown considerably (Carvalho Jr. *et al.* 1999; Lopes & Ferrari 2000; Carvalho 2002; Veiga 2006; Silva & Ferrari 2008), and it is now clear that this inopportune forecast was based on an incomplete understanding of its tolerance of habitat disturbance, in particular. In fact, the species may only be absent from relatively small or highly disturbed fragments with a long history of hunting pressure (Lopes & Ferrari 2000). Where there is no hunting pressure, the species may persist even in comparatively tiny fragments, such as the reservoir islands monitored by Veiga (2006) and Silva and Ferrari (2008), both well under 20 ha. There is an obvious need for some form of metapopulation management in such cases, however.

The primary conservation unit within this region is the Gurupi Biological Reserve which, together with contiguous Amerindian reservations, forms a nucleus of continuous forest covering a total area of approximately one million hectares. Despite its potential importance for the conservation of *Chiropotes satanas* (not to mention other endangered endemics such as *Cebus kaapori*), this area suffers intense pressure from local ranchers and lumber companies, and its future remains uncertain. Given this, and the lack of other federal protected areas, Ferrari *et al.* (1999) have emphasized the potential importance of both privately owned reserves and Amerindian reservations for the long-term conservation of *Chiropotes satanas*.

West of the Tocantins, *C. utahickae* faces a slightly less critical situation, by virtue of its larger (but see above) and more isolated geographic range, and generally lower deforestation rates. Human colonization of the region was negligible until the 1970s, when three "major projects" were initiated: the Trans-Amazon highway (BR-230), which bisects the interfluvium from east to west, the Grande Carajás mining complex, and the Tucuruí hydroelectric scheme on the Tocantins River.

While these projects have resulted in considerable environmental impact, primarily by facilitating human colonization, much larger and more continuous tracts of forest remain in comparison with the region to the east of the Tocantins, and there are a number of major conservation units, including the Carajás, Caxiuanã, Tapirapé and Tapirapé–Aquiri National Forests, with a total area of almost one million hectares. As for *C. satanas*, the presence of large Amerindian reservations may also be important, over the long term, for the conservation of the species (Ferrari *et al.* 1999), although it is equally important to remember that the exploitation of natural resources is legally permitted in both categories of reserve (Brasil 2000), which means that the bearded saki populations inhabiting these areas are vulnerable to varying degrees of anthropogenic pressure.

As for *C. satanas*, recent studies of the species have revealed contrasting ecological tendencies. Populations of *C. utahickae* are known to thrive in relatively small, isolated fragments (see Santos *et al.*, Chapter 23), whereas they appear to be relatively rare in much larger areas of continuous forest (Ferrari *et al.* 1999). Given its slightly more favorable situation, in comparison with *C. satanas*, the status of *C. utahickae* is currently "endangered" (Veiga *et al.* 2008g), although the ongoing expansion of human activities within the species' range may eventually annul much of the difference.

The third bearded saki species of the southern Amazon basin, *Chiropotes albinasus*, has the most ample distribution, ranging from the Xingu in the east to the Madeira–Jiparaná and Guaporé in the west, although, once again, there is almost no information on the occurrence of the species throughout most of its presumed range, in particular along most of the right bank of the Madeira and in northern Mato Grosso. The vast extension of this species's range is its primary line of defense against ongoing human colonization, although the Trans-Amazon highway bisects the area from east to west and, perhaps more importantly, the Santarém–Cuiabá highway (BR-163) makes most of the Xingu–Tapajós interfluvium easily accessible from the south. In recent years, this highway has become the main channel of colonization for soybean planters migrating northwards from Mato Grosso, although for the time being, this threat is limited to a relatively small proportion of the species' range. Despite its relatively ample range, *C. albinasus* has now been assigned "endangered" status by the IUCN (Veiga *et al.* 2008h), although it had been considered less vulnerable in the past.

One species of saki, *Pithecia irrorata*, occurs in the southern basin, west of the Tapajós River, although it also occurs west of the Madeira, and is thus even more widespread than *C. albinasus*. For the time being, this species, like other sakis, is considered to be of "least concern" (Veiga & Marsh 2008a), due primarily to its relatively ample geographical range.

While their distribution in the southern Amazon basin is less ample than that of *Chiropotes*, the titi monkeys are far more diverse, with no less than seven different species, including *Callicebus bernhardi*, described by van Roosmalen *et al.* in 2002. While some species, such as *Callicebus baptista*, have relatively small geographic ranges, at least by Amazonian standards, most are relatively well-protected by their comparative isolation from areas of human colonization, and all seven species are currently classified as "least concern" (IUCN 2008).

Ferrari *et al.* (2000) confirmed the presence of *Callicebus donacophilus* in southern Rondônia, and the species may also occur further south, in the western extreme of the state of Mato Grosso do Sul, on the right bank of the Paraguay River (Hirsch *et al.* 2002). The species may be more widespread in southwestern Mato Grosso, but once again, data are lacking. It is far more widespread in neighboring areas of Bolivia and Paraguay, and has been dealt with in more detail elsewhere (Porter *et al.*, Chapter 32).

Of the six other species (the *C. moloch* group, cf. Kobayashi & Langguth 1999; van Roosmalen *et al.* 2002), *Callicebus brunneus* is probably one of the most threatened, given its distribution in northern Rondônia. This region has suffered intense colonization and deforestation over the past 35 years, although it is also now relatively well-covered by protected areas, including state parks and Amerindian reservations. In addition, Iwanaga and Ferrari (2002) found that recent colonists, most of whom immigrated from southern Brazil, do not normally hunt primates, which may contribute to the relatively high densities recorded for many large-bodied species in isolated fragments. Despite these more positive aspects, the possible restriction of the species's range to northern Rondônia would seem to demand a careful re-assessment of its conservation status.

At the opposite extreme, while its exact limits may be uncertain (see above), *C. moloch* has by far the largest geographic range of the southern Amazonian titis, stretching between the Tapajós River, in the west, and the Tocantins in the east. While this species suffers a certain degree of impact at the local level, from highways and major projects (see *Chiropotes albinasus* and *Chiropotes utahickae*, above), it is relatively well protected from anthropogenic impact by the comparative isolation of its ample geographic range.

All other southern Amazonian titis are found between the ranges of these two species, i.e. between the Tapajós and Madeira/Jiparaná rivers. While this area is still relatively isolated from anthropogenic impact, a number of questions remain with regard to the zoogeography of this group, not least because all four species are known from only a small number of sites, and their ranges have been defined according to the standard practice of "filling in" interfluvia, although the ranges of three species – *C. bernhardi*, *Callicebus cinerascens* and *Callicebus hoffmannsi* – may intergrade in some way in the region of the headwaters of the Juruena, Aripuanã and Roosevelt rivers. Van Roosmalen *et al.* (2002) also propose an unusual, disjunct range for *C. baptista*. While there are no major concerns with regard to anthropogenic impact, for the time being, more detailed knowledge of distribution patterns within this area will be essential for the eventual development of effective conservation strategies.

Northern Amazon basin

Despite the fact that all four pitheciid genera occur in the northern Amazon basin, the region is characterized by relatively low species diversity in comparison with the rest of the biome. In marked contrast with other regions, only one of the species occurring in the northern Amazon basin are endemic to Brazil.

Apart from the presence of a corridor of colonization along the BR-174 highway, which links Manaus to Venezuela via the state of Roraima (and includes the Balbina hydroelectric reservoir), most of this region, to both the east and the west, remains virtually free of deforestation. In addition, all the pitheciid species found in this region also occur in neighboring countries, where they are generally at least as well protected as they are in Brazil (see Lehman *et al.*, Chapter 31 and Porter *et al.*, Chapter 32). Discussion of the sakis (*Pithecia*) and uacaris (*Cacajao*) necessarily shifts to the question of subspecific diversity (see Silva Jr. *et al.*, Chapter 4). While subspecies have been assessed individually by the IUCN in the past, the most recent update (Veiga *et al.* 2008i) provides only a species-level classification.

Chiropotes chiropotes has the smallest geographic range of the northern basin pitheciids, located between the Branco, Negro and Orinoco rivers, although it is at least as large as that of *C. satanas*, and is far more isolated, especially in the region of the Venezuelan border, where it occurs in the Pico da Neblina National Park (Boubli 2002). The status of *C. chiropotes*, as defined here, is unclear, given that all IUCN evaluations to date (Baillie & Groombridge 1996; IUCN 2008; Veiga *et al.* 2008j) have used the traditional (Hershkovitz 1985) classification. As a species distinct from *C. sagulatus*, the conservation status of *C. chiropotes* may require careful re-assessment, although it seems unlikely to be threatened at the present time, given the remoteness of its range.

Chiropotes sagulatus occurs east of the Branco River, in an ample range which includes almost the whole of the Guyanas, the Brazilian state of Amapá, the whole of northern Pará, and adjacent areas of Amazonas and Roraima. In addition to being one of the least populated parts of the Amazon basin, this is also one of the best protected, given the recent implementation of conservation units such as the Montanhas do Tumucumaque National Park, the Maicuru Biological Reserve, and the Grão Pará Ecological Station, which are contiguous with protected areas in French Guyana (Lehman *et al.*, Chapter 31). Together, they form a continuous protected area of more than 100,000 km^2. While the conservation status of this species has yet to be assessed officially, it seems reasonable to assume that it is one of the least endangered pitheciids.

Pithecia pithecia is distributed throughout the northern Amazon basin, and may be even more widespread than *Chiropotes* in Venezuela and the Guyanas (Hirsch *et al.* 2002), justifying its IUCN classification as "least concern" (Veiga & Marsh 2008b). This same status is probably equally applicable to both subspecies – *Pithecia p. pithecia* and *Pithecia p. chrysoleuca* – although the exact limits of their ranges remain unclear (Hershkovitz 1987a; Silva Jr. *et al.*, Chapter 4).

The northern or black-headed uacaris have a similarly ample distribution between the Negro–Branco and Solimões–Japurá, although the current division into three species raises some concerns. While *Cacajao ouakary* is amply distributed between the Solimões and Negro rivers and is thus of "least concern" (Barnett *et al.* 2008), both *Cacajao melanocephalus* and *Cacajao ayresi* are considered to be vulnerable on the basis of their relatively restricted ranges, north of the Negro (Boubli & Veiga 2008a, 2008b). Little is known of the newly discovered *Cacajao ayresi* (Boubli *et al.* 2008), but it does appear to be endemic to Brazil, possibly to an unusually small range. While it has a much longer history, *C. melanocephalus* is only slightly better known (see e.g. Boubli 1999), so more detailed studies of both species would appear to be essential for a more reliable evaluation of their conservation status and needs. Careful re-assessment of the taxonomic status of all three forms of black uacari would also seem to be necessary (Figueiredo *et al.*, Chapter 3).

The only titi monkey that occurs in this region, *Callicebus lugens*, is the most widely distributed of the *Callicebus torquatus* group, and occurs between the Branco and Negro rivers in Brazil, as far north as the Orinoco in Venezuela, and the upper Caquetá, in Colombia (see Lehman *et al.*, Chapter 31 and Porter *et al.*, Chapter 32). Once again, this species is protected from immediate threats by its relatively large range and isolation from areas of human colonization.

Western Amazon basin

The western Amazon basin is renowned for its biological diversity, and the overall trend is upheld by the pitheciids, in particular *Callicebus*. According to van Roosmalen *et al.* (2002), there are no less than 11 Brazilian species of titi monkeys west of the Madeira/Negro. This region is also the domain of the uacaris (at least five taxa) and the sakis (three species). Many of these species also occur in neighboring countries (Bolivia, Colombia and Peru).

The region suffers from much the same intrinsic problems associated with the understanding of species-level diversity and distribution as the southern basin, although, like the northern basin, it has yet to bear the full impact of human colonization, except in some local areas, in particular the state of Acre, but also the Madeira–Purús interfluvium, in the southeastern extreme of the region (Ferreira *et al.* 2005). Like Rondônia, Acre is bisected by the BR-364 highway, which forms the main channel of colonization from the south of the country, and has been subject to similar levels of deforestation. However, this relatively small state traverses only a limited portion of the ranges of some of the pitheciid species that occur between the Madeira and Juruá rivers, and its overall impact on their populations is still negligible.

The Madeira–Purús interfluvium is located strategically between Acre and Rondônia, to the west and south, and the major city of Manaus to the north, and is bisected longitudinally by the BR-319 highway. Fortunately, from the conservationist's point of view, the road was virtually abandoned a few years after its construction, in the mid 1970s, which averted a major influx of colonists similar to that which occurred further south in Rondônia. In 2005, however, the federal government announced plans to reconstruct the highway, which almost instantaneously resulted in incursions by speculators and potential colonists (Fearnside & Graça 2006).

It is still unclear how this will affect the colonization of the interfluvium, but it seems almost inevitable that the region will eventually be occupied in much the same way as the other major highways of the southern Amazon basin. This would have deleterious consequences for two species in particular – *Callicebus caligatus*, which is endemic to the area traversed by the highway, and the adjacent *Callicebus stephennashi*, newly described by van Roosmalen *et al.* (2002). For the time being, the former is of "least concern" (Veiga 2008a) and the latter, "data deficient" (Veiga 2008b), but in the light of recent events, the collection of more detailed information on their distribution and abundance would be recommended.

A third species, *Callicebus dubius*, occurs further upriver in this same interfluvium, but has a relatively ample range, encompassing parts of northern Bolivia, southeastern Peru and three Brazilian states. Two other species of the *Callicebus cupreus* group are also found in this region, with *C. cupreus* occurring between the Purús and Solimões, and *Callicebus discolor* in the Solimões–Içá (and over a wider area in Peru and Ecuador). None of these species are considered to be under any immediate threat of extinction, due primarily to their relatively ample and isolated ranges.

Four species of the *C. torquatus* group also occur in the region. Two of these – *Callicebus purinus* and *Callicebus regulus* – are sympatric with *C. cupreus* in the Purús–Juruá and Juruá–Solimões interfluvia, respectively, and a third (*Callicebus lucifer*) is sympatric with *C. discolor* in the southern half of its range. *Callicebus torquatus* is the only titi found in the Solimões–Negro interfluvium. As for the *C. cupreus* group, none of these species are considered to be threatened in any way at the present time.

As a species, *Cacajao calvus* is classified as "vunerable" by the IUCN (Veiga *et al.* 2008i), although some of the different subspecies (*Cacajao calvus calvus*, *Cacajao c. novaesi*, *Cacajao c. ucayalii* and *Cacajao c. rubicundus*) have been considered endangered in previous assessments (Groombridge 1994). Found south of the Solimões–Japurá, and west of the Purús, the distribution pattern of these pitheciines is somewhat unusual, in the sense that they tend to range along rivers rather than within interfluvia, like most other Amazonian platyrrhines.

With the exception of *C. c. ucayalii*, which may occur in Brazil only along the Javari river (for comments, see Porter *et al.*, Chapter 32), the subspecies have relatively small ranges, limited primarily to white-water *várzea* habitats. In addition to these ecological constraints, the distribution of populations along rivers makes them relatively more vulnerable to human impact than other species in the same area, that are found predominantly in *terra firme* habitats. The only positive factor here is that in many areas, hunters do not target these uacaris because their appearance is considered too human-like (Ayres 1986), although this is not the case for all populations.

While all three Brazilian subspecies have been classified as endangered in the past, recent developments, such as range extensions and the implementation of protected areas, have contributed to a more favorable overall situation. In fact, *C. c. calvus* is among the best-protected of all platyrrhines, considering that a large proportion of its known range is located within the area of the Mamirauá State Sustainable Development Reserve. This reserve, which grew out of the pioneering study of Ayres (1986), encompasses a considerable infrastructure for wildlife management and ongoing research, although it does also contemplate the planned exploitation of natural resources, which implies a certain level of impact on its uacari population.

Cacajao c. rubicundus is also relatively well-protected by the Jutaí–Solimões Ecological Station, decreed in 2001. This category of conservation unit is the most restrictive and, theoretically, offers the best possible protection for the taxon, although it remains unclear how effective it is in practice. Finally, *C. c. novaesi* is now known to occur along much of the Juruá River, a much larger area than first thought (Hershkovitz 1987b). This subspecies is offered some degree of protection from two sustainable-use protected areas, the Middle Juruá and the Upper Juruá extractive reserves. In contrast with the ecological station, these areas are earmarked for the managed exploitation of natural resources, which includes hunting, although, as demonstrated by the Mamirauá reserve (Queiroz 2005), participative involvement on the part of the local community may be far more effective than strict isolation, especially where resources for infrastructure and monitoring are limited.

As before, an additional question here is the validity of each subspecies, and in particular, whether at least some of them should be given full species status, as indicated by Figueiredo *et al.* (Chapter 3). Depending on this, careful re-assessment of the conservation status of the different taxa may be necessary.

Pithecia irrorata has been mentioned above with reference to the southern basin, although in addition to the widespread nominal subspecies, *Pithecia irrorata vanzolinii* is found in the western basin, on the upper Juruá River. While this subspecies appears to have a relatively small range, encompassing parts of Acre and Amazonas states, it is, once again, relatively isolated from human impact, and is probably not under any significant threat, at least for the time being. However, there are a number of considerations here. One is the location of its range, considering the intensity of human colonization in Acre, although the western part of the state, which straddles the Juruá, is much less affected than the eastern half (Ferreira *et al.* 2005). Additional studies are also required in order to better define the distribution of the subspecies, and to assess its taxonomic status (see Marsh 2006; Silva Jr. *et al.*, Chapter 4).

Pithecia monachus ranges west of the Juruá, and is relatively widespread in neighbouring countries, in particular Peru (see Porter *et al.*, Chapter 32). The only subspecies that occurs in Brazil is *P. m. monachus*, which has the most ample and cosmopolitan distribution, and is thus not in any immediate risk of extinction.

The current conservation status of the third saki species – *Pithecia albicans* – is "vulnerable" (Veiga *et al.* 2008k), mainly because of its relatively restricted range on the right bank of the Solimões, between the Tefé and Purús rivers. On the basis of current knowledge, this range is smaller than those of many Amazonian titis, and almost all other Brazilian pitheciines. However, there is an ample zone of potential parapatry with *P. irrorata* to the south and east, which needs to be examined in order to better define the ranges of both species, and specific aspects of the biological relationship between them.

Overview

Like the country itself, the pitheciid fauna of Brazil is ample and varied, and presents a correspondingly diverse set of problems for its long-term conservation. The number of taxa alone constitutes a major challenge, as does the need for the definition of species-level diversity and, in many cases, the exact limits of geographic ranges (or the identification of remnant populations). In addition, while good data are now available for some species – as demonstrated elsewhere in this volume – most ecological parameters are still poorly defined or understood.

Brazilian pitheciids include some of the World's most endangered primate species, as well as some of the least threatened with extinction. Up to a point, the region to which a species is endemic – rather than its ecological characteristics – is a good predictor of its conservation status. Given this, the most endangered species are the Atlantic Forest (and *Caatinga*) titi monkeys, which would otherwise be expected to be among the least vulnerable to most forms of anthropogenic pressure on the basis of their body size and ecological characteristics.

This regional trend becomes slightly weaker, however, if subspecific diversity is taken into account. In this case, the uacaris gain prominence by virtue of their relatively small geographic ranges, determined in part by their unique ecological characteristics. This raises a number of important questions with regard to the definition of species-level diversity, which is fundamental to the development of effective conservation strategies for any group of organisms.

There are two related tendencies here. One is regional, once again, considering that species tend to be relatively more numerous, with correspondingly smaller ranges, in southern Amazonia and the Atlantic Forest. The other is methodological, given the contrast between the current classifications of *Callicebus* and *Chiropotes*, on the one hand, and *Cacajao* and *Pithecia*, on the other, with their differing emphases on specific and subspecific diversity, respectively. Coincidentally, the latter two genera are also more widespread in northern and western Brazil, which are also the least well explored, in scientific terms. Clearly, then, more work is still needed in order to better define the exact dimensions of the diversity of Brazilian pitheciids, which will be essential for the development of effective conservation strategies.

Overall, Brazil is not only the major pitheciid habitat country, it is also at the forefront of South American Primatology, and has undergone steady improvement in conservation policy at all levels of government over the past 20 years or so. Hopefully, this will guarantee the survival of all Brazilian pitheciids over the short term, and the consolidation of their conservation and management over the longer term.

Acknowledgments

The first author is grateful to CNPq (process no. 302747/2008–7) for supporting his pitheciid research.

References

Ayres, J.M. (1981). Observações sobre a ecologia e o comportamento dos cuxius (*Chiropotes albinasus* e *Chiropotes satanas*, Cebidae: Primates). Masters dissertation, Universidade do Amazonas.

Ayres, J.M.C. (1986). Uakaris and Amazonian flooded forests. PhD thesis, Cambridge University.

Baillie, J. & Groombridge, B. (1996). *1996 IUCN List of Threatened Animals*. Gland: IUCN.

Barnett, A.A., Boubli, J.-P., Veiga, L.M., *et al.* (2008). *Cacajao melanocephalus*. In *2008 IUCN Red List of Threatened Species*, ed. IUCN. Gland: IUCN, http://www.iucnredlist.org/details/3417.

Bonvicino, C.R., Boubli, J.P., Otazu, I.B., *et al.* (2003). Morphologic, karyotypic, and molecular evidence of a new form of *Chiropotes* (Primates, Pitheciinae). *American Journal of Primatology*, **61**, 123–133.

Boubli, J.P. (1999). Feeding ecology of black-headed uacaris (*Cacajao melanocephalus melanocephalus*) in Pico da Neblina National Park, Brazil. *International Journal of Primatology*, **20**, 719–749.

Boubli, J.P. (2002). Western extension of the range of bearded sakis: a possible new taxon of *Chiropotes* sympatric with *Cacajao* in the Pico da Neblina National Park, Brazil. *Neotropical Primates*, **10**, 1–4.

Boubli, J.-P. & Veiga, L.M. (2008a). *Cacajao hosomi*. In *2008 IUCN Red List of Threatened Species, ed. IUCN*. Gland: IUCN, http://www.iucnredlist.org/details/136640.

Boubli, J.-P. & Veiga, L.M. (2008b). *Cacajao ayresi. In 2008 IUCN Red List of Threatened Species, ed. IUCN*. Gland: IUCN, http://www.iucnredlist.org/details/136419.

Boubli, J.P., da Silva, M.N.F., Amado, M.V., *et al.* (2008). A taxonomic reassessment of *Cacajao melanocephalus* Humboldt (1811), with the description of two new species. *International Journal of Primatology*, **29**, 723–741.

Brasil (1965). *Lei n. 4.771, de 15 de setembro de 1965: Código Florestal*. Brasília: Federal Government.

Brasil (1967). *Lei n. 5.197, de 3 de janeiro de 1967: Lei de Proteção à Fauna*. Brasília: Federal Government.

Brasil (1998). *Lei n. 9.605, de 12 de fevereiro de 1998: Lei de Crimes Ambientais*. Brasília: Federal Government.

Brasil (2000). *Lei n. 9.985, de 18 de julho de 2000: Sistema Nacional de Unidades de Conservação*. Brasília: Federal Government.

Brasil (2005). Portaria n. 82, de 29 de novembro de 2005: Comitê primatas amazônicos. Brasília, IBAMA. *Diário Oficial da União - Seção 1*, **229**, 126–127.

Brasil (2006a). Portaria n. 26, de 9 de março de 2006: Comitê primatas do norte da Mata Atlântica e Caatinga. *Diário Oficial da União - Seção 1*, **48**, 78.

Brasil (2006b). *Lei n. 11.428, de 22 de dezembro de 2006: Lei da Mata Atlântica*. Brasília: Federal Government.

Carvalho, M.P. (2002). Avaliação da tolerância à perturbação de hábitat das populações de cuxiú-preto *Chiropotes satanas satanas* (Primates: Pitheciinae) em fragmentos florestais da região Tocantina, na Amazônia oriental. Master's dissertation, Universidade Federal do Pará.

Carvalho Jr., O., Pinto, A.C.B. & Galetti, M. (1999). New observations on *Cebus kaapori* Queiroz, 1992, in eastern Brazilian Amazonia. *Neotropical Primates*, **7**, 41–43.

Dean, W. (1995). *With Broadax and Firebrand - the Destruction of the Brazilian Atlantic Forest*. Berkeley, CA: University of California Press.

Fearnside, P.M. & Graça, P.M.L.A. (2006). BR-319: Brazil's Manaus–Porto Velho highway and the potential impact of linking the Arc of Deforestation to central Amazonia. *Environmental Management*, **38**, 705–716.

Ferrari, S.F. (1995). Observations on *Chiropotes albinasus* from the Rio dos Marmelos, Amazonas, Brazil. *Primates*, **36**, 289–293.

Ferrari, S.F. (2004). Biogeography of Amazonian primates. In *A Primatologia no Brasil - 8*, ed. S.L. Mendes & A.G. Chiarello. Santa Teresa: Sociedade Brasileira de Primatologia, pp. 101–122.

Ferrari, S.F. & Lopes, M.A. 1996. Primate populations in eastern Amazonia. In *Adaptive Radiations of Neotropical Primates*, ed. M.A. Norconk, A.L. Rosenberger & P.A. Garber. New York, NY: Plenum Press, pp. 53–67.

Ferrari, S.F., Bobadilla, U.L. & Emídio-Silva, C. (2007). Where have all the titis gone? The heterogeneous distribution of *Callicebus moloch* in eastern Amazonia, and its implications for the conservation of Amazonian primates. *Prmate Conservation*, **22**, 49–54.

Ferrari, S.F., Emidio-Silva, C., Lopes, M.A., *et al.* (1999). Bearded sakis in south-eastern Amazonia - back from the brink? *Oryx*, **33**, 346–351.

Ferrari, S.F., Iwanaga, S., Messias, M.R., *et al.* (2000). Titi monkeys (*Callicebus* spp., Atelidae: Platyrrhini) in the Brazilian State of Rondônia. *Primates*, **41**, 229–234.

Ferreira, L.V., Venticinque, E. & Almeida, S. (2005). O desmatamento na Amazônia e a importância das áreas protegidas. *Estudos Avançados*, **19**, 157–166.

Groombridge, B. (1994). *1994 IUCN Red List Of Threatened Animals*. Cambridge: IUCN.

Heiduck, S. (2002). The use of disturbed and undisturbed forest by masked titi monkeys *Callicebus personatus melanochir* is proportional to food availability. *Oryx*, **36**, 133–139.

Hershkovitz, P. (1985). A preliminary taxonomic review of the South American bearded saki monkeys genus *Chiropotes* (Cebidae, Platyrrhini), with description of a new subspecies. *Fieldiana Zoology*, **27**, 1–45.

Hershkovitz, P. (1987a). The taxonomy of South American sakis, genus *Pithecia* (Cebidae, Platyrrhini): a preliminary report and critical review with the description of a new species and a new subspecies. *American Journal of Primatology*, **12**, 386–468.

Hershkovitz, P. (1987b). Uakaris, New World monkeys of the genus *Cacajao* (Cebidae, Platyrrhini): a preliminary review with the description of a new subspecies. *American Journal of Primatology*, **12**, 1–57.

Hershkovitz, P. (1988). Origin, speciation, dispersal of South American titi monkeys, genus *Callicebus* (Cebidae, Platyrrhini). *Proceedings of the Academy of Natural Sciences of Philadelphia*, **140**, 240–272.

Hershkovitz, P. (1990). Titis, New World monkeys of the genus *Callicebus* (Cebidae, Platyrrhini): a preliminary taxonomic review. *Fieldiana Zoology, New Series*, **55**, 1–109.

Hirsch, A., Dias, L.G., Martins, L.O., *et al.* (2002). BDGEOPRIM - database of geo-referenced localities of Neotropical primates. *Neotropical Primates*, **10**, 79–84.

IUCN (2008). *2008 IUCN Red List of Threatened Species*. Gland: IUCN. http://www.iucnredlist.org

Iwanaga, S. & Ferrari, S.F. (2002). Geographic distribution and abundance of woolly (*Lagothrix cana*) and spider (*Ateles chamek*) monkeys in southwestern Brazilian Amazonia. *American Journal of Primatology*, **56**, 57–64.

Jerusalinsky, L., Oliveira, M.M., Pereira, R.F., *et al.* (2006). Preliminary evaluation of the conservation status of *Callicebus coimbrai* Kobayashi & Langguth, 1999 in the Brazilian state of Sergipe. *Primate Conservation*, **21**, 25–32.

Johns, A.D. & Ayres, J.M. (1987). Southern bearded sakis beyond the brink. *Oryx*, **21**, 164–167.

Kinzey, W.G. & Becker, M. (1983). Activity pattern of the masked titi monkey, *Callicebus personatus*. *Primates*, **24**, 337–343.

Kleiman, D.G. & Rylands, A.B. (2002). *Lion Tamarins: Biology and Conservation*. Washington, DC: Smithsonian Institution Press.

Kobayashi, S. & Langguth, A. (1999). A new species of titi monkey, *Callicebus* Thomas, from north-eastern Brazil (Primates, Cebidae). *Revista Brasileira de Biologia*, **16**, 531–551.

Lopes, M.A. & Ferrari, S.F. (2000). Effects of human colonization on the abundance and diversity of mammals in eastern Brazilian Amazonia. *Conservation Biology*, **14**, 1658–1665.

Marsh, L.K. (2006). Identification and conservation of a new species of *Pithecia* in Amazonian Ecuador. *International Journal of Primatology*, **27**(Suppl 1), #509.

Müller, K.-H. (1996). Diet and feeding ecology of masked titis (*Callicebus personatus*). In *Adaptive Radiations of Neotropical Primates*, ed. M.A. Norconk, A.L. Rosenberger & P.A. Garber. New York, NY: Plenum Press, pp. 383–401.

Oliveira, F.F., Ferrari, S.F. & Silva, S.D.B. (2006). Mamíferos não-voadores. In *Parque Nacional Serra de Itabaiana Levantamento da Biota*, ed. C.M. Carvalho & J.C. Vilar. Aracaju: IBAMA/UFS, pp. 77–91.

Peres, C.A. (1997). Effects of habitat quality and hunting pressure on arboreal folivore densities in Neotropical forests: a case study of howler monkeys (*Alouatta* spp.). *Folia Primatologica*, **68**, 199–222.

Price, E.C. & Piedade, H.M. (2001). Ranging behavior and intraspecific relationships of masked titi monkeys (*Callicebus personatus personatus*). *American Journal of Primatology*, **53**, 87–92.

Queiroz, H.L. (2005). A reserva de desenvolvimento sustentável Mamirauá. *Estudos Avançados*, **19**, 183–203.

Silva Jr., J.S. & Figueiredo, W.M.B. (2002). Revisão sistemática dos cuxiús, gênero *Chiropotes* Lesson, 1840 (Primates, Pitheciidae). *Livro de resumos do X Congresso Brasileiro de Primatologia*, 21.

Silva, S.S.B. & Ferrari, S.F. (2008). Behavior patterns of southern bearded sakis (*Chiropotes satanas*) in the fragmented landscape of eastern Brazilian Amazonia. *American Journal of Primatology*, **70**, 1–7.

Souza-Alves, J.P. (2010). Ecologia alimentar de um grupo de Guigó-de-Coimbra-Filho (*Calliebus coimbrai* Kobayashi & Langguth, 1999): perspectivas para a conservação da espécie na paisagem fragmentada do sul de Sergipe, Brasil. Master's dissertation, Universidade Federal de Sergipe.

Trevelin, L.C., Port-Carvalho, M., Silveira, M., *et al.* (2007). Abundance, habitat use and diet of *Callicebus nigrifrons* Spix (Primates, Pitheciidae) in Cantareira State Park, Sao Paulo, Brazil. *Revista Brasileira de Zoología*, **24**, 1071–1077.

van Roosmalen, M.G.M., van Roosmalen, T. & Mittermeier, R.A. (2002). A taxonomic review of the titi monkeys, genus *Callicebus* Thomas, 1903, with the description of two new species, *Callicebus bernhardi* and *Callicebus stephennashi*, from Brazilian Amazonia. *Neotropical Primates*, **10** (Suppl.), 1–52.

Veiga, L.M. (2006). Ecologia e organização social de cuxiús, *Chiropotes satanas satanas*, em Tucuruí, Pará. PhD thesis, Universidade Federal do Pará.

Veiga, L.M. (2008a). *Callicebus caligatus*. In *2008 IUCN Red List of Threatened Species*, ed. IUCN. Gland: IUCN, http://www.iucnredlist.org/details/41555.

Veiga, L.M. (2008b). *Callicebus stephennashi*. In *2008 IUCN Red List of Threatened Species*, ed. IUCN. Gland: IUCN, http://www.iucnredlist.org/details/41552.

Veiga, L.M. & Marsh, L. (2008a). *Pithecia irrorata*. In *2008 IUCN Red List of Threatened Species*, ed. IUCN. Gland: IUCN, http://www.iucnredlist.org/details/41568.

Veiga, L.M. & Marsh, L. (2008b). *Pithecia pithecia*. In *2008 IUCN Red List of Threatened Species*, ed. IUCN. Gland: IUCN, http://www.iucnredlist.org/details/41569.

Veiga, L.M., Bowler, M., Silva Jr., J.S., *et al.* (2008i). *Cacajao calvus*. In *2008 IUCN Red List of Threatened Species*, ed. IUCN. Gland: IUCN, http://www.iucnredlist.org/details/3416.

Veiga, L.M., Ferrari, S.F., Kierulff, C.M., *et al.* (2008d). *Callicebus personatus*. In *2008 IUCN Red List of Threatened Species*, ed. IUCN. Gland: IUCN, http://www.iucnredlist.org/details/3555.

Veiga, L.M., Kierulff, C.M., Oliveira, M.M., *et al.* (2008e). *Callicebus nigrifrons*. In *2008 IUCN Red List of Threatened Species*, ed. IUCN. Gland: IUCN, http://www.iucnredlist.org/details/39943.

Veiga, L.M., Mittermeier R.A. & Marsh, L. (2008k). *Pithecia albicans*. In *2008 IUCN Red List of Threatened Species*, ed. IUCN. Gland: IUCN, http://www.iucnredlist.org/details/41567.

Veiga, L.M., Pinto, L.P., Ferrari, S.F., *et al.* (2008h). *Chiropotes albinasus*. In *2008 IUCN Red List of Threatened Species*, ed. IUCN. Gland: IUCN, http://www.iucnredlist.org/details/4685.

Veiga, L.M., Printes, R.C., Ferrari, S.F., *et al.* (2008c). *Callicebus melanochir*. In *2008 IUCN Red List of Threatened Species*, ed. IUCN. Gland: IUCN, http://www.iucnredlist.org/details/39930.

Veiga, L.M., Printes, R.C., Rylands, A.B., *et al.* (2008a). *Callicebus barbarabrownae*. In *2008 IUCN Red List of Threatened Species*, ed. IUCN. Gland: IUCN, http://www.iucnredlist.org/details/39929.

Veiga, L.M., Silva Jr., J.S., Ferrari, S.F., *et al.* (2008f). *Chiropotes satanas*. In *2008 IUCN Red List of Threatened Species*, ed. IUCN. Gland: IUCN, http://www.iucnredlist.org/details/39956.

Veiga, L.M., Silva Jr., J.S., Ferrari, S.F., *et al.* (2008g). *Chiropotes utahickae*. In

2008 IUCN Red List of Threatened Species, ed. IUCN. Gland: IUCN, http://www.iucnredlist.org/details/43892.

Veiga, L.M., Silva Jr., J.S., Mittermeier, R.A., *et al.* (2008j). *Chiropotes chiropotes.* In *2008 IUCN Red List of Threatened Species*, ed. IUCN. Gland: IUCN, http://www.iucnredlist.org/details/43891.

Veiga, L.M., Sousa, M.C., Jerusalinsky, L., *et al.* (2008b). *Callicebus coimbrai.* In *2008 IUCN Red List of Threatened Species*, ed. IUCN. Gland: IUCN, http://www.iucnredlist.org/details/39954.

Wallace, A.R. (1852). On the monkeys of the Amazon. *Proceedings of the Zoological Society of London*, **20**, 107–110.

Chapter 34

Pitheciines in captivity: challenges and opportunities, past, present and future

Jennie Becker, Andrew J. Baker, Tracy Frampton, P. Kirsten Pullen, Karen L. Bales, Sally P. Mendoza & William A. Mason

History of pitheciines in captivity

Callicebus

According to the 1996 North American Titi Monkey Studbook (Kaemmerer & Stevens 1996), *Callicebus* first appeared in zoological institutions in 1841, with a *C. "moloch"* in Paris and a *C. "personatus"* in Amsterdam. London Zoo and various other European zoos exhibited several forms of titis through the early part of the twentieth century.

The first significant breeding program for the genus, of *C. cupreus*, was initiated at the Delta Regional Primate Center (Covington, Louisiana, USA) in 1965 (Lorenz & Mason 1971). This colony was moved to the California Regional Primate Center, in Davis, in 1971, with some of the colony transferred to Göttingen, Germany. The colony at the Davis facility, now called the California National Primate Research Center (CNPRC) was supplemented by additional imports in 1990, and now numbers approximately 65 individuals, with a few more held in North American zoos. At the end of 2005, there were 44 individuals held in Europe.

A number of Bolivian gray titis (*C. donacophilus*) were imported in to North America in the 1970s and early 1980s (Kaemmerer & Stevens 1996). As of 2006, there were 44 individuals of this species in US zoos. This population has numbered in the 40s for some time, despite efforts to maximize reproduction (International Species Information System (ISIS): http://app.isis.org/abstracts/abs.asp).

Pithecia

Among the five species currently recognized within the genus *Pithecia* (IUCN Red List: http://www.iucnredlist.org), only the white-faced saki (*Pithecia pithecia*) has a significant history or current population in captive collections, although there are past records of successful reproduction of monk sakis (*Pithecia monachus*) in zoos (Lucas 1970, p. 252; Clevenger 1981) and other forms have been held in range countries. The first documented captive white-faced saki was a short-lived individual at the London Zoo in 1866 (Willis 1984). The first documented captive births for this genus were to a pair identified as golden-headed sakis (*P. p. chrysocephala*) at the San Diego Zoo, in 1950, 1951 and 1952 (Dolan 1995). The current population, all thought to be *P. p. pithecia*, is descended from animals imported to Europe and North America from home range countries from the late 1960s to the 1980s. As of 2006, there were approximately 195 individuals in European zoos and 135 individuals in North American zoos (ISIS).

Chiropotes

The bearded saki, *Chiropotes*, has a very limited history in zoos and other captive settings when compared to *Pithecia* and *Callicebus*. There were imports into Europe as early as 1881 (Hick 1968a) and into North America as early as 1903 (ISIS 2006), but the total number imported was small and a breeding population was never created in either region. The São Paulo Zoo reported both *C. albinasus* and *C. satanas* in their collection in the 1970s (Olney 1979, p. 404, and other volumes of International Zoo Yearbook) and since the 1980s, numbers of *C. satanas* have been held in at least one Guyanan facility (A. Shoemaker, pers. commun. to Baker, 2007) and several Brazilian facilities. Some of the Brazilian animals were rescued from rising waters resulting from hydroelectric projects (e.g. Kingston 1985; Malacco & Fernandes 1989). The first reported captive birth for this genus in captivity was at the Köln (Cologne) Zoo in 1968 (Hick 1968b). The offspring was a hybrid born to a female *C. albinasus* and a male *C. satanas*. Several more hybrid offspring were produced at Köln (Duplaix-Hall 1975, p. 360; Olney 1977, p. 301). Subsequently, there have been reports of captive reproduction of *C. satanas* at several Brazilian institutions (e.g. Kingston 1987; Malacco & Fernandes 1989; Javorouski *et al.* 1999; Fichtner Gomes & Bicca-Marques 2003). *Chiropotes albinasus* seems to have been held less frequently, with no published accounts other than those from the Köln Zoo (e.g. Hick 1968a).

Cacajao

Uacaris, the genus *Cacajao*, like *Chiropotes*, have a limited history in captivity, although several forms have been held in European and North American collections (e.g. Crandall 1964). The first captive birth of this species was a red uacari

Evolutionary Biology and Conservation of Titis, Sakis and Uacaris, eds. Liza M. Veiga, Adrian A. Barnett, Stephen F. Ferrari and Marilyn A. Norconk. Published by Cambridge University Press.

Figure 34.1 Grooming between *Cacajao calvus rubicundus* at Los Angeles Zoo. Photo: Tad Motoyama.

(*C. calvus rubincundus*) at the Cheyenne Mountain Zoo (Colorado Springs, Colorado, USA) in 1961 (ISIS). Several other institutions had repeated reproduction, notably Monkey Jungle (Miami, Florida, USA), with 13 births (Fontaine & Dumond 1977; Fontaine & Hench 1982), but this early success was not sustained. Small group size at most locations may have contributed to breeding difficulties (Fontaine & Hench 1982); those groups that contained two or more males appeared to be more likely to have reproductive success (C. Cox, pers. commun. to McNary, 2007). At the time of writing, outside of range countries, this genus is represented by only three *C. calvus rubicundus* at the Los Angeles Zoo (see Figure 34.1) and one reported at a private facility in France (C. Brack, pers. commun. to McNary, 2005). Additionally, there is a single *Cacajao ouakary* and another single *C. melanocephalus* at the Centro de Primatologia do Rio de Janeiro.

Captive management

Callicebus cupreus

On the basis of data collected on captive *C. cupreus* at the CNPRC and wild observations, titi monkeys are monogamous and display biparental care of offspring (Mendoza & Mason 1986b; and Bicca-Marques and Heymann, Chapter 17; Heiduck, Chapter 18; Wright, Chapter 21). Males and females form a strong bond with a pair-mate, and separation is characterized by activation of the hypothalamic–pituitary–adrenal axis (Mendoza & Mason 1986a), perhaps indicating stress or distress. The typical social structure in the wild, and the most successful for captive housing, consists of a pair-bonded male and female and their offspring of successive births (Mason 1966, 1974). Fathers may carry infants up to 90% of the time (Mendoza & Mason 1986b), although there is considerable interpair variation, with some mothers being the primary care-givers for their infants. Titi monkey infants form a selective attachment to their father, and show little distress upon removal of their mother if their father is still present (Hoffman *et al.* 1995).

Both male and female titi monkeys are reproductively mature between 18 months and 2 years of age and can be successfully paired at that time (Valeggia *et al.* 1999). However, mature offspring can be maintained within the family unit almost indefinitely without breeding and with minimal aggression. Pairing of animals is almost always without incident or aggression (Hoffman 1998), but is associated with behavioral arousal (locomotion, arching, tail-lashing), olfactory investigation and attempts by the male to gain proximity to the female (Hoffman 1998). Time spent in proximity to the pair-mate increases gradually over the next few weeks (Hoffman 1998). The average time from pairing to the birth of the first offspring is 18.5 ± 13.1 months for nulliparous females, shorter for multiparous females, with interbirth intervals of approximately 1 year following the birth of a surviving offspring (Valeggia *et al.* 1999).

In captivity, births are not confined to one season, but the majority occur between December and March (Valeggia *et al.* 1999). Infant titi monkeys are almost always singletons, although rare cases of twinning have been observed even in the wild (Knogge & Heymann 1995). The rate of twinning in 148 births at CNPRC was 1.4% (Valeggia *et al.* 1999). Infant mortality can be relatively high in titi monkeys, even in laboratory colonies with ad libitum feeding. Mothers tend to be intolerant of infants (Mendoza & Mason 1986b), handling them primarily for nursing, but there is some evidence that tolerance changes with age in a sex-specific manner. Experienced fathers spend more time in proximity to their infants and experienced mothers spend less time in proximity to and carrying their infants (Reeder 2001). Because captive titi parents show relatively low motivation to retrieve infants, they can go without nursing for extended periods if they get "stranded" on older siblings. For this reason, it is standard practice at CNPRC to remove older siblings from the group and house them separately for the first week after the birth of new infants.

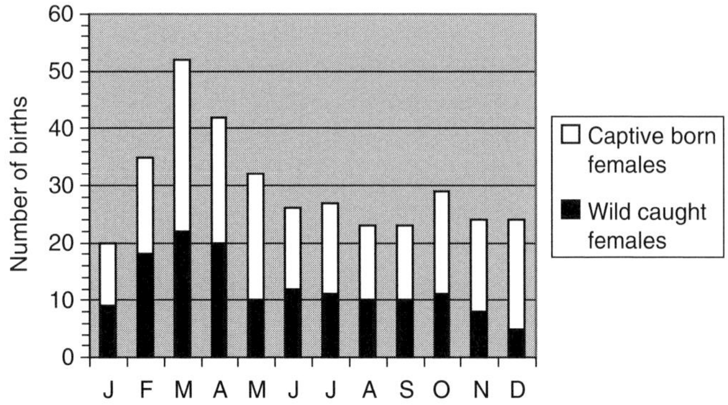

Figure 34.2 Births by month to captive born and wild caught female white-faced sakis in North American zoos current to August 2006.

Pithecia

The success of the *P. pithecia* captive population allows a general description of effective husbandry parameters and has created the opportunity for research on some aspects of the biology of the species. Unless otherwise indicated, data are based on analyses of European and/or North American studbooks (North American studbook current to August 2006; European studbook current through 2005).

Captive sakis in North America and Europe have been maintained as extended families comprised of a breeding pair and offspring, and this appears to be the most stable group composition. Most new pairings are established with little difficulty, but there have been occasional reports of attacks, one fatal (K. Kazanowski, pers. commun. to Pullen, 2006), within newly formed pairs. Although observations in the wild (e.g. Norconk 2006) demonstrate that there is flexibility in the *P. pithecia* social system and that not all groups are extended monogamous families, anecdotal reports suggest that attempts to house captive *Pithecia* in groups of unrelated individuals are likely to result in aggression and injury (Waters 1995). The current regional studbook keepers (chapter co-authors Pullen and Frampton) note that attempts to establish single male, multifemale groups using mother–daughter duos have also often failed, with the mother sometimes ousted from the group (P. Moisson, pers. commun. to Pullen, 2004). Some attempts have been successful, however, including one case in North America in which both females produced multiple offspring. In the European population, a male-biased sex ratio has led to efforts to house males together as "bachelor" groups. To date, these groups have proved successful particularly when the group is composed of full siblings (Waters 1995; Pullen, unpubl. data).

Among captive-born (and thus known-age) individuals in the North American population, earliest documented age at first reproduction is approximately 26 months for both sexes (approximately 21 months at time of fertilization/conception). In the European captive population, one male was only 21 months old at first reproduction, and thus only 16 months old at the time of fertilization (Waters 1995). However, most individuals do not reproduce until substantially later, and in one female in which hormonal status was tracked, cycling did not begin until 32 months of age (Savage *et al.* 1995). It appears that sexual maturation for females ranges at least from 21 months to 32 months of age and can occur occasionally as early as 15 months of age for males. To date, the oldest documented dam was a female imported to North America in 1974 who gave birth in 2006, at > 32 years of age. Among males, the oldest sire to date was a male imported to North America in 1967 who sired an offspring born in 1995, when he was known to be > 28 years and estimated to be 33 years of age.

Although births are distributed across the entire year, there appears to be some degree of birth seasonality in the North American population (Figure 34.2). There have been more births in each month from February to May than in any other month, with March the peak month for both wild-caught and captive-born dams.

Almost all births of *P. pithecia* are singletons. The incidence of twinning in the European population is < 2%. Only two twin births (both mixed-sex) have been documented in the North American population, out of 365 reported pregnancies. Parents successfully reared both twins in one of these cases, while in the second case, one twin was successfully parent-reared and the other hand-reared. Dams are the main caregivers with little paternal interest shown until the infants are older (Waters 1995), except in rare cases (e.g. Hick 1973, cited in Waters 1995). Older siblings of both sexes have been reported to carry infants, and it has been suggested that exposure to younger siblings may be important in the development of appropriate parental skills, particularly for females (Waters 1995).

Sexually mature offspring may remain with their natal group without incident, but may move to the periphery of the group and begin to sleep separately from other group members. Youngsters may separate from the family with no evidence of aggression from either parent, but in some cases low levels of aggression are seen. Waters (1995) reported that male offspring were often evicted from their natal group at about 2.5 years of age. Waters (1995) recommended that female offspring be maintained in their natal groups until at least 36 months of age, citing the benefits of sibling experience noted above and Haydon (2004) noted that females removed from their natal group younger than 4 years of age and those producing their first offspring before the age of 4 had reduced longevity.

Chiropotes

Relatively little can be concluded from the information published to date on the genus *Chiropotes* in captivity. On the basis of wild observations, multimale and/or multifemale groups have been successfully maintained in captivity for *C. satanas* (e.g. Malacco & Fernandes 1989; Javorouski *et al.* 1999), although one male died following within-group aggression in the latter report. Although sample size is small, reproduction does not appear to be strongly seasonal, at least for *C. satanas* (Malacco & Fernandes 1989; Fichtner Gomes & Bicca-Marques 2003). Details of birth and development for a single hybrid infant are reported by Hick (1968b, cited in van Roosmalen *et al.* 1981). One captive-born hybrid at the Köln Zoo died at a little over 26 years of age, and two others survived to approximately 15 years of age (ISIS 2006), indicating that longevity for this genus may be comparable to *Pithecia*.

Cacajao

The work published by Fontaine and DuMond (1977) can still be considered the definitive study of the captive management of what were called *C. calvus rubicundus* (but given the likely Peruvian provenance of the animals probably belong to *C. c. ucayalii*). They reported that most births occurred from May to October, mirroring an August to November birth season reported for wild populations of *C. c. ucayalii* (Bowler & Bodmer 2009). Fontaine and DuMond (1977) documented age at first viable birth at 3 years 7 months and a female at the Frankfurt Zoo gave birth as early as 4 years of age (ISIS). Available birth records suggest a 2-year interbirth interval.

Life expectancy based on the age of imported animals was estimated to be 15–20+ years (Fontaine & DuMond 1977). However, a male born at the Frankfurt Zoo in 1968 lived for 38 years and a wild-born male died at the Los Angeles Zoo at an estimated age of 31. A wild-born female died at the age of 27 years at the Los Angeles Zoo and a female of unknown origin lived to the age of approximately 31 years at the Twycross Zoo. Finally, a wild-born male *C. c. calvus* from the Cologne Zoo lived to the age of 37 years (A. Sliwa, pers. commun. to McNary, 2008). These records indicate that with proper husbandry the life expectancy could reach 20–30+ years in captivity.

Contraception

Savage *et al.* (2002) documented effective contraception using levonorgestrel implants and the Contraceptive Advisory Group of the Association of Zoos and Aquariums (AZA) currently recommends the use of melengesterol (MGA) or deslorelin implants, matching a general recommendation for all non-callitrichid New World primates (http://www.stlzoo.org/animals/scienceresearch/contraceptioncenter/). Neem oil has been tested as a potential contraceptive agent in titi monkeys (Reeder *et al.* 1998). It was found to be an effective short-term contraceptive agent and to induce abortion reliably. Short-term use did not appear to affect future fertility.

Captive diets

The development of captive titi and saki diets followed information collected on wild congeners. Wild titi monkeys primarily consume fruits, berries, tender leaves and insects (Mason 1966). Seeds constitute more than half the diet of white-faced sakis studied in the wild, with smaller quantities of fruit pulp, young leaves, other plant parts, and insects (e.g. Norconk & Conklin-Brittain 2004). Zoo diets typically include a variety of fruits, vegetables, seeds and nuts, and insects, reflecting broad similarities to the wild diet. However, there is probably little to no overlap in food species between wild and captive diets. Most North American institutions rely on a base of a prepared hard or soft commercial "chow" (various brands) or, for the titi monkey and marmoset, jelly (Mazuri Inc., Richmond, IN) with powdered vitamins added on top, while many European zoos give a multivitamin or Vitamin D3 and calcium supplements. Both regional approaches appear to meet nutritional requirements and support maintenance and reproduction. Additional food given as enrichment may includes grapes, unshelled peanuts and other novel food items given in small enough proportions that they do not distract from the nutritional balance of the daily diet. These can often be used as treats that are scattered or hidden throughout the animal's living space to encourage foraging behavior, or used as rewards in the process of training behaviors used for husbandry or medical management (e.g. crating for transportation, shifting animals between enclosures, and collecting blood and urine).

Medical management

By far, the most widespread and serious health problems seen historically in captive titi monkeys involve respiratory ailments and diarrhoea (Lorenz & Mason 1971). Because of the very high parasympathetic activity of the titi monkey autonomic nervous system, i.e. a tendency to stress or frighten easily (Mendoza & Mason 1997), gastrointestinal motility is high and extensive periods of diarrhoea can become fatal rapidly. Following a stressful experience that activates the sympathetic nervous system, the subsequent parasympathetic response can lead to a psychological shutdown similar to a panic attack. The low and irregular heartbeat in *Callicebus* caused by high levels of parasympathetic activity can be alarming to those used to working with other primate taxa (Mendoza & Mason 1997; Tardif *et al.* 2006). These tendencies should be considered when working with this genus in any captive setting. Training to enter a box for transport, as well as regular training for other procedures (such as blood draws or urine collection), can reduce the stress of participating in research and perhaps even improve reproduction (Tardif *et al.* 2006; Mendoza & Mason, unpubl. data).

There have been no population-level surveys of morbidity or mortality for *Pithecia*, and there are no medical issues perceived as specific concerns by the captive care community. There have been case reports of a variety of illnesses, infections, and parasites (e.g. Heard *et al.* 1997; Lloyd *et al.* 1995;

Klaver *et al.* 1994; Gamble *et al.* 1998). Perhaps of greatest relevance for captive caretakers is a report of a fatal multi-individual herpes simplex outbreak (Schrenzel *et al.* 2003).

A thorough review of autopsy records from all institutions may reveal a more complete understanding of the medical and dietary issues that have affected the survival of *C. c. ucayalii* in captivity, but Fontaine and Hench (1982) observed no underlying similarities among deaths at Monkey Jungle. A rare occurrence of Chagas' myocarditis was noted at the San Diego Zoo (Lasry & Sheridan 1965). Individuals at the Los Angeles Zoo have periodically suffered from poor coat quality and a slight paleness to their red face despite behavior and feeding activities that appeared to be normal. A review of 10 necropsy reports at the Los Angeles Zoo revealed that 5 had some degree of hepatic hemosiderosis (liver iron overload) and all 3 of the current captive group at Los Angeles (1 male and 2 females) have been diagnosed with the condition. Although hepatic hemosiderosis may not have been the cause of death, the finding that 8/13 uacaris have some degree of excess iron storage may be an issue in this genus in a captive environment (J. Sykes, pers. commun. to McNary, 2007).

Conclusion

The four genera of pitheciines have all been maintained in captive settings, but only two species of titi monkey and one saki species have been managed with sustained success. It seems possible and even likely that this pattern largely reflects opportunity and circumstance rather than biological differences among the genera. The three successful species came into zoos or other captive facilities in larger numbers or in more recent decades when husbandry expertise was better developed. In the case of *Callicebus cupreus*, they were the subjects of sustained attention by a single facility that developed substantial expertise.

Given the successes with *Chiropotes*, notably in Brazilian institutions, as well as the longevities achieved at Köln and even among some early imports (e.g. one *C. satanas* received at the San Diego Zoo in 1938 that survived until 1953: Dolan 1995), *Chiropotes* would appear to be at least as hardy as *Pithecia* in a captive environment. Successful captive breeding programs for *Chiropotes* would seem to be possible. Given an adequate founder stock and the history of successful reproduction and longevity of *Cacajao* at several institutions, it would appear that they too could establish a successful captive population (Fontaine & Hench 1982).

The development of breeding programs as well as husbandry and management strategies that best suit the needs of each species has been a focus of for those who work with the four genera of pitheciines in captivity. However, commitment to conservation of these species goes beyond the institutions themselves. Contributions to *in situ* research to support those individuals who are working in the field in the countries of origin, like those that have presented their work in this book, is also a very important role that zoos play in the conservation of any species.

References

Bowler, M. & Bodmer, R. (2009). Social behavior in fission–fusion groups of red uakari monkeys (*Cacajao calvus ucayalii*). *American Journal of Primatology*, **71**, 976–987.

Clevenger, M. (1981). Hand-rearing a silver-backed saki *Pithecia monachus* at Oklahoma City Zoo. *International Zoo Yearbook*, **21**, 221–223.

Crandall, L.S. (1964). Family Cebidae. In *Management of Wild Animals in Captivity*. Chicago and London: University of Chicago Press, pp. 85–101.

Dolan, J.M. Jr. (1995). The mammal collection of the Zoological Society of San Diego, a historical perspective, part XI: Cebidae. *Der Zoologische Garten*, **65**(6), 365–390.

Duplaix-Hall, N. (ed.) (1975). *International Zoo Yearbook*, vol. **15**. London: Zoological Society of London.

Fichtner Gomes, D. & Bicca-Marques, J.C. (2003). A note on the births of bearded saki and woolly monkey in Brazilian zoos. *International Zoo News*, **50**(8), 487–488.

Fontaine, R. & DuMond, F.V. (1977). The red ouakari in a seminatural environment: potentials for propagation and study. In *Primate Conservation*, ed. Prince Rainier III & G.H. Bourne. New York, San Francisco and London: Academic Press, pp. 167–236.

Fontaine, R. & Hench, M. (1982). Breeding new world monkeys at Miami's Monkey Jungle. *International Zoo Yearbook*, **22**, 77–84.

Gamble, K.C., Fried, J.J. & Rubin, G.J. (1998). Presumptive dirofilariasis in a pale-headed saki monkey (*Pithecia pithecia*). *Journal of Zoo and Wildlife Medicine*, **29**(1), 50–54.

Haydon, L. (2004). The relationship between husbandry techniques and breeding success of the white-faced saki monkey (*Pithecia pithecia*). BSc thesis, Chichester University College.

Heard, D.J., Ginn, P.E. & Neuwirth, L. (1997). *Mycobacterium avium-intracellularae* infection in a white-faced saki (*Pithecia pithecia*). *Journal of Zoo and Wildlife Medicine*, **28**(2), 185–188.

Hick, U. (1968a). The collection of saki monkeys at Cologne Zoo. *International Zoo Yearbook*, **8**, 192–194.

Hick, U. (1968b). Erstmaig gelungene Zucht eines Bartsakis (Vater: Rotrückensaki, *Chiropotes chiropotes* (Humboldt, 1811), Mutter: Weissenasensaki, *Chiropotes albinasus* (Geoffroy et Deville, 1848) im Kölner Zoo. *Freunde des Kölner Zoo*, **11**(2), 35–41.

Hick, U. (1973). Wir sind umgezogen. *Zeitschrift des Kölner Zoo*, **16**, 127–145.

Hoffman, K.A. (1998). Transition from juvenile to adult stages of development in titi monkeys (*Callicebus moloch*). Dissertation, University of California, Davis.

Hoffman, K.A., Mendoza, S.P., Hennessy, M.B., *et al.* (1995). Responses of infant titi monkeys, *Callicebus moloch*, to removal of one or both parents: evidence for paternal attachment. *Developmental Psychobiology*, **28**, 399–407.

International Species Information System (ISIS), Eagan, MN. Species Holdings. 28 Dec 2006: http://app.isis.org/abstracts/abs.asp.

Javorouski, M.L., Gomes, M.L.F., Francisco, L.R., *et al.* (1999). Manejo e reprodução de macaco-cuxiú (*Chiropotes satanas*) no Zoológico de Curitiba. In *Livro de Resumos IX Congresso Brasileiro de Primatologia*, ed. S.L. Mendes. Santa Teresa, Brazil: Congresso Brasileiro de Primatologia, p. 84.

Kaemmerer, K.R. & Stevens, A.M. (1996). *The North American Regional Studbook for Titi Monkeys (Genus* Callicebus*)*. Dallas, TX: Dallas Zoo.

Kingston, W.R. (1985). Note on the behavior of the "difficult" neo-tropical primate genera in captivity. *Laboratory Primate Newsletter*, **24**(1), 10–11.

Kingston, W.R. (1987). Captive breeding of *Alouatta belzebul* and *Chiropotes sanatus utahicki*. *Laboratory Primate Newsletter*, **26**(3), 8.

Klaver, P.S.J., Hobbelink, M.E., Erken, A.H.M., *et al.* (1994). *Gongylonema pulchrum* infection and therapy in a pale-headed saki (*Pithecia pithecia pithecia*): a case report. *Erkrankungen der Zootiere*, **36**, 409–414.

Knogge, C. & Heymann, E.W. (1995). Field observation of twinning in the dusky titi monkey, *Callicebus cupreus*. *Folia Primatologica*, **65**, 118–120.

Lasry, J.E. & Sheridan, B. (1965). Chagas' myocarditis and heart failure in the red uakari. *International Zoo Yearbook*, **5**, 140–141.

Lloyd, M.L., Susa, J.B., Pelto, J.A., *et al.* (1995). Gestational diabetes mellitus in a white-faced saki (*Pithecia pithecia*). *Journal of Zoo and Wildlife Medicine*, **26**(1), 76–81.

Lorenz, R. & Mason, W.A. (1971). Establishment of a colony of titi monkeys. *International Zoo Yearbook*, **11**, 168–175.

Lucas, J. (ed.) (1970). *International Zoo Yearbook*, vol. **10**. London: Zoological Society of London.

Malacco, A.F. & Fernandes, M.E.B. (1989). Captive colony of brown bearded sakis in Pará, Brazil. *Primate Conservation*, **10**, 34–36.

Mason, W.A. (1966). Social organization of the South American monkey, *Callicebus moloch*: a preliminary report. *Tulane Studies in Zoology*, **13**, 23–28.

Mason, W.A. (1974). Comparative studies of *Callicebus* and *Saimiri*: behaviour of male–female pairs. *Folia Primatologica*, **22**, 1–8.

Mendoza, S.P. & Mason, W.A. (1986a). Contrasting responses to intruders and to involuntary separation by monogamous and polygynous New World monkeys. *Physiology & Behavior*, **38**, 795–801.

Mendoza, S.P. & Mason, W.A. (1986b). Parental division of labour and differentiation of attachments in a monogamous primate (*Callicebus cupreus*). *Animal Behaviour*, **34**, 1336–1347.

Mendoza, S.P. & Mason, W.A. (1997). Autonomic balance in *Saimiri sciureus* and *Callicebus moloch*: relation to life-style. *Folia Primatologica*, **68**, 307–318.

Norconk, M.A. (2006). Long-term study of group dynamics and female reproduction in Venezuelan *Pithecia pithecia*. *International Journal of Primatology*, **27**(3), 653–674.

Norconk, M.A. & Conklin-Brittain, N.L. (2004). Variation on frugivory: the diet of Venezuelan white-faced sakis. *International Journal of Primatology*, **25**, 1–26.

Olney, P.J.S. (ed.) (1977). *International Zoo Yearbook*, vol. **17**. London: Zoological Society of London.

Olney, P.J.S. (ed.) (1979). *International Zoo Yearbook*, vol. **19**. London: Zoological Society of London.

Pullen, P.K. (2006). *The 9th EEP Studbook for the White-Faced Saki Monkey (Pithecia pithecia)*. Paignton: Paignton Zoo Environmental Park.

Reeder, D.M. (2001). The biology of parenting in the monogamous titi monkey (*Callicebus moloch*). Dissertation, University of California, Davis.

Reeder, D.M., Valverde, C.R. & Mendoza, S.P. (1998). Assessing the anti-fertility properties of intrauterine neem treatment in titi monkeys (*Callicebus moloch*). *American Journal of Primatology*, **45**, 203.

Savage, A., Lasley, B.L., Vecchio, A.J., *et al.* (1995). Selected aspects of female white-faced saki *(Pithecia pithecia*) reproductive biology in captivity. *Zoo Biology*, **14**, 441–452.

Savage, A., Zirofsky, D.S., Shideler, S.E., *et al.* (2002). Use of levonorgestrel as an effective means of contraception in the white-faced saki (*Pithecia pithecia*). *Zoo Biology*, **21**, 49–57.

Schrenzel, M.D., Osborn, K.G., Shima, A., *et al.* (2003). Naturally occurring fatal herpes simplex virus 1 infection in a family of white-faced saki monkeys (*Pithecia pithecia pithecia*). *Journal of Medical Primatology*, **32**(1), 7–14.

Tardif, S.D., Bales, K.L., Williams, L., *et al.* (2006). Preparing New World monkeys for laboratory research. *I.L.A.R. Journal*, **47**(4), 307–315.

Valeggia, C.R., Mendoza, S.P., Fernandez-Duque, E., *et al.* (1999) Reproductive biology of female titi monkeys (*Callicebus moloch*) in captivity. *American Journal of Primatology*, **47**, 183–195.

van Roosmalen, M.G.M., Mittermeier, R.A. & Milton, K. (1981). The bearded sakis, genus *Chiropotes*. In *Ecology and Behavior of Neotropical Primates*, Vol. 1, ed. A.F. Coimbra-Filho & R.A. Mittermeier. Rio de Janeiro: Academia Brasiliera de Ciencias, pp. 419–441.

Waters, S.S. (1995). A review of the parameters which influence breeding in white-faced saki (*Pithecia pithecia*) in captivity. *International Zoo Yearbook*, **34**, 147–153.

Willis, R. (1984). Management of white-faced saki monkeys (*Pithecia pithecia*) at The Zoological Society of London. In *Management of Prosimians and New World Primates, Symposium Proceedings* **8**, ed. J. Partridge. Bristol: Association of British Wild Animal Keepers, pp. 24–28.

Chapter 35

The challenge of living in fragments

Stephen F. Ferrari, Sarah A. Boyle, Laura K. Marsh, Marcio Port-Carvalho, Ricardo R. Santos, Suleima S.B. Silva, Tatiana M. Vieira & Liza M. Veiga

Introduction

All ecosystems are dynamic, and will vary through time and space, selecting for ecological flexibility in their component species. All species are able to adapt to certain changes in their environment, but primates tend to be more flexible than most. Not all primates are equally tolerant of habitat change, however, depending on their morphological and physiological specializations and behavioral characteristics. Among the platyrrhines, traits such as body size have important implications for the ecology of different taxa, and in broad terms, smaller-bodied species tend to be more tolerant of habitat disturbance. In fact, the smallest platyrrhines – the marmosets and tamarins (Callitrichidae) – can be considered pioneer species, in ecological terms, and may often attain much higher densities in fragmented habitat in comparison with continuous forest.

In this sense, the pitheciids represent a median subsample of the variation presented by the platyrrhines. The smallest pitheciids – the titi monkeys – are considerably larger than the largest callitrichids, but some species may be at least as tolerant of habitat disturbance as many tamarins (e.g. Jerusalinsky *et al.* 2006). At the opposite extreme, bearded sakis and uacaris are only smaller in size than the Atelidae, and appear to be more tolerant of habitat disturbance than most atelines (woolly and spider monkeys), although not necessarily the howlers, *Alouatta*, which are specialized for the dietary exploitation of leaves.

As highly specialized frugivores, both *Cacajao* and *Chiropotes* have traditionally been seen as being dependent on relatively large tracts of undisturbed forest, not least because most of the early research on these two genera was conducted in relatively pristine habitats (Mittermeier 1977; Ayres 1981, 1986; Frazão 1988). This idea was perpetuated by Johns and Ayres (1987), who predicted that *Chiropotes satanas* would be extinct by the end of the twentieth century, due primarily to ongoing deforestation in the eastern extreme of the Amazon basin. This prediction was based on the assumption that the species was dependent on large tracts of undisturbed forest in order to guarantee access to an uninterrupted supply of fruit and seeds during the course of the year.

Other, more recent studies of bearded sakis have reconfirmed very low densities (Bobadilla & Ferrari 2000) or very large home ranges (Pinto 2008; Boyle 2008) in continuous forest. However, a growing body of fieldwork (Peetz 2001; Port-Carvalho & Ferrari 2004; Veiga 2006; Silva & Ferrari 2009; Boyle *et al.*, Chapter 24 and Santos *et al.*, Chapter 23) has provided a very different perspective on their ecology, in particular their ability to exploit alternative resources such as flowers and non-reproductive plant parts in response to habitat fragmentation.

Unfortunately, our understanding of the ecology of *Cacajao* is still incipient, but there are two main reasons for expecting that uacaris may be at least as efficient as bearded sakis in this sense. One is the apparent ecological equivalence of the two genera, which are allopatric throughout virtually the whole of the Amazon basin (see Ayres & Prance, Chapter 12). The other reason is the apparent ecological specialization of the uacaris for the occupation of flooded forest ecosystems (*várzeas* and *igapós*), where natural conditions imitate some aspects of anthropogenic habitat fragmentation, such as the reduced diversity of most groups of organisms, a high proportion of edge habitat, and a highly dynamic topology (Padoch *et al.* 1999). Whether this translates into an effective tolerance of other forms of habitat disturbance remains unclear, however.

The smaller-bodied pitheciids, *Callicebus* and *Pithecia*, have always been perceived as being more flexible, in ecological terms, not only by virtue of their size, but also because of their less pronounced morphological specialization for the predation of seeds (see Kay *et al.*, Chapter 1). This is reflected in the composition of their diets, with a much greater emphasis on small, fleshy fruits, insects and non-reproductive plant parts (see Bicca-Marques & Heymann, Chapter 17 and Norconk & Setz, Chapter 25). These broad differences are also reflected in the intergeneric variation in social organization, with these smaller pitheciids typically forming family nuclei in contrast with the much larger, multimale–multifemale bands of *Cacajao* and *Chiropotes*.

Evolutionary Biology and Conservation of Titis, Sakis and Uacaris, eds. Liza M. Veiga, Adrian A. Barnett, Stephen F. Ferrari and Marilyn A. Norconk. Published by Cambridge University Press.

Photo 35.1 *Callicebus caquetensis* moving between forest fragments. Photo: Javier Garcia. (See color plate section.)

Despite the importance of understanding the natural ecological context in which each genus has evolved, a considerable proportion of pitheciid field studies have been conducted in habitats that have suffered at least some degree of anthropogenic disturbance. In some cases, such as the titi monkeys of the Atlantic Forest, there may be little alternative (see Printes *et al.*, Chapter 5). In others, the relative accessibility of sites, and the comparative visibility of animals in more peripheral forest or isolated fragments has almost certainly influenced data collection, given the renowned difficulties of observing free-ranging pitheciids (Pinto *et al.*, Chapter 13). These data nevertheless provide important insights into the challenges facing wild populations of pitheciids into the twentieth century and beyond.

Habitat fragmentation, past and present

Historical patterns

On a geological scale of time, habitat fragmentation has played a major role in the diversification of the platyrrhines, and many authors attribute present-day species-level diversity to the major fluctuations in the distribution of Neotropical forests during the Pleistocene (Kinzey 1982). During the warmer, wetter periods of this epoch, forests expanded to form a continuous bloc covering virtually the whole of the Neotropics, whereas in cooler, drier periods, they retracted into a series of smaller tracts, or "refugia", centered on areas with relatively favorable climatic, hydrological and topological characteristics. These refugia were isolated from one another by a matrix of open habitat, similar to that of the Cerrado of central Brazil.

The present-day division of South American tropical forests into two principal biomes – the Brazilian Atlantic Forest and the Amazon–Orinoco complex – appears to represent an intermediate stage between these two extremes of climatic variation. Essentially, the two biomes represent a pair of "mega-refugia", at least for organisms such as the pitheciids (but not all other platyrrhines, e.g. *Alouatta* and *Cebus*), which appear unable to colonize or disperse effectively through the savanna-like Cerrado (Printes *et al.*, Chapter 5).

The contemporary distribution of pitheciids reflects a number of different processes. The most interesting pattern is the contrast between the Amazon-centric pitheciines, and the presence of *Callicebus* in both biomes. Perhaps even more intriguing is the absence of pitheciids from Central America, given both the apparent age and extent of their radiation (Kay *et al.*, Chapter 1) and the presence of representatives of all the other major platyrrhine groups west of the Andes.

On a more local scale, the ranges of pitheciid genera are influenced considerably by major river systems, which have a fundamental role in platyrrhine zoogeography (Wallace 1852; Ayres & Clutton-Brock 1992; Ferrari 2004). Once again, present-day ranges may provide insights into both historical processes and ecological characteristics. In the case of the Atlantic Forest, for example, the northern and southern limits of the titis of the *personatus* group are formed by the São Francisco and Tietê rivers, respectively, which are less significant barriers to other small-bodied genera, such as *Callithrix* and *Leontopithecus*. This suggests that distinct processes have determined the present-day distributions of these genera. Understanding these processes may provide important insights into the ecological flexibility of the different species, and their potential tolerance of habitat fragmentation.

Anthropogenic patterns

Up until the arrival of European colonists, the inhabitants of South America's tropical forests typically practised shifting agriculture, cultivating small, temporary plots within the forest matrix. This system imitates the natural dynamic of the forest, in which small clearings appear as the result of events such as treefall and riverbank erosion, and tends to minimize the overall impact of the human presence.

European colonization of the Americas was founded on the traditions of permanent pasture and monocultures that may be efficient at temperate latitudes, but are far less appropriate to the tropics. Standard practice in this case is to clear-cut the forest, leaving only small tracts, often in relatively inaccessible locations such as hilltops, or on inadequate soils. The resulting landscape is the inverse of that created by traditional local practices, with the man-made matrix surrounding patches of forest.

From the specific viewpoint of the pitheciids, the European colonization of the Neotropics can be divided into the "old frontier" of the Brazilian Atlantic Forest, and the "new frontier" of the Amazon–Orinoco and Guianas (Ferrari 2008). Most of the old frontier is characterized by deforestation rates of over 90%, and within many areas, remaining fragments of habitat are typically of no more than a few hundred hectares. For some pitheciid species, most notably *Callicebus coimbrai*, a majority of present-day populations are found in small tracts of forest (Figure 35.1) capable of supporting little more than a few dozen individuals (Jerusalinsky *et al.* 2006).

By contrast, the new frontier is still largely intact, except for "hotspots" of colonization, such as the southern rim of the Amazon basin. The principal problem here is that some of these hotspots coincide with the geographic distribution of species such as *Chiropotes satanas* and *Chiropotes utahickae* (see Ferrari *et al.*, Chapter 33). Nevertheless, deforestation rates within these areas are still of the order of 50%, and tracts of forest of thousands of hectares are still common. Given this, most populations are still made up of hundreds of individuals, although smaller fragments are present in many areas.

Perhaps the most distinctive manifestation of present-day deforestation is the "herringbone" pattern formed by settlement roads, which run perpendicular to major highways, such as the Trans-Amazon, which now crisscross much of southern Brazilian Amazonia (Figure 35.2). These roads are distributed at regular intervals, forming standard plots of 50 or 100 ha. Even where settlers respect their legal obligation to preserve original habitats, the size and configuration of plots almost invariably results in the formation of irregular and elongated fragments, which are especially vulnerable to edge effects and eventual isolation.

One other modern pattern – which is of particular interest here – is the flooding of hydroelectric reservoirs which, in the Brazilian Amazon and Venezuelan Orinoco, typically cover

Figure 35.1 A fragment of Atlantic Forest on a ranch in Eastern Sergipe, Brazil, within the range of *Callicebus coimbrai*. Photo: Leandro Jerusalinsky.

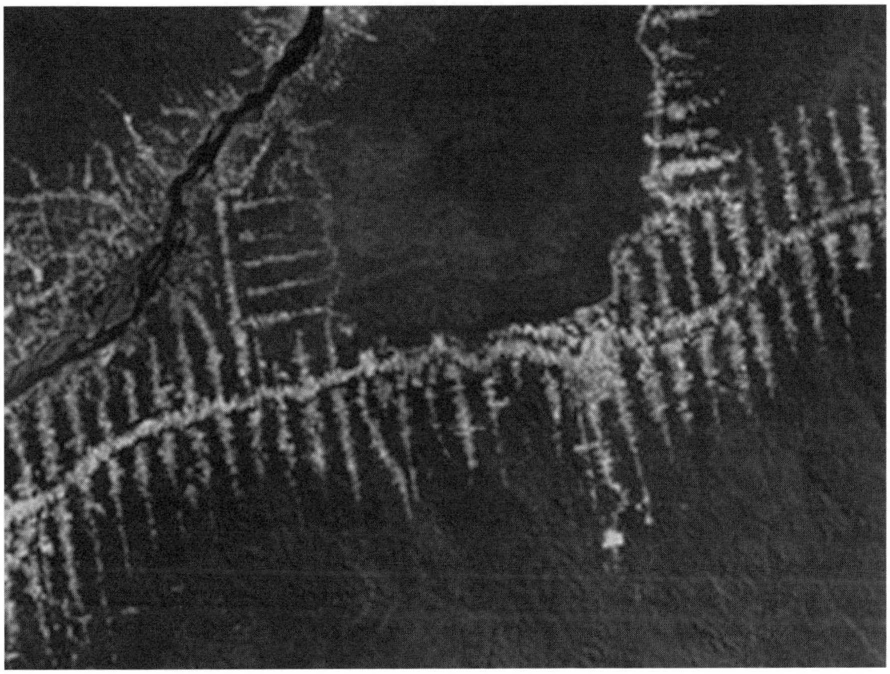

Figure 35.2 Landsat satellite image of the Trans-Amazon highway in the southern Amazon basin, south of the Tapajós River (upper left-hand corner). The lightly colored areas represent deforestation along the highway and perpendicular access roads, while the darker areas represent primary forest.

Figure 35.3 Submerged forest and islands within the Tucuruí reservoir, 20 years after the area was flooded. Photo Liza Veiga.

hundreds or thousands of square kilometers of forest. In this case, "fragments" of habitat – or, literally, islands (Figure 35.3) – are formed by the isolation of promontories of higher ground. These fragments generally encompass a specific type of upland habitat, often well-preserved, but with well-marked borders and a relatively impenetrable matrix, in the form of the reservoir itself.

Challenges to survival

Habitat fragmentation is a mainstream topic in Conservation Biology, although in many cases, the process and its effects are still poorly understood, depending on the organism in question. In the specific case of the pitheciids, a critical factor is the degree of specialization of these monkeys for an arboreal way of life, which virtually eliminates any possibility of occupying the open habitats which typically replace the original forest cover. Given this, a primary determinant of abundance in impacted populations is the extent of remaining forest cover, although other factors will also contribute to the presence or absence of remnant populations and their size and density.

The general effects of habitat fragmentation on pitheciid populations can be summarized into three main, synergistic categories:

(a) demography: the transformation of a continuous population into a series of subpopulations results in critical modifications of normal population dynamics, including reduced migration, local extinctions, distortions of the sex ratio or age structure, and increased vulnerability to catastrophic events;
(b) ecology: over the short term, fragmentation may result in increased population density due to a "refugee effect", but over the long term, alterations in the presence and relative abundance of other species – in particular dietary resources, predators and competitors – may have different, often contrasting effects on pitheciid populations; and
(c) genetics: the interruption of normal demographic processes, in particular, non-panmitic mating, may result in shifts in allele frequencies, loss of Hardy–Weinberg equilibrium and, especially in small, isolated subpopulations, random drift and inbreeding depression. All these processes may represent significant threats to long-term survival.

The influence of these effects on incumbent populations will depend primarily on two variables – fragment size and isolation – fundamental to the classic Theory of Island Biogeography (MacArthur & Wilson 1967), although other factors, such as the characteristics of the anthropogenic matrix, will be more or less important. The time scale is also fundamental here, as the influence of different factors will shift over time. Whereas genetic variability will normally be eroded progressively, for example, even a negligible level of migration among fragments may revert such loss.

Pitheciids in fragments

Demographic patterns

Many pitheciids occupy relatively large home ranges within continuous forest, although the evidence suggests that most species are able to adjust to the limitations of isolated fragments, even of relatively small size. In *Chiropotes*, at least, occupation of such small fragments generally involves a reduction in group size, although in sakis and especially titis, there may be little or no potential for such adjustment.

In fact, the social organization of the titis – specifically, their obligate monogamy – may represent a unique problem for subpopulations isolated in small fragments. Behavioral intolerance between same-sex adults results in the emigration of maturing offspring from their natal group. Under normal conditions (i.e. populations in continuous forest, or even large fragments), these migrants will usually be absorbed into the population. Under advanced habitat fragmentation, however, there may be few, if any opportunities for successful dispersal (Figure 35.4). In the extreme case of a single group inhabiting a small, isolated fragment, the only option for maturing group members may be immediate dispersal, which is likely to be a very hazardous process, with a high risk of mortality, and reduced chances of success (i.e. inclusion in the breeding metapopulation).

While the social organization of the pitheciines is more flexible, population density will inevitably increase, and the problem of dispersal will merely be postponed. Alternatively, isolated populations will face an increasing probability of inbreeding, which is generally less likely in *Callicebus*, as both sexes disperse from their natal groups. Little is known of the population dynamics of the pitheciines, however. While they may tolerate maturing offspring, sakis may be relatively territorial (see Norconk *et al.* 2003; Norconk 2011) in comparison with bearded sakis and uacaris, and this may be reflected in population dynamics, although it is unclear how this might affect migration patterns under habitat fragmentation.

Like *Callicebus*, both sexes disperse in *Pithecia*, but it is still unclear what the pattern is in *Cacajao* or *Chiropotes*. It may be premature to assume that these genera are similar to the other pitheciids, not least because of the major differences in their social organization. The formation of large, flexible groups as observed in the bearded sakis–uacaris is also typical of specialized frugivores such as the spider monkeys (*Ateles* spp.) and chimpanzees (*Pan troglodytes*), in which male philopatry is the norm (Di Fiore & Campbell 2007; Stumpf 2007). In *Ateles*, moreover, a highly biased adult sex ratio in favor of females appears to be common (Nunes & Chapman 1997). This would have important implications for fragmented populations of *Cacajao* and *Chiropotes*, although there is no evidence of any major bias in the sex ratio of free-ranging groups.

There is little direct evidence on between-fragment migrations in any pitheciid, although Ferrari (unpubl.) has observed an adult *Callicebus moloch* traveling across open pasture more than a kilometer from the nearest forest in western Pará, Brazil. In the state of Sergipe, L. Jerusalinsky (pers. commun.) recorded a number of reports of *C. coimbrai* crossing open ground between fragments. Boyle *et al.* (Chapter 24) recorded the immigration of a subadult *Chiropotes sagulatus* into an isolated fragment in central Amazonia, but did not observe the process. Pitheciids may be able to migrate between isolated fragments, then, but it remains unclear how frequent and, more importantly, how successful this process is.

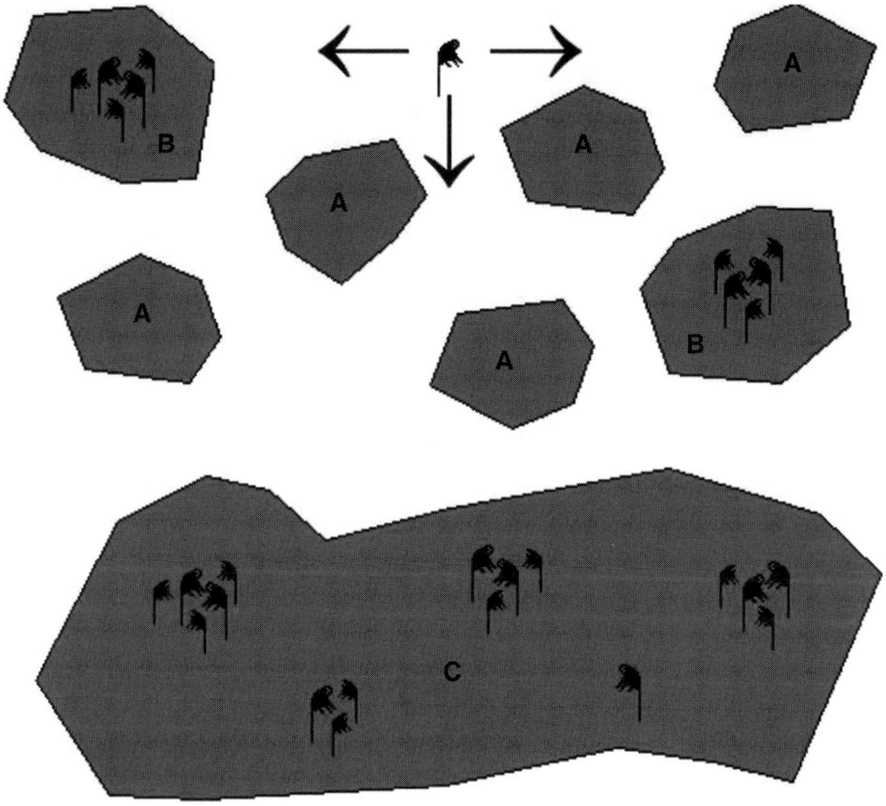

Figure 35.4 Schematic representation of the situation facing a dispersing titi monkey (top middle) in highly fragmented habitat. Most of the nearest fragments will be relatively small and either unoccupied (A) or containing resident groups (B), with no potential reproductive partners. Potential partners are more likely to exist in larger fragment (C), which are rarer and more widely spaced, and thus more difficult to reach, given the predation risk in the anthropogenic matrix.

Ecological patterns

Ecologically, the pitheciids can be divided into the smaller-bodied, more frugivorous titis and sakis, and the larger-bodied, specialized seed predators, the bearded sakis and uacaris. In addition to the exploitation of a different resource base, the larger-bodied pitheciids form larger groups and range over much wider areas of forest. These characteristics led to the early misconceptions with regard to their tolerance of anthropogenic disturbance (Johns & Ayres 1987), but nevertheless imply fundamental differences in their ability to adjust to the effects of habitat fragmentation.

While there is now a growing body of data on the ecology of pitheciids (except *Cacajao*) in fragmented habitat, little information is available on the fragmentation process itself, or the interrelationship of ecological factors. However, one common pattern is the absence of subpopulations from a certain proportion of the fragments present within a given landscape. There is a strong tendency for the vacant fragments to be of smaller size, although size is not a reliable predictor of occupancy, given the apparent influence of factors ranging from habitat characteristics through hunting pressure to random events. Obviously, the absence of species is also likely to be more persistent where fragments are more isolated. By contrast, density tends to be higher in smaller fragments, which is probably at least partly related to the problems of dispersal, as discussed above.

Interestingly, Ferrari *et al.* (2003) found contrasting patterns in the distribution of *Callicebus moloch* and *Chriopotes albinasus* in the fragmented landscape east of the Tapajós River in central Brazilian Amazonia. Whereas fragment size and hunting pressure appeared to be the primary factors determining the absence of subpopulations of *Chiropotes*, competition with the marmoset *Mico argentatus* may have been more important for *Callicebus*. In fact, *Callicebus moloch* appears to have a very patchy distribution in continuous forest, which may contribute to its absence from even very large fragments (Ferrari *et al.* 2007).

The minimum size of inhabited fragments is surprisingly small, however, especially for *Chiropotes* (Silva 2003; Veiga 2006; Boyle *et al.*, Chapter 24), but also *Pithecia* (Setz 1993; Vié *et al.* 2001). In all three genera, groups are known to inhabit isolated fragments of less than 20 ha, and in some cases, of 10 ha or less. In most cases, residence is known to be long term, i.e. covering two or more generations (e.g. Gilbert & Setz 2001; Veiga 2006; Silva & Ferrari 2009; Santos *et al.*, Chapter 23), rather than temporary.

Perhaps even more surprisingly, these monkeys appear able to maintain a diet relatively rich in fruit or seeds, even in fragments of small size. This contrasts with other platyrrhines, such as howlers (*Alouatta* spp.) and marmosets (*Callithrix* spp.), which depend on alternative resources such as leaves and gums, respectively. There is some evidence, however, that fragment-dwelling pitheciids may rely heavily on resources such as flowers, and *Callicebus*, in particular, may also become relatively folivorous (Müller 1996; Souza *et al.* 1996; Price & Piedade 2001; Souza-Alves 2010).

Some studies of bearded sakis in fragments have recorded relatively high levels of feeding on flowers (i.e. more than half the diet in some months: see Silva 2003; Santos *et al.*, Chapter 23), especially during critical periods of resource scarcity. In continuous forest, by contrast, flowers are a minor category on a par with leaves or arthropods (Ayres 1981; van Roosmalen *et al.* 1981; Kinzey & Norconk 1990; Frazão 1991; Pinto 2008). However, flowers were a negligible component of the diets of *Chiropotes chiropotes* on reservoir islands in Venezuela (Kinzey & Norconk 1993; Peetz 2001).

The moot question here is the extent to which the variation observed among studies represents differences among species or ecological variables. A priori, the relative lack of morphological variation within the genus suggests the latter, although there are just too few studies at the moment to support reliable conclusions. A similar pattern is nevertheless seen in *Pithecia*, with fragment-dwelling *Pithecia chrysocephala* consuming relatively large amounts of flowers (Setz 1993) in comparison with sakis in continuous forest (Soini 1986; Peres 1993). Once again, however, *Pithecia pithecia* on a reservoir island in Venezuela consumed even fewer flowers than these sakis (Homburg 1998).

One intriguing contrast between bearded saki species (or ecosystems) is that between the groups of *C. sagulatus* inhabiting 100-ha fragments at the Biological Dynamics of Forest Fragments Project (BDFFP) near Manaus (Boyle *et al.*, Chapter 24) and the mainland *Chiropotes satanas* group T4 at Tucuruí (Veiga 2006; Silva & Ferrari 2009). Whereas the *C. sagulatus* groups range into surrounding areas of continuous forest, the much larger *C. satanas* group, with up to 39 members, has consistently occupied a home range smaller than 100 ha, despite having access to a much larger area of forest.

One potentially important difference between biomes is that, in addition to being much smaller, in general, than their Amazonian counterparts, Atlantic Forest fragments tend to be relatively more degraded, due to their much longer history of exploitation. In fact, a characteristic of most of the fragments occupied by Amazonian pitheciids – in particular the reservoir islands and the BDFFP sites – is the general lack of anthropogenic interference other than the isolation of the fragment. This maintenance of the original forest structure may explain, at least in part, why the diets recorded for most fragment-dwelling pitheciines are relatively rich in fruit and seeds, or at least, reproductive plant parts. This is likely to be a short-term phenomenon, however, and may change significantly as habitat characteristics evolve over time. But once again, unfortunately, the complexities of the process overshadow the limitations of the available evidence.

Genetics

The effects of fragmentation on the genetic structure of pitheciid populations are unknown, not least because of the scarcity of genetic data of any kind for these primates. Most of the available studies have been phylogenetic analyses (e.g. Figueiredo *et al.*, Chapter 3), although Menescal *et al.* (2009) demonstrate relatively low heterozygosity in four microsatellite markers in populations of *C. moloch* from eastern Amazonia. While this may reflect specific ecological and demographic characteristics of the titis, including their tolerance of habitat disturbance, the evidence is still far too incipient to draw any reliable conclusions.

The lack of genetic studies of pitheciids reflects the general difficulties of studying any aspect of the biology of these elusive primates. One specific problem here is the rarity of studies that involve the capture of free-ranging animals, and hence opportunities not only for the collection of specimens, but of an adequate sample for the analysis of genetic variability at the population level.

Challenges to conservation

This chapter focuses on pitheciids, but it is important to remember that the most effective conservation programs generally adhere to an ecosystem-level approach. A priori, it seems unlikely that measures directed at the conservation of pitheciid populations will have negative effects on other elements of the ecosystem. However, as specialized seed predators, *Cacajao* and *Chiropotes*, in particular, may exert specific pressures on impacted ecosystems, quite unlike those presented by other platyrrhines. Whereas shifts in the relative abundance of different frugivores may do little more than alter patterns of seed dispersal, albeit negatively in some cases, an increase in the density of seed (or flower) predators is likely to have highly deleterious effects on the population structure of resource plant species. While seen as an encouraging trend, then, the relatively high densities of pitheciids observed in many fragments may represent an ecological "time bomb", which will erode the scant resources available within a fragment progressively over the long term.

But perhaps the main challenge to the development of effective conservation strategies is the scarcity of data on many important parameters, some of which have been highlighted above. Despite the considerable recent advances in our knowledge of the Pitheciidae, most species are still known from only one or two studies – if that – and analyses of the effects of environmental variables are still tentative, at best. While certain patterns are evident, it remains unclear how representative the data are, or to what extent they may reflect sampling choices, such as the selection of sites with specific ecological characteristics. It would be especially interesting to understand, for example, exactly which factors determine the absence of species from certain reservoir islands or BDFFP fragments.

Certainly, a growing number of studies have revealed that pitheciids may be far more tolerant of habitat fragmentation than was originally thought. However, this should be considered as no more than an unexpectedly positive first step in a very long and arduous process. Understanding the factors that determine the potential for survival in a fragment of a given size or quality will be essential for the planning of management strategies, as will the eventual evaluation of long-term processes.

Ultimately, the main challenge to conservation is likely to be the human element, ranging from legislators and landowners to the local residents who exploit the natural resources – including the pitheciids themselves – offered by the forest. While Conservation Biology offers a range of potential solutions for the problems of habitat fragmentation, they depend on the cooperation of local communities. Obviously, strategies such as the consolidation or expansion of fragments, and the establishment of corridors of habitat will be immaterial if the survival of populations is not guaranteed. In addition to its other effects, habitat fragmentation facilitates hunting by making the interior of the forest more easily accessible. While pitheciids are not preferred game species, they are hunted in many areas, not only for their meat, but in the case of *Chiropotes* and *Pithecia* for their bushy tails. Clearly, then, the integral participation of local human populations will be an essential element of any successful conservation program.

Acknowledgments

We are grateful to CNPq (process no. 302747/2008-7) and CAPES, Eletronorte S.A., Kapok Foundation, WWF-Brazil, IBAMA, INPA, BDFFP, Smithsonian Tropical Research Institute, Arizona State University, and USAID, in addition to the many people who have helped with the collection of data in the field.

References

Ayres, J.M. (1981). Observações sobre a ecologia e o comportamento dos cuxius (*Chiropotes albinasus* e *Chiropotes satanas*, Cebidae: Primates). Unpublished Master's dissertation, Universidade do Amazonas.

Ayres, J.M.C. (1986). Uakaris and Amazonian flooded forests. Unpublished PhD thesis, Cambridge University.

Ayres, J.M. & Clutton-Brock, T.H. (1992). River boundaries and species range size in Amazonian primates. *American Naturalist*, **140**, 531–537.

Bobadilla, U.L. & Ferrari, S.F. (2000). Habitat use by *Chiropotes satanas utahicki* and syntopic platyrrhines in eastern Amazonia. *American Journal of Primatology*, **50**, 215–224.

Boyle, S.A. (2008). The effects of forest fragmentation on primates in the Brazilian Amazon. Unpublished PhD thesis, Arizona State University.

Di Fiore, A. & Campbell, C.J. (2007). The atelines: variations in ecology, behaviour, and social organization. In *Primates in Perspective*, ed. C.J. Campbell, A. Fuentes, K.C. MacKinnon, M. Panger & S.K. Bearder. New York, NY: Oxford University Press, pp. 155–185.

Ferrari, S.F. (2004). Biogeography of Amazonian primates. In *A Primatologia no Brasil – 8*, ed. S.L. Mendes & A.G. Chiarello. Santa Teresa: Sociedade Brasileira de Primatologia, pp. 101–122.

Ferrari, S.F. (2008). The impact of forest fragmentation on populations of New World primates. In *Tropical Biology and Natural Resources (Encyclopedia of Life Support Systems)*, ed. K. Del Claro, P.S. Oliveira, V. Rico-Gray; A. Ramirez, A.A.A. Barbosa, A. Bonet, F.R. Scarano, F.L. Consoli, F.J.M. Garzon, J.N. Nakajima, J.A. Costello, M.V. Sampaio, M. Quesada, R. Morris, M.P. Rios, N. Ramirez, O. Marcal Junior, R.H.F. Macedo & R.J. Marq. Oxford: UNESCO, EOLSS.

Ferrari, S.F., Bobadilla, U.L. & Emídio-Silva, C. (2007). Where have all the titis gone? The heterogeneous distribution of *Callicebus moloch* in eastern Amazonia, and its implications for the conservation of Amazonian primates. *Primate Conservation*, **22**, 49–54.

Ferrari, S.F., Iwanaga, S., Ravetta, A.L., *et al.* (2003). Dynamics of primate communities along the Santarém–Cuiabá highway in southern central Brazilian Amazônia. In *Primates in Fragments*, ed. L.K. Marsh. New York, NY: Kluwer Academic, pp. 123–144.

Frazão, E.R. (1988). Dieta e estratégia de forragear de *Chiropotes satanas chiropotes* (Cebidae: Primates) na Amazônia central brasileira. Unpublished Master's dissertation, Instituto Nacional de Pesuisa da Amazônia.

Frazão, E.R. (1991). Insectivory in free-ranging bearded sakis (*Chiropotes satanas chiropotes*). *Primates*, **32**, 243–245.

Gilbert, K.A. & Setz, E.Z. (2001). Primates in a fragmented landscape. Six species in central Amazonia. In *Lessons from Amazonia: the Ecology and Conservation of a Fragmented Forest*, ed. R.O. Bierregaard Jr., C. Gascon, T.E. Lovejoy & R. Mesquita. New Haven, CT: Yale University Press, pp. 262–270.

Homburg, I. (1998). Ökologie und Sozialverhalten einer Gruppe von Weiβgesicht-sakis (*Pithecia pithecia pithecia Linnaeus 1766*) im Estado Bolívar, Venezuela. Unpublished PhD thesis, Universität Bielefeld.

Jerusalinsky, L., Oliveira, M.M., Pereira, R.F., *et al.* (2006). Preliminary evaluation of the conservation status of *Callicebus coimbrai* Kobayashi & Langguth, 1999 in the Brazilian state of Sergipe. *Primate Conservation*, **21**, 25–32.

Johns, A.D. & Ayres, J.M. (1987). Southern bearded sakis beyond the brink. *Oryx*, **21**, 164–167.

Kinzey, W.G. (1982). Distribution of primates and forest refuges. In *Biological Diversification in the Tropics*, ed. G.T. Prance. New York, NY: Columbia University Press, pp. 455–482.

Kinzey, W.G. & Norconk, M.A. (1990). Hardness as a basis of fruit choice in two sympatric primates. *American Journal of Physical Anthropology*, **81**, 5–15.

Kinzey, W.G. & Norconk, M.A. (1993). Physical and chemical composition of fruit and seeds eaten by *Pithecia* and *Chiropotes* in Surinam and Venezuela. *International Journal of Primatology*, **14**, 207–227.

MacArthur, R.H. & Wilson, E.O. (1967). *The Theory of Island Biogeography*. Princeton, NJ: Princeton University Press.

Menescal L.A., Gonçalves, E.C., Silva, A., *et al.* (2009). Genetic diversity of red-bellied titis (*Callicebus moloch*) from Eastern Amazonia based on microsatellite markers. *Biochemical Genetics*, **47**, 235–240.

Mittermeier, R.A. (1977). Distribution, synecology and conservation of Surinam monkeys. Unpublished PhD thesis, Harvard University.

Müller, K.-H. (1996). Diet and feeding ecology of masked titis (*Callicebus personatus*). In *Adaptive Radiations of Neotropical Primates*, ed. M.A. Norconk,

A.L. Rosenberger & P.A. Garber. New York, NY: Plenum Press, pp. 383–401.

Norconk, M.A. (2011). Sakis, uakaris, and titi monkeys: behavioral diversity in a radiation of primate seed predators. In *Primates in Perspective* (2nd edn), ed. C.J. Campbell, A. Fuentes, K.C. MacKinnon, S.K. Bearder & R.M. Stumpf. New York, NY: Oxford University Press, pp. 122–139.

Norconk, M.A., Raghanti, M.A., Martin, S.K., *et al.* (2003). Primates of Brownsberg Natuurpark, Surinam, with particular attention to the pitheciins. *Neotropical Primates*, **11**, 94–100.

Nunes, A. & Chapman, C.A. (1997). A re-evaluation of factors influencing the sex ratio of spider monkey populations with new data from Maracá Island, Brazil. *Folia Primatologica*, **68**, 31–33.

Padoch, C., Ayres, J.M., Piñedo-Vásquez, M., *et al.* (1999). *Várzea: Diversity, Development, and Conservation of Amazonia's Whitewater Floodplains.* New York, NY: The New York Botanical Press.

Peetz, A. (2001). Ecology and social organization of the bearded saki *Chiropotes satanas chiropotes* (Primates: Pitheciinae) in Venezuela. *Ecotropical Monographs*, **1**, 1–170.

Peres, C.A. (1993). Notes on the ecology of buffy saki monkeys (*Pithecia albicans*, Gray, 1860): a canopy seed predator. *American Journal of Primatology*, **31**, 129–140.

Pinto, L.P. (2008). Ecologia alimentar de um grupo de cuxiús-de-nariz-vermelho *Chiropotes albinasus* (Primates: Pitheciidae) na Floresta Nacional do Tapajós, Pará. Unpublished PhD thesis, Universidade Estadual de Campinas.

Port-Carvalho, M. & Ferrari, S.F. (2004). Occurrence and diet of the black bearded saki (*Chiropotes satanas satanas*) in the fragmented landscape of western Maranhão, Brazil. *Neotropical Primates*, **12**, 17–21.

Price, E.C. & Piedade, H.M. (2001). Ranging behavior and intraspecific relationships of masked titi monkeys (*Callicebus personatus personatus*). *American Journal of Primatology*, **53**, 87–92.

Setz, E.Z.F. (1993). Ecologia alimentar de um grupo de parauacus (*Pithecia pithecia chrysocephala*) em um fragmento florestal na Amazônia central. Unpublished PhD thesis, Universidade Estadual de Campinas.

Silva, S.S.B. (2003). Comportamento alimentar de cuxiú-preto (*Chiropotes* satanas) na área de influência do reservatório da usina hidrelétrica de Tucuruí-PA. Unpublished Master's dissertation, Museu Paraense Emílio Goeldi.

Silva, S.S.B. & Ferrari, S.F. (2009). Behavior patterns of southern bearded sakis (*Chiropotes satanas*) in the fragmented landscape of eastern Brazilian Amazonia. *American Journal of Primatology*, **70**, 1–7.

Soini, P. (1986). A synecological study of a primate community in the Pacaya-Samiria Reserve, Peru. *Primate Conservation*, 7, 63–71.

Souza, S.B., Martins, M.M. & Setz, E.Z. (1996). Activity pattern and feeding ecology of sympatric masked titi monkeys and buffy tufted-ear marmosets. *IPS/ASP Congress Abstracts*, #155.

Souza-Alves, J.P. (2010). Ecologia alimentar de um grupo de Guigó-de-Coimbra-Filho (*Calliebus coimbrai* Kobayashi & Langguth, 1999): perspectivas para a conservação da espécie na paisagem fragmentada do sul de Sergipe, Brasil. Unpublished Master's dissertation, Universidade Federal de Sergipe.

Stumpf, R. (2007). Chimpanzees and bonobos: diversity within and between species. In *Primates in Perspective*, ed. C.J. Campbell, A. Fuentes, K.C. MacKinnon, M. Panger & S.K. Bearder. New York, NY: Oxford University Press, pp. 321–344.

van Roosmalen, M.G.M., Mittermeier, R.A. & Milton, K. (1981). The bearded sakis, genus *Chiropotes.* In *Ecology and Behavior of Neotropical Primates*, Volume **1**, ed. A.F. Coimbra-Filho & R.A. Mittermeier. Rio de Janeiro: Academia Brasileira de Ciências, pp. 419–441.

Veiga, L.M. (2006). Ecologia e comportamento do cuxiú-preto, *Chiropotes satanas* na paisagem fragmentada da Amazônia oriental. Unpublished PhD thesis, Universidade Federal do Pará.

Vié, J.C., Richard-Hansen, C. & Fournier-Chambrillon, C. (2001). Abundance, use of space, and activity patterns of white-faced sakis (*Pithecia pithecia*) in French Guiana. *American Journal of Primatology*, **55**, 203–221.

Wallace, A.R. (1852). On the monkeys of the Amazon. *Proceedings of the Zoological Society of London.*

Chapter 36

Communities and uacaris: conservation initiatives in Brazil and Peru

Mark Bowler, João Valsecchi, Helder L. Queiroz, Richard Bodmer & Pablo Puertas

Introduction

Uacari monkeys (*Cacajao*) are some of the rarest primates in lowland Amazonia. *Cacajao melanocephalus* is listed as Least Concern and *C. calvus* as Near Threatened (IUCN 2008), but with their restricted geographical ranges (Hershkovitz 1987) and low population densities many subspecies of *Cacajao* are particularly vulnerable to threats such as logging, non-timber resource extraction and hunting (Ayres & Johns 1987; Aquino & Encarnación 1999; Bodmer *et al.* 2003). Specialist habitat and feeding requirements (Setz *et al.*, Chapter 7; Barnett *et al.*, Chapter 14) could mean that *Cacajao* is more sensitive to habitat modification than other taxa (Plumptre & Johns 2001), and the persistence of uacari monkeys in many parts of their range may require conservation action beyond that necessary for other primates.

Brazil harbours at least five, and possibly all, described subspecies of *Cacajao* (Hershkovitz 1987): *Cacajao melanocephalus melanocephalus* and *C. melanocephalus ouakary*, which also occur in Venezuela and Colombia, respectively, and *Cacajao calvus calvus*, *C. calvus rubicundus* and *C. calvus novaesi*, which are endemic to Brazil. Brazil is therefore very important for the conservation of the genus. Only *C. calvus ucayalii* may be absent or virtually absent from Brazil. This subspecies occurs in Peru, where populations represent the largest of *C. calvus*.

Few conservation initiatives specifically target *Cacajao*. This chapter gives an overview of the most prominent projects concerning *Cacajao* in Brazil and Peru. In Brazil, we review successful strategies employed by the Instituto de Desenvolvimento Sustentável Mamirauá (IDSM) in the Mamirauá Sustainable Development Reserve (MSDR), and outline newer initiatives in the Amanã Sustainable Development Reserve (ASDR). In Peru we review the conservation activities of WCS-DICE in the Ucayali–Yavarí corridor between the Rios Ucayalii and Yavarí. These projects were chosen as case studies for the conservation of *Cacajao* because the projects have been

Photo 36.1 Pet uacari. Photo: Mark Bowler.

Evolutionary Biology and Conservation of Titis, Sakis and Uacaris, eds. Liza M. Veiga, Adrian A. Barnett, Stephen F. Ferrari and Marilyn A. Norconk. Published by Cambridge University Press.

Photo 36.2 Juvenile *Cacajao calvus novaesi* kept as a pet by members of the Morro Alto community. Uacari Sustainable Development Reserve on the Rio Jurua, Brazil. Photo: Waldener Endo.

running for several years, and represent a range of conservation problems. Although the threats faced by these different populations of *Cacajao* are varied in both type and intensity, similarities are drawn in the approach adopted by the projects discussed.

Uacari conservation in Brazil

The status of *Cacajao* in the ASDR and MSDR, Brazil

Cacajao melanocephalus and *C. calvus* are both found on the Middle Rio Solimões and Lower Rio Japurá, in the western Brazilian Amazon. *Cacajao melanocephalus* occurs west of the Rio Negro, north of the Rio Solimões and northeast of the Rio Japurá into Colombia and Venezuela. *Cacajao calvus* occurs southwest of the Rio Japurá, south of Rio Solimões and west of the Rio Juruá into Peru (Hershkovitz 1987). However, the ranges of these species are poorly known and new populations are being discovered (see appendix).

The MSDR and ASDR (Figure 36.1) were created in 1990 and 1998 respectively, by the government of Amazonas State, Brazil (Queiroz & Peralta 2006). These protected areas form a block of over 6,500,000 ha with the contiguous Jaú National Park and are internationally recognized as the Central Amazon Biosphere Reserve, and the Central Amazon Protected Areas World Heritage Site (UNESCO).

The MSDR and ASDR harbor diverse primate communities, with at least nine species in Mamirauá and eight in Amanã, and the MSDR was originally created in 1982 as an Ecological Station of the Brazilian Federal Government to protect the population of *C. calvus calvus*. This reserve encompasses the major part of the total population of this subspecies (Ayres *et al.* 1999). Similarly, populations of *C. melanocephalus ouakary* living in Amanã were among a number of vertebrate species that were important in the creation of this reserve (Ayres *et al.* 1997). However, this reserve does not encompass the entire range of *C. melanocephalus ouakary*, and there are many unsurveyed areas that could hold undiscovered populations that will prove important to future conservation strategies.

Threats to *Cacajao* in the ASDR and MSDR

Anthropogenic pressures on the Middle Solimões affect each subspecies of *Cacajao* to different degrees. *Cacajao calvus calvus* is rarely hunted, and there is only one record of removal for international traffic during the last 5 years. This subspecies has a restricted geographic distribution almost entirely covered by the MSDR. In 1996 it was downlisted from Endangered to Vulnerable, and since 2001 it has been considered Near Threatened, because of the high integrity of its habitat, and the effective conservation projects in practice (IUCN 2008). Historically, this subspecies was threatened by the reduction and transformation of its habitat, because changes in plant species composition may affect the uacari's ecology and ultimately their populations.

In contrast, hunting in the ASDR is the major threat for *C. melanocephalus ouakary*, which is listed as Least Concern (IUCN 2008). Investigations in the ASDR by the IDSM, carried out with the cooperation of local people in four villages in the ASDR, have revealed important aspects about the hunting of this subspecies. These villages were selected to represent the 28 villages found inside the ASDR. Two of the villages are located in *várzea* forests (flooded by white water), and two in *terra firme* forests (tall and non-flooded) associated with *igapó* forests (flooded by black water). Although small numbers of hunted animals are traded in villages or the city of Tefé (population ≈ 71,000), most kills are consumed locally. *Cacajao melanocephalus ouakary* is the third most hunted primate species in the ASDR (Table 36.1). Between 2002 and 2006, 35 uacaris were killed, and a further 5 injured. These numbers, extrapolated to include all the villages in the ASDR, suggest a probable annual kill between 85 and 90 individuals. As hunters usually kill more than one individual during each

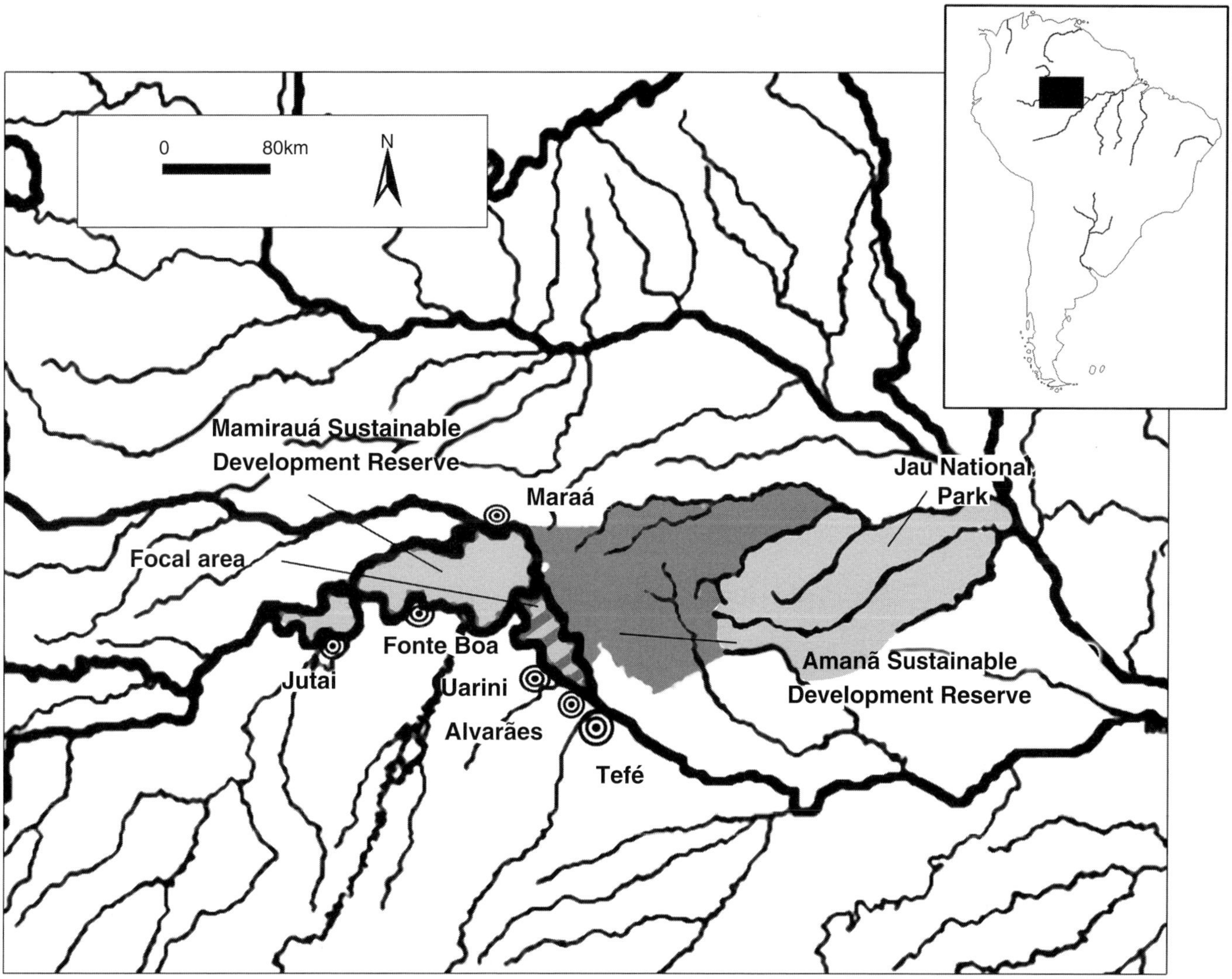

Figure 36.1 Location of Mamirauá and Amanã Sustainable Development Reserves, Amazonas State, Brazil.

hunting trip, establishing the *catch per unit of effort* (CPUE) is not straightforward. Short trips (less than 24 h) accounted for almost all uacari kills, while longer trips (lasting more than 24 h) accounted for only three. As hunting trips that resulted in uacari kills did not change in length during the study period, we can tentatively suggest that the availability of these animals did not decrease between July 2002 and June 2006. However, more data on uacari densities and hunting effort in villages in the ASDR are required in order to effectively model the sustainability of this hunting.

Conservation in the ASDR and MSDR

Participatory management of the environment and its natural resources has been implicit in the conservation strategy implemented in the MSDR since its creation. This management is based on the involvement of local people, combining traditional knowledge with the scientific management techniques of the IDSM (Queiroz & Peralta 2006). Consequently, all actions to alleviate threats to *C. calvus calvus* have been implemented with the participation of local communities. Because *C. calvus calvus* is not directly threatened by hunting, and has a restricted geographical range, the main threats to this primate are the reduction and transformation of available natural habitat. To halt habitat transformation, it was crucial to control the main forms of deforestation in the area; small-scale agriculture and logging. Since management plans were first implemented in the MSDR, an intensive program has occurred introducing new agricultural practices and change to patterns of land use. The practices of cultivating new and old fallows, sandy beaches and mud banks were introduced in 1994/95, and as a result, rates of habitat transformation have decreased considerably over the last 10 years. Monitoring of agricultural activities inside the reserve showed that in 1994 an average family in the MSDR deforested 12 ha of primary flooded forest in order to make new gardens. In 2004, this value was virtually reduced to zero, as almost all new agricultural plots were built in

Table 36.1 Game animals killed in four villages monitored in the Amanã Sustainable Development Reserve (ASDR), Brazil for three consecutive years (2002–2005).

Animal groups			Number hunted
Mammals			**2490**
Reptiles			670
Birds			344
	Ungulates		1384
	Rodents		844
	Primates		**138**
	Edentates		76
	Carnivores		21
	Sirens		13
	Cetaceans		13
	Marsupials		1
		Alouatta seniculus	46
		Cebus macrocephalus	43
		Cacajao melanocephalus	**35**
		Saimiri cassiquiarensis	4
		Ateles chamek	3
		Cebus albifrons	2
		Saguinus inustus	2
		Callicebus torquatus	2
		Lagothrix lagotricha	1

fallows, beaches and banks. As a result, less than 2.5% of all focal areas (Figure 36.1) in Mamirauá are now modified, and the risk that rates of habitat transformation will increase is low (IDSM 2003; Queiroz & Peralta 2006).

Ayres (1986) found that a few tree species make up the bulk of the diet of *C. calvus calvus*. The third most important species in the diet is "envira-vassourinha" *Xylopia frutescens* (Annonaceae). This species is also of economic importance in the Middle Solimões region, being used intensively for firewood, mainly by the brick industry. Illegal timber extraction occurred in the MSDR in the early 1990s, and almost 8000 trees were logged annually in the focal area. In 1998 a program was launched to promote sustainable timber extraction, with great success. In 2006 fewer than 50 trees were removed illegally from the area (Figure 36.2; Queiroz & Peralta 2006). At the same time, the forestry program promoted the sustainable exploitation of a reduced number of tree species. To allow this, an agreement was made with the local population, banning the logging of some threatened tree species and those species that are important for the ecological functioning of the forest. As a result logging of *X. frutescens* was halted, and in 2005 this tree species was no longer recorded (legally or illegally) in the monitoring of timber extraction in the MSDR. As a result of the measures implemented since 1990, the population of *C. calvus* in the MSDR appears to be stable, with densities recorded in 2005 similar to or greater than those recorded at virtually the same sites by Ayres (1986) in 1982 (Paim 2005).

In contrast to the MSDR situation, where threats are low and conservation measures have been implemented and consolidated, conservation work in the Amanã Sustainable Development Reserve began relatively recently (in 2000). Two main conservation strategies have been adopted. First, drawing on the experiences of the MSDR, the IDSM has involved local inhabitants in the establishment of participatory councils aiming to conserve biological diversity within the reserve. Second, the investigation on the hunting of *C. melanocephalus* described above has begun to examine the viability of this primate population under current conditions. In addition to monitoring kills, the IDSM are determining the natural abundance of *C. melanocephalus* and of other game species at Amanã. A system of line transects are censused frequently using DISTANCE sampling methods (Buckland *et al.* 1993). This is an ongoing investigation, but preliminary results will be used to model the viability of *C. melanocephalus* populations in Amanã. Once the main trends in the hunting of *C. melanocephalus* are known, efforts will be made to adapt the current management to ensure that animals are not hunted unsustainably at Amanã.

Concepts of sustainable management by local villages and hunters are currently restricted by the Brazilian legal framework, but this situation is changing. The environmental authorities are now involved in updating legislation that will legalize the sustainable management of game animals in Brazil, as long as it is appropriately organized and controlled with management plans. This action will provide local hunters with the legal capacity to plan the use of their resources, providing the means to conserve uacaris and other game animals in the reserve. As new governmental legislation on the use of wildlife by traditional populations is under preparation, one of the most important activities of the IDSM is the process of convincing policy makers and authorities to produce legislation that supports wildlife management. Wildlife management by traditional Amazonian communities should soon be a reality in Brazil. If done properly, this will provide the means to conserve *C. melanocephalus ouakary* and other wildlife species.

Uacari conservation in Peru

The status of *Cacajao calvus ucayalii* in the "Ucayali–Yavarí corridor", Peru

While Brazil contains most, and possibly all, *Cacajao* taxa, only *C. calvus ucayalii* occurs in Peru (Hershkovitz 1987). *Cacajao calvus ucayalii* is currently listed as Vulnerable (IUCN 2008), and the population in Peru represents the largest of

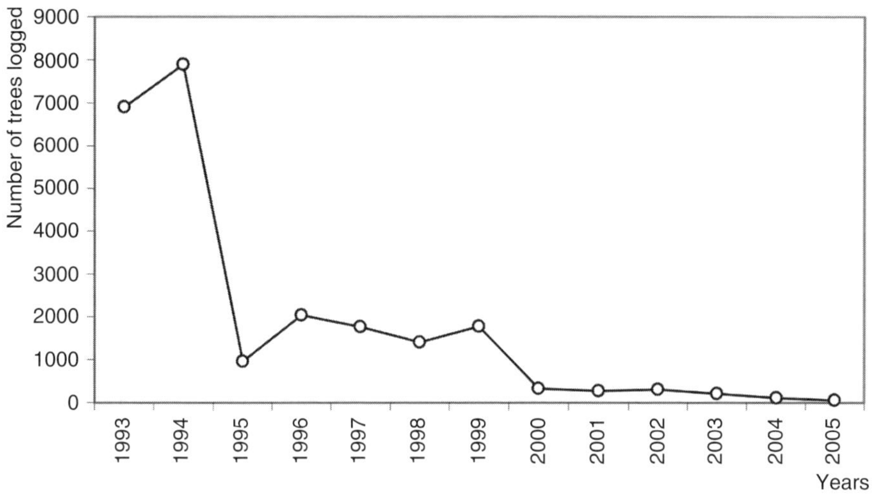

Figure 36.2 Annual number of trees illegally removed from the study area in Mamirauá Reserve, Brazil (Program of Community Forestry – IDSM).

C. calvus. The geographical range of *C. calvus ucayalii* is limited by the Rio Amazonas to the north, the Rio Yavarí to the east and the Rio Ucayali to the west, although rumours suggest that small populations also occur north of the Amazon and west of the Ucayali. Historically the range extended as far as the Rio Urubamba in the South (Hershkovitz 1987), but censuses in the 1980s suggested that range had been reduced (Aquino 1988). Aquino (1988) attributed these reductions in uacari populations to hunting and habitat disturbance, and populations have probably continued to decline (Figure 36.3).

Sites on the Yavarí and Yavarí–Mirín have the highest densities of *C. calvus ucayalii* (Puertas & Bodmer 1993; Salovaara *et al.* 2003), and several key areas for the conservation of the subspecies are in this area. These include the Tamshiyacu–Tahuayo Communal Reserve (TTCR), the Lago Preto Conservation Concession, the "Iquitos–Yavarí" Logging Concessions and the proposed Greater Yavarí Reserve (Figure 36.4). Census work by Salovaara *et al.* (2003) revealed a patchy distribution for *C. calvus ucayalii* and the distribution within the geographical range is not completely known. The subspecies is highly abundant in the Lago Preto area, but elsewhere groups are encountered more rarely. *Cacajao calvus ucayalii* is absent from the Brazilian side of the Yavarí, and from a long stretch on the Peruvian side, upriver from the mouth of the Yavarí–Mirín. On the Yavarí–Mirín, the subspecies is present on the north bank, but is absent from the south bank, except in the upper reaches (Salovaara *et al.* 2003).

Figure 36.3 Hunting of *Cacajao calvus ucayalii*. Photo taken by Mike James in the vicinity of Jenaro Herrera, "Ucayali–Yavarí corridor", Peru.

Threats to *Cacajao calvus ucayalii* in the "Ucayali–Yavarí corridor", Peru

In the Ucayali–Yavarí corridor, hunting pressure is high in areas close to the Rios Ucayali and Amazonas (Robinson & Bodmer 1999; Bodmer & Lozano 2001), and low on the Yavarí side of the forest block, where local human populations were

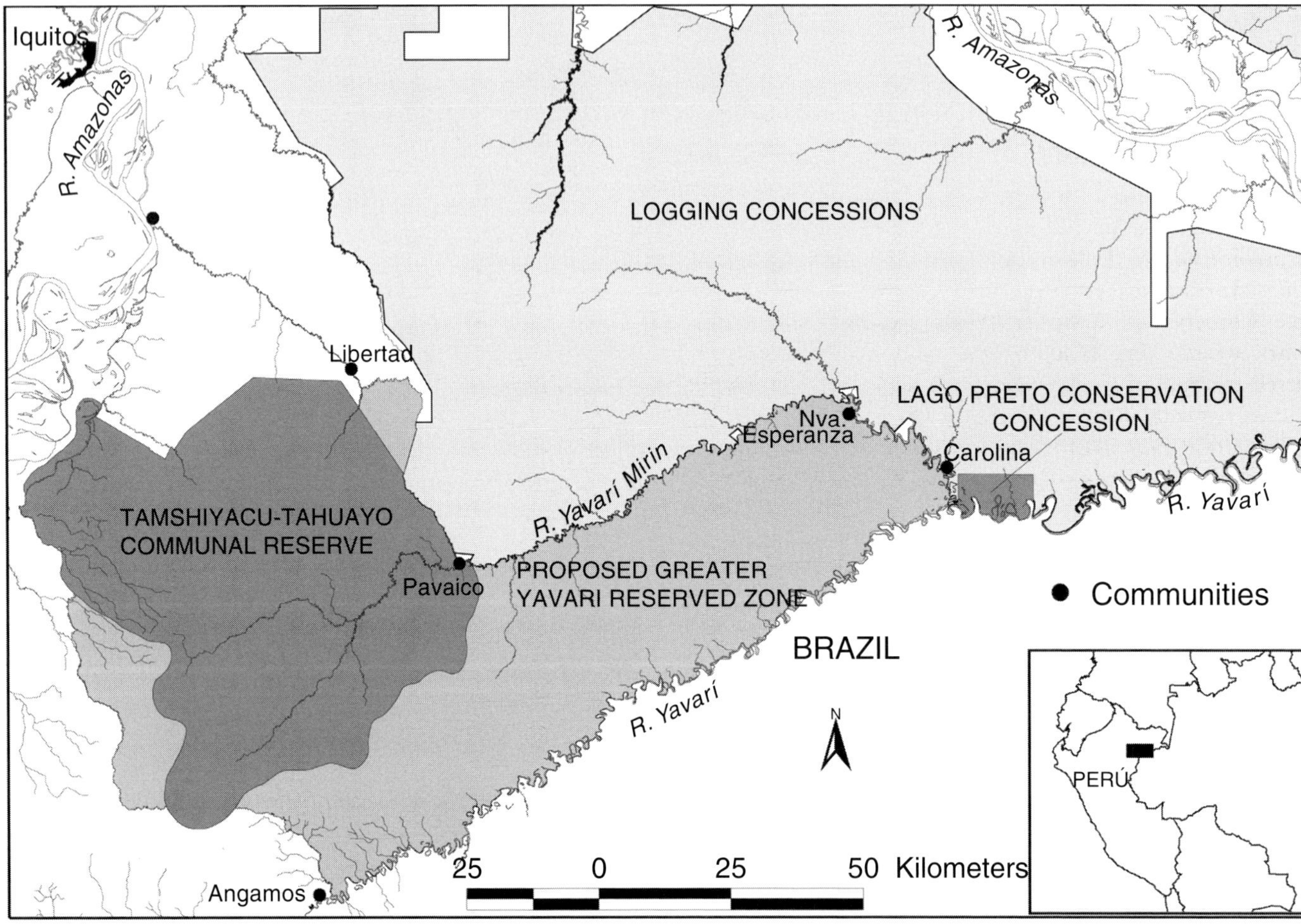

Figure 36.4 Conservation areas and logging concessions in the Ucayali–Yavarí corridor, Peru (map modified from INRENA).

probably at their lowest levels in recorded history in 2003; around 1–2 people/1000 ha (Bodmer *et al.* 2003; Bodmer & Puertas 2003; Del Campo *et al.* 2003). However, the human population on the Yavarí has swelled since 2004 with the incursion of several timber companies. More wild meat was consumed per capita in a commercial forest concession on the Yavarí than in nearby rural communities (Bodmer *et al.* 2006). Logging workers also consumed more small game species (Bodmer *et al.* 2006). Because hunting primates is relatively easy compared to hunting larger terrestrial species, it is likely to have increased and local uacari populations are under threat while logging operations are active in the area.

Logging on the Yavarí is likely to have an economic return exceeding that of existing commercial activities in the area, such as the sale of wild meat, peccary skins and other non-timber forest products (Bodmer *et al.* 2006). As timber extraction on the Yavarí is selective, impacts on uacaris will depend partially on the overlap between the species logged, and those in the diet of the monkeys (Bowler 2007). Valuable hardwood species are not important in the diet of *C. calvus ucayalii*, but some softwood timber species, such as those in the family Myristicaceae, are (Bowler 2007). On the upper Yavarí and Yavarí–Mirín, the species harvested are restricted by cost of extraction from this remote area. Mainly valuable hardwood species are removed, with only limited extraction of softwoods (Bowler 2007). Current levels of extraction are unlikely to have a great ecological impact on uacari populations on the upper Yavarí, but softwood extraction occurs on the lower Yavarí and on the Rio Amazonas side of the forest block. Softwood extraction may also occur on the Yavarí in the future, as reserves are depleted in more accessible areas (Bowler 2007).

Aquino and Encarnación (1999) and Bowler (2007) considered the effects of non-timber resource extraction on *C. calvus ucayalii. Mauritia flexuosa* (Arecaceae) fruit is the most important non-timber plant resource harvested from the forest by people in Loreto (Bodmer *et al.* 1990; Castillo *et al.* 2006). It is also the most important species for *C. calvus ucayalii* at Lago Preto, constituting 20% of the annual diet (Bowler 2007). Close by Iquitos and the Tamshiyacu Tahuayo Communal Reserve (TTCR), large quantities of *M. flexuosa* fruits are obtained simply by felling the palms (Bodmer *et al.* 1999; Meyer & Penn 2003). Such unsustainable resource extraction may negatively impact some uacari populations. On the Yavarí, the human population is low, and markets are too distant to make commercial extraction of *M. flexuosa* viable (Bowler 2007).

The conservation of *Cacajao calvus ucayalii* in the Yavarí–Ucayali corridor

Until recently the only protected area known to contain *C. calvus ucayalii* was the TTCR, but the presence of this species has become part of the justification for three newly created or proposed areas; the Sierra del Divisor Reserved Zone, the Lago Preto Conservation Concession and the proposed Greater Yavarí Reserve. The Wildlife Conservation Society and the Durrell Institute of Conservation and Ecology (WCS–DICE) are involved with all except Sierra del Divisor.

The TTCR (322,500 ha) was created in 1991 as a result of strong alliances between conservationists and local people who wanted to protect their resources from outside pressures (Newing & Bodmer 2004). It is situated close to the city of Iquitos, between the Rio Amazonas and the Yavarí, and contains 14 primate species (Puertas & Bodmer 1993). Economic and subsistence activities by rural inhabitants, including hunting and extraction of non-timber plant products, are permitted within the reserve.

Lago Preto is situated on the Rio Yavarí 175 km southeast from Iquitos. It is unique among Peruvian protected areas in that its location was determined by its densities of *C. calvus ucayalii*, which are considerably higher than those in any other area (Salovaara *et al.* 2003). However, Lago Preto is very small (9926 ha) and surrounded by logging concessions, so cannot protect the region's uacari populations on its own (Bowler *et al.*, Chapter 15).

WCS–DICE provide technical assistance for communities looking to manage their hunting, and have developed guidelines for the sustainable management of wildlife (Table 36.2). As *Cacajao* and other primates are vulnerable to overhunting (Bodmer *et al.* 1997), the guidelines require that they are not hunted. WCS–DICE work with remote communities on the Yavarí and Yavarí–Mirín, and use Lago Preto as a base for research and conservation work, enabling them to maintain a presence in the area and monitor hunting. After several years of data collection and liaison with communities, WCS–DICE began implementing community-based wildlife management in 2005, and objectives include stopping or greatly reducing hunting of primates. The establishment of source areas to replenish animals hunted in sink areas is a key part of the conservation strategy proposed by WCS–DICE for Loreto, and the Yavarí–Mirín area acts as a source area for the TTCR and surrounding areas (Bodmer 2000; Bodmer *et al.* 2003). Maintaining the current low levels of hunting on the Yavarí–Mirín could ensure healthy populations of non-primate game animals in the TTCR and therefore reduce hunting pressure on more vulnerable primate species like *C. calvus ucayalii.*

Communities working towards sustainable use are advised to manage their wildlife habitats as well as their hunting, including the extraction of timber and non-timber plant products. In more populous areas, felling of *M. flexuosa* palms clearly threatens *C. calvus ucayalii*, and its sustainable use is an important requirement in community management plans. In the TTCR, a community-run agroforestry project plants *M. flexuosa* in settlement zones (Meyer & Penn 2003). In open areas the palms grow relatively shorter, so they do not have to be cut down to retrieve the fruit. The goal is to plant enough palms in people's agricultural plots so they will not have to enter the reserve and destroy naturally occurring palms.

Community-based wildlife conservation in the TTCR has been practised since 1991, and management plans that stipulate a reduction in the hunting of monkeys succeeded in reducing the number of primates killed (Bodmer & Puertas 2000). It is hoped that the success of projects in the TTCR can be transferred to communities on the Yavarí and Yavarí–Mirín, but the situation here is more complicated. Logging concessions cover areas that greatly exceed the size of protected areas (Figure 36.4), and hunting by concession workers is the main threat. In the absence of long-term relationships between conservationists and concession workers, like those developed in community conservation projects, managing hunting is a much more difficult proposition. Despite regulations on hunting imposed by the Peruvian National Institute of Natural

Table 36.2 WCS–DICE guidelines for the sustainable management of wildlife in Loreto, Peru.

- Rural communities should have community-based wildlife management plans that set limits on harvested species that are not vulnerable to overhunting and halt or greatly reduce hunting of species vulnerable to overharvesting.
- Hunting limits should be within sustainable levels. Hunting limits will be determined through sustainable hunting models, such as the unified harvest model (Bodmer & Robinson 2004).
- Rural communities should monitor and evaluate their hunting as an integral part of community-based wildlife management. This monitoring and evaluation can be in the form of hunting registers that include information on the species, number of individuals, date, and location hunted. Monitoring wildlife hunting should also include information on catch-per-unit-effort (CPUE). CPUE can be used to evaluate trends in wildlife abundance and be used to evaluate sustainability (Puertas & Bodmer 2004). The advantage of using CPUE is that can be done by the communities with technical support.
- Rural communities will need to manage their wildlife habitats. This will require sound forestry management and sustainable use of non-timber plant products, since these plants provide food and shelter for wildlife.
- Rural communities will need to set up source-sink areas as part of their management plans. Areas with no hunting (source) should be set up near hunted areas (sink). Non-hunted areas will buffer against any unexpected fluctuations in wildlife populations or socioeconomic demands and help guarantee sustainability of wildlife hunting in the long term.

Resources (INRENA), uncontrolled hunting is likely to increase in the logging concessions and adjoining areas. WCS–DICE have proposed a management plan for the Yavarí that includes guidelines for hunting in logging concessions, but it remains to be seen whether the value of timber extraction can give INRENA sufficient leverage in enforcing restrictions on hunting, or if the logistics and political will are sufficient to implement conservation strategies in such a remote area.

Conclusions

The conservation initiatives for the populations of *Cacajao* considered in this chapter have needed to consider very different threats. However, in all the projects, community-based methods are key elements in conserving uacari populations. While this type of management obviously depends on the will of the local people involved to conserve wildlife, only the participatory nature of the management would allow the conservation successes achieved in the MSDR and TTCR. The reserves both contained a number of villages on their declaration, and it is difficult to envisage an alternative management solution that did not include these communities, indeed the TTCR was set up purely *because* of the will of local people.

Excluding local people from protected areas has proven ineffective in conserving mammal populations in several reserves in Amazonia (Schwartzman *et al.* 2000). The displacement of local people from the Pacaya–Samiria National Reserve in Peru led to high levels of illegal hunting within the reserve, and considerable conflict between local people and reserve officials. This eventually led to the resettlement of communities within the reserve and a move over to community-based conservation, similar to that practiced in the TTCR (Bodmer & Puertas 2007). The MSDR and TTCR have allowed local people to live sustainably off the natural resources, while conserving uacaris and other wildlife. Additionally, the reserves have facilitated alternative livelihoods, such as guiding tourists or monitoring wildlife, for local people who might have otherwise engaged in hunting. The MSDR and TTCR models of community management are now being used in other areas, including the ASDR and on the Yavarí, and the long-term goals of these conservation projects include the cessation of hunting of *Cacajao*.

Experiences in the TTCR and the MSDR have shown that species conservation can be achieved though community projects which allow the managed hunting of wildlife. Managing the hunting of all game species can facilitate the conservation of specific species, or those plant species on which they depend. Long-term planning and monitoring of wildlife populations is an essential part of this process, and the involvement of national and regional governments is of key importance. In Brazil, convincing authorities to produce new community-inclusive wildlife management legislation will allow community projects to proceed legally, enabling biologists to implement effective conservation for *C. melanocephalus*. In Peru, where communities adjoin logging concessions, managing hunting with participatory methods is complicated by the presence of transient timber concession workers. Here government enforcement of hunting regulations may be necessary alongside community projects to ensure the persistence of *C. calvus ucayalii* in core parts of its range. While a full discussion of the pros and cons of community-based approaches to conservation is not possible in a chapter of this size, the IDSM and WCS–DICE have converged on similar approaches to conservation in Amazonia, and it appears that, at least in the areas discussed here, the local people that share the geographical range of the uacaris will play a key role in their conservation.

Acknowledgments

We are grateful for the assistance provided by research, logistical and administrative staff and students from IDSM, WCS, DICE, UNAP, INRENA and WWF. The local people of the Yavarí, TTCR, MSDR and ASDR are thanked for their hospitality and their dedication for a sustainable future. The Peruvian project was supported by WCS, and part of the research on *C. c. ucayalii* was funded by Rufford Small Grants, the LA Zoo, Conservation International and Primate Conservation Inc. Dr. José Márcio Ayres continues to be an inspiration to us all.

References

Aquino, R. (1988). Preliminary survey on the population densities of *Cacajao calvus ucayalii*. *Primate Conservation*, **9**, 24–26.

Aquino, R. & Encarnación, F. (1999) Observaciones preliminares sobre la dieta de *Cacajao calvus ucayalii* en el Nor-Oriente Peruano. *Neotropical Primates*, 7(1), 1–5.

Ayres, J.M.C. (1986). Uakaris and Amazonian flooded forest. PhD thesis, University of Cambridge.

Ayres, J.M. & Johns, A.D. (1987). Conservation of white uacaries in Amazonian varzea. *Oryx*, **21**, 74–80.

Ayres, J.M., Alves, A.R., Queiroz, H.L., *et al.* (1999). Mamirauá: the conservation of biodiversity in the Amazonian flooded forests. In *Várzea: Diversity, Development, and Conservation of Amazonia's Floodplains*, ed. C. Padock, J.M. Ayres, M. Pinedo-Vasquez & A. Henderson, pp. 203–216. New York, NY: The New York Botanical Garden Press.

Ayres, J.M., Silva, V.F. & Nelson, B. (1997). Proposta de Criação da Reserva de Desenvolvimento Sustentável Amanã. IPAAM. Unpublished Report, Amazonas State Government, SCM, INPA, CNPq, MCT, WCS and EU, Brazil.

Bodmer, R.E. (2000). Integrating hunting and protected areas in the Amazon. In *Future Priorities for the Conservation of Mammals: Has the Panda had its Day?* ed. N. Dunstone & A. Entwistle, pp. 277–290. Cambridge: Cambridge University Press.

Bodmer, R.E. & Lozano, E. (2001). Rural development and sustainable wildlife use in the tropics. *Conservation Biology*, **15**, 1163–1170.

Bodmer, R.E. & Puertas, P.E. (2000). Community-based comanagement of wildlife in the Peruvian Amazon. In *Hunting for Sustainability in Tropical Forests*, ed. J.G. Robinson & E.L. Bennett.

New York, NY: Columbia University Press, pp. 395–409.

Bodmer, R.E. & Puertas, P. (2003). A brief history of the Yavarí Valley. In *Peru: Yavari. Rapid Biological Inventories Report 11*, ed. N. Pitman, C. Vriesendorp & D. Moskovits. Chicago, IL: The Field Museum.

Bodmer, R.E. & Puertas, P. (2007). Impacts of displacement in the Pacaya–Samiria National Reserve, Peru. In *Working Paper no. 29. Protected Areas and Human Displacement: a Conservation Perspective*, ed. K.H. Redford & E. Fearn. New York, NY: WCS Institute, Wildlife Conservation Society.

Bodmer, R. & Robinson, J. (2004). Evaluating the sustainability of hunting in the Neotropics. In *People in Nature: Wildlife Conservation in South and Central America*, ed. K. Silvius, R. Bodmer & J. Fragoso, pp. 299–323. New York, NY: Columbia University Press.

Bodmer, R.E., Allen, C.M., Penn, J.W., *et al.* (1999). Evaluating the sustainable use of wildlife in the Pacaya–Samiria National Reserve, Peru. *America Verde Working Paper No. 4 Latin America and Caribean Region*. The Nature Conservancy.

Bodmer, R.E., Eisenberg, J.F. & Redford K.H. (1997). Hunting and the likelihood of extinction of Amazonian mammals. *Conservation Biology*, **11**, 460–466.

Bodmer, R.E., Fang T.G. & Moya L.I. (1990). Fruits of the forest. *Nature*, **343**, 109.

Bodmer, R.E., Puertas, P. & Antúnez, M. (2003). Use and sustainability of wildlife hunting in and around the proposed Yavarí Reserved Zone. In *Peru: Yavari. Rapid Biological Inventories Report 11*, ed. N. Pitman, C. Vriesendorp & D. Moskovits. Chicago, IL: The Field Museum.

Bodmer, R.E., Puertas, P., Rios, C., *et al.* (2006). Investigación Sobre la Facilibilidad de Implementar un Manejo de Fauna Silvestre en Concesiones Forestales en la Amazonia Peruana. Unpublished Report, DICE–WCS, Peru.

Bowler, M. (2007). The ecology and conservation of the red uacari monkey on the Yavarí River, Peru. PhD thesis, University of Kent, Canterbury.

Buckland, S.T., Anderson, D.R., Burnham, K.P., *et al.* (1993). *Distance Sampling: Estimating Abundance of Biological Populations*. London: Chapman and Hall.

Castillo, D. del, Otárola, E. & Alvarado, L. (2006). *Aguaje: The Amazing Palm Tree of the Amazon*. Instituto de investigaciones de la Amazonía peruana – Comisión europea, Iquitos, Peru.

Del Campo, H., Valverde, Z., Calle, A., *et al.* (2003). Human communities. In *Peru: Yavari. Rapid Biological Inventories Report 11*, ed. N. Pitman, C. Vriesendorp & D. Moskovits. Chicago, IL: The Field Museum.

Hershkovitz, P. (1987). Uacaries, New World monkeys of the genus *Cacajao* (Cebidae, Platyrrhini): a preliminary taxonomic review with the description of a new subspecies. *American Journal of Primatology*, **12**, 1–53.

IDSM. (2003). *Relatório Anual do Contrato de Gestão. Instituto de Desenvolvimento Sustentável Mamirauá e MCT*. Tefé, Brazil.

IUCN. (2008). IUCN Red List of Threatened Species, http://www.iucnredlist.org/ [accessed 22 May 2009].

Meyer, D. & Penn, J. (2003). An overview of the Tamshiyacu–Tahuayo Communal Reserve. In *Peru: Yavari. Rapid Biological Inventories Report 11*, ed. N. Pitman, C. Vriesendorp & D. Moskovits. Chicago, IL: The Field Museum.

Newing, H. & Bodmer, R. (2004). Collaborative wildlife management and adaptation to change: the Tamshiyacu Tahuayo Communal Reserve, Peru. *Nomadic Peoples*, **7**, 110–122.

Paim, F.P. (2005). Relatório Final – Monitoramento Ambiental de Vertebrados Arborícolas nas Trilhas do Ecoturismo da RDSM. Unpublished manuscript, Instituto de Desenvolvimento Sustentável Mamirauá.

Plumptre, A.J. & Johns, A. (2001). Changes in primate communities following logging disturbance. In *The Cutting Edge: Conserving Wildlife in Logged Tropical Forests*, ed. R.A. Fimbel, A. Grajal & J.G. Robinson, pp. 71–92. New York, NY: Columbia University Press.

Puertas, P. & Bodmer, R.E. (1993). Conservation of a high diversity primate assemblage. *Biodiversity and Conservation*, **2**, 586–593.

Puertas, P. & Bodmer, R. (2004). Hunting effort as a tool for wildlife management in Amazonia. In *People in Nature: Wildlife Conservation in South and Central America*, ed. K. Silvius, R. Bodmer & J. Fragoso, pp. 123–138. New York, NY: Columbia University Press.

Queiroz, H.L. & Peralta, N. (2006). Reserva de desenvolvimento sustentável: Manejo integrado de recursos naturais e gestão participativa. In *Dimensões Humanas da Biodiversidade*, ed. I. Garay & B.K. Becker, pp. 447–476. Brazil: Petrópolis.

Robinson, J.G. & Bodmer R.E. (1999). Towards wildlife management in tropical forests. *Journal of Wildlife Management*, **63**, 1–13.

Salovaara, K., Bodmer, R.E., Recharte, M. & Reyes, F. (2003). Diversity and abundance of mammals. In *Peru: Yavari. Rapid Biological Inventories Report 11*, ed. N. Pitman, C. Vriesendorp & D. Moskovits. Chicago, IL: The Field Museum.

Schwartzman, S., Moreira, A. & Nepstad, D. (2000). Rethinking tropical conservation: peril in parks. *Conservation Biology*, **14**, 1351–1357.

Appendix A: Conservation Fact Sheet: Bolivia

Robert B. Wallace, Nohelia Mercado & Jesus Martinez

Taxon list

Rio Tapajós saki monkey (*Pithecia irrorata*), Bolivia: Parabacú, saki. Madidi titi monkey (*Callicebus aureipalatii*), Bolivia: Lucachi. Brown titi monkey (*Callicebus brunneus*), Bolivia: Lucachi. Bolivian gray titi (*Callicebus donacophilus*), Bolivia: Facafaca. Beni titi monkey (*Callicebus modestus*), Bolivia: Lucachi. Olalla's titi monkey (*Callicebus olallae*), Bolivia: Lucachi. Chacoan titi monkey or white-coated titi monkey (*Callicebus pallescens*), Bolivia: Ururó (Anderson 1997; Felton *et al.* 2006; Wallace *et al.* 2006; Martínez & Wallace 2007; Ayala, pers. commun. to R. Wallace, 2007).

Synonyms

Pithecia irrorata mentioned as *Pithecia monachus* or *Pithecia hirsuta* (Anderson 1997). No synonym for *Callicebus aureipalatii*. *Callicebus brunneus* cited as *Callicebus brunea*, *Callicebus moloch brunneus*, *Callicebus moloch* and *Callicebus cupreus* (Anderson 1997). *Callicebus donacophilus* mentioned as *Callithrix donacophilus*, *Callithrix donacophila*, *Callithrix cinerascens*, *Callicebus gigot donacophilus*, *Callicebus moloch*, *Callicebus moloch donacephilus* and *Callicebus donacophilus donacophilus* (Anderson 1997). *Callicebus modestus* mentioned as *Callicebus moloch brunneus* and *Callicebus moloch modestus* (Anderson 1997). *Callicebus olallae* mentioned as *Callicebus moloch brunneus* and *Callicebus moloch olallae*. *Callicebus pallescens* cited as *Callicebus moloch* and *Callicebus donacophilus pallescens* (Anderson 1997).

Protected reserves and distribution of pitheciine taxa

Pithecia irrorata observation locations are found in true Amazonian forest in northern Bolivia, but its presence is not currently reported for any national protected area. *Callicebus aureipalatii* is present in the Madidi National Park and Natural Area of Integrated Management in Bolivia. The distribution ranges from the west bank of the Beni River extending west and north into southern Peru south of the Madre de Díos River. *Callicebus brunneus* presence is reported for Pando Department and the northern part of Beni Department, with some observations made in Manuripi Heath National Wildlife Reserve. *Callicebus donacophilus* has the broadest distribution for this genus in Bolivia found from the Maniqui River of the Beni Department east to the Brazilian border (Ferrari *et al.* 2000) and south to the central–southern part of Santa Cruz Department. In Bolivia, *C. donacophilus* is present in the Beni Biosphere Reserve and Amboró National Park. *Callicebus pallescens* locations correspond to the southern part of Santa Cruz Department, exclusively in the Chaco and Pantanal regions. This Species has been registered in Kaa-Iya National Park. *Callicebus modestus* is one of the endemic Bolivian primate species and inhabits southwestern Beni Department from the Beni River to Maniqui Rivers and from the Rurrenabaque-Yucumo road north to the limits of the Amazonian forest in the northern parts of the Beni Department (Martinez & Wallace 2007). A very small fraction of this distribution is found in the western part of the Beni Biosphere Reserve. *Callicebus olallae* is the most restricted Bolivian primate species, that is found almost exclusively in riverine forest along the Yacuma River, with one record in similar forest on the Maniqui River (Martinez & Wallace 2007). This species is not found in any national protected areas, but both Bolivian *Callicebus* endemics are found in the recently created Santa Rosa Municipal Reserve.

Conservation status

In December 2007 a conservation status review of Neotropical Primates by the IUCN Neotropical Primate Specialist Group in Orlando, Florida, listed *Pithecia irrorata*, *Callicebus aureipalatii*, *Callicebus brunneus*, *Callicebus donacophilus* and *Callicebus pallescens* as Least Concern, although for the majority of these species, further studies are required to evaluate their status more definitively. At the same meeting, *Callicebus modestus* conservation status was reclassified as Endangered due to its restricted distribution and endemism, and the fact that the majority of the population is not protected at the national level. *Callicebus olallae* was reclassified as Critically Endangered due to its extremely restricted distribution at essentially one locality and again the fact that the entire population occurs outside of national protected areas.

Key studies

Distributional information for *Pithecia irrorata* and *Callicebus brunneus* is found in Anderson's (1997) Bolivian mammal summary and specific primate studies in Pando Department

Evolutionary Biology and Conservation of Titis, Sakis and Uacaris, eds. Liza M. Veiga, Adrian A. Barnett, Stephen F. Ferrari and Marilyn A. Norconk. Published by Cambridge University Press.

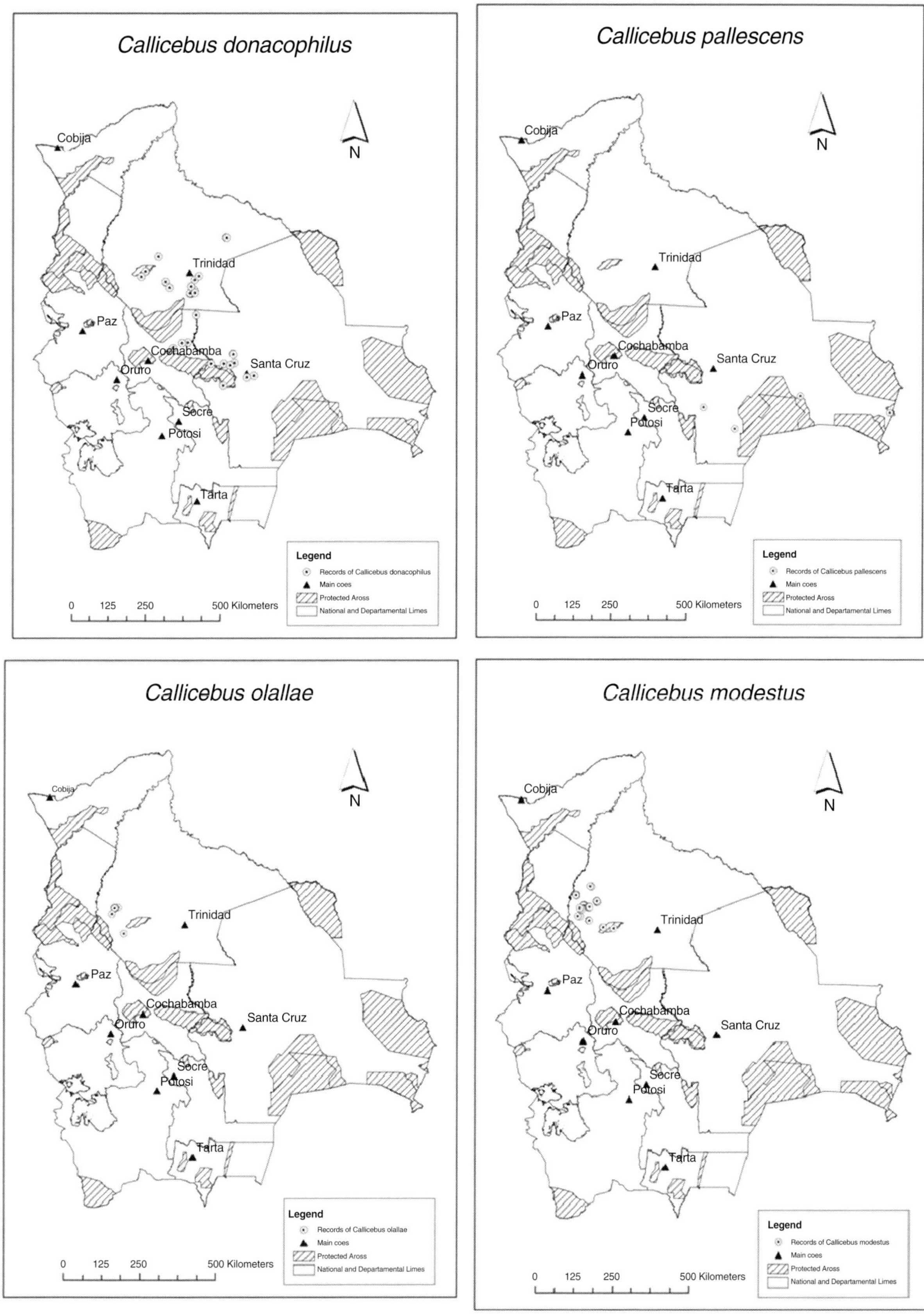

Figure A.1 Known localities of Pitheciid species and protected areas.

Figure A.1 *(cont.)*

(Kohlhaas 1988; Cameron & Buchanan-Smith 1991; Christen & Geissmann 1994; Christen 1999; Alverson *et al.* 2000, 2003; Alverson 2003; Rowe & Martínez 2003). *Callicebus aureipalatii* was recently described as a new species after 2 years of data acquisition (Wallace *et al.* 2006). The same year, a behavior field study was conducted by De la Torre and is currently in preparation. *Callicebus donacophilus*, although identified as *C. moloch*, was reported from the Beni Biosphere Reserve (Garcia & Tarifa 1988). Subsequently, information on its distribution was documented by Langstroth (1996), and additional data were obtained by Martínez and Wallace (2007). A behavioral study is currently under way at two locations on the outskirts of Santa Cruz de la Sierra (Dingess, pers. commun. to R. Wallace, 2007). For decades, the only information on the Bolivian endemics *Callicebus modestus* and *C. olallae* consisted of the original species description (Lönnberg 1939). Since 2002, distribution data were gathered by Felton *et al.* (2006) and Martinez and Wallace (2007), and Lopez-Strauss (2008) provided population density estimations for both endemic species. For *Callicebus olallae*, Martinez and Lopez are currently studying the behavioral ecology and determining feeding patterns and diet. Information about *Callicebus pallescens* is scarce.

Regarding more general information on these species, in the mid 1970s Freese and colleagues (1982) conducted studies on primate distribution and relative abundance at several sites across Bolivia. Later, Brown and Rumiz (1985) summarized existing knowledge regarding primate distribution in Bolivia. Meanwhile, Sydney Anderson (1997) began work on the classic volume on Bolivian mammal taxonomy and distribution and completed this in the mid 1990s. In 2003, an updated list of mammalian species for Bolivia was published (Salazar-Bravo *et al.* 2003). In 2007, as part of a broader systemization of distribution data, updated distribution maps for Bolivia's 23 primate species were produced (Wallace & Mercado 2007; Mercado 2008).

List of conservation programs

These programs are engaging in specific activities related to primate research and conservation.

- Wildlife Conservation Society (Greater Madidi Landscape Conservation Program in northern La Paz and southwestern Beni Departments).
- Wildlife Conservation Society (Kaa-Iya Landscape Conservation Program in Santa Cruz Department).
- Universidad Amazonica de Pando and CIPA are working to legally establish the Estacion Biologica Tahuamanu in northwestern Pando Department.
- Chicago Field Museum, Herencia and CIPA are working to legally establish a protected area, Frederico Roman, in northeastern Pando Department.
- Asociación Boliviana para la Conservación & Conservation International in the southwestern Beni Department.

References

Alverson, W. (ed.). (2003). *Rapid biological inventories: 05, Bolivia: Pando, Madre de Dios.* Chicago, IL: The Field Museum.

Alverson, W., Moskovitz, D. & Halm, I. (eds). (2003). *Rapid biological inventories: 06, Bolivia: Pando, Federico Román.* Chicago, IL: The Field Museum.

Alverson, W., Moskovitz, D. & Shopland, J.M. (eds). (2000). *Rapid biological inventories: 01, Bolivia: Pando, Río Tahuamanu.* Chicago, IL: The Field Museum.

Anderson, S. (1997). Mammals of Bolivia: Taxonomy and Distribution. *Bulletin of the American Museum of Natural History*, **231**, 1–652.

Brown, A.D. & Rumiz, D.I. (1985). Distribución y conservación de los primates en Bolivia. Estado actual de su conocimiento. Unpublished report to New York Zoological Society.

Cameron, R. & Buchanan-Smith, H. (1991). Primates of the Pando, Bolivia. *Primate Conservation*, **12**, 11–14.

Christen, A. (1999). Survey of Goeldi's monkeys (*Callimico goeldii*) in northern Bolivia. *Folia Primatologia*, **70**, 107–111.

Christen, A. & Geissmann, T. (1994). A primate survey in northern Bolivia, with special reference to Goeldi's monkey, *Callimico goeldii. International Journal of Primatology*, **15**, 239–274.

Felton, A., Felton, A.M., Wallace, R.B., *et al.* (2006). Identification, distribution and behavioural observations of the titi monkeys *Callicebus modestus* Lönnberg 1939, and *Callicebus olallae* Lönnberg 1939. *Primate Conservation*, **20**, 40–46.

Ferrari, S.F., Iwanaga, S., Messias, M.R., *et al.* (2000). Titi monkeys (*Callicebus* spp., Atelidae: Platyrrinhi) in the Brazilian State of Rondonia. *Primates*, **41**, 229–234.

Freese, C.H., Heltne, P.G., Castro, R.N., *et al.* (1982). Patterns and determinants of monkey densities in Peru and Bolivia, with notes on distributions. *International Journal of Primatology*, **3**, 53–90.

Garcia, J.E. & Tarifa, T. (1988). Primate survey of the Estación Biológica Beni, Bolivia. *Primate Conservation*, **9**, 97–100.

Kohlhaas, A.K. (1988). Primate populations in northern Bolivia. *Primate Conservation*, **9**, 93–97.

Langstroth, R. (1996) Forest islands in an Amazonian savanna of northeastern Bolivia. PhD thesis, University of Wisconsin.

Lönnberg, E. (1939). Notes on some members of the genus *Callicebus. Arkiv för Zoologi*, **31**A(13), 1–18.

Lopez-Strauss, H. (2008). Estimación de densidad y composición de grupos de dos primates, *Callicebus olallae y Callicebus modestus,* especies endémicas del sud-oeste del Departamento del Beni – Bolivia. *Undergraduate thesis,* Universidad Mayor de San Andrés, La Paz, Bolivia.

Martinez, J. & Wallace, R.B. (2007). Further notes on the distribution of the Bolivian endemic Titi Monkeys, *Callicebus modestus and Callicebus olallae. Neotropical Primates*, **14**(2), 47–54.

Mercado, N. (2008). Determinación de áreas prioritarias para la conservación de primates en Bolivia. Undergraduate thesis, Universidad Mayor de San Andrés, La Paz, Bolivia.

Rowe, N. & Martinez, W. (2003). *Callicebus* sightings in Bolivia, Peru and Ecuador. *Neotropical Primates*, **11**, 32–35.

Salazar-Bravo, J., Tarifa, T., Aguirre, L.F., *et al.* (2003). Revised checklist of Bolivian Mammals. *Museum of Texas Technical University Occasional Papers*, **220**, 1–27.

Wallace, R.B. & Mercado, N. (2007). La diversidad, distribución y abundancia de primates en Bolivia: Recomendaciones preliminares para su conservación. Abstract and Presentation at the V National Biology Congress, Santa Cruz, Bolivia, March 2007.

Wallace, R.B., Gomez, H., Felton, A., *et al.* (2006). On a new species of titi monkey, genus *Callicebus* Thomas, from western Bolivia (Primates, Cebidae) with preliminary notes on distribution and abundance. *Primate Conservation*, **20**, 29–39.

Appendix B: Conservation Fact Sheet: Brazil

Stephen F. Ferrari, José de Sousa e Silva Júnior, Manuella A. de Souza & Ana Luisa K. Albernaz

Taxon list

Titi monkeys

Twenty-three species of titi monkeys are found in Brazil, including 14 endemics (almost half of all titi species). While the vernacular names used in the English literature reflect specific morphological or geographic characteristics, local names are more homogeneous. Titis are known as "zogue-zogues" throughout the Brazilian Amazon basin, and as "guigós" or "sauás" in the Atlantic Forest region.

The endemic species include all five from the Atlantic Forest: the masked titi (*Callicebus personatus*), the black-fronted titi (*Callicebus nigrifrons*), the southern Bahian masked titi (*Callicebus melanochir*), Barbara Brown's titi (*Callicebus barbarabrownae*) and Coimbra-Filho's titi (*Callicebus coimbrai*). All the other endemics are from the southern Amazon basin, and include two species described by van Roosmalen *et al.* (2002) – Prince Bernhard's titi (*Callicebus bernhardi*) and Stephen Nash's titi (*Callicebus stephennashi*). The others are the booted titi (*Callicebus caligatus*), the dusky or red-bellied titi (*Callicebus moloch*), the Baptista Lake titi (*Callicebus baptista*), Hoffmann's titi (*Callicebus hoffmannsi*), the ashy black titi (*Callicebus cinerascens*), the brown titi (*Callicebus brunneus*), and the widow monkey, *Callicebus purinus*.

Of the remaining nine Amazonian species, only four have more ample ranges outside Brazil. These include the Bolivian gray (*Callicebus donacophilus*) and the white-coated titi (*Callicebus pallescens*), found mainly in Bolivia and Paraguay, respectively, the discolored titi (*Callicebus discolor*), which ranges west as far as Ecuador, and the collared titi, *Callicebus lugens*, found as far north as the Orinoco River in Venezuela. The other species – the coppery titi (*Callicebus cupreus*), the doubtful titi (*Callicebus dubius*), the collared titi (*Callicebus torquatus*), and the widow monkeys, *Callicebus lucifer* and *Callicebus regulus* – are all found in the western Amazon basin and neighboring areas of Peru, Colombia and Ecuador.

Sakis

Six of the eight currently recognized saki taxa are found in Brazil, and half of these are endemic. The Brazilian vernacular names are "parauacú", "macaco-voador" (flying monkey), and "macaco-velho" (old monkey). The latter refers to the grayish pelage typical of some morphotypes.

The one endemic species is the buffy saki, *Pithecia albicans*. The bald-faced saki (*Pithecia irrorata*) is represented by two subspecies, only one of which – Vanzolini's (*Pithecia irrorata vanzolinii*) – is endemic to Brazil. The other subspecies, *Pithecia irrorata irrorata* (Gray's bald-faced saki), ranges west into Peru and Bolivia. Similarly, in the northern Amazon basin, whereas the gold-faced saki (*Pithecia pithecia chrysocephala*) is endemic to Brazil, the white-faced saki (*Pithecia pithecia pithecia*) is more widespread. By contrast, the monk saki is represented in Brazil by a single taxon, Geoffroy's monk saki, *Pithecia monachus monachus*.

Bearded sakis

Of all the pitheciid genera, *Chiropotes* is by far the most Brazilian in its distribution. Three of the five species are endemic, and the two others are at least as common in Brazil as in neighboring countries. The local vernacular is "cuxiú", although some people refer to the animal as "macaco judeu", the Jewish monkey, based on its apparent resemblance to an orthodox Jew, with its long dark beard and hat-like bulbs of pelage on the top of the head. This is reflected in one synonym, *Chiropotes israelita* (see below).

The endemics are all found in the southern Amazon basin. They are the black-bearded saki (*Chiropotes satanas*), Uta Hick's bearded saki (*Chiropotes utahickae*) and the white-nosed bearded saki (*Chiropotes albinasus*). The two other species are found north of the Amazon. The tawny-olive bearded saki (*Chiropotes chiropotes*) ranges north into Venezuela as far as the Orinoco, whereas the reddish-brown bearded saki (*Chiropotes sagulatus*) is also found in the Guyanas.

Uacaris

While all seven uacari taxa appear to occur in Brazil, only three are endemic. The local vernacular is "uacari", although the bald uacari is also referred to as "macaco ingles" due to its similarity to a stereotypical Englishman, with its bald head and red complexion.

The only endemic species is the newly described Ayres' black-headed uacari (*Cacajao ayresi*). Two bald uacari subspecies are also endemic to Brazil, the white bald-headed uacari, *Cacajao calvus calvus*, and Novaes' bald-headed uacari, *Cacajao calvus novaesi*. The two other black-headed species, Humboldt's (*Cacajao melanocephalus*) and Spix's (*Cacajao

Evolutionary Biology and Conservation of Titis, Sakis and Uacaris, eds. Liza M. Veiga, Adrian A. Barnett, Stephen F. Ferrari and Marilyn A. Norconk. Published by Cambridge University Press.

ouakary) range into Venezuela and Colombia, respectively. The red bald-headed uacari (*Cacajao calvus rubicundus*) also occurs in some neighboring areas of Colombia, although the distribution of the Ucayali bald-headed uacari, *Cacajao calvus ucayalii*, is located almost entirely within Peru.

Synonyms

Up until the taxonomic review of van Roosmalen *et al.* (2002), many of the recognized forms of titi monkeys were classified at the subspecific level, in particular those of the Atlantic Forest (*personatus* group) and the *torquatus* group (subgenus *Torquatus*, cf. Groves 2005). The taxonomy of the Brazilian sakis is well-established, at the present time, at least.

In the case of the bearded sakis, only *Chiropotes albinasus* has remained unchanged since the taxonomic review of Hershkovitz (1985), which recognized only one other species, *Chiropotes satanas*, with three subspecies, corresponding to the *satanas*, *utahickae*, and *chiropotes/sagulatus* forms. In this original description of the *utahickae* form, the nomen was spelt *utahicki*, which Silva Jr. and Figueiredo (2002) corrected on the basis of the rules of Latin concordance, given that the name honors a female scientist.

The bearded sakis of the northern Amazon basin were allocated to a single taxon – *Chiropotes satanas chiropotes* – by Hershkovitz (1985), although two distinct geographic forms, separated by the Branco River, were recognized by Bonvicino *et al.* (2003). These authors nominated the eastern form *Chiropotes israelita*, and the western taxon *Chiropotes chiropotes*. However, Silva Jr. and Figueiredo (2002; see also Silva Jr. *et al.*, Chapter 4) corrected the names to *Chiropotes chiropotes* and *Chiropotes sagulatus*, respectively, based on the priority rule of the International Code of Zoological Nomenclature (1999).

The current classification of the bald uacaris is unchanged from Hershkovitz (1987), although that of the black-headed forms has recently been revised by Boubli *et al.* (2008), who proposed a new name – *Cacajao hosomi* – for *Cacajao melanocephalus*, and considered *Cacajao ouakary* a junior synonym of *C. melanocephalus*. This arrangement was not accepted here, however (see Silva Jr. *et al.*, Chapter 4).

Conservation status

With the exception of *Callicebus stephennashi*, which is listed as data deficient, the titis of the Brazilian Amazon basin are all listed as least concern by the IUCN (2008). By contrast, the *personatus* group of the Atlantic Forest includes some of the most threatened pitheciids, ranging from near threatened, in the case of *Callicebus nigrifrons*, to endangered (*Callicebus coimbrai*) and critically endangered (*Callicebus barbarabrownae*). The other two species – *Callicebus melanochir* and *Callicebus personatus* – are listed as vulnerable.

Most Brazilian sakis are relatively well protected by their ample ranges in western and northern Amazonia, and *P. pithecia*, *P. irrorata* and *P. monachus* are all considered to be of least concern by IUCN (2008). However, the endemic *P. albicans* is restricted to a comparatively small area between the lower Purus and Juruá rivers, which has led to its classification by IUCN as vulnerable (Veiga *et al.* 2008a). While the most recent IUCN listing does not assess subspecies separately, *P.i. vanzolinii* has a similarly small range, although the relative isolation of both taxa from major channels of human colonization probably minimizes risks at the present time.

By contrast, the distribution of the bearded sakis in the southern Amazon basin coincides with the "Arc of Deforestation". With a relatively small geographic range located in the most densely populated corner of the basin, *Chiropotes satanas* is the most threatened Amazonian pitheciid, and the only one to be classified as critically endangered by the IUCN (Veiga *et al.* 2008b). While they have much larger ranges, both *Chiropotes albinasus* and *Chiropotes utahickae* are also considered to be endangered (IUCN 2008), based primarily on projections of population decline.

The situation of the northern species is much less preoccupying, however, based principally on the size and isolation of their geographic ranges. Unfortunately, IUCN (2008) provides an assessment only for *C. chiropotes sensu* Hershkovitz (1985), but this is "least concern", and it seems to apply equally to the two currently recognized species, *C. chiropotes* and *C. sagultaus*.

In the case of the uacaris, only the widely distributed *Cacajao ouakary* is considered to be of "least concern", whereas the other three species have been classified as vulnerable by IUCN (2008). While little is known of *Cacajao ayresi* as yet (Boubli *et al.* 2008), its status is based on its apparently diminutive geographic range, which may be one of the smallest of any pitheciid. *Cacajao melanocephalus* also has a relatively reduced range, with a population thought to be at risk of decline.

The species *Cacajao calvus* presents a more complex scenario, considering factors such as the considerable variation in range size among the different subspecies. *Cacajao c. ucayalii*, for example, has a relatively ample distribution in the Ucayali River basin (albeit mostly outside Brazil), whereas *Cacajao c. novaesi* has a small range on the upper Juruá. *Cacajao c. calvus* also has a relatively small range, and has been considered endangered in the past (Groombridge 1994), although it is relatively well-protected in the present day (see below).

Protected reserves and distribution of pitheciid taxa

Brazilian pitheciids vary considerably in terms of both the size of their geographic distribution and the degree of protection of their populations. At one extreme, some species, such as *Cacajao ayresi* and *Callicebus coimbrai*, have unusually small ranges, with few or no protected areas. At the opposite extreme, species like *Callicebus moloch*, *Chiropotes sagulatus* and *Pithecia irrorata* have extremely ample ranges, which encompass substantial tracts of protected habitat, in some cases, equivalent to an area larger than that of the smallest geographic ranges. *Cacajao c. calvus* represents another extreme, where a large part of the subspecies' range is protected by the 1,120,000-ha Mamirauá Sustainable Development Reserve in western Amazonia. Until recently, in fact, it was thought that this reserve encompassed the whole of the geographic distribution of *C. c. calvus*, although Silva Jr. & Martins (1999) located populations further south of this area.

Between these extremes, each taxon obviously presents a uniquely different scenario, although there are three principal regional trends, in the Atlantic Forest, southern Amazon basin (east of the Madeira River), and northern Amazon basin, north of the Amazon–Madeira. The Atlantic Forest is characterized by high deforestation rates, significant levels of habitat fragmentation, and relatively small and isolated conservation units. Within this region, the least-threatened pitheciid – *Callicebus nigrifrons* – has the largest and best-protected geographic range, in conservation units such as those of the Serra do Mar complex, and even the Cantareira State Park in the city of São Paulo (Trevelin *et al.* 2007).

At the opposite extreme, the species with the smallest ranges – *Callicebus barbarabrownae* and *Callicebus coimbrai* – are the most endangered and the least well-protected. In fact, the critically endangered *C. barbarabrownae* does not appear to occur in any conservation unit whatsoever. While *C. coimbrai* is found only in the 600-ha Mata do Junco state wildlife refuge, it is protected, at least in theory, by the 2006 Atlantic Forest Law (Brasil 2006a), which prohibits deforestation of primary habitat within the ranges of endangered species.

The southern Amazon basin presents a more heterogeneous scenario, varying from vast tracts of pristine forest to areas of intense human colonization characterized by high deforestation rates, approaching those of the Atlantic Forest. However, while *Chiropotes satanas* faces the most critical levels of deforestation within its geographic range, east of the Tocantins River, it is also found in a few relatively large protected areas, such as the 340,000-ha Gurupi Biological Reserve in Maranhão, and the 570,000-ha Tucuruí Lake Environmental Protection Area. These areas alone compare more than favorably with most protected areas in the Atlantic Forest. The region's numerous Amerindian reservations may also play an important role in the conservation of this species (Lopes & Ferrari 2000), and there are a number of important areas on private properties, although in all these cases, hunting, squatting and selective logging represent significant pressures.

West of the Tocantins, *Chiropotes utahickae* faces only slightly lower levels of anthropogenic impact, although, once again, there are a number of relatively large protected areas, including the 330,000-ha Caxiuanã National Forest, the 100,000-ha Tapirapé Biological Reserve within the Carajás complex, and the western half of the Tucuruí Lake area. Amerindian reservations are also important here, especially as the local tribes such as the Kayapó and the Parakanã do not usually hunt primates (Ferrari *et al.* 1999).

Further west, where anthropogenic pressures are generally lower (except in some critical areas such as the state of Rondônia), the protected area network can be considered to be at least minimally satisfactory, especially as most conservation units cover a number of hundred thousand hectares. There is also a growing number of sustainable development reserves, which offer local pitheciid populations at least some degree of protection.

While there are some localized areas of impact, such as the state of Acre and the region of Manaus, the northern Amazon basin encompasses Brazil's largest and best-preserved tracts of native habitat. This is reflected in the conservation status of the region's pitheciids, none of which are currently classified in a category higher than vulnerable by IUCN (2008). *Chiropotes sagulatus* and *P. p. pithecia* are arguably among the best protected of Brazilian pitheciids, given the relative isolation of their geographic ranges, and the existence within this area of a number of substantial protected areas, including the 3,870,000-ha Montanhas do Tumucumaque National Park, one of the world's largest conservation units.

Whereas most other pitheciid taxa in this region are also found in substantial conservation units, such as the Pico da Neblina (*Cacajao melanocephalus*, *Callicebus lugens*) and Jaú (*Cacajao ouakary*, *P. p. chrysocephala*, *Callicebus torquatus*) national parks, and the Abunã Sustainable Development Reserve (*Cacajao c. rubicundus*, *Callicebus lucifer*), a number of species with restricted ranges may lack protection altogether, based on current knowledge. They include the poorly known *Cacajao ayresi* and *Callicebus stephennashi*, and *P. albicans*, which may, however, occur in the 290,000-ha Abufari Biological Reserve, pending confirmation.

Key studies

Ayres (1981, 1986) and Kinzey and Becker (1983) conducted the first detailed ecological studies of Brazilian pitheciids, which have served as the inspiration and methodological standard for most subsequent research. Despite these initial efforts, subsequent research has been rather patchy in its coverage of the different species, and few detailed ecological data are available for the majority of taxa.

However, two areas in the Brazilian Amazon have accumulated a certain tradition of pitheciid field studies. One is the area north of Manaus, centred on the Biological Dynamics of Forest Fragments Project (BDFF) near Manaus, which has a long history of individual projects, starting with that of Ayres (1981), and continuing to the present day (Boyle *et al.*, Chapter 24). While most research has concentrated on *Chiropotes chiropotes*, Setz (1993) also carried out one of the few studies of Brazilian sakis at this site.

The second area is the Tucuruí reservoir on the Tocantins River in southeastern Pará, where *Chiropotes satanas* and *Chiropotes utahickae* have been the subjects of a series of recent studies (Veiga 2006; Silva & Ferrari 2008; Santos *et al.*, Chapter 23). The effects of habitat fragmentation on populations of both these species have also been analyzed at a number of other sites in southeastern Amazonia (Ferrari *et al.* 1999; Lopes & Ferrari 2000; Port-Carvalho & Ferrari 2004).

Otherwise, ecological studies of Amazonian pitheciids are few and far between. Peres (1993) provides some data on *P. albicans*, the only Brazilian saki considered vulnerable to extinction, and Pinto (2008) has recently completed a study of *Chiropotes albinasus*. Boubli (1999; Boubli *et al.* 2008) and

Barnett *et al.* (2004) have also carried out extensive fieldwork on *Cacajao meanocephalus* and *Cacajao ouakary*, respectively.

The situation is little different in the Atlantic Forest, although studies are available for *Callicebus personatus* (Price & Piedade 2001), *Callicebus melanochir* (Müller 1996; Heiduck 2002) and *Callicebus nigrifrons* (Neri 1997; Trevelin *et al.* 2007). The most endangered Atlantic Forest titis are also the least well-known, although recent surveys by Jerusalinsky *et al.* (2006) and Printes (2007) have provided important data for the assessment of conservation status of *Callicebus coimbrai* and *Callicebus barbarabrownae*, respectively (see Printes *et al.*, Chapter 5). However, a number of more detailed studies of *C. coimbrai* are currently under way (Souza-Alves 2008; Chagas 2009).

Conservation programs

Brazil has relatively effective environmental legislation and a well-organized system of conservation units (Brasil 2000), both of which potentially benefit all the country's pitheciid species to some degree. The federal environment institute's Primate Protection Centre (ICMBIO/CPB) represents an extremely important initiative for all Brazilian primates, in particular endangered species. Endangered pitheciids are now included in two national committees, established to coordinate research and conservation initiatives in Brazilian Amazonia (Brasil 2005), and the northern Atlantic Forest and Caatinga (Brasil 2006b). While not specific to Brazil, the Pitheciine Action Group (www.pitheciineactiongroup.org) has a potentially important role in the coordination of conservation efforts.

In general, however, most initiatives are centered on individual studies, with specific objectives and limited long-term perspectives. Even though the pioneering efforts of the late Márcio Ayres resulted in the establishment of the Mamirauá reserve, which protects a significant proportion of the population of *Cacajao c. calvus*, there has been no further research into the ecology of this uacari at the site. Similarly, while fieldwork at BDFF and Tucuruí has also provided important data, they have not resulted in the development of specific pitheciid-oriented conservation strategies at these sites. Nevertheless, the consolidation of an effective network of protected areas, reinforced by recent changes in legislation and the establishment of specific entities such as the ICMBIO/CPB represent important advances, which will provide a sound baseline for the development of specific conservation programs.

References

Ayres, J.M. (1981). Observações sobre a ecologia e o comportamento dos cuxius (*Chiropotes albinasus* e *Chiropotes satanas*, Cebidae: Primates). Unpublished Master's dissertation, Universidade do Amazonas.

Ayres, J.M.C. (1986). Uakaris and Amazonian flooded forests. Unpublished PhD thesis, Cambridge University.

Barnett, A.A., Castilho, C.V., Shapley, R.L., *et al.* (2004). Diet, habitat selection and natural history of *Cacajao melanocephalus ouakary* in Jaú National Park, Brazil. *International Journal of Primatology*, **26**, 949–969.

Bonvicino, C.R., Boubli, J.P., Otazu, I.B., *et al.* (2003). Morphologic, karyotypic, and molecular evidence of a new form of *Chiropotes* (Primates, Pitheciinae). *American Journal of Primatology*, **61**, 123–133.

Boubli, J.-P. (1999). Feeding ecology of black-headed uacaris (*Cacajao melanocephalus melanocephalus*) in Pico da Neblina National Park, Brazil. *International Journal of Primatology*, **20**, 719–749.

Boubli, J.-P., da Silva, M.N.F., Amado, M.V., *et al.* (2008). A taxonomic reassessment of *Cacajao melanocephalus* Humboldt (1811), with the description of two new species. *International Journal of Primatology*, **29**, 723–741.

Brasil (2000). *Lei n. 9.985, de 18 de julho de 2000: Sistema Nacional de Unidades de Conservação*. Brasília: Federal Government.

Brasil (2005). Portaria n. 82, de 29 de novembro de 2005: Comitê primatas amazônicos. Brasília, IBAMA. *Diário Oficial da União – Seção 1*, **229**, 126–127.

Brasil (2006a). *Lei n. 11.428, de 22 de dezembro de 2006: Lei da Mata Atlântica*. Brasília: Federal Government.

Brasil (2006b). Portaria n. 26, de 9 de março de 2006: Comitê primatas do norte da Mata Atlântica e Caatinga. *Diário Oficial da União – Seção 1*, **48**, 78.

Chagas, R.R.D. (2009). Levantamento das populações de *Callicebus coimbrai* Kobayashi & Langguth, 1999 em fragmentos de Mata Atlântica no Sul do Estado de Sergipe, Brasil. Unpublished Master's dissertation, Universidade Federal de Sergipe.

Ferrari, S.F., Emidio-Silva, C., Lopes, M.A., *et al.* (1999). Bearded sakis in south-eastern Amazonia – back from the brink? *Oryx*, **33**, 346–351.

Groombridge, B. (1994). *1994 IUCN Red List Of Threatened Animals*. Cambridge: IUCN.

Groves, C.P. (2005). Order Primates. In *Mammal Species of the World: A Taxonomic and Geographic Reference, 3rd edition*, ed. D.E. Wilson & D.M. Reeder. Baltimore, MD: The Johns Hopkins University Press, pp. 111–184.

Heiduck, S. (2002). The use of disturbed and undisturbed forest by masked titi monkeys *Callicebus personatus melanochir* is proportional to food availability. *Oryx*, **36**, 133–139.

Hershkovitz, P. (1985). A preliminary taxonomic review of the South American bearded saki monkeys genus *Chiropotes* (Cebidae, Platyrrhini), with description of a new subspecies. *Fieldiana Zoology*, **27**, 1–45.

Hershkovitz, P. (1987). Uakaris, New World monkeys of the genus *Cacajao* (Cebidae, Platyrrhini): a preliminary review with the description of a new subspecies. *American Journal of Primatology*, **12**, 1–57.

International Commission on Zoological Nomenclature (1999). *International Code of Zoological Nomenclature*. London: British Museum of Natural History. http://www.iczn.org/iczn/index.jsp

IUCN (2008). *2008 IUCN Red List of Threatened Species*. Gland: IUCN, http://www.iucnredlist.org

Jerusalinsky, L., Oliveira, M.M., Pereira, R.F., *et al.* (2006). Preliminary evaluation of the conservation status of *Callicebus coimbrai* Kobayashi & Langguth, 1999 in the Brazilian state of Sergipe. *Primate Conservation*, **21**, 25–32.

Kinzey, W.G. & Becker, M. (1983). Activity pattern of the masked titi monkey, *Callicebus personatus*. *Primates*, **24**, 337–343.

Lopes, M.A. & Ferrari, S.F. (2000). Effects of human colonization on the abundance and diversity of mammals in eastern Brazilian Amazonia. *Conservation Biology*, **14**, 1658–1665.

Müller, K.-H. (1996). Diet and feeding ecology of masked titis (*Callicebus personatus*). In *Adaptive Radiations of Neotropical Primates*, ed. M.A. Norconk, A.L. Rosenberger & P.A. Garber. New York, NY: Plenum Press, pp. 383–401.

Neri, F.M. (1997). Manejo de *Callicebus personatus*, Geoffroy 1812, resgatados: uma tentativa de reintrodução e estudos ecológicos de um grupo silvestre na Reserva do Patrimônio Natural Galheiro – Minas Gerais. Unpublished Master's dissertation, Universidade Federal de Minas Gerais.

Peres, C.A. (1993). Notes on the ecology of buffy saki monkeys (*Pithecia albicans*, Gray, 1860): a canopy seed predator. *American Journal of Primatology*, **31**, 129–140.

Pinto, L.P. (2008). Ecologia alimentar de um grupo de cuxiús-de-nariz-vermelho *Chiropotes albinasus* (Primates: Pitheciidae) na Floresta Nacional do Tapajós, Pará. Unpublished PhD thesis, Universidade Estadual de Campinas.

Port-Carvalho, M. & Ferrari, S.F. (2004). Occurrence and diet of the black bearded saki (*Chiropotes satanas satanas*) in the fragmented landscape of western Maranhão, Brazil. *Neotropical Primates*, **12**, 17–21.

Price, E.C. & Piedade, H.M. (2001). Ranging behavior and intraspecific relationships of masked titi monkeys (*Callicebus personatus personatus*). *American Journal of Primatology*, **53**, 87–92.

Printes, R.C. (2007). Urban Monkeys Program: fourteen years of research and conservation of the brown howler (Alouatta clamitans, Cabrera, 1940) in Porto Alegre. [Abstract] CD-ROM De Resumos XII Congresso Brasileiro de Primatologia, ed. F.R. Melo, A. Hirsch, C.G. Costa, L.G. Dias, I.M.C. Mourthe, F. P. Tabacow, L.M. Scoss. 263 pages.

Setz, E.Z.F. (1993). Ecologia alimentar de um grupo de parauacus (*Pithecia pithecia chrysocephala*) em um fragmento florestal na Amazônia central. Unpublished PhD thesis, Universidade Estadual de Campinas.

Silva, S.S.B. & Ferrari, S.F. (2008). Behavior patterns of southern bearded sakis (*Chiropotes satanas*) in the fragmented landscape of eastern Brazilian Amazonia. *American Journal of Primatology*, **70**, 1–7.

Silva Jr., J.S. & Figueiredo, W.M.B. (2002). Revisão sistemática dos cuxiús, gênero *Chiropotes* Lesson, 1840 (Primates, Pitheciidae). *Livro de resumos do X Congresso Brasileiro de Primatologia*, 21.

Silva Jr., J.S. & Martins, E.S. (1999). On a new white bald uakari population in southwestern Brazilian Amazonia. *Neotropical Primates*, **7**, 119–121.

Souza-Alves, J.P. (2008). Comportamento de forrageio e dieta de *Callicebus coimbrai* (Kobayashi & Langguth, 1999) no Sul de Sergipe, Brasil. Unpublished Master's project, Universidade Federal de Sergipe.

Trevelin, L.C., Port-Carvalho, M., Silveira, M., *et al.* (2007). Abundance, habitat use and diet of *Callicebus nigrifrons* Spix (Primates, Pitheciidae) in Cantareira State Park, Sao Paulo, Brazil. *Revista Brasileira de Zoología*, **24**, 1071–1077.

van Roosmalen, M.G.M., van Roosmalen, T. & Mittermeier, R.A. (2002). A taxonomic review of the titi monkeys, genus *Callicebus* Thomas, 1903, with the description of two new species, *Callicebus bernhardi* and *Callicebus stephennashi*, from Brazilian Amazonia. *Neotropical Primates*, **10** (Suppl.), 1–52.

Veiga, L.M. (2006). Ecologia e organização social de cuxiús, *Chiropotes satanas satanas*, em Tucuruí, Pará. PhD thesis, Universidade Federal do Pará.

Veiga, L.M., Mittermeier R.A. & Marsh, L. (2008a). *Pithecia albicans*. In *2008 IUCN Red List of Threatened Species*, ed. IUCN. Gland: IUCN, http://www.iucnredlist.org/details/41567.

Veiga, L.M., Silva Jr., J.S., Ferrari, S.F., *et al.* (2008b). *Chiropotes satanas*. In *2008 IUCN Red List of Threatened Species*, ed. IUCN. Gland: IUCN, http://www.iucnredlist.org/details/39956.

Appendix C: Conservation Fact Sheet: Ecuador

Stella de la Torre

Taxon list

Coppery Titi monkey (*Callicebus discolor*), Ecuador: cotoncillo rojo, cotoncillo colorado, songo songo, sucali (Quichua), ma hua'o (Secoya, Siona).

Yellow-handed Titi monkey (*Callicebus lucifer*), Ecuador: cotoncillo negro, cotoncillo de manos amarillas, toconcillo, viudita, yana sucali (Quichua), nea hua'o (Secoya, Siona).

Monk Saki (*Pithecia monachus*), Ecuador: mono volador, parahuaque negro, parahuaco (Quichua), hua'o su'tu (Secoya, Siona).

Equatorial Saki (*Pithecia aequatorialis*), Ecuador: mono volador, parahuaque negro, parahuaco (Quichua).

Synonyms

Callicebus discolor (van Roosmalen *et al.* 2002) called *Callicebus cupreus discolor* (Groves 2001).

Callicebus lucifer (van Roosmalen *et al.* 2002) called *Callicebus torquatus lucifer* (Groves 2001).

No synonyms for the *Pithecia* species.

Protected reserves and distribution of Pitheciid taxa

Callicebus discolor occurs in the eastern lowlands of Ecuador, between 200 and 500 m a.s.l. in areas with dense vegetation, usually close to rivers and creeks. Populations of this species are found in the Sumaco-Napo Galeras and Yasuni National Parks, and in the Limoncocha, Cuyabeno and Cayambe-Coca Reserves.

Callicebus lucifer occurs in the eastern lowlands below 350 m a.s.l., usually in non-flooded forests, north of the Aguarico River. Populations of this species are found in the Cuyabeno Reserve.

Pithecia monachus occurs in the eastern lowlands below 500 m a.s.l., both in flooded and non-flooded forests. Populations of this species are found in the Yasuni National Park and the Cuyabeno Reserve.

Pithecia aequatorialis occurs in the eastern lowlands below 300 m a.s.l., south of the Napo River. There are few confirmed records of this species in flooded and non-flooded forests of the Yasuni National Park.

Conservation status

Callicebus discolor, *C. lucifer* and *Pithecia monachus* are considered Least Concern. *Pithecia aequatorialis* is in the Ecuadorian Data Deficient category because of the scarcity of records.

Key studies

Callicebus discolor has been studied since the late 1990s at two sites of the Yasuni Biosphere Reserve (Carrillo-Bilbao *et al.* 2005; Di Fiore & Schwindt 2004; Sendall *et al.* 2006, 2007; Youlatos & Rivera 1999).

Callicebus lucifer was studied in the early 1990s in the Cuyabeno Reserve (Campos *et al.* 1992; de Vries *et al.* 1993).

Pithecia monachus was studied in the early 1990s in the Cuyabeno Reserve (de Vries *et al.* 1993).

Pithecia aequatorialis has been studied in Yasuni since the early 2000s (Di Fiore *et al.* 2007; Schmitt *et al.* 2005). The taxonomy of the Ecuadorian species of the genus *Pithecia* has been revised by Marsh (unpubl. data).

List of conservation programs

There are no specific conservation programs for Ecuadorian Pitheciids and no studies in population ecology have been carried out yet for any of the species. The conservation of Ecuadorian Pitheciids in protected areas is a result of the politics and regulations for the protected areas. An evaluation of the effectiveness of protected areas on the conservation of these, and all, Ecuadorian primate species is needed.

References

Campos, F., de la Torre, S. & de Vries, T. (1992). Territorial behaviour and home range establishment of *Callicebus torquatus* (Primates: Cebidae) in Amazonian Ecuador. In Abstracts of the XIVth Congress of the International Primatological Society, Strasbourg.

Carrillo-Bilbao, G.A., Di Fiore, A. & Fernandez-Duque, E. (2005). Dieta, forrajeo y presupuesto de tiempo en cotoncillos (*Callicebus discolor*) del Parque Nacional Yasuní en la Amazonía Ecuatoriana. *Neotropical Primates*, **13**, 7–11.

de Vries, T., Campos, F., de la Torre, S., *et al.* (1993). La investigación para la

Evolutionary Biology and Conservation of Titis, Sakis and Uacaris, eds. Liza M. Veiga, Adrian A. Barnett, Stephen F. Ferrari and Marilyn A. Norconk. Published by Cambridge University Press.

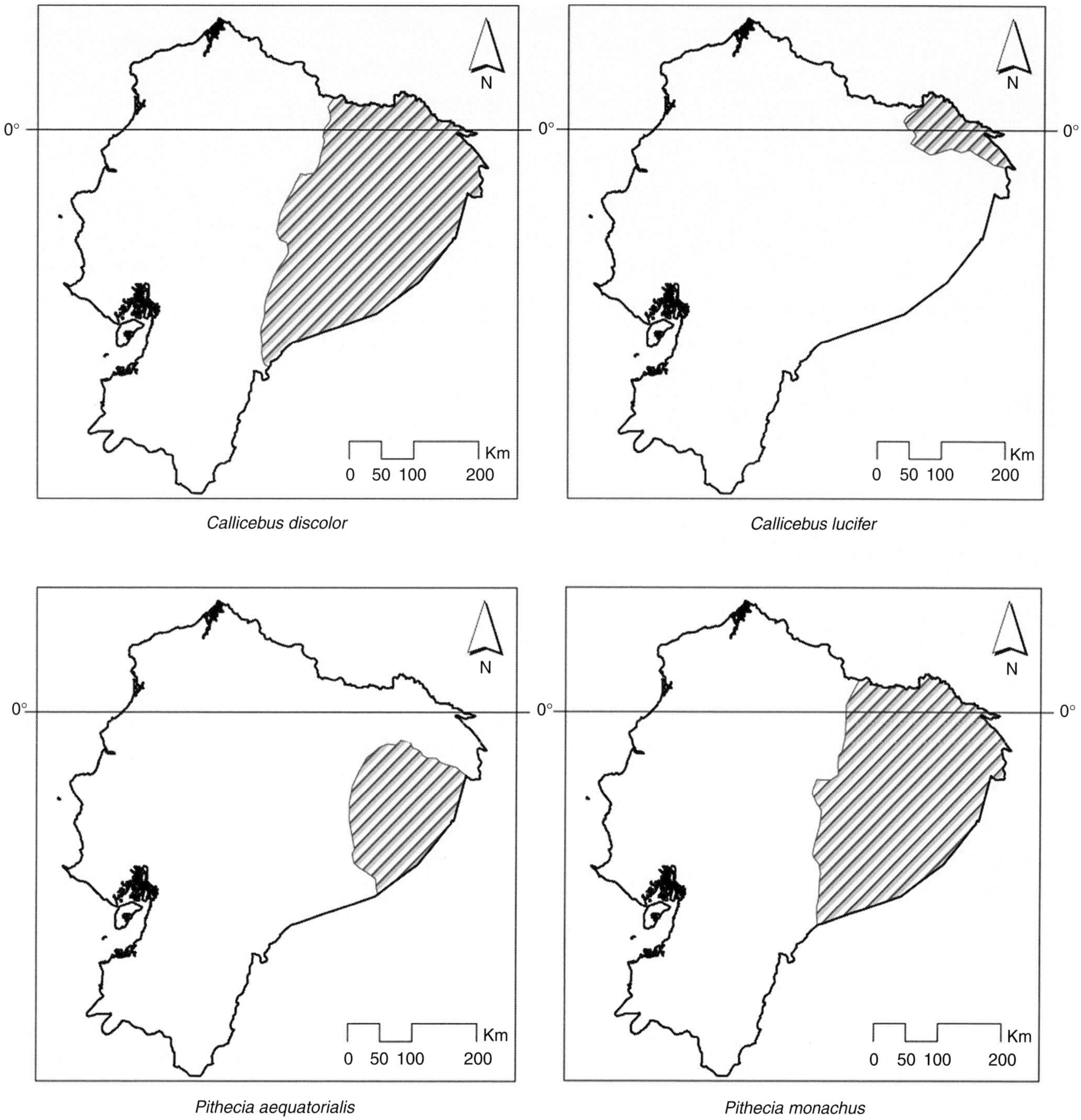

Figure C.1 Distribution maps.

conservación de la diversidad biológica en el Ecuador. In: *La investigación para la conservación de la diversidad biológica en el Ecuador*, ed. P.A. Mena & L. Suárez. Quito: Ecociencia, pp. 167–221.

Di Fiore, A. & Schwindt, D.M. (2004). A preliminary study of social behavior and pair-bonding in wild titi monkeys (*Callicebus discolor*) in Amazonian Ecuador. 73rd Annual Meeting of the American Association of Physical Anthropologists. *American Journal of Physical Anthropology*, **38**(*Suppl.*), 87.

Di Fiore, A., Fernandez-Duque, E. & Hurst, D. (2007). Adult male replacement in socially monogamous equatorial saki monkeys (*Pithecia aequatorialis*). *Folia Primatologica*, **78**, 88–98 .

Groves, C. (2001). *Primate Taxonomy*. Washington, DC: Smithsonian Institution Press.

Ministerio del Ambiente del Ecuador. Sistema Nacional de Áreas Protegidas. www.ambiente.gov.ec (accessed September 15, 2008).

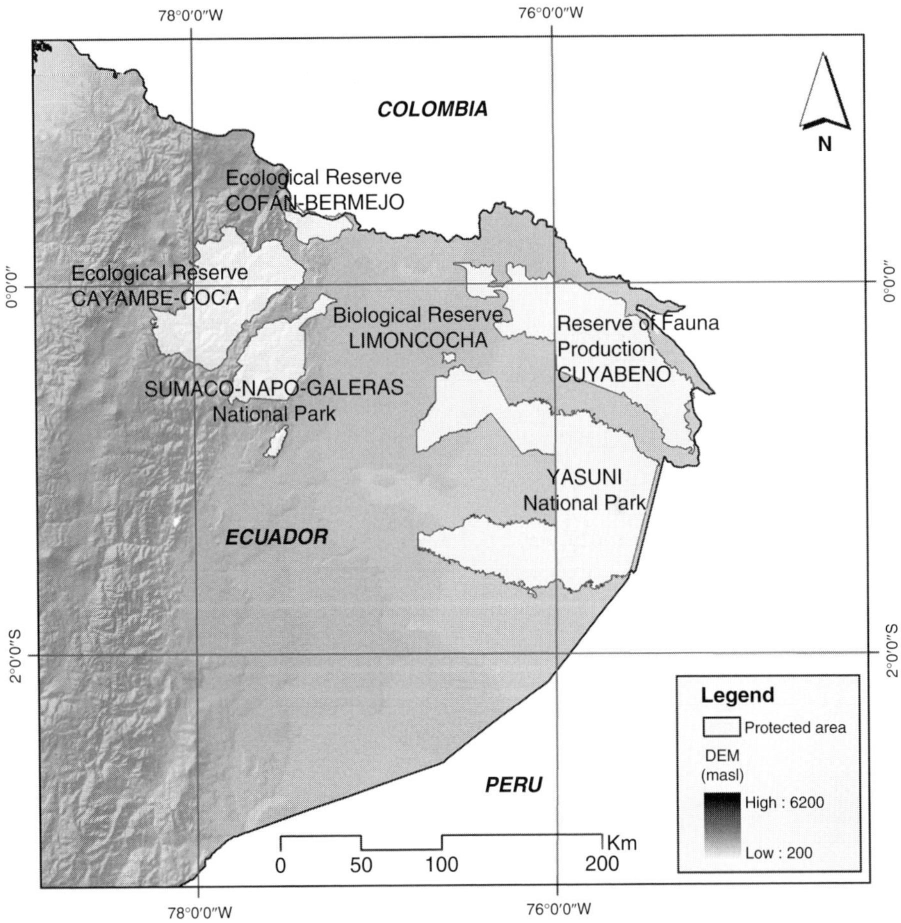

Figure C.2 Protected areas in eastern Ecuador (Ministerio del Ambiente del Ecuador www.ambiente.gov.ec)

Schmitt, C., Di Fiore, A., Hurst, D., *et al.* (2005). Maternally-initiated babysitting by wild adult male equatorial sakis (*Pithecia aequatorialis*) in Yasuni National Park, Ecuador. In 1st Annual NYCEP Symposium "Monkeys: Old and New".

Sendall, C., Fernandez-Duque, E. & Di Fiore, A. (2006). A brief investigation into the maintenance of proximity during estrus by titi monkeys (*Callicebus discolor*). In 34th Annual Meeting of the Canadian Association for Physical Anthropology.

Sendall, C., Fernandez-Duque, E. & Di Fiore, A. (2007). A preliminary study of mate-guarding in wild titi monkeys (*Callicebus discolor*). 76th Annual Meeting of the American Association of Physical Anthropologists.*American Journal of Physical Anthropology*, **44**(*Suppl.*), 214–215.

van Roosmalen, M.G., van Roosmalen, T. & Mittermeier, R.A. (2002). A taxonomic review of the titi monkeys, genus *Callicebus* Thomas, 1903, with the description of two new species, *Callicebus bernhardi* and *Callicebus stephennashi*, from Brazilian Amazonia. *Neotropical Primates*, **10**(Suppl.), 1–52.

Youlatos, D. & Rivera, W.P. (1999). Preliminary observations on the songo songo (dusky titi monkey: *Callicebus moloch*) of northeastern Ecuador. *Neotropical Primates*, **7**, 45–46.

Appendix D: Conservation Fact Sheet: French Guiana

Jean-Christophe Vié

Taxon list

White-faced saki (*Pithecia pithecia pithecia*), ver: saki à face pâle, maman guinan; and brown bearded saki (*Chiropotes sagulatus*), ver. saki satan.

Synonyms

Chiropotes sagulatus was previously named *C. satanas*. Abnormal coloration in male *P. pithecia* has led to the misidentification of some specimens. One adult male *P. pithecia* captured along the Sinnamary River possessed the female coloration.

Protected areas and distribution of pitheciine taxa

White-faced sakis occur throughout the country and are found in most forested protected areas. Brown-bearded sakis have only been observed in the south of the country, which has recently been declared a National Park. Several recent sightings have confirmed the presence of the species as far north as 03°03′N (P. Gaucher, pers. commun.). It is likely that its entire range in French Guiana lies within the Park. Both species are protected by law, but poaching is widespread including in protected areas mostly linked to illegal gold mining.

Conservation status

Pithecia pithecia is widespread. It is currently listed as Least Concern on the IUCN Red List of Threatened Species. No regional evaluation has been conducted for French Guiana, but it would certainly qualify as Least Concern given the extent of remaining habitat and its cryptic behavior, which makes it a difficult species to hunt. The taxon *Chiropotes sagulatus* is not recognized currently by the IUCN. Bearded sakis in French Guiana are classified as Least Concern, but information on their status in French Guiana is too scarce to evaluate trends in the country.

Key studies

There is no specific study on *C. sagulatus* in French Guiana and a limited number of reports. Vié (1998; Vié *et al.* 2001) conducted the first complete study of *P. pithecia* in the Petit Saut area, in the basin of the Sinnamary River. This study covered utilization of space, activity budget, diet and responses to anthropogenic disturbances. White-faced sakis have been studied in undisturbed habitat and in a forest fragment, which revealed changes in ecology and behavior as a response to the modification of the environment. Six animals were captured during the flooding of the Petit Saut dam. Blood parameters (de Thoisy *et al.* 2001), immobilization, body measurements, abnormal male coloration and behavior after translocation (Vié 1998) were also documented.

List of conservation programs

Various surveys have been conducted in protected areas (Trinité, National Park, Nouragues, Kaw) but were limited to reporting the occurrence of pitheciines. There is no specific project focusing on sakis in French Guiana. Over the past decade, the NGO Kwata has been conducting surveys in various parts of the country, looking at the impact of human disturbances (such as logging and hunting) on wildlife with the abundance of primates as a key parameter. The abundance of *P. pithecia* has been documented at a limited number of sites (de Thoisy *et al.* 2005).

References

de Thoisy B., Vogel, I., Reynes, J.-M., *et al.* (2001). Health evaluation of translocated free-ranging primates in French Guiana. *American Journal of Primatology*, **54**(1), 1–16.

de Thoisy, B., Renoux, F. & Julliot, C. (2005). Hunting in northern French Guiana and its impact on primate communities. *Oryx*, **39**(2), 149–157.

Vié, J.-C. 1998. Les effets d'une perturbation majeure de l'habitat sur deux espèces de primates en Guyane Française: translocation de singes hurleurs roux (*Alouatta seniculus*), et translocation et insularisation de sakis à face pâle (*Pithecia pithecia*). PhD thesis, Université Montpellier II, France.

Vié J.-C., Richard-Hansen, C. & Fournier-Chambrillon, C. (2001). Use of space, demography and activity patterns of white-faced sakis (*Pithecia pithecia*) in French Guiana. *American Journal of Primatology*, **55**, 203–221.

Evolutionary Biology and Conservation of Titis, Sakis and Uacaris, eds. Liza M. Veiga, Adrian A. Barnett, Stephen F. Ferrari and Marilyn A. Norconk. Published by Cambridge University Press.

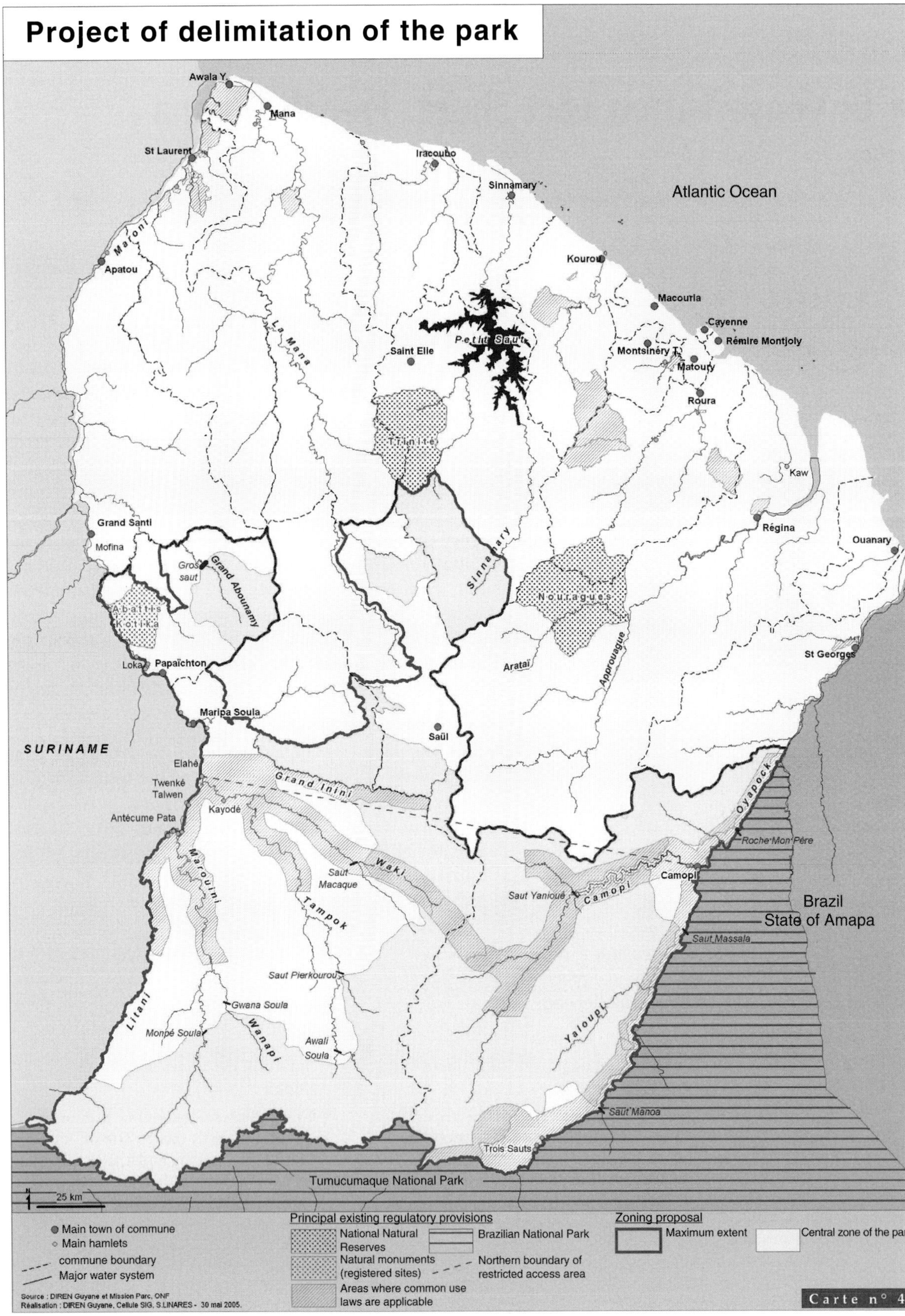

Figure D.1 Distribution of protected areas and habitat types in French Guiana.

Appendix E: Conservation Fact Sheet: Suriname

Marilyn A. Norconk

Taxon list

White-faced saki, wanaku (Surinaams), witkop-aap (Dutch) (*Pithecia pithecia pithecia* Linnaeus) and northern bearded sakis or Guianan bearded saki, bisa (Surinaams), satans-aap (Dutch) (*Chiropotes satanas chiropotes* Humboldt).

Synonyms

Chiropotes satanas chiropotes Humboldt a.k.a. *Chiropotes sagulata* Traill (Hershkovitz 1985), also known as *Chiropotes sagulatus* (Silva Jr. and Figueiredo, 2002). No synonyms for *P. pithecia pithecia* Linnaeus.

Protected reserves and distribution of pitheciine taxa

Both saki species are found in the Central Suriname Nature Reserve (CSNR) (4°0′N, 56°30′W), a 1.6 million ha reserve in central Suriname (Figure E.1). The CSNR was declared a UNESCO World Heritage Site in 2000. It is accessible only by air and river, and currently has no hunting, habitat threats, or large human population centers within its boundaries. Both species of sakis are also present at Brownsberg Nature Park (5° 01′N, 55°34′W), an 8400-ha area in eastern Suriname. The Brownsberg (500 m a.s.l.) was designated as a nature park in 1969 and is a major tourist attraction for both Surinamers and international visitors. The park is accessible by road from the capital, Paramaribo, and forms the western border of Brokopondo Lake. The long, narrow park has been under considerable stress from illegal gold mining since 2003 and is now experiencing an increase in hunting.

According to Mittermeier (1977, p. 170), who conducted the only widespread primate survey in Suriname, "*Pithecia* is found throughout the country, but occurs at low densities almost everywhere." There are few density estimates of sakis, but Norconk *et al.* (2003) found densities of known groups to be relatively high at Brownsberg Nature Park (e.g. group encounter rate/10 km surveyed = 6.1). While this could be a local artifact, Ford and Boinski (2007) and Boinski (unpubl.) found a relatively high density of saki remains at a harpy eagle nest that also suggested that actual abundance is higher than the calculated abundance assessed through censuses.

Bearded sakis are found primarily in the interior of Suriname and rarely in the old coastal plain (Mittermeier 1977) (Figure E.1). The vegetation of Suriname ranges in the north from coastal mangroves and a geologically young coastal plain replaced by an older coastal plain, deposits dating to the Cenozoic (Hammond 2005). Most of the country south of the coastal plain is referred to as "interior" and consists of a variety of forest types on the ancient Guiana Shield. Much of the interior of Suriname still supports pristine forests.

Conservation status

Pithecia pithecia is listed as Least Concern (or LC) by the IUCN. The northern bearded saki is not rated by the IUCN (http://www.iucnredlist.org).

Key studies

Mittermeier's (1977) synecology study was the first primate-related study and remains the most comprehensive assessment of distribution and conservation of monkeys in the Suriname community. It was followed by three additional studies published in the 1980s and early 1990s (Fleagle & Mittermeier 1980; van Roosmalen *et al.* 1988; Kinzey & Norconk 1990). Most of this work was conducted at Raleighvallen-Voltzberg (currently CSNR); however, there is no current research under way on sakis in the CSNR. Both species are being studied currently at the Brownsberg (Norconk *et al.* 2003; Gregory & Norconk, Chapter 28; Thompson and Norconk, Chapter 27; Norconk and Setz, Chapter 25; Gregory 2011; Thompson & Norconk 2011; Thompson *et al.* 2011; Smith & Thompson 2011). In addition, sakis were mentioned in two recent wildlife assessment surveys, both conducted at Brownsberg Nature Park (Fitzgerald *et al.* 2002; Lim *et al.* 2005). Boinski (2002) published a very accessible popular book recently, reviewing behavioral and ecological studies of all eight primate species.

List of conservation programs

The Suriname Forest Service (LBB) is responsible for 11 nature reserves (including CSNR), 1 nature park (Brownsberg), 4 multiple-use areas, and 2 new nature reserves proposed

Evolutionary Biology and Conservation of Titis, Sakis and Uacaris, eds. Liza M. Veiga, Adrian A. Barnett, Stephen F. Ferrari and Marilyn A. Norconk. Published by Cambridge University Press.

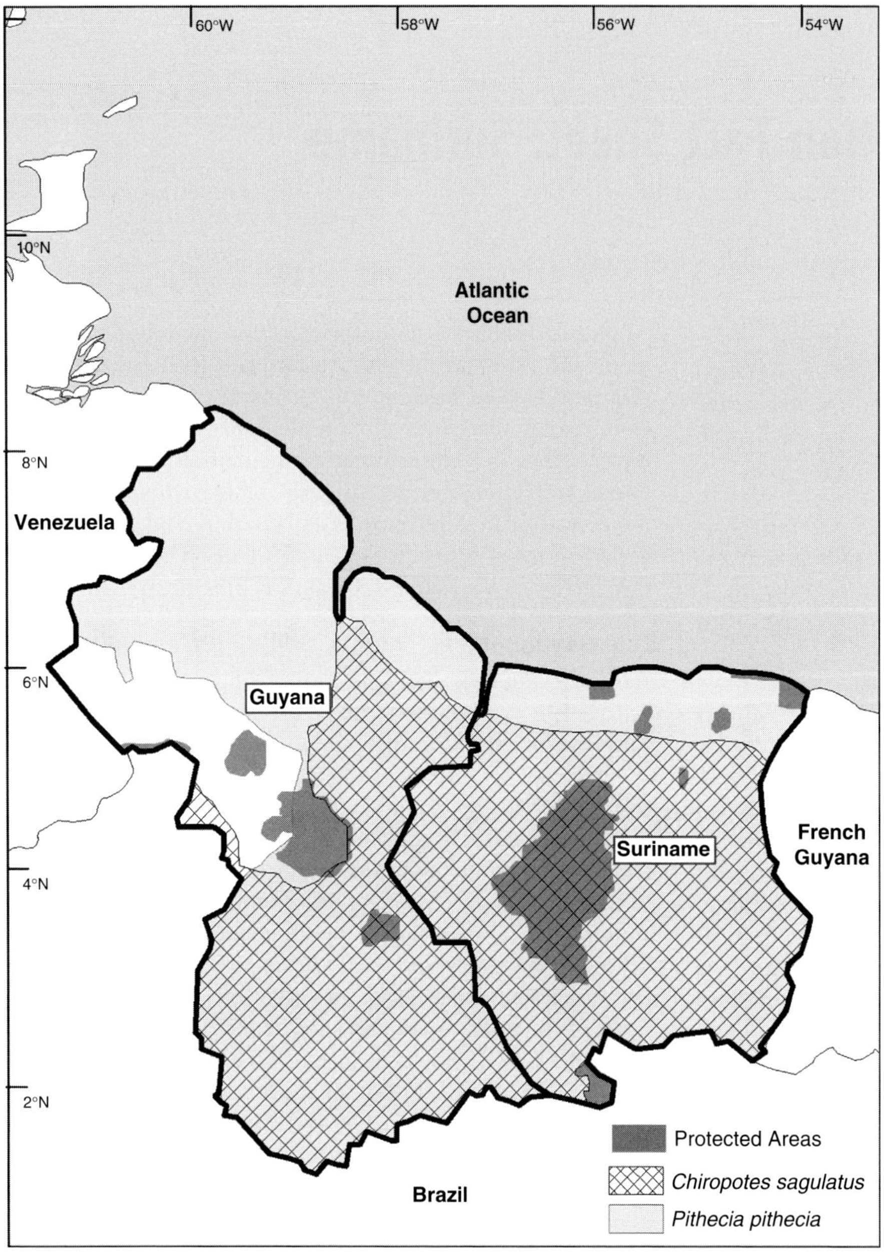

Figure E.1 Widespread distribution of *Chiropotes sagulatus* and *Pithecia pithecia* in Suriname and Guyana.

(Foundation for Nature Conservation in Suriname website: http://www.stinasu.com/nature_reserves.html). Most of these protected areas are related to sea turtle, giant otter, bird or fish conservation. Sakis may exist in more areas than the two highlighted above, but no assessment on their presence or abundance has been made. Conservation International is active in the CSNR, although their projects are not specifically related to saki conservation.

References

Boinski, S. (2002). *De apen van Suriname/ The monkeys of Suriname*. Paramaribo, Suriname: Stichting Natuurbehoud Suriname.

Fitzgerald, K.A., De Dijn, B.P.E. & Mitro, S. (2002). *Brownsberg Nature Park: Ecological Research and Monitoring Program 2001–2006*. Paramaribo, Suriname: STINASU – Foundation for Nature Conservation in Suriname.

Fleagle, J.G. & Mittermeier, R.A. (1980). Locomotor behavior, body size, and comparative ecology of seven Surinam monkeys. *Amercan Journal of Physical Anthropology*, **52**, 301–314.

Ford, S.M. & Boinski, S. (2007). Primate predation by harpy eagles in the Central Suriname Nature Reserve. *American Journal of Physical Anthropology*, **Suppl. 44**, 109.

Gregory, L.T. (2011). Socioecology of the Guianan bearded saki, *Chiropotes sagulatus*. Unpublished doctoral dissertation, Kent State University, Kent OH, USA.

Hammond, D.S. (2005). Biophysical features of the Guiana Shield. In *Tropical Forests of the Guiana Shield: Ancient Forests in a Modern World*, ed. D.S. Hammond. Wallingford: CAB International Publishing, pp. 15–194.

Hershkovitz, P. (1985). A preliminary taxonomic review of the South American Bearded Saki Monkeys genus *Chiropotes* (Cebidae, Platyrrhini), with the description of a new subspecies. *Fieldiana*, **27**[1363], 1–46. Chicago, IL: Field Museum of Natural History.

Kinzey, W.G. & Norconk, M.A. (1990). Hardness as a basis of fruit choice in two sympatric primates. *American Journal of Physical Anthropology*, **81**, 5–15.

Lim, B.K., Engstrom, M.D., Genoways, H.H., *et al.* (2005). Results of the Alcoa Foundation Suriname Expeditions. XIV. Mammals of the Brownsberg Nature Park, Suriname. *Annals of the Carnegie Museum*, **74**, 225–274.

Mittermeier, R.A. (1977). Distribution, synecology and conservation of Surinam monkeys. PhD, Harvard University.

Mittermeier, R.A., Rylands, A.B., van Roosmalen, M.G.M., *et al.* (2008). *Monkeys of the Guianas: Guyana, Suriname, French Guiana*. Washington, DC: Conservation International Tropical Pocket Guide Series.

Norconk, M.A., Raghanti, M.A., Martin, S.K., *et al.* (2003). Primates of Brownsberg Natuurpark, Suriname, with particular attention to the Pitheciins. *Neotropical Primates*, **11**, 94–100.

Silva Jr., J.S. & Figueiredo, W.M.B. (2002). Revisão sistemática dos cuxiús, gênero *Chiropotes* Lesson, 1840 (Primates Pithecidae). *Livro de Resumos do X Congresso da Socoiedade Brasileira de Primatologia, Amazônia – A Última Fronteira*, p. 21.

Smith, H.M. & Thompson, C.L. (2011). Observations of hand preference in wild groups of white-faced sakis (*Pithecia pithecia*) in Suriname. *American Journal of Primatology*, **73**, 655–664.

Thompson, C.L. & Norconk, M.A. (2011). Within-group social bonds in white-faced saki monkeys (*Pithecia pithecia*) display male–female pair preference. *American Journal of Primatology*, **73**, 1051–1061.

Thompson, C.L., Whitten, P.L. & Norconk, M.A. (2011). Can male white-faced saki monkeys (*Pithecia pithecia*) detect female reproductive state? *Behaviour*, **148**, 1313–1331.

van Roosmalen, M.G.M., Mittermeier, R.A. & Fleagle, J.G. (1988). Diet of the northern bearded saki (*Chiropotes satanas chiropotes*): a neotropical seed predator. *American Journal of Primatology*, **14**, 11–35.

Appendix F: Conservation Fact Sheet: Venezuela

Bernardo Urbani & Carlos Portillo-Quintero

Taxon list

Black-headed uacari (*Cacajao hosomi*), Ven: mono chucuto, mono rabo corto, mono feo, mono rabón, mono negro. Collared titi (*Callicebus lugens*), Ven: viudita, mono viudita, viuda de luto, mono tití. Bearded saki (*Chiropotes sagulatus*), Ven: mono capuchino lomo amarillo, mono barbudo, mono capuchino del Orinoco, mono volador. White-faced saki (*Pithecia pithecia pithecia*), Ven: mono viudo, viejito, mono viuda, mono negro cariblanco, viudito, mono barbudo cariblanco (Bodini 1983; Linares 1998; Urbani 2006).

Synonyms

Cacajao hosomi = *C. melanocephalus melanocephalus*. *Callicebus lugens* often referred to as *Callicebus torquatus lugens* (taxonomic status under discussion). *Chiropotes sagulatus* = *C. satanas chiropotes* has been referred for some Venezuelan specimens as *Pithecia chiropotes*. No synonym for *P. p. pithecia* (Bodini 1983; Bodini & Pérez-Hernández 1987; Linares 1998).

Protected reserves and distribution of pitheciid taxa

Cacajao (~130 m a.s.l.): along the Casiquiare branch and the Negro, Atabapo, Pasimoni, Yatua, Baria, Siapa, Guainía and southern upper Orinoco rivers basins. *Callicebus* (100–650 m a.s.l.): west of Paragua River and along Caura and Ventuari rivers, the Casiquiare branch and the middle and upper Orinoco River. *Chiropotes* (25–560 m a.s.l.): west of Caroní River and overlapping the channel and rivers indicated for *Callicebus*. *Pithecia* (15–650 m a.s.l.): south of Orinoco River delta (Grande River), along Cuyuní and Aguirre rivers, and east of lower Caroní River (Bodini 1983; Bodini & Pérez-Hernández 1987; Linares 1998; Aymard, pers. commun., 2006; Urbani 2006). Access to the 29 legally protected areas located in the Venezuelan Guayana are highly restricted for conservation purposes. Nonetheless, most of the known range of the pitheciids remains unprotected (59% for *Cacajao*, 75% for *Callicebus*, 75% for *Chiropotes* and 100% for *Pithecia*). Table F.1 and Figures F.1 and F.2 provide additional information on pitheciid distribution and habitat use as well as vegetation types and pitheciid taxa.

Conservation status

Cacajao and *Chiropotes* are classified as Vulnerable while *Callicebus* and *Pithecia* are considered to be under no significant threats (Rodriguez & Rojas-Suárez 1995, 2003). However, the expansion of human activities (mining, hunting, logging and cattle ranching) represents serious threats to pitheciids in the region. The recent construction of new roads has contributed with the expansion of human settlements. Because almost the entire range of white-faced sakis is within logging plots, and no part of its distribution is strictly protected, this genus seems to be at higher risk. Table F.2 presents species-by-species data on their conservation status.

Key studies

Cacajao has been surveyed in southern Venezuela (Lehman & Robertson 1994). *Chiropotes* had been studied in the field by Peetz (2001). This genus was also intensively studied from a morphological perspective (Bodini & Ferreira 1991). No systematic field work has been conducted on *Callicebus*, but the morphology of these primates was studied by Bodini (1981). *Pithecia* is the most studied Venezuelan pitheciine. The project coordinated by M.A. Norconk and W.C. Kinzey has focused on this genus for more than a decade. In addition, this project supports the research of students from Venezuela (Ceballos 1996; Riveros 1996; Urbani 2002) and abroad (Walter 1993; Homburg 1997, Gleason 2002; Cunningham 2003). Kinzey *et al.* (1988) and Urbani (2006) surveyed white-faced sakis.

List of conservation programs

There are no conservation programs that focus specifically on pitheciids. Nevertheless, there has been improved scientific coordination among local scientists and NGOs as well as an increase of field exploration in this region. Norconk and Kinzey's project in coordination with a national company produced extensive field data on *Pithecia* and *Chiropotes*. The future of Venezuelan pitheciids can only be ensured by improving the quality of life of local inhabitants, expanding local awareness and education, strengthening protected areas, incrementing official and private research support and confronting corruption (Carrillo & Perera 1995; Bevilacqua *et al.* 2002; Wright *et al.* 2007; Lehman *et al.*, Chapter 31).

Evolutionary Biology and Conservation of Titis, Sakis and Uacaris, eds. Liza M. Veiga, Adrian A. Barnett, Stephen F. Ferrari and Marilyn A. Norconk. Published by Cambridge University Press.

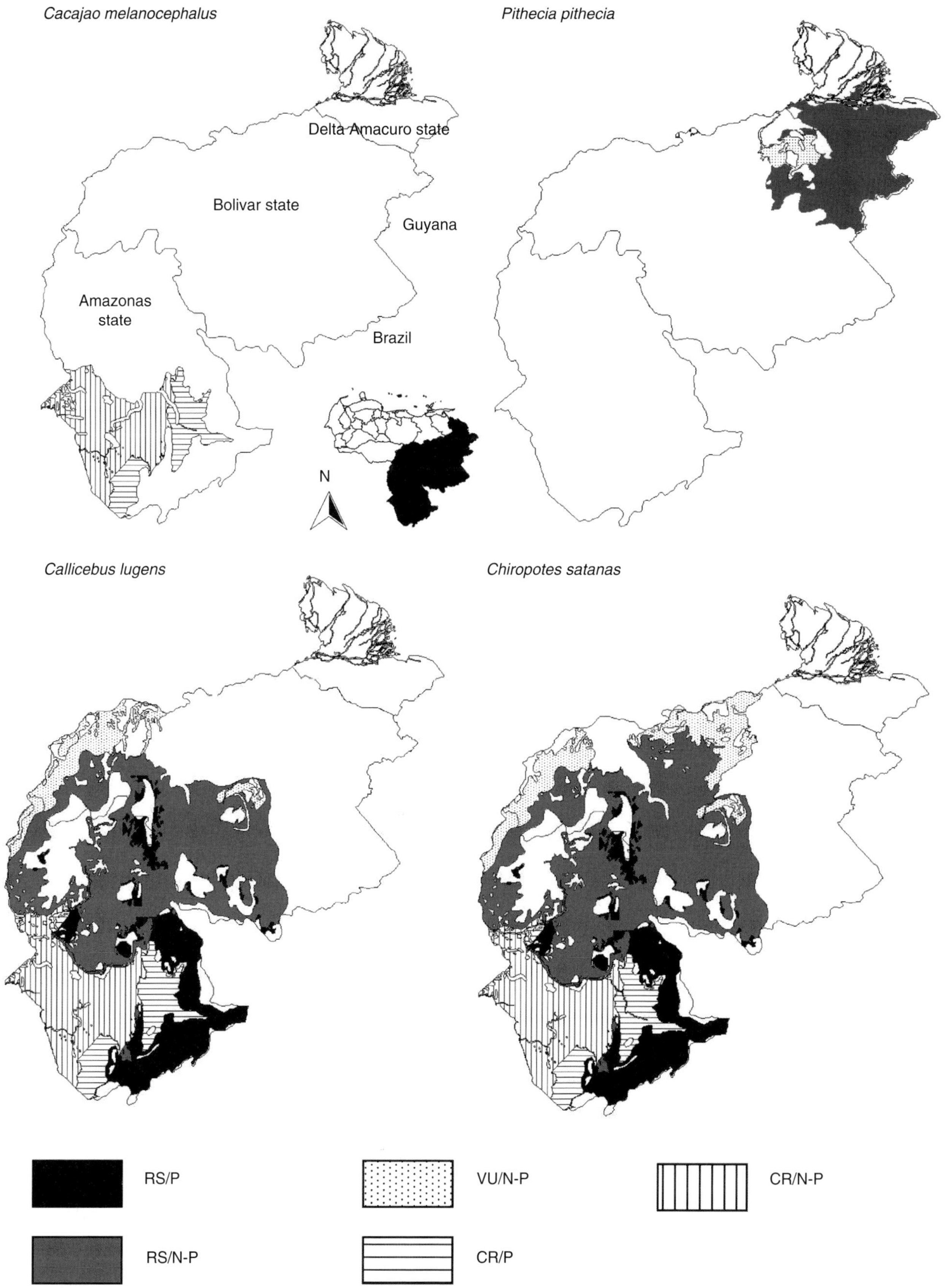

Figure F.1 Potential distribution of Venezuelan Pitheciines and the conservation status of their habitats.

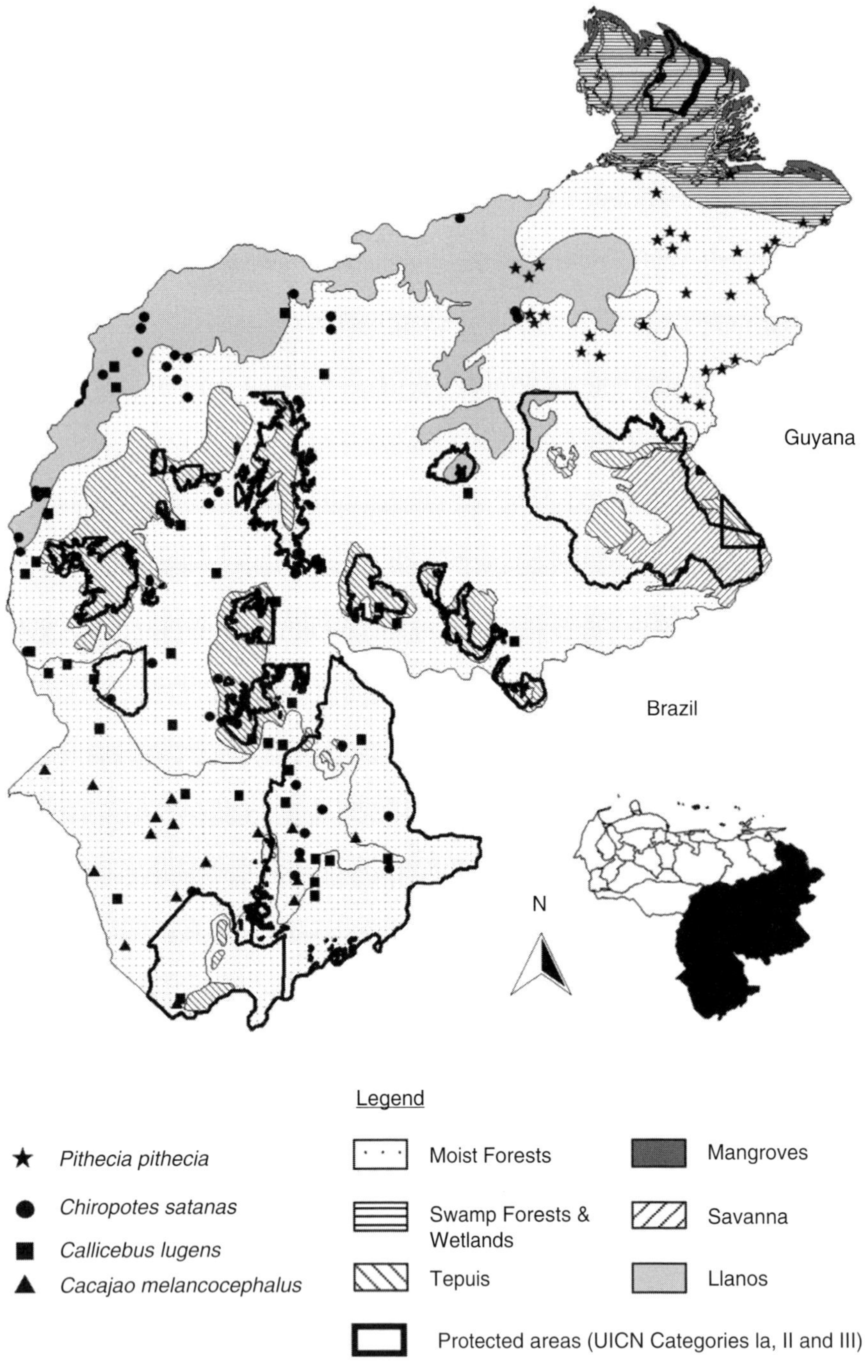

Figure F.2 Major habitat types, protected areas and the potential distribution of Venezuelan pitheciid taxa.

Table F.1 Vegetation types used by Venezuelan pitheciid taxa.

Species	Vegetation types
Cacajao melanocephalus	Ombrophilous evergreen, and partially flooded forests Complex of transitional forests between ombrophilous forests and *caatinga amazonica* Ombrophilous, sclerophyllus, evergreen forests (*caatinga amazonica*) Ombrophilous low forests, flooded, with palms
Pithecia pithecia pithecia	Ombrophilous, high, semideciduous forests (Cuyuní River basin) Ombrophilous, high, evergreen forests from south delta Ombrophilous foothill subevergreen forests Ombrophilous submontane evergreen forests Ombrophilous and swamp forests of the mid-delta
Callicebus lugens	Ombrophilous low forests, flooded, with palms. Ombrophilous foothill subevergreen forests Ombrophilous mid and high evergreen forests Ombrophilous montane evergreen forests (including low forests in tepuis' bases) Ombrophilous submontane evergreen forests Ombrophilous, sclerophyllus and evergreen forests (*caatinga amazonica*) Ombrophilous evergreen and partially flooded forests Riparian forests, seasonally flooded (Vegas del Orinoco) Riparian evergreen forests (Aguas claras y blancas) Tropophilous foothill low forests, semideciduous Complex of transitional forests between ombrophilous forests and *caatinga amazonica*
Chiropotes sagulatus	Ombrophilous foothill subevergreen forests Ombrophilous submontane evergreen forests Ombrophilous submontane subevergreen forests Ombrophilus evergreen and partially flooded forests Tropophilous foothill low forests, semideciduous Tropophilous low forests, deciduous, on rocky hills Riparian semideciduous forests with "moriche" palms Complex of transitional forests between ombrophilous forests and *caatinga amazonica*

Table F.2 Distribution of Venezuelan pitheciids within ecoregions by level of threat and protection status.

Species	RS/PR (%)	RS/N-PR (%)	VU/PR (%)	VU/N-PR (%)	CR-EN/PR (%)	CR-EN/N-PR (%)
Cacajao melanocephalus	16	6	0	0	25	53
Pithecia pithecia	0	90	0	10	0	0
Callicebus lugens	16	48	0	8	8	19
Chiropotes satanas	17	47	0	11	7	17

Abbreviations: RS (Relatively Stable), VU (Vulnerable), CR-EN (Critical or Endangered), PR (protected in parks UICN Ia-III), N-PR (Non-protected in parks UICN Ia–III).
Source: Ecoregion data from Olson *et al.* (2001).

References

Bevilacqua, M., Cárdenas, L., Flores, A.L., *et al.* (2002). *The State of Venezuela's Forest: A Case Study of the Guayana Region.* Caracas: Global Forest Watch/WRI/ACOANA/Universidad Nacional Experimental de Guayana/Provita/Fundación Polar.

Bodini, R. (1981). Musculatura locomotora de la viudita (*Callicebus torquatus*). Sus implicaciones funcionales y filogeneticas. *Memorias de la Sociedad de Ciencias Naturales La Salle*, **41**, 9–163.

Bodini, R. (1983). Distribución geográfica y conservación de primates sub-humanos en Venezuela. In *La Primatología en Latinoamérica*, ed. C.J. Saavedra, R.A. Mittermeier & I. Bastos-Santos. Bairro Cincão (Brazil): WWF/Editora Littera Maciel Ltda, pp. 101–113.

Bodini, R. & Ferreira, C. (1991). Morfología de la capsula glenohumeral y musculatura de *Chiropotes satanas chiropotes*. *Acta Biologica Venezuelica*, **13**, 51–57.

Bodini, R. & Pérez-Hernández, R. (1987). Distribution of the species and subspecies of cebids in Venezuela. *Fieldiana Zoology ns*, **39** 231–244.

Carrillo, A. & Perera, M.A. (1995). *Amazonas: Modernidad en Tradición. Contribuciones al Desarrollo Sustentable en el Estado Amazonas, Venezuela.* Caracas: SADA-Amazonas/ORPIA/CAIAH/GTZ-Venezuela.

Ceballos, N. (1996). Comportamiento social de una tropa de mono viudo, *Pithecia pithecia* (Cebidae: Primates), en una isla del Embase de Guri (Estado Bolívar). Unpublished BA thesis, Universidad Central de Venezuela.

Cunningham, E. (2003). The use of memory in *Pithecia pithecia's* foraging strategy. Unpublished PhD thesis, City University of New York.

Gleason, T. (2002). Predation risk and anti-predator adaptations in white-faced sakis, *Pithecia pithecia*. In *Eat or Be Eaten: Predation-Sensitive Foraging in Primates*, ed. L. Millar. Cambridge: Cambridge University Press, pp. 169–184.

Homburg, I. (1997). Ökologie in Sozialverhalten einer Gruppe von Weißgesicht-sakis (*Pithecia pithecia pithecia* Linnaeus 1766) im Estado Bolívar, Venezuela. Unpublished PhD thesis, Universität Bielefed.

Kinzey, W.G., Norconk, M.A. & Alvarez-Cordero, E. (1988). Primate survey of eastern Bolívar, Venezuela. *Primate Conservation*, **9**, 66–70.

Lehman, S.M. & Robertson, K.L. (1994). Preliminary survey of *Cacajao melanocephalus melanocephalus* in southern Venezuela. *International Journal of Primatology*, **15**, 927–934.

Linares, O.J. (1998). *Mamíferos de Venezuela*. Caracas: Sociedad Conservacionista Audubon de Venezuela/BP-Venezuela.

Olson, D.M., Dinerstein, E., Wikramanayake, E.D., *et al.* (2001). Terrestrial ecoregions of the world: a new map of life on Earth. *BioScience*, **51**, 933–938.

Peetz, A. (2001). Ecology and social organization of the bearded saki *Chiropotes satanas chiropotes* (Primates: Pitheciinae) in Venezuela. *Ecotropical Monographs*, **1**, 1–170.

Riveros, M. (1996). Dieta y comportamiento alimentario de una tropa de *Pithecia pithecia* (mono saki cara blanca), en una isla del Embalse de Guri, Edo. Bolívar. Unpublished BA thesis, Universidad Central de Venezuela.

Rodríguez, J.P. & Rojas-Suárez F. (1995). *Libro Rojo de la Fauna Venezolana*, 1st edition. Caracas: Provita/Fundación Polar/Wildlife Conservation Society/Profauna (MARNR)/UICN.

Rodríguez, J.P. & Rojas-Suárez F. (2003). *Libro Rojo de la Fauna Venezolana*, 2nd edition (corrected and expanded). Caracas: Provita/Fundación Polar/Wildlife Conservation Society/Profauna (MARNR)/UICN.

Urbani, B. (2002). A field observation on color selection by New World sympatric primates, *Pithecia pithecia* and *Alouatta seniculus*. *Primates*, **43**, 95–101.

Urbani, B. (2006). A survey of primate populations in northeastern Venezuelan Guayana. *Primate Conservation*, **20**, 47–52.

Walter, S. (1993). Positional adaptations and ecology of the Pitheciinae. Unpublished PhD thesis, City University of New York.

Wright, S.J., Sanchez-Azofeifa, G.A., Portillo-Quintero, C., *et al.* (2007). Poverty and corruption compromise tropical forest reserves. *Ecological Applications*, **17**(5), 1259–1266.

Index